THE INTERNATIONAL SERIES OF MONOGRAPHS ON CHEMISTRY

THE INTERNATIONAL SERIES OF
MONOGRAPHS ON CHEMISTRY

J. D. Lambert: *Vibrational and rotational relaxation in gases*

N. G. Parsonage and L. A. K. Staveley: *Disorder in crystals*

G. C. Maitland, M. Rigby, E. B. Smith, and W. A. Wakeham: *Intermolecular forces: their origin and determination*

W. G. Richards, H. P. Trivedi, and D. L. Cooper: *Spin-orbit coupling in molecules*

C. F. Cullis and M. M. Hirschler: *The combustion of organic polymers*

R. T. Bailey, A. M. North, and R. A. Pethrick: *Molecular motion in high polymers*

INTERMOLECULAR FORCES

Their Origin and Determination

By

GEOFFREY C. MAITLAND
Lecturer in the Department of Chemical Engineering and Chemical Technology
Imperial College of Science and Technology, University of London

MAURICE RIGBY
Lecturer in the Department of Chemistry
Queen Elizabeth College, University of London

E. BRIAN SMITH
Fellow of St. Catherine's College, Oxford and Lecturer in Physical Chemistry, University of Oxford

and

WILLIAM A. WAKEHAM
Reader in Chemical Physics of Fluids
Imperial College of Science and Technology, University of London

CLARENDON PRESS · OXFORD · 1981

Oxford University Press, Walton Street, Oxford OX2 6DP
London Glasgow New York Toronto
Delhi Bombay Calcutta Madras Karachi
Kuala Lumpur Singapore Hong Kong Tokyo
Nairobi Dar es Salaam Cape Town Salisbury
Melbourne Auckland
and associate companies in
Beirut Berlin Ibadan Mexico City

Published in the United States by Oxford University Press, New York

British Library Cataloguing in Publication Data

Intermolecular forces. – (International series of monographs on chemistry).
1. Molecular theory
I. Maitland, Geoffrey C   II. Series
541.2′26   QD461   80-41333

ISBN 0-19-855611-X

Filmset in Northern Ireland at The Universities Press (Belfast) Ltd
Printed in Great Britain
at the University Press, Oxford
by Eric Buckley
Printer to the University

PREFACE

The forces between molecules are of interest to scientists in a wide range of disciplines as these interactions control the progress of molecular collisions and determine the bulk properties of matter. Some twenty-five years ago, the subject was treated comprehensively in the classic text by Hirschfelder, Curtiss, and Bird: *The molecular theory of gases and liquids.* In addition to giving a thorough account of the molecular theory of matter, much of which is still relevant at the present time, this book also attempted to reconcile the properties of matter in terms of simple intermolecular potential energy functions† such as those due to Lennard-Jones. This aspect of their book has not stood the test of time, although it was not until the 1970s that a firmer foundation to the subject could be laid. During this period a quantitative understanding of the intermolecular forces of a number of monatomic species—in particular the inert gases—was achieved and found to bear little relation to the simple functions widely employed previously. These advances have been based on a number of factors such as the development of new types of experimental measurement, new attitudes to data analysis and the availability of high-speed computers, leading to a subject very different from that described by Hirschfelder, Curtiss, and Bird. It is now important to reassess the subject of molecular interactions in the light of this transformation.

This book provides a wide-ranging account of the determination of intermolecular forces. It describes both the techniques that have been developed in recent years and the potentials that have resulted from their application. Particular attention has been paid to molecular beam scattering and the spectroscopy of van der Waals dimers, which can provide extremely detailed information about certain aspects of intermolecular potentials. A comprehensive account is also given of macroscopic, thermally averaged properties such as second virial coefficients and dilute gas transport properties, which have recently become re-established as important sources of information about intermolecular forces. Among the factors which have contributed significantly to the progress of the past ten years are an increased appreciation of the complementary nature of the information contained in these different properties and the development of more direct routes from properties to potentials. Both these aspects are emphasized strongly in this book.

In writing about intermolecular forces a certain arbitrariness in defining the boundaries of the subject is necessary. We have, for example,

† Throughout this book, these will be referred to quite simply as 'potentials'.

not included the topic of hydrogen bonding and have referred only briefly to interactions involving ionic species. Naturally, we have given most attention to those systems (mainly monatomic) where the potential is known with high accuracy and have not devoted much coverage to the many superficial attempts to deduce potential functions from either limited data or inadequate techniques. However, in the case of polyatomic molecules, even though little quantitative information is at present available, we have tried to indicate those results which should provide a basis for future progress. Traditionally, many attempts have been made to deduce pair potentials from condensed phase properties. However, as some uncertainty still surrounds the role of non-pairwise additive contributions to intermolecular energy, pair potentials are most soundly deduced from those properties which involve only pair interactions. For this reason we have given only a very general account of the use of condensed phase properties in determining pair potential functions.

It is impossible to separate the determination of intermolecular forces from an understanding of their relationship to molecular and thermophysical properties, and any book of this sort must of necessity include a considerable amount of statistical mechanics and scattering theory. To give a comprehensive account of these subjects would lead to a work of excessive length. Our practice has been to include sufficient coverage of only those aspects that are essential to an understanding of the elucidation of intermolecular forces. It should be emphasized that this is not a treatise on the fundamental theory of intermolecular forces, although one chapter has been assigned to this topic.

It is our intention that this book should prove useful to a wide variety of scientists. Some may be senior undergraduates or new graduate students who require a modern account of the fundamentals of the subject, some will be research workers directly concerned with investigating the nature of the forces between molecules, whereas others will simply wish to use the best current knowledge of the force laws to calculate the thermophysical properties of substances in which they are interested. In order to cater for such diverse interests, different sections of the book have been consciously written at several levels. A key indicating these levels is attached to the table of contents. This identifies those sections which, taken together, provide a suitable introduction for those approaching the subject for the first time, and those sections, largely theoretical, which may not be necessary for those readers more concerned with the application of the results contained in the book rather than their derivation. To this end we have felt it advisable to cover some subjects more than once to provide a fairly comprehensive treatment of the subject at each level. It is inevitable that such a format should lead to inhomogeneities in the text which may trouble those who wish to read the book through. However, we believe that the benefits of this approach far outweigh the disadvantages.

In addition to its pedagogical role, the book is intended to be a work of reference. With this in mind, we have included a large quantity of numerical material in the text and appendices. This comprises tabulations, both of useful modern potential functions, and of the equilibrium and transport properties calculated using them. We have also given copies of computer programs which enable equilibrium and transport properties to be calculated for any given potential function. In addition, a quantity of selected experimental data is tabulated for a limited number of reference substances. After decades of relative stagnation, the field of intermolecular forces is now progressing at a rapid rate. In compiling the material in the book we have sought to emphasize those aspects of the subject which we believe to be the least ephemeral. To those building on the foundations described in this book, we would make a strong plea for extreme caution. The history of research into intermolecular forces shows that the most common pitfall has been overconfidence. Again and again workers have been too easily satisfied by superficial answers, and have chosen to ignore the limitations of their approach. We now appear to have a solidly based understanding of the forces in many monatomic systems and it is important that, as we turn to more complicated molecules, we do not expect easy results. Progress will only be made by an uncompromisingly critical approach.

The authors wish to thank their friends and colleagues for the enormous help which they have received in the preparation of this book. Their contributions are acknowledged below. We would be most grateful if readers would draw any mistakes to our attention and would be most happy to receive general comments and criticisms.

London and Oxford
May 1980

G. C. M.
M. R.
E. B. S.
W. A. W.

Acknowledgements

We have been most fortunate in the advice and assistance offered to us while preparing this book. We are particularly grateful to Professor J. S. Rowlinson who, as editor of the Series, gave us much encouragement and the benefit of his wide knowledge and deep scholarship. He has read the book through twice in its entirety—a feat rarely likely to be repeated! In the preparation of Chapter 7 we were privileged to have the advice of Professor R. J. LeRoy and access to his unpublished review manuscripts which proved of considerable assistance. Much of the assessment of the potential functions for the inert gases in Chapter 9 is based on the work of Professor R. A. Aziz who also kindly povided infomation prior to publication.

Professor R. B. Bernstein and Dr D. J. Tildesley have also read the manuscript and we appreciate their many important suggestions for its improvement. We are also grateful to Dr A. A. Clifford, Dr B. J. Howard, Professor N. March, Professor E. A. Mason, and Dr W. G. Richards for their advice on specific sections of the book. Any errors or omissions which remain are, of course, entirely our responsibility.

Our thanks also go to the many people who have helped with the preparation of the book. The arduous task of typing the text from an often illegible manuscript was ably performed by Miss N. Ganatra, Miss S. Hartman, Miss H. D. Hughes, Mrs M. Long, Miss M. McNeile, Miss S. M. Oakes, and Mrs E. Price. The equally laborious task of compiling many of the references for Chapter 9 was carefully carried out by Mr B. H. Wells and Mr T. Maberley. Miss H. D. Hughes, Mr R. Hainsworth, and Mr A. R. Tindell provided valuable assistance in preparing the computational material included in the appendices. We also thank Dr J. A. Barker for this permission to include in Appendix 12 a program for collision integral evaluation based upon his original computational method.

We are grateful to the publishers and authors for permission to use the following figures: American Institute of Physics: 4.11, 4.12, 4.14, 4.15, 4.25, 4.32, 4.35, 7.5, 7.10, 7.14, 7.15, 7.19; The Chemical Society: 4.34, 7.17; John Wiley & Sons, Ltd: 4.33, 7.18, 9.2; Taylor and Francis, Ltd: 4.37, 4.39, 4.39, 4.40, 7.9, 7.11.

We are happy to acknowledge the assistance given at all stages of the production of this book by the staff of Oxford University Press. Finally we thank all those close to us for their encouragement, support and forebearance during the considerable gestation period of the book.

CONTENTS

* These sections provide a suitable introduction for readers approaching the subject for the first time.

† These sections may not be necessary for all users on a first reading.

1
INTRODUCTION

1.1. Historical background

The molecular theory of matter has its roots in the philosophical speculations of the atomists of the fifth century BC, Leucippus and Democritus, who suggested that all matter was composed of small, rapidly moving particles they called atoms. This idea, somewhat modified to avoid the determinism which atomists saw as a natural corollary, was incorporated by Epicurus into his wider philosophical scheme and was an important influence on Greek and Roman thought. It was expounded in a long poem *On the nature of the universe* by Lucretius,[1] a Roman Epicurean of the first century BC. Many of the features of the molecular theory described by Lucretius are in surprising accord with modern views. Matter was conserved in all processes, there was a large, but finite, number of species of atom, and the atoms moved at high speeds. The deterministic consquences of the model were avoided by allowing the atoms to 'swerve ever so little from their course' at random times and places—an ingenious precursor of the uncertainty principle. The concept of molecular interactions was also developed, but the distinction between chemical and physical forces between the atoms was of course missing.[1]

> Again, things that seem to be hard and stiff must be composed of deeply indented and hooked atoms and held firm by their intertangling branches. In the front of this class stand diamonds... Liquids on the other hand must owe their fluid consistency to component atoms that are smooth and round.

The relationship between molecular properties and the viscosity of fluids was also considered.[1]

> We see that wine flows through a strainer as fast as it is poured in; but sluggish oil loiters. This is no doubt either because oil consists of larger atoms or because they are hooked and intertangled and therefore cannot separate so rapidly, so as to trickle through the holes one by one.

These ideas represented the limit to which molecular theory could be developed. Further advances required new experimental information that could not be obtained from a casual inspection of the universe. Two millenia were to pass before the techniques were developed and the necessary experiments performed. From the fifteenth century new observations laid the foundation for a scientifically based dynamic molecular theory. But despite many important theoretical developments, starting with Bernoulli[2] in 1738, the main tenets of the modern molecular theory of matter, in particular the idea of molecular motion in gases, were not generally accepted until the latter half of the nineteenth century. The powerful mathematical theories of Clausius, Maxwell, and Boltzmann between 1850 and 1890 brought to fruition the kinetic theory of gases.[3–9] After this period it was universally recognized that temperature and pressure are both related to the motion of the molecules. The pressure is due to the force that molecules exert on the walls of a container by virtue of their collisions with the walls and temperature is a measure of the average kinetic energy of the molecules.

The idea of intermolecular forces also has a long history. The concept of molecules as point sources of attractive and repulsive forces was first formulated by Boscovich (1783). He recognized that molecules must repel each other at very small separations but assumed that the attraction and repulsion alternate a number of times as the separation increased. By the early nineteenth century the fact that molecules could repel each other was acceptable even to the advocates of the phlogiston theory of heat who regarded gaseous molecules as stationary objects held in their positions (on a lattice) by mutual repulsion. Later, the modern view that molecules repel each other at small separations and attract each other at long range became established. It was clearly stated by Clausius in 1857.

Maxwell incorporated into the description of the kinetic behaviour of gases the idea that molecules exerted forces on one another. He presumed that the forces were entirely repulsive in character and that they decayed as the separation between the interacting molecules increased. He was able to deduce the temperature dependence of the viscosity of a gas of such molecules. Boltzmann, attempting a similar calculation, invoked a series of intermolecular force laws all of which contained attractive components. It is interesting to note that his results provided an equally acceptable description of the behaviour of the viscosity of gases to that of Maxwell despite the gross differences in the assumed force law. In 1873 van der Waals developed an equation of state for a gas whose molecules were supposed to be impenetrable rigid spheres surrounded by an attractive force field. This concept enabled van der Waals[10] to show that the pressure exerted by such a gas lay below that for a gas whose molecules were non-interacting points (the perfect gas) owing to the retardation effect of the attractive forces on molecules colliding with the wall. In addition, the non-zero volume of the gas molecules reduced the volume available for motion of molecular centres, below that for non-interacting points. Thus he deduced the van der Waals equation of state

$$\left(P+\frac{a}{\tilde{V}^2}\right)(\tilde{V}-b)=RT$$

in which a is a parameter characterizing the strength of the intermolecular attraction, b represents the reduction in the available volume owing to the finite molecular size, R is the universal gas constant, and $\tilde{V}$ is the molar volume of the gas. The importance of this equation of state lies in its prediction of a gas–liquid transition and a critical point for a pure substance, neither of which occur for the perfect gas. The work of van der Waals demonstrated that the very existence of condensed phases of matter stems from the attractive forces between molecules, and at the same time that the small compressibility of these condensed phases arises from the repulsive forces which act at short range. At this time the origins of such intermolecular forces were not understood. Nevertheless, the period saw the establishment of the fundamental connection between the macroscopic properties of matter and the forces between the constituent molecules—statistical mechanics—which is the cornerstone of modern work and the guideline of the present book.

Since the end of the nineteenth century a considerable amount of work has been devoted to the exact formulation of the connection between the properties of matter in bulk and intermolecular forces. Such a formulation represents the ultimate aim of the molecular theory of matter since, when a theory of this kind is established, a knowledge of the intermolecular forces is sufficient for the evaluation of all the properties of the bulk materials. In the cases of dilute gases and nearly perfect crystals this link was completed early in this century. However, the use of these theories for the evaluation of bulk properties was inhibited by ignorance of the exact form of the intermolecular forces. Consequently attention became concentrated on the search for the origins and nature of intermolecular forces. Even when the origins of the forces became clear around 1930 the difficulties of the *ab initio* evaluation of the forces from quantum mechanics provided a further obstacle to quantitative progress. Thus at about the same time a heuristic method of obtaining information about intermolecular forces was developed. This approach involved assuming an algebraic form for the dependence of the intermolecular force upon intermolecular separation, calculating bulk physical properties for the material through the appropriate molecular theory and finally comparing these calculations with experimental data for the same physical property. Agreement between the two sets of data was supposed to indicate the correctness of the assumed intermolecular force law. It is now recognized that this procedure yields little more than a crude estimate of intermolecular forces. However the method represents the first attempt to answer the question, 'Can the measurements of the bulk properties of materials themselves serve to determine the intermolecular forces on which they depend?' Despite much effort a unique solution to this inverse problem of molecular theory seemed impossible until the last decade.

However, recent advances in the measurement of bulk properties of materials, in the measurement of microscopic properties† of interacting molecules, and in techniques for the analysis of such experimental information have radically altered this situation. It is now possible to describe with some assurance the intermolecular forces between the simplest molecules.

It is the purpose of this book to provide a comprehensive treatment of the methods which have been employed to determine intermolecular forces both experimentally and theoretically as well as to present a summary of our current knowledge of them. Since this involves the solution of the inverse problem referred to earlier, we shall discuss the molecular theories that relate the intermolecular forces to both the properties of matter in bulk and the observable microscopic behaviour of pairs of molecules. Attention will be concentrated upon those topics which have proved of greatest value in the determination of intermolecular forces and upon those molecular systems for which most information is available. Thus, among the bulk properties, we treat in greatest depth the dilute gas state in which only interactions between pairs of molecules are important and, among the microscopic properties, the spectroscopy of dimers and the scattering of molecules. The condensed phases of matter have so far played a smaller, but not insignificant, role in the determination of intermolecular forces. If the current growth in knowledge about intermolecular forces is maintained it is likely that studies of these condensed places will be increasingly significant. Up to the present time the greatest effort has been devoted to the interaction of two monatomic species, and the interactions between more complex molecules are still but poorly understood.

1.2. Intermolecular energy

The present chapter provides an overview of the subject in a way that draws together the many, apparently disparate, aspects of the field of intermolecular forces. The opportunity is taken to establish the physical origins of intermolecular forces in a way which lends substance to the more mathematical treatment in later chapters, as well as to introduce concepts and terms in common usage in the field.

The most important concept concerns the representation of the force between two molecules by means of a potential energy function. We first consider the simplest possible situation in which two atoms, a and b, each composed of a positively charged nucleus surrounded by a negatively-charged, spherically-symmetric electron cloud, interact. When the two

† We use the term macroscopic to describe the bulk properties of matter. By contrast microscopic properties are those which are investigated at a molecular level. In this category we include for instance spectroscopic measurements and direct observation of molecular collisions.

atoms are infinitely separated they do not interact at all and the total energy of the two-atom system, E_{tot}, is just the sum of the energies of the individual atoms,

$$E_{tot}(\infty) = E_a + E_b.$$

If the two atoms are separated by a finite distance, r, the interaction between them provides an extra contribution to the total energy of the system. For the simple case we are concerned with the symmetry of the atoms requires that this energy depend only on the separation of the two atoms and not upon their relative orientation. Thus the total energy is now

$$E_{tot}(r) = E_a + E_b + U(r),$$

and the contribution to the total energy arising from the interaction, which is known as the intermolecular pair potential energy function, is

$$U(r) = E_{tot}(r) - E_a - E_b = E_{tot}(r) - E_{tot}(\infty).$$

The intermolecular pair potential thus describes the departure of the total energy of the two-atom system from its value when the two atoms are infinitely separated. This energy difference is numerically equal to the work done in bringing the two atoms from infinite separation to the separation r and is given by

$$U(r) = \int_r^\infty F(r)\,\mathrm{d}r, \quad \text{so that} \quad F(r) = -\frac{\mathrm{d}U}{\mathrm{d}r}$$

where $F(r)$ is the force acting between the two atoms at the separation r. By convention the force F is positive when it is repulsive and negative when attractive.

We shall see later that the energy of interaction between the two atoms arises from electrical forces between the charged entities of which they are made up. In the semi-classical picture of an atom the electrons are supposed to be in continuous orbit about the nucleus so that these electrical forces can be expected to fluctuate on a time scale related to the velocity of the electronic motion. If the two atoms are held stationary at a separation, r, as we supposed in the development of the intermolecular potential energy, then the net force between them is the average of these rapidly fluctuating forces. However the molecular theory of matter tells us that atoms in matter are in continuous motion so that we must re-examine the concept of the potential energy functions in this context. If the time scale on which two atoms approach each other is large compared with the time scale of the electron motion then as before the forces between the atoms will be a time average of the fluctuating values and our earlier concept of the potential energy function remains valid. It is fortunate that, because the velocities of atoms in most practical circumstances are very much less than those of the electrons about the nucleus, this is generally the case. Thus we may treat the encounters of atoms moving with a finite velocity with the aid of a unique, velocity-independent, intermolecular potential. This important idea is known as the *Born–Oppenheimer approximation* and it has the useful corollary that in the discussion of such an encounter only the motion of the nuclei needs to be considered to describe completely the dynamic events.

The work of van der Waals suggested, and later developments have confirmed, that the general form of the intermolecular potential energy function for spherically symmetric atoms $U(r)$ is as shown in Fig. 1.1. which also contains a plot of the corresponding intermolecular force, $F(r)$.

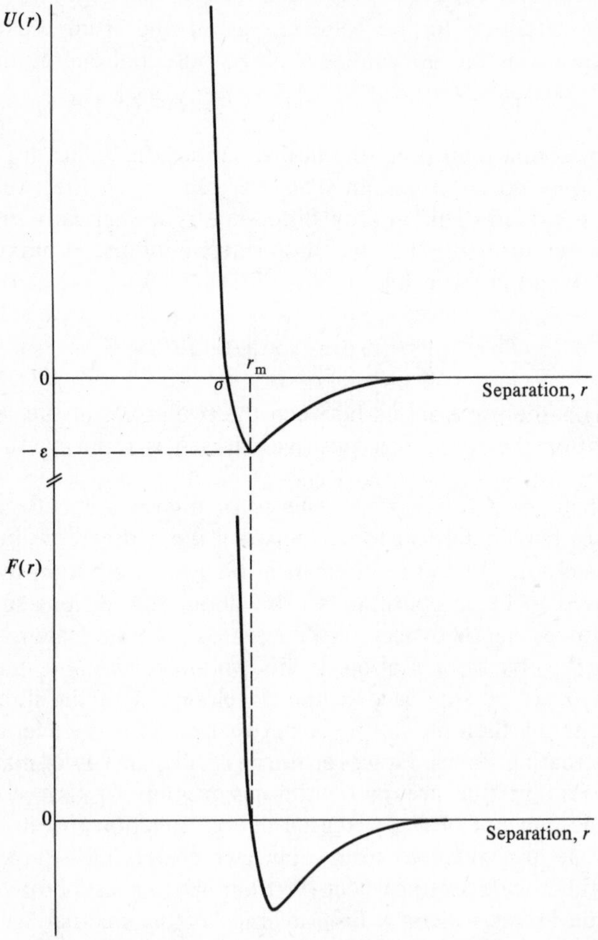

FIG. 1.1. The intermolecular pair potential energy function (schematic).

There is a strong repulsive force at short range and an attractive force at long range. In terms of the potential energy $U(r)$ this behaviour corresponds to large, positive energies at small separations and negative energies at long range, the two extreme regions being joined by a function with a single negative minimum. It is conventional to work with the intermolecular potential for convenience and Fig. 1.1 defines those parameters of the potential usually employed to characterize it. These are the separation at which the potential energy is zero, σ, the separation at which the energy attains its minimum value, r_m, and the minimum energy itself, $-\varepsilon$; ε is known as the *well depth.*

In the more general problem of the interaction of two molecules which lack spherical symmetry, we expect the intermolecular energy to depend not only upon the intermolecular separation, but also upon the relative orientation. The interactions between two nitrogen molecules or two hydrogen chloride molecules, for which each molecule possesses cylindrical symmetry, provide examples of this type of potential energy function. For such molecules there are additional contributions to the total energy of the interacting two-molecule system which arise from the internal energy of the molecules themselves, that is vibrational and rotational energy. In such cases the intermolecular potential energy function may depend on the internal state of each molecule as well as intermolecular separation and relative orientation. It is then frequently useful and necessary to consider distinct intermolecular potential energy functions for each internal energy state of the two molecules. Such interactions are as yet less well understood than those between monatomic species.

1.3. Origins of intermolecular forces

Our present knowledge allows us to classify the forces which arise between particles into four catagories.

(a) Gravitational;
(b) Electromagnetic;
(c) Strong nuclear forces;
(d) Weak nuclear forces.

The strong nuclear forces are responsible for the binding of neutrons and protons inside the nucleus, and the range over which they are significant is consequently of the order of 10^{-4} nm. The weak nuclear force, now known to be electromagnetic in origin, is of similar short range. Since molecular dimensions are typically 5×10^{-1} nm these nuclear forces cannot contribute to the intermolecular force. Conversely, the gravitational force is extremely long-range and might be thought to be the source of intermolecular attraction. However, a simple calculation shows that the gravitational potential energy of two argon atoms at a separation of 0·4 nm is only 7×10^{-52} J, which is some thirty orders of magnitude smaller than intermolecular forces. Consequently, intermolecular forces must have an electromagnetic origin.

Accepting that intermolecular forces have this origin, the source of the interaction can be seen to be due to the charged particles, electrons and protons, which make up an atom or molecule. Since the intermolecular forces are repulsive at short range and attractive at long range we can discern that there must be at least two contributions to the total force and to the corresponding intermolecular potential. Qualitatively, the origin of the repulsive forces at short range is the simpler to explain. When the electron clouds of two molecules approach each other sufficiently closely that they overlap, the Pauli exclusion principle prohibits some electrons from occupying the overlap region and so reduces the electron density in this region. The positively charged nuclei of the molecules are thus incompletely shielded from each other and therefore exert a repulsive force on each other. Such short-range forces are referred to as overlap forces.

The long-range attractive component of the intermolecular force is significant when the overlap of the electron clouds is small and arises in a completely different manner. In fact there are three possible contributions to the attractive force depending on the nature of the interacting molecules. Only one of these contributions is present in all molecular interactions and it is known as the London dispersion force. For simple molecules it is nearly always the greatest contribution to the intermolecular attractive force but, as we shall see, arises from more subtle considerations than the remainder. It is convenient to introduce the physical origins of these three contributions in the order of conceptual ease rather than in the order of their magnitude.

(1) *Electrostatic contributions.* It is well known that some molecules such as HCl possess permanent dipole moments by virtue of the electric charge distribution in the molecule. One component of the interaction energy for two such molecules at long range therefore arises from the electrostatic interaction between their dipole moments. Such an interaction takes place with no distortion of the electron distribution on either molecule and the resulting energy is for this reason termed a first-order energy. Since the electrostatic energy between two dipoles is a strong function of their relative orientation it is sometimes referred to as the orientation energy. Some non-dipolar molecules such as CO_2 possess an electric quadrupole moment which contributes to the electrostatic energy in a similar manner.

(2) *Induction contributions.* If we consider an interaction between one molecule with a permanent dipole moment and a molecule which is non-polar the electric field of the dipolar molecule distorts the electron charge distribution of the other molecule producing an induced dipole moment within it. This induced dipole then interacts with the inducing dipole to produce an attractive force. This type of interaction relies for its existence upon distortion of electron clouds and is therefore termed a second-order energy. This induction contribution is simultaneously present with the electrostatic contribution in the case of the interaction of two polar molecules.

(3) *Dispersion contributions.* If we consider the interaction of two molecules neither of which has a permanent dipole moment the source of an attractive energy is more difficult to discern. However we recognize that although a molecule may possess no permanent dipole moment, its electrons are in continuous motion so that the electron density in a molecule oscillates continuously in time and space. Thus at any instant any molecule possesses an instantaneous electric dipole which fluctuates as the electron density fluctuates. This instantaneous dipole in one molecule induces an instantaneous dipole in a second molecule. The induced dipole in the second molecule and the inducing dipole in the first interact to produce an attractive energy called the dispersion energy. In other words, the dispersion energy is a result of the correlations between the electron density fluctuations in the two molecules. The dispersion energy is second-order in the sense defined earlier, since a distortion of the electron density of the molecule is involved. In the case of interactions between two neutral, non-polar molecules it is the only contribution to the long-range energy. The name dispersion energy is applied to this long-range contribution because the physical mechanism which gives rise to the energy is that responsible for the dispersion of light in a gas.†

In the next sections a brief outline of a more quantitative description of molecular forces is given. These sections are intended to serve two functions. First, they provide an introduction to those aspects of the theory of intermolecular forces important to the remainder of the book, which are given a more complete discussion in the following chapter. Secondly, they are intended for use by a reader who wishes to gain some knowledge of intermolecular forces without undue mathematical effort. To achieve these aims the treatment given here is neither rigorous nor complete, and simplified arguments are employed wherever necessary to retain clarity.

† This relationship is discussed in Chapter 2.

1.4. Long-range energy

1.4.1. Electrostatic energy

Here we consider that part of the energy of interaction of two molecules which arises from permanent non-uniform electric charge densities within each molecule. We consider a simple case in which each molecule can be represented by a linear charge distribution as shown in Fig. 1.2.

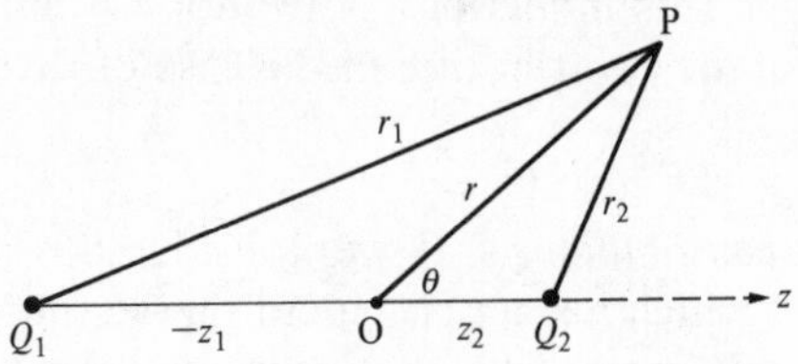

FIG. 1.2. A linear charge distribution.

The distribution consists of two charges arranged about an origin O which we take to lie at the centre of mass of the molecule. The axis of the distribution coincides with the z-axis of the coordinate system; charge Q_1 is located at $-z_1$ and charge Q_2 at z_2. Such a model might for example represent the HCl molecule. We first evaluate the electrostatic potential Φ due to one molecule at the point P of Fig. 1.2, which in terms of the symbols of the figure may be written

$$\Phi = \frac{1}{4\pi\varepsilon_0}\left\{\frac{Q_1}{r_1}+\frac{Q_2}{r_2}\right\} \tag{1.1}$$

where ε_0 is the permittivity of free space. In terms of the coordinates r, θ of the point P and the axial separations of the charges from O this may be written

$$\Phi = \frac{1}{4\pi\varepsilon_0}\left\{\frac{Q_1}{[r^2+z_1^2+2z_1 r\cos\theta]^{\frac{1}{2}}}+\frac{Q_2}{[r^2+z_2^2-2z_2 r\cos\theta]^{\frac{1}{2}}}\right\}. \tag{1.2}$$

Although (1.2) defines the potential completely, a more convenient form for our purposes can be obtained when the distance r is greater than either z_1 or z_2. In this situation the denominators of the two terms in (1.2) can be expanded in powers of z_1/r and z_2/r to yield

$$\Phi = \frac{1}{4\pi\varepsilon_0}\left\{\frac{Q_1+Q_2}{r}+\frac{(Q_2 z_2-Q_1 z_1)\cos\theta}{r^2}+\frac{(Q_1 z_1^2+Q_2 z_2^2)(3\cos^2\theta-1)}{2r^3} + \dots\right\} \tag{1.3}$$

which may be written as

$$\Phi = \frac{1}{4\pi\varepsilon_0}\left\{\frac{Q}{r}+\frac{\mu\cos\theta}{r^2}+\frac{\Theta}{2}\frac{(3\cos^2\theta-1)}{r^3}+\dots\right\}. \tag{1.4}$$

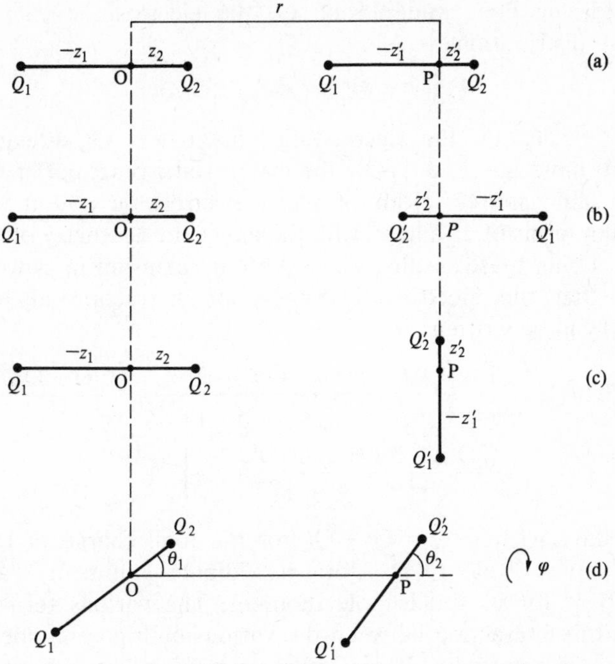

FIG. 1.3. The interaction of two linear charge distributions.

Here we have written $Q = Q_1 + Q_2$, for the total charge of the molecule, which is also known as the zeroth moment of the charge distribution; $\mu = Q_2 z_2 - Q_1 z_1$ which is the dipole moment or first moment of the charge distribution; and $\Theta = Q_1 z_1^2 + Q_2 z_2^2$ which is the quadrupole moment of the charge distribution.†

It must be emphasized that (1.4) is valid only for $r \gg z_1, z_2$. However since we are interested in the energy of two charge distributions at long-range these are the conditions for which we require the electrostatic potential.

We now consider the energy of interaction of two such charge distributions. The centre of mass of the second molecule is located at P, a distance r from that of the first, and for simplicity we consider that the axes of the two charge distributions lie in the page and restrict ourselves to the three relative orientations sketched in Fig. 1.3(a–c). The second molecule has charges Q'_1 and Q'_2 located at positions $-z'_1$ and z'_2 along its axis from P.

† More complete definitions of these quantities are given in Chapter 2.

Considering first configuration (a) the electrostatic energy of the two charge distributions is

$$U^{\mathrm{a}}_{\mathrm{el}}(r) = Q_2'\Phi(Q_2') + Q_1'\Phi(Q_1') \tag{1.5}$$

where $\Phi(Q_2')$ denotes the electrostatic potential at Q_2' due to the first (unprimed) molecule, and $\Phi(Q_1')$ the electrostatic potential at Q_1' arising from the same source. Both of these electrostatic potentials can be expressed in terms of z_2' and z_1' with the aid of the geometry of Fig. 1.3(a) and (1.4). Using these results, we find after expansion in powers of z_1'/r and z_2'/r that the electrostatic energy of interaction at long-range $(r \gg z_1', z_2')$ can be written,

$$U^{\mathrm{a}}_{\mathrm{el}}(r) = \frac{1}{4\pi\varepsilon_0}\left\{\frac{QQ'}{r} + \frac{(\mu Q' - Q\mu')}{r^2} - \frac{2\mu\mu'}{r^3} + \frac{(Q\Theta' + \Theta Q')}{r^3} + \frac{(3\Theta'\mu - 3\mu'\Theta)}{r^4} + \frac{6\Theta\Theta'}{r^5} + \dots\right\}. \tag{1.6}$$

Here we have written $Q' = Q_1' + Q_2'$ for the total charge of the second molecule, $\mu' = Q_2'z_2' - Q_1'z_1'$ for its dipole moment, and $\Theta' = Q_2'z_2'^2 + Q_1'z_1'^2$ for its quadrupole moment. The various terms of (1.6) correspond to interaction between the various multipole moments of the two charge distributions. If we consider interactions between neutral molecules $(Q = Q' = 0)$ with Q_2' and Q_2 positive then for configuration (a) the leading term in the electronic energy is the dipole–dipole interaction which varies as r^{-3} and is given by

$$U^{\mathrm{a}}_{\mu\mu'}(r) = \frac{-2\mu\mu'}{4\pi\varepsilon_0 r^3}. \tag{1.7}$$

Thus in this configuration the dipole–dipole contribution to the interaction energy is negative and there is an attractive force between the neutral molecules.

A similar development can be carried out for configurations (b) and (c) of Fig. 1.3; for the dipole–dipole interaction of two neutral molecules we find

$$U^{\mathrm{b}}_{\mu\mu'}(r) = \frac{+2\mu\mu'}{4\pi\varepsilon_0 r^3}$$

which corresponds to a repulsive force, and

$$U^{\mathrm{c}}_{\mu\mu'}(r) = 0,$$

that is the dipole–dipole interaction energy is zero.

These illustrations show that the dipole–dipole interaction energy is proportional to r^{-3} for a fixed configuration of the molecules, and that it is strongly dependent on orientation varying from attractive to repulsive as one molecule is rotated. An analysis of the interaction of two linear charge distributions in the general configuration of Fig. 1.3(d) wherein ϕ denotes rotation of the second dipole about the line joining them is presented in Chapter 2. It leads to the result for the dipole–dipole interaction energy that

$$U_{\mu\mu'}(r, \theta_1, \theta_2, \phi) = \frac{-\mu\mu'}{4\pi\varepsilon_0 r^3}\zeta(\theta_1, \theta_2, \phi) \qquad (1.8)$$

where

$$\zeta(\theta_1, \theta_2, \phi) = (2\cos\theta_1\cos\theta_2 - \sin\theta_1\sin\theta_2\cos\phi) \qquad (1.9)$$

describes the dependence of the energy on orientation.

If the two molecules are free to rotate as they are in the gas phase it is often the average of the dipole–dipole energy over all possible orientations $\langle U_{el}\rangle_{\mu\mu'}$ which is required. If all relative orientations were equally probable this average would be zero. However, the probability of observing a configuration of energy U is proportional to the Boltzmann factor, $\exp(-U/kT)$, which consequently leads to a preferential weighting for configurations of negative energy. When the Boltzmann weighted averaging is carried out we find, at sufficiently high temperatures,

$$\langle U_{el}\rangle_{\mu\mu'} = -\frac{2}{3}\frac{\mu^2\mu'^2}{r^6 kT(4\pi\varepsilon_0)^2} + \ldots \qquad (1.10)$$

The leading term of the orientation-averaged dipole–dipole contribution to the electrostatic energy of interaction of two neutral molecules is therefore attractive and inversely proportional to the sixth power of their separation. The temperature dependence of $\langle U_{el}\rangle_{\mu\mu'}$ is a result solely of the orientational averaging with the Boltzmann weighting.

Eqn (1.6) shows that there are contributions to the electrostatic energy from the dipole–quadrupole interactions which vary as $1/r^4$ for a fixed orientation and from quadrupole–quadrupole interactions which vary as $1/r^5$ for a fixed orientation. Naturally the proportionality factor depends upon the relative orientation of the two molecular axes. These contributions can be significant for some interactions and they are given further attention in Chapter 2. When the orientational averaging procedure described above is carried out for these two contributions the results are

$$\langle U_{el}\rangle_{\mu\Theta} = -\frac{1}{r^8 kT(4\pi\varepsilon_0)^2}\{\mu^2\Theta'^2 + \mu'^2\Theta^2\} + \ldots$$

$$\langle U_{el}\rangle_{\Theta\Theta'} = -\frac{14}{5}\frac{\Theta^2\Theta'^2}{r^{10} kT(4\pi\varepsilon_0)^2} + \ldots$$

These expressions show that these two contributions to the orientation-averaged electrostatic interaction energy are also attractive. This is a feature common to all higher terms in the expansion.

1.4.2. *Induction energy*

When any molecule is placed in a uniform, static electric field $\mathscr{E}$ there is a polarization of its charge distribution as shown schematically in Fig. 1.4. This polarization can be expressed in terms of induced multipole moments in the molecule, according to the ideas of the previous section.

FIG. 1.4. The process of induction.

If the electric field is small then it is found that the dipole moment induced in a molecule, $\boldsymbol{\mu}_{\text{ind}}$, is proportional to the field and we write

$$\boldsymbol{\mu}_{\text{ind}} = \alpha \boldsymbol{\mathscr{E}} \tag{1.11}$$

where α is known as the static polarizability of the molecule.†

The energy of a neutral dipolar molecule (dipole moment $\boldsymbol{\mu}$) in an electric field, $\boldsymbol{\mathscr{E}}$ is

$$U = -\int_0^{\mathscr{E}} \boldsymbol{\mu} \cdot \mathrm{d}\boldsymbol{\mathscr{E}}. \tag{1.12}$$

Thus the energy of the dipole induced in a molecule by an electric field is

$$\begin{aligned} U_{\text{ind}} &= -\int_0^{\mathscr{E}} \alpha \mathscr{E} \, \mathrm{d}\mathscr{E} \\ &= -\tfrac{1}{2}\alpha \mathscr{E}^2 \end{aligned} \tag{1.13}$$

since $\boldsymbol{\mu}$ and $\boldsymbol{\mathscr{E}}$ are parallel.

We have already seen that a neutral molecule possessing a permanent dipole moment, $\boldsymbol{\mu}$ gives rise to an electrostatic potential Φ and consequently to an electric field, $\boldsymbol{\mathscr{E}} = -(\partial\Phi/\partial\mathbf{r})$.‡ This electric field therefore induces a dipole moment in a second molecule which is nearby whether the second molecule is itself polar or not. We consider here an

† In general the polarizability of molecules is anisotropic and frequency-dependent, as will be discussed in Chapter 2.

‡ $\partial/\partial\mathbf{r}$ represents the vector operator $\mathbf{i}(\partial/\partial x)+\mathbf{j}(\partial/\partial y)+\mathbf{k}(\partial/\partial z)$ where x, y, z form a Cartesian coordinate system or $\mathbf{i}(\partial/\partial r)+\mathbf{j}(1/r)(\partial/\partial\theta)+\mathbf{k}(1/r\sin\theta)(\partial/\partial\phi)$, where r, θ, ϕ are spherical polar coordinates.

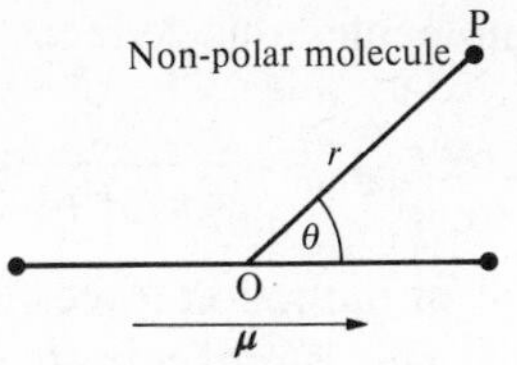

FIG. 1.5. Interaction of a dipolar molecule with a non-polar molecule.

illustrative example of the induction energy for the interaction between a molecule with a permanent dipole moment and a non-polar molecule. We consider the configuration shown in Fig. 1.5 where the non-polar molecule with a static polarizability α' lies at P.

The electrostatic potential due to the dipole at P is from (1.4)

$$\Phi = \frac{1}{4\pi\varepsilon_0}\frac{\mu\cos\theta}{r^2}$$

$$\mathscr{E} = |\mathscr{E}| = \left[\left(\frac{\partial\Phi}{\partial r}\right)^2 + \left(\frac{1}{r}\frac{\partial\Phi}{\partial\theta}\right)^2\right]^{\frac{1}{2}} = \frac{\mu}{4\pi\varepsilon_0 r^3}[4\cos^2\theta + \sin^2\theta]^{\frac{1}{2}}. \quad (1.14)$$

The magnitude of the dipole induced in the molecule at P is thus $\alpha'\mathscr{E}$ and the energy is $-\frac{1}{2}\alpha'\mathscr{E}^2$ so that the interaction energy of induction is

$$U_{\text{ind}} = -\tfrac{1}{2}\alpha'\frac{\mu^2}{r^6}\frac{(3\cos^2\theta + 1)}{(4\pi\varepsilon_0)^2} \quad (1.15)$$

which is attractive for all configurations and inversely proportional to the sixth power of the intermolecular separation for a fixed orientation. If the potential energy of (1.15) is averaged over all possible orientations, giving each orientation the Boltzmann weight, $e^{-U_{\text{ind}}/kT}$, the average induction energy between a dipolar molecule and a non-polar molecule is, at sufficiently high temperatures,

$$\langle U_{\text{ind}}\rangle = \frac{-\mu^2\alpha'}{(4\pi\varepsilon_0)^2 r^6}. \quad (1.16)$$

This leading term in the orientation-averaged energy is not temperature-dependent unlike the corresponding term for the electrostatic energy.†

In the case of the interaction of two polar molecules each molecule induces a dipole moment in the other so that there are two contributions to the total interaction energy of the form of (1.15). The orientationally-averaged induction energy for two dipolar molecules is thus

$$\langle U_{\text{ind}}\rangle_{\mu\mu'} = -\frac{1}{(4\pi\varepsilon_0)^2 r^6}\{\mu^2\alpha' + \mu'^2\alpha\} + \ldots \quad (1.17)$$

† The origins of this difference in behaviour are described in Chapter 2.

For two identical polar molecules this reduces to

$$\langle U_{\text{ind}}\rangle_{\mu\mu} = \frac{-2\alpha\mu^2}{(4\pi\varepsilon_0)^2 r^6} + \dots$$

The general treatment of induction forces is very much more complicated than that given here. This is partly because for many molecules the polarizability is not isotropic but is tensorial in character. In addition, higher-order multipoles in polar molecules can induce higher-order multipoles in other molecules and there can be other contributions to the induced dipole moment. Thus interactions such as permanent quadrupole-induced dipole and permanent quadrupole-induced quadrupole contribute to the induction energy. A brief discussion of these effects is given in Chapter 2.

1.4.3. Dispersion energy

So far we have been able to analyse the contributions to the long-range energy by means of classical electrostatics and classical mechanics. For the interaction of two molecules possessing no permanent electric dipole or higher-order moments, electrostatic and induction contributions are absent and the interaction arises solely from the dispersion energy. The dispersion energy cannot be analysed by classical mechanics since as we shall see its origins are purely quantum mechanical.

In § 1.3 it was argued that the dispersion energy arises from the interaction of instantaneous dipoles in the molecules. Here we therefore take the simplest possible model of a molecule which allows for the occurrence of such dipoles. The model we employ is a simplified version of that due originally to Drude.[11] We suppose that each molecule is composed of two charges $+Q$ and $-Q$. We imagine the charge $+Q$ to be stationary and that the negative charge oscillates about the positive charge with an angular frequency ω_0 in the z-direction which is along the line joining the positive charges of the two molecules as shown in Fig. 1.6. If we denote the displacement of the negative charge of molecule a from its positive charge by z_a and the corresponding displacement for molecule b by z_b, we note that at any time t the molecules possess instantaneous dipole moments $\mu_a = Qz_a(t)$ and $\mu_b = Qz_b(t)$. On average, of course, the molecules possess no dipole moment. We see then that the model, though

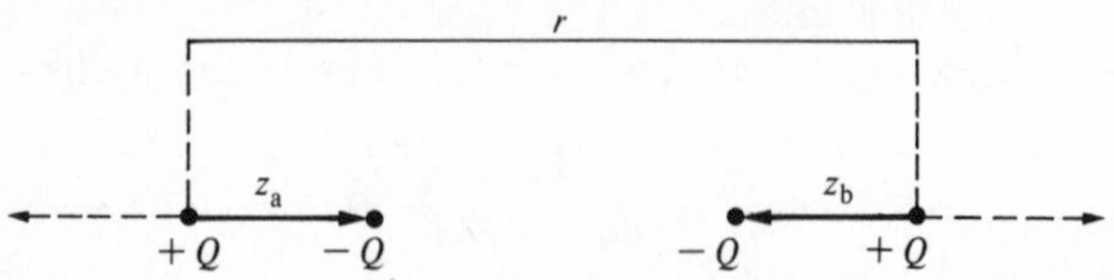

FIG. 1.6. The one-dimensional Drude model of the dispersion interaction.

an obvious approximation, has the essential features we require for a discussion of dispersion forces, despite its simplicity. Finally we denote the force constant of the harmonic oscillator by k and the mass of the oscillating charge by M; the frequency of the oscillation is thus

$$\omega_0 = (k/M)^{\frac{1}{2}}. \tag{1.18}$$

When the two molecules are infinitely separated, the Schrödinger wave equation for molecule a is

$$\frac{1}{M}\frac{\partial^2 \Psi_a}{\partial z_a^2} + \frac{2}{\hbar^2}(E_a - \tfrac{1}{2}kz_a^2)\Psi_a = 0$$

which is the equation for a simple harmonic oscillator,[12] where $\frac{1}{2}kz_a^2$ is the potential energy of the oscillator. The eigenvalues of the energy for molecules a and b are given by

$$E_a = (n_a + \tfrac{1}{2})\hbar\omega_0, \qquad E_b = (n_b + \tfrac{1}{2})\hbar\omega_0. \tag{1.19}$$

Thus, when the two molecules are infinitely separated and both are in their ground states the total energy of the two-molecule system is

$$E(\infty) = E_a + E_b = \hbar\omega_0. \tag{1.20}$$

When the molecules are separated by a finite distance r, which is still considered to be large by comparison with the dimensions of the molecules, there is an energy of interaction between the two dipoles ($\mu = z_a Q$ and $\mu' = z_b Q$) at any instant given by (1.7). Thus the Schrödinger wave equation for the two-molecule system reads

$$\frac{1}{M}\frac{\partial^2 \Psi}{\partial z_a^2} + \frac{1}{M}\frac{\partial^2 \Psi}{\partial z_b^2} + \frac{2}{\hbar^2}\left(E - \frac{1}{2}kz_a^2 - \frac{1}{2}kz_b^2 - \frac{2z_a z_b Q^2}{4\pi\varepsilon_0 r^3}\right)\Psi = 0 \tag{1.21}$$

where Ψ is the wave function for the two-molecule system. If we make the transformations

$$Z_1 = \frac{z_a + z_b}{\sqrt{2}}, \qquad Z_2 = \frac{z_a - z_b}{\sqrt{2}},$$

(1.21) can be written in the form

$$\frac{1}{M}\frac{\partial^2 \Psi}{\partial Z_1^2} + \frac{1}{M}\frac{\partial^2 \Psi}{\partial Z_2^2} + \frac{2}{\hbar^2}\left[E - \frac{1}{2}k_1 Z_1^2 - \frac{1}{2}k_2 Z_2^2\right]\Psi = 0 \tag{1.22}$$

where

$$k_1 = k - \frac{2Q^2}{4\pi\varepsilon_0 r^3} \quad \text{and} \quad k_2 = k + \frac{2Q^2}{4\pi\varepsilon_0 r^3}. \tag{1.23}$$

Equation (1.22) is the wave equation for two independent simple harmonic oscillators in the coordinates Z_1 and Z_2. Thus the eigenvalues for the total energy of the system are

$$E(r)=(n_1+\tfrac{1}{2})\hbar\omega_1+(n_2+\tfrac{1}{2})\hbar\omega_2,$$

and, if we consider the two molecules to be in their ground states,

$$E(r)=\tfrac{1}{2}\hbar(\omega_1+\omega_2). \tag{1.24}$$

Here

$$\omega_1=(k_1/M)^{\frac{1}{2}}, \qquad \omega_2=(k_2/M)^{\frac{1}{2}}$$

or

$$\omega_1=\omega_0\left\{1-\frac{2Q^2}{4\pi\varepsilon_0 r^3 k}\right\}^{\frac{1}{2}}, \qquad \omega_2=\omega_0\left\{1+\frac{2Q^2}{4\pi\varepsilon_0 r^3 k}\right\}^{\frac{1}{2}} \tag{1.25}$$

from (1.23). Now we are interested in the long-range interaction for which the perturbation potential is small so we can expand (1.25) by the binomial theorem which allows us to write (1.24) as

$$E(r)=\hbar\omega_0-\frac{Q^4\hbar\omega_0}{2(4\pi\varepsilon_0)^2 r^6 k^2}+\dots. \tag{1.26}$$

The energy of interaction of the two molecules for our model is thus

$$U_{\text{disp}}=E(r)-E(\infty)=-\frac{Q^4\hbar\omega_0}{2(4\pi\varepsilon_0)^2 r^6 k^2}+\dots. \tag{1.27}$$

A better, but still not exact, model of each molecule can be constructed by assuming that the motion of the oscillating charge can be resolved into three oscillations of identical frequency along three Cartesian coordinates centred on the positive charge. In this case (1.27) is modified to read

$$U_{\text{disp}}=\frac{-3Q^4\hbar\omega_0}{4(4\pi\varepsilon_0)^2 r^6 k^2}+\dots. \tag{1.28}$$

The force constant, k, can be simply related to the polarizability of the molecules. If we expose a single Drude molecule to an external electric field $\mathscr{E}$ a force of magnitude $Q\mathscr{E}$ acts on each charge to produce a displacement z'_a which attains a static value when the restoring force kz'_a is equal to the imposed electrical force. Then

$$Q=kz'_\text{a}/\mathscr{E}$$

so that the static dipole moment induced in the molecule by the field is $\mu_{\text{ind}}=Qz'_\text{a}=Q^2\mathscr{E}/k$.

From (1.11) we can therefore identify the polarizability α as

$$\alpha = Q^2/k.$$

so that we can write the dispersion energy for two identical molecules in their ground states for this simple model as

$$U_{\text{disp}} = C_6/r^6 \tag{1.29}$$

where

$$C_6 = -\frac{3}{4}\frac{\alpha^2\hbar\omega_0}{(4\pi\varepsilon_0)^2}. \tag{1.30}$$

The dipole–dipole dispersion energy for two molecules in their ground states is therefore attractive and inversely proportional to the sixth power of the intermolecular separation. It is worthwhile noting here that had we treated the same problem classically we should have obtained zero interaction energy. This is because the ground-state energy of a classical simple harmonic oscillator is zero so that both $E(r)$ and $E(\infty)$ of (1.20) and (1.24) would have been zero. The existence of the dispersion energy is therefore a consequence of the zero-point energy of the oscillators, a purely quantum mechanical concept. For non-spherical molecules the dispersion energy is a function of relative orientation as well as of separation.

As we might expect in view of our consideration of electrostatic and induction interactions there are additional contributions to the dispersion energy arising from instantaneous dipole–quadrupole interactions, quadrupole–quadrupole interactions, etc. The dispersion energy is therefore more completely written as

$$U_{\text{disp}} = \frac{C_6}{r^6} + \frac{C_8}{r^8} + \frac{C_{10}}{r^{10}} + \ldots \tag{1.31}$$

where, for interactions between ground-state molecules, each coefficient is negative so that each contribution is attractive.

London[13] first estimated the magnitude of the coefficient C_6 by means of the oscillator model we have described. He assumed that the frequency ω_0 of the oscillator model was that corresponding to the energy required for ionization of the molecule from its ground state, $E_{\text{I}} = \hbar\omega_0$. This assignment was supported by the fact that an identical assumption allowed a reasonable description of the variation of refractive index of a gas with the frequency of electromagnetic radiation (the effect of dispersion). With this assignment we can write (1.30) for C_6 in the form

$$C_6 = -\frac{3}{4}\frac{\alpha^2 E_{\text{I}}}{(4\pi\varepsilon_0)^2} \tag{1.32}$$

which contains only experimentally accessible quantities. In this way London obtained a value of C_6 for the interaction of two argon atoms as -5×10^{-78} J m^6 which is only some 30 per cent smaller than the value obtained by the most refined calculations of recent years.

The most recent evaluations of C_6 are based upon a full quantum mechanical treatment of dispersion energy which will be discussed in Chapter 2. This analysis confirms the form of (1.31) but leads to different expressions for the coefficients C_6, C_8, C_{10} from those of our simple model. For example, for the interaction of two ground-state atoms the coefficient C_6 depends upon the energies of transition from the ground electronic state to all excited states and upon the probabilities for these transitions, the so-called *oscillator strengths.* The former can be determined by spectroscopic methods and the latter can in principle be obtained from photoabsorption studies. Consequently C_6 can again be computed with the aid of experimental information. However, in practice, the available photoabsorption data are insufficient for an accurate determination of C_6 and supplementary experimental data must be employed. All of these techniques are discussed in Chapter 2.

1.4.4. *Long-range energy: summary*

(a) *Order of magnitude.* For the most general case of the interaction of two polar molecules the total long-range intermolecular energy is

$$U = U_{\text{el}} + U_{\text{ind}} + U_{\text{disp}}.$$

For two identical, neutral molecules, free to rotate with a dipole moment $\boldsymbol{\mu}$ and static polarizability α, the leading contributions are

$$U = -\frac{1}{(4\pi\varepsilon_0)^2}\left\{\frac{2}{3}\frac{\mu^4}{kT} + 2\mu^2\alpha + \frac{3}{4}\alpha^2\hbar\omega_0\right\}r^{-6}$$

where for simplicity we have used the Drude result for the dispersion contribution. Table 1.1 contains a list of the magnitudes of the three coefficients of the r^{-6} term for the different contributions to the interaction of like pairs of simple molecules for a temperature of 300 K. The table also provides typical values of dipole moments and polarizabilities of molecules.†

With the exception of water, a small, highly polar molecule, the dispersion contribution to the energy is seen to be the major contribution to the long-range energy. Of the other two contributions the induction energy is always small.

In order to compare the magnitude of the various attractive contributions to the intermolecular energy we have calculated them for 1 mole of

† Dipole moments are traditionally measured in Debye units; here, for consistency we quote them in coulomb metres, the unit of the rationalized SI system. 1 Debye = $3{\cdot}33\times10^{-30}$ C m.

Table 1.1

Molecule	$10^{30}\mu$/C m	$10^{30}\alpha(4\pi\varepsilon_0)^{-1}$/m^3	−Coefficient of $r^{-6}/10^{-79}$ J m^6 Electrostatic	Induction	Dispersion
Ar	0	1·63	0	0	50
Xe	0	4·0	0	0	209
CO	0·4	1·95	0·003	0·06	97
HCl	3·4	2·63	17	6	150
NH_3	4·7	2·26	64	9	133
H_2O	6·13	1·48	184	10	61

a gas on the assumption that the molecules interact in pairs, at a separation, σ, where the *total* potential energy $U(\sigma)=0$. These values for several species are tabulated in Table 1.2 where they are compared with the bond energies of the polyatomic species and the enthalpy of sublimation. Also, it is worthwhile pointing out for comparison that the kinetic energy of one mole of gas ($\frac{3}{2}RT$) at 300 K is 3·7 kJ mol^{-1}, and that the hydrogen bond familiar to chemists has a typical magnitude of 20 kJ mol^{-1}. ΔH_s° is an approximate value for the heat of sublimation of a crystal. Table 1.2 shows that the attractive component of the intermolecular pair potential at σ is very much smaller than the energy required to rupture chemical bonds. However, the same energy is comparable with the kinetic energy of molecules near room temperature. The hydrogen bond has an energy intermediate between that typical of chemical bonds and the attractive component of the intermolecular potential. It can also be seen that the heat of sublimation of the crystals increases as the total attractive contribution to the pair potential function increases, indicating the relationship of intermolecular potential energies to solid-state properties.

(b) *Limitations.* To conclude our discussion of the long-range energy we should emphasize that we have considered only a very special

Table 1.2

Molecule	σ/nm	Attractive energy contributions/kJ mol^{-1} U_{el}	U_{ind}	U_{disp}	Single bond energy/ kJ mol^{-1}	ΔH_s°/ kJ mol^{-1}
Ar	0·33	0	0	−1·2	—	7·6
Xe	0·38	0	0	−1·9	—	16
CO	0·36	-4×10^{-5}	-8×10^{-4}	−1·4	343	6·9
HCl	0·37	−0·2	−0·07	−1·8	431	18
NH_3	0·260	−6·3	−0·9	−13·0	389	29
H_2O	0.265	−16·0	−0·9	−5·3	464	47

set of conditions in our treatment. The description of the electrostatic energy has been confined to axially-symmetric charge distributions; the discussion of induction energy has been restricted to molecules with an isotropic polarizability; and the treatment of the dispersion energy limited to that for the interaction of spherically symmetric atoms in their ground states. A complete treatment of the general case is obviously more complicated. For example, as was mentioned earlier, for the interaction of two atoms not in their ground states the dispersion energy is not always attractive. Furthermore, for interactions between monatomic species not in S-states, it is possible that high-order permanent multipole interactions provide the dominant contribution to the long-range energy. For instance for the interaction of two chlorine atoms in their ^2P ground state the leading contribution to the long-range energy is the first-order electrostatic quadrupole–quadrupole term which varies as $1/r^5$. These complications will not concern us here, but it is necessary to recognize that not all systems are of the simplest type.

Finally, one other limitation implicit in our discussion of the long-range dispersion energy must be mentioned. The electromagnetic nature of the forces between molecules implies that the transmission of the force from one molecule to another takes place at the speed of light. If we consider the Drude model of the dispersion interaction, we see that it will take a non-zero time for the oscillating electric field of one dipole to be transmitted to the second molecule. Thus, by the time the field acts at the second molecule the dipole in the first molecule will have changed. This so-called retardation effect upon the dispersion energy reduces the intermolecular energy below that given by our treatment which neglected this time lag. Fortunately however the retardation effect is only significant for intermolecular separations of the order of 30 nm, that is about 100 molecular diameters. At this separation the intermolecular forces are too small to influence any of the properties discussed in this book. However, retardation effects are important in the treatment of intermolecular interactions between macroscopic bodies, between surfaces, and for the interaction of a single particle with a solid surface.[14]

(c) *Extensions.* In the preceding section we have only considered the interaction between pairs of molecules. Inasmuch as the majority of the book is devoted to the methods employed to determine this pair-interaction potential for various molecular systems this is justified. However, it is worthwhile pointing out that the total interaction energy of a group of more than two molecules is not equal to the sum of all the pairwise interaction energies. In the case of the long-range dispersion energy it has been possible to evaluate the non-additive contribution for the interaction of three atoms and to show that it is frequently significant.

The dispersion energy of three atoms a, b, and c can in fact be written as

$$U_{\text{disp}}^{\text{abc}}(r_{\text{ab}}, r_{\text{ac}}, r_{\text{ac}}) = \frac{C_6^{\text{ab}}}{r_{\text{ab}}^6} + \frac{C_6^{\text{bc}}}{r_{\text{bc}}^6} + \frac{C_6^{\text{ac}}}{r_{\text{ac}}^6} + \frac{\nu_{\text{abc}}(3\cos\theta_{\text{a}}\cos\theta_{\text{b}}\cos\theta_{\text{c}}+1)}{(r_{\text{ab}}r_{\text{bc}}r_{\text{ac}})^3}$$

when dipole–dipole interactions only are included. Here, r_{ab}, r_{bc}, and r_{ac} represent the distances between the atoms, while θ_{a}, θ_{b}, and θ_{c} are the internal angles of the triangle they form. The first three terms in this energy correspond to the addition of the intermolecular energies of pairs of molecules. The extra, non-additive term is known as the *Axilrod–Teller triple-dipole term.* This non-additive contribution is positive for equilateral triangular arrangements of atoms, and is negative for linear arrangements which are therefore stabilized compared to the pairwise additive energy. The Axilrod–Teller energy is merely the leading term of a series, the terms of which will be discussed in more detail in Chapter 8.

1.5. Short-range energy

The theoretical treatment of the short-range repulsive forces between molecules has proved much more difficult than the analysis of the long-range forces. The physical origin of these forces described in § 1.3 makes it clear why this is so. For short-range forces we are concerned with the situation when the electron clouds of two molecules overlap. The repulsive force between the molecules arises partly from the incompletely screened electrostatic repulsion of the nuclei and partly from the repulsion between the electrons of the molecules. Thus on the one hand the perturbation to the electron distribution of the separated molecules is large and on the other hand the multipole expansion employed for the long-range forces is not valid. For both of these reasons an initial unperturbed state made up from the completely separated molecules is not satisfactory for the treatment of these forces. As a consequence a variety of procedures have been developed to attempt the evaluation of the short-range forces. Until now no method has proved completely successful so that our treatment of these forces here must be much less than complete. Here we give a brief description of one approach to the evaluation of the short-range forces known as the Heitler–London or valence-bond method.[12] This method, although it has not proved the most successful method in practice, has the advantage that the physical origins of the repulsive forces are made clear. We shall employ the method to discuss the interaction of the simplest molecular system involving the interaction of two hydrogen atoms.

Fig. 1.7 shows the two hydrogen atoms at a separation r and defines the coordinates of the system. We assume that when the two atoms are

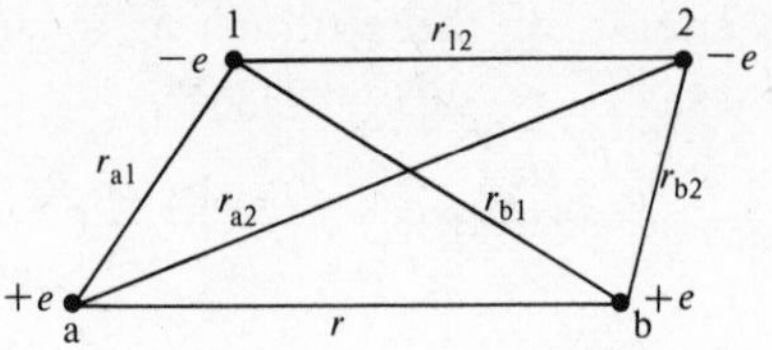

FIG. 1.7. The interaction of two hydrogen atoms.

infinitely separated electron 1 is associated with nucleus a. The Hamiltonian, H for the system can be written

$$H = H_a + H_b + V_e$$

where V_e is the electrostatic energy which arises from the interaction of the two atoms and is given by

$$V_e = -\frac{e^2}{4\pi\varepsilon_0}\left\{\frac{1}{r_{a2}} + \frac{1}{r_{b1}} - \frac{1}{r_{12}} - \frac{1}{r}\right\} \tag{1.33}$$

and H_a and H_b are the Hamiltonians of the isolated atoms. If the wave function for the interacting system is Ψ then the energy will be given by

$$\frac{\iint \Psi^* H \Psi \, d\tau_1 \, d\tau_2}{\iint \Psi^* \Psi \, d\tau_1 \, d\tau_2} = E_a + E_b + U \tag{1.34}$$

where the integrations extend over the spatial coordinates of both electrons. Here E_a and E_b are the energies of the isolated atoms and U is the intermolecular potential energy. For the hydrogen atoms in their ground state the energies E_a and E_b are known, as are the corresponding wave functions which we denote by $A(1)$ and $B(2)$. Thus, in order to evaluate the intermolecular energy, U, it is necessary only to obtain the wave function Ψ and carry out the integrations in (1.34). Unfortunately, it is not possible to solve the Schrödinger wave equation to yield the wave function Ψ for the system. Consequently it is necessary to generate suitable approximate wave functions.

When the two atoms are infinitely separated a suitable wave function Ψ would be $\Psi = A(1)B(2)$. However when the atoms are separated by a finite distance r this function is inadequate since it contains no provision for electron 1 to be associated with nucleus b and electron 2 to be associated with nucleus a. Furthermore, this wave function does not possess the correct symmetry required by the Pauli principle for the exchange of electrons. The simplest two wave functions which do possess the proper characteristics are

$$\Psi_+ = A(1)B(2) + A(2)B(1) \tag{1.35}$$

and

$$\Psi_- = A(1)B(2) - A(2)B(1). \tag{1.36}$$

The wave function Ψ_+ is appropriate to the situation where the two electrons have different spins and Ψ_- is the wave function for the case when the electrons have the same spin. It should be noted that because of electron spin the wave function Ψ_- is triply degenerate, whereas Ψ_+ is a singlet.

We use these two wave functions as approximations to the two-atom wave function Ψ in (1.34) and obtain for the intermolecular potential energy corresponding to the two wave functions,

$$\left.\begin{aligned} U_+ &= \frac{e^2}{4\pi\varepsilon_0 r} + \frac{J' + K' - 2(J + SK)}{1 + S^2} \\ U_- &= \frac{e^2}{4\pi\varepsilon_0 r} - \frac{J' - K' - 2(J - SK)}{1 + S^2} \end{aligned}\right\} \tag{1.37}$$

where

$$\left.\begin{aligned} S &= \int A(1)B(1)\, d\tau_1 \\ J &= \frac{e^2}{4\pi\varepsilon_0} \int A(1)\left(\frac{1}{r_{b1}}\right)A(1)\, d\tau_1 \\ J' &= \frac{e^2}{4\pi\varepsilon_0} \iint A^2(1)\left(\frac{1}{r_{12}}\right)B^2(2)\, d\tau_1\, d\tau_2 \\ K &= \frac{e^2}{4\pi\varepsilon_0} \int A(1)\left(\frac{1}{r_{b1}}\right)B(1)\, d\tau_1 \\ K' &= \frac{e^2}{4\pi\varepsilon_0} \iint A(1)B(2)\left(\frac{1}{r_{12}}\right)B(1)A(2)\, d\tau_1\, d\tau_2 \end{aligned}\right\} \tag{1.38}$$

These integrals all involve the wave functions of the isolated hydrogen atoms so that each can be evaluated. The integral S, the overlap integral, measures the degree of overlap of the wave functions of the two atoms. The integral J describes the coulombic interaction between the electron 1 in the orbital $A(1)$ with nucleus b, whereas J' describes the Coulombic interaction between the two electrons. J and J' are known as Coulombic integrals. The integrals K and K' are the so-called exchange integrals. K represents a Coulombic interaction between electron 1 and nucleus b involving the wave functions of both atoms and K' represents a similar interaction between the two electrons. The existence of these two terms has led to the name exchange forces for the short-range interaction. This name is somewhat misleading since it implies that the short-range energy is due to some new kind of force. It must be emphasized that the short-range forces are still electrostatic in nature which we can see in terms of the Hellmann–Feynman theorem.[15] This states that if the Schrödinger wave equation for two interacting atoms can be solved exactly to yield a wave function for the electrons, the forces on the nuclei can be calculated by applying classical electrostatics to the resulting charge distributions. The wave functions Ψ_+ and Ψ_- here provide us with approximations to the charge distribution for the interacting hydrogen atoms, and their form is essentially determined by exchange and spin effects. The forces between the two atoms then arise from Coulombic interactions between these two charge distributions. The 'exchange' terms in the energies U_+ and U_- still represent electrostatic energies of interaction. It is therefore preferable to use the name *overlap forces* for the short-range forces between molecules.

We can use the same idea to determine the qualitative features of the interaction energies U_+ and U_-. When the wave function Ψ_+ is evaluated it corresponds to an increased electron density between the two nuclei compared with that for a direct superposition as shown in Fig. 1.8(a). In this situation the two positively charged protons are attracted towards this negatively charged region and hence there is a net attractive force on them according to classical electrostatics. This wave function therefore corresponds to that of a bonding orbital for a hydrogen molecule. The wave function Ψ_-, on the other hand, leads to a decreased electron

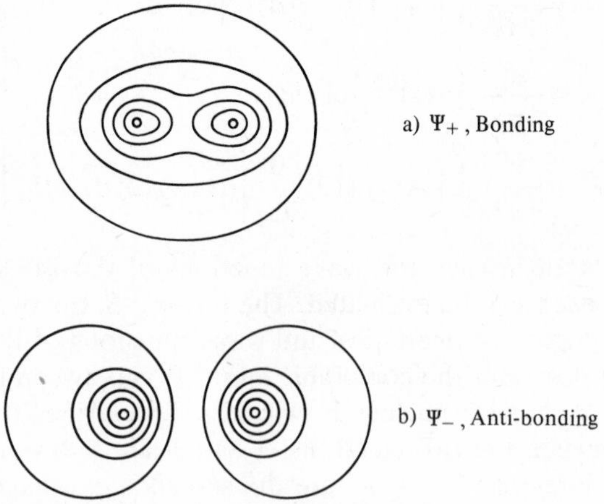

FIG. 1.8. Contours of equal electron density, in a plane containing the two nuclei, for the interaction of two hydrogen atoms.

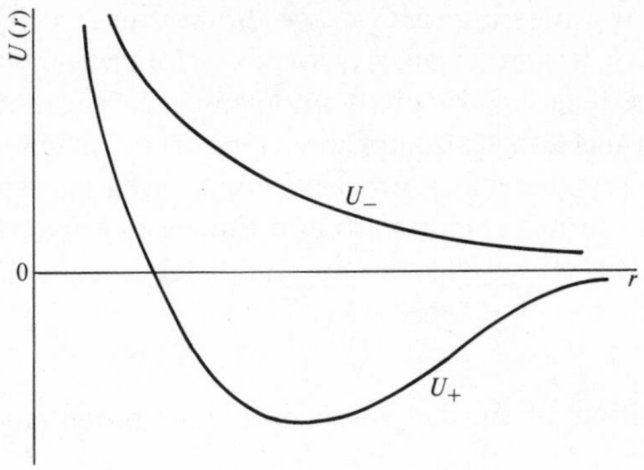

FIG. 1.9. The potential energy for the interaction of two hydrogen atoms.

density between the two nuclei as shown in Fig. 1.8(b). Thus the two nuclei are incompletely shielded from each other and an electrostatic repulsion results. The wave function Ψ_- is therefore anti-bonding.

Since the wave functions $A(1)$ and $B(2)$ for isolated hydrogen atoms are known[12] the integrals (1.38) can be evaluated and the resulting energies U_+ and U_- are sketched in Fig. 1.9 and confirm the qualitative discussion above. The energy U_+ possesses an attractive minimum region and corresponds to a stable hydrogen molecule. The energy U_- on the other hand is everywhere repulsive and because of the spin symmetry of the wave function is similar to the interaction energy for closed shell atoms. At extremely short range the energy U_- varies as $1/r$ owing to the internuclear repulsion; however, at larger separations the energy decays as e^{-2r/a_0}, where a_0 is the radius of the first Bohr orbit of the hydrogen atom. This final exponential form has been used as an analytic representation of the behaviour of short-range forces.

The Heitler–London method has been extended to the interaction of atoms containing more than one electron. However, the technique suffers from a number of disadvantages which will be discussed in Chapter 2. It is worth noting here though that the method always predicts zero dispersion energy since in the construction of the approximate wave functions $\Psi_\pm$ the atomic orbitals are not allowed to distort, a fundamental requirement for the dispersion energy. Other methods for evaluating short-range energies have been developed. Some, such as molecular orbital theory, aim at better ways of constructing the wave function Ψ for the two-atom system, whereas others use entirely different methods. These other methods are discussed in Chapter 2; however none of them is yet entirely satisfactory.

The Heitler–London method also serves to illustrate the difficulty of calculating intermolecular forces in the intermediate range of separations.

In discussing long-range interactions we did not need to include exchange effects because of the small overlap of wave functions, whereas we have seen that such effects are of great importance at short range. At intermediate separations some attempt must therefore be made to bridge this gap. No generally satisfactory procedure yet exists and so often the best that can be done in this region is to add the short-range repulsive energy to the long-range energy to obtain the total intermolecular potential energy function.

1.6. Representation of the intermolecular pair potential energy function

As mentioned in § 1.1 the difficulties of the theoretical evaluation of intermolecular pair potential energy functions led to a heuristic approach to their determination. The procedure is begun with the assumption of an analytic form for the relationship between the pair potential energy and intermolecular separation for a particular molecular species. Using this assumed form, macroscopic properties of a material composed of such molecules are calculated with the aid of the appropriate molecular theory and compared with experimental data. Agreement between calculated and experimental values for a physical property is construed as evidence for the correctness of the assumed potential energy function. It is now accepted that this procedure usually yields little more than a crude estimate of the intermolecular potential energy. However the method does provide a means of estimating physical properties of the same material for which no experimental information is available, and it is frequently used in this way today.[16,17] Consequently the analytic forms assumed for the intermolecular potential energy have not only a historical significance, but also a practical value in some instances. In general, these analytic forms for the relationship of intermolecular potential energy to separation have been formulated with some regard to the theoretical description of intermolecular forces. Normally the analytic functions are written with a number of parameters whose values are to be determined by imposing the condition that the calculated and experimental data for a particular physical property should be in agreement. Two of these parameters are almost always the depth of the potential energy well, ε and a characteristic separation, either σ or r_m of Fig. 1.1. Since these particular parameters determine the scale of the potential energy function, rather than its shape, they are distinct from other variable parameters.

1.6.1. Simple functions

The earliest and simplest representation of this kind viewed a molecule as a hard sphere of diameter σ so that the intermolecular potential energy is written

$$U(r)=\infty \qquad r\leq\sigma$$
$$U(r)=0 \qquad r>\sigma.$$

This form of potential has the single disposable parameter σ. In view of our description of the nature of intermolecular forces this model is evidently unrealistic. Nevertheless its simplicity makes it an attractive proposition for complex problems, so that it has an important place in some current efforts to describe the liquid state.

Perhaps the most frequently used model potential is that due to Lennard-Jones (LJ),[18] which we sometimes refer to as the $n-6$ potential (Appendix 1)

$$U(r)=\varepsilon\left\{\frac{6}{n-6}\left(\frac{r_{\mathrm{m}}}{r}\right)^{n}-\frac{n}{n-6}\left(\frac{r_{\mathrm{m}}}{r}\right)^{6}\right\}.$$

This function possesses the general features of the true intermolecular potential energy in that it has a repulsive short-range region joined to a long-range attractive region by a single minimum which occurs at r_{m} where the energy is $-\varepsilon$. The attractive component of the function is theoretically based on the dispersion energy contribution, but the form of the repulsive term has no theoretical justification. In this form the LJ potential has one disposable parameter n in addition to ε and r_{m}. Most often the repulsive exponent has been given the value $n=12$ and the potential is then written

$$U(r)=\varepsilon\left\{\left(\frac{r_{\mathrm{m}}}{r}\right)^{12}-2\left(\frac{r_{\mathrm{m}}}{r}\right)^{6}\right\}$$

or, equivalently,

$$U(r)=4\varepsilon\left\{\left(\frac{\sigma}{r}\right)^{12}-\left(\frac{\sigma}{r}\right)^{6}\right\}.$$

where σ $(=2^{-\frac{1}{6}}r_{\mathrm{m}})$ is the intermolecular separation for which the energy is zero. The Lennard-Jones $(12-6)$ potential has no adjustable parameters other than σ and ε, whose values can be determined by forcing agreement between experimental data for a physical property and calculated values for the potential model.

Another pair potential function occasionally employed is a modified form of that originally devised by Buckingham[19,20]

$$U(r)=\varepsilon\left\{\frac{6}{\alpha-6}\,\mathrm{e}^{-\alpha(r/r_{\mathrm{m}}-1)}-\frac{\alpha}{\alpha-6}\left(\frac{r_{\mathrm{m}}}{r}\right)^{6}\right\}.$$

This model potential also contains the leading term of the dispersion energy contribution at long range, but the repulsive contribution has the physically more realistic exponential form. In its general form the model possesses the one disposable parameter α (not to be confused with the polarizability) in addition to ε and r_m. Frequently α has been assigned a value in the range 12–15 in which case the functional form of the potential energy is fixed and only ε and r_m are allowed to vary.

We saw in § 1.4 that for the interaction of two polar molecules there was an additional electrostatic contribution to the long-range attractive energy. This was incorporated into a model potential function by Stockmayer.[21] The addition of the appropriate term to the Lennard-Jones (12−6) function yields

$$U(r, \theta_1, \theta_2, \phi) = 4\varepsilon'\left[\left(\frac{\sigma'}{r}\right)^{12} - \left(\frac{\sigma'}{r}\right)^{6}\right] - \frac{\mu^2}{4\pi\varepsilon_0 r^3}\zeta(\theta_1, \theta_2, \phi)$$

where $\zeta(\theta_1, \theta_2, \phi)$ is the function of (1.9) and Fig. 1.3(d) provides the definition of the angles of the two dipoles with respect to the line joining the centres of mass of the two molecules. The dipole moment, μ, is not a disposable parameter since the experimental value must be employed. Consequently the model has only two adjustable parameters ε' and σ' which, although they do not correspond to the minimum energy and the separation for zero energy are still characteristic of the interaction.

Many other simple functional forms for the intermolecular pair potential energy have been employed and a list of the more important ones is given in Appendix 1. None of them has been able to satisfactorily reproduce the experimental data for all of the properties of any material over a wide temperature range, which demonstrates that they are not good descriptions of the true intermolecular pair potential energy function.

1.6.2. *More flexible analytic functions*

Further analytic functions for the intermolecular pair potential energy have been employed, which incorporate a great deal more flexibility than those described above. This flexibility has been achieved by including disposable parameters whose values are determined by the requirement that the intermolecular pair potential should permit the representation of all the available experimental data for the properties of a particular material. An example of this type of function is that employed by Barker and Pompe,[22]

$$\frac{U(\bar{r})}{\varepsilon} = \exp[\alpha(1-\bar{r})]\sum_{i=0}^{5} A_i(\bar{r}-1)^i - \sum_{j=0}^{2} C_{2j+6}/(\delta+\bar{r})^{2j+6}$$

where $\bar{r} = r/r_m$.

Apart from the *scaling* parameters ε and r_m this function contains eleven disposable *form* parameters. The use of such a function clearly frees the fitting procedure from many of the constraints imposed by the choice of a simple analytic function, since the form of the potential can now be changed considerably in addition to its scale. Indeed, such functions have been employed with some success to determine the intermolecular pair potential for the interaction of simple species. Further examples can be found in Appendix 1.

1.6.3. Representational functions

A final group of functions have been devised solely to provide a convenient analytic representation of directly determined pair potential energy-separation data. Of these the simplest is the $n(\bar{r})-6$ function[23]

$$\frac{U(\bar{r})}{\varepsilon}=\left\{\left(\frac{6}{n-6}\right)\bar{r}^{-n}-\left(\frac{n}{n-6}\right)\bar{r}^{-6}\right\}$$

where n is a function of $\bar{r}$. Most commonly $n=13{\cdot}0+\gamma(\bar{r}-1)$, $\bar{r}=r/r_m$.

Although it contains only one adjustable form parameter, γ, in addition to the scale parameters ε and r_m the function provides an adequate description of the intermolecular potential energy-separation data for interactions among the monatomic gases.

1.6.4. Non-spherical molecules

The representation of the intermolecular pair potential energy function for non-spherical molecules presents additional problems since the energy now depends both on the separation and on the relative orientation of the pair of molecules. The multidimensional potential energy function for such systems may be represented in several ways.[24] A large number of pair functions might be used, one representing the energy as a function of separation at fixed relative orientations, but this does not provide a very convenient basis for calculating the pair energy at an arbitrary relative orientation and separation.

Two other approaches have been widely used. In the first, the orientation-dependent function, $U(r, \boldsymbol{\omega}_1, \boldsymbol{\omega}_2)$ is expanded as a series

$$U(r, \boldsymbol{\omega}_1, \boldsymbol{\omega}_2)=U_0(r)+\sum_{n=1}^{\infty} U_n(r) f_n(\boldsymbol{\omega}_1, \boldsymbol{\omega}_2).$$

Here the angular terms $f_n(\boldsymbol{\omega}_1, \boldsymbol{\omega}_2)$ are known standard functions of the molecular orientations $\boldsymbol{\omega}_1$, $\boldsymbol{\omega}_2$ such as Legendre polynomials or spherical harmonics and their 'coefficients' $U_n(r)$ are functions only of the separation of the molecules. The first term, $U_0(r)$, is the unweighted orientation average of the full potential. Expansions of this type are in principle exact, but their practical utility relies on the rapid convergence of the series, which does not always occur.

In the second, approximate, approach an attempt is made to decompose the energy for a pair of molecules into a sum of interactions between 'interaction sites', often coinciding with the atomic nuclei, on the different molecules. This method requires the characterization of site–site potential functions, which depend only on the separation of the sites for each type of pair interaction which may occur. The conceptual simplifications of this approach are evident, but the decomposition is approximate and consequently the determination of the site–site potential functions is difficult. One of the simplest of these site–site models, which has been used to describe diatomic molecule interactions is the diatomic Lennard-Jones 12−6 model. Here each atom of the diatom is a site for a L-J 12−6 interaction with all other sites. The potential function for the two-molecule interaction is written

$$U = 4\varepsilon \sum_{l=1}^{2} \sum_{m=1}^{2} \left\{ \left(\frac{\sigma}{r_{lm}}\right)^{12} - \left(\frac{\sigma}{r_{lm}}\right)^{6} \right\}$$

where r_{lm} is the separation of the lth and mth interaction sites.

1.7. Sources of information about intermolecular forces

Earlier in this chapter we established, in general terms, that there was a relationship between the macroscopic properties of matter, as well as any observable microscopic properties, and the intermolecular potential energy function. In this section we make the relationship explicit for those properties which have proved of the greatest value in the determination of intermolecular forces between simple molecules. We also illustrate the ways in which experimental information on these properties can be used to determine the pair potential energy function. In particular we emphasize the type of information about the intermolecular potential which each property contains in principle and the range of intermolecular separations to which available experimental data are relevant. All observable properties depend upon the entire pair potential energy function over the complete range of separation, so that these connections, although useful, are not exact.

1.7.1. Gas imperfection

The theory of equilibrium statistical mechanics leads to the result that the equation of state for a gas can be written in the form of a virial expansion

$$\frac{P\tilde{V}}{RT} = 1 + \frac{B(T)}{V} + \frac{C(T)}{\tilde{V}^2} + \ldots$$

where P is the pressure of the gas, T its temperature, and $\tilde{V}$ its molar volume. In the limit of an infinitely dilute gas ($\tilde{V} \to \infty$) this reduces to the perfect gas equation of state which corresponds to a gas whose molecules are point masses. The higher-order terms in the expansion therefore represent the effect of the intermolecular interactions. The second virial coefficient B expresses the effect of interactions of pairs of molecules upon the pressure of the gas, the third virial coefficient C the effects of interactions of groups of three molecules, etc. The second virial coefficient is therefore related to the intermolecular pair potential energy function and for molecules which interact through a spherically symmetric potential is given by (see § 3.4)

$$B(T) = -2\pi N_A \int_0^{\infty} [e^{-U(r)/kT} - 1]r^2 \, dr. \tag{1.39}$$

Thus, given $U(r)$ the calculation of $B(T)$ is straightforward, and (1.39) illustrates how the traditional method of testing guessed intermolecular potential functions outlined in the previous section could be applied. The form of the variation of $B(T)$ with temperature for a typical intermolecular pair potential energy function is sketched in Fig. 1.10. Eqn (1.39) makes it clear that the second virial coefficient at a particular temperature depends, in principle at least, upon the entire potential energy function. The second virial coefficient of a gas such as argon can be measured over a range of temperature from approximately $kT/\varepsilon = 0{\cdot}5$ to $kT/\varepsilon = 5$, where ε is the depth of the potential, by determining its low-density (PVT) properties, so that an experimental curve such as that of Fig. 1.10 can be obtained. To see the type of information which such data contain about the intermolecular pair potential we can write (1.39) in the equivalent form (see § 3.5)

$$B(T) = \frac{2\pi N_A e^{\varepsilon/kT}}{3T} \int_0^{\varepsilon} \Delta e^{-\phi/T} \, d\phi + \frac{2\pi N_A e^{\varepsilon/kT}}{3T} \int_{\varepsilon}^{\infty} r^3 e^{-\phi/T} \, d\phi$$

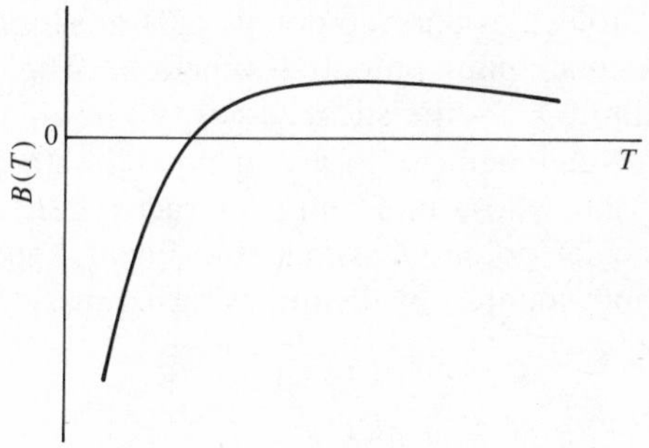

FIG. 1.10. The second virial coefficient, $B(T)$ as a function of temperature.

where $\phi(r)=(U(r)+\varepsilon)/k$ and $\Delta=r_L^3-r_R^3$, r_L and r_R being the separation coordinates of the inner and outer walls of the potential energy well at the energy ϕ. The first term corresponds to the contribution of the potential well to the virial coefficient at a temperature T whereas the second term arises from the repulsive forces for $r<\sigma$. At high temperatures the first term is negligible by comparison with the second and $B(T)$ depends only on the repulsive region of the potential $U(r)\gg 0$. Measurements of $B(T)$ in this region can be used to determine $U(r)$ explicitly. Unfortunately the temperature range of the measurements restricts this method to the low-energy repulsive region of the potential only. At low temperatures $B(T)$ is dominated by the first term and measurements provide information about the width of the potential well as a function of energy. However this limited examination does not take into account the continuity of the intermolecular potential energy function and its derivatives through the separation at which the potential changes sign. More recent treatments indicate that because of this continuity $B(T)$ is capable of yielding even more explicit information about $U(r)$ in the well region, as will be shown in Chapter 3.

For the interaction of polyatomic molecules which generally have a non-spherically symmetric potential energy function, $B(T)$ is given by an expression which also involves integration over the orientations of the two molecules. For this reason less information has so far been obtained about the interaction of polyatomic species from their second virial coefficients.

1.7.2. Transport properties of dilute gases

The transport of momentum, energy, and mass through a dilute gas under the influence of a gradient of velocity, temperature, and concentration are effected by molecular motion. The ease with which this motion takes place is determined by the scattering of molecules by others in the gas. Because the gas is dilute this scattering occurs as a result of collisions between pairs of molecules and the details of it are therefore determined by the intermolecular pair potential function. The transport coefficients of a pure dilute gas—the shear viscosity, the thermal conductivity, and the self-diffusion coefficient—are hence directly related to $U(r)$. For a monatomic gas whose molecules interact according to the general form of intermolecular potential sketched in Fig. 1.1 the transport coefficients are given approximately by the following expressions (see Chapter 5): For viscosity

$$\eta=\frac{5}{16}\left(\frac{mkT}{\pi}\right)^{\frac{1}{2}}\frac{1}{\sigma^2\Omega^{(2,2)*}(T^*)},$$

For thermal conductivity

$$\lambda = \frac{75}{64}\left(\frac{k^3T}{m\pi}\right)^{\frac{1}{2}} \frac{1}{\sigma^2\Omega^{(2,2)*}(T^*)},$$

For self-diffusion

$$D = \frac{3}{16}\left(\frac{kT}{m\pi}\right)^{\frac{1}{2}} \frac{1}{n\,.\,\sigma^2\Omega^{(1,1)*}(T^*)},$$

where m is the mass of the molecules, n their number density, and $T^* = kT/\varepsilon$. In the case of rigid spherical molecules of diameter σ, the 'reduced collision integrals' $\Omega^{(1,1)*}$ and $\Omega^{(2,2)*}$ are unity and independent of temperature. The transport coefficients then reduce to the familiar forms resulting from the simple kinetic theory of gases. For a realistic intermolecular potential energy function the reduced collision integrals are determined by the dynamics of binary collisions between molecules and hence depend upon the intermolecular potential function itself. All of the information about the intermolecular potential is therefore contained in the collision integrals. Of the transport coefficients the viscosity has proved the easiest to measure accurately either by determining the resistance of the gas to flow through a capillary tube or by observing its damping effect upon the oscillations of a torsionally oscillating disc. The temperature range covered by such measurements is 70–2000 K, and the variation of viscosity with temperature is sketched in Fig. 1.11 for a typical gas.

The information about intermolecular forces contained in viscosity data is most easily seen by examining the viscosity expression for the case where the potential energy function is monotonic. For an intermolecular

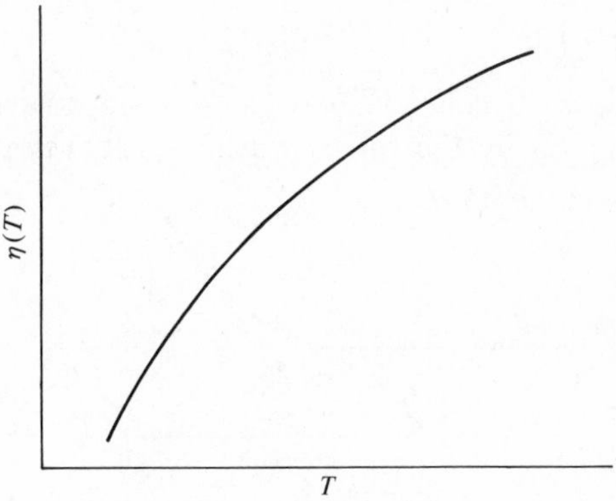

FIG. 1.11. Viscosity η as a function of temperature.

potential of the form $U(r) = C_n/r^n$ the viscosity coefficient varies with temperature as,

$$\eta = AT^{(\frac{1}{2}+\frac{2}{n})}$$

where A depends upon the sign and magnitude of C_n. Thus the temperature dependence of the viscosity contains information about the index of the inverse power potential and its coefficient. For a realistic intermolecular potential energy function such as that sketched in Fig. 1.1 the viscosity of a gas at low temperatures is dominated by the attractive dispersion energy contribution, which varies as C_6/r^6. Hence low-temperature viscosity measurements can be used to probe the intermolecular potential energy function at long-range (beyond r_m). Indeed it has been possible to estimate C_6 from experimental viscosity data in this way. At high temperatures the viscosity is dominated by the low-energy repulsive region of the potential and a similar analysis allows $U(r)$ to be determined in this region. Recently methods for the direct inversion of viscosity data to yield $U(r)$ have been developed which bridge the gap between the repulsive region and the long-range tail. These methods seem to indicate there is a close correspondence between a viscosity–temperature datum and a point on the intermolecular pair potential energy-separation curve. These new techniques are discussed in Chapter 6.

The other two transport coefficients yield similar sorts of information about the intermolecular potential but they have proved more difficult to measure accurately. The expressions given above for the transport coefficients are only rigorously applicable to monatomic species and modifications are necessary to handle the angle-dependent potentials and internal energy characteristic of polyatomic molecules. This forms an area of current research activity and little has yet been derived about the intermolecular pair potential for polyatomic species from gas transport coefficient data.

1.7.3. *Molecular beams*

The outcome of a collision between two approaching molecules is obviously determined by the forces which act between them. Fig. 1.12

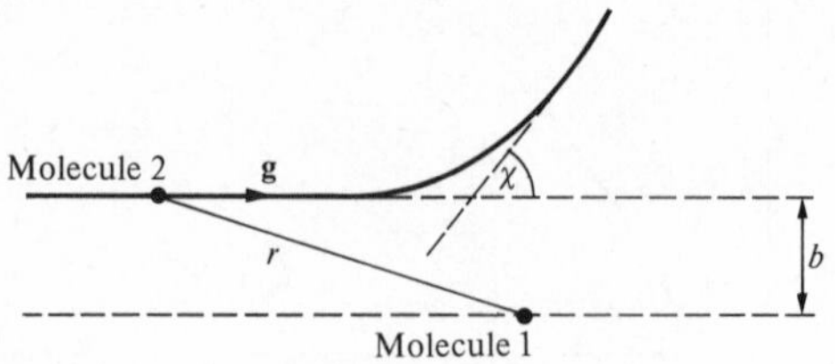

FIG. 1.12. The scattering of molecule 2 by molecule 1 (which is regarded as stationary).

shows an encounter between two molecules which interact through an intermolecular pair potential $U(r)$. We can suppose that molecule 2 approaches molecule 1 with a relative velocity $\mathbf{g}$ at an impact parameter b which represents the distance by which the molecular centres would miss if there were no interaction. As a result of the collision the trajectory of molecule 2 is deflected through an angle χ and if we assume the collision is elastic then the energy is unchanged. The deflection angle χ depends then upon the force acting between the two molecules at all points along the trajectory, upon the impact parameter b, and upon the initial relative velocity or kinetic energy of the molecules.

In practice it is not possible to observe the collision of just two molecules. However, recent technical advances in the production of monoenergetic supersonic beams of molecules and their detection have allowed a study of the dynamics of collisions for groups of molecules. Statistically such molecular beam experiments are equivalent to carrying out a large number of collisions between just two molecules over a range of impact parameters. There are two ways of looking at the scattering of molecular beams. Either one can consider the total amount of scattering in all directions or the fraction of scattering in a particular direction. The fraction of molecules in an incident beam of an energy E which suffer any scattering, which is also the probability for scattering, is determined by the integral scattering cross-section, $Q(E)$. The fraction of molecules in a beam with the same energy E which are scattered into a unit solid angle about the deflection angle χ is determined by the differential scattering cross-section, denoted by $\sigma(\chi, E)$. This differential scattering cross-section is also the probability for scattering into the unit solid angle around χ.

A full quantum mechanical analysis of the dynamics of a two-molecule encounter can relate both $\sigma(\chi, E)$ and $Q(E)$ to the intermolecular pair potential. To illustrate how the intermolecular potential influences the energy dependence of these cross-sections we consider the integral cross-section $Q(E)$. Fig. 1.13 shows a typical observed integral scattering cross-section for elastic collisions as a function of energy. The general trend is for the scattering cross-section to decrease as the energy of the colliding molecules is increased. This is to be expected on classical grounds because, as the energy is increased, the colliding particles are less influenced by the intermolecular potential energy and so the amount of scattering decreases. This argument serves to explain the behaviour of the integral cross-section at high energies completely since at such energies only the repulsive wall of the potential is significant, and then only in the low impact-parameter collisions. Measurements of $Q(E)$ at high energies thus provide a probe of the repulsive region of the potential.

At much lower energies the integral cross-section displays a series of

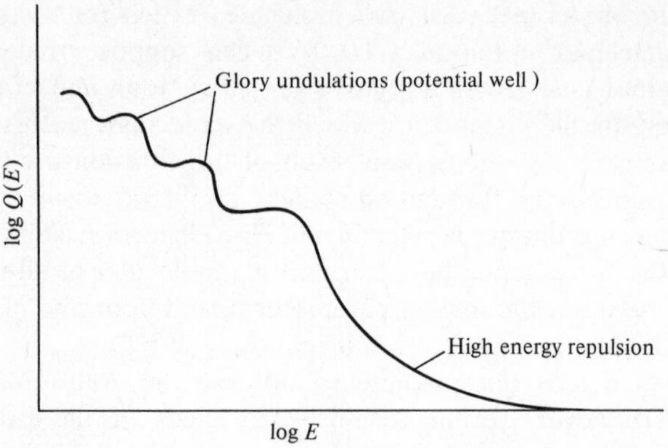

FIG. 1.13. The integral scattering cross-section $Q(E)$ as a function of relative kinetic energy E.

undulations superimposed on the classical downward trend which are known as *glory oscillations.* These oscillations arise from a quantum mechanical interference effect. Fig. 1.14 shows two trajectories for the same collision energy for which there is no net deflection ($\chi = 0$). For the trajectory at the smaller impact parameter, b_1, the incoming molecule is first attracted towards the scattering molecule and then repelled as it encounters the repulsive wall of the intermolecular potential. The path length of this trajectory is therefore longer than that for the trajectory with the larger impact parameter even though neither suffers a deflection. Because of this path-length difference the wave functions associated with the scattered molecules for the two trajectories can interfere in the same way that interference of light waves produces a diffraction pattern. Since the path-length difference between the two trajectories depends upon energy the effect of the interference and consequently the scattered wave function amplitude varies in an oscillatory fashion with energy. Because the glory trajectory, with impact parameter b_1, samples the attractive region of the potential energy function as well as the repulsive wall the significant region of the potential for the glory oscillations is the well. A detailed analysis confirms that the spacing of these oscillations is indeed sensitive to the shape of the potential well.

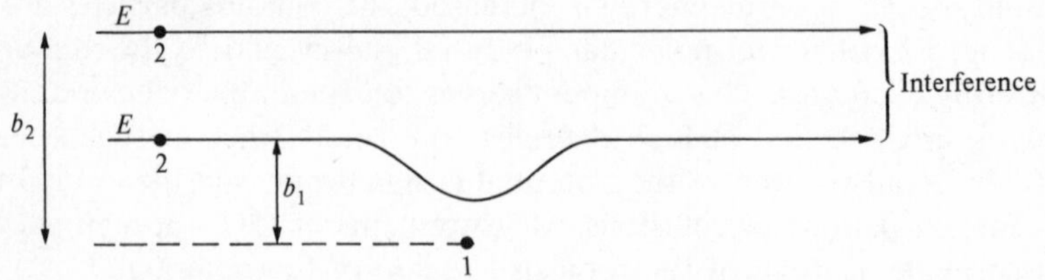

FIG. 1.14. Trajectories contributing to the glory effect.

The differential scattering cross-section provides even more detailed information about the intermolecular potential energy function in the low-energy repulsive region and the well. The exact type of information and the ways in which it can be extracted from molecular beam measurements are fully described in Chapter 4.

1.7.4. *Spectra*

Despite the fact that atoms such as the inert gases and molecules such as nitrogen do not form chemically bound molecules, they can form physically bound dimers under appropriate conditions. These physical dimers can arise when, for example, three molecules approach each other at low energy. One of these molecules may remove kinetic energy from the other two which themselves become trapped in their intermolecular potential well. If a system of two structureless molecules has an angular momentum L then the effective potential well $V(r)$ for the dimer appears as shown in Fig. 1.15, which is a combination of the intermolecular pair potential energy function $U(r)$ and the centrifugal energy $L^2/2\mu r^2$, where μ is here the reduced mass of the molecular pair ($\mu = m_1m_2/(m_1+m_2)$) and m_1 and m_2 are molecular masses. For this effective intermolecular potential the pair of molecules can exist in three types of state. First, for a pair of molecules which have an energy less than the dissociation energy, they can exist as bound dimers. Secondly if their energy is greater than the dissociation energy but less than V_{max}, the energy of the outer maximum in the effective potential, they form metastable dimers which

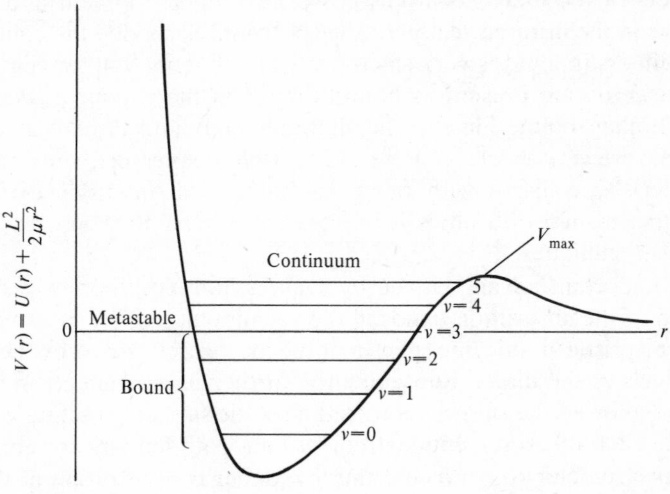

FIG. 1.15. The effective potential energy for van der Waals dimers $V(r) = U(r)+L^2/2\mu r^2$.

can dissociate by tunnelling through the potential barrier. Finally, if the energy of the molecular pair is greater than V_{max}, they are in the continuum region corresponding to a colliding pair of molecules. Within the bound and metastable dimer regions a number of discrete vibration–rotation states exist as for a chemically bound molecule, whose energies and spacing are characteristic of the potential energy well.

The physically bound, two-molecule system is often called a *van der Waals dimer* and has some similarities to a chemically bound molecule. In particular the method developed for the analysis of data for the vibration–rotation absorption spectrum of chemically bound molecules can be carried over, essentially unchanged, to the analysis of the equivalent spectrum of van der Waals molecules. The analytical method is known as the Rydberg–Klein–Rees technique; it was originally developed some fifty years ago and in recent years has been applied successfully to determine the intramolecular potential energy function of chemically bound molecules. If only the vibrational absorption bands in the spectrum can be observed then the Rydberg–Klein–Rees analysis yields the width of the potential well as a function of energy. However, if the rotational fine structure of the absorption spectrum can also be resolved, the entire potential function in the well region can be deduced from it. Of course the dissociation energy of the van der Waals complex yields the potential energy well depth immediately.

The reasons why this technique was not applied to the determination of the intermolecular potential energy function for the van der Waals dimers of simple molecules until recently are entirely experimental. First the dimers of the inert gas atoms possess no dipole moment and so are not active in the infrared region of the spectrum. Secondly, the concentration of dimers in a gas is very small; for example only 0·4 per cent of the atoms in argon are present as bound dimers in the vapour phase at the normal boiling point. Finally, the lifetimes of bound dimers at normal conditions are very short, $\sim 10^{-10}$ s at atmospheric pressure, since they are dissociated by collision with other gas molecules. In order to obviate these experimental difficulties it has been necessary to devise extremely ingenious techniques.

An important advance made by Tanaka and Yoshino[25] was the observation of the absorption bands in the vacuum ultraviolet spectrum near the characteristic atomic line spectra of the inert gases caused by the vibrational levels of the dimer. Subsequently, further improvements in resolution have allowed the observation of the rotational fine structure of these bands. In such measurements extremely long path lengths are employed in the spectrometer to overcome the low dimer concentration at the low pressures needed to provide a relatively long-lived species. Other spectroscopic techniques have also been of value in the determination of energy level spacings for dimers. For example in the case of the dimers of diatomic molecules such as N_2 and O_2 the quadrupole moment of one molecule of the pair induces a dipole moment in the other. The vibration and rotation of the dimer modulates the dipole moment and so absorption in the infrared region is possible and has been studied. A more refined technique is the use of the dimers produced in the expansion of a gas through a supersonic nozzle to produce a molecular beam. Since such molecular beams are relatively collision-free, the lifetime of the dimers is increased and this facilitates their spectroscopic study. In this way for example, the Ar–HCl complex has been studied by microwave and radiofrequency techniques. These important experimental developments and the analysis of the results are discussed in Chapter 7.

1.7.5. Solid-state properties

The four preceding topics have illustrated the wide variety of techniques and properties which have been especially useful in the determination of intermolecular forces. The properties of solids have been of less value in the determination of pair potentials since by their very nature they involve the simultaneous interaction of all molecules in the solid. It was mentioned earlier that the total energy of interaction of many molecules is not just the sum of the energy of interaction of all pairs. Consequently solid-state properties provide one example of how a knowledge of the intermolecular pair potential can be used to provide insight into the problem of many molecule interactions.

For molecular crystals their lattice energy, dimensions, and properties such as their compressibility can be directly related to intermolecular forces. This is because the stable arrangement of molecules in the lattice represents a balance of the repulsive and attractive forces between the molecules. To illustrate this we may choose as an example a solid in which the molecules interact according to the Lennard-Jones (12 − 6) pair potential function. We first presume that the total energy of a crystal of N_A molecules is given by the sum of the potential energies of interaction of all possible pairs in the lattice, and we also suppose each of the molecules is at rest on a lattice site. In this case the potential energy of the crystal becomes

$$\Phi = 2N_A\varepsilon\left\{\sum_{i=1}\left(\frac{\sigma}{r_i}\right)^{12} - \sum_i\left(\frac{\sigma}{r_i}\right)^6\right\}$$

where r_i denotes the separation of the ith molecule from a particular reference molecule and the summations extend over all the molecules around this selected molecule. For a particular crystalline structure the separation of every molecule from a particular molecule is known in terms of the nearest-neighbour distance, a, so that, if we write

$$L_n = a^n \sum_i r_i^{-n},$$

the sums L_n can be evaluated. For example, for the face-centred cubic structure of the inert gas crystals the sums $L_{12} = 12{\cdot}13$ and $L_6 = 14{\cdot}45$ have been evaluated. Then we can write for Φ

$$\Phi = 2N_A\varepsilon\left\{L_{12}\left(\frac{\sigma}{a}\right)^{12} - L_6\left(\frac{\sigma}{a}\right)^{6}\right\}.$$

At equilibrium the nearest-neighbour distance must be such that the potential energy of the crystal is a minimum ($\mathrm{d}\Phi/\mathrm{d}a = 0$); a condition met when

$$\frac{a}{\sigma} = \left\{\frac{2L_{12}}{L_6}\right\}^{\frac{1}{6}} \tag{1.40}$$

when the potential energy is $\Phi = -(N_A\varepsilon L_6^2/2L_{12})$. Including the zero point energy of the lattice Φ_0, calculated from the Debye theory of a solid the total energy is

$$\Phi_E = \frac{-N_A\varepsilon}{2}\frac{L_6^2}{L_{12}} + \Phi_0 \tag{1.41}$$

Eqns (1.40) and (1.41) demonstrate that if the assumption of pairwise additivity of potential energies was valid then measurements of the nearest-neighbour separation and sublimation energy would serve to determine σ and ε for the assumed Lennard-Jones $(12-6)$ potential function. Such calculations can of course be repeated for any given intermolecular pair potential.

As we have already noted in § 1.4.4 the assumption of pairwise additivity of intermolecular potential energies is not strictly valid. Indeed, if the best pair potential function for argon is used to compute the energy of the crystal and the result compared with experimental data it is found that there is a discrepancy of some 10 per cent. In the case of xenon the deviation is as much as 13 per cent. These discrepancies are ascribed to the effects of non-additive contributions to the energy. If account is taken of the Axilrod–Teller triple–dipole contribution to the dispersion energy good agreement with the solid-state properties can be achieved. This suggests that the triple-dipole term is probably the major non-additive contribution. Also it seems likely that there is a fortuitous cancellation of higher-order non-additive terms.

At present therefore it seems best to regard solid-state properties on the one hand as a test of intermolecular pair potentials obtained from other sources and on the other hand as a means of investigating non-additive effects. Similar comments apply to liquid-state properties although in this case the molecular theory relating the properties to the intermolecular potential energy function is less well developed. Further details of the relationship of condensed phase properties to intermolecular potential energy functions are given in Chapter 8.

1.8. Summary

This chapter has provided a brief resumé of the field of intermolecular forces both from the viewpoint of their origins and theoretical calculation and the techniques which can be used to determine them from experimental information. All these different approaches have of course been combined to provide the sum total of our knowledge about intermolecular forces today, since the ranges of separations and the type of information they provide is often complementary. The remainder of this book is devoted to describing the various routes from measurement and theory to the intermolecular potential and to emphasizing the manner in which they complement one another. The final chapter summarizes our current knowledge of intermolecular forces. At present the only well-characterized systems are the interactions of monatomic species. However the methods that have been developed to achieve this success are now being applied to polyatomic systems and progress in this direction seems assured.

References

1. Lucretius. *On the nature of the universe* (transl. R. Latham). Penguin Books, London (1951).
2. Bernoulli, D. *Hydrodynamica, sive de vivibus et motibus fluidorum commentarii.* Section Decima 'De affectionibus atque motibus fluidorum elasticorum, praecipque autem aeris,' pp. 200–4. Argentorati, Sumptibus Johannes Reinholdi Dulseckeri (1738). English translation in Brush, S. G. *Kinetic theory* Vol. I. Pergamon, London (1965).
3. Clausius, R. *Ann. Phys.* **79,** 368, 500 (1850).
4. Clausius, R. *Ann. Phys.* **100,** 353 (1857).
5. Clausius, R. *Ann. Phys.* **105,** 239 (1858).
6. Maxwell, J. C. *Phil. Mag.* **19,** 19 (1860).
7. Maxwell, J. C. *Phil. Mag.* **20,** 21 (1860).
8. Maxwell, J. C. *Phil. Trans. roy. Soc.* **157,** 49 (1867).
9. Boltzmann, L. *Sitz. Akad. Wiss., Wien* **66,** 275 (1872).
10. van der Waals, J. D. Doctoral dissertation, Leiden (1873).
11. Drude, P. K. L. *The theory of optics.* Longman, London (1933).
12. Pauling, L. and Wilson, E. B. *Introduction to quantum mechanics.* McGraw-Hill, New York (1935).
13. London, F. *Z. phys. Chem* (*B*) **11,** 222 (1930).

14. Tabor, D. *Gases, liquids and solids*. Penguin Books, Harmondsworth (1969).
15. Feynman, R. P. *Phys. Rev.* **56,** 340 (1939).
16. Hirschfelder, J. O., Curtiss, C. F., and Bird, R. B. *Molecular theory of gases and liquids*. Wiley, New York (1954).
17. Reid, R. C., Prausnitz, J. M., and Sherwood, T. K. *Properties of gases and liquids*. Wiley, New York (1977).
18. Lennard-Jones, J. E. *Proc. roy. Soc.* **A106,** 441, 463 (1924).
19. Buckingham, R. A. *Proc. roy. Soc.* **A168,** 264, 378 (1938).
20. Mason, E. A. and Rice, W. E. *J. chem. Phys.* **22,** 843 (1956).
21. Stockmayer, W. H. *J. chem. Phys.* **9,** 398 (1941).
22. Barker, J. A. and Pompe, A. *Austral. J. Chem.* **21,** 1683 (1968).
23. Maitland, G. C. and Smith, E. B. *Chem. Phys. Lett.* **22,** 443 (1973).
24. Gubbins, K. E. and Gray, C. G. *Statistical mechanics of polyatomic fluids*, Chapter 2. Oxford University Press, (in press).
25. Tanaka, Y. and Yoshino, K. *J. chem. Phys.* **53,** 2012 (1970).

2

THEORETICAL CALCULATION OF INTERMOLECULAR FORCES

2.1. Introduction

The ultimate objective of the theory of intermolecular forces is the expression of the potential energy that arises from the interaction of any two molecules over the complete range of molecular separations, in terms of fundamental constants, such as the mass and charge of an electron and Planck's constant. For a variety of reasons the complete implementation of this approach presents insuperable difficulties at present, so that less ambitious methods have had to be adopted. It is the purpose of this chapter to provide an introduction to these techniques and some of the results of their application.

The discussion of the origins of the attractive and repulsive forces between molecules contained in Chapter 1 has shown that the forces are electromagnetic in nature. The short-range forces have their origins in the overlap of the individual molecular wave functions at small separations and the symmetry requirements of the Pauli exclusion principle. The long-range forces, on the other hand, arise from the interaction between the electric multipoles of the charge distributions of the two molecules when there is little or no overlap of wave functions. The different extent of wave function overlap which occurs at the two extremes of molecular separations, and the different origins of the two types of force have necessitated the development of different techniques for the construction of the intermolecular potentials in these two regions.

For large separations the problem of the evaluation of the interaction energy may be viewed as a perturbation on the energy of two isolated molecules since their wave function overlap is small. On the other hand, at small separations the more obvious system to use as a basis for a perturbation is that where the two molecules have essentially become one. Between these two extreme zones lies a range of intermediate separations wherein neither of the two earlier approaches forms a natural basis for a perturbation theory approach. For this reason this region has proved extremely difficult to investigate and quite different methods have had to be employed.

Within each of these three ranges of intermolecular separation a further subdivision of the methods used to evaluate intermolecular potentials can be discerned. The first, and most fundamental method remains the *ab initio* evaluation of the potential in terms of fundamental constants with the aid of a quantum mechanical description of the true physical situation. The second approach treats the same physical situation but attempts to express the intermolecular potential in terms of experimentally accessible microscopic properties of the molecules, this is termed the semi-empirical method. Finally, in some cases it is necessary to resort to calculations for an approximate model of the true physical situation in order to render the problem tractable. In this chapter we shall describe all of these approaches because each has played some part in the evaluation of intermolecular forces. In particular, the approximate models provide a convenient means of discussing the physical origins and qualitative behaviour of intermolecular forces. Consequently, wherever it is possible the discussion of a specific aspect of the intermolecular forces is prefaced by a simplified treatment. It is intended that these sections can be used by a reader who requires physical insight into the subject of intermolecular forces but does not require a detailed discussion of the most recent developments.

Throughout our discussion of the theoretical evaluation of intermolecular forces we shall presume the validity of the Born–Oppenheimer approximation mentioned in Chapter 1. That is, we shall calculate the interaction potential energy of two stationary molecules for fixed positions of their nuclei (and so fixed positions and orientations of the molecules) for a series of centre-of-mass separations. It is then assumed that the pair intermolecular potential energy function generated in this way is unique and remains the same for cases in which the two molecules approach each other at the same fixed orientation but at a non-zero velocity. Since the velocities of approach of atoms in the processes we consider in the book are small compared with the speed of change of the electron charge distribution the approximation is valid. The advantage gained by its application, in quantum mechanical terms, is that it enables us to separate the nuclear kinetic energy terms from the total Hamiltonian of the two-atom system, and to obtain a wave function equation for the electronic motion for a fixed arrangement of the nuclei. Evidently, circumstances must arise where this approximation is no longer valid such as in collisions between molecules at extremely high energies, but they will not concern us here and the reader is advised to consult ref. 1 and the literature cited therein for a treatment of such cases. We also confine the discussion to conditions for which the intermolecular potential energy is not influenced by the retardation effects mentioned in Chapter 1.

To be consistent with the emphasis in the remainder of the book we shall pay special attention to the treatment of the interactions between two neutral, monatomic species, in their ground electronic states. However, the interactions of polar and non-polar polyatomic molecules and the simultaneous interaction of three molecules will also be treated. In the case of polyatomic molecular interactions the nuclear motion, separated from the electronic motion by the Born–Oppenheimer approximation, can be further subdivided according to whether it is vibrational, rotational, or translational in character. As a result of such a separation it is possible to evaluate a potential energy function between two molecules for each of their vibrational quantum states and for any relative orientation as a function of molecular separation. This leads to a family of potential energy functions dependent upon vibrational state and orientation as well as separation.

It is important to recognize that during the course of a molecular collision the relative orientation and internal states of the molecules may change with a consequent change in the appropriate potential energy function. The ways in which potential energy functions may be incorporated into a description of inelastic collisional processes will be considered in Chapter 4.

2.2. Long-range forces

In Chapter 1 it was shown that all intermolecular forces are electromagnetic in origin and that in particular the interaction of two molecules at long range, when there is no overlap of electron clouds, can be written

$$U(\mathbf{r}, \boldsymbol{\omega}) = U_{\mathrm{el}}(\mathbf{r}, \boldsymbol{\omega}) + U_{\mathrm{ind}}(\mathbf{r}, \boldsymbol{\omega}) + U_{\mathrm{disp}}(\mathbf{r}, \boldsymbol{\omega}). \tag{2.1}$$

Here $\mathbf{r}$ represents the separation of the centres of mass of the two charge distributions which make up the two molecules and $\boldsymbol{\omega}$ stands for a particular relative orientation of the two molecules in space. $U_{\mathrm{el}}(\mathbf{r}, \boldsymbol{\omega})$ represents the contribution to the interaction energy arising from permanent electric multipoles in the molecules, $U_{\mathrm{ind}}(\mathbf{r}, \boldsymbol{\omega})$ represents the contribution from multipoles induced in one molecule by permanent multipoles in the other, and $U_{\mathrm{disp}}(\mathbf{r}, \boldsymbol{\omega})$ the dispersion energy.

As we noted in Chapter 1 the electrostatic energy is first-order in the sense that it arises with no distortion of each molecule's electron cloud; the other two contributions are second-order since they arise from just such a distortion.

The greatest contribution to the long-range energy usually arises from the second-order dispersion energy; for the interaction between molecules with no permanent electric multipoles it is the only contribution. Thus, the evaluation of the dispersion energy is the most significant task of theory. However, the electrostatic origins of the dispersion energy most naturally follow from a discussion of the two other contributions to the long-range energy which are present in all interactions involving polar molecules.

2.2.1. *Electrostatic energy*

In Chapter 1 the concept of the multipole moments of an electric charge distribution was introduced. We begin our discussion of the electrostatic interaction energy with a more careful definition of them. Given a set of charges Q_i located at vectorial positions $\mathbf{r}_i$ with respect to an arbitrary origin the zeroth moment of the charge distribution is merely the total charge of the system

$$Q = \sum_i Q_i. \tag{2.2}$$

The first moment of the set of charges is called the dipole moment and is defined as

$$\boldsymbol{\mu} = \sum_i Q_i \mathbf{r}_i \tag{2.3}$$

which is a vector.

An analogous definition of the second moment of the charge distribution would be

$$\boldsymbol{\Theta}^* = \sum_i Q_i \mathbf{r}_i \mathbf{r}_i \tag{2.4}$$

which is a tensor with nine components. However, it is more convenient to work with a slightly different traceless tensor $\boldsymbol{\Theta}$ which is called the quadrupole moment tensor, the elements of which are defined by the relation[1]

$$\Theta_{\alpha\beta} = \tfrac{1}{2} \sum_i Q_i \{3r_{i\alpha}r_{i\beta} - r_i^2 \delta_{\alpha\beta}\}. \tag{2.5}$$

Here $r_{i\alpha}$ is the x, y, or z component of the vector $\mathbf{r}_i$, which defines the position of the charge Q_i, as α takes the values 1, 2, or 3, and $\delta_{\alpha\beta}$ is the Kronecker delta ($\delta_{\alpha\beta} = 1$ if $\alpha = \beta$, $\delta_{\alpha\beta} = 0$ if $\alpha \neq \beta$). The definition (2.5) may be made clearer by writing out two elements as examples

$$\begin{aligned} \Theta_{xx} &= \Theta_{11} = \tfrac{1}{2} \sum_i Q_i (2x_i^2 - y_i^2 - z_i^2) \\ \Theta_{xy} &= \Theta_{12} = \tfrac{3}{2} \sum_i Q_i x_i y_i \end{aligned} \tag{2.6}$$

If the charge distribution is linear and lies along the z-axis which also contains the origin O ($x_i = y_i = 0$ for all i) then the off-diagonal elements of $\mathbf{\Theta}$ vanish and $\Theta_{xx} = -\frac{1}{2}\sum_i Q_i z_i^2 = \Theta_{yy} = -\frac{1}{2}\Theta_{zz}$.† For such a charge distribution therefore the quadrupole moment tensor can be specified by one scalar quantity $\Theta = \Theta_{zz} = \sum Q_i z_i^2$ which is the quadrupole moment‡ introduced in Chapter 1.

Higher-order multipole moments such as the octopole and hexadecapole can also be defined in an analogous fashion and are sometimes of significance. In general, all multipole moments except the first non-zero moment depend upon the position of the origin of coordinates. It is usual to choose the origin as either the centre of mass or the centre of charge; we shall adopt here the former origin. In addition, any molecule which possesses a permanent non-zero multipole moment above the zeroth moment is termed a polar molecule.

We now wish to consider the interaction of two polar molecules, represented by electric charge distributions. We begin by recalling from Chapter 1 the electrostatic potential due to the linear charge distribution of Fig. 2.1 at the point P.

$$\Phi = \frac{1}{4\pi\varepsilon_0}\left\{\frac{Q}{r} + \frac{\mu\cos\theta}{r^2} + \frac{\Theta}{2}\frac{(3\cos^2\theta - 1)}{r^3} + \ldots\right\} \tag{2.7}$$

where Q, μ, and Θ are the total charge, dipole, and quadrupole moments respectively defined with respect to the centre of mass O. As mentioned earlier this expansion is valid only when r is greater than the dimensions of the charge distribution.§ Since we are only concerned here with the long-range energy this restriction is not serious.

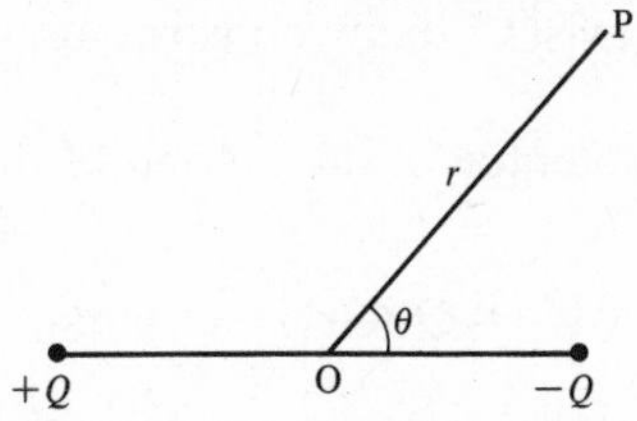

FIG. 2.1. A linear charge distribution.

We now place a similar charge distribution, $Q_1'Q_2'$, so that its centre of mass lies at the point P as shown in Fig. 2.2(a). The symbol r now denotes the separation of the centres of mass of the two distributions. The

† A similar result follows for charge distributions which are cylindrically symmetric about the z-axis.

‡ Other definitions of the quadrupole moment tensor and its scalar representation for linear charge distributions are employed and care must be exercised in their use.[1,2]

§ An alternative expansion valid for all $\mathbf{r}$ is given in ref. 2.

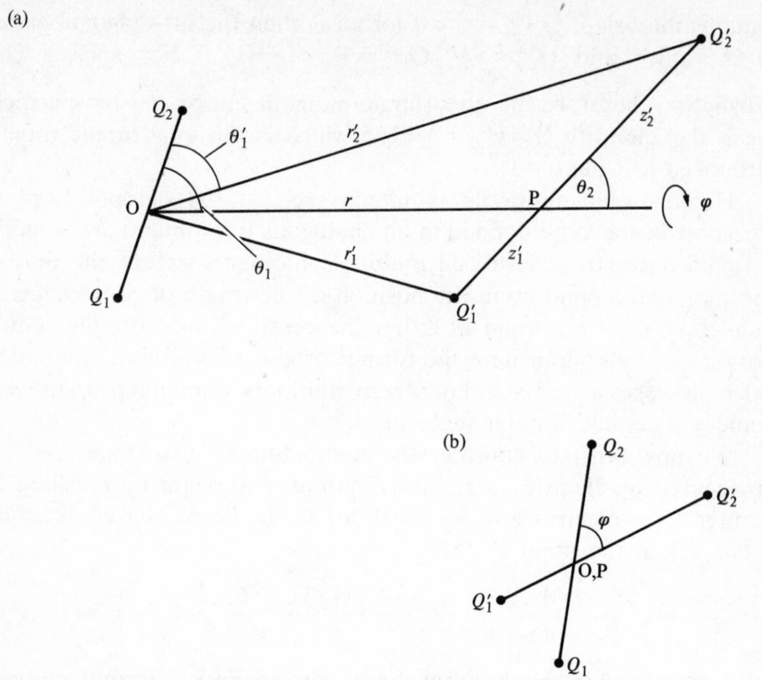

FIG. 2.2. (a) The geometry of two interacting linear charge distributions. (b) A view along OP.

first charge distribution Q_1Q_2 makes an angle θ_1 with OP, whereas the second distribution makes an angle θ_2 with the same line. The angle ϕ represents the relative twist of the two distributions about the line joining them (Fig. 2.2(b)).

The electrostatic potential of interaction of the two distributions can be written

$$U_{\text{el}} = Q_1'\Phi(\mathbf{r}_1') + Q_2'\Phi(\mathbf{r}_2')$$

where $\Phi(\mathbf{r}_1')$, $\Phi(\mathbf{r}_2')$ are the electrostatic potentials due to the first distribution Q_1Q_2 at the charges Q_1' and Q_2' respectively.

Simple trigonometry shows that the position of Q_2' with respect to O is defined by the distance

$$r_2' = r\left\{1 + \frac{z_2'^2}{r^2} + \frac{2z_2'}{r}\cos\theta_2\right\}^{\frac{1}{2}}$$

and the angle between Q_1Q_2 and r_2',

$$\theta_1' = \theta_1 - \Delta\theta_2$$

where

$$\Delta\theta_2 = \frac{z_2' \sin\theta_2 \cos\phi}{r}.$$

Then using (2.7) we may write $\Phi(\mathbf{r}_2')$ as

$$\Phi(\mathbf{r}_2') = \frac{1}{4\pi\varepsilon_0}\left\{\frac{Q}{r}\left[1+\frac{z_2'^2}{r^2}+\frac{2z_2'\cos\theta_2}{r}\right]^{-\frac{1}{2}} + \frac{\mu}{r^2}[\cos\theta_1\cos\Delta\theta_2+\sin\theta_1\sin\Delta\theta_2]\left[1+\frac{z_2'^2}{r^2}+\frac{2z_2'\cos\theta_2}{r}\right]^{-1} + \ldots\right\}. \tag{2.8}$$

A similar expression may also be written for $\Phi(\mathbf{r}_1')$. Since for the problem of interest the dimensions of the second charge distribution are small compared to r (long-range interaction), $\Phi(\mathbf{r}_2')$ and $\Phi(\mathbf{r}_1')$ can be expanded by the binomial theorem in powers of (z_2'/r), and the trigonometric functions of $\Delta\theta_2$ expanded in their appropriate power series. When these expansions are carried out and the terms in equal powers of $(1/r)$ collected we obtain for the electrostatic energy

$$(4\pi\varepsilon_0)U_{\text{el}} = \frac{Q'Q}{r}+\frac{1}{r^2}\{Q'\mu\cos\theta_1 - Q\mu'\cos\theta_2\} - \frac{\mu\mu'}{r^3}\{2\cos\theta_1\cos\theta_2 - \sin\theta_1\sin\theta_2\cos\phi\} + \frac{1}{2r^3}\{Q\Theta'(3\cos^2\theta_2-1)+Q'\Theta(3\cos^2\theta_1-1)\}+\ldots. \tag{2.9}$$

Here we have written $Q' = Q_1'+Q_2'$ for the total charge of the second distribution, $\mu' = Q_2'z_2' - Q_1'z_1'$ for its dipole moment, and $\Theta' = Q_2'z_2'^2 + Q_1'z_1'^2$ for the quadrupole moment. For the interaction of neutral species $Q = Q' = 0$, the leading term in the electrostatic energy arises from the dipole–dipole interaction and for any fixed orientation this energy is proportional to $\mu\mu'/r^3$. This result is a generalization of that obtained in Chapter 1. The angular dependent coefficient of the dipole–dipole interaction energy

$$\zeta(\theta_1, \theta_2, \phi) = (2\cos\theta_1\cos\theta_2 - \sin\theta_1\sin\theta_2\cos\phi) \tag{2.10}$$

for $\theta_1 = 0$, $\phi = 0$ varies from $+2$ to 0 to -2 as one dipole is rotated with respect to the other. The interaction can therefore be repulsive or attractive depending on the relative orientation.

The dipole–dipole interaction energy thus provides an example of a family of potential energy functions which depend upon the relative orientations of the two molecules. If the two molecules are free to rotate, as they are in a dilute gas for example, then it is frequently the average energy of interaction taken over all possible orientations which is significant.

If all relative orientations were equally probable, the average energy, $\bar{U}$, would be zero. But for freely rotating molecules the probability of observing a given relative orientation is determined by the associated Boltzmann factor, $\exp(-U(\boldsymbol{\omega}_1, \boldsymbol{\omega}_2)/kT)$, where $\boldsymbol{\omega}_1$ and $\boldsymbol{\omega}_2$ represent the orientations of molecules 1 and 2. The average energy obtained by using the Boltzmann weighting may be written†

$$\langle U_{\mathrm{el}}\rangle = \frac{\iint U(\boldsymbol{\omega}_1, \boldsymbol{\omega}_2)\exp(-U(\boldsymbol{\omega}_1, \boldsymbol{\omega}_2)/kT)\,\mathrm{d}\boldsymbol{\omega}_1\,\mathrm{d}\boldsymbol{\omega}_2}{\iint \exp(-U(\boldsymbol{\omega}_1, \boldsymbol{\omega}_2)/kT)\,\mathrm{d}\boldsymbol{\omega}_1\,\mathrm{d}\boldsymbol{\omega}_2}. \tag{2.11}$$

At temperatures sufficiently high that $U(\boldsymbol{\omega}_1, \boldsymbol{\omega}_2) \ll kT$, we may expand the exponentials and integrate term-by-term to yield

$$\langle U_{\mathrm{el}}\rangle = \frac{\iint [U(\boldsymbol{\omega}_1, \boldsymbol{\omega}_2) - U^2(\boldsymbol{\omega}_1, \boldsymbol{\omega}_2)/kT\ldots]\,\mathrm{d}\boldsymbol{\omega}_1\,\mathrm{d}\boldsymbol{\omega}_2}{\iint [1 - U(\boldsymbol{\omega}_1, \boldsymbol{\omega}_2)/kT\ldots]\,\mathrm{d}\boldsymbol{\omega}_1\,\mathrm{d}\boldsymbol{\omega}_2},$$

$$= (\bar{U} - \overline{U^2}/kT + \ldots)(1 + \bar{U}/kT + \ldots),$$

where

$$\overline{U^n} = \iint U^n(\boldsymbol{\omega}_1, \boldsymbol{\omega}_2)\,\mathrm{d}\boldsymbol{\omega}_1\,\mathrm{d}\boldsymbol{\omega}_2.$$

Since for direct interaction between permanent multipoles $\bar{U} = 0$, as noted above, the leading term in the expansion is

$$\langle U_{\mathrm{el}}\rangle = -\overline{U^2}/kT$$

and we see that this averaging process always leads to an attractive net electrostatic energy which is inversely proportional to the temperature, and having a separation dependence equal to the square of that for the fixed orientation form. For the dipole–dipole interaction, evaluation of the integral over orientation for $\overline{U^2}$ yields the result

$$\langle U_{\mathrm{el}}\rangle_{\mu\mu'} = -\frac{2}{3}\frac{\mu^2\mu'^2}{r^6 kT(4\pi\varepsilon_0)^2} \tag{2.12}$$

† The differential element $\mathrm{d}\boldsymbol{\omega}_1 = \sin\theta_1\,\mathrm{d}\theta_1\,\mathrm{d}\phi$ for a polar coordinate system having as origin the centre of mass of the dipole.

and the energy falls off as r^{-6}, in contrast to the r^{-3} dependence for fixed dipoles.

Of the higher-order terms in the expansion (2.9) for the interaction of two neutral linear molecules the dipole–quadrupole and quadrupole–quadrupole terms are generally the most significant. For a fixed relative orientation these two contributions have the form

$$U_{\mu\Theta} = \frac{3}{2(4\pi\varepsilon_0)r^4}\{\mu\Theta'[\cos\theta_1(3\cos^2\theta_2 - 1) - 2\sin\theta_1\sin\theta_2\cos\theta_2\cos\phi] - \mu'\Theta[\cos\theta_2(3\cos^2\theta_1 - 1) - 2\sin\theta_1\sin\theta_2\cos\theta_1\cos\phi]\} \quad (2.13)$$

and

$$U_{\Theta\Theta'} = \frac{3\Theta\Theta'}{4(4\pi\varepsilon_0)r^5}\{1 - 5\cos^2\theta_1 - 5\cos^2\theta_2 + 17\cos^2\theta_1\cos^2\theta_2 + 2\sin^2\theta_1\sin^2\theta_2\cos^2\phi - 16\sin\theta_1\sin\theta_2\cos\theta_1\cos\theta_2\cos\phi\}. \quad (2.14)$$

For freely rotating molecules we can again carry out the orientational averaging in an analogous fashion to that of (2.11) and we obtain, at sufficiently high temperatures

$$\langle U_{\mathrm{el}}\rangle_{\mu\Theta} = -\frac{1}{r^8 kT(4\pi\varepsilon_0)^2}\{\mu^2\Theta'^2 + \mu'^2\Theta^2\} + \ldots \quad (2.15)$$

$$\langle U_{\mathrm{el}}\rangle_{\Theta\Theta'} = -\frac{14}{5}\frac{\Theta^2\Theta'^2}{r^{10}kT(4\pi\varepsilon_0)^2} + \ldots. \quad (2.16)$$

Thus all the leading terms in the orientationally averaged electrostatic energy for neutral molecules are attractive. For a comprehensive treatment of electrostatic interactions between non-linear charge distributions the reader is advised to consult ref. 1.

2.2.2. Induction energy

We saw in Chapter 1 that when a molecule is placed in a static electric field, $\mathscr{E}$, electric multipoles may be induced in it by a distortion of its charge distribution. In the simplest case, which is the only one we consider in detail, only a dipole moment is induced and it is proportional to the field and parallel to it so that

$$\boldsymbol{\mu}_{\mathrm{ind}} = \alpha(0)\mathscr{E}. \quad (2.17)$$

The scalar quantity $\alpha(0)$ defined by this equation is the static polarizability of the molecule, which is presumed to be isotropic. The energy of a dipole $\boldsymbol{\mu}$ in an electric field $\boldsymbol{\mathscr{E}}$ is

$$U = -\int_0^{\mathscr{E}} \boldsymbol{\mu} \cdot \mathrm{d}\boldsymbol{\mathscr{E}}.$$

Thus the energy of the dipole induced in a molecule with an isotropic polarizability by the field $\boldsymbol{\mathscr{E}}$ is

$$U_{\mathrm{ind}} = -\int_0^{\mathscr{E}} \alpha(0)\boldsymbol{\mathscr{E}} \cdot \mathrm{d}\boldsymbol{\mathscr{E}}$$

$$= -\tfrac{1}{2}\alpha(0)\boldsymbol{\mathscr{E}}^2. \tag{2.18}$$

Now we have already seen that a molecule which possesses non-zero permanent multipole moments gives rise to an electrostatic potential Φ given by (2.7) and consequently to an electric field (see §1.4.2) $\boldsymbol{\mathscr{E}} = -\partial\Phi/\partial\mathbf{r}$. This electric field can therefore induce a dipole moment in a second molecule which is nearby, whether the second molecule is polar or not. We return to Fig. 2.2 to consider the induction energy between the two molecules sketched there and represented by two linear charge distributions.

The magnitude of the electric field at P due to the molecule with origin O is

$$\mathscr{E} = \left[\left(\frac{\partial\Phi}{\partial r}\right)^2 + \left(\frac{1}{r}\frac{\partial\Phi}{\partial\theta}\right)^2\right]^{\frac{1}{2}}$$

or

$$\mathscr{E} = +\frac{1}{(4\pi\varepsilon_0)}\left\{\frac{Q^2}{r^4} + \frac{4\mu Q\cos\theta_1}{r^5} + \frac{\mu^2}{r^6}(3\cos^2\theta_1 + 1) + \ldots\right\}^{\frac{1}{2}}.$$

The energy of the second molecule arising from the electric moments induced in it by this field is

$$U_{\mathrm{ind}} = -\frac{1}{2}\frac{\alpha'(0)}{(4\pi\varepsilon_0)^2}\left\{\frac{Q^2}{r^4} + \frac{4\mu Q\cos\theta_1}{r^5} + \frac{\mu^2}{r^6}(3\cos^2\theta_1 + 1) + \ldots\right\}$$

where $\alpha'(0)$ is the polarizability of the second molecule.

Of course, simultaneously the second molecule induces a dipole moment in the first which leads to a similar contribution to the induction energy. Thus the total induction energy for the two molecules with the geometry of Fig. 2.2 is

$$U_{\text{ind}} = -\frac{1}{2}\frac{(Q^2\alpha'(0)+Q'^2\alpha(0))}{r^4(4\pi\varepsilon_0)^2}$$
$$-\frac{2(\mu Q\alpha'(0)\cos\theta_1+\mu'Q'\alpha(0)\cos\theta_2)}{r^5(4\pi\varepsilon_0)^2}$$
$$-\frac{\{\mu^2\alpha'(0)(3\cos^2\theta_1+1)+\mu'^2\alpha(0)(3\cos^2\theta_2+1)\}}{2r^6(4\pi\varepsilon_0)^2}$$
$$-\ldots \tag{2.19}$$

For neutral molecules the leading term in the induction energy is the final term of (2.19) which arises from the induced dipole–permanent dipole interaction. The energy is attractive for all configurations of the molecules and for a fixed orientation is inversely proportional to the sixth power of the molecular separation. If the molecules are free to rotate then we must average the dipole-induced dipole interaction energy over all orientations using (2.11) and obtain for the leading term at high temperatures

$$\langle U_{\text{ind}}\rangle_{\mu\mu} = -\frac{1}{(4\pi\varepsilon_0)^2 r^6}\{\mu^2\alpha'(0)+\mu'^2\alpha(0)\}. \tag{2.20}$$

If the two interacting molecules are identical then this reduces to

$$\langle U_{\text{ind}}\rangle_{\mu\mu} = -\frac{2\alpha(0)\mu^2}{(4\pi\varepsilon_0)^2 r^6}. \tag{2.21}$$

In contrast to the leading direct electrostatic term this leading induction term does not depend on the temperature.

In addition, we note that for the interaction between a non-polar molecule and a neutral polar molecule there is also an induction energy whose corresponding leading term when averaged over all orientations is

$$\langle U_{\text{ind}}\rangle_{\mu} = -\frac{\mu^2\alpha'(0)}{(4\pi\varepsilon_0)^2 r^6}.$$

Here μ is the dipole moment for the polar molecule and $\alpha'(0)$ the static polarizability of the non-polar molecule.

The subsequent term in the orientation-averaged induction energy for identical neutral molecules, which arises from higher terms in the expansion (2.19), corresponds to quadrupole-induced dipole interactions and in the high temperature limit is

$$\langle U_{\text{ind}}\rangle_{\Theta\mu} = -\frac{12\alpha(0)\Theta^2}{r^8(4\pi\varepsilon_0)^2}. \tag{2.22}$$

Our discussion of induction energy has been considerably simplified. In general, the static polarizability of a molecule is not isotropic so it must be represented by a second rank tensor $\boldsymbol{\alpha}$. The elements of this tensor relate the dipole moment component induced in one direction to a field component acting in another direction. Furthermore even if the polarizability is isotropic, such as for a spherically symmetric molecule, nonlinearities between the induced dipole moment and the applied field can occur. Thus we must write

$$\mu_{\text{ind}} = \alpha\mathscr{E} + \tfrac{1}{6}\gamma\mathscr{E}^3 + \ldots$$

for such a case. The coefficient γ is called the second hyperpolarizability and in the general case is a tensor. Finally, the electric field and its gradients can induce higher order multipole moments in the molecule. Thus for example for a spherically symmetric molecule we write

$$\Theta_{\text{ind}} = \tfrac{1}{2}\beta\mathscr{E}^2 + \ldots$$

where β is the quadrupolar polarizability. Each of these factors can contribute to the induction energy between two molecules. However for the topics covered in this book these aspects are not very significant, and a comprehensive treatment can be found in ref. 1.

Two approaches may be adopted for the numerical evaluation of the electrostatic and induction contributions to the intermolecular energy. In the first of these the multipole moments and polarizabilities are obtained experimentally, in the second they are obtained from quantum mechanical calculations. If the former method is adopted the calculation of the energy contribution is semi-empirical in the sense defined earlier, whereas the latter approach constitutes an *ab initio* evaluation of the energy.

Accurate measurements of dipole moments and polarizabilities of molecules have been available for some considerable time.[3] The determinations have been carried out by a variety of techniques usually involving dielectric constant measurements. More recently measurements of the Stark-effect splittings in the rotational spectra of isotopically substituted molecules have been employed to determine the direction of the dipole moment with respect to molecular axes.[4] Higher-order multipole moments such as the octopole and hexadecapole have been derived from data for the collision-induced far-infrared spectrum of molecules such as methane.[5,6] For molecules with anisotropic polarizabilities information about the anisotropy has been obtained from studies of the depolarization of light scattered by gases.[7] The hyperpolarizabilities of molecules can be determined by making use of the anisotropy of the refractive index produced in a gas by the application of electric fields (the Kerr effect[8]).

Finally quadrupole moments can be determined by measurements of the anisotropy of the refractive index of a gas in an electric field gradient.

This wealth of experimental information makes the semi-empirical evaluation of the leading terms in the long-range energy of interaction arising from permanent multipoles the most attractive and accurate route. However, recent work[9] indicates that for higher-order terms in the interaction energy of molecules quantum mechanical evaluation of the permanent multipole moments and polarizabilities may be more accurate.

2.2.3. Dispersion energy

The final contribution to the long-range intermolecular potential, the dispersion energy, is the only one which is present in all intermolecular interactions and is the only contribution to the long-range energy for the interaction of two non-polar species. As shown in Chapter 1 even for the polar molecules the dispersion energy is generally the dominant contribution. It was also demonstrated in Chapter 1 that although the dispersion energy is a result of electromagnetic interactions it cannot be analysed by means of classical mechanics since its origins are entirely quantum mechanical. For this reason the rigorous evaluation of the dispersion energy is more complicated than that of the other long-range energy contributions. Thus we begin our discussion of the dispersion energy with a treatment of a simplified model of the interacting molecules due to Drude.[10,11]

(a) *The oscillator model.* In Chapter 1 we identified the source of the dispersion energy as the interaction between the instantaneous dipole moment in one molecule and the consequent induced dipole in the other molecule of an interacting pair. As we saw in the last section the induced dipole moment in a molecule depends upon its polarizability so we first use the simple Drude model to derive an expression for this quantity. Drude originally supposed that a molecule could be represented by a series of point charges oscillating in a simple harmonic fashion about their equilibrium positions in all three spatial directions. Here we simplify this model and assume that a molecule consists of a single negatively charged particle ($-Q$) and a charged particle of equal and opposite sign. We suppose that the negative charge of the pair moves in an oscillatory fashion about the positive charge so that its motion can be resolved into simple harmonic oscillations along three Cartesian axes, each having the same frequency. We note that although the model is crude it possesses the essential feature we require of a molecule, that it has zero average dipole moment but a non-zero instantaneous dipole moment given by

$$\boldsymbol{\mu} = Q\,\delta\mathbf{r}(t)$$

where $\delta\mathbf{r}$ represents the vectorial displacement of the negative charge from the positive. The classical equation of motion of the negative charge for the free molecule is

$$M\frac{\mathrm{d}^2(\delta\mathbf{r})}{\mathrm{d}t^2}+k\,\delta\mathbf{r}=0$$

where k is the force constant of the oscillator, and M the mass of the moving charge. The solution to this equation is

$$\delta\mathbf{r}=\delta\mathbf{r}_0\cos(\omega_0 t)$$

where ω_0 is the angular frequency of the oscillation given by $\omega_0=(k/M)^{\frac{1}{2}}$ and $\delta\mathbf{r}_0$ is the amplitude of the oscillation. If this molecule is placed in a uniform oscillating electric field of frequency ω given by $\mathscr{E}=\mathscr{E}_0\cos\omega t$ the equation of motion becomes

$$M\frac{\mathrm{d}^2(\delta\mathbf{r})}{\mathrm{d}t^2}+k(\delta\mathbf{r})=-Q\mathscr{E}_0\cos\omega t$$

which has a steady-state solution

$$\delta\mathbf{r}=\frac{Q\,\mathscr{E}_0\cos\omega t}{M\,(\omega_0^2-\omega^2)}.$$

There is, therefore, a corresponding, induced, oscillating dipole moment given by

$$\boldsymbol{\mu}_{\text{ind}}=\frac{Q^2\,\mathscr{E}_0\cos\omega t}{M\,(\omega_0^2-\omega^2)}.$$

We saw in the previous section that the proportionality factor between an induced dipole moment and a static field was the polarizability of a molecule $\alpha(0)$, so that we can identify an isotropic frequency-dependent polarizability as

$$\frac{\mu_{\text{ind}}}{\mathscr{E}}=\alpha(\omega)=\frac{Q^2}{M(\omega_0^2-\omega^2)}. \tag{2.23}$$

In the case of a static field, $\omega=0$ and we have

$$\alpha(0)=\frac{Q^2}{M\omega_0^2}. \tag{2.24}$$

The final equation also enables us to express the classical oscillator force constant in terms of the static polarizability as

$$k=M\omega_0^2=Q^2/\alpha(0). \tag{2.25}$$

An analysis of the full Drude model for a molecule consisting of many oscillating charges Q_i, masses M_i, and frequencies ω_{i0} leads to the result that the polarizability of a molecule in a static electric field is

$$\alpha(0) = \sum_i \frac{Q_i^2}{M_i \omega_{i0}^2} \tag{2.26}$$

and for a field of frequency ω it is

$$\alpha(\omega) = \sum_i \frac{Q_i^2/M_i}{(\omega_{i0}^2 - \omega^2)}. \tag{2.27}$$

Equations (2.26) and (2.27) may be written in the forms

$$\alpha(0) = \frac{e^2}{m} \sum_i \frac{f_i}{\omega_{i0}^2} \tag{2.28}$$

and

$$\alpha(\omega) = \frac{e^2}{m} \sum_i \frac{f_i}{\omega_{i0}^2 - \omega^2} \tag{2.29}$$

where

$$f_i = \frac{Q_i^2/M_i}{e^2/m} \tag{2.30}$$

and e is the charge on an electron and m its mass. The quantity f_i is called the *dipole oscillator strength.* It may be interpreted classically as the effective number of oscillators in the Drude model of the molecule having a frequency ω_{i0}.

Now the theory of the interaction of electromagnetic radiation with matter enables us to relate the polarizability of a molecule to the refractive index n of the material composed of these molecules.[2] In the case of a dilute gas, of Drude molecules, for which the number density is small, we can write[10,11]

$$n^2 - 1 = \frac{\rho e^2}{m\varepsilon_0} \sum_i \frac{f_i}{(\omega_{i0}^2 - \omega^2)} \tag{2.31}$$

where ρ is the number density. The form of this result is confirmed by a quantum-mechanical treatment of dispersion of electromagnetic radiation and as we shall see it has been of considerable value in the calculation of dispersion energies.

We can also now recognize the similarities between this situation and that when we place two Drude molecules close to each other. The oscillating dipole moment in each molecule provides the electric field to

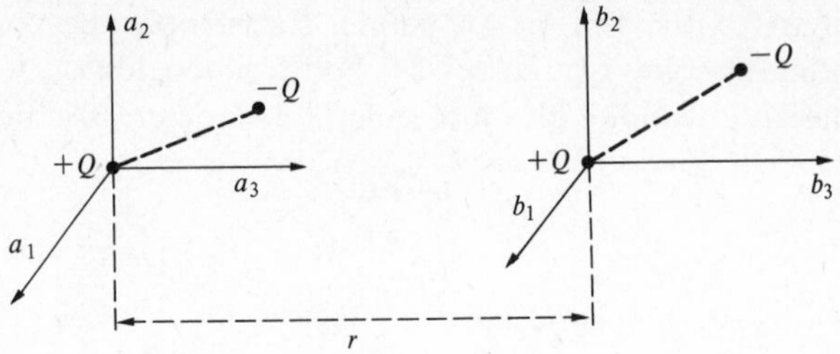

FIG. 2.3. The interaction of two Drude molecules.

induce a dipole moment in the other molecule. Each induced dipole then interacts with the field of the inducing dipole to produce an attractive energy. It is this close connection between the physical mechanism of the dispersion of electromagnetic radiation and the contribution to the long-range interaction energy of two molecules that gives rise to the name dispersion energy for the latter.

We consider the interaction of two identical molecules each represented by the simple Drude model described above. We let a_1, a_2, a_3 represent the three Cartesian coordinates of the displacement of the negative charge from the positive charge in molecule a whereas b_1, b_2, b_3 represent similar quantities for molecule b. The arrangement is shown in Fig. 2.3 where we have chosen the axes a_3 and b_3 to coincide with the line joining the centres of mass of the two molecules. We shall analyse the interaction of the two molecules quantum mechanically since as we showed in Chapter 1 a classical treatment leads to no interaction energy between molecules in their ground state.

When the two molecules are infinitely separated the wave equation for the oscillating charges of molecule a is

$$\sum_{i=1}^{3} \frac{\partial^2 \Psi_a}{\partial a_i^2} \frac{1}{M} + \frac{2}{\hbar^2}\left[E_a - \tfrac{1}{2}k \sum_{i=1}^{3} a_i^2\right]\Psi_a = 0 \tag{2.32}$$

where Ψ_a denotes the wave function for molecule a and E_a its energy. This equation is the wave equation for an isotropic three-dimensional harmonic oscillator for which the energy levels are given by[12]

$$E_a = \hbar\omega_0(n_a + \tfrac{3}{2})$$

where n_a is the total quantum number and ω_0 the frequency of the oscillation. When the molecules are in their ground states ($n_a = 0 = n_b$) the total energy of the system is

$$E(\infty) = 3\hbar\omega_0 \tag{2.33}$$

since the molecules are identical.

When the two molecules are separated by a finite distance, r, there is an electrostatic interaction between the two instantaneous dipoles. Using (2.9) and transforming to the Cartesian coordinate system of this analysis, the potential energy associated with this interaction is

$$\frac{Q^2}{4\pi\varepsilon_0 r^3}(a_1b_1+a_2b_2-2a_3b_3).$$

Thus the wave equation for the oscillating charges of both molecules is

$$\sum_{i=1}^{3}\frac{1}{M}\frac{\partial^2\Psi}{\partial a_i^2}+\frac{1}{M}\frac{\partial^2\Psi}{\partial b_i^2}$$
$$+\frac{2}{\hbar^2}\left[E-\tfrac{1}{2}k\sum_{i=1}^{3}(a_i^2+b_i^2)-\frac{Q^2}{4\pi\varepsilon_0 r^3}(a_1b_1+a_2b_2-2a_3b_3)\right]\Psi=0 \quad (2.34)$$

where Ψ is the wave function for the two molecule system.

By making the transformations

$$X_i=\frac{a_i+b_i}{\sqrt{2}} \qquad Y_i=\frac{a_i-b_i}{\sqrt{2}} \qquad (i=1 \text{ to } 3)$$

(2.34) can be put in the form

$$\sum_{i=1}^{3}\left(\frac{1}{M}\frac{\partial^2\Psi}{\partial X_i^2}+\frac{1}{M}\frac{\partial^2\Psi}{\partial Y_i^2}\right)+\frac{2}{\hbar^2}\left[E-\tfrac{1}{2}\sum_{i=1}^{3}(k_i'X_i^2+k_i''Y_i^2)\right]\Psi=0. \quad (2.35)$$

This is the wave equation for six independent harmonic oscillators in the coordinates X_i and Y_i ($i=1$ to 3), with frequencies given by

$$\left.\begin{aligned}
\omega_1=\omega_2=(k_1'/M)^{\frac{1}{2}}&=\left(\frac{k}{M}\right)^{\frac{1}{2}}\left\{1+\frac{Q^2}{4\pi\varepsilon_0 r^3 k}\right\}^{\frac{1}{2}}\\
\omega_3=(k_3'/M)^{\frac{1}{2}}&=\left(\frac{k}{M}\right)^{\frac{1}{2}}\left\{1+\frac{2Q^2}{4\pi\varepsilon_0 r^3 k}\right\}^{\frac{1}{2}}\\
\omega_4=\omega_5=(k_1''/M)^{\frac{1}{2}}&=\left(\frac{k}{M}\right)^{\frac{1}{2}}\left\{1-\frac{Q^2}{4\pi\varepsilon_0 r^3 k}\right\}^{\frac{1}{2}}\\
\text{and}\qquad\qquad&\\
\omega_6=(k_3''/M)^{\frac{1}{2}}&=\left(\frac{k}{M}\right)^{\frac{1}{2}}\left\{1-\frac{2Q^2}{4\pi\varepsilon_0 r^3 k}\right\}^{\frac{1}{2}}
\end{aligned}\right\} \quad (2.36)$$

The total energy of these six oscillators in their ground state, which is the total energy of the interacting molecules is therefore,

$$E(r)=\frac{\hbar}{2}\sum_{i=1}^{6}\omega_i.$$

The energy which arises from the interaction, denoted by U_{disp} is therefore

$$\begin{aligned} U_{\text{disp}} &= E(r) - E(\infty) \\ &= \frac{\hbar}{2} \sum_{i=1}^{6} (\omega_i - \omega_0) \end{aligned} \tag{2.37}$$

from (2.33).

Noting that $(k/M)^{\frac{1}{2}} = \omega_0$ and that $k = Q^2/\alpha(0)$ from (2.25), (2.36) can be rewritten as

$$\left.\begin{aligned} \omega_1 = \omega_2 &= \omega_0 \left\{1 + \frac{\alpha(0)}{4\pi\varepsilon_0 r^3}\right\}^{\frac{1}{2}} \\ \omega_3 &= \omega_0 \left\{1 + \frac{2\alpha(0)}{4\pi\varepsilon_0 r^3}\right\}^{\frac{1}{2}} \\ \omega_4 = \omega_5 &= \omega_0 \left\{1 - \frac{\alpha(0)}{4\pi\varepsilon_0 r^3}\right\}^{\frac{1}{2}} \\ \omega_6 &= \omega_0 \left\{1 - \frac{2\alpha(0)}{4\pi\varepsilon_0 r^3}\right\}^{\frac{1}{2}} \end{aligned}\right\}. \tag{2.38}$$

Since we are interested in the interaction only at long range $(\alpha(0)/4\pi\varepsilon_0 r^3) \ll 1$, we may expand each of these expressions by the binomial theorem. Inserting these results into (2.37) we obtain for the leading term in the dispersion energy

$$U_{\text{disp}} = C_6/r^6 \tag{2.39}$$

where

$$C_6 = -\frac{3}{4} \frac{\hbar\omega_0 \alpha^2(0)}{(4\pi\varepsilon_0)^2}. \tag{2.40}$$

C_6 may be written in an alternative form making use of the definition of the classical oscillator strength and (2.28) as

$$C_6 = -\frac{3}{4} \frac{e^4\hbar}{m^2} \frac{f_0^2}{\omega_0^3} \frac{1}{(4\pi\varepsilon_0)^2} \tag{2.41}$$

where f_0 is the classical oscillator strength for the single frequency ω_0.

The oscillator model was first employed by London in his calculations of dispersion energy[13,14] and was extended to higher multipole moment interactions by Margenau[15] and Hornig and Hirschfelder.[16] The result of this extension is that the dispersion energy can be written

$$U_{\text{disp}} = C_6/r^6 + C_8/r^8 + C_{10}/r^{10} + \ldots, \tag{2.42}$$

where all the coefficients C_6, C_8, C_{10} are negative, so that the energy contributions are all attractive.

The first evaluation of the coefficient C_6 was carried out by London according to (2.40). For the case of the single-frequency oscillator for which (2.40) was derived the corresponding expression for the refractive index of a dilute gas is

$$n^2-1=\frac{\rho e^2}{m\varepsilon_0}\frac{f_0}{(\omega_0^2-\omega^2)} \tag{2.43}$$

from (2.31). London identified the single frequency ω_0 with that corresponding to the ionization energy of a molecule, E_I, so that $\hbar\omega_0=E_I$. He found that (2.43) describes the frequency dependence of the refractive index of a gas quite well, provided that f_0 is taken as a disposable constant. Thus using this value for ω_0 (2.40) can be written

$$C_6=-\frac{3}{4}\frac{E_I\alpha^2(0)}{(4\pi\varepsilon_0)^2} \tag{2.44}$$

so that C_6 can be evaluated from entirely experimental quantities. London's calculation of C_6 for argon, despite its relative simplicity, lies within about 30 per cent of the most recent values whose calculation we now describe.

(b) *Quantum mechanical perturbation theory.* We again consider the interaction of two neutral atoms, one labelled a, the other b, which contain n_a and n_b electrons, respectively. These atoms therefore constitute charge distributions (not necessarily axi-symmetric) to which we can apply a generalization of the discussion of § 2.21 and hence deduce their electrostatic interaction energy in the form of an expansion. For the reasons mentioned there, the validity of the expansion is restricted to large intermolecular separations. The electrostatic energy is

$$U_e(4\pi\varepsilon_0)=\frac{e^2}{r^3}\sum_{i=1}^{n_a}\sum_{j=1}^{n_b}(x_ix_j+y_iy_j-2z_iz_j)$$

$$+\frac{3e^2}{2r^4}\sum_{i=1}^{n_a}\sum_{j=1}^{n_b}\{z_jr_i^2-z_ir_j^2+(z_i-z_j)(2x_ix_j+2y_iy_j-3z_iz_j)\}+\ldots \tag{2.45}$$

Here, x_i, y_i, z_i are the Cartesian coordinates of the electron i of atom a with respect to its charge or mass centre, x_j, y_j, z_j represent similar quantities for the electrons of atom b, r_i and r_j represent the radial distances of the electrons from their respective nuclei. r represents the separation of nuclei of a and b and is measured along the z-axes of the two coordinate systems.

In order to obtain the principal term in the dispersion energy for the two atoms we consider only the first term in the expansion of (2.45). To evaluate this term we use quantum-mechanical perturbation theory.† We express the Hamiltonian, H, of the interacting, two-atom system in the form

$$H = H_a(i) + H_b(j) + U_e = H_0 + U_e$$

where $H_a(i)$ is the unperturbed Hamiltonian for the electronic motion of atom a and $H_b(j)$ a similar quantity for atom b. U_e is the small perturbation potential given by (2.45). We suppose that atoms a and b are in quantum states denoted by q and q' respectively, which represent sets of quantum numbers. The state q is a particular state of the complete set of states, denoted by u, for atom a which have eigenfunctions $A_u(i)$ and energies E_{au} so that the Schrödinger equation can be written

$$H_a(i)A_u(i) = E_{au}A_u(i).$$

Similarly, denoting the eigenfunctions for atom b by $B_v(j)$ and the corresponding energies by E_{bv} we have

$$H_b(j)B_v(j) = E_{bv}B_v(j).$$

Then the complete set of wave functions for the unperturbed Hamiltonian of the two-atom system, $H_a(i) + H_b(j)$, is $\psi_{uv} = A_u(i)B_v(j)$ and the corresponding value of the energy is $E_{uv} = E_{au} + E_{bv}$. For the case when the atoms are in states q and q' the energy of the unperturbed system is therefore

$$E_{qq'} = E_{aq} + E_{bq'}$$

and the wave function, which is the zeroth-order wave function for the perturbation, is

$$\psi_0 = \psi_{qq'} = A_q(i)B_{q'}(j),$$

where the subscript 0 denotes the combination of the states q and q' of the two atoms.

We now write the wave function for the interacting (perturbed) two-atom system, Ψ_0, as an expansion about the zeroth-order wave function and expand the energy, E_0, in a similar manner so that

$$\Psi_0 = \psi_0 + \Psi_0^{(1)} + \Psi_0^{(2)} + \ldots$$
$$E_0 = E_{qq'} + E_0^{(1)} + E_0^{(2)} + E_0^{(3)} + \ldots \tag{2.46}$$

where to ensure that Ψ_0 and ψ_0 are normalized to unity for any order of

† Quantum mechanical perturbation theory is described in most elementary texts on quantum mechanics; see for example ref. 12.

expansion p we require, adopting the Dirac notation,

$$\sum_{k=0}^{p} \int \Psi_0^{(k)*} \Psi_0^{(p-k)} \, d\tau = \sum_{k=0}^{p} \langle \Psi_0^{(k)} \mid \Psi_0^{(p-k)} \rangle = \delta_{p0}$$

Here $\Psi_0^{(0)} = \psi_0$, Ψ^* is the complex conjugate of Ψ and $d\tau$ is an element of volume. Then according to the perturbation theory the terms in the energy expansion (2.46) can be obtained from the solution of the set of differential equations, arising from successive approximations to the Schrödinger wave equation for the system,

$$(H_0 - E_{qq'})\psi_0 = 0$$

$$(H_0 - E_{qq'})\Psi_0^{(1)} + (U_e - E_0^{(1)})\psi_0 = 0$$

$$(H_0 - E_{qq'})\Psi_0^{(2)} + (U_e - E_0^{(1)})\Psi_0^{(1)} = E_0^{(2)}\psi_0$$

from which it may be shown that

$$E_0^{(1)} = \langle \psi_0 | \, U_e \, | \psi_0 \rangle = \int \psi_0^* U_e \psi_0 \, d\tau$$

$$E_0^{(2)} = \langle \psi_0 | \, U_e \, | \Psi_0^{(1)} \rangle = \int \psi_0^* U_e \Psi_0^{(1)} \, d\tau.$$

These equations show that the energy of the system may be evaluated to a particular order from wave functions of a lower order. In general, if the wave function is known to kth order the energy may be calculated to order $2k+1$. If the wave functions $\Psi^{(k)}$, for $k > 0$, are now expanded in the complete set of unperturbed wave functions ψ_{uv} it is possible to show that for example

$$E_0^{(1)} = \langle A_q(i)B_{q'}(j) | \, U_e \, | A_q(i)B_{q'}(j) \rangle \tag{2.47}$$

$$E_0^{(2)} = -\sum_u{}' \sum_v{}' \frac{\langle \psi_{qq'} | \, U_e \, | \psi_{uv} \rangle \langle \psi_{uv} | \, U_e \, | \psi_{qq'} \rangle}{E_{uv} - E_{qq'}}$$

$$= -\sum_u{}' \sum_v{}' \frac{|\langle A_q(i)B_{q'}(j) | \, U_e \, | A_u(i)B_v(j) \rangle|^2}{(E_{au} - E_{aq}) + (E_{bv} - E_{bq'})} \tag{2.48}$$

where the primed summations mean a sum over all the discrete states u and v together with an integral over the continuum states of the unperturbed system, *omitting* the states with energy $E_{qq'} = E_{aq} + E_{bq'}$.

The potential energy of the two-atom system at a specified intermolecular separation, which arises from their interaction alone, is the difference between the total energy of the system at that separation and the total energy at infinite separation, $E_{qq'}$. Thus, the intermolecular potential to second order is

$$\begin{aligned} U &= (E_{qq'} + E_0^{(1)} + E_0^{(2)}) - E_{qq'} \\ &= E_0^{(1)} + E_0^{(2)}, \end{aligned} \tag{2.49}$$

with $E_0^{(1)}$ and $E_0^{(2)}$ given by (2.47) and (2.48).

In this section we wish to evaluate the contribution to the dispersion energy only so that we suppose the two atoms have zero permanent multipole moments of any order, in which case $U_e(\mathbf{r}) = U_e(r)$, and the functions $A_q(i)$ and $B_{q'}(j)$ must be spherically symmetric. Then because $E_0^{(1)}$ is merely an average of U_e weighted according to the square of the unperturbed wave functions, it will vanish. Thus, the leading term in the long-range intermolecular potential, which is the dispersion energy, arises from the second-order perturbation energy, $E_0^{(2)}$, so that

$$\mathcal{U}_{\text{disp}} = E_0^{(2)}.$$

Inserting the expansion (2.45) for U_e into (2.49) we find that the first term in the expression for the dispersion energy is

$$\mathcal{U}_{\text{disp}}(4\pi\varepsilon_0)^2 = -\frac{e^4}{r^6} \sum_u{}' \sum_v{}' \frac{\left\{ \begin{array}{c} \langle A_q(i)|\, x_a \,|A_u(i)\rangle\langle B_{q'}(j)|\, x_b \,|B_v(j)\rangle \\ +\langle A_q(i)|\, y_a \,|A_u(i)\rangle\langle B_{q'}(j)|\, y_b \,|B_v(j)\rangle \\ -2\langle A_q(i)|\, z_a \,|A_u(i)\rangle\langle B_{q'}(j)|\, z_b \,|B_v(j)\rangle \end{array} \right\}^2}{(E_{au} - E_{aq}) + (E_{bv} - E_{bq'})} \tag{2.50}$$

where

$$x_a = \sum_{i=1}^{n_a} x_i, \qquad y_a = \sum_{i=1}^{n_a} y_i, \qquad z_a = \sum_{i=1}^{n_a} z_i,$$

with a similar definition for x_b, y_b, and z_b. We note here that if q and q' represent the ground states of atoms a and b then the denominator of each term in (2.50) is necessarily positive, as is the numerator. Thus for atoms in their ground states the dispersion energy is always attractive, as for two inert gas atoms in their $^1S_0$ ground states. However, for atoms in excited states, the denominator of some terms in the summation of (2.50) is negative and the dispersion energy is therefore not necessarily attractive. The interaction of two inert gas atoms in their first excited states, $^3P_1$, is an example of this.

We now restrict our discussion to the interaction of atoms in their ground states, which we denote by the subscript 0. The dispersion energy for such an interaction takes the form

$$\mathcal{U}_{\text{disp}}(4\pi\varepsilon_0)^2 = \frac{-e^4}{r^6} \sum_u{}' \sum_v{}' \frac{\left\{ \begin{array}{c} \langle A_0(i)|\, x_a \,|A_u(i)\rangle\langle B_0(j)|\, x_b \,|B_v(j)\rangle \\ +\langle A_0(i)|\, y_a \,|A_u(i)\rangle\langle B_0(j)|\, y_b \,|B_v(j)\rangle \\ -2\langle A_0(i)|\, z_a \,|A_u(i)\rangle\langle B_0(j)|\, z_b \,|B_v(j)\rangle \end{array} \right\}^2}{(E_{au} - E_{a0}) + (E_{bv} - E_{b0})}.$$

The subscripts 0, u, and v represent quantum states for the isolated atoms. In general, these states will be characterized by a set of quantum numbers, including a magnetic quantum number, in which the energies are degenerate, in the absence of any external field. We denote the magnetic quantum numbers of atoms a and b in their ground states 0 by $m_0^{\rm a}$ and $m_0^{\rm b}$ respectively.

The corresponding angular momentum quantum numbers are represented by $l_0^{\rm a}$ and $l_0^{\rm b}$, while 0 still denotes the entire set of quantum numbers for the ground state excluding only the magnetic quantum numbers. For the states u and v of atoms a and b, we introduce magnetic quantum numbers $m_u^{\rm a}$, $m_v^{\rm b}$ and represent the remaining quantum numbers by $k_u^{\rm a}$ and $k_v^{\rm b}$. We then average the dispersion energy over all the magnetic quantum numbers of the state of the two interacting atoms according to the formula

$$U_{\rm disp} = \{(2l_0^{\rm a}+1)(2l_0^{\rm b}+1)\}^{-1} \sum_{m_0^{\rm a}} \sum_{m_0^{\rm b}} \mathcal{U}_{\rm disp} \tag{2.52}$$

where the factors $(2l_0^{\rm a}+1)$ and $(2l_0^{\rm b}+1)$ arise from the degeneracy associated with the angular momentum quantum numbers.

We now define an 'oscillator strength' for the dipole transition from the state 0 to a state u by the equation

$$f_{0u}^{(x)} = \frac{2}{3}\frac{m}{\hbar^2} |\langle A_0(i)|\, x_{\rm a}\, |A_u(i)\rangle|^2 (E_{{\rm a}u} - E_{\rm a0})$$

and two similar equations for $f_{0u}^{(y)}$ and $f_{0u}^{(z)}$ involving $y_{\rm a}$ and $z_{\rm a}$. Decomposing these equations to account for the existence of magnetic quantum numbers we have, for example

$$f_{0m_0^{\rm a}k_u^{\rm a}m_u^{\rm a}}^{(x)} = \frac{2}{3}\frac{m}{\hbar^2} |\langle A_{0m_{\rm a}^0}(i)|\, x_{\rm a}\, |A_{k_u^{\rm a}m_u^{\rm a}}(i)\rangle|^2 (E_{{\rm a}k_u^{\rm a}} - E_{\rm a0}).$$

This expression can then be summed over all degenerate final stages $m_u^{\rm a}$ and averaged over all initial states $m_0^{\rm a}$ to obtain

$$f_{0k_u^{\rm a}}^{(x)} = \frac{1}{(2l_0^{\rm a}+1)} \sum_{m_u^{\rm a}} \sum_{m_0} f_{0m_0^{\rm a}k_u^{\rm a}m_u^{\rm a}}^{(x)}$$

with similar expressions for $f_{0k_u^{\rm a}}^{(y)}$ and $f_{0k_u^{\rm a}}^{(z)}$.

When these summations are carried out it can be shown that

$$f_{0k_u^{\rm a}}^{(x)} = f_{0k_u^{\rm a}}^{(y)} = f_{0k_u^{\rm a}}^{(z)} = f_{0k_u^{\rm a}}/3$$

with

$$f_{0k_u^{\rm a}} = \frac{2}{3}\frac{m}{\hbar^2} |\langle A_0(i)|\, \mathbf{r}_{\rm a}\, |A_{k_u^{\rm a}}(i)\rangle|^2 (E_{{\rm a}k_u^{\rm a}} - E_{\rm a0}) \tag{2.53}$$

where

$$\mathbf{r}_a = \sum_i \mathbf{r}_i.$$

If these results and the analogous relations for atom b are employed in (2.52) for U_{disp} we obtain

$$U_{\text{disp}} = C_6/r^6 \tag{2.54}$$

where

$$C_6 = \frac{3}{2} \frac{e^4\hbar^4}{m^2(4\pi\varepsilon_0)^2} \sum_{k_u^a}{}' \sum_{k_v^b}{}' \frac{f_{0k_u^a} f_{0k_v^b}}{\{E_{a0} - E_{ak_u^a}\}\{E_{b0} - E_{bk_v^b}\}\{(E_{a0} - E_{ak_u^a}) + (E_{b0} - E_{bk_v^b})\}}. \tag{2.55}$$

The rigorous quantum mechanical perturbation theory therefore confirms the functional dependence of the dispersion energy on intermolecular separation obtained with the oscillator model ((2.39)–(2.41)) but the coefficient C_6 is now defined differently.

We also see that further terms in the expression for the dispersion energy would be obtained by the use of higher terms of the expansion for U_e ((2.45)) in (2.49). It is clear that inclusion of these terms will lead to a series for U_{disp} of the form

$$U_{\text{disp}} = \frac{C_6}{r^6} + \frac{C_8}{r^8} + \frac{C_{10}}{r^{10}} + \ldots.$$

The higher-order coefficients, C_8, C_{10}, etc. cannot however be expressed in terms of dipole oscillator strengths.

The evaluation of the coefficient C_6 according to (2.55) requires much more information than that for the oscillator model. In particular, it requires a knowledge of all the dipole oscillator strengths for transitions from the ground state to excited states and the corresponding energy levels. The oscillator strengths $f_{0k_u^a}$ describe the intensity of light absorption by an atom at a frequency ω_u given by

$$\omega_u = E_{qk_u^a} - E_{a0}.$$

Thus if the entire absorption spectrum of the atom were known experimentally or theoretically throughout the visible, ultraviolet, and X-ray regions, then this distribution of f values could be used in conjunction with measurements of the frequencies ω_u to evaluate the constant C_6. The theoretical evaluation of f values (as can be seen from their definition) requires a complete, exact knowledge of the wave functions for the atoms in their ground and excited states. Exact wave functions are known only for the hydrogen atom so that this route is not generally available.

Experimental determinations of the photoabsorption spectrum over the entire spectral range are also unavailable, although the frequencies of the transitions ω_u are usually well known. For these reasons it has been necessary to adopt less direct methods for the evaluation of C_6. In the next section methods utilizing a variety of experimental information are described; subsequently several *ab initio* calculations are discussed.

(c) *Semi-empirical methods of evaluation of C_6.* Although the complete photoabsorption spectrum of atoms is not known experimentally various facts about the total distribution of f values are known. These facts may be used to supplement or replace the photoabsorption information using methods originated by Dalgarno.

(1) The dipole-oscillator strengths are positive, as follows from their definition, thus

$$f_{0k_u^a} \geqslant 0.$$

(2) There is an energy gap between the ground state E_{a0} and the first excited state E_{a1}. Transition to energies in this gap is impossible

$$f_{0k_u^a} = 0 \quad \text{for} \quad E_{a0} < E < E_{a1}.$$

(3) There is a series of sum rules[17] which provide further constraints on the distribution of oscillator strengths.

The first sum rule, known as the Thomas–Reiche–Kuhn sum rule relates the total oscillator strength to n_a the number of electrons in the atom so that

$$S(0) = n_a = \sum_{k_u^a=1}^{\infty} f_{0k_u^a}.$$

A second sum rule can be obtained from the low-frequency limit of the polarizability of an atom. According to a quantum mechanical treatment the polarizability at a frequency ω is related to the dipole oscillator strengths by the relation

$$\alpha(\omega) = \frac{e^2}{m} \sum_{k_u^a=1}^{\infty} \frac{f_{0k_u^a}}{\omega_{k_u^a}^2 - \omega^2} \tag{2.56}$$

where $\omega_{k_u^a}$ is the frequency corresponding to a transition to the state k_u^a from the ground state

$$\hbar\omega_{k_u^a} = E_{ak_u^a} - E_{a0}.$$

Eqn (2.56) is thus the quantum mechanical version of (2.29) derived on the basis of the oscillator model. The low-frequency limit of the polarizability is therefore given by

$$\frac{m\alpha(0)}{e^2}=S(-2)=\sum_{k_u^a=1}^{\infty}\omega_{k_u^a}^{-2}f_{0k_u^a}.$$

Further sum rules follow from the derivatives of (2.56) with respect to ω evaluated at zero frequency.

$$S(-2l-2)=\sum_{k_u^a=1}^{\infty}\omega_{k_u^a}^{-2-2l}f_{0k_u^a}$$

$$=\frac{m}{e^2}\frac{1}{2l!}\left[\frac{d^{2l}\alpha(\omega)}{d\omega^{2l}}\right]_{\omega=0}.$$

Values of $S(-2)$, $S(-4)$, and $S(-6)$ have also been obtained from experimental measurements of the refractive index of gases. This is possible because the refractive index, n, of gas for radiation of frequency ω is related to the dipole oscillator strengths as[18]

$$n^2-1=K_0\sum_{k_u^a=1}^{\infty}\frac{f_{0k_u^a}}{\omega_{k_u^a}^2-\omega^2}. \tag{2.57}$$

It will be noted that this is the quantum mechanical representation of (2.31) quoted earlier in connection with the oscillator model. Here K_0 is a known factor whose value depends on the number density of atoms in the gas and on the system of units employed as indicated in (2.31). Provided that $\omega<\omega_{k_u^a}$ for all k_u^a (which means that $\omega<\omega_1$ the frequency for a transition from the ground state to the first excited state) (2.57) can be expanded to yield

$$n^2-1=K_0\left\{\sum_{k_u^a=1}^{\infty}\frac{f_{0k_u^a}}{\omega_{k_u^a}^2}+\omega^2K'\sum_{k_u^a=1}^{\infty}\frac{f_{0k_u^a}}{\omega_{k_u^a}^4}+\omega^4K''\sum_{k_u^a=1}^{\infty}\frac{f_{0k_u^a}}{\omega_{k_u^a}^6}+\ldots\right\} \tag{2.58}$$

$$=K_0\{S(-2)+\omega^2K_1S(-4)+\omega^4K_2S(-6)+\ldots\}$$

where K_1 and K_2 are further known constants. Thus by fitting experimental values of the refractive index as a function of frequency at low frequencies to an equation of the form (2.58) it is possible to extract values of $S(-2)$, $S(-4)$, $S(-6)$. The extension to higher-order sums is inhibited by uncertainties in the experimental data.

A further source of experimental information for sums is afforded by measurements of the Verdet constant,[18] $V(\omega)$, which describes the angle of rotation of the plane of polarization of polarized light in a gas placed in a magnetic field. It is defined by

$$\phi=VLH$$

where ϕ is the angle of rotation of the plane of polarization, L the length of the light path in the gas, and H the strength of the magnetic field. It may be shown that the Verdet constant for a gas is related to its refractive index as,[18]

$$V = V_0 \omega \frac{dn}{d\omega}.$$

where V_0 is a constant.

Repeating the development employed for the refractive index the Verdet constant too may be expressed in terms of sums at low frequencies as

$$V = K_0' \omega^2 \{S(-4) + \omega^2 K_1' S(-6) + \omega^4 K_2' S(-8) + \ldots\}$$

where K_0', K_1', K_2' are known constants under prescribed conditions.

Thus from measurements of the Verdet constant at low frequencies as a function of frequency values of $S(-4)$, $S(-6)$, and $S(-8)$ may be obtained. Again experimental errors prevent the extension to higher sum rule values. Sums for indices other than negative, even values can also be defined; however, these are not accessible to experiment and must be calculated from atomic wave functions.

In early work these sums for the distribution of oscillator strengths were used to augment the available experimental, photoabsorption data, by extrapolation and interpolation using the sum rules as checks. However, later developments have allowed the sum rules to be used directly in the evaluation of C_6. By making use of the identity

$$[ab(a+b)]^{-1} = \frac{2}{\pi} \int_0^\infty \frac{du}{(a^2+u^2)(b^2+u^2)} \quad \text{for} \quad a>0,\ b>0,$$

it is possible to transform (2.55) for C_6 into the form

$$C_6 = \frac{-3\hbar}{\pi(4\pi\varepsilon_0)^2} \int_0^\infty \alpha_a(i\omega)\alpha_b(i\omega)\, d\omega \qquad (2.59)$$

where $\alpha_a(i\omega)$ is the dipole polarizability of atom a at the imaginary frequency $i\omega$. That is from (2.56)

$$\alpha_a(i\omega) = \frac{e^2}{m} \sum_{k_u^a=1}^{\infty} \frac{f_{0k_u^a}}{\omega_{k_u^a}^2 + \omega^2}$$

with an equivalent expression for $\alpha_b(i\omega)$.

The advantage of (2.59) is that it incorporates separated information about the individual atoms a and b and does not contain sums and products of quantities relating to the two atoms, as (2.55) does. The imaginary frequencies $i\omega$ mean that the magnitude of the external polarizing field changes exponentially with time.

For real frequencies the polarizability can be written as a power series in ω^2 for small ω, so that

$$\alpha_a(\omega) = \frac{e^2}{m} \sum_{k'=0}^{\infty} S(-2k'-2)\omega^{2k'} \tag{2.60}$$

However, this series is only convergent for $\omega < \omega_1$, so that it cannot be used directly to obtain $\alpha(i\omega)$ for all values of ω. Consequently, it is necessary to find an analytic approximation for $\alpha(\omega)$ for all frequencies. This process is called analytic continuation.[19] Denoting the approximate value of α_a by $\tilde{\alpha}_a(\omega)$ we write, in the form of a Padé approximant,

$$\tilde{\alpha}_a(\omega) = \frac{a_0 + a_1\omega^2 + \ldots a_{N-2}\omega^{2N-2}}{1 + b_1\omega^2 + \ldots b_N\omega^{2N}}. \tag{2.61}$$

This expression contains $2N$ coefficients, a_p and b_p; the coefficients may be chosen so that the power series expansion of (2.61) equals that of (2.60) term-by-term up to an order $2N-1$ which is determined by the number of sums $S(-2k'-2)$ available. This procedure allows evaluation of a_p and b_p in terms of the sums themselves. The resulting expression for $\tilde{\alpha}(i\omega)$ can, for convenience be written in the form

$$\tilde{\alpha}_a(i\omega) = \frac{e^2}{m} \sum_{p=1}^{N} \tilde{f}_p^a(\tilde{\omega}_p^{a2} + \omega^2)^{-1} \tag{2.62}$$

by means of a partial fraction reduction. The constants $\tilde{f}_p^a$ and $\tilde{\omega}_p^a$ are determined from the coefficients a_p and b_p and hence, are related to the sums $S(-2k'-2)$. Inserting (2.62) and a similar one for atom b into (2.59) for C_6 and carrying out the integration we find

$$C_6 = -\frac{3e^4\hbar}{2m^2(4\pi\varepsilon_0)^2} \sum_{p,q}^{N} \frac{\tilde{f}_p^a \tilde{f}_q^b}{[\tilde{\omega}_q^a \tilde{\omega}_q^b(\tilde{\omega}_q^a + \tilde{\omega}_q^b)]}. \tag{2.63}$$

Thus, the coefficient of the dispersion energy can be expressed in terms of sums only, without the direct use of any photoabsorption data for oscillator strengths.

This procedure has been developed to the stage where methods are available for choosing the coefficients a_p and b_p of (2.61) so that they lead to both upper and lower bounds on the calculated values of C_6. When allowance is made for uncertainties in the sums these bounds are merely widened. In addition techniques have been designed to use sum rules other than the negative even values and to incorporate photoabsorption data into the calculations where they are available and reliable.[20]

The methods outlined above have been employed to evaluate the coefficient C_6 for the inert gases and some simple polyatomic systems.[20] The best available estimates for C_6 are given in Table 2.1. Similar

Table 2.1

The dipole–dipole interaction coefficients[20]

Interaction	$-C_6 \times 10^{79}$/J m^6	Interaction	$-C_6 \times 10^{79}$/J m^6
He–He	1·3989±0·0004	Ar–N_2O	111±12
He–Ne	2·93±0·084	Ar–H_2	26·9±0·8
He–Ar	9·44±0.15	Kr–Kr	127±1
He–Kr	13·07±0·06	Kr–Xe	190±5
He–Xe	19·44±0·56	Kr–O_2	97±2
He–O_2	10·37±0·19	Kr–N_2O	154±14
He–N_2O	15·9±1·5	Kr–H_2	38·0±0·8
He–H_2	3·834±0·043	Xe–Xe	288±13
Ne–Ne	6·27±0·83	Xe–O_2	143±7
Ne–Ar	19·59±0·75	Xe–N_2O	231±25
Ne–Kr	26·93±0·71	Xe–H_2	57·2±2·0
Ne–Xe	40·2±1·3	O_2–O_2	77·1±2·8
Ne-O_2	21·8±1·1	O_2–N_2O	118±15
Ne–N_2O	33·3±4·8	O_2–H_2	28·1±1·1
Ne–H_2	7·75±0·30	N_2O–N_2O	190±34
Ar–Ar	64·8±1·9	N_2O–H_2	45·9±4·2
Ar–Kr	90·6±1·6	H_2–H_2	11·56±0·32
Ar–Xe	135±5		
Ar–O_2	69·8±2·5		

C_6 is often quoted in atomic units; 1 a.u. = 0.957 15 × 10^{-79} J m^6.

techniques have been employed to estimate C_8 coefficients from quadrupole sum rules[21] and the anisotropy in C_6 for interaction between asymmetric molecules.[22] In addition, the long-range contribution to the energy of three atoms which we shall mention later, has been computed.[23]

(d) *Ab initio calculations.* Eqn (2.53) for the dipole oscillator strengths shows that the theoretical evaluation of the coefficient C_6 through (2.55) or (2.59) requires a knowledge of the wave functions of the atoms in their ground and all excited states as well as the eigenvalues of the energy. In general, these wave functions are not known exactly for many-electron atoms, owing to the impossibility of obtaining an exact solution of the Schrödinger equation for these systems. Consequently, approximate wave functions for many-electron atoms have been obtained by a variety of techniques.

A theorem of fundamental importance in all of the calculations is the Ritz variational theorem. According to this, a quantity

$$\bar{H} = \frac{\int \phi^* H \phi \, d\tau}{\int \phi^* \phi \, d\tau} \tag{2.64}$$

where H is the Hamiltonian for the system, is never smaller than the eigenvalue of the energy, E, of the Schrödinger equation $H\psi = E\psi$ (where ψ is the exact wave function of the system) for any function which has the proper symmetry and satisfies the boundary conditions imposed on ψ. Thus, if we assume for ϕ a function with arbitrary parameters, compute $\bar{H}$ and then vary the parameters to make $\bar{H}$ a minimum, it is expected that this minimum is close to E and ϕ is an approximation to the wave function ψ. This procedure can evidently be used directly to evaluate the total energy of two interacting atoms where the Hamiltonian is that of the two-atom system. However, such calculations have the disadvantage that the intermolecular potential energy must then be expressed as the difference between the variationally calculated energy of the interacting molecules and the energy of the two isolated molecules. Since this is a small difference between two very large quantities the uncertainties in the variational energy, which depend on the quality of the trial wave functions, may contribute to large errors in the interaction energy. This method, which requires wave functions centred on both molecules, has been successfully applied only to the long-range energies of simple atomic systems such as helium.[24]

Rather more success has been achieved in the theoretical calculation of long-range intermolecular forces by combining the variational procedure with the multipole expansion of the perturbation energy. This is partly because in the form of (2.59) the evaluation of the coefficient C_6 is only a one-centre problem. The essential task of the theoretical calculations in this case is the evaluation of the polarizability of the individual atoms at imaginary frequencies $\alpha(i\omega)$, that is for time-varying fields. Initial efforts in this direction employed uncoupled Hartree–Fock perturbation theory for isolated atoms. The Hartree–Fock approach involves the assumption that any one electron moves in a potential that is a spherical average of the potential due to all the other electrons. The Schrödinger equation is then solved numerically for this potential. Of course in order to construct the spherical average of the potential due to all the other electrons their wave functions must be known, so initial trial functions are assumed for them and a mean potential then calculated. The wave function for the chosen electron is then used to construct a new mean potential experienced by a second electron. The wave function for the second electron in this field is then obtained and used again to generate the mean potential for a third electron. This process is repeated for all the electrons in the atom so that first iterate wave functions of all the electrons are obtained. The entire calculation cycle is then performed several times beginning with the original electron each time until the solutions for all electrons are unchanged on successive cycles.

When the Hamiltonian for the isolated atom is perturbed by a static electric field the same principles of the Hartree–Fock method can be employed to deduce the appropriate wave functions and hence the static polarizability. If this process is carried out with the assumption that electron orbitals are affected independently by the field then the uncoupled version of the theory results. If this assumption is not applied the more accurate coupled Hartree–Fock theory results.[25,26] The extensions of the analysis to the application of a time-varying electric field results in the time-dependent Hartree–Fock theory.[27,28] Calculations based upon this latter scheme have been carried out for He–He interactions.[29] A second method of introducing the correlation between the electrons into the uncoupled Hartree–Fock theory has been developed. In this scheme the approximate wave function of the atom, obtained by neglecting electron correlation, is supposed to be a perturbation of the true wave function, and is analysed by standard perturbation theory. Then, in order to evaluate, for example, the frequency-dependent polarizability of an atom, the external field is viewed as a perturbation to the approximate wave function, or equivalently as a second perturbation to the true wave function. The method which is called double perturbation, has the advantage that explicit evaluation of the wave functions of correlated electrons is unnecessary. On the other hand the calculations involve the coordinates of two electrons simultaneously, whereas the Hartree–Fock theory deals only with single electrons. Calculations of the He–He interaction have been carried out by this method.[30]

Theoretical calculations have also been used as an auxiliary source of sum rule values for semi-empirical calculations. In general the calculations have been performed to yield sum rules not accessible from experimental information such as $S(2)$, which is related to the electron density at the nucleus for the ground-state atom.[31] These extra sum rules can be incorporated into the general semi-empirical scheme outlined earlier.[20]

The most accurate scheme for the evaluation of the coefficient C_6 in the dispersion forces for many atomic and molecular systems is undoubtedly the semi-empirical method. The theoretical calculations are likely to prove of greater value for the higher-order contributions to the dispersion forces.

(e) *Summary of the long-range pair potential energy function.* In the previous sections we have described the evaluation of each contribution to the long-range interaction of pairs of molecules in the expression

$$U(\mathbf{r}, \boldsymbol{\omega}) = U_{\text{el}}(\mathbf{r}, \boldsymbol{\omega}) + U_{\text{ind}}(\mathbf{r}, \boldsymbol{\omega}) + U_{\text{disp}}(\mathbf{r}, \boldsymbol{\omega}).$$

Here, the potential energy function depends upon the relative orientation of the two molecules specified by $\boldsymbol{\omega}$. For the interaction between two polar molecules all three terms are present, for the interaction between a polar and non-polar molecule only the last two terms contribute, whereas for the interaction between two non-polar molecules only the final term remains. If the orientationally-averaged intermolecular potential is required it may be constructed in an analogous fashion to that described in § 2.2.1 and we find

$$\langle U\rangle = \langle U_{\mathrm{el}}\rangle + \langle U_{\mathrm{ind}}\rangle + \langle U_{\mathrm{disp}}\rangle$$

since to a high degree of approximation the cross-terms in the averaging procedure can be neglected.[32] Thus, for the orientationally-averaged potential energy function for the interaction of two neutral, linear, dipolar molecules we find, collecting our earlier results

$$\langle U\rangle = -\frac{1}{r^6}\left\{\frac{2\mu^4}{3(4\pi\varepsilon_0)^2 kT} + \frac{2\mu^2\alpha(0)}{(4\pi\varepsilon_0)^2} - C_6\right\}. \tag{2.65}$$

2.2.4. Long-range three-atom potential

A further result of the theory of long-range dispersion forces which will be referred to in later parts of this book concerns the simultaneous interaction of three molecules. Here we shall consider the interaction energy for three inert gas atoms a, b, and c separated by distances r_{ab}, r_{bc}, and r_{ac} where θ_{a}, θ_{b}, θ_{c} are the internal angles of the triangle abc. By carrying the perturbation theory analysis of § 2.2.3 as far as the third-order energy $E_0^{(3)}$, it can be shown[33] that the leading terms in the expression for the dispersion energy of the three-atom system are

$$U_{\mathrm{disp}}^{\mathrm{abc}}(r_{\mathrm{ab}}, r_{\mathrm{bc}}, r_{\mathrm{ac}}) = +\frac{C_6^{\mathrm{ab}}}{r_{\mathrm{ab}}^6} + \frac{C_6^{\mathrm{bc}}}{r_{\mathrm{bc}}^6} + \frac{C_6^{\mathrm{ac}}}{r_{\mathrm{ac}}^6} + \frac{\nu_{\mathrm{abc}}(3\cos\theta_{\mathrm{a}}\cos\theta_{\mathrm{b}}\cos\theta_{\mathrm{c}} + 1)}{(r_{\mathrm{ab}}r_{\mathrm{bc}}r_{\mathrm{ac}})^3}. \tag{2.66}$$

Here, the first three terms represent the interaction energies of the molecules taken in pairs with each C_6 given by our earlier equations. If the long-range potentials were pair-wise additive these would be the only contributions. However, the fourth term on the right-hand side of (2.66) comprises a non-additive term known as the Axilrod-Teller triple-dipole term. Following the developments applied to the coefficient C_6 it is possible to show that

$$\nu_{\mathrm{abc}} = \frac{3\hbar}{\pi(4\pi\varepsilon_0)^2}\int_0^\infty \alpha_{\mathrm{a}}(\mathrm{i}\omega)\alpha_{\mathrm{b}}(\mathrm{i}\omega)\alpha_{\mathrm{c}}(\mathrm{i}\omega)\,\mathrm{d}\omega \tag{2.67}$$

where $\alpha_{\mathrm{a}}(\mathrm{i}\omega)$ is the dipole polarizability of atom a at the imaginary frequency $\mathrm{i}\omega$. Using the techniques outlined earlier it is then possible to evaluate the triple-dipole constant ν_{abc}. This has been carried out by the coupled, time-dependent Hartree–Fock method for helium and neon interactions.[34] ν_{abc} is approximately equal to $-\frac{3}{4}\alpha C_6$.

2.2.5. Combination rules

A further result of interest for some of the later material in the book is concerned with the relation between the interaction potential for two unlike atoms and the interaction potential of pairs of like atoms. An approximate relationship of this kind for the long-range dispersion potential can be obtained. We shall obtain the result from our discussion of the oscillator model although it can also be deduced by approximation of the full quantum mechanical treatment. As we saw for the interaction of a pair of molecules of species a the London result for the first dispersion energy coefficient can be approximated by

$$C_6^{\mathrm{aa}} = -\frac{3}{4}\frac{E_{\mathrm{I_a}}\alpha_{\mathrm{a}}^2(0)}{(4\pi\varepsilon_0)^2}$$

and for the interaction of a pair of molecules of species b as

$$C_6^{\mathrm{bb}} = -\frac{3}{4}\frac{E_{\mathrm{I_b}}\alpha_{\mathrm{b}}^2(0)}{(4\pi\varepsilon_0)^2}.$$

Here $E_{\mathrm{I_a}}$, $E_{\mathrm{I_b}}$ are the ionization energies of the molecules a and b whereas $\alpha_{\mathrm{a}}(0)$, $\alpha_{\mathrm{b}}(0)$ are their static polarizabilities. For the interaction between two dissimilar molecules a and b it is easily shown by a generalization of the argument of § 2.2.3a that the corresponding expression for C_6^{ab} is

$$C_6^{\mathrm{ab}} = -\frac{3}{2}\frac{E_{\mathrm{I_a}}E_{\mathrm{I_b}}}{E_{\mathrm{I_a}}+E_{\mathrm{I_b}}}\frac{\alpha_{\mathrm{a}}(0)\alpha_{\mathrm{b}}(0)}{(4\pi\varepsilon_0)^2}.$$

Now because these ionization energies vary but slightly from molecule to molecule ($E_{\mathrm{I_a}} \simeq E_{\mathrm{I_b}}$) we find that

$$C_6^{\mathrm{ab}} = (C_6^{\mathrm{aa}}C_6^{\mathrm{bb}})^{\frac{1}{2}}. \tag{2.68}$$

This relation forms the basis of one of the empirical combining rules discussed later in the book.

2.3. Short-range forces

The theoretical calculation of the short-range forces between atoms or molecules cannot be performed by an extension of the perturbation theory used for long-range forces to small separations for a number of reasons. First, the multipole expansion of the electrostatic energy (2.45) does not converge when the electron clouds of the two atoms overlap appreciably. Secondly, the wave function used for the two-atom system in the evaluation of long-range forces does not have the symmetry required by the Pauli exclusion principle as discussed in Chapter 1. Finally, the perturbation of the energy of the two isolated systems is no longer small.

These observations lead to the conclusion that the separated atomic system does not form the best starting point for any perturbation theory. Accordingly perturbation theory approaches to the calculation of short-range intermolecular forces have begun from the opposite extreme as a basis in which the two nuclei are supposed so close together that the entire system constitutes a 'united atom'. This treatment has the immediate advantage that the conditions of the Pauli principle which govern the exchange contribution to the intermolecular potential are incorporated in the wave function for the united atom. A second method which has been employed for the calculation of short range forces is an extension of the Heitler–London technique briefly discussed in Chapter 1. More recently, molecular orbital calculations have been increasingly used in conjunction with variational methods to obtain short range intermolecular potentials. Finally, a few model calculations of short-range forces have met with some success. In this section we shall briefly discuss each type of procedure.

2.3.1. Heitler–London methods

In § 1.5 we provided a preliminary example of the application of the Heitler–London method to the evaluation of the interaction of two hydrogen atoms at short range. The procedures described there have been generalized[35,36] to treat the interaction of an atom a with n_a electrons and an atom b with n_b electrons.

It will be recalled that in the Heitler–London procedure the properly anti-symmetrized wave functions for the separated atoms are combined to yield a total wave function for the interacting system. If we denote by A the unperturbed wave function for atom a and by B wave function for atom b then the general anti-symmetrized wave function for the two atoms can be written

$$\psi_a = \sum_{\lambda_a} (-1)^{\lambda_a} P_{\lambda_a} A = \mathscr{A}^a A$$
$$\psi_b = \sum_{\lambda_b} (-1)^{\lambda_b} P_{\lambda_b} B = \mathscr{A}^b B. \tag{2.69}$$

Here P_{λ_a} and P_{λ_b} represent the permutations of electrons in the atoms a and b excluding any permutations of electrons between atoms a and b. These latter permutations we denote by $P_{\lambda_{ab}}$; $\mathscr{A}$ is called the permutation operator. We then write the total wave function for the two-atom system as

$$\Psi = \sum_{\lambda_{ab}} (-1)^{\lambda_{ab}} P_{\lambda_{ab}} \psi_a \psi_b = \mathscr{A}^{ab} \mathscr{A}^b \mathscr{A}^a AB = \mathscr{A}(AB) \tag{2.70}$$

if we presume that the electrons in one molecule move independently of those in the other except for the operation of the Pauli principle.

Now, if A and B are exact eigenfunctions for the respective atoms H_a, H_b their Hamiltonians and E_a, E_b the corresponding energies, then

$$H_aA = E_aA \quad \text{and} \quad H_bB = E_bB.$$

Thus if H is the total Hamiltonian for the interacting system, excluding nuclear motion,

$$H = H_a + H_b + V_e \tag{2.71}$$

where V_e is the electrostatic interaction energy. Now,

$$H\Psi = H\mathscr{A}(AB) = \mathscr{A}HAB$$

because H is invariant to electron permutation. Thus invoking (2.69), (2.70), and (2.71)

$$H\Psi = (E_a + E_b)\Psi + \mathscr{A}(V_eAB)$$

from which it may be deduced that

$$\frac{\langle\Psi|\, H\,|\Psi\rangle}{\langle\Psi\,|\,\Psi\rangle} = E_a + E_b + U$$

where

$$U = \frac{\langle AB\,|\,\mathscr{A}(V_eAB)\rangle}{\langle AB\,|\,\mathscr{A}(AB)\rangle} \tag{2.72}$$

and represents the intermolecular potential energy.

Eqn (2.72) therefore represents the generalization of (1.37) for the interaction of atoms containing many electrons. For systems other than hydrogen atoms the exact wave functions of the isolated atoms are not known, and then (2.72) for the intermolecular potential is strictly not valid. Nevertheless, calculations based on the equation have been carried out. The first step in such a procedure is the choice of suitable approximate atomic wave functions which are generally obtained by the Hartree–Fock self-consistent field theory outlined in §2.2.3. Subsequently these atomic wave functions are combined according to (2.70) to yield total wave functions for two interacting atoms. These wave functions are then employed in the evaluation of U through (2.72). Such calculations have been carried out[37,38] for the systems Ne–Ne and Ar–Ar.

Because (2.72) is not valid if inexact wave functions are used a more reliable approach to the evaluation of the interaction energy uses a variational method. The variational technique was described earlier in § 2.2.3 and in the present context approximate atomic wave functions are combined through (2.70) to provide a total wave function which is used as the variational function of the method. By evaluating the energy at infinite nuclear separation $E(\infty)$ and at a series of finite nuclear separations $E(r)$ the intermolecular potential is determined as

$$U = E(r) - E(\infty).$$

Such calculations have been carried out in this form only for helium.[39] Heitler–London methods have the disadvantage that the atomic wave functions employed are supposed undistorted by the interaction which cannot be exactly true and for this reason the Heitler–London methods predict zero dispersion energy. The method also tends to overemphasize the localization of electrons on particular atoms.

2.3.2. *Molecular orbital theory*

Molecular orbital theory provides a further means of constructing wave functions for single molecules or for two interacting molecules. When such wave functions have been constructed the energy of the molecules or interacting pair of molecules may be evaluated. Thus, this theory and Heitler–London theory have in common the search for wave functions for the interacting molecule system which conform to the requirements of the Pauli Principle. However, in the Heitler–London theory of the interaction of two H atoms the system wave function did not allow the possibility that both electrons might be associated with the same atom, so that terms such as $A(1)A(2)$ and $B(1)B(2)$ did not enter the expression for Ψ. In fact, such an arrangement is possible and molecular orbital theory allows for this by the construction of molecular orbitals for the two-atom system which may have up to two electrons in each orbital.

The wave function for the interacting system of two identical atoms each containing N electrons is written

$$\Psi = \mathscr{A}\left(\prod_{i=1}^{2N} \Phi_i\right) \tag{2.73}$$

where Φ_i is the ith molecular orbital for an electron, and $\mathscr{A}$ is the permutation operator defined in the last section. The molecular orbitals are then written as linear combinations of atomic wave functions $\phi_{\mathrm{a}j}$ and $\phi_{\mathrm{b}j}$ centred on the two nuclei a and b.

Thus,

$$\Phi_i = \sum_j c_{ji} \left\{\frac{\phi_{\mathrm{a}j} + \lambda_i \phi_{\mathrm{b}j}}{\sqrt{2}}\right\} S$$

where S is the spin function α or β and λ_i is ± 1, chosen to satisfy the symmetry of the molecular orbital. The functions $\phi_{\mathrm{a}j}$ can be any atomic wave functions but they are usually Slater orbitals,[40] occasionally

Gaussian orbitals,[41] or for the best results the Hartree–Fock orbitals mentioned earlier.

The molecular orbital functions Φ_i are represented in the following way.[42] A set of functions X_j is chosen so that Φ_i can be written

$$\Phi_i = \sum_j c_{ji} X_j S.$$

The functions X_j are then essentially sums or differences of atomic wave functions centred on each atom. The minimum set of functions X_j is chosen to represent adequately the orbital Φ_i. The coefficients c_{ij} are then found by application of the variational method to the expectation value of the energy for the resulting wave function Ψ and by a self-consistent field technique. When the state function Ψ is known the energy of the system can be evaluated.

In this form the molecular orbital theory has been employed to evaluate the short-range interaction of many systems, for example, Li^+–H_2,[43] Cl–H_2,[44] as well as interactions among the inert gases[45] He–He, Ne–Ne, and Ar–Ar.

The deficiency of the molecular orbital theory outlined above is its tendency to overemphasize the delocalization of electrons. According to the simple form of theory each electron spends half of its time about each nucleus of an interacting pair of atoms. At a very large separation this leads to the unrealistic result that for the interaction of two hydrogen atoms for example they are equally likely to separate into $H^+ + H^-$ as into H+H. In contrast the Heitler–London theory never allows separation to H^+ and H^-. A true description of the physics of the problem evidently lies somewhere between the complete delocalization of electrons implicit in the molecular orbital theory and the complete localization of the Heitler–London theory. Either theory may be adapted to correct these defects and if this were completely achieved both methods would attain the same result. It has been more common to adjust the molecular orbital theory by means of configuration interaction calculations which allow for the construction of several different molecular orbitals from the same atomic orbitals into which the electrons of the atoms can be placed in a variety of ways. Thus the expression for the state function of the two-atom system becomes

$$\Psi = \sum_k C_k \mathscr{A}\left(\prod_i \Phi_{ki}\right)$$

where Φ_{ki} are the molecular orbitals constructed as before. Using this technique it becomes possible to introduce a greater degree of localization into the molecular orbital theory. Phillipson[46] performed a calculation of this type for helium atom interactions, using a state function Ψ

containing 64 terms, for the short-range region. Other calculations have been carried out for systems containing only a few electrons such as $Li+F_2$[47] and $He+H_2$.[48] No calculations of the type have yet been performed for the heavier inert gases owing to the complexity of the problem.

2.3.3. *United atom perturbation theory*

In the limit where the internuclear separation of two atoms (a and b) approaches zero the two-atom system may be regarded as a single atom having an atomic number, Z, equal to the sum of the values for each atom $Z = Z_a + Z_b$. The electrons can then be supposed to occupy orbitals characteristic of this united atom. This can be made the basis of a perturbation theory by presuming that when the nuclei of the interacting atoms are separated by a small distance r, the energy and wave function of the system can be calculated as perturbations on the values for the united atom.

The first developments of the united atom perturbation theory were analogous to those of the long-range perturbation theory. In order to avoid the singularity at zero nuclear separation we consider the electronic energy $W(r)$ defined by

$$W(r) = E - Z_a Z_b / r$$

where E is the total energy of the system. $W(r)$ is then expanded to lead to a series for W in powers of r about the electronic energy of the united atom,[49] $E_0 = W(0)$.

$$W = E_0 + r^2 W_2 + r^3 W_3 + r^4 W_4 + r^5 \ln r W_5' + r^5 W_5 + \ldots. \tag{2.74}$$

The remaining unspecified parameter of this theory is then the position of the centre of the united atom and its wave function. In early work this position was chosen as the centre of nuclear charge which has the advantage that it simplifies the analysis and improves the convergence of the perturbation expansion. Using this assignment an explicit expression for the coefficient W_2 has been derived. The coefficients W_2, W_3, etc. are analogous to the coefficients C_6, C_8 of the long-range perturbation theory. Unfortunately, unlike the long-range perturbation expansion, the series of (2.74) contains higher-order non-analytic terms such as the $r^5 \ln r$ term and is not accurate when truncated. Consequently, including the coefficient W_2 only yields a poor approximation to the intermolecular potential.

It has been demonstrated[50] that the essential reason why the united atom perturbation expansion given above is not satisfactory lies in the fact that no single united atom wave function, no matter where it is situated, can reproduce the behaviour for the true two-atom system wave function in the neighbourhood of the two nuclei because of the Coulombic singularities in the Hamiltonian at these points. Consequently, the zeroth-order wave function in the perturbation is a poor approximation and the burden of correcting this is thrown onto higher terms in the expansion which are not easily able to cope with the correction. Byers Brown[50] has developed an alternative perturbation expansion for the united atom theory by using as zeroth-order wave function a linear combination of two united atom wave functions one centred on each nucleus. This wave function satisfies the conditions imposed at the two nuclei by the automatic singularity so that the strain of doing this is removed from higher-order terms in any perturbation expansion. In the analysis of Byers Brown an expansion of the electronic energy in terms of the parameter $\mu = Z_a Z_b/Z^2$ emerged and the expansion takes the form

$$W(r) = E_0 + \mu W^{\mathrm{I}} + \mu^2 W^{\mathrm{II}} + \dots$$

The advantages of this expansion are that it corresponds to the natural perturbation expansion for the exactly soluble one-electron case and it converges quite rapidly so that quite accurate calculations may be performed by evaluating the first-order perturbation W^{I}. This first-order term can be written

$$W^{\mathrm{I}}(r) = Z \int \left(\frac{1}{R} - \frac{1}{|\mathbf{R} - \mathbf{r}|} \right) \rho(\mathbf{R})\, \mathrm{d}\mathbf{R} \tag{2.75}$$

where the integration extends over all space and $\rho(\mathbf{R})$ is the electron density of the united atom at the position $\mathbf{R}$ from its centre. Provided that $\rho(\mathbf{R})$, the electron density of the united atom appropriate to a particular pair of nuclei, is available, the first-order $W^{\mathrm{I}}(r)$ is easily calculated by quadrature from (2.75). For the interaction of a system such as the proton and a hydrogen atom, which contain one electron, the electron density for the one-electron united atom can be treated exactly and analytically. In the cases of systems containing more than one electron such as He–He, which contains four electrons in the united atom (which corresponds to beryllium) the Hartree–Fock self-consistent field approach is used to find approximate wave function and electron densities. The united atom theory has so far been used to determine intermolecular potentials only for small systems and then only in the high-energy repulsive region.[50]

2.3.4. *Model calculations*

Two model calculations have been attempted for short-range interactions of two atoms and both are based on a statistical model of the atom. They are the Thomas–Fermi–Dirac model and the electron gas model of Gordon and Kim. Both models have the same foundation in that they presume that each atom possesses a distribution of charge density that can be treated as a continuous variable in space about the nucleus in the same way that the density of a molecular gas is considered to be a continuous variable. That is the models consider an atom as an electron gas surrounding the nucleus whose properties are determined by application of Fermi–Dirac statistics. Such models are more reliable for interactions between atoms containing many electrons than are most of the methods described above. Thus such model calculations often represent the best that can be done at present for the heavier inert gases.

Since the two model calculation schemes have many features in common we describe these first before turning to the specific details of each model. Each atom of an interacting pair a and b is supposed to consist of a nucleus surrounded by an electron gas with densities ρ_a and ρ_b respectively. The models further presume that when the two atoms interact there is no re-arrangement of the distribution of the individual atom electron densities so that the total electron density at a point, ρ, is the sum of the contributions from both atoms at that point, i.e.

$$\rho = \rho_a + \rho_b. \tag{2.76}$$

In addition, it is supposed that the electron density in the region important to the interaction energy (usually the outer region of the atoms) is slowly varying with distance from the nucleus. In this case the energy density of the electron gas at a point in this region can be computed as if the point were in a uniform electron gas of the same electron density as exists at the point.

With these assumptions it is possible to write the (Coulombic) total energy of the two-atom system as

$$\begin{aligned}\frac{4\pi\varepsilon_0}{e^2} V_{c,tot} = \frac{Z_a Z_b}{r} + \frac{1}{2}\iint \frac{\{\rho_a(\mathbf{r}_1)+\rho_b(\mathbf{r}_1)\}\{\rho_a(\mathbf{r}_2)+\rho_b(\mathbf{r}_2)\}}{r_{12}}\,\mathrm{d}\mathbf{r}_1\,\mathrm{d}\mathbf{r}_2 \\ - Z_a\int \frac{\rho_a(\mathbf{r}_1)+\rho_b(\mathbf{r}_1)}{r_{1a}}\,\mathrm{d}\mathbf{r}_1 \\ - Z_b\int \frac{\rho_a(\mathbf{r}_1)+\rho_b(\mathbf{r}_1)}{r_{1b}}\,\mathrm{d}\mathbf{r}_1\end{aligned} \tag{2.77}$$

where the first term represents the internuclear repulsion, the second term the inter-electron repulsion, and the other two terms describe electron–nucleus attractions. Z_a and Z_b are the atomic numbers of the atoms a and b, r is the distance between their nuclei, r_{12} is the distance between two electrons, and r_{1a} and r_{1b} are the electron–nucleus distances.

The Coulombic energies of the separated atoms can be written

$$\frac{4\pi\varepsilon_0 V_{c,a}}{e^2}=\frac{1}{2}\int\frac{\{\rho_a(\mathbf{r}_1)\rho_a(\mathbf{r}_2)\}}{r_{12}}\,d\mathbf{r}_1\,d\mathbf{r}_2-Z_a\int\frac{\rho_a(\mathbf{r}_1)}{r_{1a}}\,d\mathbf{r}_1$$
$$\frac{4\pi\varepsilon_0 V_{c,b}}{e^2}=\frac{1}{2}\int\frac{\{\rho_b(\mathbf{r}_1)\rho_b(\mathbf{r}_2)\}}{r_{12}}\,d\mathbf{r}_1\,d\mathbf{r}_2-Z_b\int\frac{\rho_b(\mathbf{r}_1)}{r_{1b}}\,d\mathbf{r}_1. \tag{2.78}$$

Thus the Coulombic energy of interaction can be obtained by subtracting (2.78) from (2.77) to yield

$$\frac{4\pi\varepsilon_0 V_c}{e^2}=\iint\rho_a(\mathbf{r}_1)\rho_b(\mathbf{r}_2)\left[\frac{1}{r}+\frac{1}{r_{12}}-\frac{1}{r_{1a}}-\frac{1}{r_{2a}}\right]d\mathbf{r}_1\,d\mathbf{r}_2. \tag{2.79}$$

The Coulombic energy is not the only contribution to the interaction energy. We must also account for the exchange energy and the zero-point kinetic energy density of the electron gas.[51] Denoting the sum of these two contributions to the total energy density by v_e it may be shown that for an electron gas

$$v_e(\rho)=C_k\rho^{\frac{2}{3}}+C_e\rho^{\frac{1}{3}} \tag{2.80}$$

where C_k and C_e are universal constants.[51] If we integrate this energy density over the extent of the electron gas for the two-atom system and subtract from it the contributions from the separated atoms we obtain the net contribution to the interaction energy from exchange and zero-point kinetic energy as

$$V_g=\int\{[\rho_a(\mathbf{r}_a)+\rho_b(\mathbf{r}_b)]v_e(\rho_a+\rho_b)-\rho_a(\mathbf{r}_a)v_e(\rho_a)-\rho_b(\mathbf{r}_b)v_e(\rho_b)\}\,d\mathbf{r}. \tag{2.81}$$

The total interaction potential energy is then given by

$$U=V_c+V_g. \tag{2.82}$$

Thus $U(r)$ can be obtained, given the knowledge of the electron density distribution, by evaluation of the integrals (2.79) and (2.81) by quadrature.

In the Thomas–Fermi–Dirac model[52,53] based upon the foregoing analysis the electron density is computed from Poisson's equation for the electrostatic potential in the atom in the presence of a space charge. Given the appropriate boundary conditions this differential equation can be solved for the electrostatic potential and from it the electron density distribution can be deduced.[54,55] This density distribution is then employed in (2.79) and (2.81) to compute $U(r)$. The Thomas–Fermi–Dirac model only crudely approximates the electron density distribution in the gas and neglects the effects of correlations between electron motions.

Nevertheless recent calculations have shown that the model is capable of yielding intermolecular potential energies in substantial agreement with other model calculations which employ better electron densities and include correlation effects.[55]

The electron gas model of Gordon and Kim[56] attempts to rectify the two faults of the Thomas–Fermi–Dirac model by accounting for the effects of the correlation of electron motion in the electron gas and by employing more realistic electron density distributions. The electron densities themselves in this model are computed from Hartree–Fock wave functions obtained as described earlier. The contribution to the energy of an electron gas from the correlation of electron motions is not known exactly. However an approximate result which yields the contribution as a function of electron density is available.[56] Denoting this contribution to the energy density by v_{corr} we rewrite (2.80) as

$$v_{\text{e}}(\rho) = C_{\text{k}}\rho^{\frac{2}{3}} + C_{\text{e}}\rho^{\frac{1}{3}} + v_{\text{corr}} \tag{2.83}$$

which is to be used in (2.81). The Hartree–Fock electron densities then allow evaluation of $U(r)$ through (2.79), (2.81), and (2.82) as before. Using this model Gordon and Kim[56] have determined the intermolecular potential energies of several closed shell atoms, He–He, Ne–Ne, Ar–Ar, and Kr–Kr as well as their unlike interactions and some ionic pair systems such as K^+–Cl^-.

2.4. Intermediate-range energy

2.4.1. *Exchange perturbation theories*

The intermediate range of intermolecular forces represents the most difficult problem theoretically for a number of reasons. We have noted that the long-range dispersion forces arise in a second-order perturbation theory with unsymmetrized wave functions, that is without account being taken of the Pauli Principle. On the other hand, short-range forces appear in a first-order calculation with symmetrized wave functions. Obviously at some intermediate separations the two calculated energies will be of comparable magnitude and it would be a suspect procedure simply to add them since they are of different order (although this is frequently the best that can be done). Ideally, one would prefer to use a second-order perturbation calculation throughout for a properly symmetrized wave function. Such a process would introduce new interaction energy terms which are called second-order exchange terms. At short range these terms would be small in comparison with the first-order terms and at long range they would be small compared to the dispersion forces. However, in the intermediate range they would be significant.

There have been many attempts to develop exchange perturbation theories based upon properly symmetrized wave functions and they have been comprehensively reviewed recently.[19] In this review it is pointed out there is no unique perturbation treatment, so many different approaches have been adopted. Since it is by no means clear which methods will eventually prove most useful, and because they are very complicated, we mention here only those of Murrell and Shaw[57] and Szalewicz and Jeziorski[58] both of which have been applied to the interaction of two helium atoms.[59,60] In the perturbation scheme of Murrell and Shaw the zero-order state function of the two-atom system is chosen to be properly symmetrized whereas the higher-order terms in the expansion are left unsymmetrized. Szalewicz and Jeziorski[58] have recently formulated a double-perturbation procedure for dealing with intramolecular and intra-atomic electron correlation. The latter procedure leads to a well-depth for the He–He interaction in substantial agreement with experiment.[58]

2.4.2. *Molecular orbital theory*

The molecular orbital theory discussed earlier with the inclusion of configuration interaction calculations has been extended into the intermediate range by several authors for the helium interaction.[60–2]

This extension was made possible by the adoption of a sufficiently large basis set of atomic orbitals and sufficient configurations of the molecular orbitals so that the excitation of electrons giving rise to dispersion forces could be reproduced. Calculations of this type have rarely been carried out for more complicated systems.

2.4.3. Model calculations

A recent extension of the electron gas model to larger separations has been carried out by Kim and Gordon.[63] The original electron gas model was unable to describe inductive and dispersion forces between atoms because it presumed that there was no rearrangement of the electron density on either atom during the interaction. In order to introduce these features into the electron gas model the original version is combined with the three-dimensional Drude representation of the atom discussed in § 2.2.3. It is presumed that under the influence of the interaction at a particular instant of time the outer electron shells of the two atoms a and b are displaced by small distances $\delta\mathbf{r}_a$ and $\delta\mathbf{r}_b$ with respect to each nucleus. The total potential energy of the two-atom system is then written

$$V(\delta\mathbf{r}_a, \delta\mathbf{r}_b, \mathbf{r}) = \tfrac{1}{2}k_a(\delta\mathbf{r}_a)^2 + \tfrac{1}{2}k_b(\delta\mathbf{r}_b)^2 + V^0(|\mathbf{r}+\delta\mathbf{r}_a+\delta\mathbf{r}_b|) + V_c$$

where the first two terms present the harmonic oscillator energies for the two atoms, the third term the short-range force arising from the electron gas model, but excluding correlation effects, and V_c is the Coulombic interaction between the two atoms. Here $\mathbf{r}$ is the separation of the two nuclei and k_a and k_b are the force constants of the Drude model. By expanding V^0 and V_c about $\mathbf{r}$ and analysing the resulting oscillatory motion of the outer electrons it is possible to show that the intermolecular potential is in general the sum of three terms.

$$U(r) = V^0(r) + U_{\text{ind}}(r) + U_{\text{disp}}(r).$$

Here $U_{\text{ind}}(r)$ and $U_{\text{disp}}(r)$ are the usual induction and dispersion forces respectively. Thus, for non-polar neutral atom interactions $U_{\text{disp}} = C_6/r^6$.

Kim and Gordon have employed this model to calculate the interatomic potential of the like and unlike interactions of all of the inert gases. In view of the simplicity of the model and the relative ease of the calculations the results for all these systems except those involving helium are not unreasonable.[63]

It is clear from the foregoing discussion that for many systems there are no reliable techniques for the evaluation of intermolecular potentials in the intermediate range. In such cases the best that can be done is to add together the available short- and long-range intermolecular potentials, thereby making the assumption that even when the electronic distributions of two atoms overlap appreciably, the exchange forces have little effect on the dispersion forces. At least in the case of helium this seems to be a reasonable approximation.[57]

Finally, we mention a recent semi-empirical scheme for the evaluation of the intermolecular forces between closed shell atoms developed by Ng, Meath, and Allnatt.[64] In their procedure the short-range exchange energy of interaction is expressed as a function of the first-order Coulombic energy for the system, which is itself obtained from a self-consistent field calculation. The long-range contribution to the energy is constructed from the known dispersion energy (§ 2.2.3) combined with a suitable damping function to account for the neglect of charge overlap effects at small separations. The procedure leaves one parameter free, to be determined by forcing agreement with a known intermolecular potential for the system at one point. The results of the calculations provide reasonable agreement with the available intermolecular potential energy functions for the monatomic gases over a wide range of separations.[64]

2.5. Results

In this section we present a selection of the results which have been obtained by theoretical and model evaluations of intermolecular potentials. We have chosen to display the results for two systems, He–He and

Ar–Ar. The former system is a case where the theoretical treatments have been more exhaustive and are likely to be more accurate, whereas the latter is a more typical case of the interactions of inert gases. In order to provide a common basis for this comparison for each system we include what may be regarded as the 'best' experimental interatomic potential.

Fig. 2.4 shows the repulsive section of the pair interaction potential for helium. Included in the figure are the configuration interaction calculations,[46,65] which may be regarded as the most reliable theoretical results; they compare favourably with the potential obtained from high-energy molecular beam data.[66,67] The molecular orbital results of Gilbert and Wahl[45] predict a higher repulsion energy than the configuration interaction calculations. The Thomas–Fermi–Dirac model and the electron gas model both yield potentials in reasonable agreement with experiment despite their simplicity. Fig. 2.5 presents the attractive region for

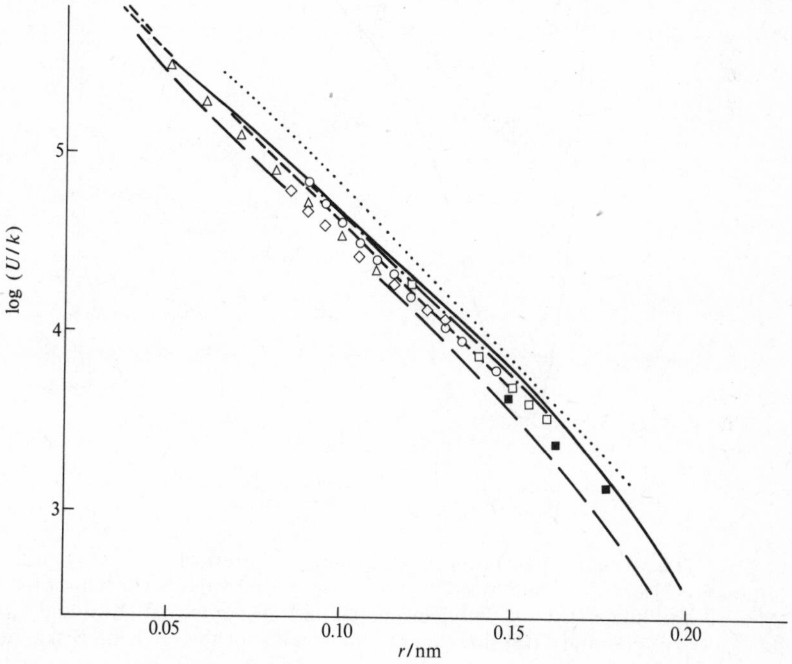

FIG. 2.4. The repulsive pair potential of helium. Theoretical: (———) configuration interaction (Phillipson[46]); (–––) configuration interaction (Matsumoto, Bender, and Davidson[65]); (–·–·–·–) Thomas–Fermi–Dirac model (Abrahamson[54]); (----) electron gas (Gordon and Kim[56]); (....) molecular orbital (Gilbert and Wahl[45]). Experimental: (△) high-energy molecular beams (Jordan and Amdur[66]); (◇) high-energy molecular beams (Kamnev and Leonas[67]); (○) high-energy molecular beams (Amdur, Jordan, and Colgate[68]); (□) high-energy molecular beams (Amdur and Harkness[69]); (■) low-energy molecular beams (Chen, Siska, and Lee[70]).

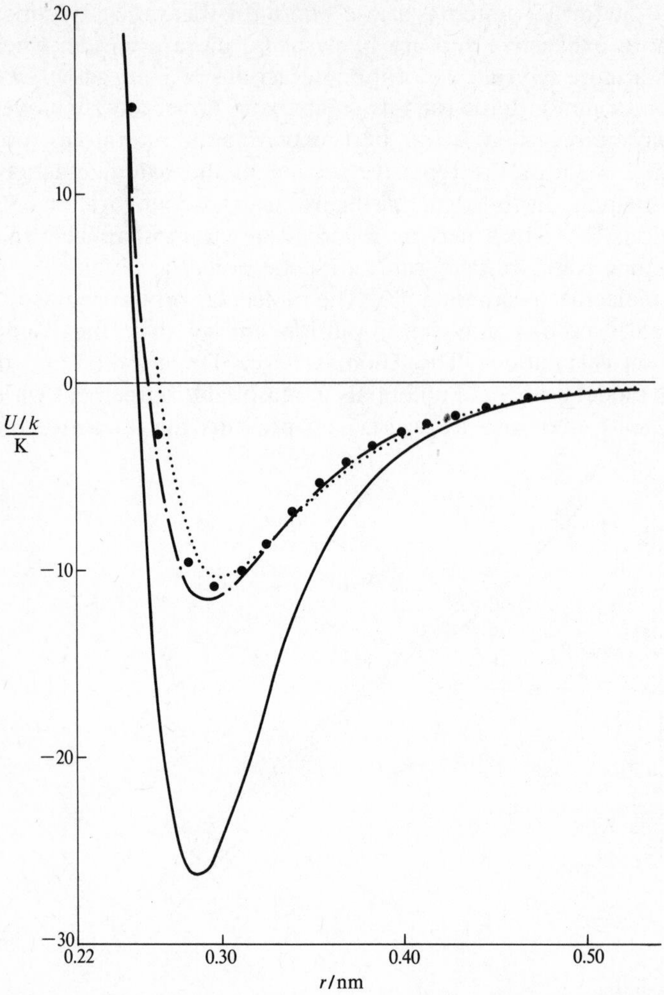

FIG. 2.5. The attractive pair potential of helium. Theoretical: (....) configuration interaction (Bertoncini and Wahl[60,61]); (–·–·–·–) exchange perturbation theory (Murrell and Shaw[57]); (———) electron gas and Drude model (Kim and Gordon[63]). Experimental: (●) low-energy molecular beams (Chen, Siska, and Lee[70]).

the helium interaction where it can be seen that the configuration interaction calculations[60,61] and the exchange perturbation theory results essentially bracket the potential obtained from low-energy beam scattering data.[70] The electron gas and Drude model[63] yields a substantially greater well depth for the interaction than is observed experimentally. These comparisons indicate that theory and experiment are in quite good accord for the helium interaction.

In the case of argon, for which the repulsive region of the interaction is shown in Fig. 2.6, there are fewer theoretical treatments and only that of Gilbert and Wahl[45] does not make use of a model. The theoretical results taken together do encompass the experimental data from molecular beams and macroscopic properties but no one calculation represents

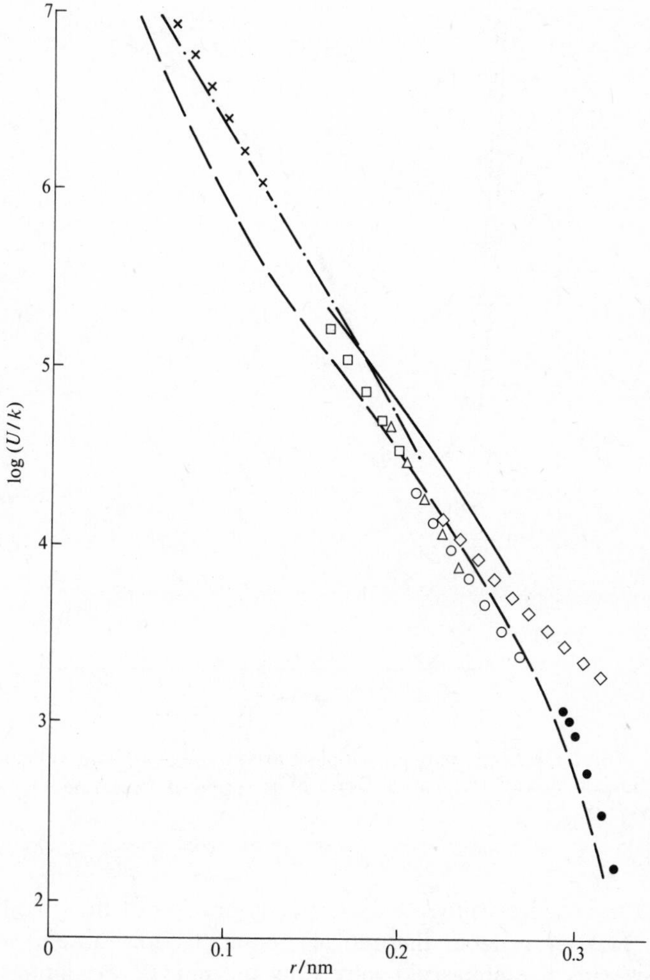

FIG. 2.6. The repulsive pair potential of argon. Theoretical: (———) molecular orbital (Gilbert and Wahl[45]); (–·–·–·–) Thomas–Fermi–Dirac model (Abrahamson[54]); (––––) electron gas model (Gordon and Kim[56]). Experimental, high-energy molecular beams: (×) Berry,[71,72] (□) Amdur, Jordan, and Bertrand;[73] (○) Amdur and Mason;[74] (△) Colgate, Jordan, Amdur, and Mason;[75] (◇) Kamnev and Leonas.[67] Experimental, combined bulk properties: (●) Aziz.[76]

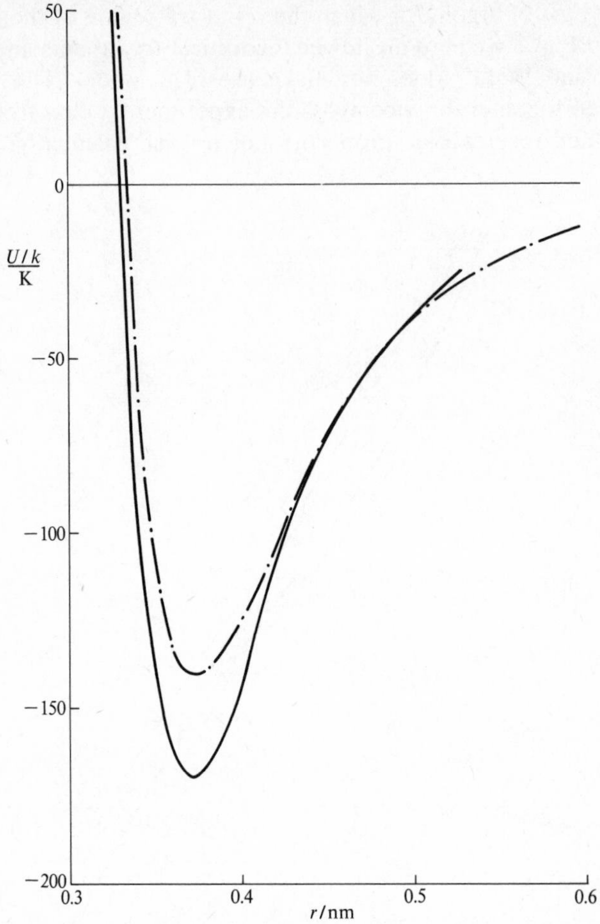

FIG. 2.7. The interatomic pair potential of argon. (———) Theoretical, electron gas and Drude model (Kim and Gordon[63]); (-·-·-·-) experimental, combined bulk properties (Aziz[76]).

all of the data successfully. In the attractive region of the potential (Fig. 2.7) the lack of reliable theoretical treatments for more complicated atomic systems is emphasized since only the model calculations of Kim and Gordon are available. Again the electron gas–Drude model yields a substantially greater well depth for the interaction than is observed experimentally.

Future theoretical evaluation of intermolecular forces is likely to involve the extension of the methods successfully applied to systems containing only a few electrons to the interactions of larger atoms and molecules. In the case of short- and long-range forces, the problems posed are in the main computational rather than problems in principle. However, in the intermediate range there remain problems of a more fundamental nature.

References

1. Buckingham, A. D. *Advan. chem. Phys.*, **12,** 107 (1967).
2. Hirschfelder, J. O., Curtiss, R. F., and Bird, R. B. *Molecular theory of gases and liquids*, Chapter 12. Wiley, New York (1954).
3. Le Fèvre, R. J. W. *Dipole moments.* Methuen, London (1953); McClellan, A. L. *Table of experimental dipole moments,* Heineman and Co., San Francisco (1963); Nelson, R. D., Lide, D. R. and Mayolt, A. A. *Selected values of dipole moments,* N.B.S. Data Reference Series **10,** (1967).
4. Townes, C. H., Dousmanis, G. C., White, R. L., and Schwarz, R. F. *Discuss. Faraday Soc.* **19,** 56 (1955).
5. Akhmodzhanov, R., Gransky, P. V., and Bulanin, M. O. *Can. J. Phys.* **54,** 519 (1976).
6. Cohen, E. R. and Birnbaum, G. *J. chem. Phys.* **66,** 2443 (1977).
7. Bridge, N. J. and Buckingham, A. D. *Proc. roy. Soc.* **295A,** 334 (1966).
8. Boyle, L. L., Buckingham, A. D., Disch, R. L., and Dunmur, D. A. *J. chem. Phys.* **45,** 1318 (1966).
9. Amos, R. D. *Mol. Phys.* **38,** 33 (1979).
10. Drude, P. K. L. *The theory of optics.* Longman, London (1933).
11. Ditchburn, R. W. *Light.* Blackie, London (1952).
12. Pauling, L. and Wilson, E. B. *Introduction to quantum mechanics,* Chapter 4. McGraw-Hill, New York (1935).
13. London, F. *Z. Phys.* **63,** 245 (1930).
14. London, F. *Z. phys. Chem.* (*B*) **11,** 222 (1930).
15. Margenau, H. *J. chem. Phys.* **6,** 896 (1938).
16. Hornig, J. F. and Hirschfelder, J. O. *J. chem. Phys.* **20,** 1812 (1952).
17. Bethe, H. A. and Salpeter, E. E. *The quantum mechanics of one- and two-electron atoms,* p. 357. Springer Verlag, Berlin, New York (1957).
18. Van Vleck, J. H. *The theory of electric and magnetic susceptibilities.* Clarendon Press, Oxford (1932).
19. Certain, D. R. and Bruch, L. W. *Intermolecular forces.* MTP International Review of Science, Physical Chemistry Series One, Theoretical Chemistry Volume. Medical and Technical Publishing Co., Oxford (1972).
20. Starkschall, G. and Gordon, R. G. *J. chem. Phys.* **54,** 663 (1971).
21. McQuarrie, D. A., Terebey, J., and Shire, J. J. *J. chem. Phys.* **51,** 4683 (1969).
22. Langhoff, R. W., Gordon, R. G., and Karplus, M. *J. chem. Phys.* **55,** 2126 (1971).
23. Dalgarno, A. *Advan. chem. Phys.* **12,** 143 (1967).
24. Kaneko, S. *J. chem. Phys.* **54,** 819 (1971).
25. Dalgarno, A. *Advan. Phys.* **11,** 281 (1962).
26. Hirschfelder, J. O., Brown, W. B., and Epstein, S. T. *Advan. quantum Chem.* **1,** 256 (1964).

27. Karplus, M. and Kolker, H. *J. chem. Phys.* **41,** 3955 (1964).
28. Dalgarno, A. and Victor, G. A. *Proc. roy. Soc.* **A291,** 291 (1966).
29. Dalgarno, A. and Stewart, A. L. *Proc. roy. Soc.* **A238,** 269 (1956).
30. Deal, W. J. and Kestner, N. R. *J. chem. Phys.* **45,** 4014 (1966).
31. Gordon, R. G. *J. chem. Phys.* **48,** 3929 (1968).
32. Margenau, H. *Rev. mod. Phys.* **11,** 1 (1939).
33. Axilrod, B. M. and Teller, E. *J. chem. Phys.* **11,** 299 (1943).
34. Kaneko, S. *J. chem. Phys.* **56,** 3417 (1972).
35. Margenau, H. and Kestner, N. R. *Theory of intermolecular forces.* Pergamon, Oxford (1969).
36. Slater, J. C. *Quantum theory of molecules and solids,* Vol. 1. McGraw-Hill, New York (1963).
37. Bleick, W. E. and Meyer, J. D., *J. chem. Phys.* **2,** 252 (1934).
38. Kunimune, M. *J. chem. Phys.* **18,** 754 (1950).
39. Margenau, H. and Rosen, P. *J. chem. Phys.* **21,** 394 (1953).
40. Atkins, P. W. *Molecular quantum mechanics,* p. 259. Clarendon Press, Oxford (1970).
41. Kotani, M., Ohno, K., and Kayama, K. *Handbuch der Physik* (ed. S. Flügge), Vol. 37/1, p. 78. Springer-Verlag, Berlin (1961).
42. Roothaan, C. C. *J. chem. Phys.* **19,** 1445 (1951).
43. Lester, W. A. *J. chem. Phys.* **54,** 3171 (1971).
44. Rottenberg, S. and Schaefer, H. F. *Chem. Phys. Lett.* **10,** 565 (1971).
45. Gilbert, T. L. and Wahl, A. C. *J. chem. Phys.* **47,** 3425 (1967).
46. Phillipson, P. E. *Phys. Rev.* **125,** 1981 (1962).
47. Balint-Kurti, G. G. and Karplus, M. *Chem. Phys. Lett.* **11,** 203, (1971).
48. Gordon, M. D. and Secrest, D. *J. chem. Phys.* **52,** 120 (1970); erratum *J. chem. Phys.* **53,** 4408 (1970).
49. Byers Brown, W. *Chem. Phys. Lett.* **1,** 655 (1968).
50. Byers Brown, W. and Power, J. D. *Proc. roy. Soc.* **A317,** 545 (1970).
51. Slater, J. C. *Quantum theory of atomic structure,* Vol. II, Appendix 22. McGraw-Hill, New York (1960).
52. Gombas, P. *Die statistische Theorie des Atoms und ihre Anwendungen.* Springer Verlag, Vienna (1949).
53. Dirac, P. A. M. *Proc. Camb. Phil. Soc.* **26,** 376 (1930).
54. Abrahamson, A. A. *Phys. Rev.* **130,** 693 (1963).
55. Yonei, K. and Goodisman, J. *J. chem. Phys.* **66,** 4551 (1977).
56. Gordon, R. G. and Kim, Y. S. *J. chem. Phys.* **56,** 3122 (1972).
57. Murrell, J. N. and Shaw, G. *Mol. Phys.* **15,** 325 (1968).
58. Szalewicz, K. and Jeziorski, B. *Mol. Phys.* **38,** 191 (1979).
59. Murrell, J. N. and Shaw, G. *J. chem. Phys.* **46,** 1768 (1967).
60. Bertoncini, P. and Wahl, A. C. *Phys. Rev. Lett.* **25,** 991 (1970).
61. Bertoncini, P. and Wahl, A. C. *J. chem. Phys.* **58,** 1259 (1973).
62. Barton, P. G. *J. chem. Phys.* **67,** 4696 (1977).
63. Kim, Y. S. and Gordon, R. G. *J. chem. Phys.* **61,** 1 (1974).
64. Ng, Kin-Chue, Meath, W. J., and Allnatt, A. R. *Mol. Phys.* **37,** 237 (1979).
65. Matsumoto, G. H., Bender, G. F., and Davidson, E. R. *J. chem. Phys.* **46,** 402 (1967).
66. Jordan, J. E. and Amdur, I. *J. chem. Phys.* **46,** 165 (1967).
67. Kamnev, A. B. and Leonas, V. B. *High Temp. Phys.* **3,** 744 (1965).
68. Amdur, I., Jordan, J. E., and Colgate, S. O. *J. chem. Phys.* **34,** 1525 (1961).
69. Amdur, I. and Harkness, A. L. *J. chem. Phys.* **22,** 664 (1954).

70. Chen, C. H., Siska, P. E., and Lee, Y. T. *J. chem. Phys.* **59,** 601 (1973).
71. Berry, H. W. *Phys. Rev.* **75,** 913 (1949).
72. Berry, H. W. *Phys. Rev.* **99,** 553 (1955).
73. Amdur, I., Jordan, J. E., and Bertrand, R. B. *Atomic and molecular collision processes.* North-Holland, Amsterdam (1964).
74. Amdur, I. and Mason, E. A. *J. chem. Phys.* **22,** 670 (1954).
75. Colgate, S. O., Jordan, J. E., Amdur, I., and Mason, E. A. *J. chem. Phys.* **51,** 968 (1969).
76. Aziz, R. A. *J. chem. Phys.* **65,** 490 (1976).

3

GAS IMPERFECTIONS

3.1. Introduction

For more than a century studies of the deviations of gases from the perfect gas law have been used as a valuable source of information about the interactions between molecules.[1] Andrews's classic measurements of P–V isotherms around the critical point of carbon dioxide[2] established the nature of the gas–liquid critical point and laid the foundation for many generations of future workers. He demonstrated clearly that the equilibrium of a pure liquid and its vapour, which is observed in a closed vessel, could only occur at temperatures below a value, characteristic of the substance, called the gas–liquid critical temperature T_c. At temperatures below T_c a gas may only be compressed until a certain pressure, the saturated vapour pressure, is reached. Further reduction in volume then leads to condensation of the gas, which continues until complete conversion to liquid has occurred. When this state is reached further reduction in volume leads to a large increase in the pressure. The saturated vapour pressure depends on the temperature, and has a maximum value, called the critical pressure, P_c, at the critical temperature. The pattern of behaviour revealed by Andrews in his study of various isotherms for carbon dioxide is shown in Fig. 3.1. Studies on other gases reveal very similar behaviour. In general it is found that small, simple molecules such as N_2 or CH_4 have low values of T_c, and these substances are gaseous under normal conditions. More complex substances such as benzene have larger values of T_c, and are often liquids or solids at room temperature and pressure.

The deviations from perfect gas behaviour may also be conveniently represented by plotting the compression factor, $P\tilde{V}/RT$, often called the compressibility factor, as a function of pressure, for several temperatures, as illustrated in Fig. 3.2. In all cases the compression factor approaches the perfect-gas limiting value of unity as the pressure approaches zero. At high temperatures (say $T > 3T_c$) $P\tilde{V}/RT$ increases steadily as the pressure is raised. At lower temperatures the compression factor first falls below the perfect gas value, rising again at higher pressures. At temperatures below T_c, discontinuities are observed, corresponding to liquefaction. At one characteristic temperature, the Boyle temperature, T_B, the limiting slope of the graph is zero, and the perfect gas equation is accurately obeyed up to moderate pressures.

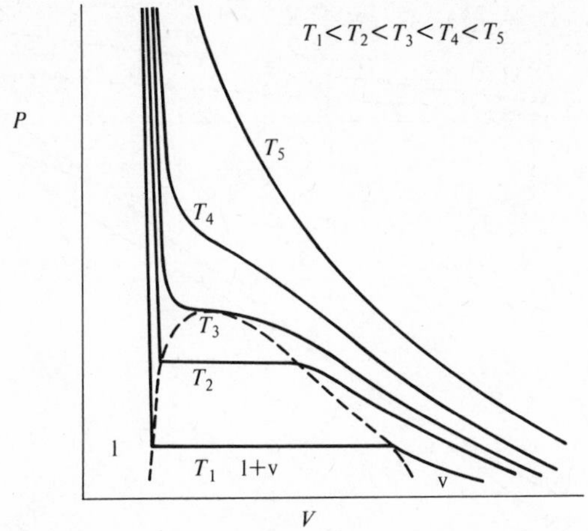

FIG. 3.1. Isotherms for an imperfect gas, near to the critical point. T_3 is the critical temperature. The coexistence curve for the liquid, l, and vapour, v, is indicated by the dotted line.

Direct evidence for intermolecular forces is to be found in the Joule–Thomson effect.[3] When a gas expands through a porous plug or nozzle, a change in temperature, usually a decrease, results. (Fig. 3.3). The temperature change depends on the drop in pressure and the Joule–Thomson coefficient, μ_{JT}, may be defined

$$\mu_{JT} = \lim_{P_2 \to P_1} \left(\frac{T_2 - T_1}{P_2 - P_1}\right)_H = \left(\frac{\partial T}{\partial P}\right)_H.$$

The process takes place under conditions of constant enthalpy, H, and by the application of standard thermodynamic manipulations the Joule–Thomson coefficient may be expressed in terms of the molar heat capacity, C_P, and the equation of state of the gas

$$\mu_{JT} = \frac{1}{C_P}\left\{T\left(\frac{\partial \tilde{V}}{\partial T}\right)_P - \tilde{V}\right\}.$$

A related experiment may be performed in which heat is supplied to (or removed from) the low-pressure side to maintain the temperature constant. The isothermal Joule–Thomson coefficient, ϕ_{JT}, studied in this way is directly related to the isenthalpic coefficient, μ_{JT}, by the equation

$$\phi_{JT} = \left(\frac{\partial H}{\partial P}\right)_T = -C_P \mu_{JT}.$$

For a perfect gas, $T(\partial \tilde{V}/\partial T)_P = \tilde{V}$, and both coefficients are zero, so

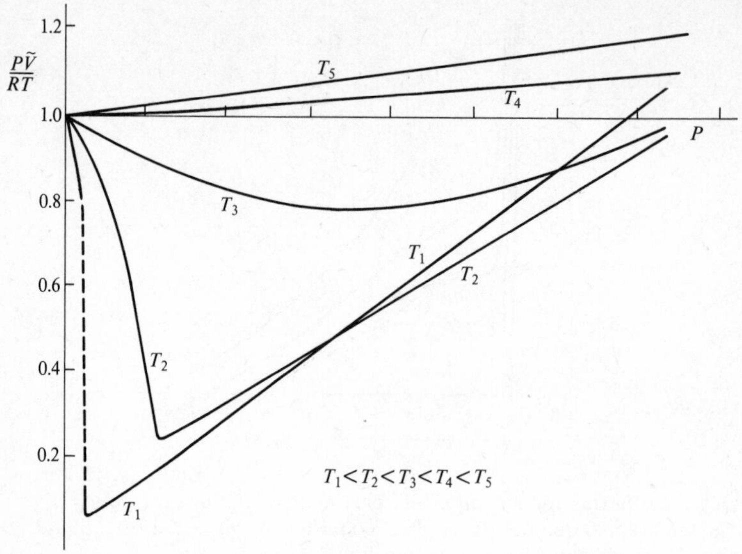

FIG. 3.2. The compression factor, $P\tilde{V}/RT$, as a function of pressure at several temperatures. T_2 is the critical temperature. T_4 is the Boyle temperature.

the occurrence of the effect is direct confirmation of the existence of intermolecular interactions. Although for most gases at low and moderate pressures μ_{JT} is positive, i.e. cooling occurs when the gas expands, there exists a temperature above which μ_{JT} is negative, and the gas warms up on expansion. The temperature at which $\mu_{JT}=0$ is called the Joule–Thomson inversion temperature, and at low pressure is approximately 1.9 times the Boyle temperature for simple substances.

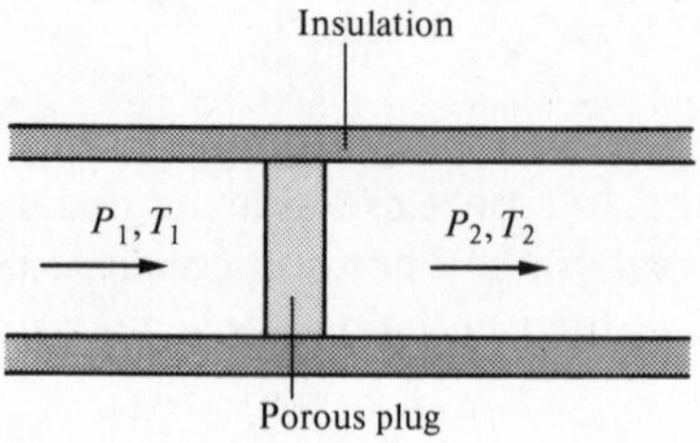

FIG. 3.3. A schematic representation of the adiabatic Joule–Thomson experiment.

3.2. Van der Waals equation of state

Andrews's measurements on carbon dioxide stimulated the Dutch physicist, J. D. van der Waals, in the development of his well-known modification of the perfect gas equation of state.[4] In this equation,

$$\left(P+\frac{a}{\tilde{V}^2}\right)(\tilde{V}-b)=RT,$$

an attempt was made to modify the perfect gas equation in such a way as to allow for the effects of intermolecular interactions. The short-range intermolecular repulsions were dealt with by treating the molecules as small spheres which could not overlap. The volume available to the molecular centres was consequently reduced by an amount b, which van der Waals calculated to be four times the actual volume of the hard spheres. A term proportional to the square of the density was introduced to take account of intermolecular attractive forces experienced at somewhat greater separations. The separation of these two factors and their effects on the pressure are clearly revealed when the equation is rewritten in the form

$$P=\frac{RT}{\tilde{V}-b}-\frac{a}{\tilde{V}^2}$$

or

$$\frac{P\tilde{V}}{RT}=\frac{\tilde{V}}{\tilde{V}-b}-\frac{a}{RT\tilde{V}}.$$

This may be expanded in powers of the reciprocal volume to give

$$\frac{P\tilde{V}}{RT}=1+\frac{b-a/RT}{\tilde{V}}+\left(\frac{b}{\tilde{V}}\right)^2+\left(\frac{b}{\tilde{V}}\right)^3+\ldots.$$

Repulsive interactions are seen to raise the pressure above the value for a perfect gas at the same density, while attractive forces have the opposite effect. Temperature and density have important roles in determining the over-all sign and size of the deviations from perfect gas behaviour, which are clearly revealed.

The shape of the van der Waals isotherms changes qualitatively with temperature, as shown in Fig. 3.4. At high temperatures, e.g. T_2, the pressure decreases monotonically with increasing volume. At low temperatures, e.g. T_1, the isotherms are more complex, and three different values of V can give rise to the same pressure, over a range of values of P. The equation can be written in a cubic form

$$\tilde{V}^3+\left(b-\frac{RT}{P}\right)\tilde{V}^2+\frac{a}{P}\cdot\tilde{V}-\frac{ab}{P}=0$$

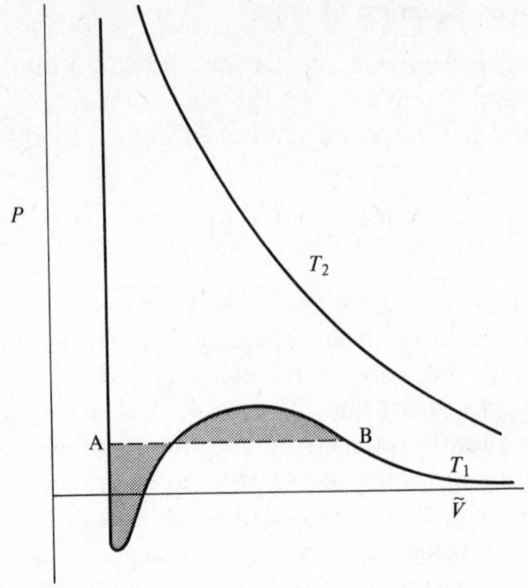

FIG. 3.4. Isotherms from the van der Waals equation. The equal areas construction is shown on the low temperature isotherm, T_1.

which has three real roots below a certain temperature. As the temperature is increased the three roots become identical and the curve shows only an inflection where $(\partial P/\partial V)_T$ and $(\partial^2 P/\partial V^2)_T$ are both zero. This temperature may be identified with the gas–liquid critical temperature, T_c. Differentiation of the equation and setting these two derivatives to zero gives

$$V_c = 3b, \qquad T_c = \frac{8a}{27Rb}, \qquad P_c = \frac{a}{27b^2}.$$

Hence the critical parameters are determined by the values of the characteristic quantities a and b, or, conversely, the experimental values of two of the critical parameters may be used to obtain suitable values of a and b for a substance.

The isotherms below T_c may be reinterpreted by means of the Maxwell equal-area construction. A line at constant pressure is drawn, cutting the isotherm in such a way that the areas above and below the line (shown shaded in Fig. 3.4) are equal. Thermodynamic arguments show that at the extreme volumes, A and B, the chemical potentials† μ^A and μ^B are equal, and these points correspond to the liquid and vapour

† The chemical potential, μ_i, of a species i is defined $\mu_i = (\partial G/\partial n_i)_{T,P,n_j}$, where G is the Gibbs free energy, and n_j represents the number of moles of other species. For a pure substance it is the Gibbs free energy per mole. The equivalent definition expressed per molecule rather than per mole is also used.

phases at the saturated vapour pressure for this temperature.[5] The modified isotherm then has the general appearance observed experimentally.

On substituting the equations for a and b in terms of the critical parameters back in van der Waals equation, it may be written in the form

$$\left[\frac{P}{P_c}+\frac{3}{(\tilde{V}/V_c)^2}\right]\left[3\left(\frac{\tilde{V}}{V_c}-1\right)\right]=8\left(\frac{T}{T_c}\right)$$

The ratios P/P_c, $\tilde{V}/V_c$, T/T_c are known as the reduced pressure, volume, and temperature. This form of the equation shows that all substances which are described by the van der Waals equation will have the same reduced pressure when they are at the same reduced temperature and volume. Alternatively, we may say that the value of $P\tilde{V}/RT$ for such substances depends only on T/T_c and either $\tilde{V}/V_c$ or P/P_c. This important observation is an example of the *principle of corresponding states* and is in fact true for all equations of state which involve only two parameters characteristic of the substance under consideration.

It is observed experimentally that the equations of state of many substances may be correlated in this way, even though their detailed behaviour is not accurately represented by the van der Waals or any other simple equation of state.[6] The principle of corresponding states is an invaluable tool in studying the relations between intermolecular forces and the properties of matter, and we shall consider its origins and applications in greater detail later.

3.3 The virial equation of state

In spite of the profound influence of the van der Waals equation on the study of molecular interactions and their consequences, it cannot represent accurately the detailed behaviour of any substance. Very many other, often quite complex, equations of state have been proposed, and these have been successful to varying degrees. However, when the relationship between intermolecular forces and the equation of state of gases is of primary interest, one equation of state, the *virial equation*, is of central importance. This equation, which represents the compression factor of a gas as an infinite series expansion in the density

$$\frac{P\tilde{V}}{RT}=1+\frac{B(T)}{\tilde{V}}+\frac{C(T)}{\tilde{V}^2}+\frac{D(T)}{\tilde{V}^3}+\dots \tag{3.1}$$

is generally associated with Kamerlingh Onnes who used a similar (though not identical) equation around the turn of this century.[7] In this equation the coefficients, B, C, $D\dots$, are called the second, third, fourth ... virial coefficients, and are functions of temperature and the nature of the material under investigation. (The alternative symbolism B_2, B_3, B_4 . . . is also used.) A knowledge of the virial coefficients and their temperature dependence then gives a complete description of the *PVT* behaviour of a substance. (It is assumed without proof that the equation is convergent.) However, the paramount importance of the virial equation, in contrast to the other equations, lies in the precisely defined relationships which exist between the virial coefficients and the intermolecular interactions.

An alternative expansion of the compressibility factor as a power series in the pressure is often encountered

$$\frac{P\tilde{V}}{RT}=1+\hat{B}(T)P+\hat{C}(T)P^2+\hat{D}(T)P^3+\dots \tag{3.2}$$

The coefficients of this expansion, $\hat{B}$, $\hat{C}$, $\hat{D}$ are closely related, for an infinite series, to those in (3.1), e.g.

$$\hat{B}=\frac{B}{RT} \qquad \hat{C}=\frac{C-B^2}{(RT)^2}\,. \tag{3.3}$$

Equations relating the higher coefficient have been derived, but are quite lengthy.[8]

Other quantities which depend on the equation of state may also be expressed in terms of the virial coefficients and their temperature derivatives. Some of the relevant equations are given in Appendix A3.1.

3.4. Statistical mechanics and the virial equation

The origins of the virial equation of state and its relationship to the interactions between molecules may be established by the methods of statistical mechanics. In statistical mechanics the macroscopic properties of matter are related to the microscopic properties of the constituent atoms or molecules. A complete treatment of statistical mechanics is beyond the scope of this book and we therefore confine ourselves to a brief description of the subject insofar as it relates to gas imperfections, and refer the reader to comprehensive texts on the subject for a fuller treatment.[9–11]

The observable, equilibrium state of a thermodynamic system may be defined in terms of a limited number of thermodynamic (macroscopic) variables such as temperature, *T*, pressure, *P*, and volume, *V*. However, since the material making up a one-component thermodynamic system consists of a very large number of identical atoms or molecules, the same observable thermodynamic state may arise from a large number of different spatial arrangements of, and energy distributions among, the molecules. In statistical mechanics one mentally constructs for the thermodynamic system of interest a very large number of replica systems, each of which has exactly the same values of the thermodynamic state variables as the original thermodynamic system but which differ in their arrangement of molecules in configuration and momentum space. This array of replicas is known as an ensemble.

When an experimental measurement of a macroscopic property of a real thermodynamic system is made, the measurement usually takes a long time compared to that characteristic of molecular motion. Consequently, in practice, one always determines a time average of the property. It is a fundamental postulate of statistical mechanics that this time-averaged value of the property is equal to the unweighted average of the property taken over all the members of the ensemble which are consistent with the thermodynamic state of the real system.

Three different ensembles are in common use; they correspond to three different types of thermodynamic system. An *isolated* thermodynamic system has a fixed number of molecules, N, a fixed volume, V, and a fixed total energy, E. The ensemble of replicas whose members have the same values of N, V, and E is known as the *microcanonical* ensemble. A *closed* thermodynamic system also contains a fixed number of molecules and has a fixed volume, but the temperature of the system, T, is defined rather than the energy. The ensemble of replicas whose members have the same values of N, V, and T is called the *canonical ensemble.* In an *open* thermodynamic system the thermodynamic state is defined in terms of the volume, temperature, and chemical potential of the molecules, μ. The corresponding ensemble of replicas with the same values of V, T, and μ is called the *grand canonical* ensemble. In this final ensemble both the energy and the number of molecules may vary among its members.

A further fundamental postulate of statistical mechanics states that for the ensemble representative of an isolated thermodynamic system (a microcanonical ensemble) the replica systems are distributed with equal probability over all the possible microscopic states which are consistent with the specified values of N, V, and E. This hypothesis, together with the earlier postulate, is assumed to be justified if the results of statistical mechanics are in agreement with observation.

The second hypothesis, for a microcanonical ensemble, is used as the basis for determining the probability, $P(E_j)$, that a member system of a canonical ensemble will be found with an energy, $E_j(N, V)$, which is characteristic of a particular microscopic state. If we restrict our attention now to a thermodynamic system which consists of a single component gas whose molecules are not distinguishable from each other by nature or localization in space, the probability $P(E_j)$ can be written

$$P(E_j) = \frac{\exp\{-E_j/kT\}}{Z_N}$$

where $Z_N(N, V, T)$ is the canonical partition function

$$Z_N = \sum_j \exp\{-E_j/kT\} \tag{3.4}$$

In (3.4) the summation extends over all distinguishable microscopic states consistent with the specified thermodynamic state (N, V, T). The energy E_j, denotes the total energy of the N molecules in the microscopic state of the member system and contains contributions from the translational kinetic energy of the molecules and the energy associated with internal degrees of freedom as well as their intermolecular potential energy.

For the grand canonical ensemble, representing a thermodynamic system specified by V, T, and the chemical potential, μ, the probability of observing a member system of the ensemble in an energy state E_j *and* containing N molecules can be shown to be

$$P(N, E_j) = \frac{\exp(-E_j/kT)\exp(N\mu/kT)}{\Xi}$$

where Ξ is the *grand canonical partition function*, which for a gas is given by the expression

$$\Xi = \sum_{N\geq 0}\sum_j \exp(-E_j/kT)\exp(N\mu/kT)$$

$$= \sum_{N\geq 0} Z_N \exp(N\mu/kT). \tag{3.5}$$

Either the canonical partition function or the grand canonical partition function may be made the basis for a derivation of the virial equation of state. Historically, the canonical partition function was used first for this purpose.[9,11] However, here we outline a development based upon the grand canonical partition function[10,12] which has the dual merits of relative simplicity and brevity.

In terms of the grand canonical partition function the ensemble average number of molecules $\bar{N}$ and the ensemble average pressure $\bar{P}$ may be written

$$\bar{N} = kT\left(\frac{\partial \ln \Xi}{\partial \mu}\right)_{V,T} \tag{3.6}$$

and

$$\bar{P} = \frac{1}{V} kT \ln \Xi. \tag{3.7}$$

According to the first postulate of statistical mechanics these ensemble averages are to be identified with the observable number of molecules in the system N_0 and the pressure of the gas P in the thermodynamic state defined by T, V, and μ.

The virial equation of state. We first assume that the molecules of the gas possess no internal degrees of freedom. This is not a necessary assumption for the development which follows, but yields a simpler treatment.

(a) *A gas composed of non-interacting molecules.* For a gas in which the molecules exert no forces on each other, the energy of a system of N particles is just the sum of their translational kinetic energies. If the quantum mechanically allowed energies for a single molecule contained in the volume V are denoted by ε_i, and if any number of molecules can occupy the same quantum state, then the canonical partition function for a system of N non-interacting particles can be written

$$Z_N = (Z_1)^N/N! \tag{3.8}$$

where

$$Z_1 = \sum \exp(-\varepsilon_i/kT) \tag{3.9}$$

and the summation extends over all allowed quantum states for the single molecule. The division by $N!$ takes account of the indistinguishability of the molecules.

Thus, for a gas in which the molecules exert no forces on each other the grand canonical partition function, Ξ^0, becomes

$$\begin{aligned} \Xi^0 &= \sum_{N \geqslant 0} (Z_1)^N \{\exp[\mu/kT]\}^N/N! \\ &= \sum_{N \geqslant 0} (Z_1\lambda)^N/N! = 1 + Z_1\lambda + \frac{(Z_1\lambda)^2}{2!} + \ldots \end{aligned} \tag{3.10}$$

where we have employed the symbol λ for the absolute activity

$$\lambda = \exp(\mu/kT). \tag{3.11}$$

Recognizing that (3.10) is an exponential series we can finally write

$$\Xi^0 = \exp(\lambda Z_1). \tag{3.12}$$

Using (3.12) and (3.11) in (3.6) and carrying out the differentiation we find that the ensemble average number of particles, or the observable number of particles in the system, is

$$\bar{N} = N_0 = \ln \Xi^0 = \lambda Z_1. \tag{3.13}$$

Thus, finally from (3.7) we obtain the observable pressure as

$$P = \bar{P} = \frac{N_0 kT}{V}$$

so that a system of non-interacting molecules conforms to the perfect gas equation of state.

(b) *A gas of interacting molecules.* In the case of a gas whose molecules interact the grand canonical partition function takes the form

$$\begin{aligned}\Xi &= \sum_{N\geq 0} Z_N \lambda^N \\ &= 1 + \sum_{N\geq 1} Z_N \lambda^N \end{aligned} \tag{3.14}$$

where the last result follows because for $N = 0$, there is only one possible value of the energy of the system, namely $E = 0$.

Eqn (3.14) is a series expansion for Ξ in powers of λ. However, in order to make use of (3.7) for the pressure it is more convenient to work with an expansion for $\ln \Xi$. Since it is an experimental fact that at low densities all gases conform to the perfect gas equation of state, we assume an expansion for $\ln \Xi$ of the form

$$\ln \Xi = \bar{N} + c_1\left(\frac{\bar{N}}{V}\right)^2 + c_2\left(\frac{\bar{N}}{V}\right)^3 + \dots \tag{3.15}$$

which yields the perfect gas result in the limit of zero density. We have already seen that for a perfect gas $\bar{N} = \lambda Z_1$, so that we can employ this relation to convert the series expansion of (3.15) into a power series expansion in λ. That is, we write

$$\ln \Xi = V \sum_{i\geq 1} b_i \left(\frac{\lambda Z_1}{V}\right)^i \tag{3.16}$$

where we have introduced new coefficients b_i, which have to be determined. We require the expansion (3.16) to be identical to that of (3.14) and we must choose the coefficients of (3.16) to ensure this. Taking the exponential of (3.16) and expanding it we obtain

$$\Xi = 1 + V \sum_{i\geq 1} b_i \left(\frac{\lambda Z_1}{V}\right)^i + \frac{V^2}{2}\left\{\sum_{i\geq 1} b_i \left(\frac{\lambda Z_1}{V}\right)^i\right\}^2 + \dots$$

which is a power series in λ as is (3.14).

To ensure the equality of these two expansions we require equality of the coefficients of various powers of λ and we find for example

$$b_1=1, \qquad b_2=\frac{Z_2V}{Z_1^2}-\frac{V}{2}, \qquad b_3=\frac{Z_3V^2}{Z_1^3}-\frac{Z_2V^2}{Z_1^3}+\frac{V^2}{3}.$$

The series for $\ln \Xi$ then reads

$$\ln \Xi = Z_1\lambda+\left(Z_2-\frac{Z_1^2}{2}\right)\lambda^2+(Z_3-Z_2Z_1+\tfrac{1}{3}Z_1^3)\lambda^3+\dots. \tag{3.17}$$

Substituting (3.17) into (3.6) and carrying out the differentiation with the aid of (3.11) we obtain an expansion for $\bar{N}$ in powers of λ which reads

$$\bar{N}=\lambda\left\{Z_1+2\left(Z_2-\frac{Z_1^2}{2}\right)\lambda+3(Z_3-Z_2Z_1+\tfrac{1}{3}Z_1^3)\lambda^2+\dots\right\}. \tag{3.18}$$

Assuming that this series is convergent, we may invert it to find a series for λ in powers of $\bar{N}$ of the form

$$\lambda=a_1\bar{N}+a_2\bar{N}^2+a_3\bar{N}^3+\dots.$$

Inserting this form into the right-hand side of (3.18), we obtain

$$\bar{N}=Z_1\{a_1\bar{N}+a_2\bar{N}^2+a_3\bar{N}^3+\dots\}$$
$$+2\left(Z_2-\frac{Z_1^2}{2}\right)\{a_1\bar{N}+a_2\bar{N}^2+a_3\bar{N}^3+\dots\}^2+\dots.$$

Equating coefficients of powers of $\bar{N}$ on both sides of the equation enables us to identify a_1, $a_2\dots$ so that we can write

$$\lambda=\frac{\bar{N}}{Z_1}-\frac{2}{Z_1^3}\left(Z_2-\frac{Z_1^2}{2}\right)\bar{N}^2+\left\{\frac{8}{Z_1^5}\left(Z_2-\frac{Z_1^2}{2}\right)^2-\frac{3}{Z_1^4}\left(Z_3-Z_2Z_1+\frac{Z_1^3}{3}\right)\right\}\bar{N}^3$$
$$+\dots.$$

Finally, using this expansion for λ in the power series expansion for $\ln \Xi$ (3.17) we obtain

$$\ln \Xi=\bar{N}-\left(\frac{Z_2}{Z_1^2}-\frac{1}{2}\right)\bar{N}^2-\left(\frac{2Z_3}{Z_1^3}-\frac{4Z_2^2}{Z_1^4}+\frac{2Z_2}{Z_1^2}-\frac{1}{3}\right)\bar{N}^3+\dots.$$

Thus, from (3.7) we obtain the virial equation of state

$$\frac{P}{kT}=\rho+B\rho^2+C\rho^3+\dots$$

where $\rho=\bar{N}/V=N_0/V$ is the number density and

$$B=-V\left(\frac{Z_2}{Z_1^2}-\frac{1}{2}\right) \tag{3.19}$$

and

$$C = -V^2\left\{\frac{2Z_3}{Z_1^3} - \frac{4Z_2^2}{Z_1^4} + \frac{2Z_2}{Z_1^2} - \frac{1}{3}\right\} \tag{3.20}$$

are the second and third virial coefficients per molecule for the gas. Expressions for higher order virial coefficients may be written down by an extension of the procedure above.[12]

Before proceeding to the evaluation of the virial coefficients themselves it is worth noting that the second virial coefficient depends only on the one- and two-molecule canonical partition functions Z_1 and Z_2, and that the third virial coefficient involves only one-, two-, and three-molecule functions. This connection of the Nth virial coefficient to the interactions of groups containing up to N molecules also applies to the higher virial coefficients.

3.5. Virial coefficients and intermolecular forces

In order to obtain the virial coefficients B, C, etc. explicitly in terms of the intermolecular potential energy we need to evaluate Z_1, $Z_2, \ldots,$ etc. For this purpose we again confine ourselves to molecules with no internal degrees of freedom. In this case the single-molecule partition function Z_1 is given by the expression

$$Z_1 = \sum \exp(-\varepsilon_i/kT)$$

where the ε_i are the translational (kinetic) energy levels of a single molecule in the volume V. The permitted quantum mechanical states for the molecule are the well-known solutions of the Schrödinger equation for the problem of a particle in a box. For a gas at experimentally accessible temperatures the spacing of these translational energy levels is generally much less than kT so that we can replace the summation for Z_1 by an integral over the kinetic energy which can be performed analytically. The resulting expression for Z_1, which is the translational molecular partition function, is

$$Z_1 = V(2\pi mkT)^{\frac{3}{2}}/h^3 \tag{3.21}$$

where m is the mass of a molecule.

In general, for Z_N, the energy of the N particles is the sum of the translational contributions plus the intermolecular potential energy U_N. These two contributions to the energy are entirely independent; furthermore, the total kinetic energy of the N molecule is just the sum of the kinetic energies of the N separate molecules. For these reasons the N-molecule canonical partition function Z_N may be factorized into components associated with the translational and intermolecular energy. The result for Z_N is then conveniently written in the form

$$Z_N = \left(\frac{2\pi mkT}{h^2}\right)^{3N/2} Q_N \tag{3.22}$$

where

$$Q_N = \frac{1}{N!}\underbrace{\int \ldots \int}_{N} \exp\{-U_N/kT\}\,\mathbf{dr}_1 \ldots \mathbf{dr}_N. \qquad (3.23)\dagger$$

Here, in passing from the summation for Z_N to the integral we have again assumed that quantum mechanical effects are negligible. The N-fold integral for Q_N is known as *the configuration integral.*

Turning to specific cases we find, for example, that

$$Q_2 = \tfrac{1}{2}\iint \exp(-U_2/kT)\,\mathbf{dr}_1\,\mathbf{dr}_2 \tag{3.24a}$$

and

$$Q_3 = \tfrac{1}{6}\iiint \exp(-U_3/kT)\,\mathbf{dr}_1\,\mathbf{dr}_2\,\mathbf{dr}_3. \tag{3.24b}$$

Using (3.22), (3.24a), (3.21) and (3.19) it may be shown that the second virial coefficient B, for one mole of gas (N_A molecules) is then given by the expression

$$B(T) = -\frac{N_A}{2V}\int\!\!\int_V [\exp(-U_{12}/kT)-1]\,\mathbf{dr}_1\,\mathbf{dr}_2 \tag{3.25}$$

where we have used the facts that $Q_1 = \int \mathbf{dr}_1 = V$ and $U_2 \equiv U_{12}$ is the intermolecular energy of a pair of molecules 1 and 2.

Eqn (3.25) can be written

$$B(T) = \frac{-N_A}{2V}\int_V\!\!\int f_{12}\,.\,\mathbf{dr}_1\,\mathbf{dr}_2 \tag{3.26}$$

where $f_{12} = \exp(-U_{12}/kT) - 1$ is *the Mayer f function.*[10] This function has the useful property of going to zero whenever the volume elements $\mathbf{dr}_1$, $\mathbf{dr}_2$ are not close together, and U_{12} is consequently small. Since f_{12} is a function only of the separation, $\mathbf{r}_2 - \mathbf{r}_1$, $(= \mathbf{r}_{12})$ of the molecular pair (and of their mutual orientation if they are non-spherical) the variables in (3.26) may be changed to $\mathbf{r}_1$ and $\mathbf{r}_{12}$, and the integration over $\mathbf{r}_1$ carried out, yielding a factor V, to give the alternative equation

$$B(T) = \frac{-N_A}{2}\int f_{12}\,.\,\mathbf{dr}_{12}.$$

† $\mathbf{dr}_1$ represents a volume element in space $dx_1\,dy_1\,dz_1$ in a Cartesian coordinate system x_1, y_1, z_1.

The integration variable, $\mathbf{r}_{12}$, may be further changed to $4\pi r_{12}^2\,\mathrm{d}r_{12}$, (for spherical molecules where f_{12} depends only on the separation of the molecules, r_{12}) and the upper limit of integration may be set to infinity, since f_{12} falls rapidly to zero for large r_{12}. Hence the second virial coefficient may be written in this case

$$B(T) = -2\pi N_{\mathrm{A}} \int_0^\infty f_{12} \cdot r_{12}^2\,\mathrm{d}r_{12}$$

$$= -2\pi N_{\mathrm{A}} \int_0^\infty [\exp(-U(r)/kT) - 1] r^2\,\mathrm{d}r \qquad (3.27)$$

and it is seen that there is a direct and simple connection between the intermolecular pair potential energy function $U(r)$ and the second virial coefficient. Eqn (3.27) may be integrated by parts to give an equivalent expression for $B(T)$ in terms of the force between the pair of molecules, $-\mathrm{d}U(r)/\mathrm{d}r$

$$\frac{B(T)}{2\pi N_{\mathrm{A}}} = \left[\frac{[1-\exp(-U(r)/kT)]r^3}{3}\right]_0^\infty - \int_0^\infty \frac{\exp(-U(r)/kT)}{kT}\frac{\mathrm{d}U(r)}{\mathrm{d}r}\frac{r^3}{3}\,\mathrm{d}r.$$

Since $U(r) \to 0$ as $r \to \infty$, the first term is zero and so

$$B(T) = \frac{-2\pi N_{\mathrm{A}}}{3kT} \int_0^\infty \exp(-U(r)/kT)\frac{\mathrm{d}U(r)}{\mathrm{d}r} r^3\,\mathrm{d}r. \qquad (3.28)$$

Yet another equivalent form, in which B is written as an integral over the intermolecular energy rather than the separation, may also be derived.[13,14] We now define an intermolecular energy by

$$\phi(r) = U(r) + \varepsilon$$

where ε is the potential energy well depth, and further split ϕ into the two monotonic branches ϕ_{L} and ϕ_{R} which join smoothly at r_m (see Fig. 3.5)

$$\phi(r) = \phi_{\mathrm{L}}(r_{\mathrm{L}}) + \phi_{\mathrm{R}}(r_{\mathrm{R}})$$

Eqn (3.28) can be re-written

$$\exp(-\varepsilon/kT)\frac{3kT}{2\pi N_{\mathrm{A}}}B(T) = -\int_0^{r_m} r^3\frac{\mathrm{d}\phi}{\mathrm{d}r}\exp(-\phi_{\mathrm{L}}/kT)\,\mathrm{d}r$$

$$-\int_{r_m}^\infty r^3\frac{\mathrm{d}\phi}{\mathrm{d}r}\exp(-\phi_{\mathrm{R}}/kT)\,\mathrm{d}r.$$

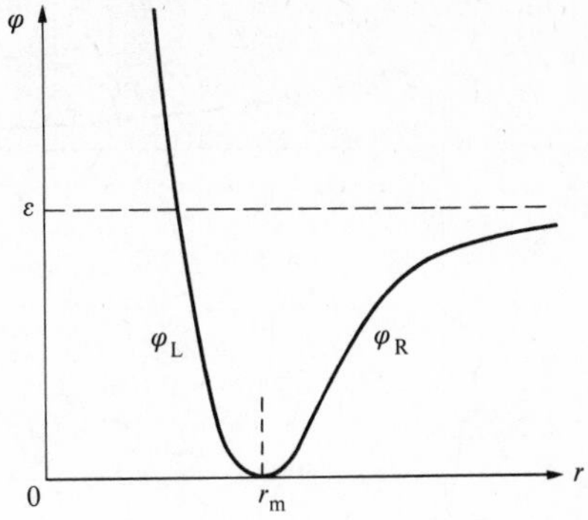

FIG. 3.5. Potential energy function in terms of $\phi = \phi_L + \phi_R = U(r) + \varepsilon$.

Changing the variable of integration to ϕ, subject to the following limits:

(1) when $r = 0$, $\phi = \phi_L = \infty$;
(2) when $r = r_m$, $\phi = \phi_L = \phi_R = 0$;
(3) when $r = \infty$, $\phi = \phi_R = \varepsilon$;

we obtain

$$\exp(-\varepsilon/kT)\frac{3kT}{2\pi N_A}B(T) = -\int_{\infty}^{0} r_L^3 \exp(-\phi_L/kT)\,d\phi_L - \int_{0}^{\varepsilon} r_R^3 \exp(-\phi_R/kT)\,d\phi_R.$$

The second integral can be extended to $\phi_R = \infty$ since the integrand has no value for $\phi_R \geqslant \varepsilon$ so that

$$\exp(-\varepsilon/kT)\frac{3kT}{2\pi N_A}B(T) = \int_{0}^{\infty} r_L^3 \exp(-\phi/kT)\,d\phi - \int_{0}^{\infty} r_R^3 \exp(-\phi/kT)\,d\phi$$

where the subscripts on ϕ_R and ϕ_L have been dropped since they refer to the same variable ϕ.

Hence we obtain a third equation for $B(T)$,

$$B(T) = \exp(\varepsilon/kT)\frac{2\pi N_A}{3kT}\int_{0}^{\infty} \Delta(\phi)\exp(-\phi/kT)\,d\phi \qquad (3.29)$$

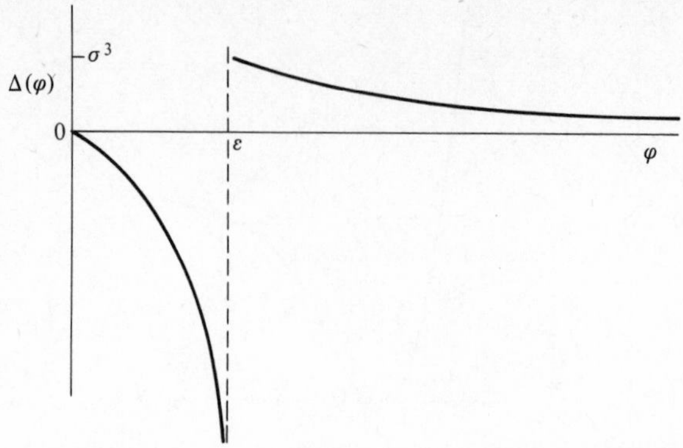

FIG. 3.6. The function $\Delta(\phi)$ for an intermolecular potential energy function of the type shown in Fig. 3.5.

where $\Delta(\phi)$ is the well-width function

$$\Delta(\phi) = r_L^3(\phi) - r_R^3(\phi), \qquad \phi \leqslant \varepsilon$$

and

$$\Delta(\phi) = r_L^3, \qquad \phi > \varepsilon.$$

r_L and r_R are the turning points of the potential well at an energy ϕ. The typical form of $\Delta(\phi)$ is illustrated in Fig. 3.6. Eqn (3.29) suggests that second virial coefficients can only yield information about the function $\Delta(\phi)$. Only for $\phi > \varepsilon$ does this lead uniquely to the pair potential energy function. In the region $\phi \leqslant \varepsilon$ the information available is in general only the width of the well as a function of its depth.

In order to calculate the third virial coefficient, $C(T)$, from (3.20)–(3.24), the integral Q_3 (3.24b) which involves U_3 must be evaluated. As has been mentioned in Chapter 1, the intermolecular energy of a group of three molecules is not usually exactly equal to the sum of the three component pair energies. In general the non-additivity of intermolecular energies must be allowed for by writing

$$\begin{aligned} U_3(r_{12}, r_{13}, r_{23}) &= U(r_{12}) + U(r_{13}) + U(r_{23}) \\ &\quad + \Delta U_3(r_{12}, r_{13}, r_{23}) \\ &\equiv U_3^{\mathrm{P}} + \Delta U_3 \end{aligned} \tag{3.30}$$

where U_3^{P} is the sum of the pair energies and the non-additivity term, ΔU_3, is generally less than U_3^{P}. If this expression is used in (3.24b) and those for Z_1 and Z_2 are also introduced, the following results for C may be obtained (after some working)

$$C(T) = C_{\text{add}} + \Delta C$$

$$C_{\text{add}} = \frac{-N_A^2}{3V} \iiint f_{12} f_{12} f_{23} \, d\mathbf{r}_1 \, d\mathbf{r}_2 \, d\mathbf{r}_3$$

$$\Delta C = \frac{-N_A^2}{3V} \iiint [\exp(-\Delta U_3/kT) - 1] \times \exp(-U_3^{\text{P}}/kT) \, d\mathbf{r}_1 \, d\mathbf{r}_2 \, d\mathbf{r}_3. \tag{3.31}$$

In these equations, C_{add} is the value of the third virial coefficient if the non-additivity were zero (i.e. in the approximation of pairwise-additivity) and ΔC represents the contribution due to this term. Since the values of U_3 depend only on the *relative* positions (and possibly orientations) of the three molecules, it is possible to transform the integration variables to the (scalar) separations of the molecules, r_{12}, r_{13}, r_{23}, using the relation[15]

$$d\mathbf{r}_1 \, d\mathbf{r}_2 \, d\mathbf{r}_3 = 8\pi^2 V r_{12} r_{13} r_{23} \, dr_{12} \, dr_{13} \, dr_{23}$$

where the integration is over those values of r_{12}, r_{13}, r_{23} which form the sides of a triangle. More convenient expressions for the calculation of third virial coefficients may then be written

$$C_{\text{add}} = \frac{-8\pi^2 N_A^2}{3} \iiint f_{12} f_{13} f_{23} r_{12} r_{13} r_{23} \, dr_{12} \, dr_{13} \, dr_{23}$$

$$\Delta C = \frac{-8\pi^2 N_A^2}{3} \iiint [\exp(-\Delta U_3/kT) - 1] \times \exp(-U_3^{\text{P}}/kT) r_{12} r_{13} r_{23} \, dr_{12} \, dr_{13} \, dr_{23}. \tag{3.32}$$

Third virial coefficients may thus be expected to provide some information about two- and three-body intermolecular interaction energies.

The expressions for the higher virial coefficients are also known.[1,16] These again involve appropriate terms for many-body interactions, and in principle the Nth virial coefficient will depend on two, three . . . N body energies. In practice there have been few reliable measurements of virial coefficients above the third, and calculations of the higher coefficients have generally been restricted to fairly simple model potential functions, and have usually assumed pairwise additivity. A few calculations of non-additive contributions have been made,[17] and these suggest that the three-body corrections dominate the total non-additive many-body effects in the higher virial coefficients.

When pairwise additivity of the intermolecular potential is assumed, the expressions for the higher virial coefficients may be conveniently written using a graphical representation.[18] The third- and higher-order coefficients are made up of integrals whose integrands are products of Mayer f-functions. These *cluster integrals* may be represented using diagrams in which a line represents a Mayer function, and the intersection of two lines represents a molecule whose coordinates are integrated over. Molecule 1 is taken to be fixed at the origin, so that $d\mathbf{r}_j \equiv d(\mathbf{r}_j - \mathbf{r}_1)$. We may then write

$$\triangle = \iint f_{12} f_{13} f_{23} \, d\mathbf{r}_2 \, d\mathbf{r}_3$$

and comparison with (3.31) shows that

$$C_{\text{add}} = \frac{-N_A^2}{3} \triangle$$

since integration of molecule 1 over the space gives a further factor, V.

Using this notation, the pairwise-additive fourth virial coefficient, D, may be written

$$D = \frac{-N_A^3}{8} [3\square - 6\boxslash + \boxtimes]$$

and the fifth virial coefficient may be expressed in terms of ten contributing diagrams. Ree and Hoover[19] introduced a modified type of diagram, in which a wiggly line stands for the function $f_{ij} + 1$. By using both types of line in the diagrams, the number of different diagrams contributing to the higher coefficients may be greatly reduced. For example, we may write

$$D = \frac{-N_A^3}{8} [3\widetilde{\boxtimes} - 2\boxtimes]$$

and the fifth virial coefficient can now be expressed in terms of five diagrams.

Although calculations of the higher virial coefficients have not generally been used directly in the study of intermolecular forces, the cluster notation described briefly above has been extensively employed in the structural analysis of liquids.

3.6. Quantum corrections

The expressions given earlier for the virial coefficients have been based entirely on classical mechanics. It is known from quantum mechanics that under some circumstances the wave-like properties of particles cannot be neglected and for light molecules at low temperatures the effects on the virial coefficients are significant.[20] Whether or not the quantum effects are important is determined by the value of the thermal de Broglie wave length, λ, associated with molecules of mass m at a temperature T, $\lambda = h/\sqrt{(3mkT)}$. Under conditions where λ is similar to the average separation of molecules, the general quantum mechanical symmetry requirement for the total wave function leads to deviations from ideal gas behaviour, even for point molecules without intermolecular interactions. For Bose–Einstein particles, this is similar to the effect of attractive forces between the molecules, while for Fermi–Dirac molecules an apparent repulsion is observed. However, for gases other than He and H_2 the magnitude of these symmetry effects is negligibly small at all accessible temperatures.

Of more general importance are effects which arise when λ is of the same order as the molecular diameter, σ. The origin of these effects may be found chiefly in the quantization of energy levels and for the second virial coefficient the evaluation of the two-molecule partition function involves summation over the two-molecule bound states, and also over the continuum of energies for the repulsive states. For substances other than He and H_2, the full quantum mechanical calculations are unnecessary, and a semi-classical approach to the quantum corrections is satisfactory.[21,22] In this approach, the virial coefficients are expressed as power series† in h^2, in which the first term is equal to the classical expression. Thus, for the second virial coefficient we may write

$$B = B_{\text{class}} + \frac{h^2}{4\pi^2 m} \cdot B_{q1} + \left(\frac{h^2}{4\pi^2 m}\right)^2 B_{q2} + \ldots \tag{3.33}$$

It has been common in the past to add to this series a statistical term, B_{perfect}, due to symmetry effects. For systems with realistic intermolecular pair potentials it has been demonstrated that it is incorrect to include this term at temperatures for which the semi-classical expansion is appropriate.[23]

The quantum corrections, B_{q1} and B_{q2}, are defined in terms of the pair potential energy function $U(r)$ and its derivatives

$$B_{q1} = \frac{\pi N_A}{6(kT)^3} \int_0^\infty \exp(-U(r)/kT)\, U'^2 r^2 \, \mathrm{d}r \tag{3.34}$$

$$B_{q2} = \frac{-\pi N_A}{6(kT)^4} \int_0^\infty \exp(-U(r)/kT) \times \left[\frac{(U'')^2}{10} + \frac{(U')^2}{5r^2} + \frac{(U')^3}{9kTr} - \frac{(U')^4}{72(kT)^2}\right] r^2 \, \mathrm{d}r \tag{3.35}$$

where $U' = \mathrm{d}U(r)/\mathrm{d}r$ and $U'' = \mathrm{d}^2U(r)/\mathrm{d}r^2$.

† For non-analytical potentials such as the hard sphere potential, terms in the odd powers of h also contribute.

The next correction term, B_{q3}, is also known.[23] It is found that the expansion converges satisfactorily, and even for ^{4}He at temperatures above 50 K the semi-classical result (including the third-order term, B_{q3}) gives results in good agreement with those obtained from the full quantum mechanical calculation.[23]

For heavier molecules the two-term correction given above is accurate at all accessible temperatures. The net effect of the quantum corrections is to make the second virial coefficient more positive. Even for molecules as light as Ne the correction is fairly small, perhaps 6 per cent of the total at 35 K.

3.7. Temperature dependence of virial coefficients

It is evident from the form of (3.27) that the second virial coefficient for a substance which has the normal type of pair potential energy function has positive and negative contributions, derived respectively from those separations at which the pair energy is positive and negative. For most values of r less than the collision diameter, σ, $U(r) \gg kT$ and $f(r) \simeq -1$. The associated positive contribution to B is relatively unaffected by changes in T, in contrast to the negative contribution derived from values of r greater than σ, for which $U(r) < 0$. The function $\exp(-U(r)/kT) - 1)$ may reach quite large values, which depend strongly on the temperature. The typical variation of $B(T)$ with temperature is shown in Fig. 3.7(a), and the changes in the Mayer function, $f(r)$, in Fig. 3.7(b).

At low temperatures the contributions from the potential well dominate, and B is negative, becoming less negative as T increases. At the Boyle temperature, T_B, B is zero, and at temperatures above T_B B is positive. Typically it is found that $T_B \simeq 2.7\ T_c$ for simple substances, so that for small light molecules such as He, H_2, Ne the value of B is positive at room temperature. For heavier substances, B is generally negative under these conditions. At temperatures above T_B the value of B first rises, reaches a maximum value and then drops slowly. For most substances this maximum occurs at an inaccessibly high temperature, and it has only been observed experimentally in the case of helium.

The temperature dependence of the third virial coefficient is illustrated in Fig. 3.8. For most substances $C(T)$ is positive in the accessible temperature range, and decreases with increasing temperature. The negative values of C expected at low temperatures have only been observed in a limited number of cases.[24,25] The experimental problems in attempting to measure the third virial coefficient at low temperatures are very considerable, since reliable measurements of C normally demand experimental work over a wide range of pressures, but at these low temperatures condensation occurs at quite low pressures.

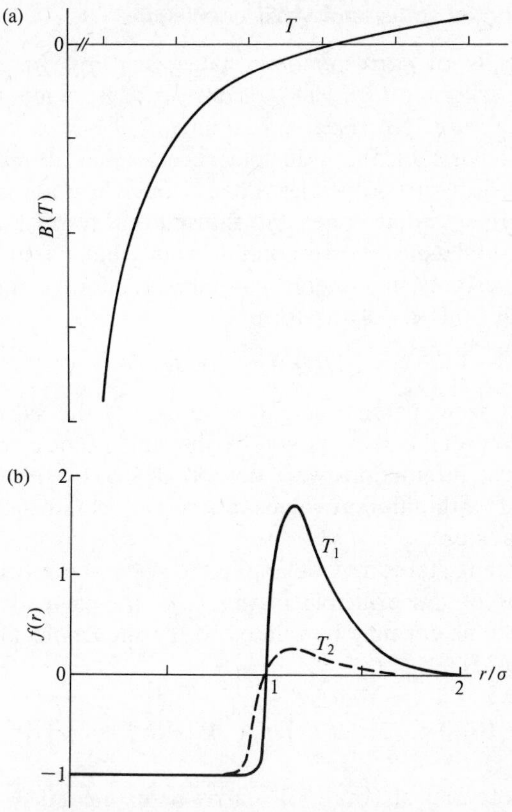

FIG. 3.7. (a) Temperature dependence of second virial coefficients. (b) Temperature dependence of the Mayer function, $f(r)$. $T_2 > T_1$.

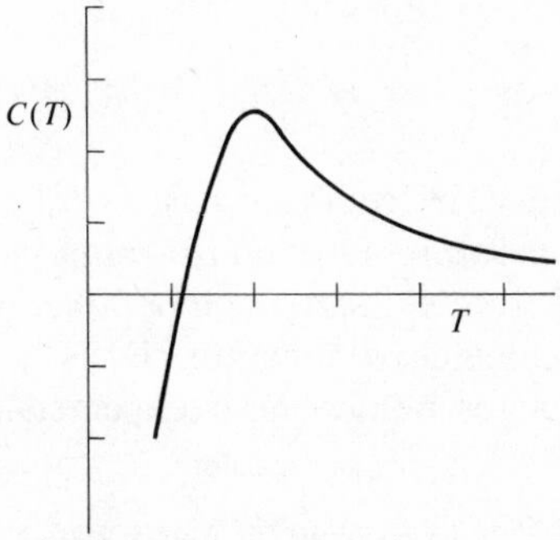

FIG. 3.8. Temperature dependence of third virial coefficients.

3.8. Corresponding states and virial coefficients

The principle of corresponding states was introduced earlier as a means of comparing the *PVT* data for different substances in terms of the reduced temperature, pressure, and volume, T/T_c, P/P_c, and $\tilde{V}/V_c$.

Further analysis of the principle reveals that it may be used to correlate many different properties. Pitzer[26] investigated the origins of the principle of corresponding states and showed that for classical substances with spherical molecules, corresponding states behaviour would be expected if the intermolecular energy was pairwise additive and derived from a pair potential function of the form

$$U(r) = \varepsilon \,.\, F(r/\sigma) \tag{3.36}$$

where ε and σ were parameters characteristic of the substance and the shape of the potential was described by the single function *F*. Hence two substances whose interactions were described, say, by the Lennard-Jones 12−6 potential, with different values of ε and σ, would show corresponding states behaviour.

Corresponding states may be applied to second virial coefficient data, and the origins of the principle in this case are particularly clear. The second virial coefficient may be related to the intermolecular pair potential $U(r)$ by (3.27)

$$B(T) = -2\pi N_A \int_0^\infty [\exp(-U(r)/kT) - 1] r^2 \, dr$$

which, for a potential of type (3.36), may be expressed

$$B(T) = -2\pi N_A \sigma^3 \int_0^\infty [\exp(-F(r/\sigma)/T^*) - 1](r/\sigma)^2 \, d(r/\sigma). \tag{3.37}$$

The value of the integral depends only on the reduced temperature, T^*, ($T^* = kT/\varepsilon$) and on the form of the function, *F*, so the second virial coefficient of a substance with potential parameters ε, σ may be written

$$B(T) = \frac{2\pi N_A}{3} \sigma^3 B^*(kT/\varepsilon) = b_0 \,.\, B^*(T^*) \qquad (3.38)\dagger$$

and the reduced second virial coefficient $B(T)/b_0$ is a function, $B^*(T^*)$, only of the reduced temperature. For other substances with the same potential function, *F*, but different characteristic parameters, ε and σ, the same reduced second virial coefficient function, $B^*(T^*)$, would be obtained. Hence corresponding states behaviour is expected for the second virial

† The characteristic molar volume $b_0 = \frac{2}{3}\pi N_A \sigma^3$ is widely used. For hard spheres it is equal to the van der Waals co-volume, *b*.

coefficients of substances which have potential energy functions of the same shape.

In order to test the validity of the principle using experimental data, it is necessary to establish the values of characteristic quantities proportional to ε and σ (or, more conveniently in this case, to σ^3). Traditionally this has been done using the critical temperature, T_c, and critical volume, V_c, although these are not strictly proportional to ε and σ^3, for reasons discussed below. When the second virial coefficients of different substances are compared in the dimensionless form, B/V_c as a function of T/T_c, it is found that a single curve is observed for many simple non-polar and slightly polar substances.[27]

Although the critical parameters have been very widely used in corresponding states reductions, they are by no means the only quantities which may be used, and indeed have some theoretical disadvantages. In particular it is not easy to compare experimental values with theoretical predictions, based on an assumed form of intermolecular potential, since it is not possible at the present time to determine the values of the critical parameters which would result from a given potential. In addition, the critical point is a dense state, at which the complications resulting from non-additive many-body interactions cannot be neglected. For many purposes these are serious impediments. An alternative form of corresponding states reduction which avoids these problems is based on the Boyle temperature, T_B, rather than the critical temperature, T_c.[28,29]

Since T_B is the temperature at which $B(T) = 0$, and this is a pair property, the value of kT_B/ε may be readily established for a model pair potential energy function. The connection between theory and experiment is thus much simpler in this case. However the experimental determination of Boyle temperatures is less convenient than that of critical temperatures and for many substances T_B lies at inconveniently high temperatures.

The use of corresponding states generally requires two characteristic parameters. The choice of a characteristic volume associated with the Boyle temperature is a little troublesome. The second virial coefficient has the units of molar volume, but is zero at this temperature. One solution which has been employed is to define a Boyle volume, V_B, by the relation

$$V_B = T_B \left(\frac{dB}{dT}\right)_{T=T_B}. \tag{3.39}$$

This is, in principle, an experimentally accessible quantity, and is also readily calculated for a model potential energy function. Its disadvantages lie in the experimental measurement of dB/dT in a region where B is very small. Nevertheless fairly reliable values of V_B have been obtained for many simple molecules, and the Boyle volume has been quite extensively used in corresponding states analyses. An alternative characteristic volume, which has some experimental advantages, is defined in terms of the value of the second virial coefficient at a temperature which is a fixed fraction of T_B.[30] We have used the characteristic volume, B_0, defined

$$B_0 = -B_{(T=0\cdot 7T_B)} \tag{3.40}$$

in our corresponding states analyses. Values of T_c, V_c, T_B, and B_0 for some simple substances are shown in Appendix 6.

The application of the principle of corresponding states to other equilibrium properties of fluids is also based on a dimensional argument.[26] Pitzer showed that the configurational integral, Q_N, (3.23) of a substance obeying the conditions noted above could be expressed as a function of the reduced volume, $V/N\sigma^3$ and temperature, kT/ε.

$$Q_N = \frac{\sigma^{3N}}{N!}\phi\left(\frac{V}{N\sigma^3}, \frac{kT}{\varepsilon}\right) \tag{3.41}$$

where ϕ is a universal function (not in general calculable from first principles) for substances with the same form of pair potential. Since the values of Q_N and its temperature and volume derivatives determine the thermodynamic properties the above relation suggests that the behaviour of all such substances will be simply related, provided that they are compared at equal values of $V/N\sigma^3$ and kT/ε. Extensive testing of the principle for many properties of dense fluids has shown that for most simple non-polar substances it is obeyed with high accuracy, and it provides an invaluable basis for the interpolation and extrapolation of results under circumstances where the present development of a molecular theory is not yet complete.[31]

It must be noted that although the occurrence of an intermolecular pair potential of a single shape is a sufficient condition for corresponding states behaviour in classical pairwise additive fluids, it has not been established that it is in practice a necessary condition. In particular, analyses of the sensitivity of various properties, including the second virial coefficient, to the form of the intermolecular potential have often revealed the futility of using experimental results in certain temperature ranges to distinguish between rival potential functions.[32] Widely differing model potentials have been found to predict behaviour which is indistinguishable when compared on a corresponding states plot. More helpfully such an analysis can indicate in what regions of temperature further experimental study may most profitably be carried out.

3.9. Virial coefficients for model potential energy functions

Although this chapter is primarily concerned with the ways in which gas imperfection data can contribute to the determination of intermolecular potential energy functions for real systems, it is of some value to consider the form of the virial coefficients which arise from a number of highly simplified model potential functions. These can provide valuable insight into the relationships between the virial coefficients and the intermolecular potential. They can in some cases be valuable for testing methods of determining intermolecular potentials from second virial coefficient data.

(a) *Hard-sphere potential.* This potential function may be represented

$$U(r) = +\infty, \qquad r \leqslant \sigma$$
$$U(r) = 0, \qquad r > \sigma$$

and corresponds to a model in which two molecules cannot approach each other to a distance less than the collision diameter, σ, but do not otherwise interact. It is the simplest representation of the non-zero size of molecules. The integration for the second virial coefficient, (3.27), is simple. For $r \leqslant \sigma$, $f(r) = -1$, and for $r > \sigma$, $f(r) = 0$.

Hence

$$B = -2\pi N_A \int_0^\infty f(r) r^2 \,dr = -2\pi N_A \int_0^\sigma (-1) r^2 \,dr$$
$$= \frac{+2\pi N_A \sigma^3}{3}.$$

B is positive and independent of temperature. The third and fourth virial coefficients may also be evaluated analytically and are listed in Table 3.1. Some higher coefficients are also given for this model, which has played an important role in the development of the theory of dense fluids.[19]

Table 3.1
Virial coefficients for the hard-sphere potential

$B = (2\pi/3) N_A \sigma^3 = b_0$
$C = \frac{5}{8} b_0^2$
$D = 0.286\,95\; b_0^3$
$E = 0.110\,40 \pm 0.000\,06\; b_0^4$
$F = 0.0386 \pm 0.0004\; b_0^5$

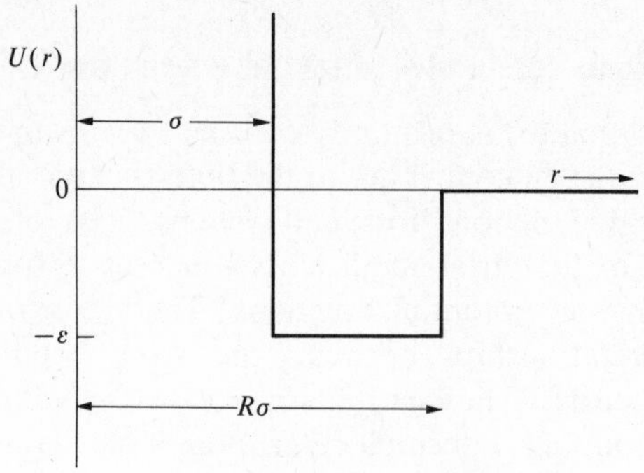

FIG. 3.9. The square-well potential.

(b) *Square-well potential.* This may be represented

$$U(r) = +\infty, \qquad r \leqslant \sigma$$
$$U(r) = -\varepsilon, \qquad \sigma < r \leqslant R\sigma$$
$$U(r) = 0, \qquad r > R\sigma$$

and is shown in Fig. 3.9. This model consists of a hard sphere to which is added an attractive well of constant depth, and a range which may be varied. Commonly studied values of R have been 1.5 and 2.0. The second virial coefficient may again be readily calculated, and can be expressed

$$\frac{B(T)}{b_0} = 1 - (R^3 - 1)\{\exp(\varepsilon/kT) - 1\}. \tag{3.42}$$

This function represents the qualitative temperature variation of experimental second virial coefficients quite well, except at very high temperatures, where it shows no maximum, but continues to rise monotonically towards the limiting hard sphere value. The third virial coefficient may also be represented analytically,[33] and again mimics the experimental data quite well. Fourth virial coefficients have also been reported[34] for this potential.

(c) *Soft-sphere potential.* An alternative representation of the repulsions between molecules replaces the hard-sphere model with an inverse power potential.[35]

$$U(r) = K/r^n = \varepsilon_s\left(\frac{\sigma}{r}\right)^n, \qquad n > 3$$

where $K(K = \varepsilon_s\sigma^n)$ is a positive constant. The hard-sphere model corresponds to the limiting case, $n = \infty$. The second virial coefficient may be expressed

$$\frac{B(T)}{b_0} = (\varepsilon_s/kT)^{3/n}\Gamma(1-3/n) \tag{3.43}$$

where Γ is the gamma function.† $B(T)$ is positive for this model, and decreases with increasing temperature.

(d) *Lennard-Jones* 12−6 *potential.* The most widely studied model intermolecular potential function is surely the Lennard-Jones 12−6 model. This is a special case of the general $n-m$ potential originally proposed by Mie[36]

$$U(r) = \frac{n\varepsilon}{n-m}\left(\frac{n}{m}\right)^{m/(n-m)}\left\{\left(\frac{\sigma}{r}\right)^n - \left(\frac{\sigma}{r}\right)^m\right\}, \qquad n > m > 3.$$

which for $n = 12$, $m = 6$ gives the familiar form

$$U(r) = 4\varepsilon\left\{\left(\frac{\sigma}{r}\right)^{12} - \left(\frac{\sigma}{r}\right)^6\right\}.$$

The extensive use of this potential in the analysis of virial coefficient data followed the demonstration by Lennard-Jones[37] of an analytical integration of B for the general $n-m$ model, which, in the 12−6 case may be written

$$B^*(T^*) = \frac{B(T^*)}{b_0} = \frac{-\sqrt{2}}{4}\sum_{j=0}^{\infty}\left\{\left(\frac{1}{T^*}\right)^{(2j+1)/4} \times \frac{2^j}{j!}\Gamma\left(\frac{2j-1}{4}\right)\right\}$$
$$= \sum_{j=0}^{\infty}\beta_j\left(\frac{1}{T^*}\right)^{(2j+1)/4} \tag{3.44}$$

where

$$\beta_j = \frac{-2^{(j+\frac{1}{2})}}{4j!}\Gamma\left(\frac{2j-1}{4}\right)$$

and

$$T^* = kT/\varepsilon.$$

It is useful to note that the β_j are related by the expression

$$\beta_{j+2} = \frac{2j-1}{(j+1)(j+2)}\beta_j$$

and that

$$\beta_0 = +1{\cdot}733\,001\,0, \qquad \beta_1 = -2{\cdot}563\,693\,4.$$

† $\Gamma(n) = \int_0^\infty x^{n-1}e^{-x}\,dx$. For positive integral n, $\Gamma(n+1) = n!$

The derivative $T^*(dB^*/dT^*)$, and the quantum corrections, B^*_{q1} and B^*_{q2} may also be expanded in a similar way.[38] Tables of the second virial coefficients and of related quantities for this potential are given in Appendix 2.

This potential function gives second virial coefficients which closely resemble experimental results in their temperature dependence, and can be used for modest interpolation. However it is not an accurate representation of the pair potential energy function for any known substance, and its use for the extrapolation of data should be carried out with some caution.

The pairwise-additive third virial coefficient may also be expressed using a series expansion,[39] and there have also been several calculations of the non-additive contribution to C, based on various approximations to the three-body energy.[40,41] Once more the qualitative behaviour of the third virial coefficient is in good general agreement with experimental results for many simple substances.

The fourth and fifth virial coefficients for this potential have been evaluated by numerical integration, assuming pairwise additivity.[45] The non-additive contributions to the fourth virial coefficient have also been reported.[17] In the absence of reliable experimental values for these quantities, it is assumed that these represent the probable experimental trends.

(e) *Realistic potential energy functions.* The potential functions discussed above must be regarded as model potentials and do not represent the interactions between any known molecules. A number of more realistic potentials are included in Appendix 1. Tables of second virial coefficients and related quantities for some of these functions are given in Appendix 2. A program for the numerical evaluation of second virial coefficients for any spherically symmetric potential function is given in Appendix 11.

3.10. Experimental measurements

The experimental determination of virial coefficients may seem at first sight to be straightforward. Simultaneous measurements of pressure, temperature, and density for a substance over a range of pressures will yield values for the compression factor as a function of density. A least squares polynomial fit to these results should then give the values of the virial coefficients and their associated uncertainties. However, a more detailed inspection reveals several problems. In measurements at low and moderate pressures the deviation from ideality is generally small, perhaps one or two per cent. Very high accuracy is thus needed in all of the component measurements if an accurate assessment of the gas imperfection, and hence the virial coefficients, is to be made. At higher pressures the extent of the deviations increases, but higher terms in the series now make substantial contributions and cannot be neglected. A further difficulty may arise because the virial equation, (3.1), is an infinite series, and the effect of approximating it by a truncated polynomial makes the equivalence of corresponding coefficients in the two series uncertain.

Nevertheless, many reliable measurements of second virial coefficients, and several of third virial coefficients have been made. In addition very many measurements of lower accuracy have been reported. An extensive compilation of experimental virial coefficient data has been published.[25]

3.10.1. Low-pressure techniques

Direct methods. Absolute measurements of P, V, T and of the amount of substance have been extensively used to determine second virial coefficients. Some typical designs are schematically represented in Fig. 3.10.

The classic apparatus (a) is that of Boyle. Gas is trapped by mercury on the left side of the J-shaped tube, whose cross-section is accurately known. The mercury level in the right side is adjusted and measurements of the heights of the two columns yield values of P and V. The amount of substance must be determined separately, perhaps by weight, or from the limiting values of PV/RT at low pressures. From a series of PV measurements at a given temperature a value of the second virial coefficient may be derived from the slope of a graph of PV against P or $1/V$. This apparatus has clear disadvantages. Its use is limited to low and moderate pressures, in order to keep the height of the mercury column to a manageable length, and it can only be used in the range of temperature in which mercury is liquid and has an acceptably low vapour pressure. The requirement of constant tube diameter is critical, and in the measurement of the heights of the mercury column due allowance must be made for capillary depression. In general measurements with this type of apparatus have been confined to organic vapours near to room temperature, and are of only moderate accuracy. Typical quoted uncertainties are ± 50 to $100\ \mathrm{cm^3\ mol^{-1}}$.

Rather improved accuracy may be achieved using a variant of the Boyle apparatus, shown in Fig. 3.10(b). In this design a given mass of gas is allowed to expand into several previously calibrated volumes, and the pressure measured at each point. Accurate calibration of the volumes may be performed during construction by filling with liquids of known density, and weighing. The pressures may either be measured directly, or

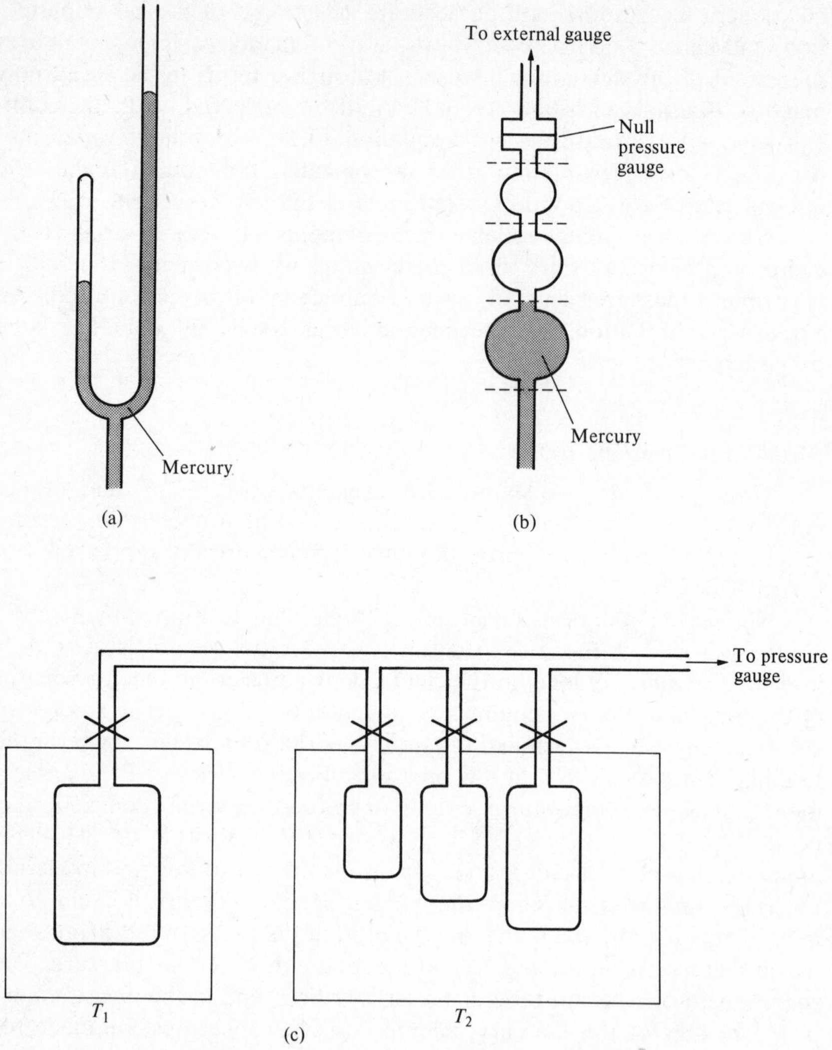

FIG 3.10. Low-pressure virial coefficient apparatus.

via a separator acting as a null gauge and permitting the use of an external pressure gauge.

An alternative form of this apparatus, suitable for use at low temperatures, below the freezing point of mercury, is shown in Fig. 3.10(c) Here the expansion volumes are contained in a thermostat near to room temperature, and only the primary gas volume vessel is held at the temperature of interest. The presence of mercury may be completely eliminated by the use of a suitable differential pressure gauge, which isolates the manometers from the apparatus. In order to achieve results of high accuracy it is vital that much attention is given to the accurate determination of the 'dead volumes' in connecting tubing, valves, etc. When this is done the apparatus can give results of high accuracy, with uncertainties which can be less than 5 $cm^3\ mol^{-1}$.[43]

Differential methods. In the methods described above, very accurate pressure measurement is essential. Uncertainties in the absolute value of the acceleration due to gravity may be the limiting source of error in such measurements! This problem may be eased by the use of differential procedures. In these, the behaviour of the gas under study is compared with that of another gas, usually chosen to have near perfect properties, which is subjected to PV changes very similar to those undergone by the subject gas. The difference in pressure resulting from similar volume changes or alternatively the additional small volume change need to restore pressure equality following a large expansion may then be used to give a more or less direct measure of the difference in the imperfection of the two gases. The second virial coefficient of the reference gas must be known if the absolute value of the second virial coefficient of the gas under study is to be determined.

These methods, and minor variants, have been quite extensively used[44] to obtain results of high accuracy, usually around room temperature, since mercury is generally used in such designs.

An alternative type of differential apparatus uses two similar vessels which are initially filled to equal pressure at room temperature, using a reference gas in one vessel and that under study in the other. The temperature of the vessels is then altered and the volume change in one vessel needed to restore equality of pressure between the vessels is determined. This may be related to the difference between the changes in second virial coefficients with temperature for the two gases. With accurate balancing of the pressures and careful initial calibration, this method can give results of high precision, and it has been successfully used in low temperature studies.[45] Uncertainties of less than 1 $cm^3\ mol^{-1}$ have frequently been reported.

Gas adsorption. Measurements of second virial coefficients at low temperatures, around the normal boiling point of a substance, are found to be particularly useful in the study of intermolecular interactions. Unfortunately under these conditions serious complications can arise as a result of the adsorption of significant amounts of gas on to the walls of the vessel.[46] This reduces the amount of substance present in the gas phase and can lead to large errors in the calculated virial coefficients. It is very desirable in such measurements to attempt a reliable estimate of the magnitude of this effect, which may be achieved by using two systems in which the surface : volume ratio is very different, and extrapolating the results to the zero surface area limit.[47] This problem has been turned to advantage in the study of the interactions of gases with surfaces, by determining the second virial coefficients appropriate to this interaction.[48]

3.10.2. High-pressure techniques

Information about the higher virial coefficients may only be reliably obtained from measurements which extend to high densities. Such results can also give valuable information about the reliability of second virial coefficients obtained from measurements at lower densities.

Direct measurements have been extensively used in high pressure studies. A design which has been quite extensively used is shown in Fig. 3.11(a). In this apparatus the gas is successively compressed into a series of smaller volumes, and the pressure measured. The design avoids the development of large pressure gradients across the calibrated volumes. Electrical contracts are used to determine the position of the mercury in the capillary tubing which joins the compression volume. The amount of substance is determined by allowing the gas to occupy a large calibrated volume at low pressure. Pressures of up to 3000 atmospheres have been used in this kind of apparatus, and it has been used in measurements on a wide range of substances.[49] The presence of mercury limits the accessible temperature range.

An alternative form of apparatus which has been widely used was devised by Burnett,[50] and uses successive expansions of gas, initially at high pressure. The basic design is shown in Fig. 3.11(b).

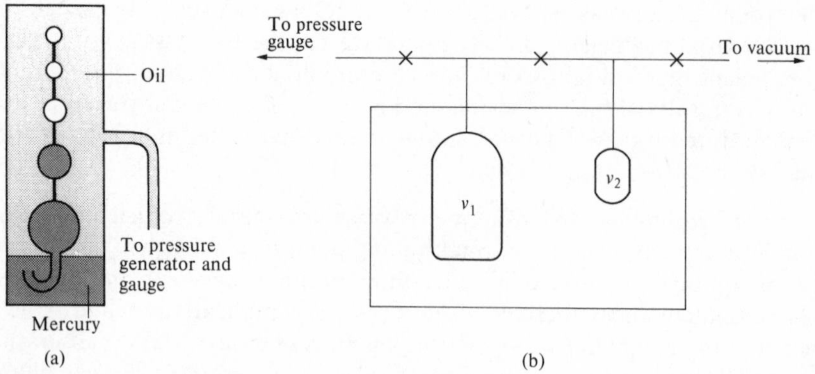

FIG. 3.11. High-pressure virial coefficient apparatus.

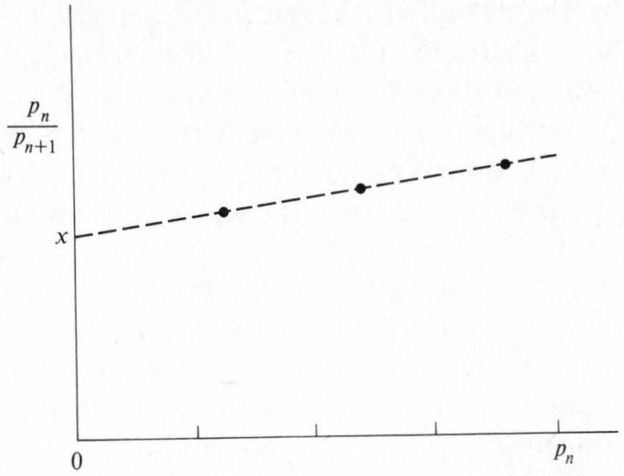

FIG. 3.12. Extrapolation of Burnett results.

Volume v_1 is initially filled to a high pressure, which is accurately measured. The smaller volume, v_2, is evacuated, and the expansion valve then opened. After equilibration, the new pressure is determined, v_2 is re-evacuated, and the procedure is repeated. It is not necessary in this apparatus to measure the amount of substance.

The data resulting from these experiments are a set of pressures, $p_0, p_1, \ldots, p_n$ from the successive expansions. In order to use these to determine the virial coefficients it is necessary to know the ratio $(v_1 + v_2)/v_1 (= x)$. This may be determined from the limiting value of the pressure ratio, p_n/p_{n+1}, as $p \to 0$, since, in general

$$\frac{p_n}{p_{n+1}} = \frac{z_n}{z_{n+1}} x$$

where z_n is the value of the compression factor $p\tilde{V}/RT$ at p_n. Since $z \to 1$ as $p \to 0$, graphical extrapolation of p_n/p_{n+1} can be used to obtain an accurate value of the volume ratio, x (Fig. 3.12).

The limiting slope of this graph at $P = 0$ may be readily shown to have the value $B(x-1)/RT$. Various alternative procedures, both graphical and directly numerical have been described which can yield second and higher virial coefficients from Burnett expansion data.[51,52]

This apparatus is suitable for use at both high and low temperatures, when it has sometimes been found convenient to maintain only v_1 at the operating temperature, and to keep v_2 near to room temperature.

3.10.3. *Calculation of virial coefficients from experimental results*

Once accurate measurements of P, $\tilde{V}$, and T have been made, the derivation of the virial coefficients may still present some problems. It might be thought that second virial coefficients could be reliably obtained from low-pressure results by plotting $P\tilde{V}/RT$ against P or $1/\tilde{V}$, and establishing the slope of this line by a least squares procedure. Uncritical use of a linear least squares calculation will indeed give results which appear to be of high precision. However, comparison between values obtained from the pressure and density series may reveal discrepancies far exceeding the declared standard errors of these results.[53] This problem appears to stem from the fact that the second virial coefficient is equal to the limiting slope of the line, at zero density, and this is less accurately determined than the mean slope over a wide density range. The use of a quadratic least squares fit, in which both second and third virial coefficients are determined appears to give results for B which are in better agreement, though with larger (presumably more realistic) errors.

High-pressure results are generally analysed by fitting high-order polynomials to the PVT values.[54] The problem of whether the polynomial coefficients accurately represent the virial coefficients is a difficult one to resolve. It seems to be necessary to carry out several fits to polynomials of varying order, and perhaps using data in different density ranges. Even then there is some evidence that the statistical uncertainty in second virial coefficients derived in this way is less than the true error. It appears to be very desirable to include low-density results in attempts to derive the second virial coefficients.

Using the procedure outlined in this section accurate values of the second virial coefficient and in some cases of the third virial coefficients may be derived, and some values of fourth virial coefficients have also been reported.[51]

3.10.4. Joule–Thomson coefficients

Virial coefficients may also be derived from measurements of other properties which may be related to the equation of state. The most important example of such a procedure involves the Joule–Thomson coefficients introduced earlier. The adiabatic coefficient, μ_{JT}, may be expressed in terms of the virial coefficients, their temperature derivatives, and the ideal gas heat capacity, C_P^0.

$$\mu_{JT}C_P^0 = -[B - TB'] + \frac{1}{\tilde{V}}\left[2B^2 - 2TBB' - 2C + TC' - \frac{RT^2}{C_P^0}(B - TB')B''\right] + \dots$$

where $B' = dB/dT$ and $B'' = d^2B/dT^2$.

The zero pressure limit is seen to be $B - T(dB/dT)$ and unusually is not the perfect gas value at this limit. Measurements of μ_{JT} can in principle provide valuable information about B and its temperature

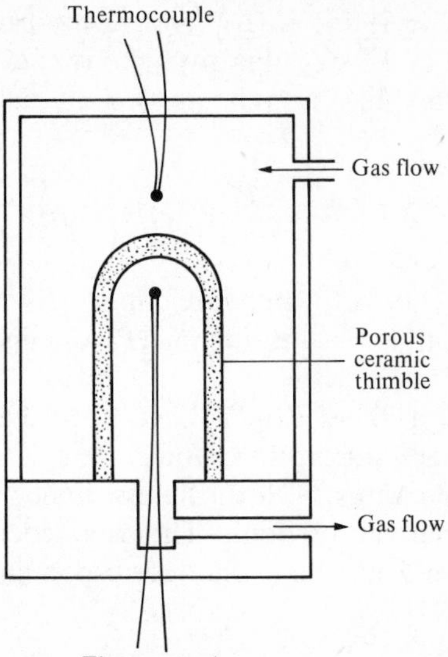

FIG. 3.13. Radial-flow Joule–Thomson apparatus.

dependence. The temperature change, ΔT, produced when a gas flows through a throttling device with a pressure gradient ΔP can yield a value of μ_{JT} provided the process is performed adiabatically. The elimination of heat leaks has proved difficult, but reliable results have been obtained using a radial-flow porous plug apparatus[55] shown schematically in Fig. 3.13.

Some recent studies have concentrated on the isothermal Joule–Thomson coefficients, ϕ_{JT}, which may be measured using a flow calorimeter, shown schematically in Fig. 3.14. In this apparatus a heater is situated beyond the throttle, and the power input, W, needed to maintain isothermal conditions at a measured flow rate, f, across the pressure gradient,

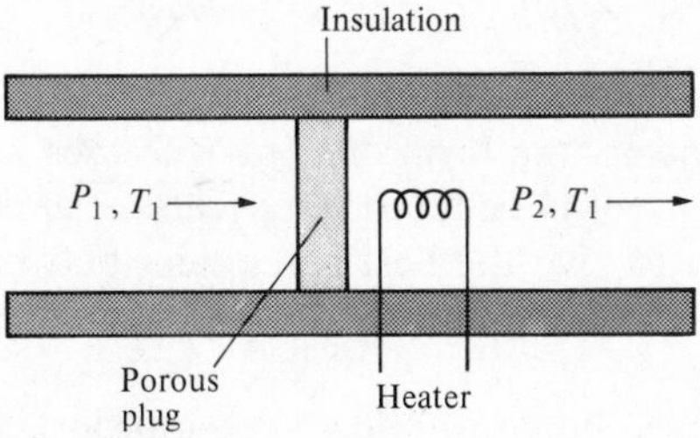

FIG. 3.14. Isothermal Joule–Thomson experiment (schematic).

ΔP, is determined. The ratio $W/f\Delta P$ is measured as a function of the mean pressure $\bar{P}$, and as $\Delta P \to 0$ this gives the isothermal Joule–Thomson coefficient, $\phi_{JT}(\bar{P})$. This may be related to the virial expansion through the equation

$$\phi_{JT}(\bar{P}) = B - TB' + \frac{2\bar{P}}{RT}\left\{C - B^2 + TBB' - \frac{TC'}{2}\right\} + \ldots.$$

where $C' = \mathrm{d}C/\mathrm{d}T$. The zero pressure limit, $B - T(\mathrm{d}B/\mathrm{d}T)$, may be obtained by extrapolating results for $\phi_{JT}(\bar{P})$ obtained over a range of pressures.

Accurate values of $B - T\,\mathrm{d}B/\mathrm{d}T$ have been measured in this way, under conditions where adsorption would be a serious problem in a conventional *PVT* apparatus.[56] Nevertheless, under most circumstances, the accurate measurement of Joule–Thomson coefficients has proved considerably more difficult than the corresponding equation-of-state studies.

3.11. Intermolecular forces from virial coefficients

3.11.1. Use of model potential functions

The traditional use of virial coefficients in the study of intermolecular forces has been based on the use of previously assumed models for the potential energy function, whose parameters have been adjusted to optimize agreement between experimental and calculated virial coefficients.[16] The calculation of second virial coefficients for a model potential is a straightforward numerical procedure, even when analytic integration methods are not available. For an intermolecular potential energy function of the general form

$$U(r) = \varepsilon F(r/\sigma)$$

the integral in (3.37) may be evaluated to yield the reduced second virial coefficient,

$$B^* = B/b_0, \qquad (b_0 = 2\pi N_A \sigma^3/3)$$

as a function of the reduced temperature, $T^*(T^* = kT/\varepsilon)$, and any parameters used to specify the form of $F(r/\sigma)$. Values of B^* may then be tabulated for a suitable set of reduced temperatures and the parameters ε and σ may be chosen to give the best agreement between calculated and experimental second virial coefficients for the substance under investigation.[57]

Many methods have been used to select appropriate values of ε and σ for a model potential.[58] One possible procedure is to take two experimental points and to choose the parameters in order to force exact agreement with calculated results for the chosen potential.[16] This is not an approach which can be recommended, however, since it neglects most of the results and is susceptible to random errors in the chosen experimental points. With the ready availability of computers the use of least squares fitting using all the data has become more common.[57] Non-linear methods must be used, but this does not present a serious problem. A disadvantage of such direct numerical methods is that they may not reveal directly a qualitative failure of a model potential, which may be unable to reproduce the experimental results with any choice of parameters. Graphical fitting is sometimes preferred for this reason.

A successful graphical method was described by Lennard-Jones,[37] in which the experimental second virial coefficients are plotted in the form $\log|B|$ vs. $\log T$. A similar plot, of $\log|B^*|$ vs. $\log T^*$, is prepared using results calculated from the model potential energy function, and the two graphs are superimposed as far as possible by translations parallel to the axes. A failure to superimpose the two plots within the experimental error indicates that the model potential is not suitable. If successful superimposition is achieved, the displacements of the axes in the x- and y-directions are readily seen to be $\log(\varepsilon/k)$ and $\log b_0$ respectively. Other graphical methods have been suggested,[59] and they generally offer a convenient, though sometimes laborious, way of choosing the characteristic parameters, although the detailed choice of the best values is necessarily rather subjective.

It was long assumed that choosing parameters for model potential functions from fits to second virial coefficients gave positive information about the intermolecular pair potential energy function. This view can no longer be accepted, since it has been established that different potential energy functions may give essentially indistinguishable second virial coefficients, particularly when a limited temperature range is studied. The ability of a particular potential function to reproduce experimental second virial coefficients is therefore not necessarily proof of its validity. This lack of specificity is clearly shown in a corresponding states analysis of second virial coefficients, using the Boyle parameters, for several model potential functions. The reduced Boyle temperature, T_B^*, ($T_B^* = kT_B/\varepsilon$) for a given potential is easily inferred from tables of reduced second virial coefficients, and the characteristic volume, B_0^*, ($B_0^* = B^*(T^* = 0{\cdot}7T_B^*)$), may also be derived. (Alternatively the Boyle volume, V_B^*, ($V_B^* = T^*(\mathrm{d}B^*/\mathrm{d}T^*)_{T^*=T_B^*}$) may be used.) A corresponding states plot of reduced second virial coefficients may then be prepared in the form $B^*(T^*)/B_0^*$ vs T^*/T_B^*. In Fig. 3.15 such graphs are shown for a number of potential functions, whose shapes differ considerably. It may be seen that the reduced plots are essentially indistinguishable over a large part of the

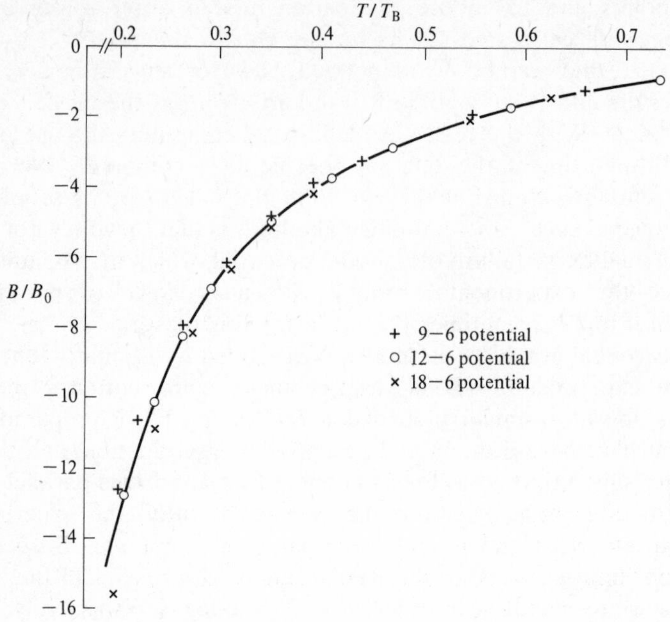

FIG. 3.15. Reduced second virial coefficients for some $n-6$ potential energy functions. The line represents the $12-6$ results. The reductions are performed using the Boyle temperature, T_B, and the characteristic volume, B_0, defined as $B_0 = -B(T = 0.7T_B)$.

temperature range—generally that region which is most easily studied experimentally. Only at rather low reduced temperatures, below $0{\cdot}4T_B$, do the second virial coefficient curves diverge in such a way as to offer a basis for choosing between these substantially different potentials.

Experimental results may be included in such a plot, using the experimental values of T_B and B_0 as reduction parameters. Comparison between the experimental and theoretical lines, which should coincide, will show clearly whether the experimental results can provide a means of choosing between the proposed model functions. The potential parameters, ε and σ, for successful models may be rapidly established by comparing the experimental and reduced values of T_B and B_0.

However, it is clear that second virial coefficient data, used in this way, can only provide very coarse information about intermolecular interactions. Recent work has shown that this lack of sensitivity is largely a consequence of the method of approach and more effective procedures for studying the intermolecular potential energy function are described in the next section.

It has been suggested that third virial coefficients may be more sensitive to the details of the intermolecular potential than second virial coefficients. For model pair potentials the integration for $C_{\rm add}$ in (3.32) may be performed numerically to give reduced pairwise-additive third virial coefficients, $C_{\rm add}/b_0^2$, as a function of the reduced temperature, T^*. When these values for different potentials are examined on a corresponding states basis, it is found that they often differ significantly, even though the second virial coefficients from the same potentials are very similar. Unfortunately, the practical value of this apparent sensitivity of third virial coefficients to the pair potential energy functions is less than might have been expected. When pairwise-additive third virial coefficients are calculated using accurate pair potentials, and are compared with experimental results, large discrepancies are observed.[60] It is believed that these result from the neglect of non-additive three-body contributions to the energy and to the third virial coefficient.

In order to evaluate the contribution to the third virial coefficient due to non-additive energy terms, a form must be assumed both for the pair potential, and for the non-additive three-body energy, ΔU_3. In most studies, the Axilrod–Teller triple-dipole contribution $U_{\rm DDD}$ has been used. (See § 1.4.4(c).)

$$U_{\rm DDD} = \frac{\nu(1+3\cos\theta_{12}\cos\theta_{13}\cos\theta_{23})}{r_{12}^3 r_{13}^3 r_{23}^3} \tag{3.42}$$

where θ_{12}, θ_{13}, θ_{23} are the interior angles of the triangle of sides r_{12}, r_{13}, r_{23}. The coefficient ν may be calculated for simple systems using sum rule methods (see Chapter 2) or may be estimated with quite good accuracy from the coefficient, C_6, of the leading term in the London dispersion energy

$$\nu \simeq -\tfrac{3}{4}\alpha C_6$$

where α is the static polarizability.

Calculations of the non-additive contribution to the third virial coefficient have been performed using the Axilrod–Teller correction, and a variety of model pair potentials.[43] In all cases it was found that the value of ΔC, the non-additive contribution to C, is large and positive. For the rare gases, at temperatures near the critical, T_c, the ratio $\Delta C/C_{\rm add}$ is around 65–90 per cent for a range of model potentials. This ratio decreases with increasing temperature, but remains significant over the experimentally accessible range, dropping to around 25–30 per cent at $2T_c$ and to about 14 per cent at $4T_c$.

The repulsive non-additive energy cannot be expressed so simply as the Axilrod–Teller term, but fairly simple theoretical models have been proposed and have been used to calculate non-additive contributions to third virial coefficients, again using model pair potentials.[41] The repulsive contributions to ΔC were found to be negative and substantial, about 45 per cent of the absolute magnitude of those derived from the Axilrod–Teller term, so that there is considerable cancellation of the two non-additive contributions.

However when an accurate pair potential for argon, due to Barker and Pompe,[60] was used to calculate third virial coefficients, good agreement with experiment was reported using only the Axilrod–Teller correction for non-additivity. It has been suggested[61] that this resulted from cancellations between the effects of repulsive non-additive terms and those due to higher-order multipole contributions to the long-range non-additive energy, which are found to give further positive contributions to ΔC. The over-all situation at the present time seems a little confused, but it is clear from the study of third virial coefficients that non-additive three-body contributions are significant, and cannot safely be neglected. This is perhaps the most important contribution to our knowledge of intermolecular forces to have come from third virial coefficients, since uncertainties in experimental values have so far limited the quantitative information which they may yield.

The methods described above represent the traditional procedures by which information about intermolecular interactions has been inferred from virial coefficients. In recent years there has been a growth of interest in alternative procedures by which intermolecular potentials may be determined directly from thermophysical properties, without using the parameter optimization techniques for model potentials described earlier. It is to the principles of these inversion procedures and their application to second virial coefficients that we now turn.

3.11.2. Inversion methods

Experimental quantities, $X(T)$, of which the second virial coefficient $B(T)$ is a specific example, are usually related to the intermolecular potential energy $U(r)$ by integration over one or more subsidiary variables. The problem is to unfold this integral, to obtain r as a function of U from X as a function of T. Whether or not this is possible will depend on a number of factors including the complexity of the functions involved.[62] In some cases a formal inversion of the integral equation is possible.[13,14] Even when this is so the sensitivity of the inversion procedure to uncertainties in the experimental data, $X(T)$, may cause it to be unreliable.

Not only are data inversion methods a convenient route from data to potential energy functions but the reformulation of the problems that they encourage can provide a direct lead into a central problem of intermolecular force determinations—that of uniqueness. For instance we find that a number of potential energy functions can often satisfy a given set of experimental second virial coefficient results and thus a potential function determined solely from such results may be essentially worthless. Although the data can provide a severe test of potential functions in that many proposed functions will prove inconsistent with them, those which are consistent are not necessarily correct.

In constructing data inversion methods, two questions should be asked about any set of physical measurements:

(a) What specific information about $U(r)$ does it contain?
(b) How best can that information be extracted?

The value of the approach and the underlying principles involved are best illustrated by means of the inversion methods that have been developed specifically for second virial coefficients.

The classical second virial coefficient of a gas can be expressed (see (3.29))

$$\frac{B(T)}{\alpha}=\frac{2\pi N_A}{3T}\int_0^\infty \Delta \exp\left(-\frac{\phi}{kT}\right)\mathrm{d}\phi \qquad (3.45)$$

where $\Delta = r_L^3 - r_R^3$, $\alpha = \exp(\varepsilon/kT)/k$ and $\phi = U+\varepsilon$. r_L and r_R are the inner and outer coordinates of the potential energy function at ϕ where ϕ is the energy measured from the bottom of the well. In the repulsive region $\Delta = r_L^3$. A formal inversion of this expression should be possible since it is apparent that $B(T)T/\frac{2}{3}\pi N_A\alpha$ is the Laplace transform[13,14] of Δ. Thus inversion of $B(T)$ can, in principle, give the repulsive branch of the potential energy function and the well width as a function of its depth. Hence $B(T)$ does not define $U(r)$ uniquely except in the repulsive region. One possible method of inverting (3.45) could be to fit $B(T)T/(\frac{2}{3}N_A\pi\alpha) = L$ to a function whose inverse Laplace transform is known. This was the first successful inversion technique[63] but it only proved practicable in the case of helium, for which results were available at sufficiently high (reduced) temperatures. The method gave the repulsive branch of $U(r)$ down to $U \approx 25\varepsilon$, but even here it was necessary to correct for the influence of attractive forces.

A formal inversion of (3.45) for a $\Delta(\phi)$ derived from a realistic intermolecular potential can be performed using numerical quadrature techniques.[64] Putting $s = 1/kT$, (3.45) may be written as

$$F(s)=\frac{3B(s)}{2\pi N_A \alpha s}=L[\Delta(\phi)]=\int_0^\infty \exp(-\phi s)\,\Delta(\phi)\,\mathrm{d}\phi.$$

We now make use of the multiplicative property of Laplace transforms

$$L[\Delta(a\phi)] = \int_0^\infty \exp(-\phi s)\,\Delta(a\phi)\,\mathrm{d}\phi = \frac{1}{a}\int_0^\infty \exp(-\phi s/a)\,\Delta(a\phi)\,\mathrm{d}\phi$$

where a is a constant, i.e.

$$\frac{F(s/a)}{a} = \int_0^\infty \exp(-\phi s)\,\Delta(a\phi)\,\mathrm{d}\phi. \tag{3.46}$$

Writing $x = \mathrm{e}^{-\phi}$ and applying an Nth-order quadrature formula to the right-hand side of (3.46) leads to

$$L = \frac{F(s/a)}{a} = \frac{3B(s/a)a}{2\pi N_A \alpha a s} = \sum_{i=1}^{N} w_i x_i^{s-1}\Delta(-a \ln x_i) \tag{3.47}$$

where w_i are suitable weighting factors at the points x_i. Knowing the left-hand side of (3.47), L, for N values of s/a, $s = 1$ to N, (corresponding to N values of $B(T)$ at the temperatures $T = a/s$, i.e. a, $a/2, \ldots, a/N$) leads to a set of N linear equations, or

$$\mathbf{A}\boldsymbol{\Delta} = \mathbf{L}$$

where $\mathbf{A}$ is the $(N \times N)$ matrix $[w_i x_i^{s-1}]$ and $\boldsymbol{\Delta}$, $\mathbf{L}$ are $(N \times 1)$ matrices. Solution of this equation to give

$$\boldsymbol{\Delta} = \mathbf{A}^{-1}\mathbf{L}$$

gives N values of the transform Δ at the appropriate values of ϕ, $-a \ln x_i$.

Unfortunately this formal inversion procedure is inherently unstable since the matrix $\mathbf{A}$ is ill-conditioned. This means that although L (or $B(T)$) is relatively insensitive to small changes in ϕ, the latter is unusually sensitive to small changes in L.

The ill-conditioning, which arises from the discontinuity in $\Delta(\phi)$ at $\phi = \varepsilon$, (see Fig. 3.6) manifests itself in solutions which oscillate about the true one; the lower the precision of the experimental data the greater the oscillations. Thus although, using specially devised programming techniques,[64] the inversion of $B(T)$ is possible for data accurate to one part in 10^4 it is not practicable for experimental data which have a precision of at best 0·1 per cent and more commonly ±1 per cent or worse.

Although the inversion of $B(T)$ by a Laplace transform is impracticable a number of alternative inversion methods[64,65] have been developed which give valuable information about $U(r)$. The principles on which the first of these methods were based provide the key to the somewhat more complex procedures developed for the inversion of transport properties, which are discussed later, in Chapter 6. They make use of the fact that one can identify certain functions of the experimental data with coordinates of the intermolecular potential energy function. A simple (but impractical) example can be constructed for high-temperature second virial coefficient data. We could write by analogy with the hard sphere expression, § 3.9(a),

$$B(T)=\frac{2N_A\pi}{3}\tilde{r}^3(T)$$

thus defining a characteristic length $\tilde{r}$. This could be thought of as the 'effective hard-sphere diameter' of the molecules at temperature T. By assuming that two molecules will approach one another until their potential energy is equal to their initial kinetic energy, which is of order kT, we postulate that

$$\tilde{r}(T)\approx r_{(U=kT)}$$

where $r_{(U=kT)}$ is the point on the potential energy function $U(r)$ at which $U=kT$. Each value of the second virial coefficient can then be used to determine a point on the potential energy function. We can test the proposed method on simulated results (second virial coefficient results calculated from a known potential energy function) and find that it can be used to define $U(r)$ with an accuracy of the order of ±5 per cent from results at temperatures greater than $kT/\varepsilon=30$, a range for which data are only available for helium. Such a method would be of little value because, for all heavier gases, the required temperature range is inaccessible but the method can be improved upon in two ways.

(i) We can improve our definition of $\tilde{r}$ by the alternative relation

$$\tilde{r}^3=\left\{B+T\frac{\mathrm{d}B}{\mathrm{d}T}\right\}\Big/\tfrac{2}{3}\pi N_A.$$

This expression has been shown[64] to lead to a superior definition of $\tilde{r}$ when the potential energy function from which the data to be inverted derives has the simple form $U(r)=A/r^n$.

(ii) Since in general $\tilde{r}(T)$ will not be identical to $r_{(U=kT)}$ we can write the more general relation

$$\tilde{r}(T)=\alpha(T)r_{(U=kT)}.$$

For a successful inversion procedure we require that $\tilde{r}(T)$ be defined so that $\alpha(T)$ is insensitive to the detailed shape of the potential function. Then a good estimate of the inversion function, $\alpha_0(T)$ can be calculated by $\alpha_0(T)=(\tilde{r}_0(T)/r_{0(U=kT)})$ using a very approximate potential function (identified by the subscript 0).

With these modifications tests using simulated second virial coefficient results have shown[64] that the inversion method allows $U(r)$ to be defined to within 1 per cent in the range $100 > U/\varepsilon > 3$ and to within 1·5 per cent down to $U/\varepsilon = 2$. The only additional information required for this inversion is an approximate potential energy function. The form of this is not critical, and a traditional function such as the Lennard-Jones 12−6 is adequate. The well depth of the approximate function is perhaps the most critical parameter and even for this a rough estimate suffices. It does not significantly affect the derived values of $U(r)$ above $U/\varepsilon = 5$ and a 10 per cent error in ε only produces an error of approximately 1 per cent in $U(r)$ at $U/\varepsilon = 2$.

The form of $\Delta(\phi)$ (see Fig. 3.6) suggests that it might be appropriate to treat the repulsive and attractive regions of the potential separately. Hence, to obtain Δ for the well region, we must separate off the contribution to $B(T)$ from the repulsive region. This may be achieved approximately by defining

$$B_{\text{attr}} = B(T) - \frac{2\pi}{3} N_A \sigma^3.$$

On the basis of this separation, functions can be devised which enable the well-width parameter Δ to be obtained by the inversion of second virial coefficient data at low temperatures ($0 < kT/\varepsilon < 1$). We can write, on the basis of an analysis that has been given in full elsewhere,[62]

$$-\Delta(\phi = kT) = \frac{B(T) - \frac{2}{3}\pi N_A \sigma^3}{P_0(T)} \tag{3.48a}$$

where $P_0(T)$ is also determined using an approximate potential energy function. It is convenient to write

$$P(T) = \frac{2\pi N_A}{3} (e^{\varepsilon/kT} - 1)\, W(T^*) \tag{3.48b}$$

where $T^* = kT/\varepsilon$ and $W(T^*)$ is a slowly varying function of T^* which is of the order of unity and is relatively insensitive to the detailed form of the potential energy function (see Table 3.2). Some insight into the physical basis of the method can be obtained by examining the equation for the second virial coefficient of the square-well potential

$$\begin{aligned} \phi &= \infty, & U &= \infty & 0 &< r \leqslant \sigma \\ \phi &= 0, & U &= -\varepsilon & \sigma &< r \leqslant R\sigma \\ \phi &= \varepsilon, & U &= 0 & r &> R\sigma \end{aligned}$$

which has the form

$$B(T) = \tfrac{2}{3}\pi N_A \sigma^3 \{1 - (e^{\varepsilon/kT} - 1)(R^3 - 1)\}.$$

Table 3.2

The well-width inversion function $W(T^*)$ for several potential functions

T^*	$W(T^*)$ BBMS[66]	12−6†	11−6−8
0.2	1.025	1.040	1.020
0.3	1.071	1.104	1.055
0.4	1.038	1.079	1.017
0.5	0.955	0.993	0.937
0.6	0.850	0.876	0.828
0.7	0.726	0.733	0.703
0.8	0.570	0.573	0.560
0.9	0.377	0.380	0.387

The potentials are given in full in Appendix 1.
† The values for the other $n-6$ potential functions where $n = 9$, 16, and 18 lie within 1 per cent of the values given for the 12−6 function.

The well-width function, Δ, is given by

$$-\Delta = \sigma^3 R^3 - \sigma^3 = \frac{B(T) - \frac{2}{3}\pi N_A \sigma^3}{\frac{2\pi N_A}{3}(e^{\varepsilon/kT} - 1)}.$$

Comparison of this equation with (3.48a) and (3.48b) indicates that $W(T^*)$ is simply a measure of the degree to which the width of the well of a realistic potential varies with energy.

This procedure requires a more reliable estimate of ε/k than the method described above for the repulsive region—a 1 per cent error in ε/k produces uncertainties in Δ of about 3 per cent at $\phi^* = \phi/\varepsilon = 0{\cdot}5$ increasing to 10 per cent as ϕ^* approaches 1·0. However only a very crude estimate of the collision diameter σ is necessary; errors of 10 per cent in σ produce only a 3 per cent uncertainty in Δ even as ϕ^* approaches unity. Thus we find that, although a formal inversion of $B(T)$ is not practicable as its execution is too demanding of experimental results, a large part of the information about $U(r)$ contained in second virial coefficient measurements can be extracted by the eductive inversion procedures described above.

These methods are novel in their approach and hard to justify formally but on the whole the results are not too surprising as the information extracted from the second virial data is that which they are known to contain. However recent work[65] on the inversion of $B(T)$ data has been more ambitious in that it seeks to obtain complete potential energy functions despite the fact that such a procedure is *in general* impossible. The separation parameter $\tilde{r}$ is again defined in terms of $B + T(\mathrm{d}B/\mathrm{d}T)$ but an alternative inversion function $F(T)$ is introduced which relates the value of U at r to kT by

$$U(\tilde{r}) = F(T)kT. \tag{3.49}$$

A first approximate value of the inversion function $F_0(T)$ is calculated from an approximate potential energy function, $U_0(r)$. When this inversion function is used in (3.49) with $\tilde{r}$ values based on experimental second virial coefficients, it leads to a series of points which define a potential function $U_1(r)$. This is found, on the basis of tests with simulated data, to be closer to the true potential function than $U_0(r)$. The improved function, $U_1(r)$, is then used to calculate $F_1(r)$ and so obtain a further improved potential, $U_2(r)$. Three iterations have been used in practice; further iterations produce insignificant variations in the potential energy function. The initial inversion function F_0 is obtained in terms of the reduced temperature, kT/ε, and in order to identify the appropriate value of T^* for a given experimental temperature, T, an estimate of the well-depth parameter is required. It is found that ε can itself be determined from second virial coefficient data by examining the consistency of the virial coefficients calculated from the potentials obtained by inversion with the original second virial coefficient data. The method has been tested on simulated data calculated for the BBMS potential function which is a good representation of the intermolecular potential for argon.[66] The $12-6$ function and four other different potential energy functions were employed as initial approximations and the convergence of the results to the true potential is illustrated in Fig. 3.16. It is seen that both branches of $U(r)$ can be determined to high accuracy despite the formal limitations of $B(T)$ data emphasized above. The factors which underly the remarkable success of this method have yet to be fully elucidated, but are presumably related to the fact that all quasi-realistic potential functions are analytic (in the mathematical sense) and have the same general shape. Thus the potential and its derivatives will be continuous at $r = \sigma$, the point at which the discontinuity in $\Delta(\phi)$ arises.

The methods described above may be readily applied to real experimental data. Suitable inversion functions, $W(T^*)$ and $F(T^*)$, have been listed in Table 3.2 and Appendix 13. Though with the highest quality data iterative methods have been applied, a single-step procedure is normally sufficient for most experimental results. The methods have so far been applied to a limited number of monatomic and pseudo-spherical systems. The extension of these methods to polyatomic systems will rest upon the degree to which the recovered potentials, which are necessarily spherical, may be unequivocally related to the true angle-dependent potential function. It is to the second virial coefficients of such non-spherical molecules that we now give our attention.

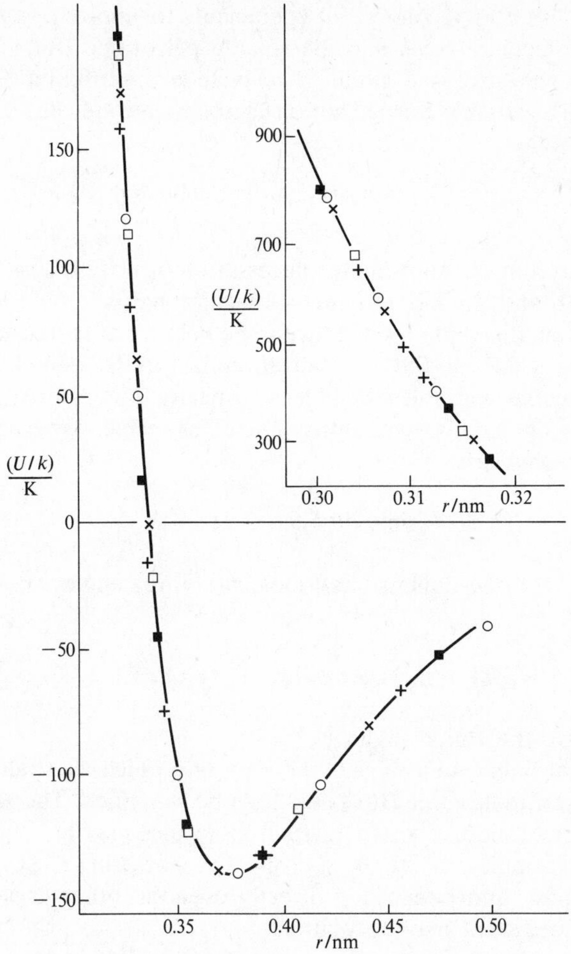

FIG. 3.16. The inversion of pseudo-experimental second virial coefficient data calculated from the BBMS potential energy function. Results after the third iteration. Initial approximate potential: (□) 9−6; (×) 11−6−8; (+) 12−6; (○) 20−6; (■) 18−6.

3.12. Virial coefficients of more complex molecules

At the present time our detailed knowledge of intermolecular forces is restricted to the simplest species, the inert gases. Future studies will undoubtedly concentrate on the interactions of more complex molecules, which lack spherical symmetry and which may have substantial electrical multipole moments.

The calculation of second virial coefficients for molecules with non-central intermolecular forces may be readily performed using (3.25), if the integration variables are modified to include the orientations of the molecules, and a suitable normalization factor is introduced.

$$B(T)=\frac{-N_{\mathrm{A}}}{2V\Omega^2}\iint[\exp(-U_{12}/kT)-1]\,\mathrm{d}\mathbf{r}_1\,\mathrm{d}\boldsymbol{\omega}_1\,\mathrm{d}\mathbf{r}_2\,\mathrm{d}\boldsymbol{\omega}_2. \qquad (3.50)$$

$\boldsymbol{\omega}_1$ and $\boldsymbol{\omega}_2$ represent the coordinates necessary to specify the orientations of molecules 1 and 2, and the intermolecular pair energy, U_{12}, now depends both on the separation and on the relative orientations of the molecules. The value of the normalization factor, Ω, depends on the number of angular variables needed to specify the orientation of a molecule. For an axially symmetric linear molecule, two angles are needed, and in this case

$$\mathrm{d}\boldsymbol{\omega}=\sin\theta\,\mathrm{d}\theta\,\mathrm{d}\phi, \qquad \Omega=4\pi.$$

For a non-linear (three-dimensional) molecule, three angles are required, and

$$\mathrm{d}\boldsymbol{\omega}=\sin\theta\,\mathrm{d}\theta\,\mathrm{d}\phi\,\mathrm{d}\psi, \qquad \Omega=8\pi^2.$$

(θ, ϕ, and ψ are the Euler angles.)

The case of linear molecules is the only one which we shall consider further, and the equation for $B(T)$ may then be simplified. The separation and relative orientation of a pair of such molecules may be expressed in terms of four variables, r_{12}, θ_1, θ_2, $\phi_1-\phi_2$. On re-writing (3.50) in terms of these variables, and integrating directly over the other variables, the second virial coefficient may be written

$$B(T)=\frac{-N_{\mathrm{A}}}{4}\int_0^\infty r_{12}^2\,\mathrm{d}r_{12}\int_0^\pi \sin\theta_1\,\mathrm{d}\theta_1\int_0^\pi \sin\theta_2\,\mathrm{d}\theta_2\int_0^{2\pi} f_{12}(r_{12},\theta_1,\theta_2,\phi_{12})\,\mathrm{d}\phi_{12} \qquad (3.51)$$

where $\phi_{12}\equiv\phi_1-\phi_2$. For model potential energy functions, the integration may be performed numerically, to give the reduced virial coefficient, B/b_0, as a function of reduced temperature kT/ε, and any shape parameters needed to define the potential function. For molecules with end-for-end symmetry, the range of angular integrations may be further reduced, with consequent savings in the calculations.

For third virial coefficients, there are similar generalizations[67] of (3.31), which give expressions applicable to non-spherical molecules.

3.12.1. *Calculations using model non-spherical potentials*

(a) *Hard convex bodies.* Various model potential energy functions have been proposed for different kinds of non-spherical shape. A simple representation of non-spherical repulsive interactions uses a generalization of the hard-sphere model to convex hard molecules of various shapes. Some of these are illustrated in Fig. 3.17. In each case, rotation about the axis AA′ gives a prolate ('cigar-shaped') body; rotation about BB′ produces an oblate ('disc-shaped') body.

For all such convex molecules, Isihara[68] showed that the second virial coefficient could be expressed in a simple form

$$B = N_A \nu_m(\gamma + 1) \qquad (3.52)$$

where ν_m is the molecular volume, and γ is a factor related to the molecular geometry which reflects the extent of the deviation from spherical shape.[69] It is defined in terms of the surface area, S, volume, ν_m, and the average value of the mean radius of curvature, $\bar{R}$.

$$\gamma = \frac{\bar{R}\,.\,S}{\nu_m} \qquad (3.53)$$

where

$$\bar{R} = \frac{1}{4\pi}\iint \tfrac{1}{2}(R_1 + R_2)\,\mathrm{d}s$$

and R_1, R_2 are the principal radii of curvature of the body at the surface element ds. The shape factors, γ, for a number of convex bodies are given in Table 3.3.

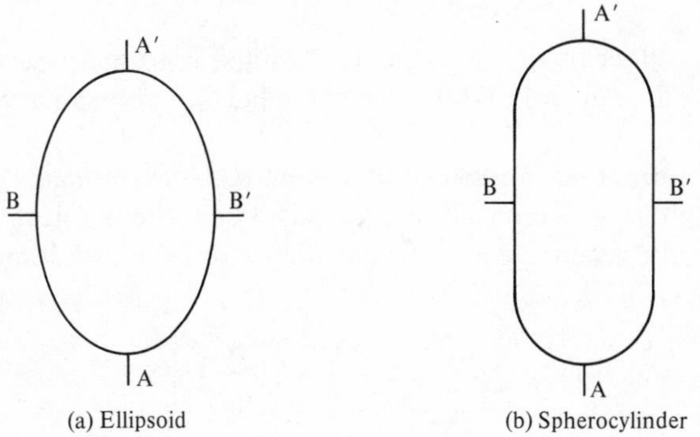

FIG. 3.17. Hard convex molecules. Rotation about AA′ generates prolate bodies. Rotation about BB′ generates oblate bodies.

Table 3.3
Shape factors, γ, for some convex bodies

Body	Shape factor, γ
Sphere	3
Regular octahedron	4.320
Cube	4.5
Regular tetrahedron	6.704
Ellipsoid of revolution, eccentricity e†	$\frac{3}{4}\left(1+\frac{\sin^{-1}(e)}{e(1-e^2)^{\frac{1}{2}}}\right)\left(1+\frac{1-e^2}{2e}\ln\frac{(1+e)}{(1-e)}\right)$
Spherocylinders, cylinder length, $R^*\sigma$, diameter, σ,	
(a) Prolate	$3+R^{*2}/(R^*+\frac{2}{3})$
(b) Oblate	$3+\frac{R^{*2}(\pi R^*+\pi^2-8)}{(4R^{*2}+2\pi R^*+\frac{8}{3})}$

† $e=(a^2-b^2)/a^2$, where a and b are the major and minor axes of the ellipse.

For hard spheres, $\gamma=3$, and for all other convex molecules $\gamma>3$. Thus the effective excluded volume of convex hard molecules is bigger than that of spheres of the same molecular volume.

Some calculations of higher virial coefficients have been reported for such molecules.[70,71] Third virial coefficients may be estimated with good accuracy using a result based on the scaled particle theory of dense fluids.[72]

$$\frac{C}{(4N_A v_m)^2}=\frac{1+2\gamma+\gamma^2/3}{16} \tag{3.54}$$

Once more the effect of the non-spherical shape is to increase the value of the virial coefficient, relative to a hard sphere of the same volume.

(b) *Two-centre and multi-centre Lennard-Jones models.* A more realistic non-spherical potential which has been the subject of fairly extensive study in recent years[73] is the diatomic Lennard-Jones model, illustrated in Fig. 3.18,

$$U_{ij}=\varepsilon\sum_{l=1}^{2}\sum_{m=1}^{2}\left\{\left(\frac{\sigma}{r_{lm}}\right)^{12}-\left(\frac{\sigma}{r_{lm}}\right)^{6}\right\}. \tag{3.55}$$

This seems to be a reasonable extension of the Lennard-Jones potential to describe the interactions of simple diatomic molecules such as N_2 or O_2. The total interaction energy between two molecules is the sum

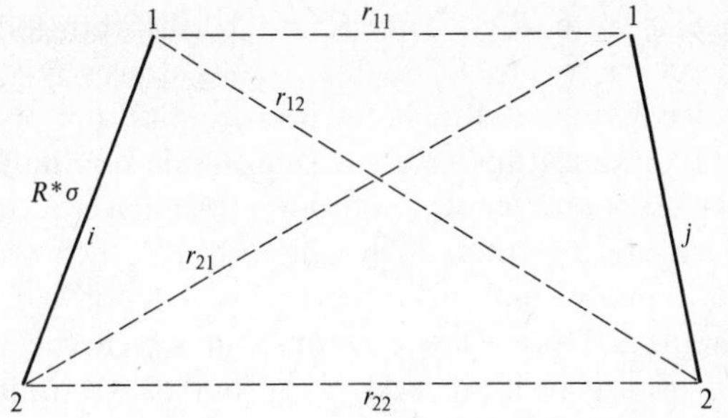

FIG. 3.18 The diatomic (site–site interaction) Lennard-Jones model.

of four atom–atom energies, each of the L-J type. The two atoms in a given molecule are separated by a distance $R^*\sigma$. Most commonly the interaction centres have been assumed to coincide with the atomic nuclei, although this is not necessary. When $R^* = 0$, this model reduces to the normal L-J form. Plausible values of R^* for molecules such as N_2 or O_2 lie in the range 0·3–0·4.

Calculations of second virial coefficients for this model have been made for a range of R^*. These are tabulated in Appendix 8 and are compared on the basis of a corresponding states plot in Fig. 3.19. It is

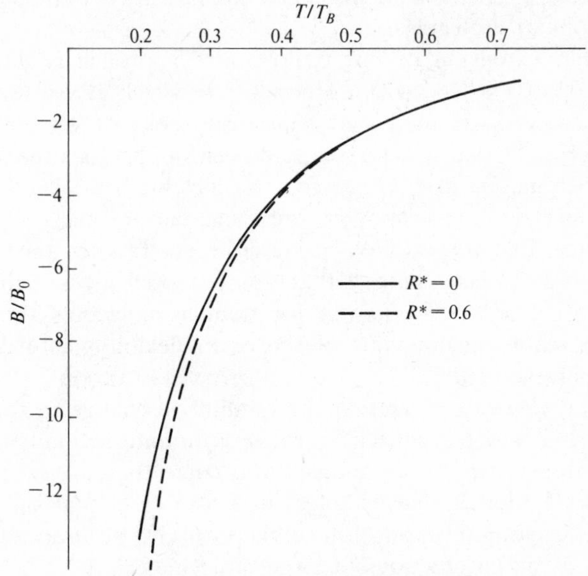

FIG. 3.19. Reduced second virial coefficients for the diatomic Lennard-Jones model, again using characteristic Boyle parameters.

seen that on going from $R^*=0$ to $R^*=0.6$, the reduced second virial coefficients fall off more rapidly at low temperatures but this change is quite small. As we shall see, this behaviour mimics that observed experimentally, and this model thus offers a reasonable qualitative description of simple non-spherical molecules, although it should not be thought of as providing an accurate intermolecular potential for such systems.

The diatomic model may be regarded as a simplified version of the four-centre potential devised by Corner,[74] in which the interactions of cylindrical molecules were mimicked using four $12-6$ interaction centres equally spaced on a line. Calculations were made of second virial coefficients and Joule–Thomson coefficients using this model, though some approximations were introduced to facilitate the numerical work. Application to experimental results for molecules such as N_2 and CO_2 gave results for the spacing of the interaction sites which were generally consistent with the known structures of these molecules.

A related multi-centre model was investigated by Hamann and Lambert[75] for globular molecules of the tetrahedral type XY_4, in which $12-6$ interactions were assumed to operate between all pairs of atoms on different molecules. The total energies of pairs of molecules were studied as functions of separation and relative orientation, and Hamann and Lambert showed that the net interaction could be adequately represented using a *central* $28-7$ potential. This model has been used to calculate second virial coefficients and has been applied to a number of quasi-spherical globular molecules.

A further extension of this multi-centre approach is the spherical shell potential of De Rocco and Hoover,[76] in which $12-6$ force centres are continuously distributed over spherical shells of diameter d. The resultant intermolecular potential is again central, and is a function of the additional parameter, d. Second virial coefficients have been calculated for this model over a wide temperature range and for many values of the shell diameter. Pairwise-additive third virial coefficients have also been reported.[77] It has been claimed that this potential gives a marked improvement over the $12-6$ potential for globular molecules, though this is perhaps not surprising, in view of the extra flexibility afforded by the adjustable parameter d.

However, it seems clear that the qualitative change in the effective central potential which results from moving the interaction centres away from the centre of mass of the molecules is correctly reproduced by all of these last three models. Nevertheless it is also clear that they can best serve as model potentials, and are unlikely to give accurate quantitative descriptions of the interactions of large molecules.

(c) *Kihara core potential.* An alternative intermolecular potential

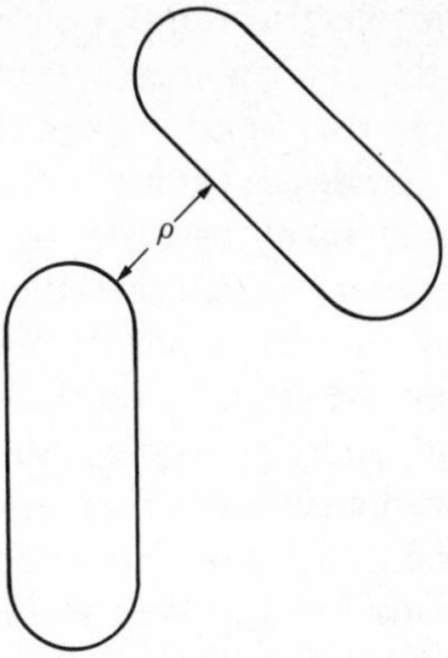

FIG. 3.20. The Kihara core model. ρ is the shortest distance between the surfaces of the cores.

for non-spherical molecules is that due to Kihara.[78] This is probably more suitable for molecules which are markedly non-spherical, and it gives second virial coefficients which may be calculated using a series expansion, similar to that described earlier for the Lennard-Jones potential. In this potential model, the molecules are regarded as having hard convex cores, which reflect the molecular geometry. The intermolecular energy of a pair of molecules depends on the shortest distance between the cores, ρ (see Fig. 3.20)

$$U(\rho)=\infty, \qquad \rho \leqslant 0$$

$$U(\rho)=\varepsilon\left[\left(\frac{\rho_m}{\rho}\right)^{12}-2\left(\frac{\rho_m}{\rho}\right)^{6}\right], \qquad \rho>0 \tag{3.56}$$

$$U(\rho_m)=-\varepsilon.$$

The energy is then calculated using a Lennard-Jones type function, in terms of this shortest separation of the pair. This model differs qualitatively from the previous models in concentrating the centres of interaction on to a single (movable) site for each molecule. It has been used extensively to describe the interaction both of simple spherical molecules (using a spherical core) and of more complex molecules, and is able to reproduce experimental second virial coefficients for many systems.[57] Once again, this must be taken as an indication of the general insensitivity of second virial coefficients to the detailed form of the potential energy function, rather than as evidence of the accuracy of the Kihara potential. Third virial coefficients, including non-additive dispersion contributions, have also been calculated for this potential,[43] using spherical cores.

(d) *Polar molecules: Stockmayer potential.* The above models may be appropriate for the description of non-spherical molecules which do not have large permanent electrical multipole moments. Deviations from spherical symmetry in the intermolecular potential then result largely from the non-spherical repulsive and dispersion interactions. Clearly any non-spherical molecule possesses some electrical multipole moments which are not zero, and in many cases these contribute substantially to the intermolecular energy. In some cases they are the major source of non-spherical interactions, since, as has been shown in Chapter 2, such multipole interactions depend strongly on the relative orientations of the molecules. Such molecules may be represented using a model potential consisting of a central, radial contribution, such as the 12−6 model, to which are added the orientationally dependent electrostatic terms. The most widely studied potential of this type is due to Stockmayer,[79] and, in its commonest form, may be written

$$U(r, \theta_1, \theta_2, \phi_{12}) = 4\varepsilon'\{(\sigma'/r)^{12} - (\sigma'/r)^6\} - \frac{\mu^2}{4\pi\varepsilon_0 r^3}\{2\cos\theta_1\cos\theta_2 - \sin\theta_1\sin\theta_2\cos\phi_{12}\} \quad (3.57)$$

where the angles are defined in Fig. 3.21. This potential includes only the direct dipole–dipole energy and neglects the effects of polarization and of higher multipoles, though the leading polarization contribution can be regarded as part of the $-4\varepsilon'(\sigma'/r)^6$ term. The integration for the second virial coefficient may be performed analytically, to give a double power series in $1/T$ and in μ. It is convenient to represent the results giving the reduced second virial coefficient, B/b_0, as a function of reduced temperature, kT/ε, and a dipole parameter, t^*, defined[80]

$$t^* = \frac{\mu^2}{\sqrt{8\varepsilon'\sigma'^3}(4\pi\varepsilon_0)} \quad (3.58)$$

Some results for this potential are shown in Fig. 3.22 and in Appendix 8. It is seen that the effect of increasing t^* is to make the second virial coefficient more negative at low temperatures, which is in accord with experiment.

In spite of the qualitative success of the Stockmayer potential, it is clear that it offers only a rather crude model for the interactions of polar substances.

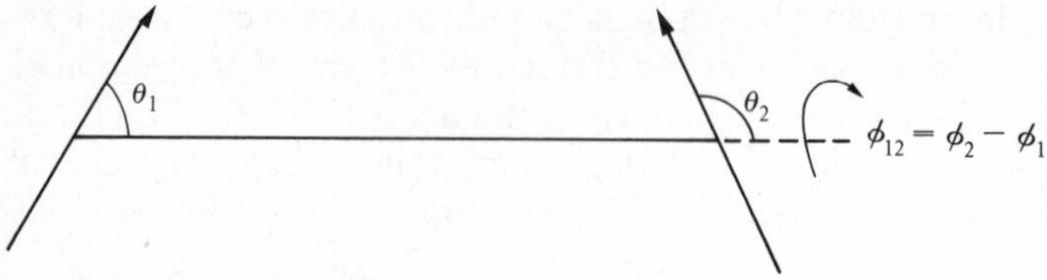

FIG. 3.21. The angles defining the relative orientation of two dipoles.

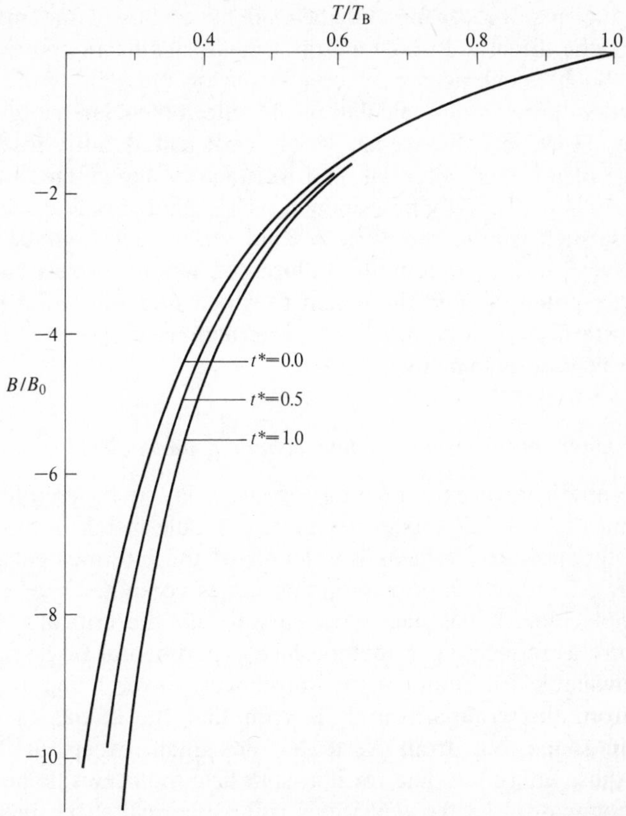

FIG. 3.22. Reduced second virial coefficients for the Stockmayer potential.

Buckingham and Pople[81] considered generalizations of (3.57) in which other orientation-dependent terms, such as those resulting from higher multipoles and from induction contributions, were included. A simple model for molecular shape was also investigated. Pople[82] developed a numerical method to calculate second virial coefficients for potential functions containing such terms in which the angle-dependent contributions were treated as perturbations of the central interaction. Buckingham and Pople tabulated the auxiliary functions needed in such calculations for the case where the central potential was a 12−6 function. They gave equations for the calculation of second virial coefficient contributions due to permanent dipole and quadrupole interactions, and those arising from their shape model. Expressions for the contributions resulting from many other non-central forces have also been derived and an extensive summary has been given by Mason and Spurling.[83] They have also demonstrated the use of this approach in calculating second virial coefficients for carbon dioxide and have shown that many terms other than the direct multipole energies make significant contributions to the total. Such calculations cast some doubt on the validity of neglecting higher order terms when calculating the intermolecular energy of polar molecules. However, the second virial coefficient is fairly insensitive to the details of the pair potential, and satisfactory agreement with experimental results can usually be obtained using quite simple models. Other properties, such as the dielectric second virial coefficient described in § 3.14, reveal serious qualitative failures of simple models such as the Stockmayer potential, and the extent to which such molecular properties may be interpreted in terms of the interactions of point multipoles is uncertain at the present time.

3.12.2. Experimental results for non-spherical molecules

A comparison of experimental second virial coefficients for a range of molecules of varying complexity gives valuable insight into the extent to which this property reflects the details of the intermolecular interactions. Fig. 3.23 shows a corresponding states comparison of results for several molecules. It has been necessary to use the critical volume and temperature as reducing parameters since experimental Boyle parameters are not available for some of the substances. Several important points emerge from this comparison. It is seen that the effects of moderate changes in shape, e.g. from Ar to N_2, are small, except at the lowest temperatures, where the data for non-spherical molecules lie below those for the inert gases. As the deviations from spherical shape increase, the effect on the second virial coefficients also increases, and the low temperature values become increasingly negative. At moderate temperatures, roughly above T_c, the effects of shape are small and good corresponding states behaviour is observed, even between the inert gases and such elongated molecules as n-C_5H_{12}. The results for polar molecules follow this same pattern, lying below the inert gas line to an extent which reflects the size of the molecular dipole moment. If large 'globular' molecules of approximately spherical shape, such as $C(CH_3)_4$, CF_4, or SF_6, are considered it is found that they too lie below the inert-gas line with deviations similar in scale to those shown above.

These observations suggest that many types of change in the intermolecular potential can lead to broadly similar effects in the second virial coefficients. In a powerful analysis of deviations from simple corresponding states behaviour, Pitzer[84] showed that almost any increase in molecular complexity, whether in the form of electical multipole moments or in non-spherical shape, has a similar effect on the angle-averaged intermolecular potential function, which dominates the form of the second

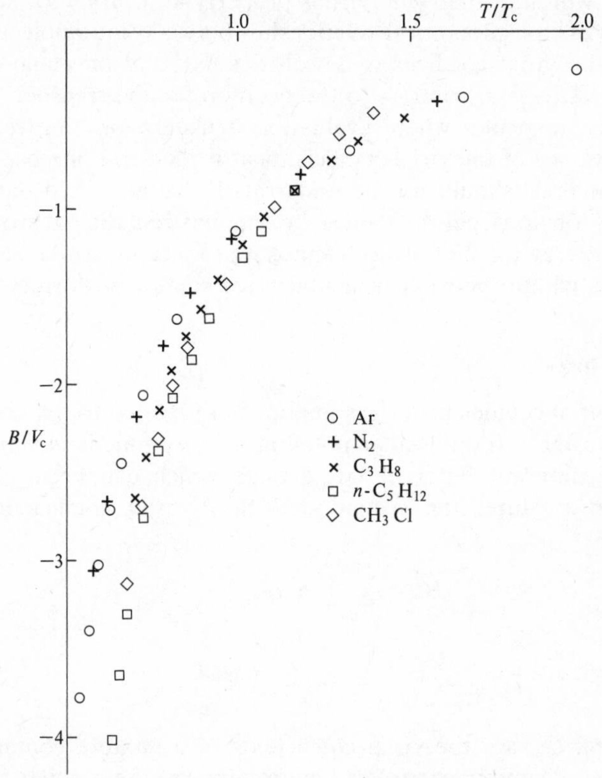

FIG. 3.23. Reduced second virial coefficients for several gases.

virial coefficient. For complex molecules, the angle-averaged interactions are characterized by steeper repulsion and narrower potential wells than those found for simple substances. This is the type of change which led to the proposal of the 28−7 potential for globular molecules, described in the previous section. Pitzer showed that a single factor, which he called *the acentric factor*, was sufficient to characterize the extent of most types of deviation from simple corresponding states behaviour. He was able to give analytical expressions which accurately represented the reduced second virial coefficients (and many other properties) of both simple and complex molecules,[85] in terms of the reduced temperature and the acentric factor. These expressions, and some acentric factors, are given in Appendices 5 and 6 and provide a useful means of correlating data for molecules of many types.

It seems possible that the insensitivity of the second virial coefficient to the specific nature of the intermolecular interaction for non-spherical molecules will limit the value of this property as a probe for such effects. Nevertheless, the theoretical relationship between intermolecular potentials and the virial coefficients is well defined, and they may be readily calculated. This is in contrast to the position for the transport properties of complex molecules where evaluation is usually prohibitively difficult. The importance of the virial coefficients as a check on proposed potential energy functions should not be underrated. At the risk of labouring an apparently obvious point, it must be emphasized that a proposed pair potential energy function which cannot reproduce measured second virial coefficients within their experimental error cannot be correct.

3.13. Mixtures

The virial coefficients of gas mixtures are related to the gas composition and to the intermolecular potential energy functions which characterize the different types of interactions which can occur. For an n-component mixture, the second and third virial coefficients may be written

$$B_{\mathrm{m}} = \sum_{i=1}^{n} \sum_{j=1}^{n} x_i x_j B_{ij}$$

$$C_{\mathrm{m}} = \sum_{i=1}^{n} \sum_{j=1}^{n} \sum_{k=1}^{n} x_i x_j x_k C_{ijk}$$

B_{m} and C_{m} are the virial coefficients of a mixture containing mole fractions x_i of component i. For binary mixtures these equations may be written explicitly

$$B_{\mathrm{m}} = x_1^2 B_{11} + 2x_1 x_2 B_{12} + x_2^2 B_{22}$$

$$C_{\mathrm{m}} = x_1^3 C_{111} + 3x_1^2 x_2 C_{112} + 3x_1 x_2^2 C_{122} + x_2^3 C_{222}.$$

In these equations B_{11} and B_{22} are the second virial coefficients of the pure components, and B_{12}, the second cross-virial coefficient, or interaction second virial coefficient, is related to the pair potential U_{12} characterizing the 1–2 interaction in just the same way that B_{11} and B_{22} depend on the potentials for 1–1 and 2–2 interactions. The third cross-coefficients C_{112} and C_{122} are also related to the different potential energy functions by obvious modifications of (3.32).

Measurements on gas mixtures may thus be used as a source of information about the interactions between unlike molecules. In practice, applications have been confined almost entirely to the second virial coefficient, B_{12}. B_{12} may be measured by using the procedures described earlier both on gas mixtures, and, separately, on the pure components. The resultant errors in derived values of B_{12} tend to be large, owing to the cumulative effects of the errors in B_{11} and B_{22}. Alternative methods of measuring B_{12} have also been used, which can give greater accuracy. For example, the change in pressure which occurs when two gases, initially at the same pressure, are mixed at constant temperature and volume may be measured, and is largely determined by the excess second virial coefficient, $B_{12}-\frac{1}{2}(B_{11}+B_{22})$. This method has been used successfully to study B_{12} for simple gas mixtures at low temperatures.[86] The related experiment, in which the volume change which is needed to achieve mixing at constant pressure is also simply related to the excess second virial coefficient, has been used in its measurement.[87]

An interesting and, at first sight, surprising source of the interaction virial coefficient, B_{12}, is measurement of the solubility of a solid[88] or liquid[89] in a compressed gas. The thermodynamic requirement for equilibrium in such an experiment is that the chemical potential of a species should have the same value in each phase. For the heavy component in the solid or liquid phase this may be readily calculated since the solid or liquid is almost pure and the effect of pressure on the chemical potential is well known. The chemical potential of the same component in the gas phase is found to be strongly dependent on B_{12}, and is fairly insensitive to the values of B_{11} and B_{22}. Hence, knowing the solubility, preferably at several pressures, the value of B_{12} leading to the best agreement with the experimental results may be established with quite a small uncertainty. Although the range of systems to which this method can be applied limits its generality, the results which can be obtained are particularly useful, since they often relate to temperatures which are below the range which is easily accessible to direct study on gas mixtures.

When values of B_{12} have been established they may be used to study the appropriate potential energy function for the unlike interaction, U_{12}. In the past, much work on unlike molecule interactions was based on the assumption that the pair potential energy functions for both pure components, and for the unlike pairs, were of the same shape, i.e. that they were conformal. Such conformality was considered in § 3.8 in connection with the principle of corresponding states, and is defined by (3.36). Attention then focused on the relationships between the parameters ε and σ characterizing the three types of interactions. Various combining rules relating the like and unlike parameters have been proposed, notably the 'Lorentz–Berthelot' rules

$$\sigma_{12}=(\sigma_{11}+\sigma_{22})/2$$
$$\varepsilon_{12}=(\varepsilon_{11}\,\varepsilon_{22})^{\frac{1}{2}}.$$

The first rule is exact for mixtures of hard spheres, and the second owes its origins to mixing rules for the dispersion energy, as discussed in Chapter 2. With the development of more direct routes to the pair potential from the measured properties it has become apparent that in many cases the assumptions of conformality and of the combining rules are incorrect. Some assessment of the practical usefulness of such rules is undertaken in Chapter 9.

3.14. Dielectric virial coefficients

Although equation-of-state data have been the major source of information about molecular interactions in gases, other equilibrium properties have also been useful, and have given valuable complementary information. The dielectric virial coefficients and the related refractivity coefficients are examples.

The density dependence of the dielectric constant, ε, of a gas may conveniently be represented by expressing the Clausius–Mossotti function, or total polarization, ${}_{T}P$, as a series expansion in the density[90,91]

$$ {}_{T}P = \frac{\varepsilon - 1}{\varepsilon + 2} \cdot \tilde{V} = A_{\varepsilon} + \frac{B_{\varepsilon}}{\tilde{V}} + \frac{C_{\varepsilon}}{\tilde{V}^2} + \dots $$

where A_{ε}, B_{ε}, C_{ε} are the first, second, and third dielectric virial coefficients. A_{ε} is the contribution to the total polarization arising from the individual molecules in the absence of intermolecular interactions, and is given by the Debye expression

$$ A_{\varepsilon} = \frac{4\pi N_{\mathrm{A}}}{3(4\pi\varepsilon_0)} (\alpha + \boldsymbol{\mu}^2/3kT) $$

where α is the mean polarizability of a molecule, and μ is its dipole moment. B_{ε} represents the initial departure from perfect gas behaviour due to contributions to the polarization, ${}_{T}P$, from pairs of interacting molecules, and may be related to the intermolecular energy of the molecular pair, U_{12}, by the equation

$$ B_{\varepsilon} = \frac{4\pi N_{\mathrm{A}}^2}{3\Omega(4\pi\varepsilon_0)} \int\int \left[\left\{ \frac{1}{2} \boldsymbol{\alpha}_{12} - \alpha \right\} + \frac{1}{3kT} \left\{ \frac{1}{2} \boldsymbol{\mu}_{12}^2 - \boldsymbol{\mu}^2 \right\} \right] \times \exp(-U_{12}/kT)\, \mathrm{d}\mathbf{r}_{12}\, \mathrm{d}\boldsymbol{\omega}_{12}. \tag{3.59} $$

$\boldsymbol{\alpha}_{12}$ and $\boldsymbol{\mu}_{12}$ are the net polarizability and net dipole moment of the pair of interacting molecules 1 and 2 (where molecules 1 and 2 are here of the same species). The integration is over all positions and orientations of molecule 2 relative to molecule 1. Ω is defined by

$$ \int \mathrm{d}\mathbf{r}_{12}\, \mathrm{d}\boldsymbol{\omega}_{12} = \Omega \tilde{V}. $$

It is seen that two factors may contribute[91] to B_{ε}. The first arises from the difference between the polarizability of the pair of molecules, $\boldsymbol{\alpha}_{12}$, and the mean polarizabilities of the two isolated molecules, 2α. The second results from the difference between the square of the vector sum of the dipole moments of the two molecules at a given orientation, $\boldsymbol{\mu}_{12}^2$, and the square of the dipole moment of an isolated molecule, $\boldsymbol{\mu}^2$.

Measurements of B_ε for pure inert gas molecules,[92] for which both $\boldsymbol{\mu}$ and $\boldsymbol{\mu}_{12}$ are zero as a result of symmetry, suggest that the contributions due to the polarizability term are generally small. This is also supported by studies of the second refractivity virial coefficient which arises in the similar density series for molar refractivity, and which depends only on this term. This is generally found to be small for both polar and non-polar molecules.[93] Since there is at present no *a priori* way of evaluating $\boldsymbol{\alpha}_{12}$, the usefulness of B_ε measurements for these systems as probes of the intermolecular potential is generally small. As the intermolecular potential energy functions for the inert gases are now reasonably well characterized, such measurements may perhaps best be regarded as giving a source of information about $\boldsymbol{\alpha}_{12}$.

Molecules which have no permanent dipole moment but possess higher multipoles, e.g. H_2, N_2, or CO_2, also have a contribution in B_ε arising from the dipole term in (3.59), since there is a fluctuating induced dipole term associated with a molecular collision. For mixed inert gases, e.g. Ar+Ne, there is a transient dipole during the collisions, which will also contribute to B_ε.

For polar gases[91] the dominant contributions to B_ε arise from the dipole term in (3.59). If all orientations were equally probable, $\langle \frac{1}{2}\boldsymbol{\mu}_{12}^2 - \boldsymbol{\mu}^2 \rangle$ would be zero. However, the relative probabilities of the orientations are determined by the Boltzmann factor, $\exp(-U_{12}/kT)$, and usually B_ε has large values for polar molecules, indicating preferred alignments. If the preferred orientation favours parallel dipoles, the value of B_ε is positive; for opposed dipoles, B_ε is negative. Since U_{12} contains contributions both from the electrostatic interactions and from the non-polar contributions, studies of B_ε can give useful information about the nature of the pair potential in such cases.

Model calculations for polar molecules using Stockmayer potentials, including induction terms, give B_ε values which are positive, since the preferred orientation at all separations is that with a head-to-tail arrangement of the dipoles. Since the experimental results for polar molecules are both positive and negative, the deficiencies of this simple potential are evident. An improvement in the model must permit the lowest energy to correspond to a head-to-head arrangement of dipoles in some circumstances. The inclusion of shape effects in the repulsive part of the potential, as proposed by Buckingham and Pople,[81] may achieve this, as shown in Fig. 3.24. If the dipole is parallel to the long axis of the

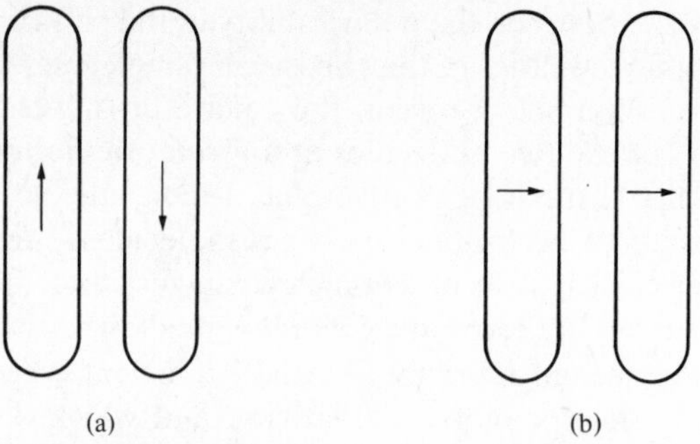

FIG. 3.24. Preferred dipole arrangements at short separations for two different types of non-spherical, polar molecules.

molecule the repulsive interactions at small separations favour an anti-parallel arrangement of dipoles. However, if the dipole is perpendicular to the long axis, both dipole–dipole interactions and the repulsive shape forces favour the head-to-tail orientation at all separations. Calculations which incorporate this factor, although an improvement on the simple Stockmayer model, have not proved to be entirely satisfactory.[94]

Measurements of dielectric virial coefficients are difficult, and the results are sparse and cover a limited range of temperature. A survey of the results available to 1971 has been given by Sutter.[91] In spite of these problems, the information contained in dielectric virial coefficients can complement equation-of-state data very usefully and it seems probable that they will be more extensively used in the future, especially as more complex molecules are investigated.

Appendix A3.1. Gas properties in terms of virial coefficients

The imperfect behaviour of gases makes contributions to the thermodynamic properties which may be conveniently expressed using virial coefficients and their temperature derivatives. The expressions below generally give the extent of the deviations from the value for a perfect gas in its standard state, denoted by the superscript ^{0}. A superior tilde (˜) represents a molar quantity.

(a) *Fugacity, f.* The fugacity of a gas is related to its free energy through the equation

$$\tilde{G} = \tilde{G}^0 + RT \ln f.$$

It is conveniently expressed in virial form in terms of the ratio f/p

$$\ln (f/p) = \frac{B}{\tilde{V}} + \frac{C + B^2}{\tilde{V}^2} + \dots$$

(b) *Internal energy, U.* The departure of the molar internal energy from the perfect gas value may be written

$$\frac{\tilde{U}-\tilde{U}^0}{RT}=-\left(\frac{B_1}{\tilde{V}}+\frac{C_1}{2\tilde{V}^2}+\ldots\right)$$

where B_n, C_n represent $T^n(\mathrm{d}^nB/\mathrm{d}T^n)$ and $T^n(\mathrm{d}^nC/\mathrm{d}T^n)$ respectively.

(c) *Enthalpy, H.* This may be related to the expression for U above, since $H=U+PV$.

$$\frac{\tilde{H}-\tilde{H}^0}{RT}=\frac{B-B_1}{\tilde{V}}+\frac{2C-C_1}{2\tilde{V}^2}$$

(d) *Entropy, S.* Combining the expressions in *a* and *c*, we obtain

$$\frac{\tilde{S}-\tilde{S}^0}{R}=-\left\{\ln P+\frac{B_1}{\tilde{V}}+\frac{B^2-C+C_1}{2\tilde{V}^2}+\ldots\right\}.$$

(e) *Heat capacities, C_P and C_V.* The contributions to these quantities due to gas imperfection may be expressed

$$\frac{\tilde{C}_\mathrm{P}-\tilde{C}^0_\mathrm{P}}{R}=-\left\{\frac{B_2}{\tilde{V}}-\frac{(B-B_1)^2-(C-C_1)-C_2/2}{\tilde{V}^2}+\ldots\right\}$$

$$\frac{\tilde{C}_\mathrm{V}-\tilde{C}^0_\mathrm{V}}{R}=-\left\{\frac{2B_1+B_2}{\tilde{V}}+\frac{2C_1+C_2}{2\tilde{V}^2}+\ldots\right\}.$$

References

1. Mason, E. A. and Spurling, T. H. *The virial equation of state.* Pergamon Press, Oxford (1969).
2. Andrews, T. *Phil. Trans. roy. Soc.* **159,** 575 (1869).
3. Lewis, G. N. and Randall, M. *Thermodynamics* (2nd edn) (ed. K. S. Pitzer and L. Brewer), Chapter 5. McGraw-Hill, New York (1961).
4. van der Waals, J. D. Dissertation, Leiden (1873); (trans. R. Threlfall and J. F. Adair) *Phys. Memoirs.* **1,** 333 (1890).
5. Ref. 3, pp. 193–6.
6. Su, G.-J. *Ind. Engng Chem.* **38,** 803 (1946).
7. Onnes, H. K. *Comm. Phys. Lab. Leiden* **71, 74** (1901).
8. Epstein, L. F. *J. chem. Phys.* **20,** 1981 (1952); **21,** 762 (1953).
9. Mayer, J. E. and Mayer, M. G. *Statistical mechanics* (2nd edn). J. Wiley, New York (1977).
10. Hill, T. L. *Introduction to statistical thermodynamics.* Addison-Wesley, Reading, Massachusetts (1960); *Statistical mechanics.* McGraw-Hill, New York (1956).
11. Rushbrooke, G. S. *Introduction to statistical mechanics,* Oxford University Press, Oxford (1949).
12. Kilpatrick, J. E. *J. chem. Phys.* **21,** 274 (1953).
13. Keller, J. B. and Zumino, B. *J. chem. Phys.* **30,** 1351 (1959).
14. Frisch, H. L. and Helfand, E. *J. chem. Phys.* **32,** 269 (1960).
15. Reed, T. M. and Gubbins, K. E. *Applied statistical mechanics,* p. 458. McGraw-Hill, New York (1973).

16. Hirschfelder, J. O., Curtiss, C. F., and Bird, R. B. *Molecular theory of gases and liquids*, Chapter 3. Wiley, New York (1954).
17. Johnson, C. H. J. and Spurling, T. H. *Aust. J. Chem.* **27,** 241 (1974).
18. Uhlenbeck, G. E. and Ford, G. W. In *Studies in statistical mechanics* (ed. J. de Boer and G. E. Uhlenbeck). North-Holland, Amsterdam (1962).
19. Ree, F. H. and Hoover, W. G. *J. chem. Phys.* **46,** 4181 (1967).
20. Ref. 16, Chapter 6.
21. Wigner, E. *Phys. Rev.* **40,** 749 (1932).
22. Kirkwood, J. G. *Phys. Rev.* **44,** 31 (1933); **45,** 116 (1934).
23. Boyd, M. E., Larsen, S. Y., and Kilpatrick, J. E. *J. chem. Phys.* **50,** 4034 (1969).
24. Hoover, A. E., Leland, T. W., and Kobayashi, R. *J. chem. Phys.* **45,** 399 (1966).
25. Dymond, J. H. and Smith, E. B. *The virial coefficients of pure gases and mixtures.* Clarendon Press, Oxford (1980).
26. Pitzer, K. S. *J. chem. Phys.* **7,** 583 (1939).
27. Guggenheim, E. A. *J. chem. Phys.* **13,** 253 (1945).
28. Kihara, T. *Rev. mod. Phys.*, **25,** 831 (1953).
29. Munn, R. J. *J. chem. Phys.* **40,** 1439 (1964).
30. Dymond, J. H., Rigby, M., and Smith E. B. *Phys. Fluids* **9,** 1222 (1966).
31. Reid, R. C., Prausnitz, J. M., and Sherwood, T. K. *The properties of gases and liquids* (2nd edn). McGraw-Hill, New York (1977).
32. Hanley, H. J. M. and Klein, M. *J. chem. Phys.* **50,** 4765 (1969).
33. Ref. 16, p. 159.
34. Barker, J. A. and Monaghan, J. J. *J. chem. Phys.* **36,** 2558 (1962).
35. Ref. 16, p. 157.
36. Mie, G. *Ann. Phys., Leipzig,* **11,** 657 (1903).
37. (Lennard-) Jones, J. E. *Proc. roy. Soc.* **A106,** 463 (1924).
38. Kihara, T., Midzuno, Y., and Shizume, T. *J. Phys. Soc. Jpn* **10,** 249 (1955).
39. Kihara, T. *J. Phys. Soc. Jpn,* **3,** 265 (1948); **6,** 184 (1951).
40. Sherwood, A. E. and Prausnitz, J. M. *J. chem. Phys.* **41,** 413 (1964).
41. Sherwood, A. E., De Rocco, A. G., and Mason, E. A. *J. chem. Phys.* **44,** 2984 (1966).
42. Barker, J. A., Leonard, P. J. and Pompe, A. *J. chem. Phys.* **44,** 4206 (1966).
43. Pool, R. A. H., Saville, G., Herrington, T. M., Shields, D. D. C. and Staveley, L. A. K. *Trans. Faraday Soc.* **58,** 1692 (1962).
44. McGlashan, M. L. and Potter, D. J. B. *Proc. roy. Soc.* **A267,** 478 (1962).
45. Schramm, B. and Hebgen, U. *Chem. Phys. Lett.* **29,** 137 (1974).
46. Fender, B. E. F. and Halsey, G. D. Jr. *J. chem. Phys.* **36,** 1881 (1962).
47. Weir, R. D., Wynn Jones, I., Rowlinson, J. S. and Saville, G. *Trans. Faraday Soc.* **63,** 1320 (1967).
48. Steele, W. A. *Interactions of gases with solid surfaces.* Pergamon Press, Oxford (1974).
49. Michels, A., Wijker, Hub. and Wijker, Hk. *Physica.* **15,** 627 (1949).
50. Burnett, E. S. *J. appl. Mech.* **A3,** 136 (1936).
51. Silberberg, I. H., Lin, D. C. K., and McKetta, J. J. *J. chem. engng Data,* **12,** 226 (1967).
52. Hoover, A. E., Canfield, F. B., Kobayashi, R., and Leland, T: W., Jr. *J. chem. engng Data.* **9,** 568 (1964).
53. Scott, R. L. and Dunlap, R. D. *J. phys. Chem.* **66,** 639 (1962).
54. Michels, A., Abels, J. C., ten Seldam, C. A., and de Graaf, W. *Physica* **26,** 381 (1960).

55. Roebuck, J. R., Murrell, T. A., and Miller, E. E. *J. Amer. chem. Soc.* **64,** 400 (1942).
56. Al-Bizreh, N. and Wormald, C. J. *J. chem. Therm.* **9,** 749 (1977).
57. Sherwood, A. E. and Prausnitz, J. M. *J. chem. Phys.* **41,** 429 (1964).
58. Ref. 1, Chapter 4.
59. Buckingham, R. A. *Proc. roy. Soc.* **A168,** 264 (1938).
60. Barker, J. A. and Pompe, A. *Aust. J. Chem.* **21,** 1683 (1968).
61. Johnson, C. H. J. and Spurling, T. H. *Aust. J. Chem.* **24,** 2205 (1971).
62. Maitland, G. C. and Smith, E. B. *Proc. Seventh Symp. on Thermophysical Properties*, p. 412. Amer. Soc. Mech. Eng. (1977).
63. Jonah, D. A. and Rowlinson, J. S. *Trans. Faraday Soc.* **62,** 1067 (1966).
64. Maitland, G. C. and Smith, E. B. *Mol. Phys.* **24,** 1185 (1972).
65. Cox, H. E., Crawford, F. W., Smith, E. B., and Tindell, A. R. *Mol. Phys.* **40,** 705 (1980).
66. Maitland, G. C. and Smith, E. B. *Mol. Phys.* **22,** 861 (1971).
67. Ref. 1, p. 38.
68. Isihara, A. *J. chem. Phys.* **18,** 1446 (1950).
69. Kihara, T. *J. Phys. Soc. Jpn.* **6,** 289 (1951); **8,** 686 (1953).
70. Nezbeda, I. *Chem. Phys. Lett.* **41,** 55 (1976).
71. Monson, P. A. and Rigby, M. *Mol. Phys.* **35,** 1337 (1978).
72. Gibbons, R. M. *Mol. Phys.* **17,** 81 (1969).
73. Sweet, J. R. and Steele, W. A. *J. chem. Phys.* **47,** 3022 (1967); **47,** 3029 (1967).
74. Corner, J. *Proc. roy. Soc.* **A192,** 275 (1948).
75. Hamann, S. D. and Lambert, J. A. *Aust. J. Chem.* **7,** 1 (1954).
76. De Rocco, A. G. and Hoover, W. G. *J. chem. Phys.* **36,** 916 (1962).
77. Storvick, T. S., Spurling, T. H., and De Rocco, A. G. *J. chem. Phys.* **46,** 1498 (1967).
78. Kihara, T. *Advan. chem. Phys.* **5,** 147 (1963).
79. Stockmayer, W. H. *J. chem. Phys.* **9,** 398 (1941).
80. Rowlinson, J. S. *Trans. Faraday Soc.* **45,** 974 (1949).
81. Buckingham, A. D. and Pople, J. A. *Trans. Faraday Soc.* **51,** 1173 (1955).
82. Pople, J. A. *Proc. roy. Soc.*, **A221,** 498, 508 (1954).
83. Ref. 1, p. 237.
84. Pitzer, K. S. *J. Amer. chem. Soc.* **77,** 3427 (1955).
85. Pitzer, K. S. and Curl, R. F., Jr. *J. Amer. chem Soc.* **79,** 2369 (1957).
86. Knobler, C. M., Beenakker, J. J. M., and Knaap, H. F. P. *Physica* **25,** 909 (1959).
87. Edwards, A. E. and Roseveare, W. E. *J. Amer. chem. Soc.* **64,** 2816 (1942).
88. Rowlinson, J. S. and Richardson, M. J. *Advan. chem. Phys.* **2,** 85 (1959).
89. Prausnitz, J. M. and Benson, P. R. *Amer. Inst. chem. Engng J.* **5,** 161 (1959).
90. Buckingham, A. D. and Pople, J. A. *Trans. Faraday Soc.* **51,** 1179 (1955).
91. Sutter, H. *Dielectric and related molecular properties*, Specialist periodical report Chem. Soc., **1,** 65 (1972).
92. Orcutt, R. H. and Cole, R. H. *J. chem. Phys.* **46,** 697 (1967).
93. Ashton, H. M. and Halberstadt, E. S. *Proc. roy. Soc.* **A245,** 373 (1958).
94. Sutter, H. and Cole, R. H. *J. chem. Phys.* **54,** 4988 (1971).

4
MOLECULAR COLLISIONS

4.1. Introduction

The preceding chapter has been concerned with the equilibrium properties of matter, that is those quantities which are independent of the time scale of molecular motions. We now turn to those properties which do depend on the variation of molecular velocities and positions with time and consider the dynamics of molecular collisions. For some of these quantities, such as dilute gas transport properties at normal temperatures, it is sufficient to use classical mechanics to obtain a quantitative understanding of the phenomena involved. Indeed, fairly crude and simple models of molecular interactions are often able to give a remarkably realistic description of these transport processes. Consequently we begin this chapter with a classical description of the collision between two simple molecules. When one comes to consider the angular distribution of the scattered molecules in more detail, a classical treatment is far from satisfactory. Many of the phenomena observed in molecular beam scattering experiments are only explained, even qualitatively, by a quantum mechanical treatment. As in most cases where quantum mechanics is used, various approximations may be applied in the quantitative treatment of scattering processes. The conditions under which some of these methods may be used, and some of their important conclusions, will be discussed.

In describing the various levels of approximation by which molecular scattering phenomena may be related to the potential energy function of the colliding pair, the aim has been to give sufficient mathematical detail to enable the reader to follow the relationships through, whilst maintaining a physical appreciation of the various steps involved. Consequently not all sections of this chapter show the same degree of rigour in the development of the working equations. The chapter begins with a rigorous classical treatment of the dynamics of binary collisions (§ 4.2). This provides an understanding of the important factors which govern the outcome of scattering processes and gives the general features of observed cross-sections. The shortcomings of such a classical approach can be anticipated on theoretical grounds (§ 4.3) but are probably highlighted most graphically by the contrast of its predictions with experiment. § 4.4 describes the main features of molecular beam-scattering equipment and gives illustrative examples of the results obtained for spherically symmetric systems.

The detail of such results can only be explained quantitatively by a quantum-mechanical approach and this is described for elastic scattering in § 4.5. This is presented at a less rigorous level than the classical treatment, giving sufficient detail for the reader to appreciate how one step follows on from the other without necessarily providing all the intermediate equations. The object is to obtain the relationships between measured experimental quantities and the intermolecular potential and to provide an understanding of the origins, both physical and mathematical, of these relationships. In § 4.6 it is shown how, using these developments, information about spherically symmetric intermolecular pair potential energy functions may be extracted at various levels of detail from the cross-sections described in § 4.4. Finally, in § 4.7 attention is turned to polyatomic systems where the potential function is anisotropic. Differences from spherical systems in experimental techniques and the nature of results are outlined first and then it is shown how the treatment of elastic scattering may be generalized through the so-called scattering matrix. After deriving in some detail a formal quantum-mechanical statement of the scattering problem in polyatomic systems, the various approximation methods available for its solution are then described in outline. These methods enable non-spherical potentials to be related to the scattered intensities and emphasis is placed on the physical approximations made in them, and the conditions under which they may be used with confidence.

It is apparent, then, that in order to describe how information about intermolecular forces can be extracted from molecular scattering studies, this chapter must cover a wide range of theoretical and experimental topics. The reader is referred to a number of specialist texts[1,2] and collections of reviews[3–6] for more detailed treatments of those areas which, of necessity, are not exhaustively treated here.

4.2. Classical dynamics of molecular collisions

4.2.1. Binary elastic collisions

The simplest situation to examine theoretically is the elastic collision between a pair of particles which interact with a spherically symmetrical potential $U(r)$, i.e. they exert an equal and opposite force of magnitude F on each other. In practice, it is usually possible to make measurements at pressures where three- and higher body interactions are so rare that their effect on collisional properties is either negligible or easily applied as a small correction. An understanding of the binary collision situation should therefore aid us in the interpretation of a wide range of experimental results.

If we consider the encounter of two masses m_a and m_b whose coordinate vectors in the laboratory frame are $\mathbf{r}_a$ and $\mathbf{r}_b$ then the vector for the centre of mass $\mathbf{r}_c$ is

$$\mathbf{r}_c = \frac{m_a\mathbf{r}_a + m_b\mathbf{r}_b}{m_a + m_b}. \tag{4.1}$$

We may introduce the separation vector $\mathbf{r}$.

$$\mathbf{r} = \mathbf{r}_a - \mathbf{r}_b. \tag{4.2}$$

The situation is as illustrated in Fig. 4.1.

Applying Newton's equations of motion to each particle we have

$$\begin{aligned} \mathbf{F} &= m_a \frac{d\mathbf{c}_a}{dt} \\ -\mathbf{F} &= m_b \frac{d\mathbf{c}_b}{dt} \end{aligned} \tag{4.3}$$

where $\mathbf{c}_a$, $\mathbf{c}_b$ are the respective velocities of a and b. Now by differentiation of (4.1) we see that

$$\frac{d\mathbf{c}_c}{dt} = \frac{m_a(d\mathbf{c}_a/dt) + m_b(d\mathbf{c}_b/dt)}{m_a + m_b} = \frac{\mathbf{F} - \mathbf{F}}{m_a + m_b} = 0, \tag{4.4}$$

that is, the centre of mass moves with constant velocity $\mathbf{c}_c$. The equations of motion also show that

$$\frac{d}{dt}\left(\frac{d\mathbf{r}}{dt}\right) = \frac{d\mathbf{c}}{dt} = \frac{d\mathbf{c}_a}{dt} - \frac{d\mathbf{c}_b}{dt} = \left(\frac{1}{m_a} + \frac{1}{m_b}\right)\mathbf{F} = \mu^{-1}\mathbf{F} \tag{4.5}$$

where μ is the reduced mass of the system. Eqn (4.5) defines the relative motion of a and b, which is confined to a plane containing both particles

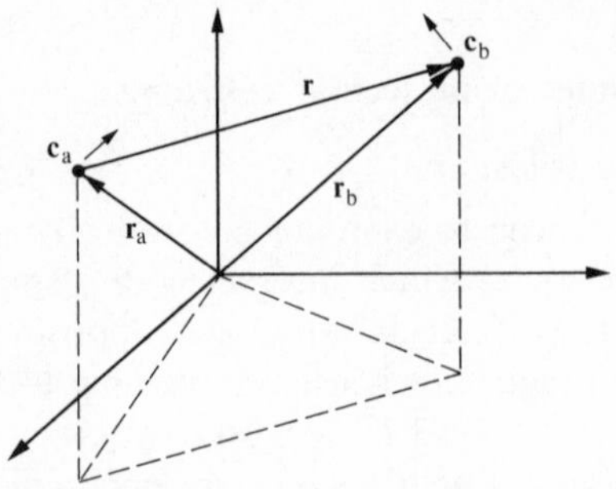

FIG. 4.1. Coordinate vectors of two colliding molecules.

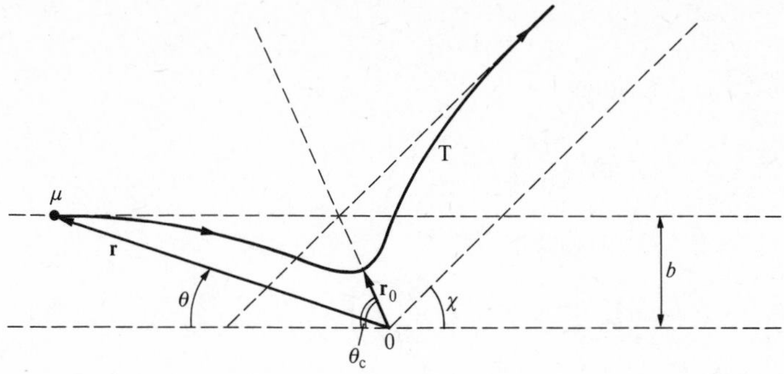

FIG. 4.2. One-body trajectory equivalent to a binary collision. b is the impact parameters, r_0 the classical turning point, χ the scattering angle, and μ the reduced mass of the colliding pair.

and their centre of mass. This plane, which is defined by the vectors $\mathbf{r}$ and $\mathbf{c}$, is therefore moving with the constant velocity of the centre of mass, $\mathbf{c}_c$. In this *centre-of-mass frame* the relative motion of the two particles is thus equivalent to the motion of a single particle of mass μ in the central force field $\mathbf{F}$, i.e. under the influence of the potential $U(r)$. We therefore need consider only the much simpler equivalent one-body problem illustrated in Fig. 4.2.

4.2.2. The equivalent one-body problem

The vector $\mathbf{r}$ now joins the centre of mass O to the position of the particle of mass μ, which follows the trajectory T. The angle between $\mathbf{r}$ and the initial direction of motion is called the *orientation angle* θ. This has the value θ_0 at the distance of closest approach, or turning point, r_0. The *impact parameter*, b, is the distance of closest approach in the absence of the potential $U(r)$ and, hence, in the absence of any deflection. The ultimate *angle of deflection* is χ, which is related to θ_0 by

$$\chi = \pi - 2\theta_0. \tag{4.6}$$

The dynamics of this collision may be described by considering the conservation of both energy and momentum in the system. The initial energy E is all kinetic, $\frac{1}{2}\mu g^2$; where $\mathbf{g}$ is the *initial relative velocity* $\{\mathbf{c}(t=0)\}$. At some point during the trajectory when the particle has Cartesian coordinates x and y relative to the origin, O

$$\begin{aligned} -x &= r\cos\theta, \qquad & -\mathrm{d}x/\mathrm{d}t &= c\cos\theta - r\frac{\mathrm{d}\theta}{\mathrm{d}t}\sin\theta \\ y &= r\sin\theta, \qquad & \mathrm{d}y/\mathrm{d}t &= c\sin\theta - r\frac{\mathrm{d}\theta}{\mathrm{d}t}\cos\theta, \end{aligned} \tag{4.7}$$

where

$$c = \mathrm{d}r/\mathrm{d}t$$

Thus the total energy E is given by

$$E = \tfrac{1}{2}\mu\left\{\left(\frac{\mathrm{d}x}{\mathrm{d}t}\right)^2 + \left(\frac{\mathrm{d}y}{\mathrm{d}t}\right)^2\right\} + U(r),$$

i.e.

$$E = \tfrac{1}{2}\mu g^2 = \tfrac{1}{2}\mu\left(c^2 + r^2\left(\frac{\mathrm{d}\theta}{\mathrm{d}t}\right)^2\right) + U(r). \tag{4.8}$$

Similarly we may equate the asymptotic value of the angular momentum ($L = \mu bg$) with its value at some point during the trajectory:

$$L = \mu bg = \mu\left(x\frac{\mathrm{d}y}{\mathrm{d}t} - y\frac{\mathrm{d}x}{\mathrm{d}t}\right) = \mu r^2\frac{\mathrm{d}\theta}{\mathrm{d}t} \tag{4.9}$$

Eqn (4.9) may be used to eliminate $\mathrm{d}\theta/\mathrm{d}t$ from (4.8), giving

$$E = \tfrac{1}{2}\mu c^2 + \tfrac{1}{2}\mu g^2\frac{b^2}{r^2} + U(r). \tag{4.10}$$

Since the first term of (4.10) is the kinetic energy of the particle, the last two terms are usually grouped together as the *effective potential energy* $V(L, r)$. It consists of the true potential energy $U(r)$ together with the *centrifugal potential energy* $U_L(r)$ which can be expressed in terms of either L or E:

$$U_L(r) = \tfrac{1}{2}\mu g^2\frac{b^2}{r^2} = \frac{L^2}{2\mu r^2} = \frac{Eb^2}{r^2}. \tag{4.11}$$

For a given potential $U(r)$ there are an infinite number of possible $V(L, r)$ depending on the values of g and b. The nature of the possible trajectories can be understood by examining the behaviour of $V(L, r)$. Fig. 4.3 shows how the shape of the effective potential energy function varies as the angular momentum increases from zero.

The effect of the centrifugal potential is to raise the effective potential energy and to introduce a *centrifugal barrier*, X, at intermediate separations as L increases from zero. Eventually this barrier disappears, when the coordinate of the maximum Y coincides with that of the potential minimum Z. Above this inflection point (I) ($\equiv L_4$) the effective potential is purely monotonic and repulsive.

(a) *The classical turning point.* Here, the velocity $c = 0$ and according to (4.10)

$$E - V(L, r) = 0 \tag{4.12}$$

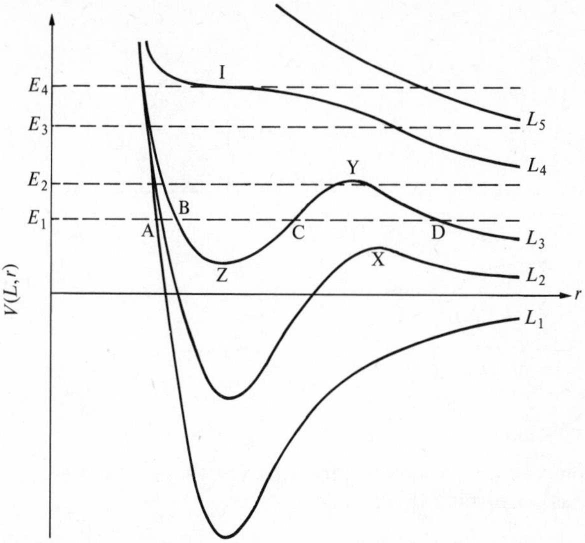

FIG. 4.3. Effective potential energy functions (schematic). $V(L, r) = U(r) + L/(2\mu r^2) = U(r) + Eb^2/r^2$, where E is the relative kinetic energy and L the angular momentum. $L_5 > L_4 > L_3 > L_2 > L_1 = 0$.

or

$$r_0 = b\left\{1 - \frac{U(r_0)}{E}\right\}^{-\frac{1}{2}}. \tag{4.13}$$

Eqn (4.12) tells us that the classical turning point will be the point at which a horizontal line plotted at an energy E on Fig. 4.3 crosses $V(L, r)$. When E is below E_4, the inflection value of $V(L, r)$, e.g. E_1, three turning points are possible but since a classical particle cannot penetrate the centrifugal barrier it is the outermost point, D, which gives r_0.

The value of r_0 and hence the form of the trajectory depends on whether the attractive or repulsive parts of the potential are effective in deflecting the particle's path. For instance consider a trajectory having an initial energy E_1. For low angular momenta (L_2) r_0 is at A, $U(r_0) > 0$ and, from (4.13), $r_0 > b$. Repulsion dominates and the path is of the form of trajectory 1 in Fig. 4.4. As the angular momentum increases, we expect the attractive part of the potential to become more effective. For L_3, r_0 is at D and $U(r_0) < 0$, so that $r_0 < b$. In this case the trajectory resembles number 2 of Fig. 4.4.

Likewise a change of impact parameter at fixed energy E will alter the effective potential. At small b, repulsion dominates and $r_0 > b$ whereas for a large b, net attraction results and $r_0 < b$. For a series of trajectories at constant angular momentum (L_3) we may vary the impact parameter

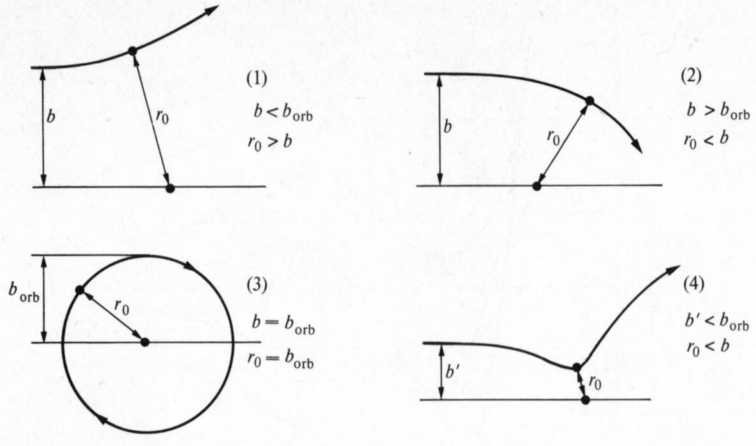

FIG. 4.4. Some typical trajectories for different values of the impact parameter b and the classical turning point r_0.

by increasing the energy E. At an energy E_1, we have seen that the net effect is attractive (trajectory 2). When $E=E_2$, $\partial V(L,r)/\partial r=0$ at the turning point Y and therefore r remains constant. The particle goes into a circular orbit about the centre of mass and $\chi\rightarrow\infty$ (trajectory 3). This phenomenon is called *orbiting* and whether it occurs or not obviously depends on the particular combination of g and b. The attractive and repulsive contributions of the potential are here exactly balanced. (As E and L increase, the radius of the orbiting trajectory decreases until above E_4, orbiting cannot occur irrespective of the values of g and b.†) As E increases above the orbiting value (to E_3), repulsion takes over and the path resembles trajectory 4.

(b) *The classical deflection function.* The argument may be made more quantitative by calculating the angle of deflection χ: (4.10) gives

$$c=\frac{\mathrm{d}r}{\mathrm{d}t}=\pm\left[\frac{2}{\mu}\left\{E-\tfrac{1}{2}\mu g^2\frac{b^2}{r}-U(r)\right\}\right]^{\frac{1}{2}}. \tag{4.14}$$

Eqn (4.9) gives:

$$\frac{\mathrm{d}\theta}{\mathrm{d}t}=\frac{L}{\mu r^2}$$

Hence

$$\frac{\mathrm{d}\theta}{\mathrm{d}r}=\frac{\mathrm{d}\theta}{\mathrm{d}t}\frac{\mathrm{d}t}{\mathrm{d}r}=\pm\frac{L}{\mu r^2}\left[\frac{2}{\mu}\left\{E-\tfrac{1}{2}\mu g^2\frac{b^2}{r}-U(r)\right\}\right]^{-\frac{1}{2}}.$$

† E_4, the critical energy for orbiting, is given by the conditions that $r=r_0$ and $(\mathrm{d}V(L,r)/\mathrm{d}r)=0$, i.e. $E_4=U(r)+\frac{1}{2}r\,\mathrm{d}U(r)/\mathrm{d}r$.

At the turning point r_0, taking the incoming trajectory,

$$\theta_0 = \int_0^{\theta_0} \mathrm{d}\theta = -\int_\infty^{r_0} \frac{L}{\mu r^2}\left[\frac{2}{\mu}\left\{E - \tfrac{1}{2}\mu g^2 \frac{b^2}{r} - U(r)\right\}\right]^{-\frac{1}{2}} \mathrm{d}r$$

which by (4.6) gives

$$\chi(b, g) = \pi - 2\int_{r_0}^{\infty} \frac{L}{\mu r^2}\left[\frac{2}{\mu}\left\{E - \tfrac{1}{2}\mu g^2 \frac{b^2}{r^2} - U(r)\right\}\right]^{-\frac{1}{2}} \mathrm{d}r.$$

Taking the asymptotic values of $E(=\tfrac{1}{2}\mu g^2)$ and $L\ (=\mu bg)$, we have

$$\chi(b, g) = \pi - 2b\int_{r_0}^{\infty} \frac{\mathrm{d}r/r^2}{\{1 - b^2/r^2 - U(r)/\tfrac{1}{2}\mu g^2\}^{\frac{1}{2}}}. \tag{4.15}$$

The variation of χ with impact parameter is termed *the classical deflection function.* Its typical behaviour for a realistic potential having both attractive and repulsive branches is illustrated in Fig. 4.5.

The trajectories of Fig. 4.4 can also be understood in terms of these classical deflection functions. Collisions with a large positive χ can be attributed to the repulsive portion of the potential acting at small impact parameters (or small L). As b increases, χ decreases until at b_1, the net force is zero and there is no deflection. For $b > b_1$, χ passes through a minimum χ_r at b_2 and converges to zero as b tends towards infinity. The angle χ_r at which $\mathrm{d}\chi/\mathrm{d}b$ is zero, is called the *rainbow angle.* As the relative velocity decreases, the depth of the minimum in χ increases and for small values of g ($<(2E_4/\mu)^{\frac{1}{2}}$) becomes infinite. $b = b_0$ is then the *orbiting value.* It should be noticed at this stage that as long as $\chi < +\chi_r$, several values (usually three) of the impact parameter can lead to the same value of $|\chi|$ at the same relative velocity. The relevance of $|\chi|$ will become apparent when we come to consider the experimental study of

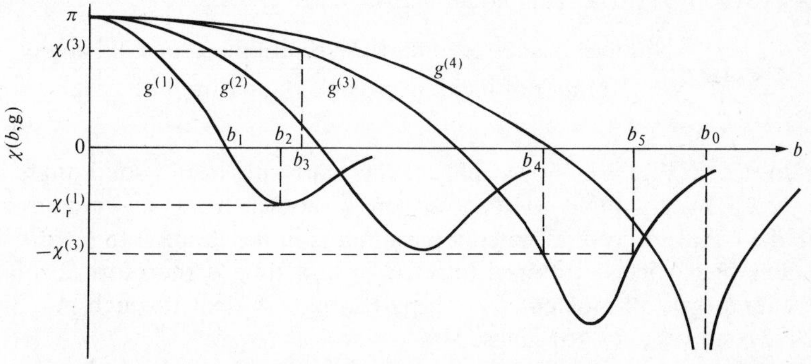

FIG. 4.5. Typical deflection functions for central force scattering, corresponding to the trajectories of Fig. 4.4. Relative initial speeds: $g^{(1)} > g^{(2)} > g^{(3)} > (2E_4/\mu)^{\frac{1}{2}} > g^{(4)}$. ($E_4$ refers to the critical orbiting energy; see Fig. 4.3.)

molecular collisions in scattering experiments where the sign of χ cannot be measured, only its magnitude. For instance, in Fig. 4.5 impact parameters b_3, b_4, and b_5 will all lead to the same angle of deflection $|\chi^{(3)}|$.

4.3. Experimental characterization of scattering: cross-sections

It is impossible to study a single molecular collision experimentally. The simplest experiment which can be devised involves the collision of two narrow beams having well-defined initial velocities and the measurement of the resultant scattering as a function of angle. So although energy and angle may both be specified, it is not possible to confine the collisions to a single impact parameter. Experimental results are therefore reported in terms of *collision cross-sections*, which are quantitative measures of the intensity of the observed scattering. Cross-sections may be defined for three types of scattering:

1. Elastic, in which there is no change in the internal states of the colliding molecules;
2. Inelastic, in which a change from one specified internal energy state to another occurs;
3. Reactive, in which there is a change in the chemical nature of the species during collision.

Since our main aim here is to study intermolecular forces and not energy transfer, we shall for the present confine our attention to elastic scattering.

4.3.1. Definition of collision cross-sections

Scattering at a particular angle χ is characterized by the differential cross-section $\sigma(\chi, E)$. This is defined by

$$\sigma(\chi, E) = \frac{\text{Number of scattered particles/unit time/unit solid angle}}{\text{Number of incident particles/unit time/unit area}}. \tag{4.16}$$

Reference to Fig. 4.6 shows that all the scattering into a solid angle $d\Omega$, projected by an angle $d\chi$ centred on χ, comes from an annular area $2\pi b\, db = A_1$, i.e. from all impact parameters in the range b to $b + db$. The number of particles scattered into $d\Omega$ in unit time is therefore $2\pi b I_0\, db$. In other words all molecules entering through A_1 exit through A_2. Since $d\Omega = 2\pi \sin\chi\, d\chi$, (4.16) gives

$$\sigma(|\chi|, E) = \left|\frac{2\pi b I_0\, db/2\pi \sin\chi\, d\chi}{I_0}\right| = \left|\frac{b}{\sin\chi(d\chi/db)}\right|. \tag{4.17}$$

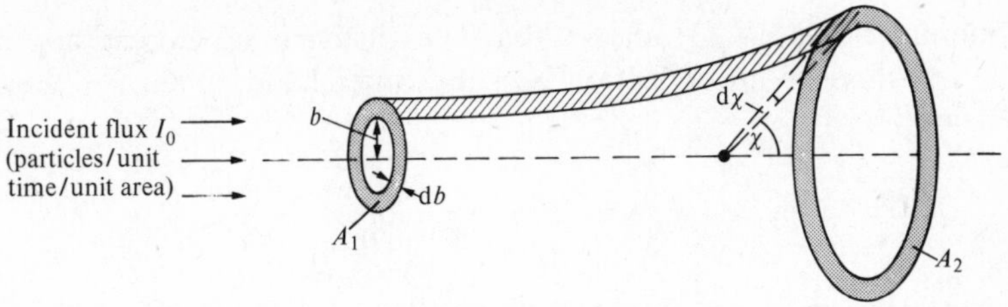

FIG. 4.6. Three-dimensional scattering: incident particles within the annular area A_1 are scattered through an angle χ into the annulus, area A_2.

In those cases where more than one impact parameter leads to the same value of $|\chi|$,

$$\sigma(\chi, E) = \sum \left| \frac{b}{\sin \chi (\mathrm{d}\chi/\mathrm{d}b)} \right|. \tag{4.18}$$

The *integral cross-section*, $Q(E)$, is a measure of all the scattering irrespective of angle, and is simply the integral of $\sigma(\chi, E)$ over all solid angles:

$$Q(E) = \int \sigma(\chi, E)\, \mathrm{d}\Omega = \int_0^{\pi} \sigma(\chi, E) 2\pi \sin \chi\, \mathrm{d}\chi. \tag{4.19}$$

We will now examine how these classical cross-sections behave for some model systems.

(a) *Hard sphere cross-sections.* For hard spheres of diameter d,

$$U(r) = \infty \qquad r \leqslant d$$
$$U(r) = 0 \qquad r > d$$

and the effective potential $V(L, r)$ is illustrated in Fig. 4.7. For impact

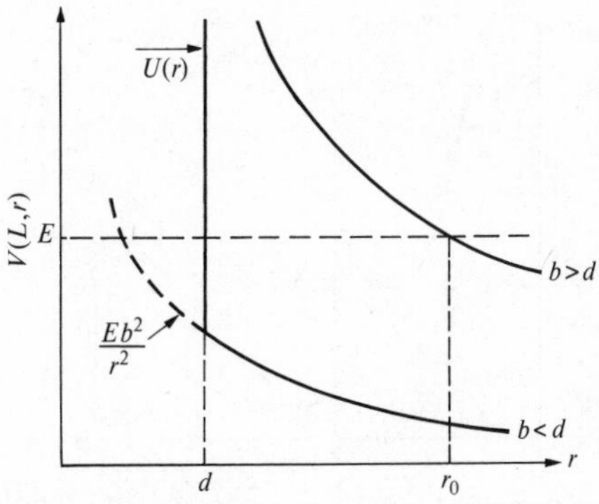

FIG. 4.7. Effective potential energy for collisions of hard-spheres, diameter d.

parameters $b > d$, (4.13) shows that the distance of closest approach $r_0 = b$, i.e. there is no deflection. On the other hand, when $b < d$, $r_0 = d$ and using (4.15):

$$\chi = \pi - 2b \int_d^\infty \frac{dr/r^2}{(1 - b^2/r^2)^{\frac{1}{2}}}$$
$$= \pi - 2\sin^{-1}(b/d)$$

Thus

$$b = d \sin \tfrac{1}{2}(\pi - \chi) = d \cos(\chi/2)$$

and

$$db/d\chi = d(-\sin(\chi/2))/2.$$

Using (4.17) gives

$$\sigma(\chi, E)_{hs} = d^2/4 \tag{4.20}$$

and (4.19) yields

$$Q_{hs} = \frac{2\pi}{4} \int_0^\pi d^2 \sin\chi \, d\chi = \pi d^2, \tag{4.21}$$

i.e. scattering is independent of both energy and angle and the integral cross-section is determined by the distance of closest approach, all as expected.

(b) *Cross-sections for realistic potential functions.* For potential functions having both repulsive and attractive branches, (4.15), (4.17), and (4.19) must in general be evaluated numerically. Typical behaviour of the classical differential cross-section is illustrated in Fig. 4.8.

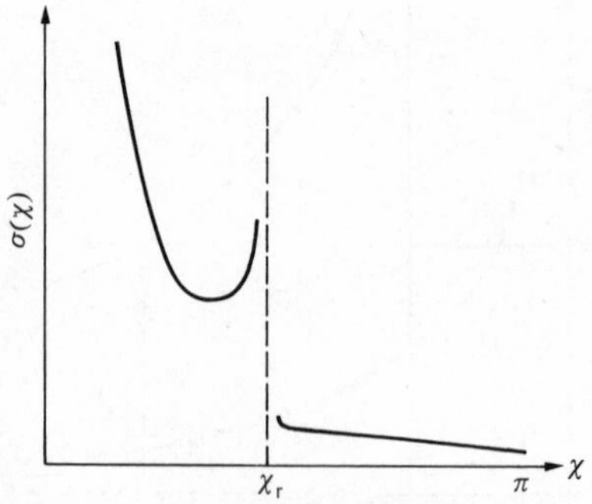

FIG. 4.8. Classical differential cross-section for a potential of the 12−6 type, showing the singularities at the rainbow angle χ_r, and as $\chi \to 0$.

It will be noticed that $\sigma(\chi)$ becomes singular on two occasions:

(i) *The glory effect* occurs when $\sin\chi$ in (4.17) can become zero and hence $\sigma(\chi)$ becomes infinite. A forward glory exists when $\chi \to 0$ and a backward glory when $\chi \to -n\pi$. No singularity occurs as $\chi \to \pi$ since b is also zero here.

(ii) *The rainbow effect* arises when $\mathrm{d}\chi/\mathrm{d}b$ becomes zero at some intermediate impact parameter, again causing $\sigma(\chi)$ to become infinite, this time at χ_r. This effect derives its name from the analogous turning point in the deflection angle of a light ray internally reflected in a water droplet, which gives rise to an optical rainbow.

The trajectories contributing to these two effects are illustrated in Fig. 4.9.

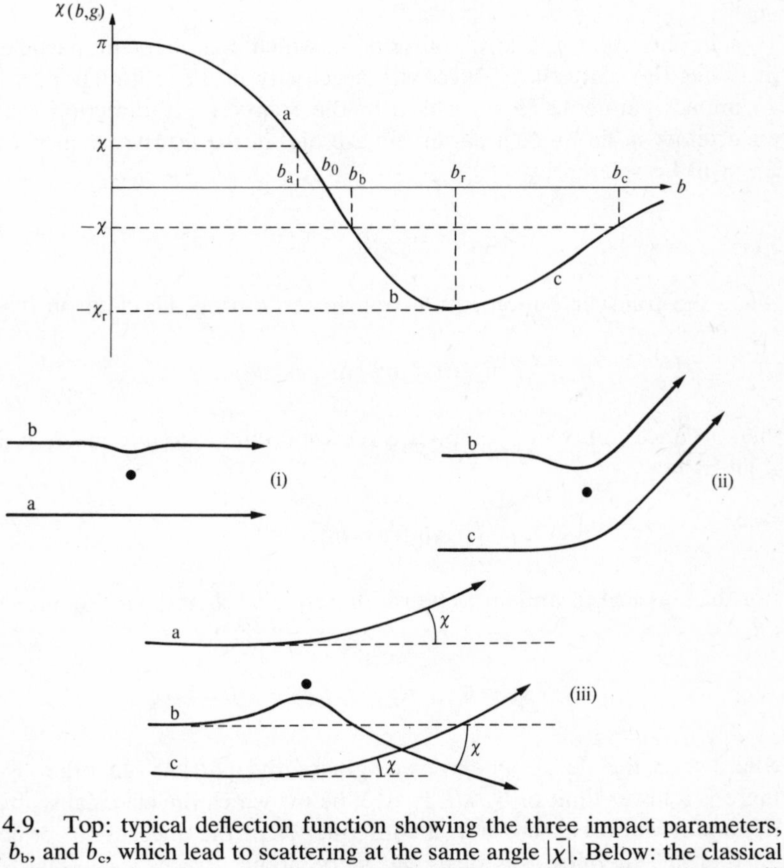

FIG. 4.9. Top: typical deflection function showing the three impact parameters, b_a, b_b, and b_c, which lead to scattering at the same angle $|\bar{\chi}|$. Below: the classical trajectories giving rise to interference effects in scattering cross-sections. (i) Glory effect, $b_a \sim b_0 \sim b_b$, $|\chi| \sim 0$; (ii) rainbow effect $b_b < b_r < b_c$, $|\chi| \sim |\chi_r|$; (iii) rapid quantum oscillations, $|\chi_r| > |\chi| > 0$.

4.3.2. Shortcomings of classical treatment

The general features of elastic scattering by atoms and molecules can be successfully described by the classical treatment given above. However, there are a number of important respects in which experimental experience differs from classical predictions; these arise because of the wave-like nature of the particles and can only be explained by a quantum approach. The major limitations of classical theory are summarized here.

The uncertainty principle prevents the simultaneous definition of the position, x, and the conjugate momentum, p_x, of a particle with a precision greater than that given by the equation

$$\Delta x\,\Delta p_x \geqslant \frac{h}{2\pi} = \hbar.$$

This implies that for any collision in which the incident particle approaches the scattering centre with a velocity in the x-direction, c_x, and an impact parameter b measured in the transverse y-direction that the uncertainty in the impact parameter Δb and in the transverse momentum Δp_y will be related by

$$\Delta b\,\Delta p_y \simeq \hbar.$$

Since the transverse momentum transferred during the collision is

$$p_y - 0 = \mu c_x \sin\chi \simeq \mu c_x \chi,$$

then an uncertainty in p_y implies a corresponding one in deflection angle χ given by

$$\Delta b\,\Delta\chi \simeq \hbar/\mu c_x.$$

For the classical treatment to be valid, both $\Delta b \ll b$ and $\Delta\chi \ll \chi$ must hold, i.e.

$$\chi \gg \Delta\chi \simeq \hbar/\mu c\,\Delta b \gg \hbar/\mu c b = \lambda/b = \hbar/L, \tag{4.22}$$

where λ is the de Broglie wavelength of the particle. In other words, there is a lower limit on χ, λ/b $(=\chi_q)$, below which the classical scattering treatment will break down as a consequence of the uncertainty principle. At thermal energies, this turns out to be about 1°.

This breakdown of classical mechanics in the case of low-angle scattering (where large impact parameters dominate) may be illustrated analytically by making the assumptions

1. That only the long-range part of the potential, $U(r) = C_n/r^n$ $(n>2)$ is important;
2. $b^2/r^2 \gg U(r)/E$ so that the classical deflection function, (4.15), may be expanded in powers of E^{-1} to first order and the integration performed analytically.

These assumptions lead to[7]

$$\chi = (n-1)C_n f(n)/Eb^n \tag{4.23}$$

where

$$f(n) = \frac{\pi^{\frac{1}{2}}\,\Gamma(n/2-1/2)}{2\Gamma(n/2)}$$

and $\Gamma(x)$ is the gamma function. Insertion in (4.17) gives

$$\sigma(\chi, E) = \{(n-1)f(n)C_n/E\}^{2/n}/\{n\chi^{1+(2/n)}\sin\chi\}. \tag{4.24}$$

For the important case of $n=6$, $\sigma(\chi, E) \propto \chi^{-\frac{7}{3}}$, and clearly diverges at low angles. This low-angle behaviour also causes $Q(E)$ to become infinite when $\sigma(\chi, E)$ is integrated over all angles. It is a general result that classical theory predicts infinite total cross-sections for potentials of infinite range. By contrast, experiments show that $\sigma(\chi, E)$ has a finite value as $\chi \to 0$ and that the integral cross-section is also finite.

We have seen that classical theory predicts a singularity at the rainbow angle χ_r. The deflection angle χ goes through a minimum at χ_r (see Fig. 4.5) and near this angle we can represent the deflection by a simple quadratic dependence on impact parameter b

$$\chi \simeq \chi_r + K(b-b_r)^2. \tag{4.25}$$

There will be two trajectories which have the same deviation $\Delta\chi$ from the rainbow angle χ_r. The corresponding difference in the impact parameters is given by

$$\Delta b = b - b_r = \left(\frac{\chi-\chi_r}{K}\right)^{\frac{1}{2}} = \left(\frac{\Delta\chi}{K}\right)^{\frac{1}{2}}. \tag{4.26}$$

As the rainbow angle is approached then these differences become smaller than the uncertainties in b and χ which are consistent with the uncertainty principle. At this stage classical mechanics fails and according to (4.22) this occurs when

$$\Delta\chi < \frac{\hbar}{\mu c\,\Delta b} = \frac{\hbar}{\mu c}\left(\frac{K}{\Delta\chi}\right)^{\frac{1}{2}}.$$

i.e.

$$\Delta\chi < \left(\frac{\hbar^2 K}{\mu^2 c^2}\right)^{\frac{1}{3}}. \tag{4.27}$$

Quantum-mechanical effects cause the rainbow singularity to be smoothed out over the range $\Delta\chi$ and the peak becomes finite, although still a dominant feature of the scattering.

In cases such as those above where more than one classical trajectory contributes to the scattering at an angle χ, the intensities no longer remain additive when the wave nature of the scattered particles is considered. Interference between the various contributions can occur and oscillations in both $\sigma(\chi, E)$ and $Q(E)$ are predicted, and observed experimentally, superimposed on the classical envelope.

The classical trajectories which give rise to these interference effects are summarized in Fig. 4.9. It can be seen that if $|\chi| < |\chi_r|$ then trajectories corresponding to three impact parameters b_a, b_b, and b_c can lead to scattering at the same angle $|\chi|$. (The sign of χ cannot be distinguished experimentally). For $|\chi| \sim 0$, trajectories a and b can have the same deflection angle, but different path lengths (Fig. 4.9(i)). These trajectories will thus arrive at the detector out-of-phase and will interfere, producing the *glory oscillations* in the differential cross-section. Similarly, trajectories of type b and c can arrive at the detector having both been scattered through an angle $|\chi|$ near $|\chi_r|$, the rainbow angle, having travelled via different paths. (Fig. 4.9(ii)). Interference between these trajectories leads to the *rainbow maximum* in the cross-section near χ_r. Higher order supernumary rainbows are also produced at lower scattering angles. For intermediate angles $|\chi_r| > |\chi| > 0$, it is possible for all three branches of the deflection function, a, b, and c, to contribute a trajectory to the net scattering at a particular angle. Interference here leads to rapid oscillations superimposed over the more widely spaced rainbow structure—the so-called *rapid quantum oscillations*. These different types of interference phenomena are a valuable source of information about the intermolecular potential and are considered in some detail in later sections. Their experimental observation is well illustrated in Fig. 4.15 (p. 185).

Finally, it was seen in Fig. 4.3 that orbiting collisions occurred when the collision energy E coincided with the height of the outer maximum in the effective potential $V(L, r)$. Quantum effects allow tunneling through this barrier, so that the singularity in the classical deflection function at b_0 in Fig. 4.5 is removed. Orbiting contributions to the cross-sections[1] arise because of the accessibility of quasi-bound states within the central region of $V(L, r)$.

It is quite clear from this discussion that scattering behaviour significantly different from the predictions of classical mechanics is observed in practice, and that this is mainly due to quantum effects. However, some differences occur simply because of the finite resolution of molecular beam instruments used to study such scattering phenomena. So before a quantum treatment is considered in more detail, we will survey briefly the experimental methods used to determine cross-sections and examine the type of results obtained.

4.4. Experimental techniques

4.4.1. Principles of molecular beam experiments

Differential and integral cross-sections for all three types of scattering—elastic, inelastic, and reactive—are measured using molecular beams. A molecular beam is a well-collimated stream of atoms or molecules which are at a sufficiently low density and moving at sufficiently low velocities relative to each other to eliminate collisions between them, hence minimizing both spatial and velocity broadening of the beam. The beam moves in a high vacuum to reduce the frequency of spurious collisions with the beam. It impinges on a target from which the scattering can be monitored by a suitable detector.

In the study of elastic collisions, two configurations can be used. A typical single-beam experiment is shown schematically in Fig. 4.10(a). Here the target is simply a small volume of gas and the attenuation of the beam as it passes through the sample is measured. Here only the integral cross-section $Q(E)$ can be obtained, using the relation

$$\ln \frac{I_0}{I} = nQ(E)l \tag{4.28}$$

where I_0, I are the incident and scattered beam intensities respectively, and n is the number density of target molecules in the collision path, of length l.

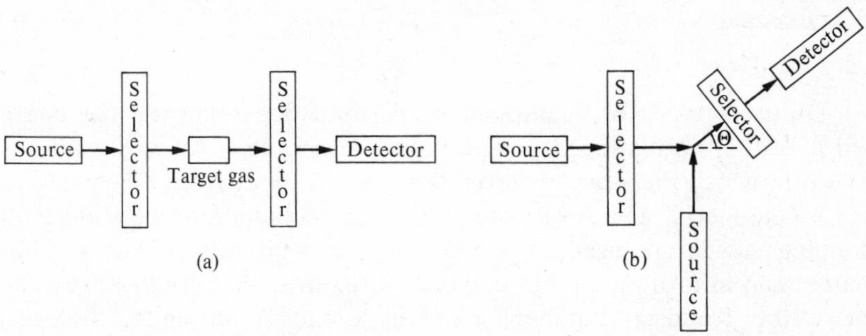

FIG. 4.10. Schematic diagram of molecular beam experiments. (a) Single beam technique; (b) cross-beam technique. The selectors can be used to select particular velocities or internal states if required.

The cross-beam method, illustrated in Fig. 4.10(b), uses another molecular beam as the target, and can be used to determine the angular dependence of the scattering, i.e. the differential cross-section, as well as $Q(E)$.

In both cases, the beams can be formed by effusion or expansion from oven sources and are collimated by knife-edge slits. A narrow range of velocities can be selected by mechanical slotted-wheel devices and, where necessary, particular internal quantum states selected by magnetic or electrical means.† Ionization detectors are usually used and in the cross-beam case these move round the scattering volume to determine the intensity as a function of laboratory scattering angle Θ. By modulating the cross-beam, the scattering due to this beam can be distinguished from that due to background. The success of a molecular beam experiment rests on maximizing scattered intensity and angular resolution.

The problems of detection can be severe. A few typical values[2] will exemplify this. A typical beam would be 1 cm high and 0·05 mm wide and for a thermal source the flux might be about 10^{16} molecules s^{-1} m^{-2}. For a beam velocity of 10 m s^{-1}, this corresponds to a density of 10^{15} molecules m^{-3} which should be compared with a good laboratory vacuum of 10^{-7} mm Hg, also equivalent to 10^{15} molecules m^{-3}, and a typical flux at the detector of 10^6–10^7 molecules s^{-1} m^{-2}. Consequently the beam density is comparable with that of the background and because the over-all path lengths are far greater than the target beam thicknesses, the number of molecules scattered from background molecules can be far in excess of those deflected by the cross-beam. Ultrahigh-vacuum techniques capable of pressures as low as 10^{-9} mm Hg are hence almost essential; particle counting techniques are then also essential to resolve the extremely low scattered fluxes found at high angles. These problems have been considerably alleviated by the advent of supersonic nozzle beams, whose intensities are considerably greater than those from thermal sources, typically 10^{19} molecules s^{-1} m^{-2}.

4.4.2. Sources

Different types of beam source are used depending on the energy range. In the thermal energy region, up to 1 eV, most sources use a flow system in which the beam material flows as a vapour from a heated oven into a chamber(s) at a lower pressure. Here collimation and differential pumping stages are used to produce a final well-defined beam. These sources can in turn operate in one of two regimes. At very low pressures, where the Knudsen number‡ (Kn) is greater than unity, molecular

† More commonly nowadays supersonic nozzle sources are used which provide a narrow velocity distribution directly.

‡ The Knudsen number is the ratio of the mean free path of the molecules to the relevant dimension of the container, in this case the slit width.

effusion occurs and the molecules pass through the slit without colliding with each other. The advantages of these sources are that they are simple to construct, have a well defined and relatively narrow velocity distribution and produce an equilibrium mixture of internal states which can be controlled simply by heating. On the other hand the fluxes produced are relatively low—$\sim 5 \times 10^{16}$ molecules $m^{-2}\,s^{-1}$.

Alternative sources operate in the *hydrodynamic flow* regime where $Kn < 1$. Here collisions are frequent and for polyatomic molecules a degree of energy transfer from internal modes to translational energy occurs. They have a great advantage of producing fluxes up to 10^3 higher than effusive sources. The beam material emerges as a supersonic jet of gas through a nozzle from a chamber in which the pressure is 0·1–2 bar. Such an adiabatic expansion produces extremely low translational (and rotational) temperatures in the beam molecules, and a narrow velocity distribution. On expansion into the nozzle chamber, the jet strikes a truncated hollow cone (the skimmer) which isolates a rapidly moving gas beam and deflects the remainder of the jet away to the pumps. The flow then rapidly changes to molecular flow and collimation and differential pumping follow as before. The availability of these *supersonic nozzle beam* sources,[8] producing typically fluxes of 10^{18}–10^{20} molecules $sr^{-1}\,s^{-1}$†, has made it possible to perform high resolution experiments for inert materials such as the rare gases, nitrogen, etc., for which there are only low efficiency detectors.

The nozzle sources described above operating in the thermal energy range are the most important for elastic scattering experiments since this energy range samples the low-energy repulsion and attractive energy regions of typical intermolecular potential functions. However, *superthermal* sources are important for the study of the higher-energy repulsive region and, incidentally, are very important for reactive scattering studies.[9–11] Three major methods are used:[3]

(a) Seeded nozzle beams in which a few per cent of the heavy beam material is mixed with a light carrier gas such as helium. These produce a high bulk flow rate due to the low mean mass of the seeded beam and carry the beam molecules along at energies well in excess of their thermal values. Typical fluxes are 10^{16} molecules $sr^{-1}\,s^{-1}$, at energies up to 4 eV. Alternative methods run nozzle beams at elevated temperatures and a combination of this and seeding techniques have produced argon beams in Ar/He mixtures of intensity 10^{19} atoms $sr^{-1}\,s^{-1}$, up to 20 eV. Shock heating in the nozzle chamber can produce even higher intensities.

† sr ≡ steradian, the unit of solid angle. One steradian is the solid angle subtended at the centre of a unit sphere by unit surface area.

(b) Sputtering sources, in which high-energy atoms are expelled from the surface of a solid target on bombardment with energetic ion beams. These sources can produce energies up to about 50 eV but have a very broad velocity distribution and low intensities.

(c) Resonant charge exchange sources, in which a well-focused ion beam having a narrow energy distribution is passed through a sample of the neutral gas to which charge transfer occurs with high probability. This produces a high energy neutral beam which retains the directional and energy distribution characteristics of the ion beam and is the major type of source used for energies greater than 20 eV. Unfortunately, intensities tend to be rather low, typically $\sim 10^{14}$ molecules sr^{-1} s^{-1}.

4.4.3. Velocity and state selection

In order to determine cross-sections at a well-defined collision energy, it is desirable to use beams as near monoenergetic as possible. Since thermal sources produce beams containing a distribution of molecular velocities, it is usually necessary to use velocity selection. A common form of velocity selector is the multi-disc mechanical system in which the beam passes through a series of slotted rotating discs so that only those molecules which are in a narrow band of velocities arrive at every slot at the correct time to be transmitted.[12]† To achieve velocity selection on detection an alternative time-of-flight method[13,14] is sometimes used. This operates on a similar principle to phase-sensitive detection: the incident and detected beams are chopped out of phase with one another and a time-of-arrival spectrum of the scattered molecules built up. For polar and paramagnetic species, rotational and magnetic state selection can also be used.[15–17] This usually involves the use of inhomogeneous fields in which the deflection of molecules is dependent upon their particular internal state.

4.4.4. Detectors

As mentioned earlier, the major problems of detection are differentiation between beam and background molecules and measurement of extremely low fluxes. The majority of detectors in current use are ionization detectors,[2] which ionize the scattered beam molecules and measure the resulting ion current. The initial ionization may be performed using a hot wire (such as tungsten), which is suitable at low energies <3 eV, cold filaments, which are effective for high energy beams,

† Nozzle beams have a distinct advantage over chopped oven sources since the production of a narrow velocity distribution without the need for selection results in no attenuation of the beam and hence higher fluxes.

electron bombardment, or field ionization.† Semi-conductor and metal strip bolometers which respond either to the kinetic energy of the beam or to some surface exothermic reaction have also been used.[19]

4.4.5. *Centre-of-mass–laboratory coordinates*

It is important to realize that the quantity determined in a molecular beam experiment is a cross-section as a function of the angle of deflection Θ in laboratory (LAB) coordinates. Calculated differential cross-sections are based on the centre-of-mass (CM) coordinate frame, in which the angle χ is the angle between the initial and final *relative* velocity vectors of the colliding molecules. Before the two sets of cross-sections may be compared, therefore, conversion to a common coordinate system must be made. This CM–LAB coordinates transformation is readily achieved by means of Newtonian velocity vector diagrams[2] and is summarized in terms of the relevant Jacobian, J[20]

$$(\mathrm{d}x\ \mathrm{d}y\ \mathrm{d}z)_{\mathrm{LAB}} = J\left(\frac{(xyz)_{\mathrm{LAB}}}{(xyz)_{\mathrm{CM}}}\right)(\mathrm{d}x\ \mathrm{d}y\ \mathrm{d}z)_{\mathrm{CM}}. \tag{4.29}$$

4.4.6. *Integral cross-section measurements*

In order that the cross-section determined using (4.28) should be the true $Q(E)$, a number of criteria must be fulfilled. The angular resolution of the instrument necessary for thermal energies is 1–10 minutes of arc.[21] The velocity or energy resolution is limited by the averaging which arises from the motion of the target molecules. This can cause errors in the absolute magnitude of $Q(E)$ and also obscure the fine oscillations. This effect is significantly greater for the single-beam experiment where the molecules in the scattering chamber have a Maxwellian distribution, but for both kinds of experiment the effect can be corrected for. For absolute values of $Q(E)$, the single-beam method is usually used since the absolute intensity is difficult to measure in the cross-beam method. Chamber pressures are typically 10^{-4} mm Hg, giving attenuations of 10–20 per cent. Under these conditions multiple scattering is rare but pressure measurement is difficult. Much of the early work was in error due to systematic errors in using McLeod gauges.[22] For the most accurate work, end effects arising from target molecules flowing out of the chamber by the entrance and exit slots must also be considered.

A typical experimental set-up based on that of Dondi, Scoles, Torello, and Pauly[23] is illustrated in Fig. 4.11. This uses a single nozzle beam operating at 77 K. The scattering chamber is cooled to 2 K to

† Many recent advances have been made by the use of high-resolution mass spectrometers incorporating ion-counting systems as detection devices.[18]

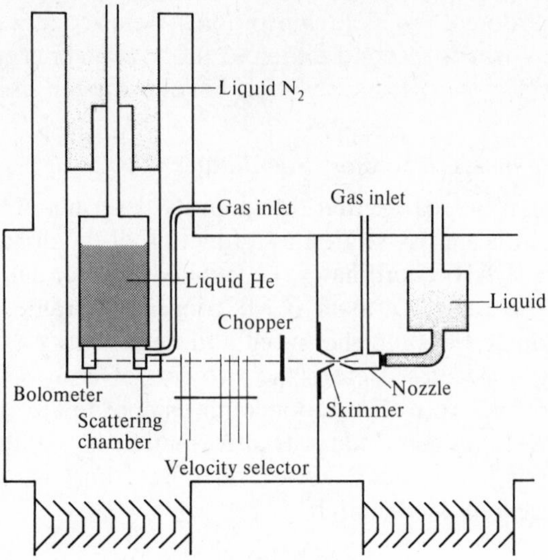

FIG. 4.11. Single-beam apparatus for measuring total cross-sections. (M. G. Dondi, G. Scoles, F. Torello, and H. Pauly.[23]).

minimize velocity-averaging and the primary beam is velocity-selected mechanically to give a resolution $\Delta v/v$ of 5 per cent. (Δv is the half-height width of the velocity distribution whose peak value is v.) The results for $Q(E)$ for He^4/He^4 scattering produced by this apparatus are shown in Fig. 4.12. The large number of oscillations observed for this system are not typical, but arise from the identical nature of the scattering particles (see later). Fig. 4.13 shows a more typical plot of $Q(E)$ vs. E, showing a number of glory modulations at low energy superimposed on a gradual

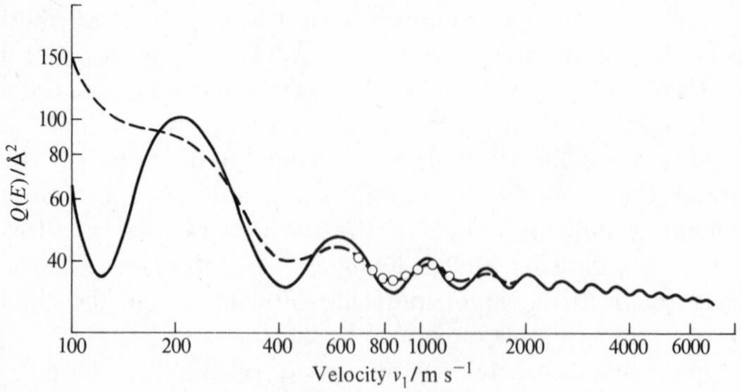

FIG. 4.12. Integral cross-section $Q(E)$ for $He^4 + He^4$ scattering. The full curve is calculated using a 12−6 potential, $r_m = 0.30$ nm, $\varepsilon = 1{\cdot}4 \times 10^{-22}$ J, assuming no velocity averaging; the dashed curve is calculated for a chamber temperature of 2 K. The experimental measurements are shown as circles. (Based on M. G. Dondi, G. Scoles, F. Torello, and H. Pauly.[23])

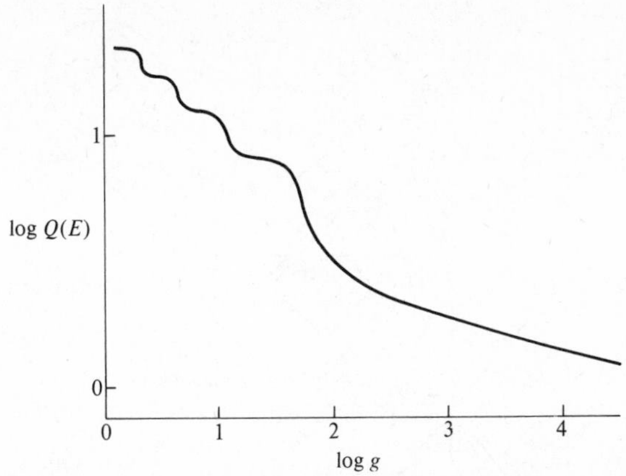

FIG. 4.13. Major features of $Q(E)$ for $12-6$ potential (arbitrary units).

decrease in Q with increasing energy. We will return to the interpretation of these data after the treatment of quantum scattering.

4.4.7. *Differential cross-section measurements*

For two cross-beams having fluxes I_1 and I_2, velocities $\mathbf{v}_1$ and $\mathbf{v}_2$, the scattered intensity, S, reaching a detector a distance R from the scattering centre in a direction Θ is related to the CM differential cross-section by the equation[2]

$$S(\Theta)=\frac{A}{R^2}\frac{I_1I_2}{v_1v_2}JV\,|\mathbf{v}_1-\mathbf{v}_2|\,\sigma(\chi, E) \tag{4.30}$$

where A is the detector area, V the scattering volume, J the CM $\rightarrow$ LAB Jacobian, and χ is the centre-of-mass scattering angle as defined previously.

The most important factor in reducing the resolution of $\sigma(\chi, E)$ is the spread of velocities in the (usually unselected) cross-beam. As well as directly affecting $S(\Theta)$, any such spread also leads to loss of resolution in the LAB $\rightarrow$ CM transformation. To reduce such effects, the heavier beam is usually used as the cross-beam and the source cooled. In order to resolve the rainbow and its supernumary peaks, a 15 per cent spread in velocity is permissible, but to resolve the high-frequency oscillations, $\Delta v/v < 10$ per cent and an angular resolution of $\sim 0{\cdot}2°$ is required.

An apparatus capable of resolving the detailed fine structure of $\sigma(\chi, E)$ for inert gas systems, where the problems of detection sensitivity are probably more severe than for any other species, is illustrated in Fig. 4.14. This has been used by Lee and co-workers[24] in a definitive series of experiments on the inert gases. It uses two nozzle beams in an ultra-high vacuum ($<10^{-10}$ mm Hg) chamber. Detection is by electron

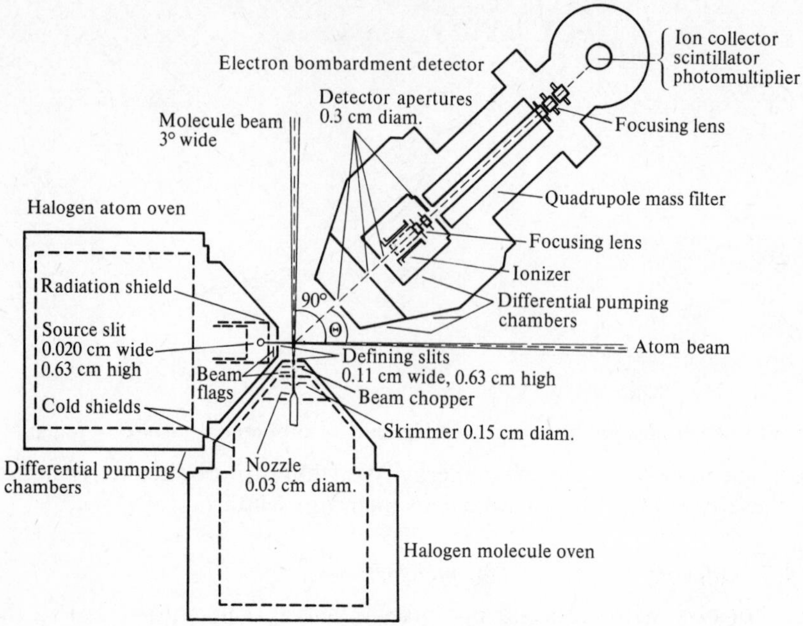

FIG. 4.14. Molecular beam apparatus for measuring differential cross-sections. (Y. T. Lee, J. D. McDonald, P. R. LeBreton, and D. R. Herschbach.[24])

bombardment mass filter using a modulated ion-counting system. Typical of the results that can be obtained with such an instrument are those illustrated in Fig. 4.15 for the inert gases.[25] This shows three main quantum effects superimposed on the classical prediction of Fig. 4.8:

(a) A finite rainbow maximum at Θ_r where Θ_r is the laboratory frame angle corresponding to the centre-of-mass rainbow angle χ_r;
(b) A series of supernumary rainbow maxima at $\Theta < \Theta_r$;
(c) High-frequency, lower-amplitude interference oscillations superimposed on the rainbow structure.

A detailed interpretation of these effects will be deferred until after the treatment of quantum scattering in the next section.

4.5. Quantum mechanical treatment of elastic scattering[1,4]

In the previous sections we have presented a rigorous and detailed classical treatment of spherical particle scattering and given some physical insight into the origins of quantum scattering phenomena. We now give a quantum mechanical treatment, which is formally correct but, in order to keep to a reasonable length and to stress the physical origins of the

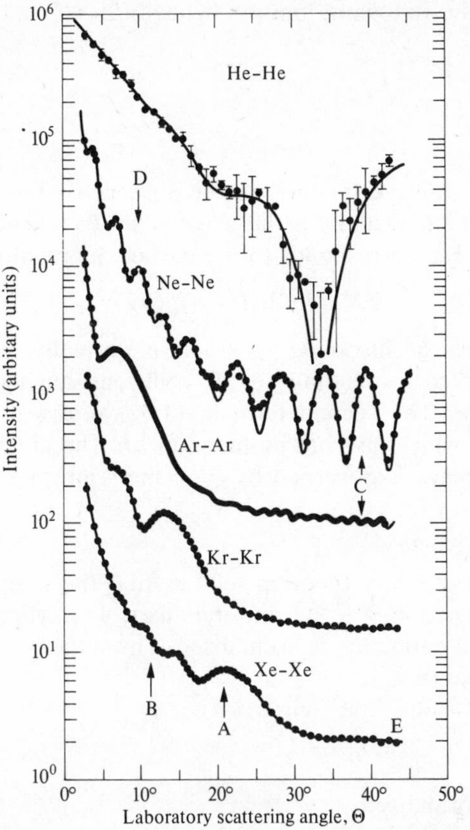

FIG. 4.15. Differential cross-sections $\sigma(\Theta, E)$ for the inert gases, He, Ne, Ar, Kr, and Xe, showing fine structure. Principal features: (A) rainbow maximum; (B) supernumary rainbow peaks; (C) symmetry (identical particle) oscillations; (D) high-frequency quantum oscillations; (E) monotonic large angle scattering. (J. M. Farrar, T. P. Schafer, and Y. T. Lee.[25])

phenomena, less detailed explanation is given of the links between various steps in the argument.

4.5.1. *The wave function*

We begin by considering the same single-particle problem dealt with classically in § 4.2. A full quantum treatment of elastic scattering for a spherically symmetric potential describes the system in terms of a wave function $\Psi(\mathbf{r})$ which depends only on the relative position $\mathbf{r}$ of the particle. $\Psi(\mathbf{r})$ is an eigenfunction of the Schrödinger equation

$$\left(-\frac{\hbar^2}{2\mu}\nabla^2 + U(r)\right)\Psi(\mathbf{r}) = E\Psi(\mathbf{r}). \tag{4.31}$$

$\Psi(\mathbf{r})$ must have the following limiting behaviour:

$$\begin{aligned} &1.\ \Psi(\mathbf{r}) = 0, \qquad r \to 0 \\ &2.\ \Psi(\mathbf{r}) = \Psi^{(0)}(\mathbf{r}) + \frac{f(\chi)e^{ikr}}{r}, \qquad r \to \infty \end{aligned} \tag{4.32}$$

where $\Psi^{(0)}(\mathbf{r})$ is the wave function in the absence of $U(r)$, $f(\chi)$ is called the *scattering amplitude*, and $\mathbf{k}$ is the *wavevector*. Its magnitude, k, known as the *wave number*, corresponds to the particle momentum $(2\mu E)^{\frac{1}{2}}$

$$k^{-1} = \hbar/(2\mu E)^{\frac{1}{2}} = \hbar/(\mu g). \tag{4.33}$$

$\Psi^{(0)}(\mathbf{r})$ is therefore the incoming wave and is normally represented by a plane wave e^{ikz}, corresponding to all collisions having a z-direction momentum of $k\hbar$. The second term in (4.32) represents the outgoing (scattered) wave, which also has momentum $k\hbar$. The angular distribution of the scattered wave is governed by $f(\chi)$, the scattering amplitude.

4.5.2. The cross-sections

The Born probability theorem tells us that the intensity of a wave having an eigenfunction Ψ is Ψ^2, and the flux is $\Psi^2 v$, where v is the wave velocity, if the wave function is normalized. The velocity v of the incident wave is $\hbar/k\mu$, from (4.33).
Hence the incident flux $= |\Psi^{(0)}(\mathbf{r})|^2\, v$

$$= \hbar k/\mu \tag{4.34}$$

since $\Psi^{(0)}(\mathbf{r})$ is normalized.

Similarly since $\Psi(\mathbf{r})$ is also normalized the scattered flux is $|f(\chi)/r|^2\, v$, using (4.32).

Now it can be seen in Fig. 4.6 that any scattering through the annular area A_2 $(= 2\pi r^2 \sin\chi\, d\chi)$ has been scattered into a solid angle $2\pi \sin\chi\, d\chi$, so that

$$\begin{aligned} \text{Rate of scattering per unit solid angle} &= \left|\frac{f(\chi)}{r}\right|^2 v \frac{2\pi r^2 \sin\chi\, d\chi}{2\pi \sin\chi\, d\chi} \\ &= |f(\chi)|^2\, v. \end{aligned} \tag{4.35}$$

Consequently, using the definition of differential-cross section, (4.16), we may write

$$\sigma(\chi, E) = \frac{|f(\chi)|^2\, v}{v} = |f(\chi)|^2. \tag{4.36}$$

It follows from the definition of the integral cross-section (4.19) that

$$Q(E) = \int \sigma(\chi, E)\, d\Omega = \int_0^{\pi} 2\pi\, |f(\chi)|^2 \sin\chi\, d\chi. \tag{4.37}$$

The crux of the quantum mechanical treatment is therefore to determine $|f(\chi)|$, just as in the classical case it was necessary to find the deflection function $\chi(b)$. In other words, the Schrödinger equation (4.31) must be solved for $\Psi(\mathbf{r})$.

4.5.3. *Solution of the wave equation: method of partial waves*

The wave equation (4.31) can be simplified to

$$(\nabla^2 + k^2 - u(r))\Psi(\mathbf{r}) = 0 \tag{4.38}$$

where $u(r)$ is simply $U(r) . 2\mu/\hbar^2$. The wave function can be expressed as the sum over *partial waves*, each of which corresponds to a particular angular momentum state of the system.

$$\Psi(\mathbf{r}) = r^{-1} \sum_{l=0}^{\infty} C_l \psi_l(r) P_l(\cos\chi) \tag{4.39}$$

where C_l are the amplitudes of the partial waves $\psi_l(r)$, $P_l(\cos\chi)$ are the Legendre polynomials,[26] and l is the angular momentum quantum number. Substituting (4.39) into (4.38) results in a set of equations for $\psi_l(r)$:

$$\left(\frac{\mathrm{d}^2}{\mathrm{d}r^2} + k^2 - u_l(r)\right)\psi_l(r) = 0 \tag{4.40}$$

where

$$u_l(r) = u(r) + \{l(l+1)/r^2\}. \tag{4.41}$$

Eqn (4.40) is called the partial wave equation and will be used extensively in the treatment that follows.

Since the angular momentum L is $\hbar\{l(l+1)\}^{\frac{1}{2}}$, comparison with the classical expression for L, μbg (4.9) using (4.33) enables us to identify

$$b \simeq \frac{\{l(l+1)\}^{\frac{1}{2}}}{k} \tag{4.42}$$

and to draw an analogy between $u_l(r)$ and the classical effective potential energy $V(L, r)$ (4.10) and between $l(l+1)/r^2$ and the classical centrifugal potential energy $U_L(r)$ (4.11).

It can be shown that,[27] providing $ru(r) \to 0$ as $r \to \infty$ the solution of the wave equation (4.40) for each partial wave has the form

$$\psi_l(r) \simeq \sin(kr + \delta_l) \tag{4.43}$$

where δ_l is the relative phase of the incoming and outgoing parts of $\psi_l(r)$. If $u(r) = 0$, then the solution to (4.40) is[1]

$$\psi_l^{(0)}(r) \simeq \sin(kr - l\pi/2) \tag{4.44}$$

where $-l\pi/2$ is $\delta_l^{(0)}$, the phase term arising solely from the centrifugal

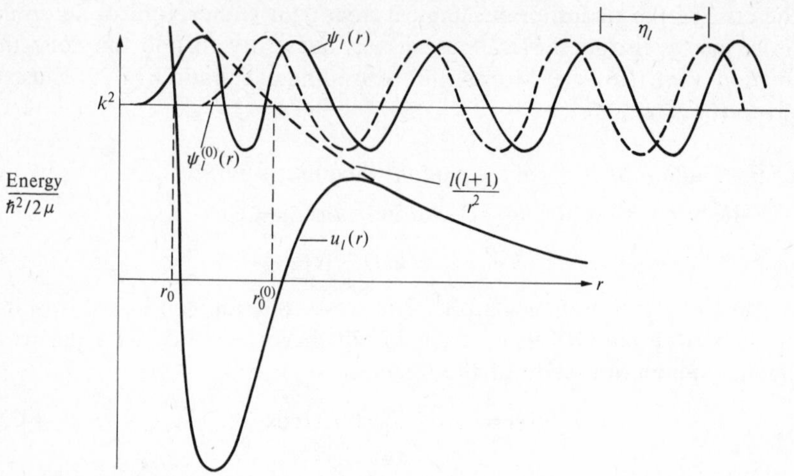

FIG. 4.16. Wave functions for a $12-6$ type potential at high l values. The phase shift η_l is indicated as the positive displacement necessary to superimpose at long range the wave function $\psi_l(r)$ and the wave function $\psi_l^{(0)}(r)$ due solely to the centrifugal term $l(l+1)/r^2$.

contribution $l(l+1)/r^2$. So the *phase shift* caused by the potential $u(r)$ is

$$\eta_l = \delta_l - \delta_l^{(0)} = \delta_l + l\pi/2. \tag{4.45}$$

The partial waves contributing to $\Psi(\mathbf{r})$ therefore have the asymptotic form

$$\psi_l(r) \simeq \sin(kr + \eta_l - l\pi/2). \tag{4.46}$$

The phase shift is simply the difference in phase between $\psi_l(r)$ and $\psi_l^{(0)}(r)$ for large r, as illustrated for a $12-6$ potential in Fig. 4.16. Note that η_l is determined by

(i) Any difference in wavelength of the oscillations of the two wave functions;
(ii) Any shift in the classical turning points, r_0 and $r_0^{(0)}$, which constitute the origins of the respective wave functions.

At intermediate and high values of l (corresponding to large impact parameters), the attractive part of the potential dominates. This has the effect of decreasing the wavelength of the scattered wave relative to the centrifugal function $\psi_l^{(0)}$, and causes r_0 to be at smaller separations than $r_0^{(0)}$ (the distance of closest approach in the absence of the potential i.e. in the presence of the centrifugal potential alone)—see Fig. 4.16. Consequently η_l is positive.

As l is increased, the amount of $U(r)$ sampled by the oscillating wave function is reduced, so that $\eta_l \to 0$ as $l \to \infty$. Conversely, at low values of

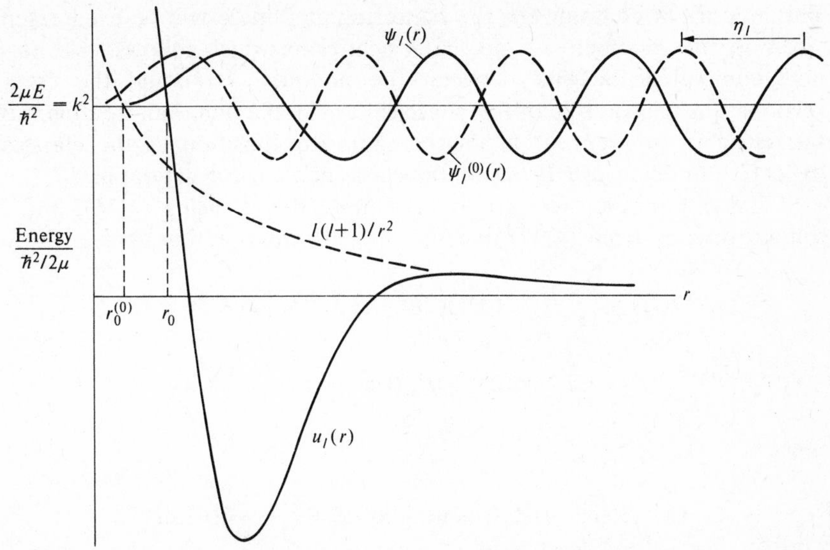

FIG. 4.17. Wave functions for a 12−6 type potential at low l values. Here the phase shift η_l required to superimpose $\psi_l(r)$ and $\psi_l^{(0)}(r)$ at long range is negative.

l, the repulsive part of $U(r)$ dominates and η_l becomes negative—see Fig. 4.17.

A typical dependence of η_l on l for a potential of the 12−6 type is shown in Fig. 4.18.

We may determine the coefficients C_l in (4.39) by writing down the normalization condition for $\Psi(\mathbf{r})$, $\Psi\Psi^* = 1$, and by making use of the orthogonality of Legendre polynomials, ensuring that the incoming terms of $\Psi(\mathbf{r})$, $\Psi^{(0)}(\mathbf{r})$, vanish when both wave functions are expressed in partial wave form.[1] Hence we find that

$$f(\chi) = \frac{1}{2ik} \sum_{l=0}^{\infty} (2l+1)(e^{2i\eta_l} - 1)P_l(\cos\chi). \tag{4.47}$$

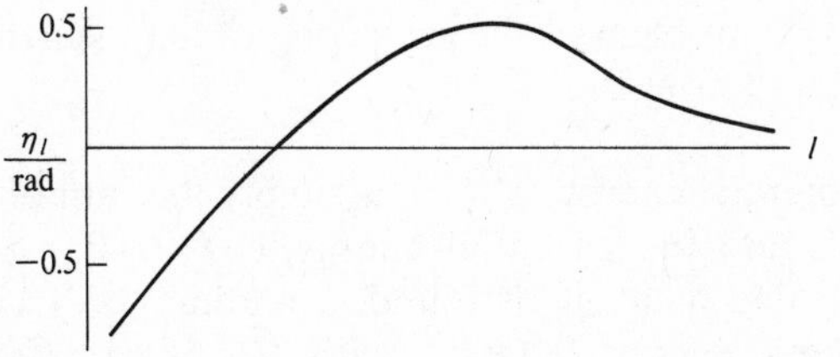

FIG. 4.18. Dependence of phase shift on angular momentum quantum number for a potential containing attractive and repulsive branches—schematic.

This important equation for the scattering amplitude will be used extensively in the treatment of molecular scattering which follows. We have now completed the link between the potential $U(r)$ and the cross-sections; the central role of the phase shift η_l in the quantum treatment of scattering is apparent. It holds the same key position as the classical deflection function $\chi(b, E)$ has in the classical approach set out in § 4.2.

The differential cross-section is simply $|f(\chi)|^2$, using (4.36), and it follows directly from (4.37) that the integral cross-section is given by

$$Q(E) = \frac{2\pi}{4k^2} \sum_{l,m} (2l+1)(2m+1)\,|(e^{2i\eta_l} - 1)(e^{2i\eta_m} - 1)| \times \int_{-1}^{1} P_l(\cos\chi) P_m(\cos\chi)\, d(\cos\chi).$$

Since[26]

$$\int_{-1}^{1} P_l(\cos\theta) P_m(\cos\theta)\, d(\cos\theta) \equiv \frac{2}{2l+1}\,\delta(l-m)$$

(where δ in this particular equation is the Kronecker delta function),

$$Q(E) = \frac{4\pi}{k^2} \sum_{l=0}^{\infty} (2l+1)\sin^2\eta_l. \tag{4.48}$$

If we compare this with (4.47) we find that $Q(E)$ may be identified with

$$\frac{4\pi}{k}\,\mathrm{Im}\{f(0)\}. \qquad (4.49)\dagger$$

This identity is termed the *optical theorem.*

Before examining how η_l can be calculated for a realistic potential, let us compare the quantum treatment of hard-sphere scattering with the classical results obtained in § 4.3.1(a).

4.5.4. *Hard-sphere scattering*

The effect of using the quantum mechanical expressions for the scattering cross-sections can most easily be demonstrated by applying them to the simple problem of the scattering of hard spheres, diameter d. Consider the situation in Fig. 4.19:

The effective potential energy, $V(l, r)$, is simply the sum of the centrifugal term, $l(l+1)/r^2$, and the potential energy $U(r)$ $(=0, r>d;\ =\infty, r\leq d)$. The classical turning point, r_0, is found, according to (4.12), by equating the relative kinetic energy E $(\equiv k^2)$ with $V(l, r)$. Hence if $l>kd$, then

† Im denotes the imaginary part.

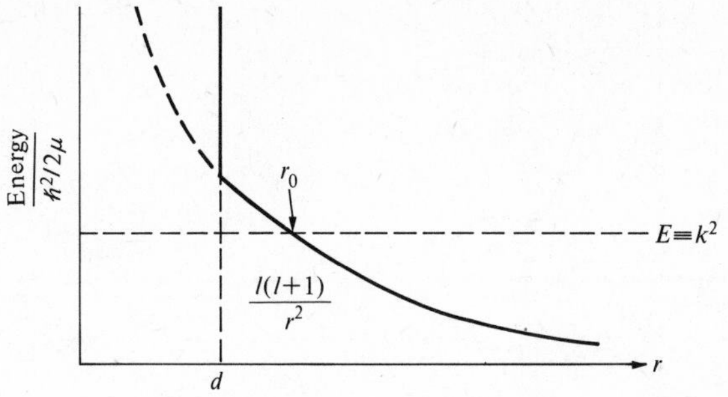

FIG. 4.19. Effective potential energy function for hard-sphere scattering.

$r_0 = k^{-1}\{l(l+1)\}^{\frac{1}{2}}$, there is essentially no effect of $U(r)$ on $\psi_l(r)$ and η_l is negligibly small. However, if $l < kd$, then $r_0 = d$ and solution of (4.40) yields

$$\eta_l = l\pi/2 - kd. \tag{4.50}$$

It follows that at low energies when only $\psi_0(r)$ contributes significantly to the cross-sections:

$$\psi_0(r) = \sin(kr - kd) \tag{4.51}$$

$$f(\chi) = \frac{1}{2ik}(e^{-2ikd} - 1) \tag{4.52}$$

$$= \frac{1}{2ik}(1 + 2ikd - 1), \qquad kd \ll 1$$

i.e.

$$f(\chi) = d.$$

So from (4.36) and (4.48)

$$\left.\begin{aligned} \sigma(\chi, E) &= d^2 \\ Q(E) &= \frac{4\pi}{k^2} \,.\, k^2 d^2 = 4\pi d^2 \end{aligned}\right\}. \tag{4.53}$$

These cross-sections are both four times the classical values (4.20) and (4.21); because $d \ll k^{-1} \sim \lambda$, diffraction effects occur to increase the effective cross-section.

For higher energies, many partial waves contribute to the cross-section. Since the phase shift varies over the whole range of values as l rises from 0 to kd, it is a reasonable approximation to replace $\sin^2\eta_l$ in (4.48) by the average value of 0.5—this is often called the *random phase*

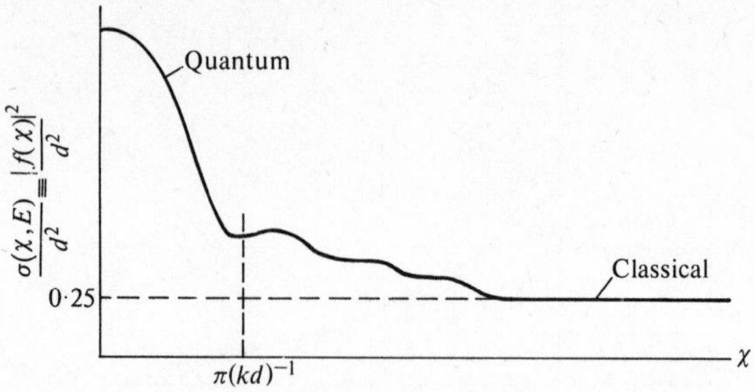

FIG. 4.20. Differential cross-section for hard-sphere scattering at high energies illustrating the additional non-classical *shadow scattering* in the region $0<\chi<\pi/kd$.

approximation

$$Q(E)=\frac{4\pi}{k^2}\sum_{l=0}^{kd}\tfrac{1}{2}(2l+1)$$
$$=2\pi d^2. \tag{4.54}$$

Thus the total cross-section is just *twice* the classical value. The 'extra' contribution of πd^2 also comes from diffraction effects which in this case are sometimes referred to as *shadow scattering.* It was seen in § 4.3.2 that there is a lower limit on χ, λ/b or $1/bk$ (see (4.22)), below which a classical treatment breaks down. Since to observe scattering at these low angles for a hard sphere, b must be close in value to d, the extra diffraction contribution to $Q(E)$ comes from $\chi \lesssim \pi(kd)^{-1}$, as illustrated in Fig. 4.20.

4.5.5. *The semi-classical phase shift*

We have seen how the phase shift η_l arises naturally in the quantum treatment of scattering. In order to emphasize the physical significance of this quantity, and to illustrate the links between a classical and quantum approach, it is instructive to consider a semi-classical interpretation of n_l.

Consider a particle of mass μ which is moving with velocity $\mathbf{c}$ in field-free space; associated with it will be a wave motion of wavelength λ given by (4.33):

$$\lambda/2\pi \equiv k^{-1} = \hbar/\mu c \equiv \hbar/(2\mu E)^{\frac{1}{2}}. \tag{4.55}$$

Now if the same particle is subjected to a force field, the wavelength will depend on the distance of the particle from the centre of force, r, and this will cause its wave function $\Psi(r)$ to be shifted in phase from the field-free case (see Fig. 4.21).

In the semi-classical approximation,[28] the phase shift is interpreted as the difference between the actual path and that in the absence of the

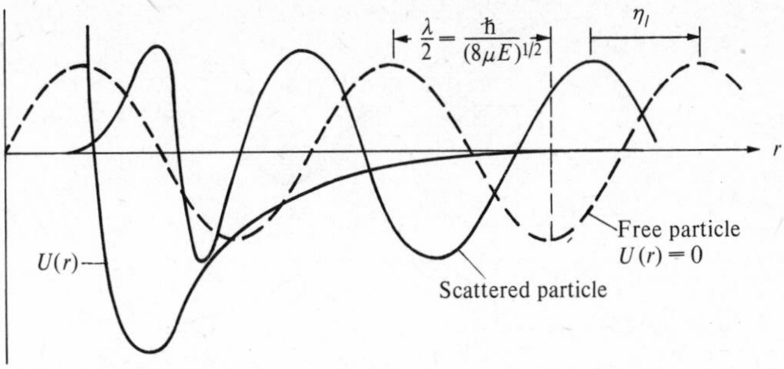

FIG. 4.21. A semi-classical interpretation of phase shift: radial wave function $\psi(r)$ for free and scattered particles.

potential. This difference will depend on the value of the impact parameter, b. At distances sufficiently far removed from the scattering centre ($r = 0$), the phase shift η becomes constant and can be defined by[29]

$$\eta(b) = \lim_{r\to\infty}\Big(\underbrace{\int_{r_0}^{r} \mathrm{d}r/\lambda}_{\text{Actual path}} - \underbrace{\int_{b}^{r} \mathrm{d}r/\lambda}_{U(r)=0}\Big). \tag{4.56}$$

Now (4.55) gives

$$\lambda = k(2\mu E)^{\frac{1}{2}}/\mu c. \tag{4.57}$$

The radial velocity c is given by (4.14) as

$$c = \pm\left\{\frac{2}{\mu}\left(E - \frac{Eb^2}{r^2} - U(r)\right)\right\}^{\frac{1}{2}}$$

so that $\eta(b)$ becomes

$$\eta(b) = k\left(\int_{r_0}^{\infty}\left(1 - \frac{b^2}{r^2} - \frac{U(r)}{E}\right)^{\frac{1}{2}} \mathrm{d}r - \int_{b}^{\infty}\left(1 - \frac{b^2}{r^2}\right)^{\frac{1}{2}} \mathrm{d}r\right) \tag{4.58}$$

which is a semi-classical expression for the phase shift since it is based on both classical and quantum ideas.

Introducing the angular momentum quantum number l, which in the semi-classical approximation is related to the angular momentum by $L = (l + \frac{1}{2})\hbar \simeq l\hbar$, and using the classical value of the angular momentum $L = \mu bc$ (4.9) we see that

$$l + \tfrac{1}{2} \simeq l = \mu cb/\hbar \equiv bk \tag{4.59}$$

where we have used the definition of k, (4.33). This is the semi-classical equivalent of the b, l relation (4.42).

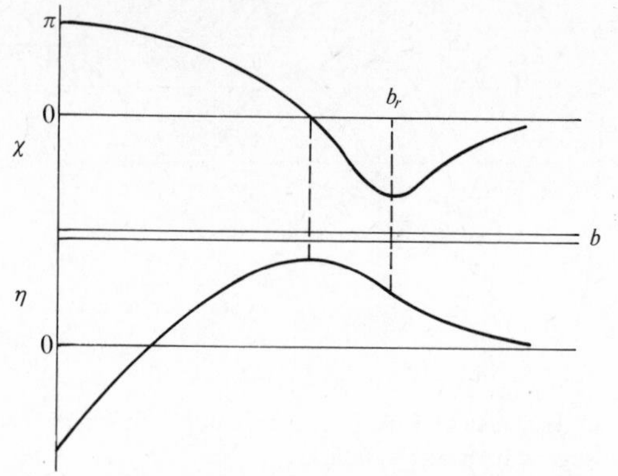

FIG. 4.22. A comparison of the classical deflection function $\chi(b)$ and the phase shift $\eta(b)$ for a potential having both attractive and repulsive branches. b_r is the rainbow impact parameter.

Differentiating (4.58) with respect to l to

$$\begin{aligned}\frac{\partial\eta(l)}{\partial l} &= -\int_{r_0}^{\infty}\frac{b\,\mathrm{d}r}{r^2(1-b^2/r^2-U(r)/E)^{\frac{1}{2}}}+\int_{b}^{\infty}\frac{b\,\mathrm{d}r}{r^2(1-b^2/r^2)^{\frac{1}{2}}}\\ &= \frac{\pi}{2}-\int_{r_0}^{\infty}\frac{b\,\mathrm{d}r}{r^2(1-b^2/r^2-U(r)/E)^{\frac{1}{2}}}\\ &= \chi/2 \qquad \text{(from (4.15))}. \end{aligned} \tag{4.60}$$

Hence there is a direct relationship between the semi-classical phase shift and the classical deflection function and we may regard η as a function of either b or l. $\eta(b)$ is compared with $\chi(b)$ for a typical potential having both attractive and repulsive branches in Fig. 4.22.

Two features worth noting are that the zero-crossing of χ(glory) corresponds to the maximum in $\eta(b)$ and the minimum in χ(rainbow) corresponds to an inflexion in the phase shift.

4.5.6. *Analytical evaluation of phase shifts*

The precise evaluation of the phase shift[30] for an arbitrary potential $U(r)$ involves numerical solution of (4.40) for $\psi_l(r)$ and the comparison of this result with the solution in the case $U(r)=0$. The latter is given by[1]

$$\psi_l^{(0)}(r)=krJ_l(kr) \tag{4.61}$$

where $J_l(kr)$ is a spherical Bessel function[26] of which (4.44) is the asymptotic form. For our purposes analytical solutions for $\psi_l(r)$ are much more useful and this section considers two widely used approximations.

The first method, called the *Born Approximation*,[31] is a perturbation treatment in which the partial wave equation (4.40) is solved for $\psi_l(r)$ to yield an expansion in powers of $u(r)$. The first-order solution has the form

$$\psi_l(r) \simeq \psi_l^{(0)}(r) + \int_0^\infty G_l(r, r')u(r')\psi_l^{(0)}(r')\,\mathrm{d}r' \tag{4.62}$$

where $\psi_l^{(0)}(r)$ is the unperturbed solution given by (4.61) and $G_l(r, r')$ is a Green's function† which also depends only on the unperturbed wave function. The solution is[1]

$$\psi_l(r) \simeq \mathrm{e}^{\mathrm{i}\eta_l} \sin(kr + \eta_l - l\pi/2) \tag{4.63}$$

where

$$\eta_l = -\frac{1}{k}\int_0^\infty u(r)(\psi_l^{(0)}(r))^2\,\mathrm{d}r. \tag{4.64}$$

Using (4.47) for $f(\chi)$ and recognizing that[26]

$$\sum_{l=0}^{\infty} (2l+1)\{J_l(kr)\}^2 P_l(\cos\chi) = \frac{\sin Kr}{Kr},$$

where $K = 2k\sin(\chi/2)$, gives

$$f(\chi) = -\frac{2\mu}{\hbar^2}\int_0^\infty \frac{\sin Kr}{Kr} U(r) r^2\,\mathrm{d}r. \tag{4.65}$$

The necessary condition for (4.64) to be valid is[1]

$$\eta_l \sim -\frac{1}{2k}\int_{l/k}^\infty u(r)\,\mathrm{d}r \ll 1 \tag{4.66}$$

which implies that, since $b \sim l/k$ (4.59), the Born approximation should hold at energies which satisfy

$$E\left(=\frac{\hbar^2 k^2}{2\mu}\right) \gg \frac{\hbar^2}{2\mu}\cdot\frac{1}{4}\cdot\frac{4\mu^2}{\hbar^4}\left(\int_b^\infty U(r)\,\mathrm{d}r\right)^2$$

i.e.

$$E \gg \frac{\mu}{2\hbar^2}\left(\int_b^\infty U(r)\,\mathrm{d}r\right)^2. \tag{4.67}$$

The higher b (or l), or the lower k, the more likely is condition (4.67) to be met.

A more generally useful approach is based on the *JWKB semi-classical approximation*,[28,29,33] an outline of which is given in Appendix A4.1. It has already been referred to in § 4.5.5, in connection with a

† The Green's function[32] is given by $\psi_a(r)\psi_b(r')/[\psi_a(\mathrm{d}\psi_b/\mathrm{d}r) - \psi_b(\mathrm{d}\psi_a/\mathrm{d}r)]$, where ψ_a and ψ_b are known unperturbed ($r \to \infty$) solutions of the homogeneous form of (4.40).

classical derivation of the phase shift. The semi-classical method assumes a particularly simple form for the wave function

$$\Psi(r) = e^{iS(r)\hbar} \tag{4.68}$$

where the function $S(r)$ is found to be, in zero order, simply the classical action integral

$$S(r) = \pm \int p(r)\,dr. \tag{4.69}$$

Higher approximations to $\Psi(r)$ involve an expansion of $S(r)$ in powers of $\hbar$ using (4.69) as the zero-order term. The first-order term, which introduces an amplitude proportional to $p(r)^{-\frac{1}{2}}$, is essential to obtain a physically meaningful solution for which the probability density, $\Psi^2(r)$, varies inversely with the velocity of the corresponding classical particle.

When this method is applied to the solution of the partial wave equation (4.40), the wave function is found to have the form[29]

$$\psi_l(r) \simeq \alpha_l(r)^{-\frac{1}{2}} \sin\left(\int_{r_0}^{r} \alpha_l(r)\,dr + \frac{\pi}{4}\right) \tag{4.70}$$

where

$$\alpha_l^2(r) = k^2 - u(r) - (l+\tfrac{1}{2})^2/r^2. \tag{4.71}$$

The semi-classical replacement of $l(l+1)$ by $(l+\frac{1}{2})^2$ in (4.71) is due to Langer.[34]

When $u(r) \to 0$, the unperturbed solution becomes

$$\begin{aligned} \psi_l^{(0)}(r) &\sim k^{-\frac{1}{2}} \sin\left(\int_{(l+\frac{1}{2})/k}^{r} \left(k^2 - \frac{(l+\frac{1}{2})^2}{r^2}\right)^{\frac{1}{2}} dr + \frac{\pi}{4}\right) \\ &\simeq k^{-\frac{1}{2}} \sin(kr - l\pi/2) \quad \text{as} \quad r \to \infty \end{aligned} \tag{4.72}$$

which has the same form as the exact solution (4.44). Hence the phase shift is given by

$$\begin{aligned} \eta_l &= \lim_{r\to\infty} \{\text{Phase}\ \psi_l(r) - \text{Phase}\ \psi_l^{(0)}(r)\} \\ &= \lim_{r\to\infty} \left\{\int_{r_l}^{r} \alpha_l(r)\,dr + \frac{\pi}{2}(l+\tfrac{1}{2}) - kr\right\} \\ &= \int_{r_l}^{\infty} \alpha_l(r)\,dr - \int_{r_l^0}^{\infty} \alpha_l^0(r)\,dr \end{aligned} \tag{4.73}$$

with

$$[\alpha_l^0(r)]^2 = k^2 - (l+\tfrac{1}{2})^2/r^2. \tag{4.74}$$

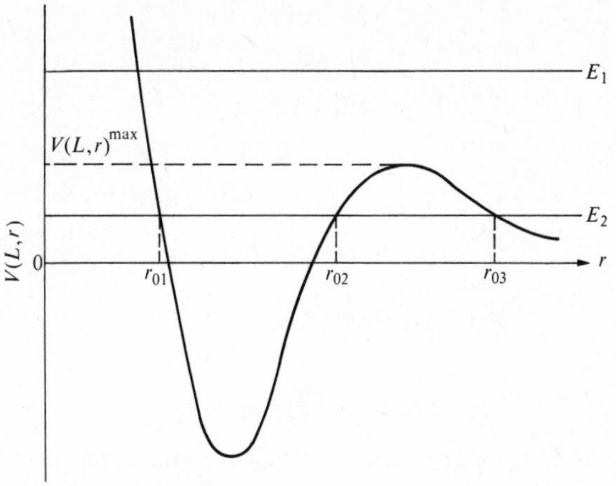

FIG. 4.23. An effective potential energy curve, $V(L, r) = u(r) + (l+\frac{1}{2})^2/r^2$, illustrating the occurrence of multiple classical turning points for energies less than $V(L, r)^{\max}$.

Here r_l, $r_l^{(0)}$ are the classical turning points in the presence and absence of the potential energy $u(r)$, respectively. The identity of (4.73) and (4.58) may be readily verified using the definition of $\alpha_l(r)$ (4.71), $\alpha_l^{(0)}(r)$ (4.74), and $b \simeq l/k$ (4.59).

The above treatment applies to the simple case of a single turning point—energy E_1 in Fig. 4.23. At energies sufficiently low that orbiting can occur (e.g. E_2 in Fig. 4.23) the particle can become temporarily trapped in a quasi-bound energy level between r_{01} and r_{02}. The problem now is that although (4.70) still applies to $\psi_l(r)$ between r_{01} and r_{02}, this must be extended outwards beyond r_{03} in order to evaluate η_l. This is done by expressing the wave function in the form[35]

$$\psi_l(r) = \alpha_l(r)^{-\frac{1}{2}}\left\{C_i \exp\left(\mathrm{i}\int_{r_{0i}}^{r} \alpha_l(r)\,\mathrm{d}r\right) + C_i' \exp\left(-\mathrm{i}\int_{r_{0i}}^{r} \alpha_l(r)\,\mathrm{d}r\right)\right\}$$

in each of the three regions:

$$\begin{aligned} i&=1 \quad (r_{01}<r<r_{02}); \\ i&=2 \quad (r_{02}<r<r_{03}); \\ i&=3 \quad (r_{03}<r), \end{aligned}$$

and using standard connection formula techniques to work outwards to C_3, C_3' from the known coefficients C_1, C_1'. The result is that[35]

$$\eta_l = \lim_{r\to\infty}\left\{\int_{r_{03}}^{r} \alpha_l(r)\,\mathrm{d}r + \frac{\pi}{2}(l+\tfrac{1}{2}) - kr + \eta_l^0 - \phi_l/2\right\}. \qquad (4.76)$$

This differs from the non-orbiting phase shift (4.73) by the two terms η_l^0

and $\phi_l/2$:

$$\eta_l^0 = \arctan\left\{\left(\frac{(e^{2\pi\varepsilon_l}+1)^{\frac{1}{2}}-1}{(e^{2\pi\varepsilon_l}+1)^{\frac{1}{2}}+1}\right)\tan(\beta_{l1}-\phi_l/2)\right\} \tag{4.77}$$

where

$$\varepsilon_l = \frac{\mu^{\frac{1}{2}}(E - V(L, r)^{\max})}{h(-\partial^2 V(L, r)/\partial r^2)^{\frac{1}{2}}}$$

$$\beta_{l1} = \int_{r_{01}}^{r_{02}} \alpha_l(r)\, dr$$

and

$$\phi_l = \arg\Gamma(i\varepsilon_l + \tfrac{1}{2}) - \varepsilon_l \ln|\varepsilon_l| + \varepsilon_l$$

where arg is the argument of the complex gamma function. In general the effects of these additional terms are small. Their most important consequence is at low scattering energies where ε_l is negative. In this case, because $e^{2\pi\varepsilon_l} \ll 1$, (4.77) simplifies to

$$\eta_l^0 = \arctan\{\tfrac{1}{2}e^{-2\pi|\varepsilon_l|}\tan(\beta_{l1}-\pi_l/2)\}. \tag{4.78}$$

When $(\beta_{l1}-\phi_l/2)$ passes through $\pi/2$, $3\pi/2$, etc., η_l^0 and hence η_l increase by π; this is known as a *resonant change* in the phase shift.[36] Physically it corresponds to the situation where the scattering energy E coincides with the energy E_n of a quasi-bound level in the range $0 < E < V(L, r)^{\max}$. Such resonance features will be reflected in the scattering cross-sections at low energies.

4.5.7. *The scattering cross-sections*

Having now obtained analytic expressions for the phase shifts, it is possible to evaluate the cross-sections. We will first discuss the differential cross-section $\sigma(\chi, E)$ for several regions of the scattering angle χ, within the JWKB approximation. These different regions arise because of the various approximations which may be used for $P_l(\cos\chi)$ in the evaluation of the scattering amplitude, (4.47).

(a) $\chi > l^{-1}$, $\neq \chi_r$. This range covers scattering at most angles, excluding very small values near $\chi = 0$ and scattering near the rainbow angle χ_r. Since, according to (4.36), $\sigma(\chi, E)$ is given by $|f(\chi)|^2$, we need to evaluate (4.47) for $f(\chi)$ subject to the above constraints on χ. Using the fact that[26]

$$P_l(\cos\chi) \simeq (\tfrac{1}{2}\pi l \sin\chi)^{-\frac{1}{2}}\cos\{(l+\tfrac{1}{2})\chi - \pi/4\}, \quad \sin\chi \gg l^{-1} \qquad (\text{i.e. } \chi > l^{-1}), \tag{4.79}$$

(4.47) can be simplified to

$$f(\chi) \simeq \frac{1}{k(2\pi\sin\chi)^{\frac{1}{2}}}\int_0^\infty (e^{i\delta_+} + e^{i\delta_-})l^{\frac{1}{2}}\, dl \tag{4.80}$$

where the summation has been replaced by an integral because the number of phase shifts involved is large. Here δ_+ and δ_- are the phases of the integrand,

$$\delta_{\pm} = 2\eta_l - \pi/2 \pm (l\chi - \pi/4). \tag{4.81}$$

(The positive signs apply to an attractive potential, or branch of a potential, the negative signs to a repulsive contribution.) Eqn (4.80) can be approximated further by expressing a particular phase, δ_- say, as

$$\delta_- \simeq \delta_-^{s} + \left(\frac{\partial^2 \delta_-}{\partial l^2}\right)_{l=l_s} \cdot \frac{(l-l_s)^2}{2} \tag{4.82}$$

where δ_-^{s} is the stationary value of δ_- (i.e. $(\partial\delta_-/\partial l)_{l=l_s} = 0$). In other words, for the purposes of evaluating (4.80), it is necessary only to know the phases accurately where they are slowly varying functions of l since where they vary rapidly with l, destructive interference occurs. This is called the *stationary phase approximation.*[1] For the stationary phase, (4.81) becomes

$$\frac{\partial \delta_{\pm}}{\partial l} = 2\frac{\partial \eta_l}{\partial l} \pm \chi = 0 \tag{4.83}$$

or the classical deflection function is given by

$$|\chi| = 2\left(\frac{\partial \eta_l}{\partial l}\right). \tag{4.84}$$

This very important relation between the classical and semi-classical quantities has already been obtained by classical arguments in (4.60). By differentiating (4.73) for η_l, using (4.71):

$$\frac{\partial \eta_l}{\partial l} = \int_{(l+\frac{1}{2})/k}^{\infty} \frac{(l+\frac{1}{2})\,\mathrm{d}r}{r^2(k^2-(l+\frac{1}{2})^2/r^2)^{\frac{1}{2}}} - \int_{r_l}^{\infty} \frac{(l+\frac{1}{2})\,\mathrm{d}r}{r^2(k^2-u(r)-(l+\frac{1}{2})^2/r^2)^{\frac{1}{2}}} \tag{4.85}$$

and using (4.33), (4.59), and the definition of $u(r)$ $(=U(r)2\mu/\hbar^2)$ we obtain

$$\chi(E, L) = \pi - 2b\int_{r_0}^{\infty} \frac{\mathrm{d}r}{r^2(1-b^2/r^2-U(r)/E)^{\frac{1}{2}}} \tag{4.86}$$

which agrees with the classical expression for χ, (4.15). Using

$$\left(\frac{\partial^2 \delta_-}{\partial l^2}\right)_{l=l_s} = 2\left(\frac{\partial^2 \eta_l}{\partial l^2}\right)_{l=l_s} = \left(\frac{\partial \chi}{\partial l}\right)_{l=l_s} \tag{4.87}$$

in (4.82) to obtain δ_-, $f(\chi)$ can be determined from (4.80) for a monotonic repulsive potential (where there will only be a contribution from δ_-) to be

$$f(\chi) = \left(\frac{l_s}{\pi k \sin \chi (\partial^2 \delta_- / \partial l^2)_{l=l_s}}\right)^{\frac{1}{2}} \exp\{i(\delta^s_- - \pi/4)\}. \tag{4.88}$$

Using (4.42) and (4.87) this can be rewritten as

$$f(\chi) = \left\{\frac{b}{\sin \chi \, |d\chi/db|}\right\}^{\frac{1}{2}} \exp\{i(\delta^s_- - \pi/4)\} \tag{4.89}$$

$$= \{\sigma(\chi, E)_{\text{class}}\}^{\frac{1}{2}} \exp\{i(\delta^s_- - \pi/4)\}. \tag{4.90}$$

(See (4.17).)

The situation when $U(r)$ has both attractive and repulsive branches can be understood by referring to Fig. 4.24.

For any given scattering angle, ($|\chi| < \chi_r$) there are in general three contributing branches to the scattering (a, b, and c). In terms of the phase shifts there are hence three stationary phase regions (corresponding to three phase shifts η_a, η_b, η_c at $l_s = l_a$, l_b, and l_c). As long as these regions do not overlap, i.e.

$$l_i - l_j \gg |\partial\chi/\partial l|^{-\frac{1}{2}}$$

(where i and j are any two of a, b, and c) then the total scattering amplitude is the sum of the contributions from the three branches, which

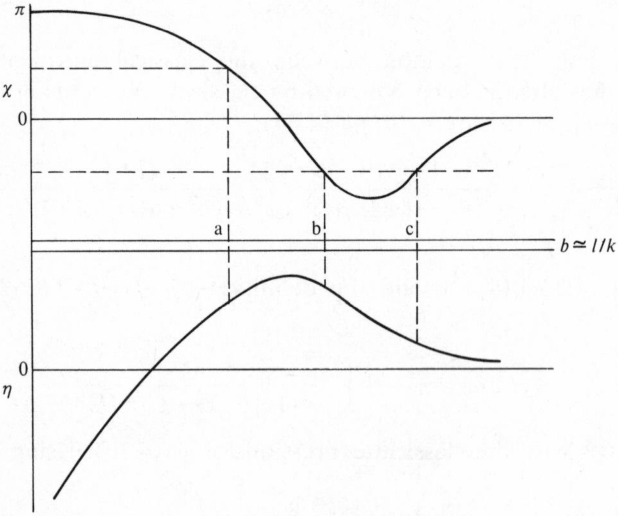

FIG. 4.24. Relation of phase shifts $\eta(l)$ to the classical deflection function $\chi(b)$. For $|\chi| < \chi_r$, the rainbow angle, three branches contribute to the scattering: the repulsive branch, a; the inner attractive branches, b; the outer attractive branch, c.

by analogy with (4.90) we may write

$$f(\chi)=\sum_{j=\mathrm{a,b,c}} \sigma_j^{\frac{1}{2}} \exp\{\mathrm{i}(\delta_j^{\mathrm{s}} \pm \pi/4)\} \tag{4.91}$$

where σ_j is the corresponding classical differential cross-section given by

$$\sigma_j=\frac{l_j}{\pi k^2 \sin\chi\, |\partial^2\delta_j/\partial l^2|_{l=l_j}} \equiv \frac{b_j}{\sin\chi\, |\mathrm{d}\chi/\mathrm{d}b|_{b=b_j}}. \tag{4.92}$$

δ_j^{s} in (4.91) is the stationary phase corresponding to the jth branch ($j=$ a, b, or c), given by (4.81), and the positive sign in the exponential term in this equation applies when $(\partial^2\eta_l/\partial l^2)_{l=l_j}$ is positive (i.e. $l>l_r$ or $b>b_r$, an outer attractive branch, c), and negative when $(\partial^2\eta_l/\partial l^2)_{l=l_j}<0$ (i.e. $l>l_r$, $b>b_r$, either an inner attractive branch, b, or a repulsive branch, a).

The semi-classical cross-section is then given by

$$\sigma(\chi, E)_{\mathrm{s/c}}=|f(\chi)|^2. \tag{4.93}$$

The important difference from the classical result (4.18) is that interference between the contributing branches occurs in (4.91), resulting in $\sigma(\chi, E)$ oscillating about the classical result. This can be seen most readily by examining the behaviour of (4.91) at low angles.

(b) *Small-angle behaviour:* $l^{-1}<\chi<\pi/4$. In this region the approximation (4.79) for $P_l(\cos\chi)$ still holds and in addition we put $\sin\chi\simeq\chi$. The two inner branches, a and b in Fig. 4.24, come closer together (and in the limit $\chi\to 0$ become identical, of course—see section (c) below). Let us write

$$f(\chi)=f_{\mathrm{c}}(\chi)+f_{0}(\chi) \tag{4.94}$$

where the contributions from the two branches a and b have been combined in the single function $f_0(\chi)$

$$f_0(\chi)=f_{\mathrm{a}}(\chi)+f_{\mathrm{b}}(\chi). \tag{4.95}$$

Using (4.90) the repulsive contribution of $f_{\mathrm{a}}(\chi)$ can be written as

$$f_{\mathrm{a}}(\chi)=\sigma^{\frac{1}{2}} \exp\{\mathrm{i}(\delta_{\mathrm{a}}^{\mathrm{s}}-\pi/4)\} \tag{4.96}$$

where

$$\delta_{\mathrm{a}}^{\mathrm{s}}\equiv\delta_{-}^{\mathrm{s}}\equiv 2\eta_{l\mathrm{a}}-\pi/4-l_{\mathrm{a}}\chi$$

and σ_{a} is the classical cross-section given by (4.92). Likewise the inner attractive contribution $f_{\mathrm{b}}(\chi)$ is

$$f_{\mathrm{b}}(\chi)=\sigma_{\mathrm{b}}^{\frac{1}{2}} \exp\{\mathrm{i}(\delta_{\mathrm{b}}^{\mathrm{s}}-\pi/4)\} \tag{4.97}$$

where

$$\delta_b^s \equiv \delta_+^s \equiv 2\eta_{lb} - 3\pi/4 + l_b\chi. \tag{4.98}$$

l_a and l_b are the stationary phase values of l for the a and b branches respectively.

Since at small angles

$$l_a \simeq l_b \simeq l_0 \quad \text{and} \quad \eta_{la} \simeq \eta_{lb} \simeq \eta_{l0},$$

(4.95) becomes

$$f_0(\chi) \simeq \sigma_0^{\frac{1}{2}}(\exp\{i(2\eta_0 - \pi/4 - l_0\chi - \pi/4)\} + \exp\{i(2\eta_0 - 3\pi/4 + l_0\chi - \pi/4)\}) \tag{4.99}$$

$$\simeq \sigma_0^{\frac{1}{2}} \exp\{i(2\eta_0 - 3\pi/4)\}(\exp\{-i(l_0\chi - \pi/4)\} + \exp\{i(l_0\chi - \pi/4)\}) \tag{4.100}$$

$$\simeq 2\sigma_0^{\frac{1}{2}} \cos(l_0\chi - \pi/4)\exp\{i(\delta_0^s - \pi/4)\} \tag{4.101}$$

where

$$\sigma_0 \simeq \sigma_a \simeq \sigma_b$$

and

$$\delta_0^s = 2\eta_0 - \pi/2. \tag{4.102}$$

The outer attractive branch c has a scattering amplitude

$$f_c(\chi) = \sigma_c^{\frac{1}{2}} \exp\{i(\delta_c^s + \pi/4)\} \tag{4.103}$$

where σ_c is given by (4.92) and

$$\delta_c^s \equiv \delta_+^s = 2\eta_{lc} - 3\pi/4 + l_c\chi. \tag{4.104}$$

Substituting these expressions for $f_c(\chi)$ and $f_0(\chi)$ into (4.94) gives for the over-all scattering amplitude

$$f(\chi) \simeq \sigma_c^{\frac{1}{2}} \exp\{i(\delta_c^s + \pi/4)\} + 2\sigma_0^{\frac{1}{2}} \cos(l_0\chi - \pi/4) \exp\{i(\delta_0^s - \pi/4)\} \tag{4.105}$$

$$\simeq \sigma_c^{\frac{1}{2}} \exp\{i(\delta_c^s + \pi/4)\}\left\{1 + 2\left(\frac{\sigma_0}{\sigma_c}\right)^{\frac{1}{2}} \cos(l_0\chi - \pi/4)\exp\{i(\delta_c^s - \delta_c^s - \pi/2)\}\right\}. \tag{4.106}$$

The differential cross-section consequently becomes

$$\sigma(\chi, E) = |f(\chi)|^2$$

$$\simeq \sigma_c\{1 + 2G \cos(\delta_0^s - \delta_c^s - \pi/2) + G^2\} \tag{4.107}$$

$$\simeq \sigma_c\{1 + 2G \cos(\delta_0^s - \delta_c^s - \pi/2)\} \tag{4.108}$$

where

$$G = (\sigma_0/\sigma_c)^{\frac{1}{2}} \cos(l_0\chi - \pi/4),$$

which is very much less than unity here. The cosine term in (4.108) may be rewritten as

$$\cos(l_0\chi - \pi/4)\cos(\delta_0^s - \delta_c^s - \pi/2) = \cos(l_0\chi - \pi/4)\cos(2(\eta_0 - \eta_{lc}) - l_c\chi - \pi/4)$$
$$= \tfrac{1}{2}[\cos\{2(\eta_0 - \eta_{lc}) + \chi(l_0 - l_c) - \pi/2\} + \cos\{\chi(l_0 + l_c) - 2(\eta_0 - \eta_{lc})\}] \quad (4.109)$$

which has a periodicity $2\pi(l_0 \pm l_c)^{-1}$ at fixed energy.

The 'shape' of the scattering is now clear. Since σ_c is the classical differential cross-section for the long-range attractive branch—where $U(r)$ is typically C_6r^{-6} and the cross-section is given by (4.24)—$\sigma(\chi, E)$ will oscillate about this classical value with an angular spacing between successive maxima of

$$\Delta\chi = 2\pi(l_0 + l_c)^{-1}. \quad (4.110)$$

The nature of these oscillations is illustrated in Fig. 4.25, which shows the complex oscillatory behaviour observed[37] for Li–Hg.

Several features of this scattering can be explained by the arguments presented so far, notably the oscillations about the classical prediction from all but the very lowest angles up to the rainbow angle χ_r. Although we have restricted ourselves to angles below about 30° in order to simplify the expressions, there would be no difficulty in generalizing (4.94)–(4.109), keeping the a and b branches separate. The angular spacing then becomes

$$\Delta\chi = 2\pi(l_b + l_c)^{-1}. \quad (4.111)$$

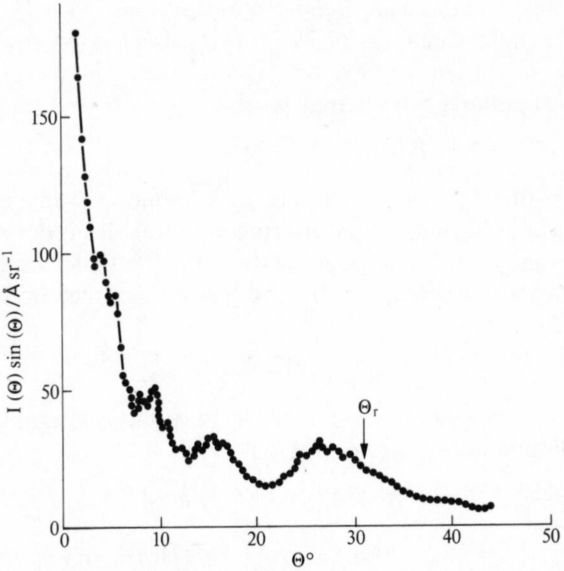

FIG. 4.25. The differential scattering cross-section of Li–Hg at an energy $E = 48{\cdot}9 \times 10^{-21}$ J showing oscillatory structure. (U. Buck, H. O. Hoppe, F. Huisken, and H. Pauly.[37])

This spacing should increase with χ, or equivalently Θ, since l_c decreases with increasing angle (see Fig. 4.24); this is borne out by Fig. 4.25. Also, because (4.59) gives

$$l \simeq bk$$

then the angular spacing should decrease with increasing wave number k for fixed energy E. In other words, using (4.33) ($k=(2\mu E)^{\frac{1}{2}}/\hbar$), 'quantum effects' should diminish with increasing mass μ or, in terms of the characteristic parameters of the interaction potential, with decreasing reduced quantum parameter Λ^* (see Appendix 2):

$$\Lambda^* = h/\sigma(2\mu\varepsilon)^{\frac{1}{2}} \equiv 2\pi E^{\frac{1}{2}}/k\sigma\varepsilon^{\frac{1}{2}}. \tag{4.112}$$

By contrast the amplitude of the oscillations, which (4.108) gives as the classical cross-section ratio

$$G_{\max} \equiv 2\left(\frac{\sigma_0}{\sigma_c}\right)^{\frac{1}{2}}. \tag{4.113}$$

does not depend on l and is therefore independent of Λ^* and, through (4.92), depends only on the relative kinetic energy for any fixed angle.

The detailed explanation of the interference effects in the vicinity of the glory (χ_g) and rainbow (χ_r) angles requires further refinement.

(c) *Low-angle scattering, monotonic potential:* $\chi < l^{-1} \simeq 0$. At very low angles, approximation (4.79) for $P_l(\cos\chi)$ is no longer valid and the above treatment cannot be used directly. However, by expanding $P_l(\cos\chi)$ about $\chi=0$ we may write[26]

$$P_l(\cos\chi) \simeq 1 - \tfrac{1}{4}l^2\chi^2. \tag{4.114}$$

Consider in the first instance scattering by a monotonic inverse power potential, $U(r)=C_n r^{-n}$. Since the scattering at low angles is dominated by large angular momentum/small phase shift terms (cf. (4.47) or Fig. 4.24) the Jeffreys–Born approximation[31] for the phase shifts may be used—see Appendix A4.2.

$$\eta_l = F(n)l^{1-n}. \tag{4.115}$$

The differential cross-section obtained as a result of evaluating the scattering amplitude according to (4.47) is[38]

$$f_n(\chi) = \{\sigma_n(\chi)\}^{\frac{1}{2}} \exp\{\mathrm{i}(\delta_n^{\mathrm{s}} - \pi/4)\} \tag{4.116}$$

$$\sigma_n(\chi) = |f_n(\chi)|^2 \simeq \left(\frac{k\sigma_c}{4\pi}\right)^2 A(n)\exp\left(\frac{-B(n)k^2\sigma_c\chi^2}{8\pi}\right) \tag{4.117}$$

where $A(n)$, $B(n)$, and δ_n^{s} are *all functions of n only*:

$$A(n) = 1 + \tan^2\{\pi/(n-1)\} \tag{4.118}$$

$$B(n) = \frac{1}{2\pi}\tan\{2\pi/(n-1)\}\frac{(\Gamma\{2/(n-1)\})^2}{\Gamma\{4/(n-1)\}} \tag{4.119}$$

$$\delta_n^s = \frac{\pi(n-3)}{2(n-1)} + \frac{\pi}{4}. \tag{4.120}$$

Hence in contrast to the classical result (4.24), which predicts a divergent cross-section at low angles, the semi-classical approach predicts a finite differential cross-section as $\chi \to 0$, and hence a finite total cross-section, in agreement with observation (see Fig. 4.26).

(d) *Glory scattering for non-monotonic potentials:* $\chi < l^{-1} \simeq 0$. Let us now consider low-angle scattering from a realistic, non-monotonic potential. As discussed earlier in § 4.3.1(b), classically the cross-section becomes infinite when χ passes through zero (or through $-n\pi$ if E is sufficiently low). Fig. 4.27 illustrates the classical trajectories which contribute to the divergence in the forward scattering or forward glory at $\chi = 0$. One of these corresponds to a large impact parameter b (or equivalently, l)—the c-branch (see Fig. 4.24); the other is the glory trajectory at small b, corresponding to the coalescence of the a and b branches of the scattering.

Since these trajectories have different path lengths, quantum mechanical interference can occur resulting in a finite cross-section and oscillations in the intensity as the relative path lengths change with energy. The mathematical description of the effect is as follows.

As $\chi \to 0$ and the a and b branches coalesce

$$f(\chi) = f_g(\chi) + f_c(\chi) \tag{4.121}$$

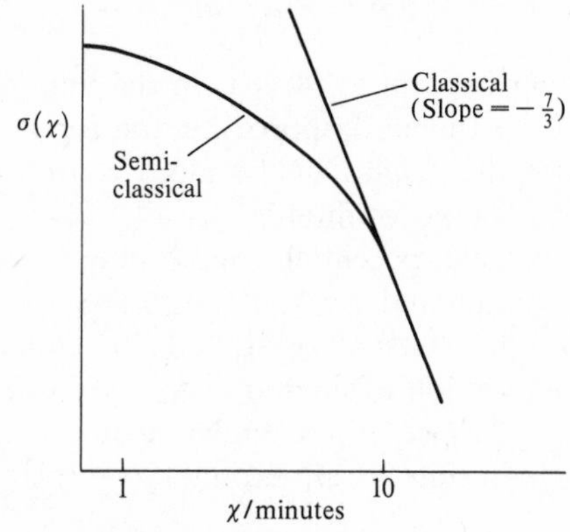

FIG. 4.26. Low-angle differential cross-sections for a potential of asymptotic form $U(r) = C_6 r^{-6}$ (schematic).

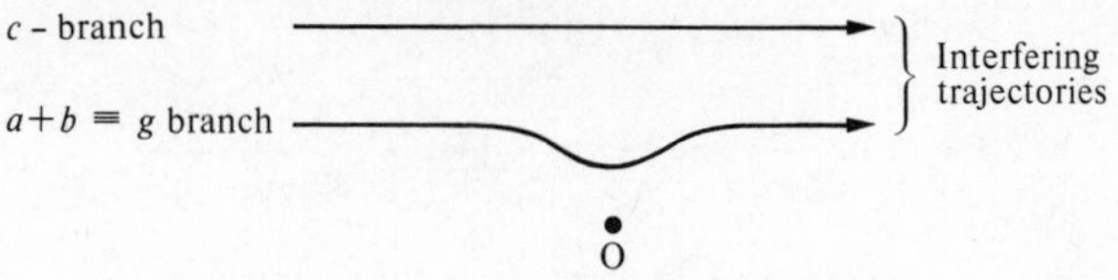

FIG. 4.27. Typical trajectories giving rise to the glory effect.

where $f_g(\chi)$ is the glory contribution to the scattering given by

$$f_g(\chi)=\lim_{a\to b}\{f_a(\chi)+f_b(\chi)\}. \tag{4.122}$$

It can be evaluated accurately by again applying the stationary phase approximation to (4.47), expanding the phase shift η_l about its maximum value, η_g, (see also Fig. 4.24)

$$\eta_l=\eta_g+\frac{1}{2}\left(\frac{d^2\eta}{dl^2}\right)_{l=l_g}(l-l_g)^2 \tag{4.123}$$

resulting in

$$f_g(\chi)\simeq\frac{e^{2i\eta_g}}{ik}l_gP_{l_g}(\cos\chi)\int_{-\infty}^{\infty}\exp\left\{\frac{i\,d^2\eta_g}{dl^2}(l-l_g)^2\right\}d(l-l_g) \tag{4.124}$$

$$\simeq\frac{P_{l_g}(\cos\chi)l_g}{k}\left(\frac{2\pi}{|d\chi/dl|}\right)^{\frac{1}{2}}\exp\{i(\delta_g^s-\pi/4)\} \tag{4.125}$$

where the phase is given by

$$\delta_g^s=2\eta_g-\pi/2. \tag{4.126}$$

Eqn (4.125) is similar in form to (4.101) in the limit of $0\to g$ and would be identical to it if the (invalid) approximation for $P_{l_g}(\cos\chi)$ (4.79) were used. This illustrates the origins of the glory contribution in the coalescence of the a and b branches though; see (4.122).

Even for a realistic potential, the c-branch contribution to the scattering will be dominated by the long-range part of the potential (where $U(r)\sim C_n r^{-n}$, $n=6$ usually). So (4.117) can be used for $f_c(\chi)$ and the differential cross-section evaluated as $\chi\to 0$ from (4.121). Recognizing that $|f_c(\chi)|\gg|f_g(\chi)|$, since at low angles scattering is dominated by the long-range c-branch, results in an expression similar to (4.109) for the low-angle region:

$$\sigma(\chi,E)_{\chi\to 0}=|f(\chi)|^2\simeq\sigma_c(\chi,E)\{1+2G_g\cos(\delta_g^s-\delta_c^s)\} \tag{4.127}$$

where

$$G_g = (\sigma_g/\sigma_c)^{\frac{1}{2}}$$

$$\sigma_g \simeq \frac{P_{l_g}(\cos\chi)l_g}{k}\left(\frac{2\pi}{|\mathrm{d}\chi/\mathrm{d}l|}\right)^{\frac{1}{2}} \qquad \text{from (4.125).}$$

As for the monotonic potential, because $(\mathrm{d}\chi/\mathrm{d}l)_{\chi\to 0} > 0$ $\sigma(\chi, E)$ clearly remains finite as $\chi \to 0$. In addition the interference between the c- and g-contributions results in oscillations about the classical value since the argument of the cosine term in (4.127) is energy-dependent through the maximum phase shift, η_g.

The *integral cross-section* may be readily evaluated from (4.121) by making use of the so-called optical theorem, (4.49):

$$Q(E) = \frac{4\pi}{k}\,\mathrm{Im}(f(0)) \tag{4.49}$$

Taking the imaginary part of (4.117) yields for a monotonic potential[39,40]

$$\mathrm{Im}\{f_c(0)\} = \xi(n)\left(\frac{C_n}{\hbar g}\right)^{2/(n-1)}\frac{k}{4\pi} \tag{4.128}$$

where

$$\xi(n) = \frac{\pi^2\{2F(n)\}^{2/(n-1)}}{\sin\{\pi/(n-1)\}\Gamma\{2/(n-1)\}} \tag{4.129}$$

and $F(n)$ is given in Appendix A4.2.

The glory contribution is, from (4.125),

$$\mathrm{Im}\{f_g(0)\} = \left(\frac{2\pi l_g^2}{k^2\,|\mathrm{d}\chi/\mathrm{d}l|}\right)^{\frac{1}{2}}\sin(2\eta_g - 3\pi/4) \tag{4.130}$$

since, from (4.114), $P_{l_g}(\cos\chi) \to 1$ as $\chi \to 0$. Consequently,

$$Q(E) = \xi(n)\left(\frac{C_n}{\hbar g}\right)^{2/(n-1)} + \frac{4\pi}{k}\,\sigma_g(0)^{\frac{1}{2}}\sin(2\eta_g - 3\pi/4). \tag{4.131}$$

Hence the integral cross-section also exhibits oscillatory behaviour since the maximum phase increases with a decrease in k, or equivalently relative velocity g. Provided the potential well is deep enough the phase η_g can pass successively through $N\pi/2$. These undulations superimposed on the monotonic c-branch term are termed *glory oscillations* and have been resolved experimentally for many systems.[41] Typical behaviour is shown in Fig. 4.28 which shows the velocity dependence of the total cross-section for scattering from a 12−6 potential.

The maxima occur when, from (4.130),

$$\eta_g = (N - \tfrac{3}{8})\pi \qquad N = 1, 2, 3, \ldots \tag{4.132}$$

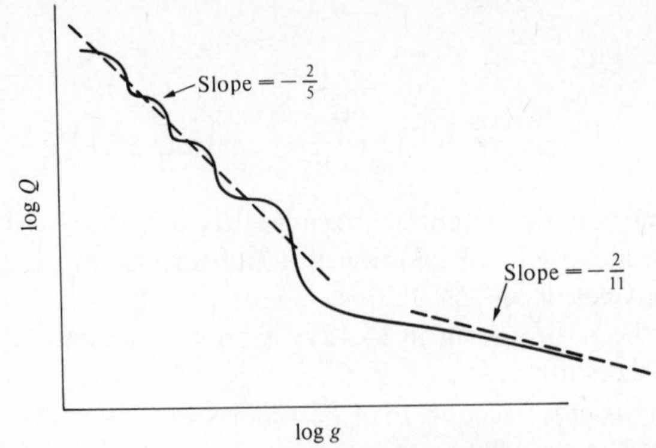

FIG. 4.28. Integral cross-section for 12−6 potential as a function of relative kinetic energy (schematic). The mean slopes at high and low energies are consistent with the first term of (4.131).

and oscillate about the low-velocity limiting slope corresponding to the r^{-6} attractive potential. The number of maxima that can be observed in such 'impact spectra' has been shown to be equal to the number of bound vibrational states in the potential well.[42] At high velocities, the repulsive r^{-12} branch of the potential dominates the scattering and, in agreement with (4.126), the slope of log Q vs. log g approaches $-2/11$. Here there are no interference effects since only one branch—the a branch—contributes to the scattering process.

Although integral cross-section measurements have been widely used to obtain information about intermolecular potential energy functions, several other contributions can occur in addition to those already considered. These include such effects as induced dipole–quadrupole and retardation effects,[40] and the possibility of orbiting resonances.[1] A treatment of these effects is beyond the level of this book—the reader is referred to the work of Child[1] and of Bernstein[40] for details—but their existence means that considerable care should be exercised in the interpretation of integral cross-section data.

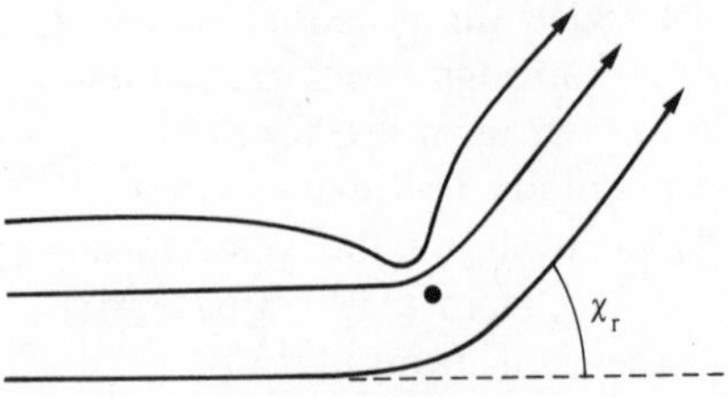

FIG. 4.29. Classical trajectories giving rise to the rainbow effect.

(e) *Rainbow scattering:* $\chi \simeq \chi_r$. The second cause of a singularity in the classical cross-section pointed out in § 4.3.1(b) is the *rainbow effect.*[43] The minimum in $\chi(b)$ at χ_r causes $\sigma(\chi)$ to become infinite in the classical limit because $d\chi/db$ becomes zero. Physically this can be thought to arise from trajectories having a wide range of impact parameters all converging on the same scattering angle, χ_r, thus leading to an infinitely large rise in intensity over a small range of angle. (See Fig. 4.29.)

It was shown earlier ((4.60), Fig. 4.22) that in the semi-classical treatment, χ_r corresponds to an inflexion in the η vs. l curve at l_r, i.e.

$$|\chi_r| = -\chi_r \equiv 2\left(\frac{d\eta}{dl}\right)_{l=l_r} < 0 \tag{4.133}$$

$$\left(\frac{d\chi}{dl}\right)_{l=l_r} = 2\left(\frac{d^2\eta}{dl^2}\right)_{l=l_r} = 0 \tag{4.134}$$

$$\left(\frac{d^2\chi}{dl^2}\right)_{l=l_r} = 2\left(\frac{d^3\eta}{dl^3}\right)_{l=l_r} > 0. \tag{4.135}$$

It is convenient to define $(d^2\chi/dl^2)_{l=l_r}$ as q. The rainbow phenomenon therefore corresponds to the coalescence of the outer attractive c-branch and the inner attractive b-branch (see Fig. 4.24) and, in a similar fashion to the glory case, interference between these contributions to the scattering results in the two characteristic quantum features of finite cross-sections accompanied by intensity oscillations.

In an analogous manner to the glory effect treatment, we write the scattering amplitude near χ_r as

$$f(\chi) = f_a(\chi) + f_r(\chi) \tag{4.136}$$

where the rainbow contribution $f_r(\chi)$ arises from the combination (and interference) of f_b and f_c. The repulsive branch contribution is given by (4.80), (4.81), and (4.92):

$$f_a(\chi) = (\sigma_a)^{\frac{1}{2}} \exp\{i(\delta_a^s - \pi/4)\} \tag{4.137}$$

where

$$\delta_a^s = 2\eta_{l_a} - l\chi - \pi/4. \tag{4.138}$$

$f_r(\chi)$ can be obtained by using the stationary phase approximation as before. Near l_r, χ can be expanded as

$$-\chi = -\chi_r + q(l - l_r)^2 \tag{4.139}$$

where $|\chi|$ is the observed scattering angle whose sign is not open to experimental determination. Integration of (4.133) using this expansion gives

$$\eta = \eta_r - \tfrac{1}{2}\chi_r(l - l_r) + \tfrac{1}{6}q(l - l_r)^3. \qquad (4.140)$$

Since f_r arises from attractive branches of the potential, we must use (4.81) with the phase δ_+, i.e.

$$\begin{aligned}\delta_+(l \sim l_r) &= 2\eta_l + l\chi - 3\pi/4 \\ &= 2\eta_r - \chi_r(l - l_r) + \tfrac{1}{3}q(l - l_r)^3 + l\chi - 3\pi/4. \end{aligned} \qquad (4.141)$$

Rewriting this as

$$\delta_+(l \sim l_r) = \delta_r^s + (\chi - \chi_r)(l - l_r) + \tfrac{1}{3}q(l - l_r)^3 \qquad (4.142)$$

where

$$\delta_r^s = 2\eta_r + l_r\chi - 3\pi/4 \qquad (4.143)$$

and substituting in (4.80) gives

$$f_r(\chi) \simeq \left(\frac{l_r}{k^2(2\pi \sin \chi)}\right)^{\frac{1}{2}} \exp(\mathrm{i}\delta_r^s) \int_0^\infty \exp(\mathrm{i}\{(\chi - \chi_r)(l - l_r) + \tfrac{1}{3}q(l - l_r)^3\})\, \mathrm{d}(l - l_r). \qquad (4.144)$$

This reduces to

$$\begin{aligned} f_r(\chi) &\simeq \frac{1}{k}\left(\frac{2\pi l_r}{\sin \chi}\right)^{\frac{1}{2}} \frac{\mathrm{Ai}(x)}{q^{\frac{1}{3}}} \exp(\mathrm{i}\delta_r^s) \qquad (4.145) \\ &\simeq \sigma_r(\chi)^{\frac{1}{2}} \exp(\mathrm{i}\delta_r^s) \qquad (4.146) \end{aligned}$$

where $\mathrm{Ai}(x)$ is the Airy function,[26] shown in Fig. 4.30 and x is $(\chi - \chi_r)q^{-\frac{1}{3}}$.

Consequently combining (4.135) and (4.144) we find that in the region of χ_r

$$f(\chi) = \sigma_a(\chi)^{\frac{1}{2}} \exp\{\mathrm{i}(\delta_a^s - \pi/4)\} + \sigma_r(\chi)^{\frac{1}{2}} \exp\{\mathrm{i}\delta_r^s\} \qquad (4.147)$$

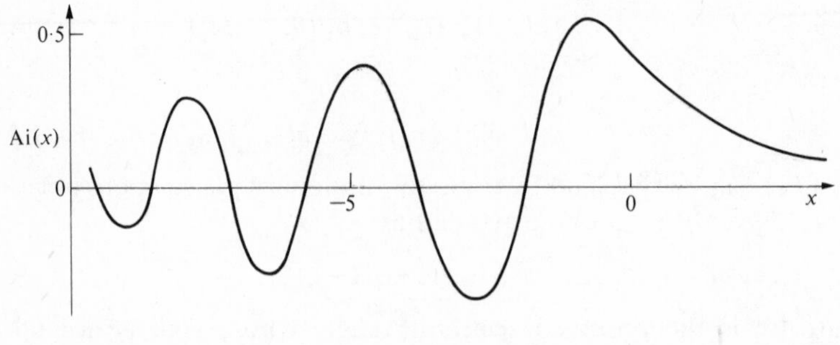

FIG. 4.30. The Airy function: $\mathrm{Ai}(x) = (1/2\pi)\int_{-\infty}^{\infty} \exp(\mathrm{i}xs + s^3/3)\,\mathrm{d}s$.

and so

$$\sigma(\chi)=|f(\chi)|^2=\sigma_a(\chi)+\sigma_r(\chi)+2(\sigma_a\sigma_r)^{\frac{1}{2}}\cos(\delta_r^s-\delta_a^s+\pi/4). \quad (4.148)$$

Eqn (4.148) therefore explains the complex structure of the differential cross-section in the region of χ_r observed in Fig. 4.25. The classical singularity at $\chi=\chi_r$ has been smoothed out and replaced at $\chi<\chi_r$ by a double oscillatory feature:

(i) The low-frequency broad oscillating envelope arises from the damped oscillations in Ai(x) for $x<0$; these are the *rainbow maxima.* The main peak occurs significantly below χ_r, where $x=-1{\cdot}019$, i.e.

$$\chi_{\max}=\chi_r-1{\cdot}019q^{\frac{1}{3}}. \quad (4.149)$$

Subsidiary *supernumerary rainbows* occur at lower angles ($x=-3{\cdot}248$, $-4{\cdot}820$, etc.).

(ii) Superimposed on these broad features are higher-frequency quantum oscillations arising from the interference cosine term in (4.148). These oscillate between $(\sigma_r^{\frac{1}{2}}\pm\sigma_a^{\frac{1}{2}})^2$ with a periodicity

$$\Delta\chi=2\pi/(l_r+l_a). \qquad \text{(c.f. (4.111))} \quad (4.150)$$

Both these features are superimposed on a rising background due to the attractive c-branch which becomes increasingly dominant as we pass to lower angles.

(f) *Large scattering:* $\chi>\chi_r$. High angle scattering, $\chi>\chi_r$, implies that the scattering is dominated by the repulsive branch, a. As χ increases beyond the rainbow angle, the influence of the c- and b-branches rapidly diminishes. Just beyond χ_r, the oscillations persist with decreasing amplitude as the angle increases, superimposed on the exponentially decaying background arising from the tail in Ai(x) for $x>1$ (see Fig. 4.30), until eventually the scattering is dominated by the single short-range repulsive a-branch. Here the cross-section decreases smoothly with angle as for a monotonic repulsive potential with

$$\sigma(\chi)=|f_a(\chi)|^2=\sigma_a(\chi) \quad (4.151)$$

where $\sigma_a(\chi)$ is given by (4.92) and becomes equivalent to the classical expression.

4.5.8. *Scattering of identical particles*

If the scattering species are indistinguishable, then the wave function describing their relative motion $\Psi(\mathbf{r})$ must either be symmetric (if the particles are bosons) or anti-symmetric (if they are fermions) with respect to interchange of the particles.[44] Consequently only even or odd partial waves will contribute to the scattering amplitude and in this case (4.47) becomes

$$f_{\text{id}}(\chi)=\frac{2}{2ik}\sum_{\substack{l\text{ even}\\ \text{or odd}}}(2l+1)(e^{2i\eta_l}-1)P_l(\cos\chi) \tag{4.152}$$

which can be rewritten as

$$f_{\text{id}}(\chi)=\frac{1}{2ik}\sum_{l=0}^{\infty}(2l+1)(e^{2i\eta_l}-1)[P_l(\cos\chi)\pm P_l\cos(\pi-\chi)] \tag{4.153}$$

where the last term takes the positive sign for the odd states (*ortho*) and the negative sign for the even summation (*para*). Hence

$$f_{\text{id}}(\chi)=f(\chi)\pm f(\pi-\chi) \tag{4.154}$$

and

$$\sigma(\chi)=|f_{\text{id}}(\chi)|^2=|f(\chi)|^2+|f(\pi-\chi)|^2\pm 2\,|f(\chi)|\,|f(\pi-\chi)|. \tag{4.155}$$

Substitution for $f(\chi)$ and $f(\pi-\chi)$ in (4.155) in the same manner as in previous sections leads to oscillations ('symmetry oscillations') in $\sigma(\chi)$ in the region of $\chi=\pi/2$ where $f(\chi)$ and $f(\pi-\chi)$ are of comparable magnitudes. Physically these arise because of interference between particles arriving at the detector, scattered through an angle χ, from the main beam and those arriving from the other beam having been scattered through the angle $\pi-\chi$ (in the centre-of-mass frame), from which they are indistinguishable.

Application of the optical theorem to (4.153) shows that similar oscillations are to be expected in the integral cross-section $Q(E)$, arising from interference between forward and back scattered trajectories from the two beams i.e. $\chi=\pi$. This effect augments the normal glory undulations and can be readily seen in Fig. (4.11) for He^4+He^4 scattering,[23] although in this particular case a full quantum treatment is necessary to account quantitatively for the data.

§§ 4.5.7 and 4.5.8 have explained the origins of the complex oscillatory structure of typical differential scattering cross-sections for elastic collision processes as illustrated in Fig. 4.25. The features described here have been observed and extensively studied experimentally for a wide range of molecular systems. We now go on to examine how information about molecular interactions can be extracted from these experimental data.

4.6. The determination of intermolecular forces from elastic scattering cross-sections

4.6.1. *Quality of data*

The treatment in § 4.5 described the structure of differential and integral cross-section measurements which would be observed in the ultimate device for complete angular resolution—the theoretical experiment! In practice, of course, such perfect resolution cannot be obtained. § 4.4 described how instruments are constructed to maximize intensities, angular and energetic resolution, the problem being that these requirements are often mutually exclusive. There is no universal solution to these problems, the instrument design and the resolution obtainable being very much dependent on the particular system being studied and the energy range of interest. Hence the plethora of molecular beam instruments described in the literature. However, careful design and improved techniques have led, in certain favourable cases, to instruments with sufficiently high angular resolution to observe all the fine details described in § 4.5. Examples of such measurements, mainly for inert gas and diatomic molecule systems, are given in § 4.4.

The precise information about intermolecular pair potential energy functions which can be extracted from molecular beam scattering cross-sections is highly dependent on the degree of resolution of the data. In general, the greater the angular or energy resolution, the more completely can $U(r)$ be obtained. Bearing in mind that data of varying quality are to be used, the analysis will be described under three headings, relevant to data where

(i) No oscillatory fine structure is resolved;
(ii) A limited number of oscillatory features are observed (where potential fitting procedures and parameter optimization must be used);
(iii) All oscillations are resolved (where inversion of cross-sections may be used to give $U(r)$ directly).

4.6.2. *Interpretation of data having no oscillatory fine structure*

This situation arises either because the data have low resolution or because the relative energy E, is so high that the scattering is dominated by the repulsive branch of the potential, i.e. $U(r)$ is effectively monotonic, and there is no fine structure to be observed.

(a) *Differential cross-sections.* In the absence of glory and rainbow oscillations, the classical equations for $\sigma(\chi, E)$ (4.17) and, as an example, (4.24) for an r^{-n} potential will apply at angles greater than some small angle χ_q below which classical mechanics breaks down (see § 4.3.2) and the small angle treatment of § 4.5.7(c) must be used. Since χ_q is typically less than 1°, this is scarcely a practical restriction. Hence at low reduced energies ($E^* = E/\varepsilon \simeq 1$) when the scattering is dominated by the long-range part of $U(r)$, where

$$U(r) \simeq C_n r^{-n}, \tag{4.156}$$

(4.24) shows that a plot of log σ vs. log χ will yield n from the slope and C_n from the intercept.[39,45] Similarly high-energy beams[46,47] may be used to determine the repulsive interactions between molecules. If it is assumed that the interaction is of the form (4.156), then plots of log σ vs. log χ at $E^* \gg 1$ will give the relevant coefficients, C_n and n (see Fig. 4.26).

(b) *Integral cross-sections.* A similar situation applies to measurements of $Q(E)$. Low-resolution data may be analysed using the first term of (4.131), i.e.

$$Q(E) \simeq \xi(n)\left(\frac{\mu C_n}{2\hbar E}\right)^{2/(n-1)} \tag{4.157}$$

where $\xi(n)$ is solely a function of n given by (4.129), and E is the relative kinetic energy of approach. As in the cases above, a plot of log Q vs. log E should be linear in the low-energy region, and again at very high energies, thus enabling the n, C_n parameters describing the long-range and short-range branches of the potential to be evaluated.[47] Even if quantum oscillations are resolved, the above analyses of $\sigma(\chi, E)$ and $Q(E)$ may still be applied to the mean background cross-sections.

4.6.3. *Interpretation of data for which some quantum fine structure is observed*

One approach which has been widely used in the literature, not only for interpreting the results of molecular beam experiments but also thermophysical properties, is to assume a functional form for $U(r)$ containing several variable parameters, including the well depth, ε, and a characteristic length, say r_m. $\sigma(\chi, E)$ and $Q(E)$ are then calculated using the equations of § (4.5) and compared with the experimental results.[48,49] The potential parameters are varied to optimize the agreement between theory and experiment. The dangers inherent in such an approach have already been outlined for the case of second virial coefficients (see Chapter 3). Although scattering cross-sections are in principle more sensitive discriminants between proposed potential functions than are virial or transport coefficients, it is still possible to fit the same experimental data satisfactorily using several quite different functional forms for $U(r)$. The most satisfactory use of this data-fitting procedure is therefore to eliminate unsuitable potential energy functions. The data may also be used in conjunction with a wide range of other data, e.g. thermophysical and spectroscopic measurements, which are each sensitive to different regions of the potential function, in multiparameter-fitting procedures to determine the intermolecular potential. Indeed the advent of high resolution $\sigma(\chi, E)$ measurements for argon,[24] and later the other inert gases,[50–53] and their use in a procedure of this type played a key role in the elucidation of $U(r)$ for this class of molecules in the early-1970s.

For a given potential function a number of procedures have been devised for determining reasonably accurate values of the parameters ε and r_m directly from the cross-sections, so adding useful constraints to the full fitting procedure or simply providing useful independently determined scaling parameters.

(a) *Differential cross-sections, rainbow structure.* The value of the rainbow angle, χ_r, can be determined experimentally as the position of the inflection in $\sigma(\chi, E)$ beyond the rainbow maximum (see § 4.5.7(e)). It has been shown[45] that for collision energies $E^* \gtrsim 1$, the value of $\chi_r(E^*)$ is nearly independent of the form of $U(r)$. Knowing χ_r for a given value of E, therefore, enables ε to be determined as E/E^*, to within about 10 per cent. χ_r has also been shown to correlate with $r_i(\mathrm{d}U/\mathrm{d}r)_{r_i}$, where r_i is the separation at the inflection point in the potential.[54]

The parameter r_m can be estimated for a given potential energy function from the position of the rainbow maximum or, more accurately, from the positions of the supernumerary rainbows.[1] The value of $(\mathrm{d}\chi/\mathrm{d}l^2)_{l=l_r}$, q, can be found

1. By comparing the experimental $\sigma(\chi, E)$ with the predictions of (4.148);
2. From the position of the rainbow maximum, χ_1, and the rainbow angle χ_r using (4.149)

$$q = \left(\frac{\chi_r - \chi_1}{1 \cdot 019}\right)^3;$$

3. From the positions of the supernumerary maxima, χ_2, χ_3, etc. and the known values of $x(=(\chi - \chi_r)q^{-\frac{1}{3}})$ for the maxima in the Airy Function, $\mathrm{Ai}(x)$.

The assumed potential function $U(r)$ can be used to calculate the reduced classical deflection function $\chi(b^*, E^*)$ using (4.15) corresponding to the experimental energy E since ε can be evaluated from χ_r. Consequently χ in the region of χ_r can be fitted to

$$\chi = \chi_r + q^*(b^* - b_r^*)^2 \tag{4.158}$$

and the parameter q^* evaluated. Since experimentally it is known that[62]

$$\chi = \chi_r + q(l - l_r)^2 \tag{4.159}$$

[see (4.139)], and $b^* = b/r_m$ can also be expressed as l/kr_m from (4.59), it follows that

$$r_m = k^{-1}(q^*/q)^{\frac{1}{2}}. \tag{4.160}$$

(b) *Integral cross-sections, glory undulations.* It was shown in § 4.5.7(d) that the extrema in the integral cross-section at low energy occur when

$$\eta_g(E_N) = \pi(N - \tfrac{3}{8}) \tag{4.161}$$

where $\eta_g(E_N)$ is the maximum phase shift at an energy E_N and $N = 1, 2, 3, \ldots$ are the indices of the extrema, labelling from high to low energy. Now it has been shown that in general for a two-parameter potential[1,42,55]

$$\eta_g = P\left(\frac{\mu g r_m}{\hbar}\right)\frac{1}{E^*} \tag{4.162}$$

where P is a constant for a given potential energy function. This equation applies at energies for which the Born approximation is valid, $E^* \gg 1$. Hence,

$$N - \frac{3}{8} = \frac{P}{\pi}\left(\frac{\mu g_N r_m}{\hbar}\right)\frac{1}{E^*} = \frac{P}{\pi}\left(\frac{\varepsilon r_m}{\hbar g_N}\right), \quad E^* \gg 1 \tag{4.163}$$

where g_N is the velocity at the Nth extremum ($\frac{1}{2}\mu g_N^2 = E_N$). By plotting $N - \frac{3}{8}$ against g_N^{-1}, therefore, the product εr_m may be found from the initial slope. The assignment can be checked by ensuring that the plot passes through the origin. The product εr_m may be separated by determining the long-range dispersion coefficient, C_6, from the energy dependence of the mean background total cross-section, as in § 4.6.2 above. Since for any assumed two-parameter potential of the type used in these analyses the relationship between C_6 and εr_m^6 will be known, r_m and ε can be separately determined. A further check on the value of r_m can be made by using (4.161) to evaluate η_g at each velocity g_N and to fit the results to the form

$$\eta_g = \alpha_1 g_N^{-1} - \alpha_2 g_N^{-2} + \ldots. \tag{4.164}$$

Bernstein and O'Brien[55] have shown that $\alpha_1 = P' C_6 r_m^{-5}$ where P' is another constant which depends only on the form of $U(r)$.

Whilst is appears here that molecular beams can give unequivocal information about C_n and n, care should be exercised that under the experimental conditions used the scattering was dominated by the potential in the truly asymptotic range.

4.6.4. *Interpretation of fully-resolved scattering data*

All the methods described so far involve some prior assumption about the functional form of the intermolecular potential. As already indicated, this can sometimes give misleading results and generally much more confirmatory evidence is required to demonstrate that a particular assumed form for $U(r)$ corresponds to the true potential.

Indeed, when most of the quantum oscillations are resolved, application of the above procedures does not make maximum use of the scattering data and cannot extract all the information about $U(r)$ contained in them. A much more satisfactory procedure is to *invert* the scattering cross-section data to give $U(r)$ directly, without making any assumptions about its shape in advance.

Such a procedure is, not unexpectedly, fraught with both theoretical and experimental difficulties. Many formal inversion schemes have been proposed, differing in the approximations made and in the range of scattering angles and energies of the data required for their implementation. Two basic types of scheme exist for differential cross-sections: those requiring σ as a function of χ at fixed E, and those based on σ as a function of E at fixed angular momentum. The interested reader is referred to the excellent reviews by Buck[49,56] for details of a variety of such methods. Here we shall restrict ourselves to a single example—the inversion of $\sigma(\chi)$, fixed E, as devised by Buck.[57] Not only is this the most common type of data measured experimentally, but this particular scheme has been implemented successfully for a number of systems, notably alkali metal–Hg.[58–60] It is therefore a practically viable scheme which can operate with data of a resolution obtainable using present-generation molecular beam apparatus.

The method consists of two stages:

1. The determination of the classical deflection function $\chi(b, E)$ from the experimental differential cross-section data, $\sigma(\chi, E)$;
2. Inversion of $\chi(b, E)$ to yield the intermolecular potential $U(r)$.

Stage 1 is difficult and only achievable by semi-empirical methods; formal procedures do exist for performing stage 2.

(a) *Determination of the classical deflection function from the cross-sections*

Monotonic potentials. In this case there is only a single contribution to $\chi(b, E)$, (4.92), which becomes equivalent to the classical expression

(4.17). This gives directly

$$b^2(\chi) = 2\int_\chi^\pi \sigma(\chi)\sin\chi\, d\chi \tag{4.165}$$

Integration over the measured (monotonic) cross-section therefore yields the monotonic function $b(\chi)$. This gives $\chi(b)$ directly, which can be inverted to $U(r)$ by the method of § (b) below.

Potentials containing an attractive minimum. The method devised by Buck[57] for potentials having a monotonic repulsive branch and a long-range attractive branch joined by a single attractive minimum is based on several premises and observations:

1. For potentials of this kind the classical deflection function and the phase shift curve have the general shape shown in Fig. (4.24), i.e. a repulsive a-branch, an inner attractive b-branch, and an outer attractive c-branch joined by a single extremum.
2. The semi-classical expressions derived in § 4.5.7 are valid descriptions of the scattering and in particular the phase shift curve $\eta(l, E)$ and $\chi(b, E)$ are related by (4.60):

$$|\chi| = 2\left|\left(\frac{d\eta(l)}{dl}\right)\right|.$$

3. As seen earlier, the oscillations in $\sigma(\chi, E)$ are determined by certain limited ranges of the deflection function, e.g. the glory oscillations by $\chi(b)$ near $b = b_0$ and the rainbow oscillations by $\chi(b)$ near $b = b_r$. Their amplitudes and the average cross-section are sensitive to the derivatives of $\chi(b)$.

Formal quantum mechanical inversions of $\sigma(\chi, E)$ to give $\eta(l)$ have been devised,[56] but none in a form which enables them to be implemented in practice. Based on the above observations Buck[57] has devised methods which are much simpler, without leading to significant loss of generality. In order to unfold the multivalued character of $b(\chi)$, the quantity which, through (4.84)–(4.88), appears in the semi-classical expressions for $\sigma(\chi, E)$, the deflection function is split up into a series of monotonic regions, each of which is readily inverted:

$$\chi(b) = \sum_i g_i(b) \tag{4.166}$$

$$b(\chi) = \sum_i g_i^{-1}(\chi). \tag{4.167}$$

The forms taken for $g_i(b)$ are those which arise naturally out of the semi-classical theory viz:

a parabola round the minimum:

$$\chi(b) = -\chi_r + q(b - b_r)^2 \tag{4.168}$$

a linear (cf. (4.123)) region near the zero point:

$$\chi(b) = a_1(b - b_0) \tag{4.169}$$

an inverse power law for the asymptotic region at high b:

$$\chi(b) = a_2 b^{-n}. \tag{4.170}$$

For angles $\chi > |\chi_r|$, $b(\chi)$ is a single-valued function and hence (4.165) may be used to evaluate $\chi(b)$ directly by integration from χ_r to π. The repulsive part of $\chi(b)$ can hence be constructed without assuming a functional form—the less general alternative would be to use an expression similar to (4.170).

To determine the rest of $\chi(b)$, (4.166) and (4.167) are used to calculate the differential cross-section and the functions g_i (i.e. the parameters in (4.168)–(4.170)) are determined by comparison with the experimental measurements. The specific data generally used are:

1. The positions of the rainbow oscillation extrema;
2. The amplitudes of the rainbow oscillations;
3. The separation of the rapid quantum oscillations;
4. The monotonic large-angle scattering in $\sigma(\chi, E)$ (see Figs. (4.15) and (4.25));
5. The position of the extrema of the glory oscillations in $Q(E)$;
6. The dispersion energy coefficient, C_6, from semi-empirical calculations (see Chapter 2).

The most reliable information is the positions of the interference oscillations since at high resolution they are not affected by any averaging process, unlike the amplitudes which might be severely modified by instrumental factors. The rainbow oscillations in $\sigma(\chi, E)$ can be used to determine q and χ_r (see § 4.5.7(e)); the number N of the Nth extremum is a measure of the difference of the interfering impact parameters:[56]

$$2\pi(N - \tfrac{3}{4}) = 2\Delta\eta_{ij} + k\chi(b_i - b_j). \tag{4.171}$$

By plotting $z(=\{0{\cdot}75(2\pi[N-\tfrac{3}{4}])\}^{\frac{2}{3}})$ vs. χ_N, q, and χ_r are evaluated from the initial linear region. The superimposed rapid quantum oscillations completely determine b_0:[56]

$$b_0 = \frac{\pi}{k(\chi_1 - \chi_2)}\{N_q(\chi_1) - N_q(\chi_2) - N_r(\chi_1) + N_r(\chi_2)\} \tag{4.172}$$

where N_r and N_q are the indices of the rainbow and rapid oscillations, respectively, each corresponding to two angles χ_1 and χ_2. From the ratio of the amplitude of the rainbow maximum to the monotonic large-angle scattering (see Figs. (4.15) and (4.25)) b_r can be found once q and χ_r have been determined.[57] The remainder of the coefficients are found by imposing certain continuity conditions and minimizing the remaining deviations between calculated and experimental quantities. The glory oscillations and C_6 are useful additional information for determing $\chi(b)$ at low angles since this region is not particularly well probed by $\sigma(\chi, E)$.

The net result of this procedure is a classical deflection curve $\chi(b)$ for the scattering energy E which may be expressed either in numerical or piece-wise analytical form. This information can now be inverted to give $U(r)$ by the procedure given in section (b) below. It is useful to determine $\chi(b)$ at several energies if possible; these may be inverted to give $U(r)$ separately and a comparison between the resulting potentials is a check on the accuracy and consistency of the procedure.

The determination of $\chi(b, E)$ from $\sigma(\chi, E)$ is summarized in Fig. (4.31), which demonstrates the physical relation of the various regions of the deflection function to the fine structure of the cross-section.

(b). *Inversion of the classical deflection function to give* $U(r)$. The classical deflection function $\chi(b, E)$ is related to $U(r)$ through (4.15)

$$\chi(b, E) = \pi - 2b\int_{r_0}^{\infty} \frac{\mathrm{d}r}{r^2\{1 - b^2/r^2 - U(r)/E\}^{\frac{1}{2}}}. \tag{4.173}$$

This may be inverted to give $U(r)$ by a method first described by Firsov.†[61] If the integration variable is changed to

$$\alpha(r) = r^2(1 - U(r)/E), \tag{4.174}$$

(4.173) becomes

$$\chi(b, E) = \pi - \int_{b^2}^{\infty} \frac{\mathrm{d}\ln r}{\mathrm{d}\alpha} \frac{2b\,\mathrm{d}\alpha}{(\alpha - b^2)^{\frac{1}{2}}}. \tag{4.175}$$

Now when $U(r) = 0$, $\alpha = r^2$ and $\chi = 0$, hence

$$\int_{b^2}^{\infty} \frac{\mathrm{d}\ln\alpha}{\mathrm{d}\alpha} \frac{b\,\mathrm{d}\alpha}{(\alpha - b^2)^{\frac{1}{2}}} = \pi \tag{4.176}$$

which enables (4.175) to be simplified to:

$$\chi(b, E) = \int_{b^2}^{\infty} \frac{\mathrm{d}\beta}{\mathrm{d}\alpha} \frac{b\,\mathrm{d}\alpha}{(\alpha - b^2)^{\frac{1}{2}}} \tag{4.177}$$

† In general JWKB relations of this form can be inverted by similar methods, e.g. the RKR inversion procedure for spectroscopic data; see Chapter 7.

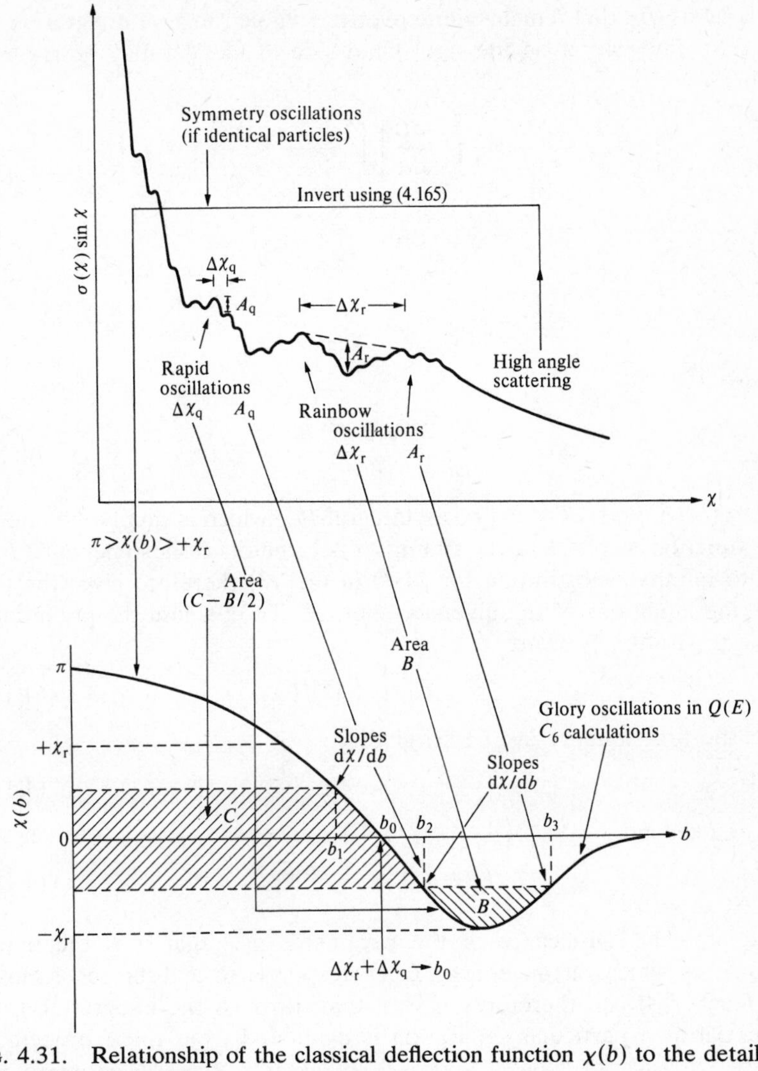

FIG. 4.31. Relationship of the classical deflection function $\chi(b)$ to the details of the differential scattering cross-section.

where

$$\beta(\alpha) = \ln(\alpha/r^2) = \ln(1 - U(r)/E). \tag{4.178}$$

In order to evaluate β at some particular value of $\alpha(=s^2)$ both sides of (4.177) must be multiplied by $(b^2 - s^2)^{-\frac{1}{2}}$ and the result integrated with respect to b from $b = s$ to $b = \infty$. This gives

$$\int_s^\infty \frac{\chi(b, E)\, \mathrm{d}b}{(b^2 - s^2)^{\frac{1}{2}}} = \frac{1}{2} \int_{s^2}^\infty \left[\int_{b^2}^\infty \frac{(\mathrm{d}\beta/\mathrm{d}\alpha)\, \mathrm{d}\alpha}{\{(\alpha - b^2)(b^2 - s^2)\}^{\frac{1}{2}}} \right] \mathrm{d}b^2. \tag{4.179}$$

As long as $(\mathrm{d}\beta/\mathrm{d}\alpha)$ remains finite over the whole range of integration, the order of integration on the right-hand side of (4.179) may be reversed, giving

$$\int_s^\infty \frac{\chi(b, E)\,\mathrm{d}b}{(b^2-s^2)^{\frac{1}{2}}} = \frac{1}{2}\int_{s^2}^\infty \frac{\mathrm{d}\beta}{\mathrm{d}\alpha}\left[\int_{s^2}^\alpha \frac{\mathrm{d}b^2}{\{(\alpha-b^2)(b^2-s^2)\}^{\frac{1}{2}}}\right]\mathrm{d}\alpha \qquad (4.180)$$

$$= \frac{\pi}{2}\int_{s^2}^0 \mathrm{d}\beta$$

$$= -\frac{\pi}{2}\beta(s^2).$$

Hence

$$\beta(s^2) = -\frac{2}{\pi}\int_s^\infty \frac{\chi(b, E)\,\mathrm{d}b}{(b^2-s^2)^{\frac{1}{2}}} \equiv -2I(s) \qquad (4.181)$$

where (4.181) serves to define an integral $I(s)$ which is simply the integral of a function of $\chi(b, E)$ over all impact parameters from some value $b = s$ out to infinity. Substitution for $\beta(s^2)$ in (4.178) therefore gives the final working equations of the inversion method. To evaluate the potential at some separation $\bar{r}$, where

$$s^2 = \bar{r}^2(1 - U(\bar{r})/E) \qquad (4.182)$$

then the first equality in (4.178) gives

$$\bar{r} = s\mathrm{e}^{I(s)} \qquad (4.183)$$

and the second

$$U(\bar{r}) = E(1 - \mathrm{e}^{-2I(s)}). \qquad (4.184)$$

(The physical significance of the parameter s is that it is the impact parameter which, at an energy E, corresponds to a distance of closest approach $\bar{r}$. It can therefore be seen that, through the integral $I(s)$, the potential at a particular separation is evaluated from those trajectories whose distance of closest approach is greater than or equal to this particular separation of interest.)

A point on the potential energy curve $(\bar{r}, U(\bar{r}))$ may therefore be determined from the experimental $\chi(b, E)$ values by evaluating $I(s)$ numerically for some value of b, s. Values of $U(\bar{r})$ are determined at increasing separations $\bar{r}$ by gradually increasing the value of the lower limit s from which $\chi(b, E)$ is integrated. Hence the c-branch of the deflection function alone serves to determine the very long-range part of the potential, as expected intuitively and in line with the more restricted methods of § (4.6.2), but the b- and, eventually, the a-branches are required in addition to determine $U(r)$ at increasingly smaller separations.

In order for this inversion procedure to be valid, the following conditions must be met:[56]

1. The interpolation of the phase shifts, which are defined at discrete values of l, must be unambiguous.

2. Since for s to remain non-complex, $E \geqslant U(r_0)$, the potential can only be evaluated up to separations r_0, the classical turning point at the energy E.

3. In order that the order of integration may be reversed in (4.179), $d\beta/d\alpha$ must remain finite over the whole domain of integration. Differentiating (4.178),

$$\frac{d\beta}{d\alpha} = \frac{1}{(U(r)/E-1)}\frac{1}{2r}\left\{\frac{dU(r)/dr}{E-U(r)-\frac{1}{2}r\,dU(r)/dr}\right\}. \qquad (4.185)$$

In order for this to remain finite:

$$\frac{dU}{dr} < \infty \qquad (4.186)$$

and

$$E - U(r) - \tfrac{1}{2}r\,dU(r)/dr \neq 0.$$

The final condition is satisfied so long as

$$E > E_c = U(r) + \frac{r}{2}\frac{dU(r)}{dr}. \qquad (4.187)$$

In other words, only data at energies greater than the critical energy for the onset of *orbiting collisions* may be inverted uniquely.†

4. Only very precise experimental data must be used. Any convolution effects due to finite energy resolution will distort the inverted results. As a guide, it has been found that Buck's procedure works adequately as long as the energy resolution is such that the position of the rainbow maximum is unaffected.

This method has been tested by Buck *et al.* on simulated data using a 12−6 potential and applied to experimental data for alkali metal–Hg systems.[58–60] The quality of the results is illustrated in Fig. 4.32. Where the quality of the experimental data permits their application, such inversion methods are the most unambiguous method of analysis of scattering data.

† See footnote on p. 168.

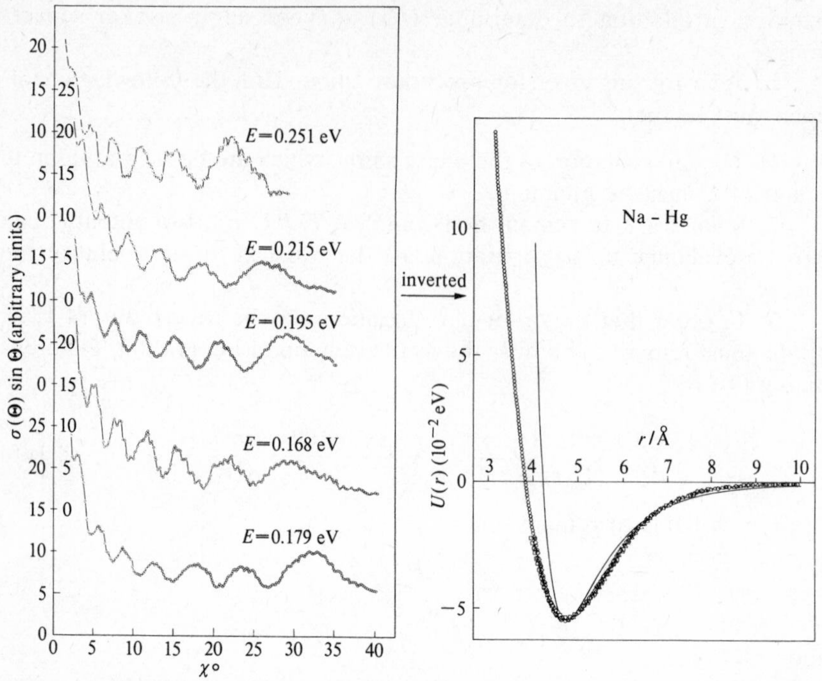

FIG. 4.32. Potential function for Na–Hg obtained by inversion of differential scattering cross-section. (U. Buck and H. Pauly.[60]) Points are obtained by inversion of $\sigma(\Theta)$ at $E = 0{\cdot}18$ eV (○); $E = 0{\cdot}19$ eV (◇); $E = 0{\cdot}20$ eV (+); $E = 0{\cdot}22$ eV (△); and $E = 0{\cdot}25$ eV (□); the solid line is a $12-6$ potential having the same r_m and ε values as the inverted potential.)

It is possible to express $I(s)$ in terms of $\eta(b)$ or $\eta(l)$, so enabling $U(r)$ to be evaluated from the phase shift curve which is obtained as an intermediate step in stage (i) of the inversion process:

$$I(s) = \frac{2}{\pi s k}\frac{\mathrm{d}}{\mathrm{d}s}\int_s^\infty \frac{\eta(b) b\, \mathrm{d}b}{(b^2 - s^2)^{\frac{1}{2}}} \tag{4.188}$$

where $b = (l + \frac{1}{2})/k$ (4.59). Partial integration gives the alternative form

$$I(s) = \frac{2}{\pi k}\int_s^\infty \frac{(\mathrm{d}\eta/\mathrm{d}b)\, \mathrm{d}b}{(b^2 - s^2)^{\frac{1}{2}}}. \tag{4.189}$$

Alternative inversion procedures which use the experimental data to determine a parametrized phase shift curve[62] (or S-matrix;[63] see § 4.7.3) via a minimization procedure have been devised and shown to be useful in certain cases. In particular the method of Klingbeil,[64] though not suited to cases where many angular momenta are involved, works in the absence of resolved rainbow structure, unlike the Buck method.

4.7. Scattering processes in polyatomic systems

When one or both of the molecules involved in a binary collision contains more than one atom, there are two important effects which make the observed scattering, and its interpretation, more complex than in the atomic case.

(i) The intermolecular potential energy is now a function not only of the separation between the molecules, r, but also of their relative orientation and the internal coordinates of the molecules R_1 and R_2. It is therefore dependent on $\mathbf{r}$, $\mathbf{R}_1$, and $\mathbf{R}_2$, $U(\mathbf{r}, \mathbf{R}_1, \mathbf{R}_2)$, and in contrast to the cases considered so far, is *anisotropic*.

(ii) The presence of internal degrees of freedom means that *inelastic processes* can occur during collisions at thermal energies. Coupling between the orbital angular momentum of the colliding pair and the angular momenta of the individual molecules, through the anisotropic part of the potential, leads to energy transfer and inelastic scattering.

From the viewpoint of intermolecular forces, the objective is to determine the potential energy surface $U(\mathbf{r}, \mathbf{R}_1, \mathbf{R}_2)$, or at least parts of it, from the experimental scattering cross-sections. Internal state selection of the molecules is usually employed so that both elastic and inelastic cross-sections can be determined for particular initial orientations of the colliding molecules with respect to their relative velocity vector. Information about the isotropic part of $U(\mathbf{r}, \mathbf{R}_1, \mathbf{R}_2)$ is contained principally within the elastic scattering component. Differences between the elastic cross-sections for different relative orientations of the molecules, and the inelastic scattering, are sensitive to the anisotropic, non-central part of the potential.

Extraction of this information from the scattering data is much more complex, and less fully developed, than for spherically symmetric systems. No inversion procedures yet exist for fully determining $U(\mathbf{r}, \mathbf{R}_1, \mathbf{R}_2)$ directly from the cross-sections and consequently trial-and-error fitting of an assumed potential to the experimental observations must be used. As in the case of spherical molecules, the postulated potential has to be based on a judicious combination of *ab initio* calculations of the full surface,[65] theoretical estimates of the asymptotic long-[66,67] and short-range[68] behaviour, experimental observation[17] and empiricism. The limitations of this type of approach have already been pointed out. In addition to the presence of many more disposable parameters for polyatomic systems, which in itself must tend to lessen the probability of obtaining a unique fit to the data, another factor renders the process less satisfactory than in the monatomic case. The essential characteristics of the scattering in polyatomic systems arise because of quantization and, although this can sometimes be incorporated into classical trajectory calculations in an *ad hoc* manner, the precise relation of $U(\mathbf{r}, \mathbf{R}_1, \mathbf{R}_2)$ to the scattering cross-sections requires a quantum approach. The exact quantum treatment requires the solution of an infinite set of close coupled wave equations even when the collision energy is below the threshold for rotation–vibration excitation. Even given high-powered computing facilities, these can only be reasonably solved by making a series of approximations. It is therefore necessary first to choose a calculation procedure in which the approximations made are sufficiently valid under the prevailing scattering conditions to give an essentially exact result for the relevant cross-sections. Only then is refinement of a proposed potential, by an iterative comparison of theory with experiment, a valid procedure.

Polyatomic molecule scattering is a rapidly expanding field of research, both experimentally[5] and theoretically.[1,6] It is not the intention of this section to provide a comprehensive treatment of either. Here we merely indicate the important kinds of experimental data, how they may be obtained, and the most useful calculation procedures which have been used, or are currently being developed. Emphasis will be placed on the relation between the intermolecular potential and the scattering cross-sections and the physical basis of the various approximation procedures. The reader is referred to several excellent reviews[1,2,5,6,17] for the detail of the individual techniques.

4.7.1. *Inelastic scattering cross-sections*

Because of the possibility of state-sensitive beam selection and detection, scattering cross-sections can in principle be measured for a large number of combinations of initial and final states. The definition of cross-sections is therefore somewhat more detailed than for elastic scattering, where there are simply two cross-sections, differential and integral (see § 4.5.2). For inelastic scattering there are several types of differential and integral cross-sections, the quantity measured in any particular experiment depending on the selectivity of the source/detection system.

(a) *Differential cross-sections.* As for elastic scattering, an inelastic differential cross-section is defined, by (4.16), as the ratio of the scattered flux per unit solid angle centred on the centre-of-mass scattering angle χ to the incident particle flux. A collision for which the initial internal quantum states of the molecules are characterized by α and the final states by β can be characterized by the initial and final values of the wave vector, $\mathbf{k}_\alpha$ and $\mathbf{k}_\beta$. The direction of these vectors is given by the centre-of-mass separation vector $\mathbf{r}$ and the corresponding values of the wave number are given by (see (4.33))

$$k_\alpha^2 = (2\mu/\hbar^2)(E - \varepsilon_\alpha) \tag{4.190}$$

$$k_\beta^2 = (2\mu/\hbar^2)(E - \varepsilon_\beta) \tag{4.191}$$

where E is the total (conserved) energy of the system and ε_α, ε_β are the initial and final internal (rotation, vibration, etc.) energies respectively. A particular differential cross-section can therefore be identified either by the scattering angle and energy (χ, E) as before or, equivalently, by $\mathbf{k}_\beta$.

To distinguish inelastic differential cross-sections from those for elastic scattering, $\sigma(\chi, E)$, we will use the symbol I. Designating particular internal rotational quantum numbers of the scattering molecules 1 and 2 by j, the corresponding azimuthal quantum numbers by m_j, and the vibrational quantum numbers by ν, we may define three levels of cross-section I_α^β.

1. At the highest level of discrimination, there is the *individual differential cross-section* which is a measure of the scattered intensity due to one particular set of (j, m_j, ν) transitions:

$${}_{\mathrm{i}}I_\alpha^\beta \equiv I(j_1' m_{j1}' \nu_1', j_2' m_{j2}' \nu_2' \leftarrow j_1 m_{j1} \nu_1, j_2 m_{j2} \nu_2 \,|\, \mathbf{k}_\beta) \tag{4.192}$$

where primed quantities correspond to the final state β.

2. Unless magnetic or electric state selection is being used, individual m_j states cannot be distinguished and the experimental scattered intensity gives the *degeneracy-averaged differential cross-section*

$${}_{\mathrm{d}}I_\alpha^\beta \equiv I(j_1' \nu_1' j_2' \nu_2' \leftarrow j_1 \nu_1 j_2 \nu_2 \,|\, \mathbf{k}_\beta) = \sum_{\substack{m_{j1}, m_{j2} \\ m_{j1}', m_{j2}'}} {}_{\mathrm{i}}I_\alpha^\beta \tag{4.193}$$

3. Many experiments measure only the scattered intensity summed over all final rotational states for state-selected primary beams to yield the *total differential cross-section*:

$${}_{\mathrm{t}}I_\alpha^\beta \equiv I(\nu_1' \nu_2' \leftarrow j_1 \nu_1 j_2 \nu_2 \,|\, \mathbf{k}_\beta) = \sum_{j_1', j_2'} {}_{\mathrm{d}}I_\alpha^\beta. \tag{4.194}$$

Here, the initial states may also be m_j selected

(b) *Integral cross-sections.* Integral cross-sections corresponding to each of these three differential cross-sections can be obtained by integrating over all angles according to the equivalent of (4.19), viz.

$$Q_\alpha^\beta = \int I_\alpha^\beta \, \mathrm{d}\Omega = \int_0^\pi I_\alpha^\beta 2\pi \sin\chi \, \mathrm{d}\chi = \int_{-1}^{1} I_\alpha^\beta \, \mathrm{d}\cos\chi. \tag{4.195}$$

The most commonly encountered integral cross-section is the *total integral cross-section*

$$ {}_{\mathrm{t}}Q^{\beta}_{\alpha}(\nu_1'\nu_2' \leftarrow j_1\nu_1 j_2\nu_2) = \int_{-1}^{1} {}_{\mathrm{t}}I^{\beta}_{\alpha}\,\mathrm{d}\cos\chi. \tag{4.196} $$

4.7.2. *Experimental methods*

The molecular beam methods used to measure these state-to-state scattering cross-sections are essentially those described in § 4.4, incorporating two major methods of state selection. The first technique resembles that used in molecular beam resonance spectroscopy[69] whereby inhomogeneous electric[70] or magnetic[71] fields are used to select particular (j, m_j) states for the primary beams or for detection. Such fields deflect a beam of polar molecules by an amount which depends on their effective dipole moment (i.e. (j, m_j) state) and their velocity. In combination with velocity selection and suitably positioned slits, specific (j, m_j) states only may be allowed to pass into the primary beam(s) initially or to the detector after scattering.

An early example of the use of this technique was the work of Bennewitz, Kramer, Paul, and Toennies[72] who studied the scattering of TlF and CsF by a range of other molecules using the single beam technique. By applying a steady electric field to the scattering chamber they were able to maintain the polarization of the beam as prepared by the inhomogeneous selection field. A field perpendicular to the relative velocity maintained the (1, 0) state whereas a parallel field maintained the (1, 1) polarization. This enabled the total integral cross-section ratio ${}_{\mathrm{t}}Q(1, 0)/{}_{\mathrm{t}}Q(1, 1)$ to be measured over a range of velocities. Such information can be used to study the short- and long-range anisotropy of the intermolecular potential. More recently, the method has been exploited by Reuss and co-workers.[17] Fig. 4.33 shows the apparatus used by Zandee and Reuss[73] to study scattering in H_2–inert gas systems.

An alternative method of detecting inelastic processes is to measure changes in the translational energy of molecules after collision. The energy transfer $\Delta\varepsilon$ is simply related to the initial and final relative velocities, g_{i} and g_{f}, by

$$ \frac{g_{\mathrm{i}}}{g_{\mathrm{f}}} = \left(\frac{E}{E \pm \Delta\varepsilon}\right)^{\frac{1}{2}} = \frac{t_{\mathrm{f}}}{t_{\mathrm{i}}} \tag{4.197} $$

where E is the initial relative kinetic energy, $\frac{1}{2}\mu g_{\mathrm{i}}^2$, and t_{i}, t_{f} are the corresponding flight times of molecules in the beam from the scattering volume to the detector. Hence either high-resolution velocity analysis (e.g. electrostatic focusing) or time-of-flight detection can be used. In

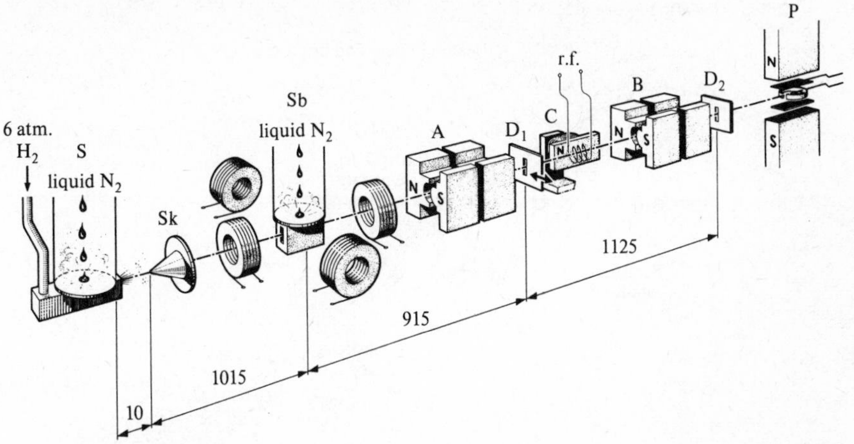

FIG. 4.33. Schematic diagram (view) of the molecular beam apparatus of Reuss and co-workers for the measurement of total collision cross-sections for oriented H_2 beam molecules scattered by various gases. The H_2 beam is formed by emission from a cooled nozzle source S through a skimmer Sk; it passes through a cooled scattering box Sb, which is surrounded by magnetic field coils to establish the orientation axis. The collimated beam is state-selected by Rabi deflection fields A and B, between which is the r.f. transition field (C). The intensity of the transmitted beam is measured by a Penning detector P. Distances are in millimetres. (J. Reuss.[17])

order to measure specific inelastic cross-sections, corresponding to small energy changes $\Delta\varepsilon$, it is necessary to have a high detection resolution over a range of scattering angles and a narrow initial spread of relative velocities. This has been achieved for ion–molecule collisions by Toennies and co-workers[74,75] using electrostatic focusing of the primary ion beam (energy spread approximately 0·4 per cent) which was electrically chopped at 10 kHz before intersecting a high-flux nozzle target beam (see Fig. 4.34(a)). The times of arrival of the ion pulses at an electron multiplier detector, some distance (0·5–1·7 m) from the scattering chamber, were recorded to give a time-of-flight spectrum which resolves specific inelastic processes. Examples of such spectra showing both vibrational transitions[74,75] and (higher resolution) rotational transitions[76] for $Li^+ + H_2$ are given in Fig. 4.34(b). It is now possible to resolve single quantum rotational transitions, and to observe oscillatory fine structure in the angular dependence of state-to-state differential cross-sections,[77] similar to that observed for elastic cross-sections in spherical systems (see Fig. 4.35). Such detail should prove to be a sensitive probe of the non-central part of the intermolecular potential energy function.

Studies of the angular dependence of vibrational cross-sections[78]

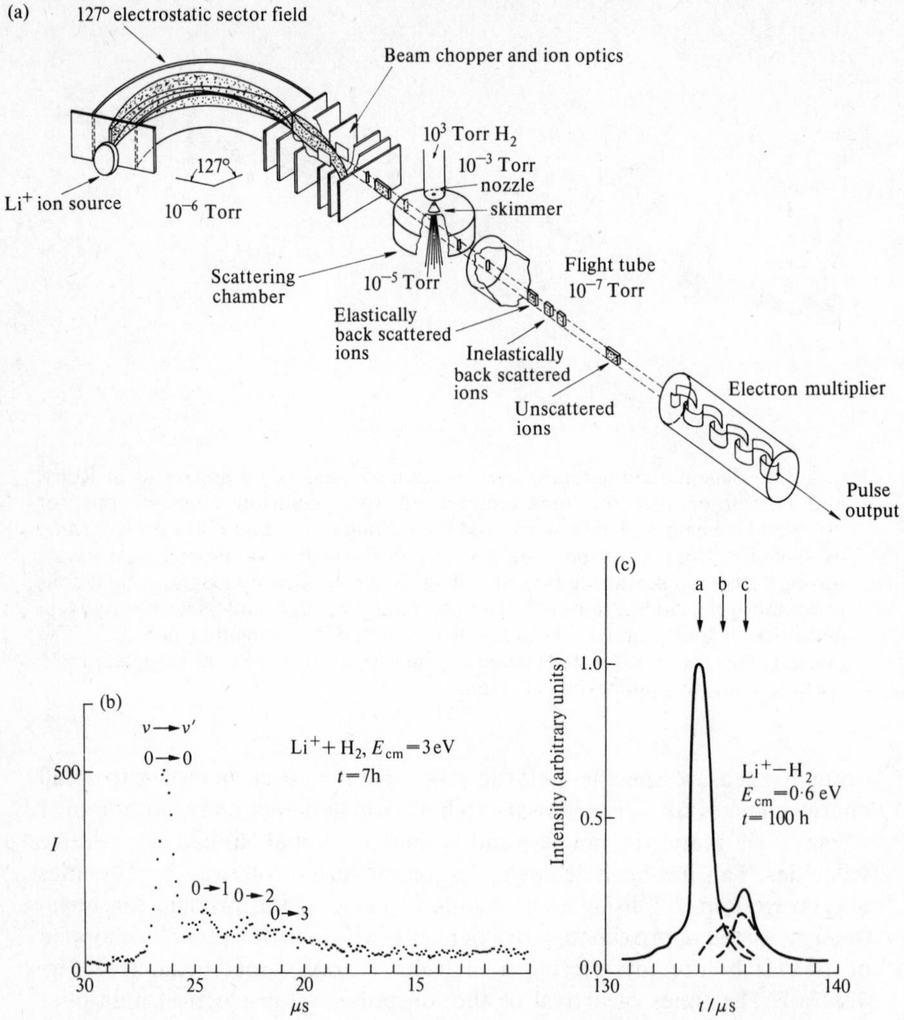

FIG. 4.34. Time-of-flight spectra for Li^+ scattered inelastically by H_2. (a) Schematic diagram of time-of-flight apparatus. The ion beam energy is selected by a 127° electrostatic sector field. The beam is then electrically chopped and passed into the scattering chamber, where it is crossed with an H_2 nozzle beam. The pulsed output is accumulated over a period of time t. (b) Vibrational inelastic scattering, $v = 0 \to v' = 0, 1, 2, 3$. (c) Rotational inelastic scattering: $j = 0 \to 0$, $1 \to 1$, a; $j = 0 \to 2$, b; $j = 1 \to 3$, c. (Based on R. David, M. Faubel, P. Marchand, and J. P. Toennies[74] and H. F. Van den Bergh, M. Faubel, and J. P. Toennies.[76])

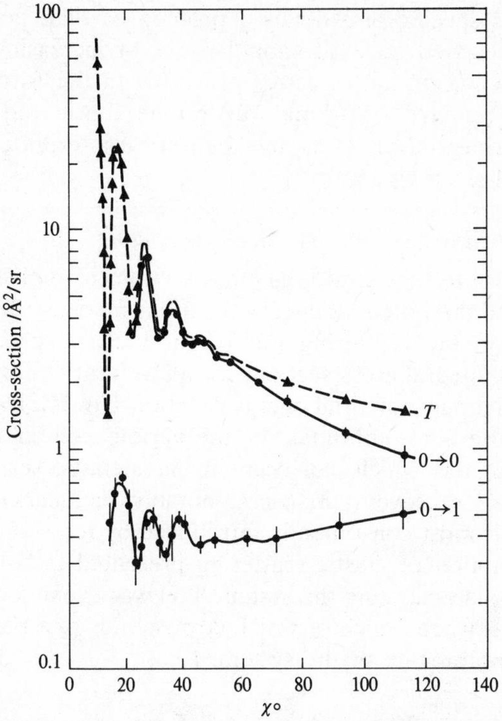

FIG. 4.35. Resolved differential scattering cross-sections for $j' \leftarrow j$, $0 \leftarrow 0$, and $1 \leftarrow 0$ transitions for HD scattered by D_2. T is the total differential cross-section. Energy, $E_{tr} = 454$ meV. (U. Buck, F. Huisken, and J. Schleusener.[77])

suggest that similar oscillatory behaviour will be resolved here. At higher energies, a host of electronically non-adiabatic collision processes have been studied ranging from simple electronic excitation and energy transfer to other modes, to collisional ionization[80] and atom transfer.[81] Measurements of cross-sections as functions of energy and scattering angle for these processes probe the relative shapes and positions of the various potential energy surfaces between which electronic energy is being transferred, especially in the region of potential curve crossing. Finally, in the field of reactive scattering,[11] velocity and state selection techniques are widely and ingeniously used to give detailed state-to-state differential reaction cross-sections as a function of the relative translational energy of the reagents. Such data shed light on the mechanisms of fundamental chemical processes and help to refine the complex potential energy surfaces suggested by *ab initio* and semi-empirical calculations for multi-atom reactive systems. Laser excitation of particular reactant internal states[82] enables specific and detailed probing of such potential energy surfaces to be carried out and should prove of increasing value in the study of both inelastic and reactive processes in the future. Electronic excitation and reactive scattering will not be considered further here; details of experimental methods and theoretical interpretation are given in several excellent reviews.[5,6,79,83,84]

4.7.3. The scattering (S-) matrix: elastic scattering

We now turn to how scattering cross-sections for polyatomic systems can be related to the potential energy functions involved. In particular we need to develop the scattering theory for inelastic processes so that differential and integral cross-sections for a given set of conditions can be calculated from a pair potential energy function $U(\mathbf{r}, \mathbf{R}_1, \mathbf{R}_2)$. The evaluation of the scattering amplitudes for the various combinations of input and output channels which can occur in an inelastic scattering process involves the use of several matrices, notably *the scattering matrix* or *S-matrix.* This is most conveniently introduced by returning briefly to the quantum description of elastic scattering presented in § 4.5. There, the wave function characterizing the system $\Psi(\mathbf{r})$ was expanded in (4.39) as a sum over partial waves, each of which corresponds to a particular orbital angular momentum state of the system, l

$$\Psi(\mathbf{r}) = r^{-1} \sum_{l=0}^{\infty} C_l \psi_l(r) P_l(\cos \chi). \qquad (4.39) \equiv (4.198)$$

Similarly the state of the system in the absence of the scattering potential may be expressed as

$$\Psi^{(0)}(\mathbf{r}) = r^{-1} \sum_{l=0}^{\infty} C_l^{(0)} \psi_l^{(0)}(r) P_l(\cos \chi). \qquad (4.199)$$

The asymptotic ($r \to \infty$) forms of the partial wave functions were shown in (4.44), (4.46) to be:

$$\psi_l^{(0)}(\mathbf{r}) = \sin(kr - l\pi/2) \equiv \frac{1}{2\mathrm{i}} (\mathrm{e}^{\mathrm{i}(kr - l\pi/2)} - \mathrm{e}^{-\mathrm{i}(kr - l\pi/2)}) \qquad (4.200)$$

$$\psi_l(\mathbf{r}) = \sin(kr + \eta_l - l\pi/2) \equiv \frac{1}{2\mathrm{i}} (\mathrm{e}^{\mathrm{i}(kr + \eta_l - l\pi/2)} - \mathrm{e}^{-\mathrm{i}(kr + \eta_l - l\pi/2)}). \qquad (4.201)$$

Now according to (4.32)

$$\Psi(\mathbf{r}) - \Psi^{(0)}(\mathbf{r}) = \frac{f(\chi)\mathrm{e}^{\mathrm{i}kr}}{r}, \qquad r \to \infty. \qquad (4.202)$$

Substituting (4.200) and (4.201) into (4.199) and (4.198) respectively, we see that

$$\Psi(\mathbf{r}) - \Psi^{(0)}(\mathbf{r}) = \frac{1}{2ir} \sum_{l=0}^{\infty} [C_l\{e^{i(kr+\eta_l - l\pi/2)} - e^{-i(kr+\eta_l - l\pi/2)}\}$$

$$- C_l^{(0)}\{e^{i(kr-l\pi/2)} - e^{-i(kr-l\pi/2)}\}]P_l(\cos\chi). \quad (4.203)$$

Eqn (4.202) shows that the coefficient of e^{-ikr} on the right-hand of (4.203) vanishes, consequently

$$C_l e^{-i(\eta_l - l\pi/2)} = C_l^{(0)} e^{-i(-l\pi/2)}$$

or

$$C_l = C_l^{(0)} e^{i\eta_l}. \quad (4.204)$$

Making use of this relation between the partial wave amplitudes in the presence and absence of the scattering potential, we may rewrite (4.203) as

$$\Psi(\mathbf{r}) \overset{r\to\infty}{=} \frac{1}{2ir} \sum_{l=0}^{\infty} C_l^{(0)} e^{i\eta_l}[e^{i(kr+\eta_l - l\pi/2)} - e^{-i(kr+\eta_l - l\pi/2)}]P_l(\cos\chi) \quad (4.205)$$

or

$$\Psi(\mathbf{r}) \overset{r\to\infty}{=} \frac{1}{2ir} \sum_{l=0}^{\infty} \{C_l^{(0)} e^{2i\eta_l} e^{i(kr-l\pi/2)} - C_l^{(0)} e^{-i(kr-l\pi/2)}\}P_l(\cos\chi). \quad (4.206)$$

We now define the diagonal elements of the scattering matrix $\mathbf{S}$, S_{ll}, as the ratios of the amplitudes of the outgoing (e^{ikr}) and incoming (e^{-ikr}) spherical waves in the lth channel in the limit of large r. From (4.206) we see that here

$$S_{ll} = \frac{C_l^{(0)} e^{2i\eta_l}}{C_l^{(0)}} = e^{2i\eta_l}. \quad (4.207)$$

In this case of *elastic scattering* with a spherically symmetric potential, the wave equations for the different angular momentum states, (4.40), are *uncoupled* and the conservation of orbital angular momentum means that the matrix $\mathbf{S}$ is diagonal i.e. only S_{ll} are non-zero. Physically the elements of the S-matrix can be thought of as transforming an incident spherical wave in one channel into an outgoing wave in the same (elastic case) or a different (inelastic scattering) channel. They contain all the information about the scattering due to $U(r)$ and are a much more flexible alternative to the phase shift notation. Using (4.207), the elastic scattering amplitude, $f(\chi)$, of (4.47) becomes

$$f(\chi) = \frac{1}{2ik} \sum_{l=0}^{\infty} (2l+1)(S_{ll} - 1)P_l(\cos\chi). \quad (4.208)$$

If inelastic scattering occurs, such that some molecules change from state α to state β, then (4.208) may still be used to calculate the *elastic* scattering amplitude, summing over diagonal elements only:

$$f_{\mathrm{el}}(\chi) = \frac{1}{2ik} \sum_{l=0}^{\infty} (2l+1)(S_{\alpha l,\alpha l} - 1) P_l(\cos \chi). \tag{4.209}$$

The elastic integral cross-section $Q(E)$ is most readily obtained in terms of the S-matrix elements by applying the optical theorem (4.49) to (4.208), leading to

$$Q(E) = \frac{2\pi}{k^2} \sum_{l=0}^{\infty} (2l+1)[1 - \mathrm{Re}(S_{ll})] \tag{4.210}$$

where Re is the real part.

It follows from (4.207) that

$$\mathbf{SS}^* = 1. \tag{4.211}$$

i.e. $\mathbf{S}$ is a unitary matrix. This is a perfectly general result as will be seen in the next section.

4.7.4. *Scattering matrices: the general case*

If a number of internal states $\alpha, \beta, \gamma \ldots$ are available to the molecules undergoing collision then the asymptotic form of the system wave function equivalent to (4.206) is

$$\Psi(\mathbf{r}) \overset{r\to\infty}{=} \psi_\alpha r^{-1} \sum_{l=0}^{\infty} C_l^{(0)} \{ e^{-i(k_\alpha r - l\pi/2)} - S_{\alpha l,\alpha l} e^{i(k_\alpha r - l\pi/2)} \} P_l(\cos \chi)$$

$$+ \psi_\beta r^{-1} \sum_{l,l'=0}^{\infty} [C_l^{(0)} S_{\alpha l,\beta l'} e^{i(k_\beta r - l\pi/2)}] P_{l'}(\cos \chi) + \ldots \tag{4.212}$$

where the initial state is α, wave function ψ_α, and the final states are β, γ, etc., wave functions $\psi_\beta \ldots$, and as before primed quantities are associated with the exit channel. Since the wave functions ψ_i are orthogonal, it follows that

$$\sum_{i=\alpha,\beta\ldots} \sum_{\bar{l}=0}^{\infty} S_{\alpha l,i\bar{l}} S^*_{\beta l',i\bar{l}} = 0, \qquad \alpha \neq \beta \quad \text{with} \quad l \neq l'. \tag{4.213}$$

Also $\Psi(\mathbf{r})$ must remain normalized to conserve the particle flux, so that

$$\sum_{i=\alpha,\beta\ldots} \sum_{\bar{l}=0}^{\infty} |S_{\alpha l,i\bar{l}}|^2 = 1. \tag{4.124}$$

Eqns (4.213) and (4.214) imply that $\mathbf{S}$ is unitary in all cases.

The physical significance of an off-diagonal element of the S-matrix $S_{\alpha l,\beta l'}$, is that the probability of any particular partial wave, l, starting in state α and emerging in state β is

$$P(\beta l' \leftarrow \alpha l) = \sum_{l'=0}^{\infty} |S_{\alpha l,\beta l'}|^2. \tag{4.215}$$

Two other matrices closely related to $\mathbf{S}$ are commonly used in scattering theory. The *transition matrix* $\mathbf{T}$ is simply $\mathbf{1}-\mathbf{S}$, where $\mathbf{1}$ is the unit matrix. Because $\mathbf{S}$ is complex, it is sometimes more convenient to solve a system of equations in real arithmetic using the $\mathbf{R}$ matrix, which is related to $\mathbf{S}$ by

$$\mathbf{S} = (\mathbf{1}+\mathrm{i}\mathbf{R})(\mathbf{1}-\mathrm{i}\mathbf{R})^{-1}. \tag{4.216}$$

The S-matrix may then be regenerated at the end of the calculation to calculate scattering amplitudes and cross-sections. It is to the relation of $\mathbf{S}$ to inelastic scattering cross-sections that we now turn.

4.7.5. Relation of the S-matrix to scattering cross-sections

Consider a general binary collision between two molecules X and Y, initially in internal states a and b respectively. Collectively this initial state will be labelled α, having associated with it a wave vector $\mathbf{k}_\alpha$,

$$\underbrace{X_a + Y_b}_{\text{State }\alpha} \rightarrow \underbrace{X_{a'} + Y_{b'}}_{\text{State }\beta}. \tag{4.217}$$

For a particular partial wave having an orbital angular momentum quantum number l, and a corresponding azimuthal quantum number m (not to be confused with j and m_j which are reserved for the internal motion of the molecules and are consequently included in a, b, etc.) we may express the associated wave function $\psi_{\alpha lm}$ as[2]

$$\psi_{\alpha lm} = r^{-1}\xi_{\alpha lm}(r)\Phi_\alpha Y_{lm}(\chi, \phi). \tag{4.218}$$

Here Y_{lm} is a spherical harmonic[26] which is closely related to the Legendre polynomials $P_l(\cos\chi)$,

$$Y_{lm}(\chi, \phi) = (-1)^m\left(\frac{2l+1}{4\pi}\right)^{\frac{1}{2}}\left(\frac{l-m}{l+m}\right)^{\frac{1}{2}} P_l(\cos\chi)\mathrm{e}^{im\phi} \tag{4.219}$$

where χ, ϕ are the polar angles relating $\mathbf{k}_\alpha$ to $\mathbf{k}_\beta$.

Φ_α is the total wave function for the isolated molecules (X_a+Y_b) and $\xi_{\alpha lm}(r)$ is the translational wave function for the system. To derive an expression for the scattering amplitude for the process $\alpha \rightarrow \beta$, we proceed as before by considering the asymptotic ($r \to \infty$) behaviour of the translational wave functions:

$$\xi_{\alpha lm}(r) \overset{r\to\infty}{=} k_\alpha^{-\frac{1}{2}}(C^{(0)}_{\alpha lm} e^{i(k_\alpha r - l\pi/2)} - C_{\alpha lm} e^{i(k_\alpha r - l\pi/2)}) \tag{4.220}$$

$$\xi_{\beta l'm'}(r) \overset{r\to\infty}{=} k_\beta^{-\frac{1}{2}}(C_{\beta l'm'} e^{i(k_\beta r - l\pi/2)}) \tag{4.221}$$

adopting a similar notation to § 4.7.3.

The amplitude $C^{(0)}_{\alpha lm}$ can be chosen by requiring that an incoming plane wave is associated solely with state α. This constraint leads to[2]

$$C^{(0)}_{\alpha lm} = -2i(i)^l \left\{\frac{4\pi(2l+1)}{k_\alpha}\right\}^{\frac{1}{2}}. \tag{4.222}$$

The other amplitudes $C_{\beta l'm'}$ follow directly from the S-matrix:

$$C_{\beta l'm'} = C^{(0)}_{\alpha lm} S_{\alpha lm, \beta l'm'}. \tag{4.223}$$

Consequently the total wave function of the system $\Psi(r)$ can be evaluated in the limit of $r \to \infty$ by summing (4.218) over all values of l and comparing the resulting expression with the boundary condition (4.32) to identify the scattering amplitude $f(\beta \leftarrow \alpha \mid \mathbf{k}_\beta)$ associated with the inelastic process $\alpha \to \beta$ as the coefficient of $e^{ik_\beta r}/r$. The result is

$$f(\beta \leftarrow \alpha \mid \mathbf{k}_\beta) = \left(\frac{4\pi}{k_\alpha k_\beta}\right)^{\frac{1}{2}} \frac{1}{2i} \sum_{l,l',m'} i^{l-l'}(2l+1)^{\frac{1}{2}} S_{\alpha lm, \beta l'm'} Y_{l'm'}(\chi, \phi). \tag{4.224}$$

This expression is formally similar to that for elastic scattering, (4.209). A flux calculation similar to that in § 4.5.2 shows that the individual differential cross-section is related to $f(\beta \leftarrow \alpha \mid \mathbf{k}_\beta)$ by

$$_iI^\beta_\alpha = \frac{k_\beta}{k_\alpha} \left| f(\beta \leftarrow \alpha \mid \mathbf{k}_\beta) \right|^2. \tag{4.225}$$

The total integral cross-section follows from application of the optical theorem:

$$\begin{aligned} {}_tQ^\beta_\alpha &= \frac{4\pi}{k_\beta} \mathrm{Im}[f(\beta \leftarrow \alpha \mid \mathbf{k}_\beta)] \\ &= \frac{\pi}{k_\alpha^2} \sum_{l'm'} \left| \sum_l (2l+1)^{\frac{1}{2}} i^l S_{\alpha l0, \beta l'm'} \right|^2. \end{aligned} \tag{4.226}$$

Other cross-sections may be readily calculated using the expressions of § 4.7.1.

4.7.6. Evaluation of the S-matrix: quantum approach

Having established relationships between the various inelastic cross-sections and the S-matrix, the problem of calculating the scattering from a given intermolecular potential $U(\mathbf{r}, \mathbf{R}_1, \mathbf{R}_2)$ and comparing with experiment has been reduced to the evaluation of the S-matrix for that potential function under the relevant scattering conditions. To do this it is necessary to solve the Schrödinger equation for the total wave function of the system. An exact solution can only be found with extreme tedium and developments in inelastic scattering theory have centred around devising approximation methods which are more direct, less time-consuming, but not appreciably less accurate than a full quantum treatment.

In order to present the alternative approximate methods, it is necessary to given an outline of the basic quantum approach.[85] To clarify the notation and amplify the equations as much as possible the particular example of the scattering between an atom and a diatomic molecule will be used throughout the remainder of this section. The inclusion of more internal degrees of freedom can usually be done by analogy and is presented in detail in specialist scattering texts.[1,85a,86] The coordinate system used is given in Fig. 4.36.

Let us designate the initial internal quantum state of the diatomic molecule BC by α, and the final state by β. The scattering process is therefore

$$A + BC_\alpha \rightarrow A + BC_\beta.$$

The Schrödinger equation is

$$(E - H)\Psi = 0. \qquad (4.227)$$

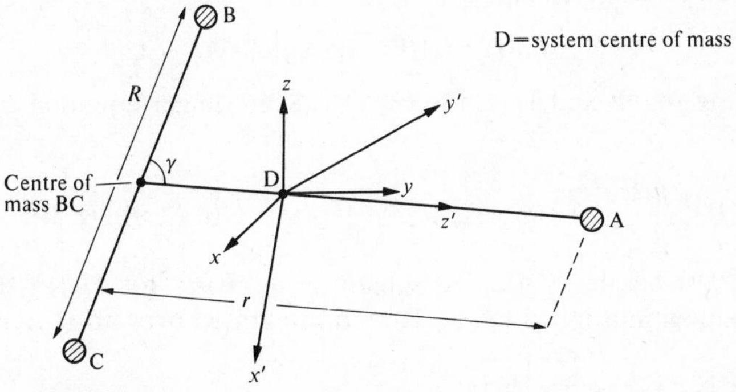

FIG. 4.36. Coordinate system for scattering of atom A with diatomic molecule BC. x, y, z are space-fixed (SF) axes; x', y', z' are body-fixed (BF) and are obtained by rotating x, y, z about D so that z lies along r and x is in the plane containing A and BC.

where the Hamiltonian in the centre-of-mass coordinate system is

$$H=\frac{-\hbar^2}{2\mu r}\frac{\partial^2}{\partial r^2}r+\frac{\mathbf{L}^2}{2\mu r^2}+H_{BC}+U(r,R,\gamma) \tag{4.228}$$

Here μ is the atom–molecule reduced mass, $\mathbf{L}$ the orbital angular momentum of the atom relative to the molecule, and H_{BC} is the Hamiltonian for the isolated molecule BC:

$$H_{BC}=\frac{-\hbar^2}{2\mu_{BC}R}\frac{\partial^2}{\partial R^2}R+\frac{\mathbf{j}^2}{2\mu_{BC}R^2}+U_{BC}(R) \tag{4.229}$$

where μ_{BC} is the molecular reduced mass and $\mathbf{j}$ is the molecular angular momentum operator. The unperturbed states of BC having energy E_β, and wave function ϕ_β are given by

$$(E_\beta-H_{BC})\phi_\beta=0. \tag{4.230}$$

Note that the intermolecular potential is a function of both internal and external coordinates and is orientation-dependent.

In order to obtain the wave function of the over-all system, $\Psi(\mathbf{r},\mathbf{R})$, it is expanded in terms of the complete set ϕ_β, including all states, both discrete and continuous

$$\Psi(\mathbf{r},\mathbf{R})=\sum_\beta \phi_\beta(\mathbf{R})F_\beta(\mathbf{r}). \tag{4.231}$$

$F_\beta(\mathbf{r})$ is the translational part of the wave function in the exit channel. For the elastic case it has the asymptotic form

$$F_\alpha(r)=e^{ik_\alpha r}+f_\alpha(\alpha\leftarrow\alpha\mid\mathbf{k}_\alpha)e^{ik_\alpha r}/r \tag{4.232}$$

and for the inelastic channels

$$F_\beta(\mathbf{r})=f_\beta(\beta\leftarrow\alpha\mid\mathbf{k}_\beta)e^{ik_\beta r}/r. \tag{4.233}$$

Substituting for Ψ and H in the over-all Schrödinger equation therefore gives

$$\sum_\beta \phi_\beta(R)\left(\frac{\hbar^2}{2\mu r}\frac{\partial^2}{\partial r^2}r+E-E_\beta-\frac{\mathbf{L}^2}{2\mu r^2}\right)F_\beta(\mathbf{r})=U(r,R,\gamma)\Psi(\mathbf{r},\mathbf{R}). \tag{4.234}$$

Here (4.230) has been used to substitute $\phi_\beta(\mathbf{R})E_\beta$ for $H_{BC}\phi_\beta(\mathbf{R})$. Eqn (4.234) is now multiplied by $\phi_\beta^*(\mathbf{R})$ and integrated over all $\mathbf{R}$ to give

$$\left\{\frac{\hbar^2}{2\mu}\frac{\partial^2}{\partial r^2}+E-E_\beta-\frac{\mathbf{L}^2}{2\mu r^2}\right\}F_\beta(r)$$
$$=\sum_\gamma F_\gamma(r)\int \phi_\beta^*(\mathbf{R})U(r,R,\gamma)\phi_\gamma(\mathbf{R})\,d\mathbf{R}. \tag{4.235}$$

Recognizing that the wave vector $\mathbf{k}_\beta$ has a length given by

$$k_\beta^2 = \frac{2\mu}{\hbar^2}(E - E_\beta) \tag{4.236}$$

and defining the matrix element $u_{\beta\gamma}(\mathbf{r})$ by

$$u_{\beta\gamma}(\mathbf{r}) = \frac{2\mu}{\hbar^2}\int \phi_\beta^*(\mathbf{R})\, u(r, R, \gamma)\phi_\gamma(\mathbf{R})\, d\mathbf{R} \tag{4.237}$$

for the special case of a spherical potential where $F_\beta(r)$ may simply be expanded in terms of partial waves $F_{\beta l}(r)$ in a similar manner to (4.39), we obtain for each partial wave the infinite set of coupled equations†

$$\left.\begin{array}{c} \left\{\dfrac{\partial^2}{\partial r^2} + k_0^2 - \dfrac{l(l+1)}{r^2}\right\} F_{0l}(\mathbf{r}) = \sum\limits_\gamma F_{\gamma l}(\mathbf{r}) u_{0\gamma}(\mathbf{r}) \\ \vdots \qquad\qquad\qquad \vdots \\ \left\{\dfrac{\partial^2}{\partial r^2} + k_\beta^2 - \dfrac{l(l+1)}{r^2}\right\} F_{\beta l}(\mathbf{r}) = \sum\limits_\gamma F_{\gamma l}(\mathbf{r}) u_{\beta\gamma}(\mathbf{r}). \\ \beta = 0 \to \infty \end{array}\right\} \tag{4.238}$$

The set of coupled differential equations (4.238) form the basis of all practicable methods for evaluating scattering cross-sections for anisotropic systems. Short of solving the quantum equations of motion numerically, which is only possible at very low energies where only a small number of rotational levels are accessible, they represent the most accurate approximation to the scattering problem and can in principle give any desired accuracy providing sufficient equations are retained. However, the large amounts of computer time involved in such procedures mean that the set must usually be severely truncated. This reduced set is referred to as the close-coupling equations. The procedure for carrying out such close-coupling calculations is outlined in the next section.

4.7.7. *Close-coupling calculations*

Although this procedure is an approximation in that the set (4.238) is truncated to some reasonable degree, the expansion has been shown to converge and by adding more terms until the results no longer change significantly, essentially exact results for scattering cross-sections can be obtained. Since many of the approximation procedures to be described in

† For the general case of an anisotropic potential, where coupling between the orbital ($\mathbf{L}$) and rotational ($\mathbf{j}$) angular momenta must be included, the expansion of $\Psi(\mathbf{r}, \mathbf{R})$ is more complex, involving eigenfunctions, $\mathcal{Y}^J_{\alpha l}$, of the total angular momentum $\mathbf{J}(=\mathbf{L}+\mathbf{j})$. However, the resulting equations[85] for the radial components of the partial waves are identical to (4.238), with $\mathcal{Y}^{J*}_{\alpha l}$, $\mathcal{Y}^J_{\gamma l}$, replacing ϕ_β^*, ϕ_γ in the definition of the matrix elements (4.237).

later sections involve solution of a similar set of coupled differential equations to (4.238), the general method for solution outlined here is not restricted to the close-coupling case.

Given that a certain number of states β must be retained in (4.238) due to the energy range of the scattering process being considered, the first task is to reduce the number of coupled equations that need to be solved by judicious choice of the wave functions used in expansion (4.231) for $\Psi(\mathbf{r}, \mathbf{R})$.[86] The wave functions of the separated molecules $\phi_\beta(\mathbf{R})$ are not necessarily the best choice, and consideration of the symmetry of the system can often lead to large computational savings. For instance, since the Hamiltonian is invariant under a rotation of the coordinate system, total angular momentum is conserved. Choosing eigenfunctions of the total angular momentum leads to a block diagonal matrix $\mathbf{U}$. The equations need only be solved for one of these blocks since they are identical. The large system of coupled equations is hence reduced to a number of smaller coupled systems.[86]

The next stage is to solve the reduced system of coupled equations subject to the boundary condition

$$F_{\beta l}(\mathbf{r}) \overset{r\to\infty}{=} \frac{1}{k_\beta^{\frac{1}{2}}}\{C\mathrm{e}^{-\mathrm{i}(k_\alpha r - l\pi/2)} - S_{\alpha\beta}\mathrm{e}^{\mathrm{i}(k_\beta r - l\pi/2)}\} \tag{4.239}$$

which enables the S-matrix elements to be determined. More practically, the real boundary condition

$$F_{\beta l}(\mathbf{r}) \overset{r\to\infty}{=} \frac{1}{k_\beta^{\frac{1}{2}}}\{C \sin(k_\beta r - l\pi/2) + R_{\alpha\beta} \cos(k_\beta r - l\pi/2)\} \tag{4.240}$$

is used, where $R_{\alpha\beta}$ are the elements of the $\mathbf{R}$-matrix, related to $\mathbf{S}$ through (4.216). The differential cross-section can then be evaluated using (4.224) and (4.225).

There are several methods for solving the set of coupled equations. In some cases they are solved numerically,[87] whereas an alternative approach is to approximate the potential matrix $\mathbf{U}$ using, for instance, a Legendre polynomial expansion and to solve the close coupled equations analytically.[88] Either of these methods may be used in conjunction with one of two techniques for obtaining the $\mathbf{S}$ (or $\mathbf{R}$) matrices. One technique starts the solution in the non-classical region where $U > E$ and proceeds to follow the solution step by step to large r. The other obtains the R-matrix corresponding to separate portions of the potential which are combined to cover larger portions until $\mathbf{R}$ for the whole potential is built up. For details of these methods, the reader is referred to an excellent review by Secrest.[86] Which combination of procedures is used for particular problems depends on the accuracy required and the number of solutions required. The numerical methods are generally more accurate for a given computer time.

The same general approach can be used to treat both vibrational and rotational processes. The major problem in the latter case is that, because of the $2j+1$ degeneracy associated with the energy levels of a rotating molecule, the number of coupled equations that need to be solved increases rapidly with the maximum value of j included in the basis set used to expand $\Psi(\mathbf{r}, \mathbf{R})$. Since the computer time required to solve the coupled equations depends on the cube of the number in the set, the problem becomes prohibitively difficult at quite low j values. Consequently a large number of approximate methods have been developed in an attempt to reduce solution time without significant loss of accuracy. These will be briefly reviewed and suggestions made about the range of conditions in which they are useful.

4.7.8. *The centrifugal sudden approximation*

The essence of sudden approximations is that they regard certain groups of quantum numbers associated with the scattering systems as degenerate. The centrifugal sudden (CS),[89] or j_z-conserving coupled states,[90] approximation simply approximates the centrifugal potential in the wave equation, $\mathbf{L}^2/2\mu r^2$, by an eigenvalue form, $\bar{l}(\bar{l}+1)\hbar^2/2\mu r^2$. The purpose of this substitution is that it enables the closure property of the complete set to be used to carry out certain sums over l, m_l. The method has been developed independently by McGuire and Kouri[90] and by Pack.[89] Physically, it assumes that the relative kinetic energy is so large that the precise value of the centrifugal potential is not important. This is a reasonable approximation as long as the classical turning point does not vary rapidly with l. This is true for a purely repulsive potential but will become less valid for potentials having attractive minima when the relative kinetic energy is below that for which three turning points exist.

A common choice for $\bar{l}$ is the initial value of l, l_0. The CS treatment produces a set of coupled equations similar to (4.238) except that all values of l are replaced by l_0. These are solved by the methods outlined in § 4.7.7 giving the following expressions for atom–diatom systems:[91]

Scattering amplitude

$$f(j'm_j' \leftarrow jm_j \mid \mathbf{r}\cdot\mathbf{k}_j) = \frac{i\,\delta_{m_j m_j'}}{2(k_j k_{j'})^{\frac{1}{2}}} \sum_{l_0} (2l_0+1) T_{l_0 m_j}(j' \leftarrow j) P_{l_0}(\mathbf{r}\cdot\mathbf{k}_j), \quad (4.241)$$

where δ is the Kronecker delta function.

Individual differential cross-section

$$_{\mathrm{i}}I_\alpha^\beta \equiv I(j'm_j' \leftarrow jm_j) = \frac{\delta_{m_j m_j'}}{4k_j^2} \sum_{l_0, \bar{l}_0} (2l_0+1)(2\bar{l}_0+1) T_{l_0 m_j}(j' \leftarrow j) P_{l_0}(\mathbf{r}\cdot\mathbf{k}_j) P_{\bar{l}_0}(\mathbf{r}\cdot\mathbf{k}_j) \quad (4.242)$$

Degeneracy-averaged cross-section

$$
{}_{\mathrm{d}}I^{\beta}_{\alpha} \equiv I(j' \leftarrow j) = \frac{1}{(2j+1)4k_j^2} \sum_{l_0, \bar{l}_0, m_j} (2l_0+1)(2\bar{l}_0+1) T_{l_0 m_j}(j' \leftarrow j) \times T^*_{\bar{l}_0 m_j}(j' \leftarrow j) P_{l_0}(\mathbf{r} \cdot \mathbf{k}_j) P_{\bar{l}_0}(\mathbf{r} \cdot \mathbf{k}_j) \tag{4.243}
$$

Integral cross-sections

$$
Q(j'm_j' \leftarrow jm_j) = \frac{\pi}{k_j^2} \delta_{m_j m_j'} \sum_{l_0} (2l_0+1) \, |T_{l_0 m_j}(j' \leftarrow j)|^2 \tag{4.244}
$$

$$
{}_{\mathrm{t}}Q(j' \leftarrow j) = \frac{\pi}{(2j+1)k_j^2} \sum_{l_0, m_j} (2l_0+1) \, |T_{l_0 m_j}(j' \leftarrow j)|^2. \tag{4.245}
$$

All these equations refer to coordinates such that the quantization direction z is along the final separation vector $\mathbf{r}$. Expressions have been given for cross-sections defined with respect to space fixed axes of arbitrary direction, the most notable changes being in the magnetic scattering.[92] However, degeneracy-averaged cross-sections are independent of the choice of quantization direction. The choice of the final exit channel orbital angular momentum quantum number for $\bar{l}$, l', also gives different magnetic transitions compared with $\bar{l} = l_0$ in body-fixed coordinates but for the standard experimental axes the choices are equivalent.[92,93]

A CS theory for molecule–molecule systems has been developed by Heil, Green, and Kouri.[94]

4.7.9. *The infinite-order sudden* (*IOS*) *approximation*

In the infinite-order sudden method, the centrifugal sudden approximation is made as before and, additionally, the energy sudden (ES) approximation is made. This involves replacing the rotational energy of the diatom in (4.229) by an effective eigenvalue form

$$
\mathbf{j}^2 \sim \bar{j}(\bar{j}+1)\hbar^2. \tag{4.246}
$$

This has the effect of replacing all the wave numbers k_j in the coupled equations by a single effective wave number $\bar{k}$. This enables the closure property of the complete set of rotor states to be used to simplify summations over j, m_j. The resulting set of coupled equations again resembles (4.238), now with both l and k_β replaced everywhere by $\bar{l}$ and $\bar{k}$.

The use of the ES and CS approximations together is termed infinite-order because all the resulting equations include the effects of all terms in the potential to all orders. Because all operators which depend on the angle γ between $\mathbf{r}$ and $\mathbf{R}$ (see Fig. 4.36) have been removed, this angle only enters the wave equation as a parameter. Hence one can solve the wave equations for fixed orientations γ and obtain the relevant scattering information by projecting the fixed orientation amplitude with the appropriate initial and final rotor states.

We first give the IOS equations for the general case in which both rotational (j, m_j) and vibrational (ν) transitions can occur.[95]

The replacement of all the l values by $\bar{l}$ and all the j values by $\bar{j}$ in the close-coupling equations results in (4.224) for the inelastic scattering amplitude becoming

$$f(j'm_{j'}\nu' \leftarrow jm_j\nu \mid \mathbf{r}) = \mathrm{i}\left(\frac{\pi}{k_{j\nu}k_{j'\nu'}}\right)^{\frac{1}{2}} \sum_{l,l',m'} (-1)^{l-\bar{l}}(2l+1)^{\frac{1}{2}} T^{\mathrm{IOS}}_{j'm_{j'}\nu'l',jm_j\nu l} Y_{l'm'}(\mathbf{r}) \quad (4.247)$$

where $T^{\mathrm{IOS}}_{j'l'\nu'jl\nu}$ is the IOS transition matrix, related to the IOS scattering matrix simply by

$$T^{\mathrm{IOS}}_{j'm_j'\nu'l',jm_j\nu,l} = \delta_{jj'}\delta_{m_jm_j'}\delta_{ll'}\delta_{\nu\nu'} - S^{\mathrm{IOS}}_{j'm_j'\nu'l',jm_j\nu l} \quad (4.248)$$

and is given in terms of the scattering elements corresponding to a fixed relative orientation γ, $T^{\bar{j}\bar{l}}_{\nu'\nu}(\gamma)$, by

$$T^{\mathrm{IOS}}_{j'm_{j'}\nu'l',jm_j\nu l} = \langle Y_{j'm_{j'}} Y_{l'm'} |\, T^{\bar{j}\bar{l}}_{\nu',\nu}(\gamma) \,| Y_{jm_j} Y_{lm} \rangle \quad (4.249)$$

where $\langle\,\rangle$ indicates integration over all angular configurations of the colliding molecules and Y_{jm_j}, Y_{lm} are the (spherical harmonic) rotational wavefunctions of the molecule and the colliding pair, respectively (see (4.219)).

Replacing $\bar{l}$ by l, the initial orbital angular momentum quantum number, enables the completeness property of $Y_{l'm}$ to be used in the summation over l', m' in (4.247) above, resulting in

$$f(j'm_j'\nu' \leftarrow jm_j\nu \mid \mathbf{r}) = \left(\frac{k_{\bar{j}\nu'}k_{\bar{j}\nu}}{k_{j'\nu'}k_{j\nu}}\right)^{\frac{1}{2}} \langle j'm_j' |\, f^{\bar{j}}(\nu' \leftarrow \nu, \gamma \,|\chi) jm_j\rangle_{\mathrm{SF}} \quad (4.250)$$

where the average is performed over space-fixed (SF) coordinates and $f^{\bar{j}}$ is the orientation-dependent scattering amplitude given by

$$f^{\bar{j}}(\nu' \leftarrow \nu, \gamma \mid \chi) = \frac{\mathrm{i}}{2(k_{\bar{j}\nu'}k_{\bar{j}\nu})^{\frac{1}{2}}} \sum_{l} (2l+1) T^{\bar{j}l}_{\nu',\nu}(\gamma) P_l(\cos\chi). \quad (4.251)$$

This expression is just that expected for inelastic scattering by spherical particles (cf. 4.209). In other words, the IOS approximation reduces to holding the relative orientation of the molecules fixed throughout the collision, calculating the vibrationally inelastic scattering for this spherically symmetric case, and then integrating over all angles to obtain the full inelastic scattering amplitude. Eqn (4.250) is not diagonal in m_j quantum numbers and so allows magnetic transitions. However it involves integration over two angular coordinates relating the orientation of the molecules to the SF axes. An equivalent form to (4.250) can be obtained, involving body-fixed coordinates, which reduces the averaging procedure to a single integral over the relative orientation γ. However, this is diagonal in m_j and so does not allow magnetic transitions, which close coupling calculations show is unrealistic.

The other common alternative choice for $\bar{l}$, l' leads to the following expression for the scattering amplitude:

$$f'(j'm_j'v' \leftarrow jm_jv\,|\chi) = (-1)^{j-j'}\left(\frac{k_{\bar{j}v'}k_{\bar{j}v}}{j_{j'v'}k_{jv}}\right)^{\frac{1}{2}}\delta_{m_jm_j'} \times \langle j'm_j|\,f^{\bar{j}}(v' \leftarrow v, \gamma\,|\chi)\,|jm_j\rangle_{\mathrm{BF}} \quad (4.252)$$

which is similar to (4.250) and has the advantage, being in the body fixed (BF) frame, of only involving a one-dimensional integral, i.e. over γ. It can be seen that here again, however, no magnetic transitions are allowed. Nevertheless it is only here that the choice of $\bar{l}$ is critical since it has been shown[93] that the use of either (4.250) or (4.252) leads to the same result for degeneracy-averaged cross-sections.

Using (4.225) and (4.193), we see that the degeneracy-averaged differential cross-section turns out to be

$$I(j' \leftarrow jv\mid\chi) = \frac{k_{j'v'}}{k_{jv}}\frac{1}{2j+1}\sum_{m_j',m_j}|f(j'm_j'v' \leftarrow jm_jv)\mid\mathbf{r})|^2 \quad (4.253)$$

which using (4.250) gives

$$I(j'v' \leftarrow jv\mid\chi) = \left(\frac{k_{\bar{j}v'}k_{\bar{j}v}}{k_{jv}^2}\right)\frac{1}{2j+1}\sum_{m_j}|\langle j'm_j'|\,f^{\bar{j}}(v' \leftarrow v, \gamma\mid\chi)\,|jm_j\rangle_{\mathrm{BF}}|^2. \quad (4.254)$$

Hence the cross-section is simply obtained by averaging over a series of fixed orientation spherically symmetric collisions. The state-to-state integral cross-section follows from (4.254) and (4.252) by integration over all scattering angles χ, or alternatively, from (4.226). It is given by

$$Q(j'v' \leftarrow jv) = \left(\frac{\pi}{k_{jv}^2}\right)\sum_l (2l+1)\mathcal{P}_l(j'v' \leftarrow jv) \quad (4.255)$$

where $\mathcal{P}_l$ are the opacity functions[96]

$$\mathcal{P}_l(j'v' \leftarrow jv) = \frac{1}{2j+1}\sum_{m_j}|\langle j'm_j|\,T^{\bar{j}l}_{v'v}(\gamma)\,|jm_j\rangle_{\mathrm{BF}}|^2. \quad (4.256)$$

The total differential cross-section can be obtained from (4.254) by summation over all final rotational states j'. At this point, $\bar{j}$ can be identified with j, the initial rotational state, which this time simplifies the summation over j', m_j' due to the completeness of $Y_{j'm_j'}$.[26] This leads to

$$ {}_{t}I(\nu' \leftarrow j\nu \mid \chi) = \frac{1}{2}\int_{-1}^{1} I^{j}(\nu' \leftarrow \nu, \gamma \mid \chi)\, d\cos\gamma \tag{4.257} $$

where

$$ I^{j}(\nu' \leftarrow j\nu, \gamma \mid \chi) = \frac{k_{j\nu'}}{k_{j\nu}} |f^{j}(\nu' \leftarrow \nu, \gamma \mid \chi)|^2. \tag{4.258} $$

$I^{j}(\nu' \leftarrow j\nu, \gamma \mid \chi)$ is simply the equivalent spherical particle differential cross-section calculated by holding the angle γ fixed throughout the collision. The total differential inelastic cross-section for the transition $\nu' \leftarrow \nu$ is simply the angle average of this quantity in the IOS approximation. The corresponding integral cross-section follows by integration over χ

$$ {}_{t}Q(\nu' \leftarrow \nu) = \frac{1}{2}\int_{-1}^{1} Q(\nu' \leftarrow \nu, \gamma)\, d\cos\gamma \tag{4.259} $$

where, from (4.255),

$$ Q(\nu' \leftarrow \nu, \gamma) = \left(\frac{\pi}{k_{j\nu}^2}\right) \sum_{l} (2l+1)\, |T^{jl}_{\nu'\nu}|^2. \tag{4.260} $$

The classical equivalent of this fixed orientation approximation introduced by Monchick and Mason,[97] has been used in transport theory for some years. The relevance of IOSA to justifying this approach and its wider applications to transport properties is discussed in Chapter 6.

For the special case of an atom colliding with a rigid rotor the close-coupled equations (4.238) reduce to a single uncoupled equation:

$$ \left[\frac{\partial^2}{\partial r^2} + \bar{k}^2 - \frac{\bar{l}(\bar{l}+1)}{r^2} - \frac{2\mu}{\hbar^2} U(r,\gamma)\right] F_{\bar{k}\bar{l}}(r) = 0 \tag{4.261} $$

with

$$ \bar{k}^2 = \frac{2\mu}{\hbar^2}\left(E - \frac{\hbar^2 \bar{j}(\bar{j}+1)}{2I_{BC}}\right), \tag{4.262} $$

where I_{BC} is the moment of inertia of the rotor.
Solution of this equation leads to the following particularly straightforward IOS expressions[95]

(a) *Scattering matrices*

S-matrix: $S^{\bar{k}\bar{l}}(\gamma) = e^{2i\eta^{\bar{k}\bar{l}}(\gamma)}$, T-matrix: $T^{\bar{k}\bar{l}}(\gamma) = 1 - S^{\bar{k}\bar{l}}(\gamma)$, where $\eta^{\bar{k}\bar{l}}(\gamma)$ are the phase shifts for a fixed value of γ. As in the case of elastic scattering, it is common practice to use the JWKB expression (4.73) for calculating these fixed orientation phase shifts:

$$\eta^{\bar{k}\bar{l}}(\gamma)=\bar{k}\int_{r_0}^{\infty}\left[\left\{1-\frac{U(r,\gamma)}{E}-\frac{(\bar{l}+\frac{1}{2})^2}{\bar{k}^2r^2}\right\}^{\frac{1}{2}}-1\right]\mathrm{d}r-\bar{k}r_0+\frac{\pi}{2}(\bar{l}+\tfrac{1}{2}). \quad (4.263)$$

(b) *Scattering amplitude*

For $\bar{l}=l'$

$$f(j'm_j'\leftarrow jm_j\mid\mathbf{r})=(-1)^{j-j'}\frac{\bar{k}^2}{(k_jk_{j'})^{\frac{1}{2}}}\,\delta_{m_jm_j'}\langle j'm_j\,|f^{\bar{k}}(\gamma\mid\chi)|\,jm_j\rangle_{\mathrm{BF}} \quad (4.264)$$

where BF indicates body-fixed coordinates and

$$f^{\bar{k}}(\gamma\mid\chi)=\frac{\mathrm{i}}{2\bar{k}}\sum_l(2l+1)(1-\mathrm{e}^{2\mathrm{i}\eta^{\bar{k}l}(\gamma)})P_l(\cos\chi) \quad (4.265)$$

which is simply the expression for elastic scattering by spherical particles.

(c) *Degeneracy-averaged differential cross-section*

$$_{\mathrm{d}}I_\alpha^\beta\equiv I(j'\leftarrow j\mid\chi)=\frac{\bar{k}^2}{(2j+1)k_j^2}\sum_{m_j}|\langle j'm_j\mid f^{\bar{k}}(\gamma\mid\chi)\,|jm_j\rangle_{\mathrm{BF}}|^2. \quad (4.266)$$

(d) *Total cross-sections*

For $\bar{j}=j$

$$I(\chi)=\tfrac{1}{2}\int_{-1}^{1}I(\gamma\mid\chi)\,\mathrm{d}\cos\gamma \quad (4.267)$$

where

$$I(\gamma\mid\chi)=|f^{\bar{k}}(\gamma\mid\chi)|^2. \quad (4.268)$$

$$_{\mathrm{t}}Q=\tfrac{1}{2}\int_{-1}^{1}Q(\gamma)\,\mathrm{d}\cos\gamma \quad (4.269)$$

where

$$Q(\gamma)=\frac{4\pi}{k^2}\sum_l(2l+1)\sin^2\eta^{kl}(\gamma). \quad (4.270)$$

All these expressions are relatively simple and represent the most convenient computational route from an anisotropic potential energy function to inelastic scattering cross-sections, where the assumptions of the IOS method are valid (see § 4.7.13). It will be seen in Chapter 6 that the expressions may be further used in the evaluation of bulk transport properties in a fairly straightforward manner.

An important additional property of the IOS expressions is that they may be factorized into sums of products of system-independent spectroscopic coefficients and system-dependent dynamical coefficients.[95,98] Hence the degeneracy-averaged differential cross-section is given by

$$I(j' \leftarrow j \mid \chi) = \frac{k_0^2}{k_j^2} \sum_{j''} C^2(j, j'', j', 000) I(j'' \leftarrow 0 \mid \chi) \tag{4.271}$$

and

$$I(j'' \leftarrow 0 \mid \chi) = \frac{\bar{k}^2}{k_0^2} |F_{j''}^{\bar{k}}(\chi)|^2 \tag{4.272}$$

where C is a Clebsch–Gordan coefficient.[99] In other words, all the relevant cross-sections can be calculated from the cross-sections for scattering out of the $j=0$ state. If the final state is not resolved

$$_{\mathrm{t}}I_\alpha^\beta \equiv I(j \mid \chi) = \frac{k_0^2}{k_{j_0}^2} \sum_{j''} I(j'' \leftarrow 0 \mid \chi) \tag{4.273}$$

which implies that the total differential cross-section is essentially independent of the initial state of the rotor. Similar expressions to (4.271)–(4.273) exist for the corresponding integral cross-sections.[95]

4.7.10. *The space-fixed orientation approximation*

It was seen in § 4.7.9 that the IOS approximation is essentially a body-fixed orientation sudden approximation in which averages over fixed relative orientations are obtained. A complementary approach has been adopted whereby the direction of the molecular axis of each molecule, **R**, is kept fixed and the cross-sections obtained by averaging over fixed *absolute*, or space-fixed, molecular orientations. We designate this the space-fixed orientation or SFO approximation.[100,101] It is assumed that the molecules move along a classical trajectory determined solely by the isotropic part of the potential, $U_0(r)$. Orientation dependent phase shifts $\eta_l(\phi)$ are calculated by

$$\eta_l(\phi) = \eta_{l_0} - \frac{1}{2\hbar} \int_{-\infty}^{\infty} U'(t)\, \mathrm{d}t \tag{4.274}$$

where ϕ is the azimuthal angle of the molecule(s) in space-fixed coordinates and η_{l_0} is the phase shift resulting solely from $U_0(r)$. The trajectory having an impact parameter $b = (l+\frac{1}{2})/k$ is determined entirely by $U_0(r)$; this gives η_{l_0} and $r(t)$. The latter is used in turn to evaluate the integral in (4.274) which is simply over the anisotropic part of the potential $U' \equiv U - U_0$.

This phase shift is then used to replace $\eta^{\bar{l}\bar{k}}(\gamma)$ in the IOS expressions for the orientation-dependent scattering amplitudes such as (4.251) or

(4.265) to give $f^{\bar{k}}(\phi \mid \chi)$. This is then averaged over all molecular orientations (ϕ) to give the various collision cross-sections as in the previous section. Because the anisotropic part of the potential is introduced into the problem as a perturbation in (4.274), and the whole potential function not used throughout, the method is not infinite-order in the potential; in fact it has been shown to be only valid to first-order.[100] Nevertheless the method is comparable in ease of computation to the IOS approximation and has been applied successfully to elastic scattering situations[6,101] and to inelastic scattering[100] where the anisotropies in the potential are not large (see § 4.7.13).

4.7.11. *The l_z-conserved energy sudden approximation*

This approximation[102,103] is complementary to the CS and IOS methods, involving an energy sudden (ES) approximation only, the centrifugal potential being treated by the close-coupling procedure. It is valid when the rotational energy level spacing is small compared with the relative kinetic energy and when the non-available levels are not strongly coupled to the states involved in the scattering process. It is also useful when the CS or IOS procedure is not valid, such as when the relative kinetic energy is small compared with the potential well depth or when an extremely long-range anisotropic potential is acting. The expressions for the various cross-sections are similar in form to the CS and IOS, but require considerably more computing time for their evaluation. They are not reproduced here; the reader is referred to ref. 91 for a detailed discussion of this method.

4.7.12. *Other approximate methods*

A method designed for use in systems where the centrifugal potential is dominant is the *decoupled l-dominant* (*DLD*) *approximation.*[104] In contrast to the CS method, where the intermolecular potential is treated exactly and the centrifugal potential is approximated, the DLD method includes **L** exactly and approximates the intermolecular potential in a series expansion. It is likely to be useful whenever there is a large contribution to the inelastic scattering coming from the long-range part of the potential since here the energy varies only slowly with r and changes in l can produce significant changes in the classical turning point.

In the simplest form of the *distorted wave approximation* (DWA)[105] (4.238) are decoupled by retaining only the first term on the right-hand side and replacing $F_{0l}(\mathbf{r})$ by a wave function $F_i(\mathbf{r})$ determined by the potential matrix element $u_{00}(\mathbf{r})$ according to the wave equation

$$\left\{\frac{\partial^2}{\partial r^2} - u_{00}(\mathbf{r}) + k_0^2 - l(l+1)/r^2\right\}F_i(\mathbf{r}) = 0 \qquad (4.275)$$

This is used to solve the decoupled equation for $F_\beta(\mathbf{r})$:

$$\left\{\frac{\partial^2}{\partial r^2} - u_{\beta\beta}(\mathbf{r}) + k_\beta^2 - l(l+1)/r^2\right\}F_\beta(\mathbf{r}) = u_{\beta 0}F_i(\mathbf{r}) \tag{4.276}$$

In the *exponential distorted wave* (EDW)[106,107] method, the distorted waves are calculated using WKB wave functions. These methods are in general less widely useful than the sudden approximations, although they are valid under some conditions in which the latter fail (see § 4.7.13).

Semi-classical calculations using time-dependent perturbation theory have also been used to calculate inelastic cross-sections.[107,108] The basis of the methods is similar to that used in the SFO approximation. The nuclear motion is calculated classically in a trajectory calculation and the internal modes treated by a series of first-order differential equations for the S-matrix elements derived from the time-dependent wave equation:

$$i\hbar\frac{\partial S}{\partial t}(\beta \leftarrow \alpha \mid t) = \sum_\gamma \langle\beta|\, U(t)\,|\gamma\rangle S(\gamma \leftarrow \beta \mid t)e^{i(E_\beta - E_\gamma)t/\hbar} \tag{4.277}$$

where the quantities α, β, γ represent particular internal quantum states of a target molecule, and U is the intermolecular potential. The integral cross-section for a particular transition is obtained using the relation

$$Q(\beta \leftarrow \alpha) = 2\pi\int_0^\infty |S(\beta \leftarrow \alpha)|^2\, b\, \mathrm{d}b \tag{4.278}$$

where b is the impact parameter of the trajectory. The time dependence of U is fed into (4.277) by determining the separation $\mathbf{r}$ as a function of time in a classical trajectory calculation. This is often simplified by using only the central part of U to calculate $\mathbf{r}(t)$ at a mean energy for the collision.

4.7.13. *Criteria for choosing approximate methods*

The close-coupling, sudden, and other approximate methods given in this chapter have been described chiefly from the point of view of rotational energy transfer since this has proved the major obstacle to be cleared in the development of any accurate inelastic scattering theory, and the area to which most theoretical effort has been devoted. However, similar methods may be applied to vibrational and electronic excitation. In particular, close-coupling calculations have been made for both cases, providing a body of accurate results against which various sudden and other approximations may be tested. The IOS method has been applied with some success to rotation–vibration excitation and the factorization properties of the cross-sections carry over to this more complex situation. Most studies of electronic effects have used WKB semi-classical approaches, but combined sudden-approximations/close-coupling approaches should prove useful in this area too.

There are no universal criteria for the suitability of particular approximation methods to scattering problems. For rotational excitation it is possible to give an indication of the range of experimental conditions where each method should be most valid. In a few cases, quantitative criteria can be given. However, there is no guarantee that such observations will still apply when vibrational and electronic excitation processes are being considered. Indeed there has been far less comparative study with accurate close coupling calculations in these latter areas anyway. Consequently approximation methods for treating vibrational and electronic inelastic scattering will not be considered further here. We will consider each of the approximation methods described earlier in turn and give a brief guide to the conditions under which they may be applied to rotational inelastic scattering processes, with particular emphasis on atom–diatom collisions as described by Kouri and co-workers.

(a) *The energy sudden approximation.* The conditions under which it is valid to replace the set of wave numbers k_j by a single value $\bar{k}$ are:

1. The relative centre-of-mass velocity of the colliding species must be substantially greater than the velocities of the atoms within the molecule for all states which are strongly coupled to the particular transition under consideration. Hence as the collision energy increases, the ESA becomes more useful and as a rough guide will be a reasonable approximation if $E_{\text{trans}} \gtrsim \varepsilon$, where ε is the potential well-depth. For a given relative kinetic energy, increasing the reduced mass μ of the colliding pair or increasing j to high values both decrease the difference between the translational and rotational velocities and so eventually lead to situations where the ES approximation is no longer valid.

2. The difference between the atom velocities within the molecule for the initial and final states, j and j', should be small compared with the molecule's mean velocity. Top and Kouri[109] have quantified this condition for the atom–diatom case

$$\left(\frac{2\mu B_e}{\mu_{\text{BC}}}\right)^{\frac{1}{2}} \frac{|[j(j+1)]^{\frac{1}{2}} - [j'(j'+1)]^{\frac{1}{2}}|}{[2E - B_e(j(j+1) + j'(j'+1))]^{\frac{1}{2}}} < 1 \tag{4.279}$$

or, for small B_e,

$$\left(\frac{B_e \bar{j}(\bar{j}+1)}{\mu_{\text{BC}} E}\right)^{\frac{1}{2}} \ll 1 \tag{4.280}$$

where B_e is the rotational constant, E the total energy, and $\bar{j}$ the higher of j, j'. In other words, the approximation is valid for small Δj values where $\Delta E_{\text{rot}} \ll E_{\text{trans}}$.

(b) *The centrifugal sudden approximation.* Approximating the centrifugal potential by the single value $\bar{l}(\bar{l}+1)\hbar^2/2\mu r^2$ is valid if the relative kinetic energy of the colliding molecules is sufficiently large that the precise value of the centrifugal potential is unimportant. Since the relative kinetic energy is a minimum at the classical turning point, r_0, it is here that the greatest breakdown of the CS approximation is likely to occur. Hence if r_0 varies significantly with l, then the CS approximation will not be a useful one whereas if r_0 is relatively insensitive to changes in l it should be successful. We may use these ideas to set out the conditions when the CS approximation is valid:

1. For an anisotropic potential energy function, the effective potential V (the sum of the potential and centrifugal energies) will depend on relative orientation as well as separation. However, the variation of V with r along any given trajectory will qualitatively resemble that for spherical systems shown in Fig. 4.3. When the asymptotic relative kinetic energy E is greater than the outer maximum in V, i.e. $E > E_{\text{orbiting}}$, then the classical turning point r_0 is determined by the repulsive wall of the potential and, because of the steepness of the curve here, is relatively insensitive to changes in l. For lower energies ($E < E_{\text{orbiting}}$), r_0 is at separations greater than that corresponding to the maximum in V. It is determined almost entirely by the centrifugal potential and is extremely sensitive to the value of l. For collisions of this type, the CS approximation breaks down.

2. This may be generalized to a statement that for those collisions in which the short-range part of the potential function dominates, the CS approximation is a good one. In contrast, whenever a collision is mainly influenced by the long-range part of the potential, a region in which the centrifugal potential is very sensitive to l (and in which r_0 is correspondingly sensitive due to the low slope of the curve), the approximation fails.

3. These criteria have been quantified by Kouri, Heil, and Shimoni[110] who showed that the CS approximation is valid when

$$r_0 > b \tag{4.281}$$

where b is the impact parameter. r_0 and b can be related to the parameters used in the CS calculation by using (4.11) and (4.12) to give

$$r_0^2 \simeq \frac{l(l+1)\hbar^2}{2\mu(E_j - U(r_0))} \tag{4.282}$$

where E_j is the relative kinetic energy for state j at large separations, and the semi-classical expression (4.59) for the impact parameter:

$$b \simeq \frac{(l+\frac{1}{2})}{k_j}. \tag{4.283}$$

4. Criterion (3) shows that at very low relative kinetic energies where $E_j \simeq \Delta E_{j' \leftarrow j}$, b is large and greater than r_0 at low l. Consequently, the CS approximation breaks down at energies near to the excitation threshold.

5. The sufficient, but not necessary, condition

$$E_j, E_{j'} > \varepsilon \tag{4.284}$$

usually ensures that criteria (1) and (4) are both satisfied.

6. As l is increased, at fixed relative kinetic energy, then the long-range part of the potential will eventually dominate the collision and the CS approximation breaks down. The point at which this occurs will be determined by the form of the potential and its anisotropy, i.e. the extent to which a particular $j' \leftarrow j$ transition is determined by regions of the potential at long and short range. However, the CS approximation remains valid up to large l for elastic collisions.

(c) *The infinite-order sudden approximation.* Since both the CS and ES approximations are made in the IOSA, this approximation is only valid when the criteria for both these simplifications are satisfied, i.e. it becomes increasingly valid as the relative kinetic energy increases, for small j transitions where the short-range part of the potential energy function dominates the scattering. It is by far the easiest of the various approximations discussed here to implement and is recommended for initial qualitative calculations even under conditions where the validity crtieria are not met. Its convenience arises because

1. The expressions for the cross-sections (§ 4.7.9) are particularly straightforward, simply requiring integration over a series of fixed relative orientation evaluations.

2. The elements of the S-matrix at each orientation γ, $S^{k\bar{l}}(\gamma)$, may be evaluated in a straightforward fashion by using the JWKB phase shift expression, (4.73), using $l = \bar{l} = l_0$ to give $\eta^{k\bar{l}}(g)$, under conditions where the semi-classical approximation is valid (4.263);

3. The factorization property of the cross-sections enables all the cross-sections to be evaluated from those for scattering out of the $j = 0$ state.

(d) *Other approximations.* The DLD approximation is designed for those situations where both the l_z-conserved ES and the CS approximations are not valid, i.e. where $r_0 > r_m$, corresponding to collisions dominated by high values of the orbital angular momentum quantum number, l.[111] The exponential distorted wave approximation is also applicable under these conditions.

The SFO approximation has been shown to give results comparable in accuracy to the DWA for elastic scattering from anisotropic potentials,[112] and in many cases to be superior to the IOS approximation for both elastic and inelastic scattering. It appears to be more accurate in situations where the long-range part of the potential dominates, particularly where the anisotropy is marked.[100] This is probably a consequence of the fact that, unlike in the IOS case, the atomic angular coordinates change along an individual trajectory in the SFO phase shift calculation.

(e) *Semi-classical calculations.* When none of the above approximations are suitable, it may be possible to apply semi-classical techniques to the internal motions.[108] For instance, we have seen that collisions in which the reduced mass of the partners is large and the rotational transitions, characterized by a small value of B_e, are to high j values cannot be treated using the ES approximation. However, these are just the conditions for which one would expect a semi-classical treatment to be useful for both the relative translational motion and the internal rotational motion, since the de Broglie wavelength in each case is likely to be small compared with the distance over which the potential energy varies significantly. One stage of approximation further, at the opposite extreme from close-coupling calculations, are classical trajectory calculations[6] in which quantization is introduced in an approximate way and interference effects are ignored completely. Such calculations are sometimes satisfactory in reproducing the qualitative features of scattering experiments but are not suitable for any rigorous test of the intermolecular potential.

4.8. The determination of intermolecular forces from inelastic cross-section measurements

In contrast to spherically symmetric systems, no formal inversion procedures exist at present for determining the intermolecular pair potential energy function directly from cross-section measurements for polyatomic molecules although some progress has been made in developing approximate methods. Consequently it is currently usually necessary to use the more traditional method of assuming a particular functional form for the potential energy and determining a limited number of variable parameters by fitting calculated cross-sections to experimental measurements. Naturally this should always be done having due regard to the region(s) of the potential to which the particular measurements are sensitive.

Although state-to-state cross-sections are likely to be most sensitive to the anisotropic part of the potential function, it is on elastic and total differential cross-sections that most attention has centred to date. The approach, and the effect of potential anisotropy on the scattering, is well illustrated by the work of Buck *et al.*[100] on the scattering of sodium atoms and tetrahedral molecules $M(CH_3)_4$. These workers used a potential function of the form

$$U(r,\gamma,\phi)=\varepsilon\left(\frac{r_m}{r}\right)^{12}(1+b_3\cos\theta_x\cos\theta_y\cos\theta_z)$$
$$-2\varepsilon\left(\frac{r_m}{r}\right)^{6}\left(1+a_3\left(\frac{r}{r_m}\right)\cos\theta_x\cos\theta_y\cos\theta_z\right)\quad(4.285)$$

where r is the Na–M separation and θ_x, θ_y, θ_z are the angles which r makes with body-fixed axes (see Fig. 4.37). This was chosen to have a long-range form in line with theory, and is a simple extension of an isotropic Lennard-Jones $12-6$ potential. It can be written in terms of spherical polar coordinates r, γ and ϕ

$$U(r,\gamma,\phi)=\varepsilon\left\{\left(\frac{r_m}{r}\right)^{12}-2\left(\frac{r_m}{r}\right)^{6}\right\}+(-i)\left(\frac{2\pi}{105}\right)^{\frac{1}{2}}\varepsilon\{Y_{23}(\gamma,\phi)-Y_{3-2}(\gamma,\phi)\}$$
$$\times\left\{b_3\left(\frac{r_m}{r}\right)^{12}-a_3\left(\frac{r_m}{r}\right)^{7}\right\}.\quad(4.286)$$

The long-range anisotropy parameter a_3 is given by

$$a_3=\frac{8A}{\alpha r_m}\quad(4.287)$$

where α and A are the dipole and dipole-quadrupole polarizabilities of the molecule respectively.

Calculations of total differential cross-sections for this potential function using the SFO method, believed to be a good approximation at

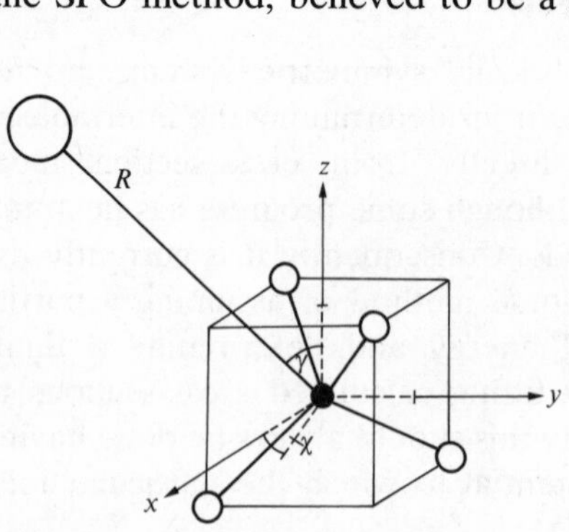

FIG. 4.37. Body-fixed coordinate system used for potential energy functions of tetrahedral molecules interacting with atoms.[100]

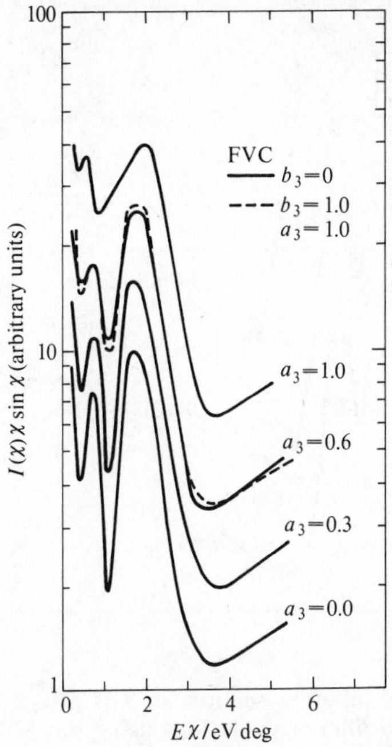

FIG. 4.38. Space-fixed orientation approximation calculations of effect of anisotropy changes on differential cross-sections in the centre-of-mass system. a_3 and b_3 are the attractive and repulsive anisotropy parameters defined by (4.286). (U. Buck, V. Khare and M. Kick.[100])

thermal energies for the slowly rotating ($\tau_{rot} \sim 5$ ps) molecules studied here, show the effects of increasing anisotropy on the scattering (Fig. 4.38). In general, anisotropic interactions tend to quench the interference structure of the cross-sections. As a_3 is increased, the main rainbow peak is broadened, the heights of the supernumary rainbows decrease and their positions are shifted. For very large long-range anisotropies, the supernumaries disappear completely. Short-range anisotropy tends to partially compensate for these effects e.g. the cross-sections for $a_3 = 0{\cdot}6$, $b_3 = 0$, and $a_3 = 1{\cdot}0$, $b_3 = 1{\cdot}0$ are very similar. Consequently only an effective anisotropy, $a_3 - xb_3$ where $x \simeq 0{\cdot}4$ can be determined by fitting to experiment for any system. This is a reflection of the fact that the rainbow structure is sensitive to a superposition of repulsive and attractive forces. Changes in the well parameters have a similar effect to that for spherically symmetric systems, ε changes producing a change in the rainbow maximum ($\Delta\varepsilon \propto \Delta\Theta_r$) and r_m changes altering the angular spacing of the supernumary rainbows. In integral cross-section measurements the glory amplitude is also quenched as the potential anisotropy increases.

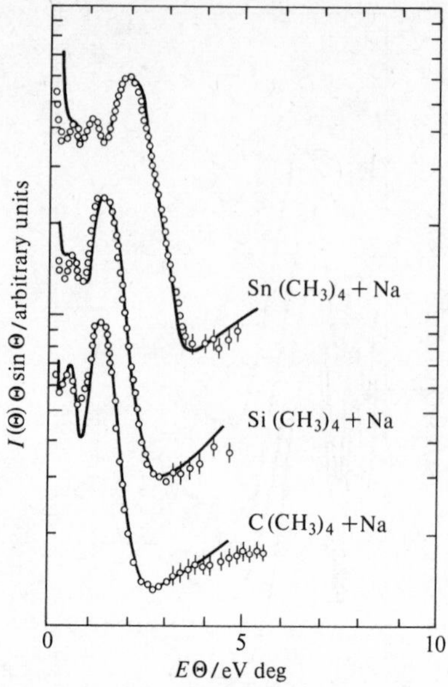

FIG. 4.39. Total differential cross-section for $M(CH_3)_4 + Na$. *Circles*: experimental measurements; *solid lines*: calculations using the SFO method based on the potential of (4.286). The base line of each curve is different. The curves are normalized at the rainbow maximum. (U. Buck, V. Khare and M. Kick.[100])

The experimental total differential cross-sections for $Na + M(CH_3)_4$, $M = C$, Si, Sn are illustrated in Fig. 4.39. It can be seen that as the mass of M increases, the rainbow maxima become progressively broader. The effective anisotropy parameters determined by fitting increase in the same order: $C = 0{\cdot}20 \pm 0{\cdot}10$, $Si = 0{\cdot}44 \pm 0{\cdot}06$, $Sn = 0{\cdot}70 \pm 0{\cdot}07$. The potential function chosen reproduces the experimental curves semi-quantitatively. Two possible combinations of a_3 and b_3 for $Na + Si(CH_3)_4$ are shown in Fig. 4.40 for two relative configurations of the molecules. The expectation that a direct approach of Na to a methyl group would have a steeper repulsive branch than a path directed between the methyl groups may be used to discriminate between possible combinations of a_3 and b_3 that are consistent with the rainbow structure.

The anisotropic 12–6 potential of (4.285) can only be regarded as a model function for systems of tetrahedral symmetry, and can at best be a semi-quantitative representation of the actual potential energy surface. However it (and its analogous forms for other symmetries) provides a valuable guide to the sensitivity of different parts of the scattering cross-sections to specific regions of the potential, and can suggest the directions in which improvements towards more sophisticated potential

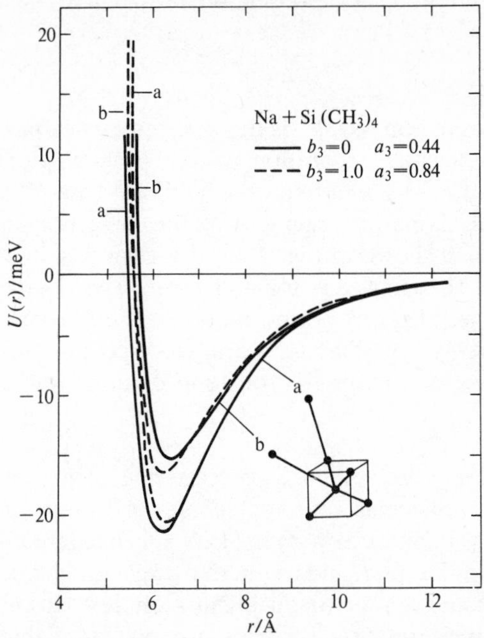

FIG. 4.40. Comparison for two relative orientations of two potential energy functions consistent with the cross-sections of Fig. 4.39 for Si $(CH_3)_4$+Na. (U. Buck, V. Khare and M. Kick.[100])

functions should be made. Increased experimental resolution combined with accurate quantum calculation procedures should eventually lead to practicable inversion procedures, for portions of the potential at least. In particular, separation of the effects of the isotropic and anisotropic parts of the potential on the height of the primary rainbow, or resolution of rapid quantum oscillations, would enable the isotropic part of the potential to be obtained by inversion methods analogous to those described in § 4.6.4 for spherically symmetric systems.

Molecules whose potential functions are more anisotropic than $M(CH_3)_4$ show more pronounced distortions of the rainbow structure. Linear molecules,[113] for instance, exhibit a double rainbow effect; similar broadening and shifting of the peaks occurs as the long-range anisotropy increases. These qualitative features are described for the limited number of systems studied to date by Reuss.[17]

The oscillatory structure of state-to-state differential cross-sections provides a sensitive test of the anisotropic parts of the potential function. It was shown in § 4.5.7 that for spherical molecules the angular spacing of the rapid oscillations superimposed on the rainbow structure is given, using (4.111) and (4.59), by

$$\Delta\chi \simeq \frac{\pi}{kb^*\sigma} \tag{4.288}$$

where k is the wave number, b^* is the reduced impact parameter b/σ, and σ is the intermolecular separation when $U(r)=0$. This expression has been shown to be also valid for the inelastic case,[114] interpreting the derivative of the S-matrix argument as the deflection function. For the $HD+D_2$ scattering shown in Fig. 4.35, this relationship was used to infer that $\sigma_{00}/\sigma_{01}=1{\cdot}0$, using $\Delta\chi$ for the $0\leftarrow 0$ and $1\leftarrow 0$ $j'\leftarrow j$ differential cross-sections. This sort of information is a useful precursor to setting up potential energy functions to use in close-coupling or more approximate calculations as a more rigorous and detailed test of the molecular interactions.

We have seen that the use of scattering techniques to determine anisotropic intermolecular potential energy functions is somewhat less developed than for spherical systems. Few approximate inversion schemes have yet been developed and even the direct calculation of scattering cross-sections from the potential is highly complex. However useful, rapid and accurate approximation schemes are now available to make such calculations feasible, even routine. Experimental techniques have developed to the extent of resolving oscillatory fine structure in inelastic cross-sections of similar detail to that which has commonly been observed for elastic cross-sections. These parallel developments in theory and experiment should contribute significantly to the synthesis of potential energy surfaces for polyatomic molecules. Indeed, some progress is being made in the development of inversion procedures for inelastic cross-sections. Gerber, Buch, and Buck[120] have proposed an approximate scheme, based on the j_z-conserved sudden and the exponential distorted wave approximations, for obtaining $U_0(r)$ and $U_2(r)$ [where $U(r, R, \gamma)=\sum_i U_i(r, R)P_i(\cos\gamma)$] for atom–diatom systems by direct inversion of rotationally elastic and inelastic differential cross-sections. Although the method is restricted to weakly coupled systems with large rotational energy spacings, such as $X+H_2$ or D_2, it provides clear guidelines for the development of more sophisticated procedures. Despite the emergence of these more direct techniques, however, it is essential to complement the scattering results by other data which probe different regions of the potential, in particular spectroscopic measurements (see Chapter 7).

Appendix A4.1. The JWKB approximation

The semi-classical JWKB approximation is associated with the names of Jeffreys,[115] Wentzel,[116] Kramers,[117] and Brillouin.[118] It is based on the idea that if the wavelength λ associated with a particle varies only slowly with position, then the wave function ψ will be very similar to that which would occur if λ was constant and equal to the mean value over the range under consideration. In its simplest form, the method draws on the result that for a particle in a one-dimensional box the wave function is given by[28]

$$\Psi = \bar{A} \sin\left(\frac{2\pi r}{\lambda}\right) \tag{A4.1.1}$$

where r is the coordinate. In the JWKB approximation, the wave function is taken to be approximately sinusoidal, with the amplitude $\bar{A}$ as well as the wavelength λ varying with position. For positive classical kinetic energies, i.e. total energy $E > U(r)$, a solution of the Schrödinger equation

$$\left\{\frac{d^2}{dr^2} + \frac{2m}{\hbar^2}(E - U(r))\right\}\Psi(r) = 0 \tag{A4.1.2}$$

having this form, which satisfies the condition $d\lambda/dr \ll 1$, is found to be

$$\Psi(r) = \frac{A}{p^{\frac{1}{2}}} \sin\left(\frac{1}{\hbar}\int p \, dr + \alpha\right) \tag{A4.1.3}$$

where A and α are arbitrary constants and $p(r)$ is the momentum, $\{2m(E-U(r))\}^{\frac{1}{2}}$. An amplitude proportional to $p^{-\frac{1}{2}}$ ensures a probability density which varies inversely as the velocity $v(=p/m)$ of the classical particle, i.e. a constant particle flux $v\,|\Psi(r)|^2$.

In regions where the classical kinetic energy is negative $(E < U(r))$, the approximate solution equivalent to (A4.1.3) is

$$\Psi = p^{-\frac{1}{2}}\left\{B \exp\left(\frac{1}{\hbar}\int p \, dr\right) + C \exp\left(-\frac{1}{\hbar}\int p \, dr\right)\right\}. \tag{A4.1.4}$$

The solutions (A4.1.3, 4) must join smoothly onto each other in the region of the classical turning point, where $E - U(r) \rightarrow 0$ and both approximate solutions diverge and become unrealistic. This joining procedure is not straightforward and the reader is referred elsewhere[1] for a detailed description of the *semi-classical connection formulae* involved.

An important result to emerge from this treatment is that the wave function decays effectively to zero within a distance $\lambda/8$ beyond the classical turning point. For the motion of a linear oscillator, which is bounded by two classical turning points, r_{01} and r_{02}, and has wave functions of the form illustrated in Fig. 4.41, this behaviour implies that the total number of wavelengths is

$$\int_{r_{01}}^{r_{02}} \frac{p}{h} \, dr + \frac{1}{4}. \tag{A4.1.5}$$

Since the wave function must contain an integer number of half-wavelengths to be well-behaved, we arrive at the condition

$$2\int_{r_{01}}^{r_{02}} \frac{p}{h} \, dr + \frac{1}{4} = \nu \tag{A.4.1.6}$$

or

$$\oint p \, dr = h(\nu + \tfrac{1}{2}) \tag{A4.1.7}$$

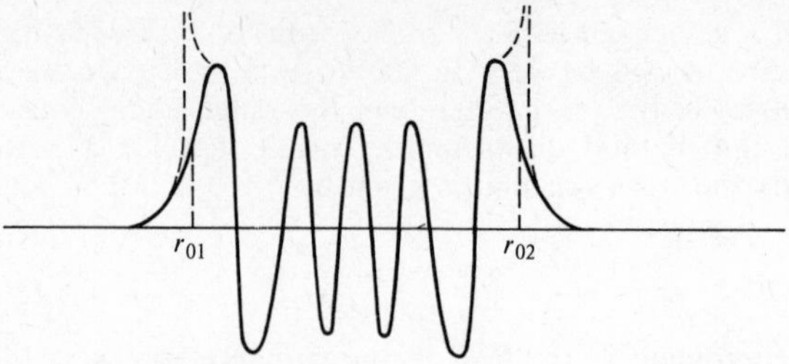

FIG. 4.41. Wave function for the $v=8$ level of a linear oscillator, illustrating the JWKB solutions (indistinguishable from exact solution except where shown dotted).

where the cyclic integral on the left-hand side of (A4.1.7) is the *action* or *phase integral*. This quantization condition is one of the fundamental results of the JWKB approximation.

A more general solution to (A4.1.2) has the form

$$\Psi = e^{iS(r)/\hbar}. \tag{A4.1.8}$$

Substitution in (A4.1.2) gives

$$i\hbar\frac{d^2S}{dr^2} - \left(\frac{dS}{dr}\right)^2 + p^2(r) = 0. \tag{A4.1.9}$$

If the term in $\hbar$ is small, to a first approximation we have

$$S_0(r) = \pm\oint p\,dr + \text{constant}. \tag{A4.1.10}$$

This applies if

$$\left|\hbar\frac{d^2S}{dr^2}\right| \ll \left(\frac{dS}{dr}\right)^2 \tag{A4.1.11}$$

or, using (A4.1.10), if

$$\left|\frac{\hbar}{p^2}\frac{dp}{dr}\right| = \left|\frac{d}{dr}\left(\frac{\hbar}{p}\right)\right| \ll 1. \tag{A4.1.12}$$

This implies that

$$\left|\frac{d\lambda(r)}{dr}\right| \ll 2\pi \tag{A4.1.13}$$

which is exactly the slowly varying wavelength criterion referred to earlier. Higher-order corrections to $S_0(r)$ are obtained by writing

$$S(r) = S_0(r) + \hbar S_1(r) + \hbar^2 S_2(r) + \ldots \tag{A4.1.14}$$

By equating the coefficients of $\hbar$ when this is substituted into the Schrödinger equation we obtain (from (A4.1.9))

$$i\left(\frac{d^2S_0}{dr^2}\right) - 2\left(\frac{dS_0}{dr}\right)\left(\frac{dS_1}{dr}\right) = 0 \tag{A4.1.15}$$

so that

$$S_1(r) = \frac{i}{2}\ln\left(\frac{dS_0}{dr}\right) + \text{constant} = \frac{i}{2}\ln p(r) + \text{constant}. \tag{A4.1.16}$$

To a second approximation, therefore,

$$\Psi(r) = \frac{A}{p^{\frac{1}{2}}}\exp\left\{\pm\frac{i}{\hbar}\int p(r)\,dr + \alpha\right\} \tag{A4.1.17}$$

where A and α are constants. This is the generalized solution corresponding to (A4.1.3) and (A4.1.4). Higher-order approximations are given in ref. 119.

Appendix A4.2. Phase shifts for potentials of form $U(r) = C_n r^{-n}$

Massey and Smith[31] have shown that at high angular momenta a good approximation to the mean value of the unperturbed wave function $\psi_l^{(0)}(r)$ in (4.44) is

$$\langle\{\psi_l^{(0)}(r)\}^2\rangle = \tfrac{1}{2}(1 - l^2/k^2r^2)^{-\frac{1}{2}}. \tag{A4.2.1}$$

Substituting this in (4.64), remembering that $u(r) = U(r)\,.\,2\mu/\hbar^2$, gives

$$\eta_l = -\frac{C_n}{k}\frac{\mu}{\hbar^2}\int_0^\infty \frac{(1 - l^2/k^2r^2)^{-\frac{1}{2}}}{r^n}\,dr \tag{A4.2.2}$$

or

$$\eta_l = \frac{F(n)k^{n-2}C_n\mu}{l^{n-1}\hbar^2} \tag{A4.2.3}$$

where

$$F(n) = \frac{\pi^{\frac{1}{2}}\Gamma(n/2 - \frac{1}{2})}{2\Gamma(n/2)} \quad \text{(cf. (4.23))}. \tag{A4.2.4}$$

(A4.2.3) is called the *Jeffreys–Born approximation.*

References

1. Child, M. S. *Molecular collision theory.* Academic Press, London (1974).
2. Fluendy, M. A. D. and Lawley, K. P. *Chemical applications of molecular beam scattering.* Chapman and Hall, London (1973).
3. Hirschfelder, J. O. (Ed.) *Intermolecular forces. Advan. chem. Phys.* **12,** (1967).
4. Ross, J. (Ed.) *Molecular beams. Advan. chem. Phys.* **10,** (1966).
5. Lawley, K. P. (Ed.) *Molecular scattering. Advan. chem. Phys.* **30,** (1975).
6. Bernstein, R. B. (Ed.) *Atom–molecule collision theory: A guide for the experimentalist.* Plenum Press, New York (1979).
7. Kennard, E. H. *Kinetic Theory of Gases.* McGraw-Hill, New York (1938); Smith, F. T. *Phys. Rev.* **150,** 79 (1966).
8. Anderson, J. B., Andres, R. P., and Fenn, J. B. *Advan. chem. Phys.* **10,** 275 (1966); Pauly, H. and Toennies, J. P. *Methods exp. Phys.* **7A**, 227 (1968).

9. Parks, E. K., Kuhry, J. G., and Wexler, S. *J. chem. Phys.* **67,** 3014 (1977).
10. Kinsey, J. L. *MTP int. Rev. Sci. phys. Chem. Ser. 1,* **9,** 173 (1972).
11. Grice, R. *Advan. chem. Phys.* **30,** 247 (1975).
12. Kinsey, J. L. *Rev. Sci. Instrum.* **37,** 61 (1966).
13. Gentry, W. R. and Giese, C. F. *J. chem. Phys.* **67,** 5389 (1977).
14. Toennies, J. P. *Chem. Soc. Rev.* **3,** 407 (1974).
15. Bromberg, E. E. A., Proctor, A. E. and Bernstein, R. B., *J. chem. Phys.* **63,** 3287 (1975); Stolte, S., Proctor, A. E., Pope, W. M., and Bernstein, R. B. *J. chem. Phys.* **66,** 3468 (1977).
16. Brash, H. M., Campbell, D. M., Farago, P. S., Rae, A. G. A., Siegman, H. C., and Wykes, J. S. *Proc. roy. Soc. Edinburgh* **A68,** 158 (1969).
17. Reuss, J. *Advan. chem. Phys.* **30,** 389 (1975).
18. Valentini, J. J., Lee, Y. T., and Auerbach, D. J. *J. chem. Phys.* **67,** 4866 (1977).
19. Cavallini, M., Gallinaro, G., and Scoles, G. *Z. Naturforsch.* **22,** 415 (1967).
20. Morse, F. A. and Bernstein, R. B. *J. chem. Phys.* **37,** 2019 (1962).
21. von Busch, F. *Z. Phys.* **193,** 412 (1966).
22. Rothe, E. W., Neynaber, R. H., and Trujillo, S. M. *J. chem. Phys.* **42,** 3310 (1965).
23. Dondi, M. G., Scoles, G., Torello, F., and Pauly, H. *J. chem. Phys.* **51,** 392 (1969).
24. Parson, J. M., Siska, P. E., and Lee, Y. T. *J. chem. Phys.* **56,** 1511 (1972); Lee, Y. T., McDonald, J. D., LeBreton, P. R., and Herschbach, D. R. *Rev. Sci. Instrum.*, **40,** 1402 (1969).
25. Farrar, J. M., Schafer T. P., and Lee, Y. T. *AIP Conference Proc.* (*USA*) **11,** 279 (1973).
26. Abramowitz, M. and Stegun, I. A. *Handbook of mathematical functions.* Dover, London (1965).
27. Mott, H. S. W. and Massey, N. F. *The theory of atomic collisions* (3rd ed), p. 23. Oxford University Press (1965).
28. Slater, J. C. *Quantum theory of atomic structure*, Vol. 1. McGraw-Hill, New York (1960).
29. Landau, L. D. and Lifshitz, E. M. *Quantum mechanics.* Pergamon Press, London (1959).
30. Bernstein, R. B. *J. chem. Phys.* **33,** 795 (1960); **34,** 361 (1961).
31. Massey, H. S. W. and Smith, R. A. *Proc. roy. Soc.* (*London*) **A142,** 142 (1933).
32. Ziman, J. M. *Elements of advanced quantum theory*, Chapter 4. Cambridge University Press (1969).
33. Fröman, N. and Fröman, P. *JWKB approximation.* North-Holland, Amsterdam (1965).
34. Langer, R. E. *Phys. Rev.* **51,** 669 (1937).
35. Connor, J. N. L. *Mol. Phys.* **15,** 621 (1969).
36. Wu, T. Y. and Ohmura, T. *Quantum theory of scattering.* Prentice Hall, New Jersey (1962).
37. Buck, U., Hoppe, H. O., Huisken, F., and Pauly, H. *J. chem. Phys.* **60,** 4925 (1974).
38. Mason, E. A., Vanderslice, J. T., and Raw, C. J. G. *J. chem. Phys.* **40,** 2153 (1964).
39. Bernstein, R. B. and Kramer, K. H. *J. chem. Phys.* **38,** 2507 (1963).
40. Bernstein, R. B. *Advan. chem. Phys.* **10,** 75 (1966).

41. For example, Linse, C. A., van den Biesen, J. J. H., van Veen, E. H., and van den Meijdenberg, C. J. N. *Physica*, **99A,** 166 (1979).
42. Bernstein, R. B. *J. chem. Phys.* **37,** 1880 (1962); **38,** 2599 (1963).
43. Ford, K. W. and Wheeler, J. A. *Ann. Phys.* (*NY*) **7,** 259, 287 (1959).
44. Ziman, J. M. *Elements of advanced quantum theory*, Chapters 1 and 2. Cambridge University Press (1969).
45. Bernstein, R. B. and Muckerman, J. T. *Advan. chem. Phys.* **12,** 389 (1967).
46. Amdur, I. and Jordan, J. E. *Advan. chem. Phys.* **10,** 29 (1967).
47. Mason, E. A. and Vanderslice, J. T. *Atomic and molecular processes* (ed. D. R. Bates), Chapter 17. Academic Press, New York (1962); Inouge, H. J. *J. chem. Phys.*, **72,** 3695 (1980).
48. Siska, P. E., Parson, J. M., Schafer, T. P., and Lee, Y. T. *J. chem. Phys.* **55,** 5762 (1971).
49. Buck, U. *Advan. chem. Phys.* **30,** 313 (1975).
50. Farrar, J. M., Lee, Y. T., Goldman, V. V., and Klein, M. L. *Chem. Phys. Lett.* **19,** 359 (1973).
51. Buck, U., Dondi, M. G., Valbusa, U., Klein, M. L., and Scoles, G. *Phys. Rev.* **A8,** 2409 (1973).
52. Barker, J. A., Watts, R. O., Lee, J. K., Schafer, T. P., and Lee, Y. T. *J. chem. Phys.* **61,** 3081 (1974).
53. Farrar, J. M. and Lee, Y. T. *J. chem. Phys.* **56,** 5801 (1972).
54. Düren, R. and Schlier, Ch. *Discuss. Faraday Soc.* **40,** 56 (1965).
55. Bernstein, R. B. and O'Brien, T., J. P. *Discuss. Faraday Soc.* **40,** 35 (1965); *J. chem. Phys.* **46,** 1208 (1967); Bernstein, R. B. and LaBudde, R. A. *J. chem. Phys.* **58,** 1109 (1973).
56. Buck, U. *Rev. mod. Phys.* **46,** 369 (1974).
57. Buck, U. *J. chem. Phys.* **54,** 1923 (1971).
58. Buck, U., Kick, M., and Pauly, H. *J. chem. Phys.* **56,** 3391 (1972).
59. Buck, U., Hoppe, H. O., Huisken, F., and Pauly, H. *J. chem. Phys.* **60,** 4925 (1974).
60. Buck, U. and Pauly, H. *J. chem. Phys.* **51,** 1929 (1971).
61. Firsov, O. B. *Zh. eksp. Teor. Fiz.* **24,** 279 (1953).
62. Vollmer, G. *Z. Phys.* **226,** 423 (1969).
63. Remler, E. A. *Phys. Rev.* **A3,** 1949 (1971).
64. Klingbeil, R. *J. chem. Phys.* **56,** 132 (1972).
65. Schaefer, H. F. In *Atom–molecule collision theory* (ed. R. B. Bernstein), Chapter 2. Plenum Press, New York (1979).
66. Buckingham, A. D. *Advan. chem. Phys.* **12,** 107 (1967).
67. Langhoff, P. W., Gordon, R. G., and Karplus, M. *J. chem. Phys.* **55,** 2126 (1971).
68. Gordon, M. D. and Secrest, D. *J. chem. Phys.* **52,** 120 (1970).
69. Dyke, T. R., Howard, B. J., and Klemperer, W. *J. chem. Phys.* **56,** 2442 (1972).
70. Stolte, A., Reuss, J., and Schwartz, H. L. *Physica* **57,** 254 (1972).
71. Moerkerken, H. C. H. A., Prior, M., and Reuss, J. *Physica* **50,** 499 (1970).
72. Bennewitz, H. G., Kramer, K. H., Paul, W., and Toennies, J. P. *Z. Phys.* **177,** 84 (1964).
73. Zandee, L. and Reuss, J. *Chem. Phys.* **26,** 327 (1977).
74. David, R., Faubel, M., Marchand, P., and Toennies, J. P. *Proc. VII IC-PEAC*, p. 252. Amsterdam, (1971).
75. Toennies, J. P. *Ann. Rev. Phys. Chem.* **27,** 225 (1976).

76. Van den Bergh, H. E., Faubel, M., and Toennies, J. P. *Discuss. Faraday Soc.* **55,** 203 (1973).
77. Buck, U., Huisken, F., and Schleusener, J. *J. chem. Phys.* **68,** 5654 (1978).
78. Held, W.-D., Schöttler, J., and Toennies, J. P. *Chem. Phys. Lett.* **6,** 304 (1970).
79. Kempter, V. *Advan. chem. Phys.* **30,** 417 (1975).
80. Aten, J. A. and Los, J. *Chem. Phys.* **25,** 47 (1977).
81. Geddes, J., Krause, H. F., and Fite, W. L. *J. chem. Phys.* **56,** 3298 (1972).
82. Lau, A. M. F. and Rhodes, C. K. *Phys. Rev.* **A15,** 1570 (1977).
83. Schlier, C. (Ed.). *Molecular beams and reaction kinetics.* Academic Press, New York (1970).
84. Miller, W. H. (Ed.). *Dynamics of molecular collisions, Part B,* Chapter 5. Plenum Press, New York (1976).
85. Arthurs, A. M. and Dalgarno, A. *Proc. roy. Soc.* (*Lond.*) **A256,** 540 (1960).
85a. Curtiss, C. F. *J. chem. Phys.* **52,** 4832 (1970) + references therein.
86. Secrest, D. In *Atom–molecule collision theory* (ed. R. B. Bernstein), Chapter 8. Plenum Press, New York (1979).
87. Sams, W. N. and Kouri, D. J. *J. chem. Phys.* **51,** 4809, 4815 (1969); Lester W. A. and Bernstein, R. B. *J. chem. Phys.* **48,** 4896 (1968).
88. Gordon, R. G. *J. chem. Phys.* **51,** 14 (1969).
89. Pack, R. T. *J. chem. Phys.* **60,** 633 (1974).
90. McGuire, P. and Kouri, D. J. *J. chem. Phys.* **60,** 2488 (1974).
91. Kouri, D. J. In *Atom–molecule collision theory* (ed. R. B. Bernstein), Chapter 9. Plenum Press, New York (1979).
92. Kouri, D. J. and Shimoni, Y. *J. chem. Phys.* **67,** 86 (1977).
93. Khare, V. *J. chem. Phys.* **67,** 3897 (1977).
94. Heil, T. G., Green, S., and Kouri, D. J. *J. chem. Phys.* **68,** 2562 (1978).
95. Parker, G. A. and Pack, R. T. *J. chem. Phys.* **68,** 1585 (1978).
96. Secrest, D. *J. chem. Phys.* **62,** 710 (1975).
97. Monchick, L. and Mason, E. A. *J. chem. Phys.* **35,** 1676 (1961).
98. Goldflam, R., Kouri, D. J., and Green, S. *J. chem. Phys.* **67,** 5661 (1977).
99. Rose, M. E. *Elementary theory of angular momentum,* p. 35. Wiley, New York (1957).
100. Buck, U., Khare, V., and Kick, M. *Mol. Phys.* **35,** 65 (1978).
101. Kuijpers, J. W. and Reuss, J. *Chem. Phys.* **4,** 277 (1974).
102. Khare, V. *J. chem. Phys.* **68,** 4631 (1978).
103. Chu, S.-I. and Dalgarno, A. *J. chem. Phys.* **63,** 2115 (1975); **64,** 3085 (1976).
104. DePristo, A. E. and Alexander, M. H. *J. chem. Phys.* **64,** 3009 (1976).
105. Mies, F. H. *J. chem. Phys.* **42,** 2709 (1965).
106. Levine, R. D. *Mol. Phys.* **22,** 497 (1971).
107. Balint-Kurti, G. G. In *International review of science, physical chemistry Series Two* (ed. A. D. Buckingham and C. A. Coulson), p. 283. Butterworths, Boston (1975).
108. Clark, A. P., Dickinson, A. S., and Richards, D. *Advan. chem. Phys.* **36,** 63 (1977).
109. Top, Z. H. and Kouri, D. J. *Chem. Phys.* **37,** 265 (1979).
110. Kouri, D. J., Heil, T. G., and Shimoni, Y. *J. chem. Phys.* **65,** 1462 (1976).
111. De Pristo, A. E. and Alexander, M. H. *Chem. Phys. Lett.* **44,** 214 (1976).
112. Klaasen, D., Thuis, H., Stolte, S., and Reuss, J. *Chem. Phys.* **27,** 107 (1978).
113. Cross, R. J. *J. chem. Phys.* **52,** 5703 (1970).

114. Collins, M. A. and Gilbert, R. G. *Chem. Phys. Lett.* **41,** 108 (1976).
115. Jeffreys, H. *Proc. London Math. Soc.* **23,** 428 (1923).
116. Wentzel, G. *Z. Phys.* **38,** 518 (1926).
117. Kramers, H. A. *Z. Phys.* **39,** 828 (1926).
118. Brillouin, L. *J. Phys. Radium* **7,** 353 (1926).
119. Landau, L. D. and Lifshitz, E. M. *Quantum mechanics, non-relativistic theory* (2nd edn). Addison-Wesley, New York (1965).
120. Gerber, R. B., Buch, V., and Buck, U. *J. chem. Phys.* **72,** 3596 (1980).

5

THE KINETIC THEORY OF NON-UNIFORM, DILUTE GASES

5.1. Introduction

The kinetic theory of gases was developed by Clausius, Maxwell, and Boltzmann in the middle of the last century. It seeks to explain the macroscopic properties of gases in terms of the motion and interaction of molecules. In the particular case of dilute gases of monatomic species this aim was formally achieved, early in this century, by the development of the theory carried out by Enskog and by Chapman. They were able to show how the transport properties (that is the viscosity, thermal conductivity, diffusion coefficient, and thermal diffusion factor) of such gases were related to the properties of their molecules. The most significant feature of their theory is that each of the transport coefficients of the gas or gas mixture can be expressed in terms of well defined integrals over the intermolecular potential for each of the possible binary encounters in the system. Thus, provided that the intermolecular pair potential for the interaction of the monatomic species is known it is possible to calculate the transport properties of a gas consisting of such molecules to any desired degree of accuracy at any temperature.

Since the experimental determination of the various transport coefficients of a gas is relatively straightforward (in principle) these observations naturally lead to an important question. If a knowledge of the intermolecular pair potential permits the evaluation of the transport properties of a dilute gas, does a knowledge of the transport properties of the gas over a finite range of temperature permit the direct determination of the intermolecular potential? For some fifty years after the work of Enskog and Chapman it was thought that the answer to this question was negative, except in special, physically unrealistic cases, because of the convoluted relation between the intermolecular potential and the transport coefficients. Within the last ten years however this situation has been radically altered by two developments. First, more accurate measurements of transport coefficients of gases and gas mixtures have been made. Second, it has been demonstrated that the inverse problem of determining the intermolecular potential from such data is indeed soluble.

In order to provide a detailed treatment of the determination of intermolecular forces from measured transport coefficients of gases, the above discussion indicates that a knowledge of the kinetic theory of dilute gases is necessary. Consequently, the present chapter is dedicated to providing the background knowledge. For the reader who is meeting the subject for the first time, we provide in § 5.2 a brief treatment of the kinetic theory in its simplest form. Though very approximate, this approach can give valuable physical insight, and together with Chapter 6 would provide a suitable introduction to the kinetic theory of transport properties in dilute gases. On the other hand those who wish to pursue the full theory directly, and in reasonable depth, could well bypass this section and proceed to § 5.3 where the framework of the rigorous kinetic theory is established for a pure, dilute gas of spherically symmetric monatomic molecules.

In § 5.3 an equation, first given by Boltzmann, which describes the evolution of the state of an initially non-uniform gas is formulated. In the subsequent section this equation is solved by a method of successive approximations and it is shown that the evolution of the macroscopic state of the gas according to the kinetic theory is consistent with that given by the equations of continuum hydrodynamics. Then, by using the thermal conductivity of a pure gas as an example, it is shown that the coefficient of thermal conductivity may be related to the dynamics of binary collisions occurring in the gas and therefore to the intermolecular pair potential. Subsequently, in § 5.6 these results are generalized to other transport coefficients of pure gases and gas mixtures. It is shown there that a knowledge of the intermolecular pair potential for all the possible binary interactions among the molecules of a dilute monatomic gas suffices to determine its transport coefficients. The kinetic theory approaches and the resulting expressions for the transport coefficients of dilute polyatomic gases are given in § 5.8 and § 5.9. The chapter ends with a description of the extensions of the kinetic theory to dense gases in § 5.10.

The following chapter then discusses the interrelation between the transport coefficients and the intermolecular pair potential and in particular how measurements of the former may be used to derive the latter.

5.2. Simple kinetic theory

In this section we present a simple approach to the derivation of the transport coefficients of a gas which makes clear the physical principles underlying the phenomena of viscosity, thermal conduction and diffusion. This naive treatment gives physical insight which is less easily perceived in the more rigorous development of subsequent sections, and despite the gross approximations employed, leads to the correct functional dependence of the transport coefficients on the appropriate variables. The resulting expressions for the transport coefficients therefore differ from those of the exact kinetic theory for hard spheres only by constant numerical factors.

We consider a gas composed of identical hard spheres of diameter d and mass m. These particles exert no forces on one another except on impact and are in continuous random motion. We assume that their speeds are distributed according to the Maxwell–Boltzmann Law (see § 5.4)

$$\mathrm{d}n_c = 4\pi N\left(\frac{m}{2\pi kT}\right)^{\frac{3}{2}} C^2 \mathrm{e}^{-mC^2/2kT}\,\mathrm{d}C$$

where $\mathrm{d}n_c$ is the number of molecules whose speeds lie in the interval C to $C+\mathrm{d}C$ and N is the total number of molecules in the gas under consideration. The mean molecular speed corresponding to this distribution is

$$\bar{C} = \left(\frac{8kT}{\pi m}\right)^{\frac{1}{2}}. \tag{5.1}$$

In order to evaluate the transport properties of gases we require an estimate of the fraction of the gas molecules moving in a particular Cartesian direction, for example, the positive x-direction (i.e. the direction of increasing x). As there are six equivalent directions in space we shall make the very crude assumption that the number of molecules moving in the positive x-direction is just $\frac{1}{6}$ of the total number of molecules.

Another quantity we need to evaluate is the mean distance a molecule travels between collisions—*the mean free path*, l. As the mean speed of the molecules is $\bar{C}$, they sweep out, in unit time, a cylinder of length $\bar{C}$ and radius d. If the centre of another molecule falls within this cylinder a collision will occur (Fig. 5.1). Thus, if there are n molecules per unit volume the number of collisions in unit time will be $\pi d^2 \bar{C} n$. The mean distance a molecule travels between such collisions, l, is therefore

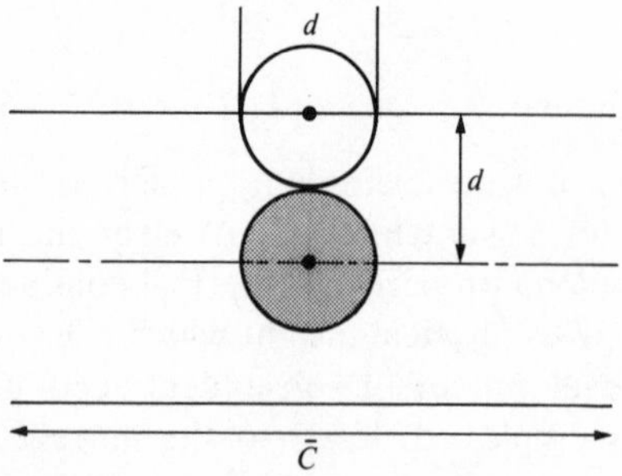

FIG. 5.1. A hard-sphere collision.

given by

$$l = \frac{\text{distance travelled in unit time}}{\text{number of collisions in unit time}} = \frac{\bar{C}}{\pi d^2 n \bar{C}}$$

$$= \frac{1}{\pi d^2 n}. \quad (5.2)$$

We may now proceed to obtain expressions for the transport properties of the gas.

5.2.1. Viscosity

In order to understand first how the viscosity of a gas arises we consider the gas contained between two plates (Fig. 5.2(a)), the upper one of which is moving with a velocity u in the positive x-direction. The gas in contact with the lower, stationary plate has no flow velocity in the positive x-direction and thus a velocity gradient exists in the gas which causes the gas to exert a drag on the upper plate. In order to maintain the velocity of the upper plate a force must be applied to overcome the viscous drag. Newton's law of viscosity defines the coefficient of shear viscosity, η, in terms of the viscous force, F, on a plate of area A, and the velocity gradient $(\mathrm{d}u/\mathrm{d}z)$ by the equation

$$F = -\eta A \frac{\mathrm{d}u}{\mathrm{d}z}. \quad (5.3)$$

The origin of viscous forces can be illuminated by a pleasing analogy. We imagine two trains moving on parallel tracks passing each other with different speeds and that parcels of mail are exchanged between the trains. The result of this exchange is that the parcels arriving on the slower train from the faster one have a higher velocity than the slow train and thus tend to speed it up. Conversely, the parcels landing on the faster

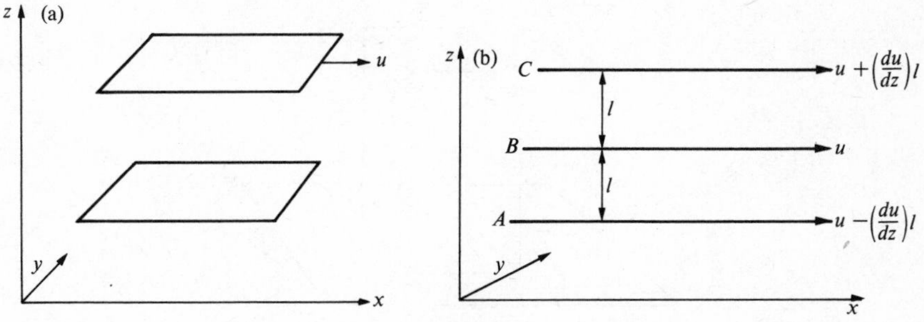

FIG. 5.2. (a) Shear in a fluid, (b) momentum transport between planes.

train tend to slow it down. The net effect will be to reduce the difference in speed of the two trains, and a distant observer would conclude that they exerted a frictional force on one another. In the case of the frictional force in a gas it is the layers in the gas which correspond to the trains and the molecules moving from one layer to another which correspond to the parcels.

We consider a reference layer in the gas lying in the xy-plane in which the gas moves with a flow velocity u in the positive x-direction only (Fig. 5.2(b)). The velocity gradient $(\mathrm{d}u/\mathrm{d}z)$ is assumed to be uniform. The viscous forces can be deduced by considering the molecules which reach the reference plane B. Those molecules which reach plane B will, on average, have travelled since their last collision a distance, l, the mean free path. Those which arrive at plane B from above (Fig. 5.2(b)) will have originated in plane C and will have a flow velocity in the positive x-direction of

$$u+\left(\frac{\mathrm{d}u}{\mathrm{d}z}\right)l$$

and a positive x-direction flow momentum of

$$m\left(u+\left(\frac{\mathrm{d}u}{\mathrm{d}z}\right)l\right)$$

where m is the mass of a molecule. As discussed earlier, we assume that $\frac{1}{6}$ of all the molecules are moving downwards in the negative z-direction. Thus the number of molecules which will cross a unit area in the xy-plane in unit time will be all those contained within a column of height $\bar{C}$ and unit cross-sectional area which are moving in the correct direction (Fig. 5.3), that is $n\bar{C}/6$.

The molecules reaching the reference plane, B, from above will therefore transport downwards an amount of positive x-direction

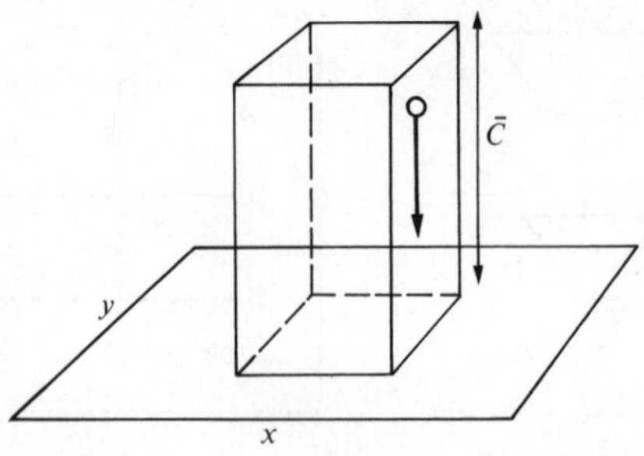

FIG. 5.3. The flux of molecules in the z-direction.

momentum

$$\tfrac{1}{6}n\bar{C}m\left\{u+\left(\frac{\mathrm{d}u}{\mathrm{d}z}\right)l\right\}$$

across unit area in unit time. The molecules that reach plane B from below (plane A of Fig. 5.2(b)) will transport upwards an amount of positive x-direction momentum

$$\tfrac{1}{6}n\bar{C}m\left\{u-\left(\frac{\mathrm{d}u}{\mathrm{d}z}\right)l\right\}$$

across unit area in unit time. The net transport of momentum across unit area of the reference plane B in unit time is the frictional force, F, on this plane in the positive x-direction

$$F=-\tfrac{1}{6}nm\bar{C}\left\{u+\left(\frac{\mathrm{d}u}{\mathrm{d}z}\right)l\right\}+\tfrac{1}{6}nm\bar{C}\left\{u-\left(\frac{\mathrm{d}u}{\mathrm{d}z}\right)l\right\}=-\tfrac{1}{3}nm\bar{C}\left(\frac{\mathrm{d}u}{\mathrm{d}z}\right)l. \quad (5.4)$$

The frictional force per unit area for a unit velocity gradient is, according to (5.3), the coefficient of viscosity, η, so that we may identify η as

$$\eta=\tfrac{1}{3}nm\bar{C}l=\tfrac{1}{3}\rho\bar{C}l \quad (5.5)$$

where ρ is the mass density of the gas. Substituting for the mean free path of the gas, l, from (5.2) and the mean molecular speed we obtain the alternative expressions

$$\eta=\frac{1}{3}\frac{m\bar{C}}{\pi d^2}=\frac{1}{3}\frac{m}{\pi d^2}\left(\frac{8kT}{\pi m}\right)^{\frac{1}{2}}. \quad (5.6)$$

This equation predicts that the viscosity of a dilute gas of hard spheres is independent of density. This surprising result was first obtained theoretically by Maxwell who subsequently tested it for real gases by measuring their viscosity as a function of pressure. He found that this prediction of the simple hard-sphere model was in keeping with the behaviour of *real* gases. This was a triumph for the newly emergent kinetic theory and did much to convince the sceptics of its validity. The physical reason for this independence is that in denser gases, though more molecules pass from one layer to another, the mean free path is correspondingly shorter and the excess momentum carried by each molecule is therefore less. More careful measurements on real gases show that the result is not in fact exact, but it is a very reasonable approximation at moderate densities. Eqn (5.6) also predicts that the viscosity of a rigid sphere gas should be proportional to $T^{\frac{1}{2}}$. This is a less satisfactory approximation for real gases whose viscosity generally varies more rapidly with temperature.

5.2.2. *Thermal conductivity*

The viscosity of a gas arises from the transport of momentum along a velocity gradient, whereas for a rigid sphere gas thermal conduction is the transport of kinetic energy along a temperature gradient. The phenomenological law of heat conduction was given by Fourier as

$$Q = -\lambda A\left(\frac{\mathrm{d}T}{\mathrm{d}z}\right) \tag{5.7}$$

where Q is the energy (heat) flow across an area A and λ is the coefficient of thermal conductivity. In order to obtain an expression for this transport coefficient we again consider a reference plane in a gas in which there is a uniform temperature (and energy) gradient in the z-direction alone. If we denote the mean energy of a molecule by E, then molecules reaching the reference plane from above (the higher temperature region) will have an energy

$$E + \frac{\mathrm{d}E}{\mathrm{d}z}l.$$

Those reaching the reference plane from below will have an energy

$$E - \frac{\mathrm{d}E}{\mathrm{d}z}l.$$

By similar arguments to those used in the case of viscosity the net transport of energy across unit area of the reference plane in unit time in the direction of the energy gradient is

$$-\tfrac{1}{3}n\bar{C}\left(\frac{\partial E}{\partial z}\right)l.$$

However,

$$\frac{\mathrm{d}E}{\mathrm{d}z} = \left(\frac{\mathrm{d}T}{\mathrm{d}z}\right)\cdot\left(\frac{\mathrm{d}E}{\mathrm{d}T}\right) = c_v\left(\frac{\mathrm{d}T}{\mathrm{d}z}\right)$$

where c_v is the heat capacity of the gas per molecule. Thus, for a unit temperature gradient the transport of kinetic energy across unit area in unit time, which is the thermal conductivity coefficient, λ, is

$$\lambda = \tfrac{1}{3}nc_v\bar{C}l$$

or, using (5.2),

$$\lambda = \frac{1}{3}\frac{c_v}{\pi d^2}\bar{C} = \frac{c_v}{3\pi d^2}\left(\frac{8kT}{\pi m}\right)^{\frac{1}{2}}. \tag{5.8}$$

The thermal conductivity of a hard-sphere gas is therefore predicted to be independent of density and proportional to $T^{\frac{1}{2}}$ as was the viscosity. In fact, our simple model leads to the conclusion that there is a proportionality between the viscosity and thermal conductivity coefficients of a rigid sphere gas, namely

$$\lambda = \eta c_v/m. \tag{5.9}$$

This proportionality turns out to be an extremely good approximation for real monatomic gases, albeit with a different proportionality constant (see § 5.6).

5.2.3. *Diffusion*

Diffusion is the transport of mass along a concentration gradient. Since the argument for two gases diffusing into each other is complicated, let us imagine a gas composed of two isotopes, whose properties, including their molecular mass are sufficiently similar so that we can treat the values of $\bar{C}$ and l as identical. We select one of the isotopes as a reference and suppose that n is its number density in a reference plane which is situated in the gas normal to a concentration gradient of the isotope $[\mathrm{d}(nm)/\mathrm{d}z]$. Then, following our earlier arguments the mass of the reference isotope reaching the reference plane, per unit area in unit time, from a plane of higher concentration a distance, l, above is

$$\tfrac{1}{6}\bar{C}\left[nm + \frac{\mathrm{d}(nm)}{\mathrm{d}z}\right]l.$$

This represents a downward flux of mass. The mass of molecules of the reference isotope reaching the plane from below is

$$\tfrac{1}{6}\bar{C}\left\{mn - \frac{\mathrm{d}(nm)}{\mathrm{d}z}\,l\right\}.$$

The net mass flow across unit area of the reference plane in the direction of the concentration gradient is therefore

$$-\tfrac{1}{3}\bar{C}l\frac{\mathrm{d}(nm)}{\mathrm{d}z}.$$

The transport coefficient for this process, known as the self-diffusion coefficient, D, is defined by the equation of Fick's Law

$$J = -AD\frac{\mathrm{d}(nm)}{\mathrm{d}z} \tag{5.10}$$

where J is the mass flow of molecules of the reference isotope. We can isolate an expression for D by considering transport of mass across unit area in Fick's Law and we find

$$D = \tfrac{1}{3}\bar{C}l$$

$$= \tfrac{1}{3}\bar{C}\frac{1}{\pi n d^2} = \frac{1}{3\pi d^2 n}\left(\frac{8kT}{\pi m}\right)^{\frac{1}{2}}. \tag{5.11}$$

Thus, the diffusion coefficient is inversely proportional to the number of molecules per unit volume. Furthermore, for a perfect gas

$$n = P/kT,$$

so that we can write

$$D = \frac{1}{3\pi d^2}\left(\frac{8kT}{\pi m}\right)^{\frac{1}{2}}\frac{kT}{P}. \tag{5.12}$$

The self-diffusion coefficient of a hard-sphere gas is therefore predicted to be inversely proportional to pressure and directly proportional to $T^{\frac{3}{2}}$.

5.2.4. *Comparison with exact kinetic theory for hard-sphere molecules*

The exact kinetic theory for hard-sphere molecules gives the same general form of the equations for the transport properties as derived above for our simple model. That is the dependencies on temperature, density, molecular mass, and diameter predicted by our simple argument, are all identical with those of the rigorous kinetic theory. However, the numerical coefficients in the expressions are quite different in the two treatments. The correct functional dependencies are a result of our use of an essentially correct physical mechanism for the transport processes. The discrepancies in the numerical coefficients arise, not surprisingly, from errors in some of our assumptions.

First, in calculating the mean free path, it is not proper to regard all molecules except the reference molecule as stationary and the correct average 'relative' molecular speed, which should be employed is $\sqrt{2}\,\bar{C}$. This factor of $\sqrt{2}$ reduces the value of the mean free path. Secondly the argument which led to an estimate of the flux of molecules in any particular direction as $n\bar{C}/6$ is incorrect. In order to obtain the correct value, $n\bar{C}/4$, an appropriate angular averaging of velocities must be carried out. Thirdly, it is an implicit assumption of the simple model that molecules reaching the reference plane acquire the characteristics of that plane following a single collision; this is an oversimplification. Piecemeal corrections to the simple kinetic theory to allow for these defects, though often employed, are not rewarding. Even in the case of rigid spheres it is more profitable to develop a rigorous kinetic theory as is done in the later sections of this chapter. The expressions for the transport coefficients of a hard-sphere gas obtained using this rigorous theory are summarized in

Table 5.1
Transport properties of hard-sphere gases

	Transport of	Simple theory coefficient ξ†	Rigorous theory coefficient ξ†	SI units of transport coefficient
Viscosity	Momentum	$\frac{2\sqrt{2}}{3\pi}=0{\cdot}300$	$\frac{5}{16}=0{\cdot}313$	$\mathrm{N\,s\,m^{-2}}$ (Pa s)
Thermal conductivity	Kinetic energy	$\frac{2\sqrt{2}}{3\pi}=0{\cdot}300$	$\frac{25}{32}=0{\cdot}781$	$\mathrm{W\,m^{-1}\,K^{-1}}$
Diffusion	Mass	$\frac{2\sqrt{2}}{3\pi}=0{\cdot}300$	$\frac{3}{8}=0{\cdot}375$	$\mathrm{m^2\,s^{-1}}$

† These are the coefficients in the equations: $\eta=\xi\{(\pi mkT)^{\frac{1}{2}}/\pi d^2\}$; $\lambda=\xi\{(\pi kT/m)^{\frac{1}{2}}/\pi d^2\}c_v$; $D=\xi\{(\pi kT/m)^{\frac{1}{2}}/n\pi d^2\}=\xi\{(\pi k^3T^3/m)^{\frac{1}{2}}/\pi d^2P\}$.

Table 5.1, and compared with the expressions obtained using the simple theory.

5.2.5. Application to real gases

Even the results of the rigorous theory given in Table 5.1 relate only to gases composed of hard-sphere molecules and each expression contains (through the mean free path, l) a cross-section πd^2 which is independent of temperature. However, for real substances, with intermolecular forces that can be both attractive and repulsive according to separation, the situation is more complicated, and the cross-sections which influence the transport properties depend on temperature. The rigorous kinetic theory for real gases therefore introduces into the expressions for the transport coefficients temperature-dependent factors which may be thought of as corrections to the hard-sphere cross-section to allow for the effects of intermolecular forces.

Thus for real gases we write

$$\eta=\frac{5(\pi mkT)^{\frac{1}{2}}}{16\pi d^2\Omega^*_\eta} \tag{5.13}$$

$$\lambda=\frac{25(\pi kT/m)^{\frac{1}{2}}c_v}{32\pi d^2\Omega^*_\lambda} \tag{5.14}$$

$$D=\frac{3(\pi k^3T^3/m)^{\frac{1}{2}}}{8P\pi d^2\Omega^*_D} \tag{5.15}$$

where the dimensionless factors Ω^*_η, Ω^*_λ, and Ω^*_D introduce into the equations the effects of the intermolecular forces on the binary collisions in the gas. It is one of the strengths of the advanced kinetic theory that it enables these factors to be calculated to any desired accuracy from a knowledge of the intermolecular pair potential.

A measure of the correction factor Ω_η^* can be obtained from the data for argon at 300 K. The collision diameter for argon is known to be approximately 0·33 nm, and the mass of an argon atom is $6{\cdot}63\times10^{-26}$ kg. Thus the mean speed of an argon atom at 300 K is 400 m s^{-1} and the mean free path in the gas at a pressure of 1 bar is 120 nm. The viscosity of argon gas under these conditions according to our simple kinetic theory for hard spheres is 26·0 μN s m^{-2}, and according to the rigorous kinetic theory of hard spheres is 27·0 μN s m^{-2}. The experimentally observed viscosity of argon at 300 K is 22·7 μN s m^{-2}, so that from (5.13) and Table 5.1 we find the correction factor Ω_η^* has the value 1·19 at 300 K. It will be seen later that this value is in keeping with the predictions of the rigorous theory.

5.3. The rigorous kinetic theory

5.3.1. The Boltzmann equation†

Our discussion of the kinetic theory will be restricted to single component gases of monatomic molecules which have spherically symmetric force fields and between which collisions are perfectly elastic and are governed by classical mechanics. In addition we shall consider those gas densities which are high enough on the one hand so that a typical molecule undergoes many more collisions with other molecules than it does with the walls of the container, but which are low enough so that the majority of collisions involve only one other molecule. The first condition implies that the mean free path of the gas molecules is much smaller than the dimensions of the vessel, whereas the second requires that the thermodynamic state of the gas should be adequately described by a virial expansion up to and including the second virial coefficient (Chapter 3).

A complete, deterministic treatment of the state of a gas containing N molecules would aim to describe the positions and velocities of all the molecules in the gas as a function of time by means of classical mechanics. Because of the extremely large number of molecules ($N\sim10^{20}$) in even a small sample of gas this treatment is not possible. Consequently, it is necessary to analyse the problem statistically and to seek the *probable* behaviour of the entire N-molecule system. For the density range we have chosen to study this permits a simpler approach. The assumption that only binary collisions occur implies that the position and velocity of a

† A more complete and rigorous derivation of the Boltzmann equation can be found in, for example, ref. 1.

typical molecule are independent of the positions and velocities of the other $(N-1)$ molecules in the system. Since a similar statement can be made about any of the N molecules it is clear that an adequate description of the probable behaviour of the N molecules is given by specifying the probable behaviour of a single molecule. This reduces the analysis to the search for a function $f(\mathbf{r}, \mathbf{c}, t)$ which is so defined that $f\,\mathrm{d}\mathbf{r}\,\mathrm{d}\mathbf{c}$ is the probable number of molecules whose centres have, at time t, position coordinates in the range† $\mathrm{d}\mathbf{r}$ about $\mathbf{r}$ and velocity coordinates in a range $\mathrm{d}\mathbf{c}$ about $\mathbf{c}$. This definition of f demonstrates that the number density of molecules in the gas, their mean velocity $\mathbf{u}$, and their mean thermal kinetic energy $\mathscr{E}$, which may all be functions of $\mathbf{r}$ and t but not $\mathbf{c}$, are given by

$$n(\mathbf{r}, t) = \int f(\mathbf{r}, \mathbf{c}, t)\,\mathrm{d}\mathbf{c} \tag{5.16}$$

$$n(\mathbf{r}, t)\mathbf{u}(\mathbf{r}, t) = \int \mathbf{c}f(\mathbf{r}, \mathbf{c}, t)\,\mathrm{d}\mathbf{c} \tag{5.17}$$

$$n(\mathbf{r}, t)\mathscr{E}(\mathbf{r}, t) = \int \tfrac{1}{2}m(\mathbf{c}-\mathbf{u})^2 f(\mathbf{r}, \mathbf{c}, t)\,\mathrm{d}\mathbf{c} \tag{5.18}$$

where the integrations extend over all molecular velocities.

For a gas whose molecules are not subject to external forces the only means whereby the velocity of a molecule can change is by collisions with other molecules. Thus in the absence of collisions, if we consider a group of molecules located near $\mathbf{r}$ with velocities near $\mathbf{c}$ at time t, at the end of an interval $\mathrm{d}t$ these same molecules will be located near $\mathbf{r}+\mathbf{c}\,\mathrm{d}t$ with unchanged velocities so that

$$f(\mathbf{r}, \mathbf{c}, t)\,\mathrm{d}\mathbf{r}\,\mathrm{d}\mathbf{c} = f(\mathbf{r}+\mathbf{c}\,\mathrm{d}t, \mathbf{c}, t+\mathrm{d}t)\,\mathrm{d}\mathbf{r}\,\mathrm{d}\mathbf{c}.$$

However, if collisions occur in the gas then in the time $\mathrm{d}t$ some of the original molecules may be lost from the group due to velocity changes upon collision whereas others not initially in the group may join it by virtue of collisions. If we denote the number of molecules joining the group in time $\mathrm{d}t$ by $\Gamma^+\,\mathrm{d}\mathbf{r}\,\mathrm{d}\mathbf{c}\,\mathrm{d}t$ and those leaving the group by $\Gamma^-\,\mathrm{d}\mathbf{r}\,\mathrm{d}\mathbf{c}\,\mathrm{d}t$ then we can write

$$f(\mathbf{r}+\mathbf{c}\,\mathrm{d}t, \mathbf{c}, t+\mathrm{d}t) - f(\mathbf{r}, \mathbf{c}, t) = (\Gamma^+ - \Gamma^-)\,\mathrm{d}t. \tag{5.19}$$

Expanding the first term on the left of this equation in a Taylor series and

† As defined in Chapter 3, $\mathrm{d}\mathbf{r}$ represents a volume element in real space, so that $\mathrm{d}\mathbf{r} = \mathrm{d}x_1\,\mathrm{d}x_2\,\mathrm{d}x_3$ in a Cartesian coordinate system (x_1, x_2, x_3). Similarly $\mathrm{d}\mathbf{c}$ represents a volume element in velocity space, $\mathrm{d}\mathbf{c} = \mathrm{d}c_{x_1}\,\mathrm{d}c_{x_2}\,\mathrm{d}c_{x_3}$ in the rectangular coordinate system or $\mathrm{d}\mathbf{c} = c^2\,\mathrm{d}c \sin\theta\,\mathrm{d}\theta\,\mathrm{d}\phi$ in spherical polar coordinates (c, θ, ϕ).

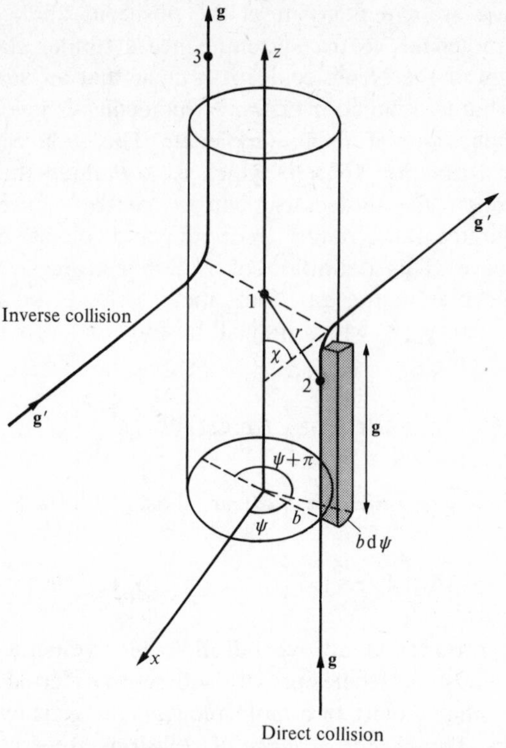

FIG. 5.4. Direct and inverse binary collisions. The collision is drawn so that molecule 1 is at rest. The angle χ is the angle of deflection of the relative velocity of molecule 2. The angle ψ defines the orientation of the plane of the collision in space.

taking the limit as $\mathrm{d}t$ tends to zero we have†

$$\left(\frac{\partial}{\partial t}+\mathbf{c}\,.\,\frac{\partial}{\partial \mathbf{r}}\right)f(\mathbf{r},\mathbf{c},t)=\Gamma^{+}-\Gamma^{-}. \tag{5.20}$$

Evidently, in order to complete this equation to yield the time and positional change of the function f we must compute explicitly the number of molecules gained and lost by our original group of molecules. Fig. 5.4 shows a collision which results in a loss of one molecule (molecule 2) from the initial group having a velocity $\mathbf{c}$. Before the collision the molecule 1 has a velocity $\mathbf{c}_1$, the velocity of molecule 2 being $\mathbf{c}$. Thus, molecule 2 is one of the initial group of molecules considered above, but molecule 1 is not. The relative velocities before and after

† Again the notation $\partial/\partial\mathbf{r}$ represents $\mathbf{i}(\partial/\partial x_1)+\mathbf{j}(\partial/\partial x_2)+\mathbf{k}(\partial/\partial x_3)$ where $\mathbf{i}$, $\mathbf{j}$, $\mathbf{k}$ are unit vectors parallel to the x_1, x_2, x_3 axes of a rectangular coordinate system.

collision are defined by

$$\mathbf{g} = \mathbf{c} - \mathbf{c}_1, \qquad \mathbf{g}' = \mathbf{c}' - \mathbf{c}_1', \tag{5.21}$$

respectively, where $\mathbf{c}_1'$ and $\mathbf{c}'$ are the velocities of molecules 1 and 2 respectively after collision. Because the collision is elastic, $|\mathbf{g}| = |\mathbf{g}'|$, and the effect of the collision is therefore merely to rotate the relative velocity through an angle χ, where $\mathbf{g} \cdot \mathbf{g}' = g^2 \cos \chi$. Furthermore, because the centre of mass of the two-body system moves at a constant velocity, $\mathbf{G}$, throughout the encounter we have

$$\mathbf{G} = \tfrac{1}{2}\mathbf{c}_1 + \tfrac{1}{2}\mathbf{c} = \tfrac{1}{2}\mathbf{c}' + \tfrac{1}{2}\mathbf{c}_1' \tag{5.22}$$

so that

$$\mathbf{c}_1 = \mathbf{G} - \tfrac{1}{2}\mathbf{g}, \qquad \mathbf{c} = \mathbf{G} + \tfrac{1}{2}\mathbf{g} \tag{5.23}$$

$$\mathbf{c}_1' = \mathbf{G} - \tfrac{1}{2}\mathbf{g}', \qquad \mathbf{c}' = \mathbf{G} + \tfrac{1}{2}\mathbf{g}'. \tag{5.24}$$

In addition to the relative velocity of the two molecules the geometry of the collision is defined by the impact parameter b and the azimuthal angle ψ which defines the orientation of the plane in which the collision takes place. This collision changes the velocity of molecule 2 so that after collision it is no larger part of the original group. The number of collisions in time $\mathrm{d}t$ of molecules with initial velocity between $\mathbf{c}$ and $\mathbf{c} + \mathrm{d}\mathbf{c}$ with a single molecule of velocity $\mathbf{c}_1$, such that the impact parameter lies between b and $b + \mathrm{d}b$ and the azimuthal angle is between ψ and $\psi + \mathrm{d}\psi$ is given by the number of molecules with velocity near $\mathbf{c}$ which lie somewhere in a 'cylinder' of volume $gb\,\mathrm{d}b\,\mathrm{d}\psi\,\mathrm{d}t$ at the beginning of this interval, and is, by definition, equal to

$$b\,\mathrm{d}b\,\mathrm{d}\psi\,g\,f(\mathbf{r}, \mathbf{c}, t)\,\mathrm{d}\mathbf{c}\,\mathrm{d}t.$$

Now the number of molecules with velocities near $\mathbf{c}_1$, in a volume $\mathrm{d}\mathbf{r}$ near $\mathbf{r}$ is $f(\mathbf{r}, \mathbf{c}, t)\,\mathrm{d}\mathbf{r}\,\mathrm{d}\mathbf{c}_1$, so that the number of collisions between molecules with a velocity of $\mathbf{c}$ and molecules with any velocity $\mathbf{c}_1$ at any impact parameter and any azimuthal angle in unit time is

$$\mathrm{d}\mathbf{r}\,\mathrm{d}\mathbf{c} \int_0^\infty \mathrm{d}\mathbf{c}_1 \int_0^\infty b\,\mathrm{d}b \int_0^{2\pi} \mathrm{d}\psi\,g\,f(\mathbf{r}, \mathbf{c}, t)\,f(\mathbf{r}, \mathbf{c}_1, t)$$

This is equal to $\Gamma^-\,\mathrm{d}\mathbf{r}\,\mathrm{d}\mathbf{c}$ since each of these collisions results in a molecule leaving the initial group we considered. In formulating this expression we have used the condition that only binary collisions occur because we have neglected the possibility that the collision cylinders of two or more molecules overlap. In addition, we have assumed that the probable number of molecules near $\mathbf{r}$ with velocities near $\mathbf{c}$ is independent of the positions and velocities of other molecules. This is the statistical assumption known as the assumption of *molecular chaos*. The assumption and its limitations have a significance far beyond the scope of this book.[2]

Fig. 5.4 also contains a diagram of a collision between molecule 3 and molecule 1 which is said to be the *inverse* of that between molecules 2 and 1 in that, after collision, molecule 3 has a velocity $\mathbf{c}$ and consequently represents a gain of a molecule to the group $f\,\mathbf{dr}\,\mathbf{dc}$. Such an *inverse collision* occurs if molecule 3 has a velocity prior to collision $\mathbf{c}'$ whereas molecule 1 has a velocity $\mathbf{c}_1'$, so that the relative velocity prior to collision is $\mathbf{g}' = (\mathbf{c}' - \mathbf{c}_1')$. Furthermore, it is necessary that the inverse collision occurs at the same impact parameter, b, and an azimuthal angle $\psi + \pi$. By a similar argument to that used for the direct collisions the number of inverse collisions occurring in a volume $\mathbf{dr}$ near $\mathbf{r}$ in unit time, each of which leads to the gain of a molecule to our original group, is

$$\Gamma^+\,\mathbf{dr}\,\mathbf{dc} = \mathbf{dr}\,\mathbf{dc}\int_0^\infty \mathbf{dc}_1 \int_0^\infty b\,\mathrm{d}b \int_0^{2\pi} \mathrm{d}\psi\, g\, f(\mathbf{r}, \mathbf{c}', t) f(\mathbf{r}, \mathbf{c}_1', t).$$

Here we have made use of the fact that the collision is perfectly elastic so that using the results of classical mechanics[3] $|\mathbf{g}| = |\mathbf{g}|' = g$ and $\mathbf{dc}\,\mathbf{dc}_1 = \mathbf{dc}'\,\mathbf{dc}_1'$.

Thus we can write a final equation of change for the function f as

$$\left\{\frac{\partial f}{\partial t}\right\} + \mathbf{c}\cdot\frac{\partial f}{\partial \mathbf{r}} = J(ff) \tag{5.25}$$

$$J(ff) = \int_0^\infty \mathbf{dc}_1 \int_0^\infty b\,\mathrm{d}b \int_0^{2\pi} \mathrm{d}\psi\, g\{f(\mathbf{r}, \mathbf{c}', t) f(\mathbf{r}, \mathbf{c}_1', t) - f(\mathbf{r}, \mathbf{c}, t) f(\mathbf{r}, \mathbf{c}_1, t)\}. \tag{5.26}$$

This non-linear integro-differential equation is known as *Boltzmann's equation* after its originator. The solution of this equation, $f(\mathbf{r}, \mathbf{c}, t)$, should enable the calculation of all the macroscopic properties of a dilute gas according to our original hypothesis. The intermolecular forces can be seen to influence the solution of the equation, and hence the calculated properties, through their determination of the relationship between $\mathbf{c}$ and $\mathbf{c}_1$, and $\mathbf{c}'$ and $\mathbf{c}_1'$ according to the laws of classical mechanics.

5.4. The Maxwell–Boltzmann equilibrium velocity distribution

In order to proceed to the solution of the Boltzmann equation for $f(\mathbf{r}, \mathbf{c}, t)$ in the general case we first consider a gas in equilibrium, that is a gas whose parameters of state and properties do not change with position or time. For such a gas we may infer that the solution of the Boltzmann equation f_E itself is not a function of position and time but only of velocity, so that the left-hand side of the Boltzmann equation is equal to zero. It is evident that a sufficient† condition for this equality is that

$$f_E(\mathbf{c})f_E(\mathbf{c}_1) = f_E(\mathbf{c}')f_E(\mathbf{c}_1') \tag{5.27}$$

for all values of $\mathbf{c}$, $\mathbf{c}_1$, b, and ψ.

Taking the logarithm of both sides of (5.27) we have

$$\ln f_E(\mathbf{c}) + \ln f_E(\mathbf{c}_1) = \ln f_E(\mathbf{c}') + \ln f_E(\mathbf{c}_1').$$

This shows that for any collision the sum of the function $\ln f_E$ for the two colliding molecules is unchanged by the collision. Since the only independent quantities in a collision for which this same result is true are the number of molecules, the three components of linear momentum and the kinetic energy of a molecule, $\ln f_E$ must be a linear combination of these five *collisional invariants.*

Thus

$$\ln f_E(\mathbf{c}) = a_1 + \mathbf{a}_2 \cdot \mathbf{c} + \tfrac{1}{2}a_3 mc^2 \tag{5.28}$$

or

$$f_E(\mathbf{c}) = a_4 \exp\{-\tfrac{1}{2}a_6(\mathbf{c} - \mathbf{a}_5)^2\} \tag{5.29}$$

where a_1–a_6 are constants. The values of these constants may be deduced following application of the equations

$$n = \int f_E(\mathbf{c})\, d\mathbf{c} \tag{5.30}$$

$$n\mathbf{u} = \int f_E(\mathbf{c})\mathbf{c}\, d\mathbf{c} \tag{5.31}$$

$$n\mathscr{E} = \tfrac{3}{2}nkT = \int f_E(\mathbf{c})\tfrac{1}{2}m(\mathbf{c} - \mathbf{u})^2\, d\mathbf{c} \tag{5.32}$$

which follow from the definition of the function f through (5.16)–(5.18).

Here, n is the number density, $\mathbf{u}$ the average velocity, and T the temperature. Eqn (5.32) defines the kinetic theory temperature, T, in terms of the average translational kinetic energy of the molecules which arises by virtue of their motion relative to the mean velocity of the gas. This definition renders the kinetic theory temperature identical with the thermodynamic temperature. The solution for $f_E(\mathbf{c})$ is

$$f_E(\mathbf{c}) = n\left(\frac{m}{2\pi kT}\right)^{\frac{3}{2}} \exp\left(\frac{-m\mathbf{C}^2}{2kT}\right), \tag{5.33}$$

† It may be shown that this condition is also necessary. (See for example ref. 4.)

where

$$\mathbf{C}=\mathbf{c}-\mathbf{u} \tag{5.34}$$

is the velocity of a molecule relative to the average velocity; it is termed the *peculiar velocity*.

For a gas at equilibrium one is often interested in the distribution of molecular speeds, $F(C)$, without regard to direction. To obtain this result we integrate (5.33) over all directions, using the results of Appendix A5.1, and find

$$F(C)=4\pi n\left(\frac{m}{2\pi kT}\right)^{\frac{3}{2}}C^2\exp\left(\frac{-mC^2}{2kT}\right). \tag{5.33a}$$

This distribution of molecular speeds is known as the Maxwell–Boltzmann Law which was mentioned in § 5.2.

5.5. The Chapman–Enskog solution for non-uniform gases

The solution of the Boltzmann equation in the more general case is now begun by examining the processes whereby an initially non-uniform gas approaches the equilibrium state. We shall restrict our attempts to find a solution to the most usual practical situation where the non-uniformities in the quantities $n(\mathbf{r}, t)$, $\mathbf{u}(\mathbf{r}, t)$ and $T(\mathbf{r}, t)$ are such that they do not change significantly over a distance equal to the mean free path of the gas l. That is, we require that

$$\delta^*=\frac{l}{x}\frac{\partial x}{\partial \mathbf{r}}\ll 1$$

where x may be n, $\mathbf{u}$, or T. The quantity δ^* is a measure of the departure of the gas from the uniform state.

It is instructive to distinguish three time scales for molecular processes in the dilute gas. The first, t_1, corresponds to the motion of a molecule through the intermolecular potential of another molecule and is typically of the order of 10^{-12} s. The second is the average time between collisions of molecules in the gas, $t_2 \sim 10^{-9}$ s, and the third is the time taken for a molecule to traverse a distance characteristic of the gas container, $t_3 \sim 10^{-4}$ s. For times $t \leqslant t_2$ the function $f(\mathbf{r}, \mathbf{c}, t)$ must be obtained from a general solution of the Boltzmann equation because it is determined by the exact dynamics of particular collisions. For $t_3 > t \gg t_2$ many collisions occur in a volume element $d\mathbf{r}$ of the gas near $\mathbf{r}$ and $f(\mathbf{r}, \mathbf{c}, t)$ changes rapidly with time because of the velocity changes at each collision. However, the total number of molecules, their total momentum, and total energy remain constant because they are conserved in each collision. Thus the function f can rapidly adjust itself to the essentially constant values of $n(\mathbf{r}, t)$, $\mathbf{u}(\mathbf{r}, t)$, and $T(\mathbf{r}, t)$ which exist in $d\mathbf{r}$ near $\mathbf{r}$. Here, the word 'constant' means that the quantity is independent of position within $d\mathbf{r}$ (because of our restriction on spatial non-uniformities) and also that it is independent of time on a time scale $t_3 > t \gg t_2$. This process rapidly establishes a local equilibrium within $d\mathbf{r}$ such that to a first approximation $f(\mathbf{r}, \mathbf{c}, t)$ is the function characteristic of the 'constant' values of $n(\mathbf{r}, t)$, $\mathbf{u}(\mathbf{r}, t)$, and $T(\mathbf{r}, t)$. On a larger time scale than t_3 the non-uniformities in $n(\mathbf{r}, t)$, $\mathbf{u}(\mathbf{r}, t)$, and $T(\mathbf{r}, t)$ decay when every molecule has traversed the container many times. This second process leading to equilibrium is much slower, so that the changes in $n(\mathbf{r}, t)$, $\mathbf{u}(\mathbf{r}, t)$, and $T(\mathbf{r}, t)$ occur at a rate such that the rapid changes in $f(\mathbf{r}, \mathbf{c}, t)$ always allow it to keep up with them and maintain the local equilibrium, at least to a first approximation.

The mathematical implications of these physical considerations lead to a simplification in our search for a solution of the Boltzmann equation for a non-uniform gas provided that we restrict ourselves to an examination of the slower equilibrating process, that is to the so-called *hydrodynamic regime.* This is because, in these circumstances, we may regard the function f as dependent on t only through the dependencies of n, $\mathbf{u}$, and T on t, since f always changes almost sufficiently rapidly to keep up with their changes. In addition we may assume that the explicit dependence of f on $\mathbf{r}$ is not greater than the indirect dependence of f on $\mathbf{r}$ through the dependence of n, $\mathbf{u}$, and T on $\mathbf{r}$. We consider first the time dependence of f. For $t \gg t_2$ we may write

$$\frac{\partial f(\mathbf{r}, \mathbf{c}, t)}{\partial t} = \frac{\partial f}{\partial n} \cdot \frac{\partial n}{\partial t} + \frac{\partial f}{\partial \mathbf{u}} \cdot \frac{\partial \mathbf{u}}{\partial t} + \frac{\partial f}{\partial T} \cdot \frac{\partial T}{\partial t}. \tag{5.35}$$

Now we have already established by physical reasoning that if the frequency of collisions is very high the solution for f would approximate to a local equilibrium solution at each point in the gas. This suggests that we seek a solution for the Boltzmann equation in terms of successive approximations which describe increasingly large departures from equilibrium. That is we shall seek a solution for the Boltzmann equation which is of the form

$$f = f_0 + \xi f_1 + \xi^2 f_2 + \ldots, \tag{5.36}$$

where $1/\xi$ measures the collision frequency in the gas. This is consistent with the fact that an infinite collision frequency would lead to the local equilibrium solution f_0, in line with the argument above. Arbitrary variation of ξ alters the total number of collisions occurring in unit volume in unit time, but does not affect the fraction of collisions with prescribed values of molecular velocities, impact parameter, and orientation. The introduction of the parameter ξ implies that we must modify the Boltzmann equation itself to accommodate a varying collision frequency.

Examination of the Boltzmann equation (5.25) shows that it is only the term on the right-hand side, $J(ff)$, which introduces the effect of molecular collisions. Since the term is proportional to the collision frequency, the parameter ξ is properly introduced in the form

$$\frac{\partial f}{\partial t}+\mathbf{c}\,.\frac{\partial f}{\partial \mathbf{r}}=\frac{1}{\xi}J(ff). \tag{5.37}$$

This formulation ensures that the total number of collisions in unit time can be varied arbitrarily without affecting the fraction of collisions of a particular type.

Substituting the expansion for f, (5.36), into the right-hand side of (5.37) and using the expression for $J(ff)$, (5.26), we obtain for the right-hand side of the modified Boltzmann equation (5.37)

$$\frac{1}{\xi}J(f_0f_0)+\{J(f_0f_1)+J(f_1f_0)\}+\dots. \tag{5.38}$$

By substituting (5.35) for $\partial f/\partial t$ into (5.37) we see that in order to find a similar expansion for the left-hand side of the modified Boltzmann equation we must first obtain expressions for $\partial n/\partial t$, $\partial \mathbf{u}/\partial t$, and $\partial T/\partial t$. Such expressions are provided by the conservation equations for molecular number, momentum and energy in the gas. These conservation equations may be obtained from the Boltzmann equation (5.37) by multiplying both sides by l, $m\mathbf{c}$, and $mc^2/2$ respectively and integrating with respect to $\mathbf{c}$. Because the collision term on the right of the Boltzmann equation automatically conserves the number of molecules, their momentum, and their energy, we find

$$\int\frac{\partial f}{\partial t}\,\mathrm{d}\mathbf{c}+\int\mathbf{c}\,.\frac{\partial f}{\partial \mathbf{r}}\,\mathrm{d}\mathbf{c}=0 \tag{5.39}$$

$$\int m\mathbf{c}\frac{\partial f}{\partial t}\,\mathrm{d}\mathbf{c}+\int m\mathbf{c}\mathbf{c}\,.\frac{\partial f}{\partial \mathbf{r}}\,\mathrm{d}\mathbf{c}=0 \tag{5.40}$$

$$\tfrac{1}{2}\int mc^2\frac{\partial f}{\partial t}\,\mathrm{d}\mathbf{c}+\tfrac{1}{2}\int mc^2\mathbf{c}\,.\frac{\partial f}{\partial \mathbf{r}}\,\mathrm{d}\mathbf{c}=0 \tag{5.41}$$

for the equations of conservation of mass, momentum, and energy. Remembering that $\mathbf{C}=\mathbf{c}-\mathbf{u}$, that $m\mathbf{c}$ and $mc^2/2$ are functions of c only and using the definitions of n, $\mathbf{u}$, and $\mathscr{E}$ given in (5.16)–(5.18), the conservation equations can be written, after some manipulation, in the more useful forms

$$\frac{\partial n}{\partial t}=-\sum_j u_j\frac{\partial n}{\partial x_j}-n\sum_j\frac{\partial u_j}{\partial x_j} \tag{5.42}$$

$$nm\frac{\partial u_i}{\partial t} = -nm\sum_j u_j\frac{\partial u_i}{\partial x_j} - \sum_j \frac{\partial p_{ij}}{\partial x_j} \qquad i = 1, 2, 3 \tag{5.43}$$

$$\tfrac{3}{2}nk\frac{\partial T}{\partial t} = -\tfrac{3}{2}nk\sum_j u_j\frac{\partial T}{\partial x_j} - \sum_i \frac{\partial q_i}{\partial x_i} - \sum_i\sum_j p_{ij}\frac{\partial u_i}{\partial x_j} \tag{5.44}$$

where

$$p_{ij} = \int f(mC_iC_j)\,\mathrm{d}\mathbf{c} \qquad i, j = 1 \text{ to } 3 \tag{5.45}$$

and

$$q_i = \int f(\tfrac{1}{2}mC^2)C_i\,\mathrm{d}\mathbf{c} \qquad i = 1 \text{ to } 3. \tag{5.46}$$

In (5.44), the time dependence of the energy at a point arises from three separate contributions: convective, conductive, and viscous dissipative processes corresponding respectively to the three terms on the right-hand side. In formulating these equations we have employed a rectangular Cartesian coordinate system x_j ($j = 1$ to 3) and corresponding coordinates for velocity components, C_j ($j = 1$ to 3), in order to emphasize the relationship between the time derivatives of n, $\mathbf{u}$, and T and spatial derivatives. The summations in the equations therefore run over the values of the indices 1 to 3. The elements p_{ij} and q_i may be given a physical significance. For example p_{11} is the x_1-direction momentum (mC_1) of all the molecules crossing a plane of unit area perpendicular to the x_1-direction in unit time. It is therefore the force exerted on this plane in the x_1-direction (the normal stress). On the other hand, p_{21} is the x_2-direction momentum (mC_2) of all the molecules crossing the same plane perpendicular to the x_1-direction in unit time. It is therefore the force per unit area in the x_2-direction on a plane perpendicular to the x_1-direction, a shear stress. The element q_1 represents the amount of thermal kinetic energy transported across unit area perpendicular to the x_1-direction in unit time. This is the heat flux in the x_1-direction.

The elements p_{ij} and q_i can evidently be evaluated to any desired degree of accuracy once a solution for f is obtained. Since it is just these elements which reveal themselves macroscopically as viscous forces in a gas under shear, and as heat flows in a gas in a temperature gradient, we may expect that their evaluation will lead to expressions for the viscosity and thermal conductivity coefficients for gases. Corresponding to successive approximations to the function f, we can define successive approximations to the elements p_{ij} and q_i, viz. to zeroth order

$$p_{ij} = p_{ij}^{(0)} = \int f_0(mC_iC_j)\,\mathrm{d}\mathbf{c} \tag{5.47}$$

$$q_i = q_i^{(0)} = \int f_0(\tfrac{1}{2}mC^2)C_i\,\mathrm{d}\mathbf{c} \tag{5.48}$$

and to first order

$$p_{ij}^{(1)} = p_{ij}^{(0)} + \xi \int f_1(mC_iC_j)\,\mathrm{d}\mathbf{c} \tag{5.49}$$

$$q_i^{(1)} = q_i^{(0)} + \xi \int f_1(\tfrac{1}{2}mC^2)C_i\,\mathrm{d}\mathbf{c}, \tag{5.50}$$

and so on for higher approximations. The elements of any particular order can be calculated from a knowledge of the distribution function to this same order.

Using the expansion (5.36) for f in terms of reciprocal collision frequency, ξ, in the equations for p_{ij} (5.45) and q_i (5.46), the conservation equations (5.42)–(5.44) can be written in the forms:

$$\frac{\partial n}{\partial t} = \left(\frac{\partial n}{\partial t}\right)^{(0)} \tag{5.51}$$

$$\frac{\partial u_i}{\partial t} = \left(\frac{\partial u_i}{\partial t}\right)^{(0)} + \xi\left(\frac{\partial u_i}{\partial t}\right)^{(1)} + \xi^2\left(\frac{\partial u_i}{\partial t}\right)^{(2)} + \dots \qquad i = 1, 2, 3 \tag{5.52}$$

$$\frac{\partial T}{\partial t} = \left(\frac{\partial T}{\partial t}\right)^{(0)} + \xi\left(\frac{\partial T}{\partial t}\right)^{(1)} + \xi^2\left(\frac{\partial T}{\partial t}\right)^{(2)} + \dots \tag{5.53}$$

where

$$\left(\frac{\partial n}{\partial t}\right)^{(0)} = -\sum_j u_j \frac{\partial n}{\partial x_j} - n\sum_j \frac{\partial u_j}{\partial x_j} \tag{5.54}$$

$$\left(\frac{\partial u_i}{\partial t}\right)^{(0)} = -\sum_j u_j \frac{\partial u_i}{\partial x_j} - \frac{1}{nm}\sum_j \frac{\partial p_{ij}^{(0)}}{\partial x_j} \qquad i = 1, 2, 3 \tag{5.55}$$

$$\left(\frac{\partial u_i}{\partial t}\right)^{(1)} = -\frac{1}{nm}\sum_j \frac{\partial p_{ij}^{(1)}}{\partial x_j} \qquad i = 1, 2, 3 \tag{5.56}$$

$$\left(\frac{\partial T}{\partial t}\right)^{(0)} = -\sum_j u_j \frac{\partial T}{\partial x_j} - \frac{2}{3nk}\left\{\sum_i \frac{\partial q_i^{(0)}}{\partial x_i} + \sum_i\sum_j p_{ij}^{(0)} \frac{\partial u_i}{\partial x_j}\right\} \tag{5.57}$$

$$\left(\frac{\partial T}{\partial t}\right)^{(1)} = -\frac{2}{3nk}\left\{\sum_i \frac{\partial q_i^{(1)}}{\partial x_i} + \sum_i\sum_j p_{ij}^{(1)} \frac{\partial u_i}{\partial x_j}\right\}. \tag{5.58}$$

We now substitute (5.51)–(5.53) into the full expression for $\partial f/\partial t$, (5.35), and use the result on the left-hand side of the modified Boltzmann equation (5.37) together with the expansion for f in terms of ξ, (5.36). Thus we obtain the desired expansion for the left-hand side of the Boltzmann equation, which reads

$$L_0 + \xi L_1 + \xi^2 L_2 \tag{5.59}$$

where, for example

$$L_0 = \left(\frac{\partial f_0}{\partial n}\right)\left(\frac{\partial n}{\partial t}\right)^{(0)} + \left(\frac{\partial f_0}{\partial \mathbf{u}}\right).\left(\frac{\partial \mathbf{u}}{\partial t}\right)^{(0)} + \left(\frac{\partial f_0}{\partial T}\right)\left(\frac{\partial T}{\partial t}\right)^{(0)} + \mathbf{c}.\frac{\partial f_0}{\partial \mathbf{r}}. \quad (5.60)$$

Since ξ may be varied arbitrarily both sides of the Boltzmann equation, (5.38) and (5.59), are equal for any value of ξ and we can equate the coefficients of different powers of ξ independently. We find for the coefficient of ξ^{-1},

$$0 = J(f_0 f_0), \quad (5.61)$$

and for the coefficient of ξ^0,

$$\left(\frac{\partial f_0}{\partial n}\right)\left(\frac{\partial n}{\partial t}\right)^{(0)} + \left(\frac{\partial f_0}{\partial \mathbf{u}}\right)\left(\frac{\partial \mathbf{u}}{\partial t}\right)^{(0)} + \left(\frac{\partial f_0}{\partial T}\right)\left(\frac{\partial T}{\partial t}\right)^{(0)} + \mathbf{c}\cdot\frac{\partial f_0}{\partial \mathbf{r}} = J(f_0 f_1) + J(f_1 f_0). \quad (5.62)$$

We now see the advantage of the particular separation of $\partial n/\partial t$, $\partial \mathbf{u}/\partial t$, and $\partial T/\partial t$ employed in the conservation equations (5.51)–(5.58). We note that once (5.61) has been solved for f_0, the zero-order solution, this completely determines the left-hand side of (5.62) so that it may be solved for f_1. This process may be continued to obtain higher-order terms in the expansion of f.

5.5.1. *The zeroth-order solution*

Eqn (5.61) is similar in form to that which we have already solved for the equilibrium case. Applying the method used there the solution can be shown to be

$$f_0(\mathbf{r}, \mathbf{c}, t) = n\left(\frac{m}{2\pi kT}\right)^{\frac{3}{2}} \exp\frac{-m[\mathbf{c}-\mathbf{u}]^2}{2kT} \quad (5.63)$$

$$= n\left(\frac{m}{2\pi kT}\right)^{\frac{3}{2}} \exp\left\{\frac{-mC^2}{2kT}\right\} \quad (5.64)$$

$$= n\left(\frac{m}{2\pi kT}\right)^{\frac{3}{2}} \mathrm{e}^{-\mathscr{C}^2} \quad (5.65)$$

where (5.64) and (5.65) follow from the definitions of the peculiar velocity $\mathbf{C} = \mathbf{c} - \mathbf{u}$, and a dimensionless velocity $\mathscr{C}$ which is

$$\mathscr{C} = \left(\frac{m}{2kT}\right)^{\frac{1}{2}} \mathbf{C} \quad (5.66)$$

In (5.63) n, T and $\mathbf{u}$ are functions of $\mathbf{r}$ and t and have been deliberately identified with the *local* number density, temperature, and mean velocity respectively, that is using equations analogous to (5.30)–(5.32)†

$$n(\mathbf{r}, t) = \int f_0 \, d\mathbf{c}, \qquad n(\mathbf{r}, t)\mathbf{u}(\mathbf{r}, t) = \int \mathbf{c} f_0 \, d\mathbf{c} \tag{5.67}$$

and

$$\tfrac{3}{2} n(\mathbf{r}, t) kT(\mathbf{r}, t) = \int \tfrac{1}{2} m (\mathbf{c} - \mathbf{u})^2 f_0 \, d\mathbf{c}. \tag{5.68}$$

As a consequence of these identifications it follows from (5.16)–(5.18) and (5.36) that for f_1, and all higher terms in the ξ expansion of f (5.36),

$$\int f_1 \, d\mathbf{c} = \int \mathbf{c} f_1 \, d\mathbf{c} = \int \tfrac{1}{2} m (\mathbf{c} - \mathbf{u})^2 f_1 \, d\mathbf{c} = 0. \tag{5.69}$$

Eqn (5.64) is a significant result and shows that our zeroth-order solution for f does indeed correspond to that for local equilibrium and that the approach to this state is via collisions, as we argued earlier since the collision term $J(f_1 f_0)$ is involved in its evolution.

We may now use f_0 to evaluate the elements $q_i^{(0)}$ of the heat flux and the elements $p_{ij}^{(0)}$, through their defining equations (5.47) and (5.48). Using the results of Appendix A5.1 it can be shown that

$$q_i^{(0)} = 0 \qquad i = 1, 2, 3 \tag{5.70}$$

$$p_{ii}^{(0)} = P = nkT \qquad i = 1, 2, 3 \tag{5.71}$$

where P is the pressure of gas. Hence the zeroth-order solution for the normal stresses reduces to the perfect gas law. Furthermore, evaluation of the off-diagonal elements $p_{ij}^{(0)}$, the zeroth-order shear stresses, with the aid of the results of Appendix A5.1 leads to the result

$$p_{ij}^{(0)} = 0 \qquad i \neq j. \tag{5.72}$$

The equations of conservation (5.51)–(5.53) in zeroth order are therefore

$$\frac{\partial n}{\partial t} = -\sum_j u_j \frac{\partial n}{\partial x_j} - n \sum_j \frac{\partial u_j}{\partial x_j} = \left(\frac{\partial n}{\partial t}\right)^{(0)} \tag{5.73}$$

$$mn \frac{\partial u_i}{\partial t} = -nm \sum_j u_j \frac{\partial u_i}{\partial x_j} - \frac{\partial P}{\partial x_i} = mn \left(\frac{\partial u_i}{\partial t}\right)^{(0)} \tag{5.74}$$

$$\frac{3nk}{2} \frac{\partial T}{\partial t} = -\frac{3nk}{2} \sum_j u_j \frac{\partial T}{\partial x_j} - P \sum_j \frac{\partial u_j}{\partial x_j} = \frac{3nk}{2} \left(\frac{\partial T}{\partial t}\right)^{(0)} \tag{5.75}$$

which are Euler's equations of hydrodynamics for an inviscid gas.

5.5.2. The first-order solution

In order to obtain the first-order solution we must solve (5.62) for f_1. To do this we eliminate the derivatives on the left-hand side using the

† It must be emphasized that f_0, the *local* equilibrium solution, is not the same as f_E, the global equilibrium solution for a *uniform* gas.

equations of conservation in zeroth-order obtained above (5.73)–(5.75) and appropriate differentiation of the known function f_0 (5.63). In this way we obtain the following equation for f_1,

$$J(f_0f_1)+J(f_1f_0)=-f_0\left\{(\mathscr{C}^2-\tfrac{5}{2})\sum_i C_i\frac{\partial \ln T}{\partial x_i}+\left(2\sum_i\sum_j b_{ij}\frac{\partial u_i}{\partial x_j}\right)\right\} \quad (5.76)$$

$$b_{ij}=\mathscr{C}_i\mathscr{C}_j \qquad i\neq j$$

and

$$b_{ii}=\mathscr{C}_i\mathscr{C}_i-\tfrac{1}{3}\mathscr{C}^2 \quad \text{for} \quad i=1,2,3. \quad (5.77)$$

Here $\mathscr{C}_i$ denotes the ith component of the dimensionless velocity $\mathscr{C}$, that is,

$$\mathscr{C}_i=C_i\left(\frac{m}{2kT}\right)^{\frac{1}{2}}$$

and

$$\mathscr{C}^2=\sum_i \mathscr{C}_i^2.$$

We now define a new function Θ by the equation

$$f_1=f_0\Theta \quad (5.78)$$

and substitute this expression for f_1 into the left-hand side of (5.76). We then obtain an integral equation for Θ.

$$n^2I(\Theta)=-f_0\left\{(\mathscr{C}^2-\tfrac{5}{2})\sum_i C_i\frac{\partial \ln T}{\partial x_i}+2\sum_i\sum_j b_{ij}\frac{\partial u_i}{\partial x_j}\right\} \quad (5.79)$$

where, using the definition of $J(ff)$, (5.26),

$$n^2I(\Theta)=\int f_0(\mathbf{c})f_0(\mathbf{c}_1)\,\mathrm{d}\mathbf{c}_1\int b\,\mathrm{d}b\int \mathrm{d}\psi\, g\{\Theta(\mathbf{c})+\Theta(\mathbf{c}_1)-\Theta(\mathbf{c}')-\Theta(\mathbf{c}_1')\}. \quad (5.80)$$

It can be shown that the solution to (5.79) is of the form

$$\Theta=-\frac{1}{n}\left(\frac{2kT}{m}\right)^{\frac{1}{2}}\sum_j A_j(\mathscr{C})\frac{\partial \ln T}{\partial x_j}-\frac{2}{n}\sum_i\sum_j B_{ij}(\mathscr{C})\frac{\partial u_i}{\partial x_j} \quad (5.81)$$

where $A_i(\mathscr{C})$ and $B_{ij}(\mathscr{C})$ are, as yet, undetermined coefficients. Substitution of this solution into (5.79) and equating the coefficients of similar gradients leads to the following integral equations for A_i and B_{ij}.

$$nI(A_i)=f_0(\mathscr{C}^2-\tfrac{5}{2})\mathscr{C}_i \qquad i=1,2,3 \quad (5.82)$$

$$nI(B_{ij})=f_0(\mathscr{C}_i\mathscr{C}_j) \qquad i\neq j \quad (5.83)$$

$$nI(B_{ii})=f_0(\mathscr{C}_i\mathscr{C}_i-\tfrac{1}{3}\mathscr{C}^2) \qquad i=1,2,3. \quad (5.84)$$

The first of these equations shows that because the $\mathscr{C}_i$ are the components of a vector, the A_i must themselves be the components of a vector. Since the only vector available is $\boldsymbol{\mathscr{C}}$, the only possible form for $A_i(\boldsymbol{\mathscr{C}})$ is

$$A_i(\boldsymbol{\mathscr{C}}) = A(\mathscr{C})\mathscr{C}_i \tag{5.85}$$

where $A(\mathscr{C})$ is a scalar. A similar argument may be applied to the B_{ij}, which are the elements of a tensor; they must be of the form

$$B_{ij}(\boldsymbol{\mathscr{C}}) = B(\mathscr{C})\mathscr{C}_i\mathscr{C}_j \qquad i \neq j \qquad i, j = 1, 2, 3 \tag{5.86}$$

$$B_{ii}(\boldsymbol{\mathscr{C}}) = B(\mathscr{C})(\mathscr{C}_i\mathscr{C}_i - \tfrac{1}{3}\mathscr{C}^2) \qquad i = 1, 2, 3 \tag{5.87}$$

where $B(\mathscr{C})$ is a scalar.

Eqns (5.69) evidently impose conditions upon the solution for f_1 given by (5.78) and (5.81). Substitution of the solution for f_1, $f_0\Theta$, into each of (5.69) in turn and use of (5.85)–(5.87) for $A_i(\boldsymbol{\mathscr{C}})$ and $B_{ij}(\boldsymbol{\mathscr{C}})$ allows the integrals to be evaluated with the aid of Appendix A5.1. It follows that the first and third of these conditions are automatically satisfied by (5.81) but that the second condition requires

$$\int f_0 A(\mathscr{C})\mathscr{C}^2 \, d\mathbf{c} = 0. \tag{5.88}$$

The solution for f_1, namely (5.78) and (5.81), allows us to evaluate the first-order approximations to the heat flux vector $\mathbf{q}$ and the elements p_{ij} through the defining equations (5.49) and (5.50). In these evaluations we assign a value of unity to the collision frequency parameter, ξ. As a result (5.37) reduces to (5.25), the original Boltzmann equation, so that our solution now refers to this equation. We are permitted to make this assignment since we may choose the parameter arbitarily according to the conditions of its original introduction. The first-order approximation to $\mathbf{q}$ is then

$$\mathbf{q}^{(1)} = \tfrac{1}{2}m \int f_0 \Theta C^2 \mathbf{C} \, d\mathbf{c} \tag{5.89}$$

because $\mathbf{q}^{(0)} = 0$. Using (5.81) for Θ and the results of Appendix A5.1, this may be written in component form as

$$q_i^{(1)} = -\frac{m}{2n}\left(\frac{2kT}{m}\right)^{\frac{1}{2}} \int f_0 C_i C^2 A(\mathscr{C}) \sum_j \mathscr{C}_j \frac{\partial \ln T}{\partial x_j} \, d\mathbf{c}$$

or in dimensionless form

$$q_i^{(1)} = -\frac{m}{2n}\left(\frac{2kT}{m}\right)^2 \int f_0 \mathscr{C}_i \mathscr{C}^2 A(\mathscr{C}) \sum_j \mathscr{C}_j \frac{\partial \ln T}{\partial x_j} \, d\mathbf{c}. \tag{5.90}$$

Adding the components vectorially and using the results of Appendix A5.1 we obtain for the first-order heat flux vector, $\mathbf{q}^{(1)}$,

$$\mathbf{q}^{(1)} = \frac{-2k^2T^2}{3mn}\frac{\partial \ln T}{\partial \mathbf{r}}\int f_0 \mathscr{C}^4 A(\mathscr{C})\,\mathrm{d}\mathbf{c}. \tag{5.91}$$

Noting the condition (5.88) we can subtract from (5.91) any multiple of the left-hand side of (5.88) without affecting it. For later convenience we may therefore write

$$\mathbf{q}^{(1)} = \frac{-2k^2T}{3mn}\int f_0 \mathscr{C}^2(\mathscr{C}^2 - \tfrac{5}{2})A(\mathscr{C})\,\mathrm{d}\mathbf{c}\,\frac{\partial T}{\partial \mathbf{r}} \tag{5.92}$$

or finally

$$\mathbf{q}^{(1)} = -\lambda \frac{\partial T}{\partial \mathbf{r}} \tag{5.93}$$

where

$$\lambda = \frac{2k^2T}{3mn}\int f_0 \mathscr{C}^2(\mathscr{C}^2 - \tfrac{5}{2})A(\mathscr{C})\,\mathrm{d}\mathbf{c} \tag{5.94}$$

$$= \frac{2k^2T}{3m}\int \mathbf{A}\,.\,I(\mathbf{A})\,\mathrm{d}\mathbf{c} \tag{5.95}$$

making use of (5.82) and (5.85) which define $I(\mathbf{A})$ and $\mathbf{A}$.

Eqn (5.93) is Fourier's law of heat conduction and λ is the thermal conductivity coefficient. Following a similar line of argument for $p_{ij}^{(1)}$ it can be shown that

$$p_{ij}^{(1)} = \frac{-2kT}{5n}\int f_0 \mathscr{C}^4 B(\mathscr{C})\,\mathrm{d}\mathbf{c}\left\{\frac{\partial u_i}{\partial x_j} + \frac{\partial u_j}{\partial x_i}\right\} \tag{5.96}$$

because $p_{ij}^{(0)} = 0$. Although the normal stress elements $p_{ii}^{(1)}$ differ from $p_{ii}^{(0)}$ we omit them here since they are not essential to the discussion which follows.

We restrict ourselves for the moment to a gas flowing in the x_1-direction only so that $u_2 = u_3 = 0$ and for which u_1 is a function of x_2 only. This is a situation identical to that described in § 5.2.1. In this case (5.96) reduces to the form

$$p_{21}^{(1)} = -\eta \frac{\partial u_1}{\partial x_2}. \tag{5.97}$$

Now p_{21} is a shear stress as indicated in § 5.5 so that (5.97) is a statement of Newton's law of viscosity, and η is the viscosity coefficient given by

$$\eta = \frac{2kT}{5n}\int f_0 \mathscr{C}^4 B(\mathscr{C})\,\mathrm{d}\mathbf{c}. \tag{5.98}$$

If the first-order expressions for the heat flux vector $\mathbf{q}^{(1)}$ and the elements of the stress tensor $p_{ij}^{(1)}$ are substituted into the general conservation equations (5.51)–(5.53) we obtain the Navier–Stokes equations of hydrodynamics. Using the definition of the viscosity coefficient of (5.98) the elements of the pressure tensor to this first-order approximation may be written

$$p_{ij}^{(1)} = P\delta_{ij} - 2\eta S_{ij} \tag{5.96a}$$

where

$$S_{ij} = \frac{1}{2}\left(\frac{\partial u_j}{\partial x_i} + \frac{\partial u_i}{\partial x_j}\right) - \frac{1}{3}\frac{\partial}{\partial \mathbf{r}} \cdot \mathbf{u}\delta_{ij}.$$

5.5.3. *Explicit evaluation of the thermal conductivity coefficient*

In order to obtain explicit expressions for the thermal conductivity and viscosity of a simple gas it is necessary to evaluate $A(\mathscr{C})$ and $B(\mathscr{C})$ using (5.82)–(5.87). To retain simplicity we discuss in detail only the thermal conductivity and hence $A(\mathscr{C})$. The analysis for the viscosity is similar, but more complicated.

We first expand the function $A(\mathscr{C})$ as an infinite series of orthogonal polynomials called Sonine polynomials. The polynomials of interest here are defined by the equation†

$$S_{\frac{3}{2}}^{(n)}(x) = \sum_{l=0}^{n} \frac{(-x)^l(\frac{3}{2}+n)!}{(\frac{3}{2}+l)!(n-l)!l!} \tag{5.99}$$

so that

$$S_{\frac{3}{2}}^{(0)}(x) = 1 \tag{5.100}$$

$$S_{\frac{3}{2}}^{(1)}(x) = \tfrac{5}{2} - x. \tag{5.101}$$

These particular polynomials are chosen because the right-hand side of (5.82) for $A_i(\mathscr{C})$ can be directly related to a Sonine polynomial in $\mathscr{C}^2$, namely $S_{\frac{3}{2}}^{(1)}(\mathscr{C}^2)$.

We expand $A(\mathscr{C})$ in the form

$$A(\mathscr{C}) = \sum_{l=0}^{\infty} a_l S_{\frac{3}{2}}^{(l)}(\mathscr{C}^2) \tag{5.102}$$

where the coefficients a_l are independent of $\mathscr{C}$.

† These polynomials have the orthogonality property

$$\int_0^\infty e^{-x} S_{\frac{3}{2}}^{(p)}(x) S_{\frac{3}{2}}^{(q)}(x) x^{\frac{3}{2}}\, dx = 0 \qquad p \neq q$$

$$= \frac{(\frac{3}{2}+p)!}{p!} \qquad p = q.$$

Using this expansion in condition (5.88) we find that

$$\int_0^\infty e^{-\mathscr{C}^2}\mathscr{C}^3 \sum_{l=0}^{\infty} a_l S_{\frac{3}{2}}^{(0)}(\mathscr{C}^2) S_{\frac{3}{2}}^{(l)}(\mathscr{C}^2)\, d\mathscr{C}^2 = 0 \tag{5.103}$$

where use has been made of the solution for f_0, (5.65), and the fact that $S^{(0)}(\mathscr{C}^2) = 1$. The integration has also been performed over all directions for $\boldsymbol{\mathscr{C}}$ (Appendix A5.1). Using the orthogonality condition for the polynomials and the results of Appendix A5.1 again, we find that (5.103) requires that $a_0 = 0$, so that

$$\mathbf{A}(\boldsymbol{\mathscr{C}}) = A(\mathscr{C})\boldsymbol{\mathscr{C}} = \sum_{l=1}^{\infty} a_l \mathbf{a}^{(l)} \tag{5.104}$$

where

$$\mathbf{a}^{(l)} = S_{\frac{3}{2}}^{(l)}(\mathscr{C}^2)\boldsymbol{\mathscr{C}}. \tag{5.104a}$$

The values of a_l are to be determined from the integral equation (5.82). Multiplying the vectorial version of this equation by $\mathbf{a}^{(l)}$ and integrating over all values of $\mathbf{c}$ we obtain

$$\int \mathbf{a}^{(l)} \,.\, I(\mathbf{A})\, d\mathbf{c} = \int \mathbf{a}^{(l)} \,.\, I\Big(\sum_{p=1}^{\infty} a_p \mathbf{a}^{(p)}\Big)\, d\mathbf{c} = \alpha_l$$

where (5.105)

$$\alpha_l = -\frac{2}{\pi^{\frac{1}{2}}}\int_0^\infty S_{\frac{3}{2}}^{(l)}(\mathscr{C}^2)\mathscr{C}^3 S_{\frac{3}{2}}^{(1)}(\mathscr{C}^2) e^{-\mathscr{C}^2}\, d\mathscr{C}^2, \qquad l = 1 \text{ to } \infty,$$

using the definition of $S_{\frac{3}{2}}^{(1)}(\mathscr{C}^2)$ and the result for f_0 (5.65). Carrying out the integration and noting the orthogonality conditions we find that

$$\alpha_l = -\tfrac{15}{4} \qquad l = 1,$$
$$\alpha_l = 0 \qquad l \neq 1. \tag{5.106}$$

Now, if we let

$$a_{pl} = \int \mathbf{a}^{(l)} \,.\, I(\mathbf{a}^{(p)})\, d\mathbf{c} \tag{5.107}$$

then (5.105) reads

$$\sum_{p=1}^{\infty} a_p a_{pl} = \alpha_l \quad \text{for} \quad l = 1 \text{ to } \infty. \tag{5.108}$$

Since the functions a_{pl} are known, in principle at least, because the functions $\mathbf{a}^{(p)}$ are the known Sonine polynomials, (5.108) constitute an infinite set of equations for the coefficients a_p.

The thermal conductivity may be related to the coefficients a_p since from (5.95) and (5.104)

$$\lambda = \frac{2k^2T}{3m}\sum_{p=1}^{\infty} a_p \int \mathbf{a}^{(p)} \cdot I(\mathbf{A})\, \mathrm{d}\mathbf{c}$$

which owing to (5.105) and (5.106) simplifies to

$$\lambda = -\frac{5k^2T}{2m}a_1. \tag{5.109}$$

Hence the thermal conductivity depends only on the coefficient a_1.

Now it can be shown that an mth-order approximation to any of the coefficients a_p (denoted by $a_p^{(m)}$) may be obtained by replacing the infinite set of equations (5.108) by a finite set of m equations. Thus, in the mth-order approximation,

$$\sum_{p=1}^{m} a_p^{(m)} a_{pl} = \alpha_l \qquad l = 1 \text{ to } m. \tag{5.110}$$

Because in practice the first approximation to a_1 differs by only 1 per cent or so from its true value and because its derivation contains the essential physics of interest we shall work here with that approximation. Thus, to first order, since $\alpha_1 = -\frac{15}{4}$

$$a_1^{(1)} = -\frac{15}{4a_{11}} \tag{5.111}$$

so that a first-order approximation to the thermal conductivity $[\lambda]_1$ may be written

$$[\lambda]_1 = +\frac{75k^2T}{8ma_{11}}. \tag{5.112}$$

Finally, we may evaluate the function a_{11}, through (5.107), (5.104a), and (5.80) for $I(\Theta)$, to obtain

$$a_{11} = \int S_{\frac{3}{2}}^{(1)}(\mathscr{C}^2)\boldsymbol{\mathscr{C}} \cdot I(S_{\frac{3}{2}}^{(1)}(\mathscr{C}^2)\boldsymbol{\mathscr{C}})\, \mathrm{d}\mathbf{c} = \frac{1}{n^2}\iiiint f_0(\mathbf{c}_1)f_0(\mathbf{c})S_{\frac{3}{2}}^{(1)}(\mathscr{C}^2)\boldsymbol{\mathscr{C}}$$
$$\times[S_{\frac{3}{2}}^{(1)}(\mathscr{C}^2)\boldsymbol{\mathscr{C}} + S_{\frac{3}{2}}^{(1)}(\mathscr{C}_1^2)\boldsymbol{\mathscr{C}}_1 - S_{\frac{3}{2}}^{(1)}(\mathscr{C}'^2)\boldsymbol{\mathscr{C}}' - S_{\frac{3}{2}}^{(1)}(\mathscr{C}_1'^2)\boldsymbol{\mathscr{C}}_1']gb\, \mathrm{d}b\, \mathrm{d}\psi\, \mathrm{d}\mathbf{c}\, \mathrm{d}\mathbf{c}_1 \tag{5.113}$$

Here, the primed variables correspond to the velocities after a binary collision. Because of symmetry in the collision we could write a similar equation for a_{11} wherein primed and unprimed variables were exchanged; also a further expression for a_{11} can be written for each permutation of primed and unprimed variables by an exchange of $\mathscr{C}$ and $\mathscr{C}_1$. By adding these four equations and using (5.65) for f_0 we obtain:

$$a_{11}=\frac{1}{4\pi^3}\iiiint e^{-(\mathscr{C}^2+\mathscr{C}_1^2)}\{S_{\frac{3}{2}}^{(1)}(\mathscr{C}'^2)\boldsymbol{\mathscr{C}}'+S_{\frac{3}{2}}^{(1)}(\mathscr{C}_1'^2)\boldsymbol{\mathscr{C}}_1'-S_{\frac{3}{2}}^{(1)}(\mathscr{C}^2)\boldsymbol{\mathscr{C}}$$

$$-S_{\frac{3}{2}}^{(1)}(\mathscr{C}_1^2)\boldsymbol{\mathscr{C}}_1\}^2 gb\, db\, d\psi\, d\boldsymbol{\mathscr{C}}d\boldsymbol{\mathscr{C}}_1 \quad (5.114)$$

where the definition of the reduced velocity (5.66) has been invoked. Noting the definition of $S_{\frac{3}{2}}^{(1)}(\mathscr{C}^2)$, $(\frac{5}{2}-\mathscr{C}^2)$, and making use of the results of Appendix A5.1 it can be shown that (5.114) reduces to

$$a_{11}=\frac{1}{4\pi^3}\iiiint e^{-(\mathscr{C}^2+\mathscr{C}_1^2)}[\mathscr{C}'^2\boldsymbol{\mathscr{C}}'+\mathscr{C}_1'^2\boldsymbol{\mathscr{C}}_1'-\mathscr{C}^2\boldsymbol{\mathscr{C}}-\mathscr{C}_1^2\boldsymbol{\mathscr{C}}_1]^2 gb\, db\, d\psi\, d\boldsymbol{\mathscr{C}}\, d\boldsymbol{\mathscr{C}}_1. \quad (5.115)$$

Using the definitions of the centre-of-mass velocity for the collision, **G**, (5.22) and the relative velocity, **g**, (5.21) and rendering both dimensionless by the transformations.

$$\left.\begin{aligned} \boldsymbol{\mathscr{G}}&=\left(\frac{m}{2kT}\right)^{\frac{1}{2}}\mathbf{G}, \qquad \boldsymbol{\gamma}=\left(\frac{m}{2kT}\right)^{\frac{1}{2}}\mathbf{g} \\ \boldsymbol{\gamma}'&=\left(\frac{m}{2kT}\right)^{\frac{1}{2}}\mathbf{g}' \end{aligned}\right\} \quad (5.116)$$

we can write

$$a_{11}=\frac{1}{4\pi^3}\iiiint e^{-2\mathscr{G}^2}e^{-\frac{1}{2}\gamma^2}[\gamma^2(\boldsymbol{\gamma}\,.\boldsymbol{\mathscr{G}})^2+\gamma'^2(\boldsymbol{\gamma}'\,.\boldsymbol{\mathscr{G}})^2$$

$$-2(\boldsymbol{\gamma}\,.\,\boldsymbol{\gamma}')(\boldsymbol{\gamma}\,.\,\boldsymbol{\mathscr{G}})(\boldsymbol{\gamma}'\,.\,\boldsymbol{\mathscr{G}})]gb\, db\, d\psi\, d\boldsymbol{\mathscr{G}}\, d\boldsymbol{\gamma}$$

The integration over $\boldsymbol{\mathscr{G}}$ can be carried out[5] and leads to

$$a_{11}=\frac{\sqrt{2}}{64\pi^{\frac{3}{2}}}\iiint e^{-\frac{1}{2}\gamma^2}\{\gamma^4+\gamma'^4-2(\boldsymbol{\gamma}\,.\,\boldsymbol{\gamma}')^2\}gb\, db\, d\psi\, d\boldsymbol{\gamma}$$

or since for an elastic collision $\gamma^2=\gamma'^2$ and $\boldsymbol{\gamma}\,.\,\boldsymbol{\gamma}'=\gamma^2\cos\chi$ where χ is the scattering angle of Fig. 5.4

$$a_{11}=\frac{\sqrt{2}(2kT/m)^{\frac{1}{2}}}{32\pi^{\frac{3}{2}}}\iiint e^{-\frac{1}{2}\gamma^2}\gamma^5(1-\cos^2\chi)b\, db\, d\psi\, d\boldsymbol{\gamma}.$$

The integration over the angular dependence of $\boldsymbol{\gamma}$ can be carried out using spherical polar coordinates and the integration over ψ is straightforward so that finally

$$a_{11}=\frac{1}{4}\left(\frac{kT}{\pi m}\right)^{\frac{1}{2}}\int_0^\infty \exp(-\tfrac{1}{2}\gamma^2)\gamma^7\left\{2\pi\int_0^\infty(1-\cos^2\chi)b\, db\right\}d\gamma$$

$$=4\left(\frac{kT}{\pi m}\right)^{\frac{1}{2}}\int_0^\infty \exp(-\gamma^2)\gamma^7\left\{2\pi\int_0^\infty(1-\cos^2\chi)b\, db\right\}d\gamma \quad (5.118)$$

$$=8\left(\frac{kT}{\pi m}\right)^{\frac{1}{2}}\bar{\Omega}^{(2,2)}(T). \quad (5.119)$$

Consequently, the first-order expression for the thermal conductivity (5.112) can be written

$$[\lambda]_1 = \frac{75}{64}\left(\frac{k^3 T\pi}{m}\right)^{\frac{1}{2}} \frac{1}{\bar{\Omega}^{(2,2)}(T)}$$

or

$$[\lambda]_1 = \frac{25}{32}\left(\frac{kT\pi}{m}\right)^{\frac{1}{2}} \frac{c_v}{\bar{\Omega}^{(2,2)}(T)} \tag{5.120}$$

where in the last expression we have invoked the definition of c_v for a monatomic gas $c_v = 3k/2$. This result may be compared with that of the simple kinetic theory obtained in § 5.2.

The quantity $\bar{\Omega}^{(2,2)}(T)$ is known as a *collision integral* and is defined by

$$\bar{\Omega}^{(2,2)}(T) = \frac{1}{6}\int_0^\infty Q^{(2)}(g)\mathrm{e}^{-\gamma^2}\gamma^6\,\mathrm{d}(\gamma^2) = \tfrac{1}{6}(kT)^{-4}\int_0^\infty Q^{(2)}(E)\mathrm{e}^{-E/kT}E^3\,\mathrm{d}E \tag{5.121}$$

Here we have introduced $E = \gamma^2 kT = \frac{1}{4}mg^2$, the relative kinetic energy of the collision, and the *transport cross-section* $Q^{(2)}(E)$ defined by

$$Q^{(2)}E = 3\pi\int_0^\infty (1-\cos^2\chi)b\,\mathrm{d}b. \tag{5.122}$$

The classical deflection angle χ has been related to the intermolecular potential energy function $U(r)$ for spherical molecules in Chapter 4; it is given by the expression (4.15)

$$\chi(E, b) = \pi - 2b\int_{r_0}^\infty \frac{\mathrm{d}r/r^2}{\sqrt{\{1 - b^2/r^2 - U(r)/E\}}} \tag{5.123}$$

where r_0 is the classical distance of closest approach in a collision. Eqns (5.120)–(5.123) show how the intermolecular pair potential for molecules of the gas influences its thermal conductivity.

Arguments analogous to those presented above can be used to obtain a first-order expression for the viscosity of a pure, monatomic gas. The resulting expression for this quantity $[\eta]_1$ is

$$[\eta]_1 = \frac{5}{16}\frac{(m\pi kT)^{\frac{1}{2}}}{\bar{\Omega}^{(2,2)}(T)}, \tag{5.124}$$

which involves the same collision integral $\bar{\Omega}^{(2,2)}(T)$ as occurs in the expression for the thermal conductivity. This arises because the transport of momentum and energy are dominated by binary molecular collisions of the same type. Thus, to a first-order approximation we find

$$[\lambda]_1 = \frac{15k}{4m}[\eta]_1$$
$$= \frac{5}{2m}[\eta]_1 c_v. \quad (5.125)$$

Thus the proportionality between the thermal conductivity and viscosity of a pure monatomic gas, noted in § 5.2, is confirmed by rigorous theory to a first-order approximation.

5.6. Transport coefficients for pure gases

For rigid spherical molecules of diameter d, it may be shown that the deflection angle χ is given by the expression $\cos\frac{1}{2}\chi = b/d$ (Chapter 4, § 4.3.1). Substitution of this result into the expressions above allows explicit evaluation of $\bar{\Omega}^{(2,2)}(T)$ for this case; we find that

$$\bar{\Omega}^{(2,2)}_{\mathrm{rs}}(T) = \pi d^2. \quad (5.126)$$

For an intermolecular potential of the form $U(r) = \varepsilon f(r/\sigma)$, where ε and σ are the characteristic energy and size parameters, it is convenient to define reduced collision integrals $\Omega^{(2,2)*}(T^*)$ by

$$\Omega^{(2,2)*}(T^*) = \bar{\Omega}^{(2,2)}/\bar{\Omega}^{(2,2)}_{\mathrm{rs}}(T) \quad (5.127)$$

where $\bar{\Omega}^{(2,2)}_{\mathrm{rs}}(T)$ is the collision integral for rigid spheres of diameter σ and the reduced temperature, T^*, is defined by $T^* = kT/\varepsilon$.

By defining additional reduced variables

$$r^* = r/\sigma, \qquad b^* = b/\sigma, \qquad U^* = U/\varepsilon, \qquad E^* = E/\varepsilon,$$

(5.121)–(5.123) become

$$\Omega^{(2,2)*}(T^*) = \frac{1}{6T^{*4}} \int_0^\infty \mathrm{e}^{-E^*/T^*} E^{*3} Q^{(2)*}(E^*)\,\mathrm{d}E^* \quad (5.128)$$

where

$$Q^{(2)*}(E^*) = 3\int_0^\infty (1 - \cos^2\chi) b^*\,\mathrm{d}b^* \quad (5.129)$$

and

$$\chi(E^*, b^*) = \pi - 2b^* \int_{r_0^*}^\infty \frac{\mathrm{d}r^*/r^{*2}}{\sqrt{\left\{1 - \frac{b^{*2}}{r^{*2}} - \frac{U^*(r^*)}{E^*}\right\}}} \quad (5.130)$$

with

$$r_0^* = r_0/\sigma.$$

In terms of this *reduced collision integral* the first-order expression for the thermal conductivity of a gas obtained in the previous section may be written

$$[\lambda]_1 = \frac{75}{64}\left(\frac{k^3T}{m\pi}\right)^{\frac{1}{2}} \frac{1}{\sigma^2\Omega^{(2,2)*}(T^*)}\,. \tag{5.131}$$

Similarly the first approximation to the viscosity is given by the expression

$$[\eta]_1 = \frac{5}{16}\left(\frac{mkT}{\pi}\right)^{\frac{1}{2}} \frac{1}{\sigma^2\Omega^{(2,2)*}(T^*)}\,. \tag{5.132}$$

To obtain higher-order approximations to the transport properties, it is necessary to solve (5.110) for higher values of m. The mathematics of the solution becomes progressively more complicated but it is found that (5.131) and (5.132) represent a very good approximation to the true values of the viscosity and thermal conductivity. We can write for the mth approximation to η and λ

$$[\eta]_m = [\eta]_1 f_\eta^{(m)} \tag{5.133}$$

and

$$[\lambda]_m = [\lambda]_1 f_\lambda^{(m)} \tag{5.134}$$

so that

$$\frac{[\lambda]_m}{[\eta]_m} = \frac{15}{4}\frac{k}{m}\frac{f_\lambda^{(m)}}{f_\eta^{(m)}} = \frac{5}{2}\,c_v\,\frac{f_\lambda^{(m)}}{f_\eta^{(m)}}\,. \tag{5.135}$$

Eqn (5.135) is the exact kinetic theory result for the relationship between the viscosity and thermal conductivity of a monatomic gas. The higher order correction factors $f_\eta^{(m)}$ and $f_\lambda^{(m)}$ differ from unity by only 1 or 2 per cent for all values of m over a wide temperature range. Thus the exact kinetic theory results confirm those of the simple theory of § 5.2, in that the viscosity and thermal conductivity for a monatomic gas are nearly proportional to each other. Expressions for the higher-order correction factors are given for $m = 2, 3$ in Appendix A5.2. They are functions of generalized collision integrals, $\bar{\Omega}^{(l,s)}(T)$, analogous to that defined by (5.121) and (5.122). Thus,

$$\bar{\Omega}^{(l,s)}(T) = [(s+1)!(kT)^{s+2}]^{-1}\int_0^\infty Q^{(l)}(E)\mathrm{e}^{-E/kT}E^{s+1}\,\mathrm{d}E \tag{5.136}$$

where the transport cross-section $Q^{(l)}(E)$ is

$$Q^{(l)}(E)=2\pi\left[1-\frac{1+(-1)^l}{2(1+l)}\right]^{-1}\int_0^\infty(1-\cos^l\chi)b\,\mathrm{d}b \tag{5.137}$$

with the deflection angle χ given by (5.123).† The corresponding reduced collision integrals are defined by the equations

$$\Omega^{(l,s)*}(T^*)=[(s+1)!T^{*s+2}]^{-1}\int_0^\infty Q^{(l)*}(E^*)\mathrm{e}^{-E^*/T^*}E^{*s+1}\,\mathrm{d}E^* \tag{5.138}$$

and

$$Q^{(l)*}(E^*)=2\left[1-\frac{1+(-1)^l}{2(1+l)}\right]^{-1}\int_0^\infty(1-\cos^l\chi)b^*\,\mathrm{d}b^* \tag{5.139}$$

with the deflection angle χ given by (5.130).

5.6.1. *Formulae for special intermolecular potential models*

Two particular simple forms of intermolecular potential have special historical significance for the transport properties of gases. If we first presume that the molecules are rigid elastic spheres of diameter d then all the reduced collision integrals $\Omega^{(l,s)*}(T^*)$ are temperature-independent and equal to unity so that according to the rigorous kinetic theory the viscosity coefficient, to a first-order approximation, is given by

$$[\eta]_1=\frac{5(\pi mkT)^{\frac{1}{2}}}{16\pi d^2} \tag{5.140}$$

and the thermal conductivity by

$$[\lambda]_1=\frac{25}{32}\left(\frac{kT\pi}{m}\right)^{\frac{1}{2}}\frac{c_v}{\pi d^2} \tag{5.141}$$

where $c_v=3k/2$ for rigid spheres. These results may be compared with the expressions derived on the basis of simple mean free path arguments

† The collision integrals $\bar{\Omega}^{(l,s)}(T)$ are defined so that for a rigid sphere of diameter d they are equal to πd^2 for all l and s. This is done to emphasize their significance as energy averaged cross-sections. They therefore differ from the collision integrals $\Omega^{(l,s)}(T)$ defined by Chapman and Cowling[6] and Hirschfelder, Curtiss, and Bird,[7] but are related to them by the equation

$$\bar{\Omega}^{(l,s)}(T)=2\left\{(s+1)!\left[1-\frac{1+(-1)^l}{2(1+l)}\right]\right\}^{-1}\left(\frac{2\pi m_i m_j}{kT(m_i+m_j)}\right)^{\frac{1}{2}}\Omega^{(l,s)}(T)$$

for the case of an interaction between two unlike species of masses m_i and m_j. The reduced collision integrals $\Omega^{(l,s)*}$ are, however, identical with those of other authors.

earlier (see § 5.2). The dependence of the transport coefficients on microscopic and macroscopic quantities is identical for the two theories although the numerical factors determining the absolute values of the coefficients differ.

A second simple potential model which has proved of great value in interpreting transport property data at both very high and very low temperatures is that which assumes the molecules interact as point centres of repulsion or attraction. To examine this potential model we suppose that the intermolecular potential is given by $U(r) = C_n/r^n$, where C_n is positive for a repulsive potential and negative if the potential is attractive.

For such potentials it may be shown that the collision integrals $\bar{\Omega}^{(2,2)}$ take the form

$$\bar{\Omega}^{(2,2)} = \frac{\pi}{2}\left[\frac{n\,|C_n|}{kT}\right]^{2/n} S^{(2)}(n, C_n)\Gamma(4-(2/n))$$

where $S^{(2)}$ is a numerical constant for a particular potential function that depends on the magnitude of n but only on the sign of C_n, and Γ is a gamma function. Values for $S^{(2)}(n, C_n)$ for several typical values of n and for the two signs of C_n, which have been evaluated by quadrature, are given in Table 5.2. It can be seen that the dependence of $S^{(2)}$ upon the magnitude of n is very weak.

For a gas composed of molecules of this type we see that the thermal conductivity and viscosity can be written in the forms

$$[\lambda]_1 = \frac{25}{16}\left(\frac{k}{m\pi}\right)^{\frac{1}{2}}\left(\frac{k}{n\,|C_n|}\right)^{2/n} c_v T^{(\frac{1}{2}+2/n)}\Big/\Gamma\left(4-\frac{2}{n}\right)S^{(2)}(n, C_n) \quad (5.142)$$

$$[\eta]_1 = \frac{5}{8}\left(\frac{km}{\pi}\right)^{\frac{1}{2}}\left(\frac{k}{n\,|C_n|}\right)^{2/n} T^{(\frac{1}{2}+2/n)}\Big/\Gamma\left(4-\frac{2}{n}\right)S^{(2)}(n, C_n). \quad (5.143)$$

These relations show that the existence of intermolecular forces modifies the temperature dependence of the transport coefficients from that of the

Table 5.2

n	$S^{(2)}(n, C_n)$	$S^{(2)}(n, C_n)$
	$C_n(+ve)$	$C_n(-ve)$
4	0·308	
6	0·283	0·328
8	0·279	
10	0·278	
12	0·279	
14	0·280	
20	0·286	
24	0·289	
∞(rigid sphere)	0·333	

rigid sphere model. It is interesting to note here that although measurements of λ or η as functions of temperature would serve to determine the exponent n in the intermolecular potential the same measurements could not be used to determine C_n uniquely since its sign could not be established. However, if the sign of C_n were known from some other source then the entire potential function could be determined uniquely.

In practice, the sign of a realistic intermolecular potential at extremely large and extremely small separations is known from the physical reasoning given in Chapter 1. At the extremes of high and low temperatures, only a monotonic repulsive potential or a monotonic attractive potential contributes significantly to the collision integral even for real intermolecular potentials. By approximating the intermolecular potential in these regions by an inverse power function it is therefore possible to determine these sections of the potential uniquely. For example the discussion in Chapters 1 and 2 has demonstrated that at large separations the intermolecular potential between two neutral atoms can be written $U = C_6/r^6$. Thus measurement of the viscosity of gases at very low temperatures, where the viscosity will vary as $T^{\frac{5}{6}}$, can in principle be used to determine the value of C_6 through the above relationships.

5.7. Transport coefficients for binary mixtures

In a binary gas mixture there are three types of two-body collisions which may occur, each characterized by a different intermolecular potential function, viz. $U_{11}(r)$, $U_{22}(r)$, and $U_{12}(r)$ having parameters ε_{11}, σ_{11}, ε_{22}, σ_{22}, ε_{12}, and σ_{12}, respectively. By arguments similar to those presented for pure gases, it is possible to deduce expressions for the viscosity and thermal conductivity of a binary mixture which are dependent on reduced collision integrals, defined by (5.130), (5.137), and (5.139) for each pair interaction. Denoting the collision integral corresponding to U_{12} by $\Omega_{12}^{(l,s)*}(T_{12}^*)$ where $T_{12}^* = kT/\varepsilon_{12}$ and those corresponding to U_{11} by $\Omega_{11}^{(l,s)*}(T_{11}^*)$, etc., the viscosity of a binary gas mixture may be written to a first-order approximation

$$[\eta_{\text{mix}}]_1 = \frac{1+Z_\eta}{X_\eta + Y_\eta} \tag{5.144}$$

where

$$X_\eta = \frac{x_1^2}{[\eta]_1} + \frac{2x_1x_2}{[\eta_{12}]_1} + \frac{x_2^2}{[\eta_2]_1}$$

$$Y_\eta = \tfrac{3}{5}A_{12}^\star\left[\frac{x_1^2}{[\eta_1]_1}\left(\frac{m_1}{m_2}\right) + \frac{2x_1x_2}{[\eta_{12}]_1}\left\{\frac{(m_1+m_2)^2}{4m_1m_2}\right\}\left(\frac{[\eta_{12}]_1^2}{[\eta_1]_1[\eta_2]_1}\right) + \frac{x_2^2}{[\eta_2]_1}\left(\frac{m_2}{m_1}\right)\right]$$

$$Z_\eta = \tfrac{3}{5}A_{12}^\star\left[x_1^2\frac{m_1}{m_2} + 2x_1x_2\left\{\frac{(m_1+m_2)^2}{4m_1m_2}\left(\frac{[\eta_{12}]_1}{[\eta_1]_1} + \frac{[\eta_{12}]_1}{[\eta_2]_1}\right) - 1\right\} + x_2^2\left(\frac{m_2}{m_1}\right)\right]$$

where

$$[\eta_1]_1 = \frac{5}{16}\left(\frac{m_1 kT}{\pi}\right)^{\frac{1}{2}} \frac{1}{\sigma_{11}^2 \Omega_{11}^{(2,2)*}(T_{11}^*)}$$

$$[\eta_2]_1 = \frac{5}{16}\left(\frac{m_2 kT}{\pi}\right)^{\frac{1}{2}} \frac{1}{\sigma_{22}^2 \Omega_{22}^{(2,2)*}(T_{22}^*)}$$

are the first-order approximations to the viscosities of the pure gases and

$$[\eta_{12}]_1 = \frac{5}{16}\left(\frac{2m_1 m_2 kT/(m_1+m_2)}{\pi}\right)^{\frac{1}{2}} \frac{1}{\sigma_{12}^2 \Omega_{12}^{(2,2)*}(T_{12}^*)} \tag{5.145}$$

is the so-called *interaction viscosity*, which corresponds to the viscosity of a hypothetical gas of molecules with a mass equal to twice the reduced mass of the two species in the mixture whose intermolecular potential is $U_{12}(r)$. In addition, x_1 and x_2, are the mole fractions of the two species. The function $A_{12}^{\star}$ is the ratio of two collision integrals,

$$A_{12}^{\star}(T_{12}^*) = \Omega_{12}^{(2,2)*}(T_{12}^*)/\Omega_{12}^{(1,1)*}(T_{12}^*). \tag{5.146}$$

An expression can also be written for the thermal conductivity of a binary gas mixture to first order and is given in Appendix A5.2. The formulae for the viscosity and thermal conductivity of binary mixtures can in principle be extended to a higher-order approximation than the first; however, the formulae rapidly become complicated. They are given in Appendix A5.2.

For a binary mixture two other transport coefficients may be defined which are of importance to the investigations of intermolecular forces. The first is the *binary diffusion coefficient*, D_{12}, defined by Fick's law

$$\mathbf{J}_1 = -D_{12}\frac{\partial n_1}{\partial \mathbf{r}}$$

where $\mathbf{J}_1$ is the number flux of species 1 in an isothermal binary mixture subject to a gradient of the number density, n_1, of species 1, that is a concentration gradient. By applying the Chapman–Enskog analysis to a binary mixture it can be shown that the diffusion coefficient is given in first order by the expression

$$[D_{12}]_1 = \frac{3}{16n}\left(\frac{2kT(m_1+m_2)}{\pi m_1 m_2}\right)^{\frac{1}{2}} \frac{1}{\sigma_{12}^2 \Omega_{12}^{(1,1)*}(T_{12}^*)}. \tag{5.147}$$

Thus, the diffusion coefficient depends on the collision integral $\bar{\Omega}^{(1,1)}$ unlike viscosity and thermal conductivity. Furthermore, within the first-order approximation the diffusion coefficient depends only upon the unlike interaction potential $U_{12}(r)$. By means of the definition of the collision integral ratio $A_{12}^{\star}$ (5.146) we can relate the diffusion coefficient to the interaction viscosity $[\eta_{12}]_1$, and write

$$[D_{12}]_1 = \frac{3}{5} A_{12}^{\star} \left(\frac{m_1 + m_2}{m_1 m_2} \right) [\eta_{12}]_1 \frac{1}{n}. \tag{5.148}$$

To a higher order of approximation we can write

$$[D_{12}]_m = [D_{12}]_1 f_D^{(m)} \tag{5.149}$$

where $f_D^{(m)}$ departs from unity by only a few per cent and depends upon the two like intermolecular potentials for the mixture as well as the unlike interaction (Appendix A5.2). An analogous transport coefficient to the binary diffusion coefficient can be defined for a pure gas; to first order it is

$$[D_{11}]_1 = \frac{3}{8n} \left(\frac{kT}{\pi m} \right)^{\frac{1}{2}} \frac{1}{\sigma_{11}^2 \Omega_{11}^{(1,1)*}(T_{11}^*)}. \tag{5.150}$$

This is the *self-diffusion coefficient* described in § 5.2.3.

The final transport property of interest here is the *thermal diffusion factor.* This factor describes the extent of the partial separation of an initially uniform gas mixture which occurs when it is subject to a temperature gradient. The effect can be expressed phenomenologically by the relation

$$\frac{\partial x_1}{\partial \mathbf{r}} = -\alpha_T x_1 x_2 \frac{\partial (\ln T)}{\partial \mathbf{r}} \tag{5.151}$$

where α_T is the thermal diffusion factor. The kinetic theory permits evaluation of the thermal diffusion factor in an analogous fashion to that for the other transport coefficients. To a second-order approximation,† within this framework (the Chapman–Cowling approximation scheme), the thermal diffusion factor may be written in the form

$$[\alpha_T]_1 = (6C_{12}^{\star} - 5) \frac{S_1 x_1 - S_2 x_2}{Q_1 x_1^2 + Q_2 x_2^2 + Q_{12} x_1 x_2} \tag{5.152}$$

where

$$C_{12}^{\star} = \frac{\Omega_{12}^{(1,2)*}}{\Omega_{12}^{(1,1)*}} \tag{5.153}$$

and S_1, S_2, Q_1, Q_2, and Q_{12} are complicated functions of molecular masses and collision integrals $\Omega^{(l,s)*}$ for all three binary interactions in the gas. Detailed formulae for these functions have been collected in Appendix A5.2.

† To a first-order approximation equivalent to that used in obtaining the thermal conductivity from (5.110) $\alpha_T = 0$ identically. For this reason thermal diffusion is often termed a second-order effect.

In the particular case of a binary mixture of isotopes of the same species, which will prove of interest later, these equations simplify considerably. This is because the two isotopes will differ only in their molecular mass and specifically $U_{11}(r) = U_{22}(r) = U_{12}(r)$, $\sigma_{11} = \sigma_{22} = \sigma_{12}$, and $\varepsilon_{11} = \varepsilon_{22} = \varepsilon_{12}$. In this case (5.152), after expansion in powers of $(m_1 - m_2)/(m_1 + m_2)$, becomes

$$[\alpha_T]_1 = \left\{\frac{15(6C^\star - 5)(2A^\star + 5)}{2A^\star(16A^\star - 12B^\star + 55)}\right\}\left(\frac{m_1 - m_2}{m_2 + m_2}\right) = [\alpha_0]_1\left(\frac{m_1 - m_2}{m_1 + m_2}\right) \tag{5.154}$$

provided that $(m_1 - m_2)/(m_1 + m_2) \ll 1$. Here, $[\alpha_0]_1$ represents the first approximation to the *isotopic thermal diffusion factor* and

$$B^\star = \frac{5\Omega^{(1,2)*} - 4\Omega^{(1,3)*}}{\Omega^{(1,1)*}} \tag{5.155}$$

where the subscripts have been dropped owing to the equivalence of the intermolecular potentials.

A different method of approximating the solutions of the equations equivalent to (5.110) for the Sonine polynomial coefficients was developed by Kihara.[7–9] This corresponds to setting $B^\star = \frac{5}{4}$ in (5.154) and leads to the result

$$[\alpha_0]_1^K = \frac{15(6C^\star - 5)}{16A^\star} \tag{5.156}$$

in the first approximation and to

$$[\alpha_0]_2^K = [\alpha_0]_1 f_\alpha^{K(2)} \tag{5.157}$$

in the second approximation. The second-order correction term, $f_\alpha^{K(2)}$ is given in Appendix A5.2 and is determined by higher-order collision integrals $\Omega^{(l,s)*}$. The correction factor $f_\alpha^{K(2)}$ is not very sensitive to the intermolecular potential used for its evaluation, and the Kihara approximation, denoted by the superscript K, is generally more accurate than its Chapman–Cowling counterpart to an equivalent order of approximation. For these reasons the Kihara approximation is more commonly used for the thermal diffusion factor, and will prove the more convenient for later purposes.

5.7.1. Formulae for special intermolecular potential models

For the rigid-sphere potential model it is easily verified that the binary diffusion coefficient and self-diffusion coefficient reduce to the forms

$$[D_{12}]_1 = \frac{3}{16n}\left(\frac{2kT(m_1+m_2)}{\pi m_1 m_2}\right)^{\frac{1}{2}} \frac{1}{d_{12}^2}$$

$$[D_{11}]_1 = \frac{3}{8n}\left(\frac{kT}{\pi m_1}\right)^{\frac{1}{2}} \frac{1}{d_{11}^2}$$

or in terms of the pressure $P = nkT$

$$[D_{12}]_1 = \frac{3}{16P}\left(\frac{2k^3T^3(m_1+m_2)}{\pi m_1 m_2}\right)^{\frac{1}{2}} \frac{1}{d_{12}^2} \tag{5.158}$$

$$[D_{11}]_1 = \frac{3}{8P}\left(\frac{k^3T^3}{\pi m_1}\right)^{\frac{1}{2}} \frac{1}{d_{11}^2} \tag{5.159}$$

where $d_{12} = \frac{1}{2}(d_{11} + d_{22})$ and d_{11} and d_{22} are the rigid-sphere diameters of the molecular species 1 and 2. The result for the self-diffusion coefficient of a gas is thus identical in functional form to that given by the simple mean free path kinetic theory, (5.12), although the numerical factor is different.

In the case when the interaction potential between the two different species in a binary mixture conforms to the equation $U_{12}(r) = C_n/r^n$, then the binary diffusion coefficient takes the form

$$[D_{12}]_1 = \frac{\frac{3}{16}\left(\frac{2k^3(m_1+m_2)}{\pi m_1 m_2}\right)^{\frac{1}{2}}\left(\frac{k}{|C_n|n}\right)^{2/n} T^{(\frac{3}{2}+2/n)}}{PS^{(1)}(n, C_n)\Gamma\left(3-\frac{2}{n}\right)}.$$

Here $S^{(1)}(n, C_n)$ is a constant for a particular intermolecular potential function evaluated by quadrature. $S^{(1)}$ depends upon the magnitude of n weakly and only upon the sign of C_n. Values of $S^{(1)}$ have been tabulated elsewhere.[10] In the limit of very low temperatures when $U_{12}(r) = C_6/r^6$ the diffusion coefficient will vary as $T^{\frac{11}{6}}$ for non-polar gas molecules.

It is possible to express the other transport coefficients of mixtures in closed form for these special intermolecular potential models. However, because such properties depend upon all possible binary interactions in the mixture, including those between like molecules, the expressions are lengthy and are not given here.

5.8. Polyatomic gases

Earlier in this chapter the behaviour of a dilute, monatomic gas was described with the aid of a distribution function, f, giving the number of molecules in the gas with positions and translational velocities in particular ranges at a given time. An equation describing the evolution of f in a non-uniform gas was obtained (the Boltzmann equation), and its solution employed to evaluate the transport coefficients of a gas in terms of the dynamics of binary collisions. For a gas containing polyatomic molecules such a distribution function is inadequate because a complete description of the behaviour of the molecules requires a knowledge of the number of molecules in particular *internal* states which are within the prescribed range of configuration and velocity space.

5.8.1. *Quantum mechanical theory*

The quantization of the internal energy of molecules implies that strictly a full quantum mechanical treatment of the kinetic theory is necessary. Such a treatment is, however, complicated by the fact that quantum mechanically it is not possible to specify the position and velocity of a molecule simultaneously. The distribution function, f, employed in the classical treatment of monatomic gases can therefore no longer be employed. Instead it is necessary to employ the *Wigner distribution function*[13] to describe the probability of finding molecules in particular regions of configuration and velocity space *and* in specified internal states.† A kinetic theory has been developed in which a Boltzmann-type equation is formulated for the Wigner distribution function in a non-uniform gas; the resulting equation is known as the *Waldmann–Snider* equation.[14,15] As in the case of classical monatomic gases the dynamics of binary molecular collisions influence the solution to this equation. However, in addition to the conservation of mass, linear momentum, and energy characteristic of monatomic encounters, a fourth conservation condition, for the angular momentum, must also be considered. The Waldmann–Snider equation has been solved subject to these constraints, and expressions for some of the transport coefficients have been given.[16–18]

The quantum mechanical kinetic theory and the resulting expressions for the transport coefficients of the polyatomic gas are exceedingly complicated. For this reason, the expressions have not yet been employed for studies of intermolecular forces. However, the theory predicts the existence of several new transport coefficients for a polyatomic gas which do not exist for monatomic species. These coefficients are mainly sensitive to the non-spherical part of the intermolecular potential and because they are likely in the future to prove of some value in the study of such interactions, they are discussed briefly in § 5.9.4.

For the more usual transport coefficients the quantum mechanical theory introduces only small corrections to formulae derived on the basis

† The Wigner distribution function has the property that it reduces to the classical distribution function in the classical limit.

of other, simpler, but less rigorous theories. Consequently, the majority of the work concerned with the influence of intermolecular forces on the transport properties of polyatomic gases has so far been carried out with expressions derived from these simpler theories. An entirely classical kinetic theory of polyatomic gases has been given by Taxman,[19] whereas Wang Chang and Uhlenbeck[20] have provided a semi-classical treatment.

5.8.2. *The semi-classical theory*

In the theory of Wang Chang and Uhlenbeck a distribution function, f_i, is defined for each of the possible quantized internal states of the molecules. Each of these semi-classical distribution functions describes the number of molecules in a particular internal state which have positions and translational velocities in prescribed ranges. By considering the dynamics of binary molecular collisions (which may be inelastic, involving exchange of translational and internal energy) a Boltzmann-like equation can be established for the evolution of the distribution functions f_i in a non-uniform gas. It is an essential part of the Wang Chang and Uhlenbeck approach that the dynamics of the binary encounters are treated quantum mechanically since this allows the concept of inverse collisions to be invoked, but that in all other respects the analysis is classical. The Wang Chang, Uhlenbeck, and de Boer equation for f_i is solved by methods similar to those employed for the Boltzmann equation earlier. In particular use is made of the conservation of mass, linear momentum, and total energy in binary collisions although conservation of angular momentum is not considered. From the solutions for f_i it is possible to construct expressions for the transport coefficients for the gas and its macroscopic conservation equations, following the earlier pattern.

One of the limitations of the semi-classical analysis stems from its neglect of angular momentum conservation in binary collisions. Furthermore, the treatment is unable to account for all of the transport coefficients of a polyatomic gas consisting of rotating molecules. This is because the semi-classical argument requires that the molecules be in definite internal states before and after a collision. According to the uncertainty principle this can only be valid if ΔE_{int}, the energy difference between two internal states of a molecule and the time between molecular collisions, t_{coll}, satisfy the inequality

$$\Delta E_{int} t_{coll} \geqslant \hbar.$$

In the case of rotational energy levels, which are degenerate, $\Delta E_{int} = 0$ and the inequality is not satisfied. Thus, strictly, the Wang Chang, Uhlenbeck, and de Boer theory cannot be applied to molecules with rotational energy states. This is unfortunate because rotational states are just the ones which are most significant at normal temperatures. It seems that the numerical errors in the transport coefficients arising from the use of the semi-classical expressions in such cases are small for the usual transport coefficients of the gas.[21] However the theory completely fails to account for the new transport coefficients of a polyatomic gas which are characteristic of the full quantum mechanical analysis.

5.8.3. The classical theory

The entirely classical kinetic theory of Taxman employs the same ideas as the Wang Chang, Uhlenbeck, and de Boer theory but treats both the dynamics of the binary collisions and the internal energy of the molecules classically. In a classical analysis of polyatomic molecular collisions inverse collisions cannot be assumed to exist. However, Taxman was able to show that a finite set of collisions exists whereby molecules can return to their states before a direct collision from those after the direct collision. Using this result a classical Boltzmann equation can be constructed and solved to yield expressions for the transport coefficients.

5.9. The transport coefficients of polyatomic gases

The semi-classical theory of Wang Chang, Uhlenbeck, and de Boer has been the most commonly used scheme for the study of the intermolecular forces of polyatomic molecules and their transport properties. Consequently in this section we present the formal results of this theory for the transport coefficients of a dilute polyatomic gas.

5.9.1. The semi-classical transport coefficients

According to the semi-classical theory the elements of the stress tensor, p_{ij}, are given to a first-order approximation by the expressions

$$p_{ii}^{(1)} = nkT - \tfrac{2}{3}\eta\left(2\frac{\partial u_i}{\partial x_i} - \frac{\partial u_j}{\partial x_j} - \frac{\partial u_k}{\partial x_k}\right) - \kappa\frac{\partial}{\partial \mathbf{r}}\cdot\mathbf{u} \qquad (5.161)$$

and

$$p_{ij}^{(1)} = -\eta\left\{\frac{\partial u_j}{\partial x_i} + \frac{\partial u_i}{\partial x_j}\right\}. \qquad (5.162)$$

Here, n is the number density of the gas, u_i, u_j, u_k are the Cartesian components of its flow velocity in the directions x_i, x_j, x_k and the coefficient η is the shear viscosity introduced earlier. The coefficient κ is the bulk viscosity and does not occur in the corresponding expression for the elements of the stress tensor for a dilute monatomic gas (5.96a). It therefore represents a new quantity which arises here because of the polyatomic nature of the gas.

If a small volume element of a dilute polyatomic gas is suddenly compressed the energy is initially associated with the translational motion of the molecules which is then not in equilibrium with the internal motion. After a period of time, through inelastic collisions, the energy is distributed to the internal degrees of freedom and a new equilibrium is established. Immediately after the compression the pressure, which is due to translational motion only, will be slightly higher than it would be if equilibrium between internal and translational motion were established instantaneously. Thus the phenomenon of bulk viscosity in a dilute polyatomic gas is related to the finite rate of occurrence of inelastic collisions in the gas, a fact we shall find useful later.

The heat flux vector, $\mathbf{q}$, in a polyatomic gas is, to a first-order approximation, given by an expression identical to that for a monatomic gas and

$$\mathbf{q} = -\lambda \frac{\partial T}{\partial \mathbf{r}}, \tag{5.163}$$

where λ is the thermal conductivity coefficient.

To an approximation equivalent to the first-order Chapman and Cowling scheme the shear viscosity of a dilute polyatomic gas may be written[22]

$$[\eta]_1 = \frac{5}{16} \frac{(m\pi kT)^{\frac{1}{2}}}{Z(T)}. \tag{5.164}$$

A comparison with (5.124) indicates that the quantity $Z(T)$ replaces the collision integral $\bar{\Omega}^{(2,2)}(T)$ for the monatomic case. The term $Z(T)$ does indeed incorporate all the information about binary molecular collisions which occur in the gas at a particular temperature. However, the dynamics of binary collisions between two polyatomic molecules are very much more complicated than those between atoms and the expression for $Z(T)$ is correspondingly more complicated than that for $\bar{\Omega}^{(2,2)}(T)$.

We consider a binary molecular encounter between two polyatomic molecules initially in internal quantum states denoted by i and j with an initial relative velocity represented by $\mathbf{g}$. As a result of the encounter the internal quantum states of the molecules may be changed by exchange of internal energy and transfer of translational energy to internal energy. Furthermore, because polyatomic molecules do not interact through central forces the molecules may not only be scattered through an angle χ in the initial plane of the encounter but also through an azimuthal angle ψ out of the initial plane. If we denote the energy of the ith internal quantum state of a molecule by ε_i' then the factor $Z(T)$ can be written in the form

$$Z(T)=\frac{1}{2}\left[\sum_i \exp(-\varepsilon_i)\right]^{-2}\sum_{ijkl}\exp(-\varepsilon_i-\varepsilon_j)$$
$$\times\int_0^\infty\int_0^{2\pi}\int_0^\pi[\gamma^4\sin^2\chi+\tfrac{1}{3}(\Delta\varepsilon)^2-\tfrac{1}{2}(\Delta\varepsilon)^2\sin^2\chi]\gamma^3$$
$$\times\exp(-\gamma^2)I_{ij}^{kl}\sin\chi\,\mathrm{d}\chi\,\mathrm{d}\psi\,\mathrm{d}\gamma. \tag{5.165}$$

In this expression we have written for convenience

$$\boldsymbol{\gamma}=\left(\frac{m}{4kT}\right)^{\frac{1}{2}}\mathbf{g}$$
$$\varepsilon_i=\varepsilon_i'/kT \tag{5.166}$$

and

$$\Delta\varepsilon=\varepsilon_k+\varepsilon_l-\varepsilon_i-\varepsilon_j. \tag{5.167}$$

In addition, we have employed the degeneracy-averaged differential scattering cross-section I_{ij}^{kl} defined in Chapter 4. In general the differential scattering cross-section depends upon the relative kinetic energy of the collision, and the scattering angles χ and ψ so that I_{ik}^{kl}, which is determined by the intermolecular pair potential, should be written $I_{ij}^{kl}=I_{ij}^{kl}(g,\chi,\psi)$. If the molecules possess no internal energy $\varepsilon_i=0$, and if we replace $I_{ij}^{kl}\sin\chi\,\mathrm{d}\chi$ by its classical counterpart $b\,\mathrm{d}b$ then the expression for $Z(T)$ reduces exactly to that for the collision integral $\bar{\Omega}^{(2,2)}(T)$ given by (5.121) and (5.122) for the monatomic case. The bulk viscosity, κ, may also be expressed in terms of the dynamics of binary collisions and it may be shown that if there is only one mode of internal motion

$$\kappa=\tau(nk^2Tc_{v,\mathrm{int}}/c_v^2). \tag{5.168}$$

Here, $c_{v,\mathrm{int}}$ is the constant volume internal heat capacity per molecule, c_v the total constant volume heat capacity of the gas per molecule, and n is the number density in the gas. The parameter, τ, contains the dynamical information and may be written

$$\tau^{-1}=\left(\frac{2nk}{c_{v,\mathrm{int}}}\right)\left(\frac{kT}{\pi m}\right)^{\frac{1}{2}}\left[\sum_i\exp(-\varepsilon_i)\right]^{-2}\sum_{ijkl}\exp(-\varepsilon_i-\varepsilon_j)$$
$$\times\int_0^\infty\int_0^{2\pi}\int_0^\pi(\Delta\varepsilon)^2\gamma^3\exp(-\gamma^2)I_{ij}^{kl}\sin\chi\,\mathrm{d}\chi\,\mathrm{d}\psi\,\mathrm{d}\gamma. \tag{5.169}$$

It may be thought of as the relaxation time for the approach of an initially non-equilibrium distribution of internal energy in the gas towards equilibrium. Eqn (5.168) may be generalized for the case of more than one internal mode of motion.

A quantity related to the relaxation time τ is also frequently used and is denoted by ζ_{rot}. The quantity ζ_{rot} may be thought of as the number of collisions which are required for equilibration of rotational and translational energy in the gas. This rotational collision number is defined by the equation

$$\zeta_{\text{rot}} = \left(\frac{4}{\pi}\right)\left(\frac{P\tau}{\eta}\right) \tag{5.170}$$

where $P = nkT$ is the pressure.

The thermal conductivity of a polyatomic gas may be written, to the same order of approximation, as the sum of two contributions, one from the flux of translational energy and one from the flux of internal energy.[22]

$$[\lambda]_1 = [\lambda_{\text{trans}}]_1 + [\lambda_{\text{int}}]_1 \tag{5.171}$$

The expressions for the two contributions are

$$[\lambda_{\text{trans}}]_1 = \frac{75k^2T}{8m\alpha_{11}}\left\{\frac{1-\frac{2}{5}(\alpha_{12}/\alpha_{22})(c_{v,\text{int}}/k)}{1-\alpha_{12}^2/\alpha_{11}\alpha_{22}}\right\} \tag{5.172}$$

and

$$[\lambda_{\text{int}}]_1 = \frac{3}{2\alpha_{22}}\frac{c_{v,\text{int}}^2 T}{m}\left\{\frac{1-\frac{5}{2}(\alpha_{12}/\alpha_{11})(k/c_{v,\text{int}})}{1-\alpha_{12}^2/\alpha_{11}\alpha_{22}}\right\}. \tag{5.173}$$

In these expressions the collision integrals α_{ij} may be written

$$\begin{aligned}\alpha_{11} = 4\left(\frac{kT}{\pi m}\right)^{\frac{1}{2}}&\left[\sum_i \exp(-\varepsilon_i)\right]^{-2}\sum_{ijkl}\exp(-\varepsilon_i-\varepsilon_j)\\ &\times\int_0^\infty\int_0^{2\pi}\int_0^\pi [\gamma^4\sin^2\chi + \tfrac{11}{8}(\Delta\varepsilon)^2 - \tfrac{1}{2}(\Delta\varepsilon)^2\sin^2\chi]\\ &\times\gamma^3\exp(-\gamma^2)I_{ij}^{kl}\sin\chi\,d\chi\,d\psi\,d\gamma\end{aligned} \tag{5.174}$$

$$\begin{aligned}\alpha_{12} = -\frac{5}{2}\left(\frac{kT}{\pi m}\right)^{\frac{1}{2}}&\left[\sum_i \exp(-\varepsilon_i)\right]^{-2}\sum_{ijkl}\exp(-\varepsilon_i-\varepsilon_j)\\ &\times\int_0^\infty\int_0^{2\pi}\int_0^\pi (\Delta\varepsilon)^2\gamma^3\exp(-\gamma^2)I_{ij}^{kl}\sin\chi\,d\chi\,d\psi\,d\gamma\end{aligned} \tag{5.175}$$

and

$$\begin{aligned}\alpha_{22} = \left(\frac{kT}{\pi m}\right)^{\frac{1}{2}}&\left[\sum_i \exp(-\varepsilon_i)\right]^{-2}\sum_{ijkl}\exp(-\varepsilon_i-\varepsilon_j)\\ &\times\int_0^\infty\int_0^{2\pi}\int_0^\pi [|\boldsymbol{\gamma}(\varepsilon_i-\varepsilon_j)-\boldsymbol{\gamma}'(\varepsilon_k-\varepsilon_l)|^2+\tfrac{3}{2}(\Delta\varepsilon)^2]\\ &\times\gamma^3\exp(-\gamma^2)I_{ij}^{kl}\sin\chi\,d\chi\,d\psi\,d\gamma\end{aligned} \tag{5.176}$$

where $\gamma'^2 = \gamma^2 - \Delta\varepsilon$, and $\boldsymbol{\gamma}'$ is the relative velocity after the collision. Again if the molecules possess no internal energy the results for the thermal conductivity reduce to those for the monatomic gas

$$[\lambda_{\text{trans}}]_1 = \frac{75}{64}\left(\frac{k^3 T\pi}{m}\right)^{\frac{1}{2}} \frac{1}{\bar{\Omega}^{(2,2)}(T)} \tag{5.177}$$

and

$$[\lambda_{\text{int}}]_1 = 0. \tag{5.178}$$

The coefficient of self-diffusion for a pure polyatomic gas may also be obtained from the semi-classical theory and is written to a first approximation

$$[D_{11}]_1 = \frac{3}{8n}\left(\frac{\pi k T}{m}\right)^{\frac{1}{2}} \frac{1}{\beta_{11}(T)} \tag{5.179}$$

where the collision integral β_{11} is given by

$$\beta_{11}(T) = \left[\sum_i \exp(-\varepsilon_i)\right]^{-2} \sum_{ijkl} \exp(-\varepsilon_i - \varepsilon_j)$$
$$\times \int_0^\infty \int_0^{2\pi} \int_0^\pi \gamma^3(\gamma^2 - \gamma\gamma' \cos\chi)\exp(-\gamma^2) I_{ij}^{kl} \sin\chi \, d\chi \, d\psi \, d\gamma. \tag{5.180}$$

For the case of no internal energy this reduces to the collision integral $\bar{\Omega}^{(1,1)}(T)$.

It is also possible to define a diffusion coefficient for internal energy in the polyatomic gas. This describes the process whereby internal energy is transported in a gas and is given by the expression

$$D_{\text{int}} = \frac{3}{8n}\left(\frac{\pi k T}{m}\right)^{\frac{1}{2}} \frac{1}{\beta_{\text{int}}(T)} \tag{5.181}$$

where the collision integral β_{int} is given by

$$\left(\frac{c_{v,\text{int}}}{k}\right)\beta_{\text{int}}(T) = \left[\sum_i \exp(-\varepsilon_i)\right]^{-2} \sum_{ijkl} \exp(-\varepsilon_i - \varepsilon_j)$$
$$\times \int_0^\infty \int_0^{2\pi} \int_0^\pi \gamma^3 [(\varepsilon_i - \bar{\varepsilon})\{(\varepsilon_i - \varepsilon_j)\gamma^2 - (\varepsilon_k - \varepsilon_l)\gamma\gamma' \cos\chi\}]$$
$$\times \exp(-\gamma^2) I_{ij}^{kl} \sin\chi \, d\chi \, d\psi \, d\gamma \tag{5.182}$$

and

$$\bar{\varepsilon} = \sum_i \varepsilon_i \exp(-\varepsilon_i) \Big/ \sum_i \exp(-\varepsilon_i)$$

is the mean reduced internal energy of the molecules.

It will be noted that the expression for the diffusion coefficient for internal energy differs from that for the self-diffusion coefficient for mass, a difference which implies that internal energy may be transported at a different rate from the molecules themselves by means of inelastic collisions. Normally this difference is small.

The formal expressions of the semi-classical kinetic theory pose severe problems for the evaluation of the transport coefficients of a polyatomic gas for a realistic intermolecular potential function. This is because the intermolecular potential enters into the evaluation of the inelastic differential scattering cross-section I_{ij}^{kl} and the accurate calculation of these cross-sections requires an immense computational effort as indicated in Chapter 4. Consequently, in order to provide useful kinetic theory formulae it has been necessary to adopt a less ambitious route and either to make further approximations to the formal kinetic theory expressions or to obtain relationships between the various transport coefficients. Monchick, Pereira, and Mason[23] have been able to follow the latter route without introducing any approximations beyond those implicit in Wang Chang, Uhlenbeck, and de Boer theory. By eliminating the various contributions to the integrals of (5.174)–(5.176) in favour of experimentally accessible, or at least quite accurately estimated, quantities, they have shown that to the equivalent of a first-order Chapman–Cowling approximation the thermal conductivity and the viscosity of a polyatomic gas are related by the equation,

$$\frac{m[\lambda]_1}{k[\eta]_1} = \tfrac{5}{2}(\tfrac{3}{2} - \Delta) + \frac{\rho D_{\text{int}}}{[n]_1}\left(\frac{c_{v,\text{int}}}{k} + \Delta\right) \tag{5.183}$$

where

$$\Delta = \left(\frac{2c_{v,\text{int}}}{k\pi\zeta_{\text{rot}}}\right)\left[\frac{5}{2} - \frac{\rho D_{\text{int}}}{[\eta]_1}\right] \times \left[1 + \left(\frac{2}{\pi\zeta_{\text{rot}}}\right)\left(\frac{5}{3}\frac{c_{v,\text{int}}}{k} + \frac{\rho D_{\text{int}}}{[\eta]_1}\right)\right]^{-1}, \tag{5.184}$$

$\rho = nm$ is the mass density of the gas and ζ_{rot} is the collision number introduced in (5.170).

If both the shear viscosity and the thermal conductivity are available from accurate measurements and the rotational collision number can be obtained from experiment or estimation, (5.183) may be employed to determine D_{int} which is not available from any other source.[24] More frequently only accurate viscosity data are available and then they may be used in conjunction with estimates of ζ_{rot} and D_{int} to calculate the thermal conductivity. For this purpose it is usual to equate D_{int} with D_{11}, the self-diffusion coefficient of the gas which can itself be calculated from the viscosity.[23–4] However, other methods of estimating D_{int} which are based on calculations for simple molecular models may also be employed.[25]

The form of (5.183) indicates that the relationship between the viscosity and the thermal conductivity is not very sensitive to the intermolecular pair potential which enters only through D_{int} and ζ_{rot}. Thus for the study of intermolecular forces we require an explicit equation for individual transport coefficients. In order to obtain such relations it is necessary to introduce approximations to the semi-classical formulae beyond those of the Wang Chang, Uhlenbeck, and de Boer theory. These approximations have only proved useful for the viscosity and diffusion coefficient and the viscosity is the only transport property which has been measured accurately over a wide range of temperature so that we confine ourselves to this coefficient here.

5.9.2. The Mason–Monchick approximation[22]

The starting point for the simplification of the semi-classical formulae is the recognition that the change of internal energy, $\Delta\varepsilon$, in a binary encounter usually involves only a few quanta of internal energy and that at moderate temperatures the significant internal energy is almost entirely restricted to rotational modes. The quanta of internal energy for rotation are generally very much less than kT so that $\Delta\varepsilon \ll 1$. Now the reduced, relative kinetic energy for the collision, γ^2, is of order unity so that to a first approximation we may neglect $\Delta\varepsilon$ compared to γ^2.

It also follows from this argument that the trajectories of the colliding molecules are insignificantly distorted by the exchange of $\Delta\varepsilon$ between translational and internal energies so that we also replace the inelastic differential cross-section I_{ij}^{kl} by its value for elastic scattering I_{el}. Using these results in (5.165) and carrying out the summation over internal states we find that the shear viscosity may be written as[22]

$$[\eta] = \frac{5}{16} \frac{(mkT\pi)^{\frac{1}{2}}}{\langle \bar{\Omega}^{(2,2)}(T) \rangle} \tag{5.185}$$

Here, $\langle \bar{\Omega}^{(2,2)}(T) \rangle$ is similar to the collision integral defined by (5.130), (5.136), and (5.137) for the monatomic case. However, it is not identical to this collision integral because the intermolecular pair potential for non-spherical polyatomic molecules is dependent upon their relative orientation. This implies that scattering out of the initial plane of a binary collision may occur. The final deflection angle, χ, in a binary collision therefore depends not only on the energy and impact parameter but also upon the initial relative orientation of the molecules. Thus, we must write

$$\langle \bar{\Omega}^{(2,2)}(T) \rangle = \tfrac{1}{4} \int_0^\infty \int_0^{2\pi} \int_0^\infty \{[1 - \cos^2 \chi(b, \gamma, \psi)] b \, \mathrm{d}b \, \mathrm{d}\psi\} \exp(-\gamma^2) \gamma^6 \, \mathrm{d}\gamma^2 \tag{5.185a}$$

where we have used the result that the classical equivalent of $I_{el} \sin \chi \, d\chi$ is $b \, db$. We postpone a discussion of the evaluation of such collision integrals until the next chapter, but note now that in this approximation the expression for the viscosity of a polyatomic gas is *formally* identical to that for a monatomic gas.

A better approximation to the viscosity of a polyatomic gas which does not entirely neglect inelastic effects in collisions has also been given by Mason and Monchick.[22] In (5.165) the most difficult term to handle is that involving the product $(\Delta\varepsilon)^2 \sin^2 \chi$ which couples the internal and translational motions; consequently we approximate $\sin^2\chi$ in this term by giving it its mean value on a unit sphere, that is $\frac{2}{3}$. This corresponds to using a rigid-sphere model for the evaluation of this contribution. With this approximation it may be seen that the terms involving $(\Delta\varepsilon)^2$ again disappear from (5.165) and, using I_{el} in place of I_{ij}^{kl} again, we recover the formal expression for the viscosity given by (5.184). Thus, even to this order of approximation, the viscosity of a polyatomic gas is unaffected by inelastic collisions.

5.9.3. Polyatomic gas mixtures

Monchick, Yun, and Mason have extended the analysis of Wang Chang, Uhlenbeck, and de Boer to systems of polyatomic gas mixtures.[26] According to this extension the *form* of the relationship between the viscosity of a gas mixture and that of its components is exactly the same as that for a monatomic gas mixture in the first-order approximation of a Chapman–Cowling scheme. The differences which occur are only in the definitions of the collision integrals which arise in the resulting expressions. They are generalizations of those introduced earlier for a pure gas and, in principle, incorporate the effects of inelastic collisions. However, the argument of Mason and Monchick can be repeated for the case of mixtures to demonstrate that the viscosity is essentially independent of the occurrence of inelastic collisions. Thus, within this approximation, even the collision integrals for a polyatomic gas mixture have the same form as for a monatomic mixture provided account is taken of the non-spherical nature of the intermolecular pair potential for the former. There is a similar result for diffusion coefficients in polyatomic systems.

For the thermal conductivity of polyatomic mixtures the situation is more complicated since further approximations have to be introduced to obtain a useful formula which even then merely relates the thermal conductivity of the mixture to its viscosity and other macroscopic quantities. The resulting expressions are quite complicated and are given in Appendix 5.3. In the same appendix methods for evaluating the quantities necessary to calculate the thermal conductivity of a polyatomic gas mixture from its viscosity are also given. So far as intermolecular forces are concerned the viscosity is of major interest since it has been more accurately measured than other properties and the formal analysis is essentially the same as that for monatomic gases which will be described in the next chapter.

In summary, it should be emphasized that the available kinetic theory formulae for the transport coefficients of polyatomic gases are inherently less accurate than those for monatomic gases. First even the Wang Chang and Uhlenbeck results are only first-order approximations in the equivalent of the Chapman–Cowling scheme. Second, even these expressions have required further approximations to render calculations practicable.

5.9.4. *Quantum mechanical effects*

As mentioned earlier the semi-classical theory is not strictly applicable to molecules with degenerate rotational energy states. For this reason it fails to account for a number of interesting phenomena which are likely to prove of value in the study of the intermolecular forces of polyatomic molecules. In this section we consider the modifications to the earlier semi-classical results introduced by the quantum mechanical theory and several new phenomena which it is able to describe.

(a) *The transport coefficients.* The solution to the Waldmann–Snider kinetic equation yields the following expression for the elements of the stress tensor in a polyatomic gas.[16]

$$p_{ij} = \left(nkT - \kappa\frac{\partial}{\partial \mathbf{r}}\cdot\mathbf{u} - \kappa_1 \boldsymbol{\alpha}\cdot\frac{\partial}{\partial \mathbf{r}}\times\mathbf{u}\right)\delta_{ij} - 2\eta S_{ij} - 2\eta_1 S^{(1)}_{ij} - 2\eta_2 S^{(2)}_{ij}. \tag{5.186}$$

Here, $\boldsymbol{\alpha}$ corresponds classically to the local mean angular velocity in the gas and the elements S_{ij} are those of the usual rate-of-strain tensor

$$S_{ij} = \frac{1}{2}\left[\frac{\partial u_j}{\partial x_i} + \frac{\partial u_i}{\partial x_j}\right] - \frac{1}{3}\frac{\partial}{\partial \mathbf{r}}\cdot\mathbf{u}\delta_{ij}$$

The remaining elements are†

$$S^{(1)}_{ij} = \frac{1}{2}\left\{\alpha_j\left(\frac{\partial u_k}{\partial x_j} - \frac{\partial u_j}{\partial x_k}\right) - \alpha_i\left(\frac{\partial u_k}{\partial x_i} - \frac{\partial u_i}{\partial x_k}\right)\right\} - \frac{1}{3}\boldsymbol{\alpha}\cdot\frac{\partial}{\partial \mathbf{r}}\times\mathbf{u}\delta_{ij}$$

† i, j, k are cyclic indices. ε_{ijk} is defined so that

$$\varepsilon_{ijk} = \begin{cases} 0 & \text{if any } i, j, k \text{ are the same} \\ +1 & 123, 231, 312 \\ -1 & 132, 321, 213. \end{cases}$$

and

$$S_{ij}^{(2)} = \varepsilon_{ijk}[\alpha_j S_{ik} - \alpha_i S_{ik} + \alpha_k(S_{ii} - S_{jj})] \qquad i \neq j$$

$$S_{ii}^{(2)} = (2\alpha_j S_{ik} - \alpha_k S_{ij}).$$

From (5.186) we see that there are contributions to the elements of the stress tensor which did not arise in the semi-classical treatment and which occur because of the existence of a non-zero local mean angular velocity. Five transport coefficients are now required to relate the elements of the stress tensor to macroscopic gradients. They are η, the shear viscosity, κ the bulk viscosity, and three new coefficients, κ_1, η_1, and η_2 which do not occur in the Wang Chang and Uhlenbeck expressions.

In essence, the new coefficients arise because, in a shear field, rotating non-spherically symmetric molecules tend to achieve a preferential alignment of their angular momentum vectors (spin-polarization) owing to the different cross-sections for collision in different orientations. Because the Wang Chang and Uhlenbeck semi-classical theory does not treat rotational internal energy states properly it fails to account for these effects.

The various transport coefficients may be evaluated by generalizations of the methods employed earlier but the details are too complicated to be given here.

In the quantum mechanical theory the components of the heat flux vector are given by the expression

$$\begin{aligned} q_i = -\lambda \frac{\partial T}{\partial x_i} - \lambda'\left(\alpha_j \frac{\partial T}{\partial x_k} - \alpha_k \frac{\partial T}{\partial x_j}\right) - \nu\left(\frac{\partial \alpha_k}{\partial x_j} - \frac{\partial \alpha_j}{\partial x_k}\right) \\ - 2\nu_1(\Omega_{ii}^{(2)}\alpha_i + \Omega_{ij}^{(2)}\alpha_j + \Omega_{ik}^{(2)}\alpha_k) \\ - 2\nu_2(\Omega_{ii}^{(1)}\alpha_i + \Omega_{ij}^{(1)}\alpha_j + \Omega_{ik}^{(1)}\alpha_k) \\ - \nu_3\alpha_i\left(\frac{\partial}{\partial \mathbf{r}} \cdot \boldsymbol{\alpha}\right) \end{aligned} \tag{5.187}$$

where

$$\Omega_{ij}^{(2)} = \frac{1}{2}\left[\frac{\partial \alpha_j}{\partial x_i} + \frac{\partial \alpha_i}{\partial x_j}\right] - \frac{1}{3}\frac{\partial}{\partial \mathbf{r}} \cdot \boldsymbol{\alpha}\delta_{ij}$$

and

$$\Omega_{ij}^{(1)} = \frac{1}{2}\left[\frac{\partial \alpha_j}{\partial x_i} - \frac{\partial \alpha_i}{\partial x_j}\right].$$

Again there are additional contributions to the heat flux, and new transport coefficients, λ', ν, ν_1, ν_2, and ν_3 for a gas of rotating non-spherical molecules which arise from the spin polarization in addition to those of the Wang Chang and Uhlenbeck theory. The new transport coefficients characteristic of the quantum mechanical theory can be expressed in terms of the dynamics of binary molecular encounters. However, no calculations have yet been performed of these coefficients for realistic potentials owing to the complexity of the expressions, so that we omit details here.

(b) *The effects of magnetic and electric fields.* In 1930 Senftleben[27] observed that the transport properties of a paramagnetic gas were affected by the application of a magnetic field. However, no great interest was displayed in the effect until Beenakker and his collaborators[28,29] began a systematic study of it in the early-1960s. It was found that the effect occurred in diamagnetic as well as paramagnetic gases and furthermore that an analogous phenomenon was associated with the application of an electric field to polar gases.[28]

The origin of the effect is now well understood qualitatively and recent developments indicate that a quantitative description of the phenomenon may shortly be available.[30,31] Because the effects only occur in non-spherical molecules and depend primarily on the non-spherical part of the intermolecular potential they are likely to prove of considerable value to the study of intermolecular forces for polyatomic systems. The development of the theory is very complicated and not yet complete; we therefore give here only a brief outline of the description of the simplest phenomenon in the case of a magnetic field.

As was mentioned earlier the imposition of a velocity or temperature gradient produces a preferential alignment of the angular momentum vectors of non-spherically symmetric molecules. If a molecule possesses a permanent magnetic dipole moment then application of a magnetic field to the gas causes the magnetic moment to precess about the field direction. This precession partially destroys the preferential alignment of molecules established by the coupling between the macroscopic gradients and molecular collisions. The extent of this destruction of the polarization depends upon how fast the molecule precesses relative to the time between collisions in the gas which tend to re-establish the alignment. If we denote the precession frequency by ω_p and the time between collisions by τ_{coll}, then the destruction of the polarization will increase as $\omega_p \tau_{coll}$ increases. The precession frequency of a magnetic dipole is proportional to H, the applied magnetic field, whereas the collision frequency is inversely proportional to the gas pressure P. Hence the degree of destruction of the polarization varies with the ratio H/P. Thus, the transport properties of such a gas, which as we have seen depend upon the existence of spin polarization, vary according to the value of H/P. For large values of $\omega_p \tau_{coll}$ the spin polarization is completely destroyed and further increase, by increasing H or lowering P, will produce no further effect. This saturation corresponds to complete splitting of the degenerate rotational energy states of the molecule by the applied field (the Zeeman effect). A similar argument can be given for the effect of an electric field on the transport properties of a polar gas.[28]

This qualitative description of field effects on transport properties has been supported by the development of a quantitative kinetic theory. The theory begins by the formulation of a generalized Boltzmann-type equation which incorporates the effects of the applied field upon the molecular distribution function. As before the equation has been solved to yield expressions for the stress tensor and heat flux vector and more recently to yield explicit expressions for the various transport coefficients.

It turns out that the transport coefficients depend upon the orientation of the magnetic field relative to the gradient of temperature or velocity. As an example we write the components of the heat flux vector in the form

$$q_i = -\lambda_{ii}\frac{\partial T}{\partial x_i} - \lambda_{ij}\frac{\partial T}{\partial x_j} - \lambda_{ik}\frac{\partial T}{\partial x_k}. \tag{5.188}$$

Table 5.3 lists the elements of the thermal conductivity tensor λ_{ij} in standard notation for the case where a magnetic field is applied in the x_1-direction. It will be noted that if the field and the temperature gradient are applied in the same direction the relevant thermal conductivity is $\lambda_{\parallel}$. On the other hand, if the temperature gradient is applied perpendicular to the field and the heat flux is measured in the direction of the temperature gradient the thermal conductivity is $\lambda_{\perp}$ and is different from $\lambda_{\parallel}$. More surprisingly, if the temperature gradient and the field are perpendicular it is predicted that there is a heat flux perpendicular to both of them associated with a characteristic thermal conductivity λ_{tr}. A similar description may be given for the viscosity where the stress tensor depends on seven independent transport coefficients.

The magnitude of the changes in the thermal conductivity and viscosity in a polyatomic gas brought about by application of magnetic

Table 5.3

Elements of the thermal conductivity tensor for a gas in a magnetic field

	$\partial T/\partial x_1$	$\partial T/\partial x_2$	$\partial T/\partial x_3$
q_1	$-\lambda_{\parallel}$	0	0
q_2	0	$-\lambda_{\perp}$	λ_{tr}
q_3	0	$-\lambda_{\mathrm{tr}}$	$-\lambda_{\perp}$

If $H=0$, $\lambda_{\parallel}=\lambda_{\perp}$, $\lambda_{\mathrm{tr}}=0$.

fields is small. Even at saturation the relative change in the thermal conductivity parallel to the field, $(\Delta\lambda_{\parallel}/\lambda)_{\text{sat}}$, amounts to only 0·1–1 per cent. Nevertheless, it has been possible to measure the effect in specially designed experiments.[28,29]

(c) *Corrections to the semi-classical expressions.* As noted earlier the Wang Chang and Uhlenbeck theory leads to a relatively simple and useful relationship between the viscosity and thermal conductivity of a polyatomic gas (5.183). However, because the semi-classical theory omits the effects due to spin polarization the expression is less accurate than is necessary to describe the best available measurements. In order to introduce the effects of spin polarization into the semi-classical expression in an approximate manner Viehland, Mason, and Sandler[32] have used the fact that the magnetic field effects on the transport coefficients essentially measure the influence of the polarization in the gas. With the aid of the existing quantum mechanical kinetic theory for diatomic molecules they showed that an improved relation between the viscosity and thermal conductivity of such a gas is[32,33]

$$\frac{m[\lambda]_1}{k[\eta]_1}=\tfrac{5}{2}(\tfrac{3}{2}-\Delta)+\frac{\rho D_{\text{int}}}{[\eta]_1}\left(\frac{C_{v,\text{int}}}{k}+\Delta\right)\left\{1-\frac{5}{3}\left(1+\frac{\lambda_{\text{trans}}}{\lambda_{\text{int}}}\right)\left(\frac{\Delta\lambda_{\parallel}}{\lambda}\right)_{\text{sat}}\right\} \tag{5.189}$$

where Δ is given by (5.184) and

$$\frac{m\lambda_{\text{trans}}}{k[\eta]_1}=\frac{5}{2}\left(\frac{3}{2}-\Delta\right)$$

and

$$\frac{m[\lambda_{\text{int}}]_1}{k[\eta]_1}=\frac{\rho D_{\text{int}}}{[\eta]_1}\left(\frac{C_{v,\text{int}}}{k}+\Delta\right)\left\{1-\frac{5}{3}\left(1+\frac{\lambda_{\text{trans}}}{\lambda_{\text{int}}}\right)\left(\frac{\Delta\lambda_{\parallel}}{\lambda}\right)_{\text{sat}}\right\}$$

Here, $(\Delta\lambda_{\parallel}/\lambda)_{\text{sat}}$ is the fractional change in the thermal conductivity measured parallel to a magnetic field at saturation, an experimentally accessible quantity.

5.10. The kinetic theory of dense gases

Finally in our discussion of the kinetic theory of gases we consider briefly the theory of dense monatomic gases. In our treatment of the Boltzmann equation for dilute gases we assumed that the density was such that only binary collisions needed to be considered and that the mean distance between molecules was large compared with their dimensions. In dense gases these conditions are not fulfilled and the kinetic theory must therefore be examined afresh.

5.10.1. *The Enskog theory*

The first attempt towards a theory of dense gases was made by Enskog[34] and was developed on the assumption that the molecules of a gas are rigid spheres. Enskog merely extended the dilute-gas analysis by introducing corrections which account for the non-zero size of the molecules. First, the centres of two rigid-sphere molecules of diameter d cannot approach each other closer than d. Consequently in evaluating the collision term in the Boltzmann equation for a dense gas it is not possible to evaluate the two distribution functions $f(\mathbf{c})$ and $f(\mathbf{c}_1)$ for two colliding molecules at the same point as was done earlier for the dilute gas (§ 5.3.1). The use of a more correct form for the collision term provides a new mechanism for the transport of momentum and energy in the gas because these quantitites can now be transported over the diameter of a molecule in a collision and not merely by free molecular motion. This *collisional transfer* of momentum and energy is in fact the dominant mechanism of transport in dense fluids such as liquids. A second result of the non-zero volume of the molecules is to increase their collision frequency above that for molecules of infinitesimal size. Enskog supposed the collision frequency was increased by a factor $\tilde{\chi}$ which may depend on density.

By making these modifications to the Boltzmann equation, and following essentially the same method of solution as for the dilute gas, expressions may be derived for the viscosity of a dense hard-sphere gas. The results are[35]

$$\eta(n)=\frac{1}{\tilde{\chi}}[1+\tfrac{4}{15}\pi nd^3\tilde{\chi}]^2\eta(0)+\tfrac{4}{15}n^2d^4\tilde{\chi}(\pi mkT)^{\frac{1}{2}} \qquad (5.190)$$

and

$$\lambda(n)=\frac{1}{\tilde{\chi}}[1+\tfrac{2}{5}\pi nd^3\tilde{\chi}]^2\lambda(0)+\tfrac{2}{3}n^2d^4\tilde{\chi}\left(\frac{\pi k^3T}{m}\right)^{\frac{1}{2}} \qquad (5.191)$$

where n is the molecular number density and $\eta(0)$ and $\lambda(0)$ are the low-density limiting values of the transport coefficients. The factor $\tilde{\chi}$, for rigid spheres can be written in the form

$$\tilde{\chi}=1+0{\cdot}6250(\tfrac{2}{3}\pi nd^3)+0{\cdot}28695(\tfrac{2}{3}\pi nd^3)^2+\ldots \qquad (5.192)$$

where the numerical coefficients are related to the virial coefficients of hard-sphere gases (§ 3.9). When this expression is inserted into (5.190) and (5.191) and an expansion in density performed a polynomial expansion for the transport coefficients results in the forms

$$\eta(n)=\eta(0)+a_1n+a_2n^2+\ldots \qquad (5.193)$$

and

$$\lambda(n)=\lambda(0)+b_1n+b_2n^2+\ldots \qquad (5.194)$$

where according to the Enskog theory

$$a_1 = \tfrac{7}{40}(\tfrac{2}{3}\pi n d^3)\eta(0)$$

and

$$b_1 = \tfrac{23}{40}(\tfrac{2}{3}\pi n d^3)\lambda(0).$$

Enskog's analysis has been extended to mixtures of monatomic gases by Thorne[36,37] and less rigorously to polyatomic gases and gas mixtures. The Enskog theory and its extensions provide a useful means of correlating experimental data especially if semi-empirical modifications to account for the real forces between molecules are included.[38,39]

5.10.2. The general kinetic theory of dense gases

In order to develop a rigorous kinetic theory of dense gases it is necessary to return to first principles and to rederive a new Boltzmann equation for dense gases. This development is beyond the scope of this work and a recent review has been given by Dorfman and van Beijeren.[40] Although the development of the theory is not yet complete one of the conclusions which can be drawn so far is that the transport coefficients of a dense gas are not simply polynomial expansions in the density.

Instead the expansion for the viscosity takes the form

$$\frac{\eta(n)}{\eta(0)} = 1 + c_1(nd^3) - c_2'(nd^3)^2 \log(nd^3) + c_2''(nd^3)^2 + \ldots \qquad (5.195)$$

with a similar expansion for the thermal conductivity. The logarithmic term in this density expansion arises essentially because of correlations in the velocities of colliding particles over the distance of a mean free path which are specifically not included in the Enskog theory. There is no convincing experimental evidence for the existence of the logarithmic term although computer simulation studies of dense gases strongly indicate its presence.

The first correction to the low-density viscosity contains contributions from collisional momentum transfer in binary collisions and from effects of three-body encounters which may include correlated binary collisions as well as genuine three-body collisions where three molecules collide simultaneously. Explicit expressions for c_1 do exist[40] and a calculation of the coefficient for a dense gas of hard spheres[41] is remarkably close (±5 per cent) to that predicted by the Enskog theory.

There remain considerable formal difficulties in the kinetic theory of dense gases. Furthermore the calculation of the coefficients of the density expansion of the transport properties has not yet been carried out for realistic intermolecular potential functions. It is therefore too early to speculate on the role of such coefficients in an understanding of intermolecular forces.

Appendix A5.1 Integrals occurring in the kinetic theory

(A) *Angular integration in spherical polar coordinates C, θ, ϕ*

Let $F(C)$ be any scalar function of C,

$$\int F(C)\,\mathrm{d}\mathbf{C} = \int_0^{2\pi}\int_0^{\pi}\int_0^{\infty} F(C)C^2\,\mathrm{d}C \sin\theta\,\mathrm{d}\theta\,\mathrm{d}\phi \tag{A5.1.1}$$

from the definition of the elemental volume $\mathrm{d}\mathbf{C}$ in polar coordinates.

$$\int F(C)\,\mathrm{d}\mathbf{C} = 4\pi\int_0^{\infty} F(C)C^2\,\mathrm{d}C. \tag{A5.1.2}$$

(B) *Integrals involving exponentials*

$$\int_0^{\infty} \mathrm{e}^{-\alpha C^2} C^r\,\mathrm{d}C = \frac{\sqrt{\pi}}{2}\cdot\frac{1}{2}\cdot\frac{3}{2}\cdot\frac{5}{2}\cdots\frac{r-1}{2}\,\alpha^{-\frac{1}{2}(r+1)} \qquad r\text{, even,} \qquad r>0 \tag{A5.1.3}$$

$$\int_0^{\infty} \mathrm{e}^{-\alpha C^2} C^r\,\mathrm{d}C = \tfrac{1}{2}\alpha^{-\frac{1}{2}(r+1)}\left(\frac{r-1}{2}\right)! \qquad r\text{, odd} \qquad r>0. \tag{A5.1.4}$$

(C) *Integrals involving powers of components of a vector* $\mathbf{C}$

Let C_1, C_2, C_3 be the components of $\mathbf{C}$. Then

$$\int C_1^r F(C)\,\mathrm{d}\mathbf{C} = \int_{-\infty}^{\infty}\int_{-\infty}^{\infty}\int_{-\infty}^{\infty} C_1^r F(C)\,\mathrm{d}C_1\,\mathrm{d}C_2\,\mathrm{d}C_3 \tag{A5.1.5}$$

from the definition of the elemental volume $\mathrm{d}\mathbf{C}$ in Cartesian coordinates.

Case 1: r odd. In the integration over C_1 the contribution to the integral from negative values of C_1 is exactly cancelled by the contribution from positive values of C_1. Thus the integral vanishes.

Case 2: $r=2$

$$\int C_1^2 F(C)\,\mathrm{d}\mathbf{C} = \int C_2^2 F(C)\,\mathrm{d}\mathbf{C} = \int C_3^2 F(C)\,\mathrm{d}\mathbf{C}$$

by symmetry. Thus

$$\begin{aligned}\int C_1^2 F(C)\,\mathrm{d}\mathbf{C} &= \tfrac{1}{3}\int (C_1^2+C_2^2+C_3^2)F(C)\,\mathrm{d}\mathbf{C}\\ &= \tfrac{1}{3}\int C^2 F(C)\,\mathrm{d}\mathbf{C}.\end{aligned} \tag{A5.1.6}$$

Case 3: $r=4$

$$\int C_1^4 F(C)\,\mathrm{d}\mathbf{C} = \tfrac{1}{5}\int F(C)C^4\,\mathrm{d}\mathbf{C}. \tag{A5.1.7}$$

Case 4.

$$\int C_1^2 C_2^2 F(C)\,d\mathbf{C} = \tfrac{1}{15}\int F(C)C^4\,d\mathbf{C}. \tag{A5.1.8}$$

(D) **A** *is a constant vector with components* A_1, A_2, A_3

Consider the integral

$$\int F(C)(\mathbf{A}\,.\,\mathbf{C})C_1\,d\mathbf{C} = \int F(C)[A_1C_1 + A_2C_2 + A_3C_3]C_1\,d\mathbf{C}$$

$$= A_1\int F(C)C_1^2\,d\mathbf{C}$$

$$= \frac{A_1}{3}\int F(C)C^2\,d\mathbf{C}. \tag{A5.1.9}$$

Appendix A5.2 Kinetic theory formulae for the transport properties of monatomic gases and gas mixtures

This appendix lists the kinetic theory expressions for the transport coefficients for dilute monatomic gases. The list includes all the expressions necessary for an understanding of the material in Chapter 6. In addition, formulae are given for several higher-order approximations to the transport coefficients of both pure gases and mixtures. These higher-order approximations are given for two schemes where it is appropriate and where they have been derived. The first scheme, discussed in the text, is the Chapman–Cowling method; the alternative method, developed by Kihara, is discussed elsewhere.[8,9] In general, the latter scheme leads to somewhat simpler formulae, which to an equal order of approximation are more accurate than the corresponding Chapman–Cowling expressions. However, the Kihara approximations are not available for all the transport properties of binary gas mixtures.

A5.2.1. The reduced collision integrals

For an intermolecular potential of the form

$$U_{ij}^*(r^*) = \frac{U_{ij}(r/\sigma_{ij})}{\varepsilon_{ij}}$$

characteristic of the interaction between two monatomic species i and j the deflection angle in a binary collision is

$$\chi(E^*, b^*) = \pi - 2b^*\int_{r_0^*}^{\infty} \frac{dr^*/r^{*2}}{\sqrt{1 - \dfrac{b^{*2}}{r^{*2}} - \dfrac{U_{ij}^*(r^*)}{E^*}}} \tag{A5.2.1}$$

where $E^* = E/\varepsilon_{ij}$ is the reduced relative kinetic energy of the collision, $b^* = b/\sigma_{ij}$ is the reduced impact parameter, $r^* = r/\sigma_{ij}$ is the reduced intermolecular separation, and $r_0^* = r_0/\sigma_{ij}$ is the reduced closest distance of approach. The reduced transport cross-sections are defined by the equations

$$Q_{ij}^{(l)*}(E^*) = 2\left[1 - \frac{1+(-1)^l}{2(1+l)}\right]^{-1} \int_0^\infty (1-\cos^l\chi) b^* \, db^* \qquad \text{(A5.2.2)}$$

and the collision integrals by

$$\Omega_{ij}^{(l,s)*}(T_{ij}^*) = [(s+1)!T_{ij}^{*s+2}]^{-1} \int_0^\infty Q_{ij}^{(l)*}(E^*) e^{-E^*/T_{ij}^*} E^{*s+1} \, dE^* \qquad \text{(A5.2.3)}$$

where $T_{ij}^* = kT/\varepsilon_{ij}$.

It is useful to define some combinations of reduced collision integrals which occur in kinetic theory expressions for the transport coefficients.†

$$\left.\begin{aligned}
&A_{ij}^\star = \Omega_{ij}^{(2,2)*}/\Omega_{ij}^{(1,1)*} \qquad B_{ij}^\star = \frac{5\Omega_{ij}^{(1,2)*} - 4\Omega_{ij}^{(1,3)*}}{\Omega_{ij}^{(1,1)*}} \\
&C_{ij}^\star = \Omega_{ij}^{(1,2)*}/\Omega_{ij}^{(1,1)*} \qquad E_{ij}^\star = \Omega_{ij}^{(2,3)*}/\Omega_{ij}^{(2,2)*} \\
&F_{ij}^\star = \Omega_{ij}^{(3,3)*}/\Omega_{ij}^{(1,1)*} \qquad G_{ij}^\star = \Omega_{ij}^{(1,4)*}/\Omega_{ij}^{(1,1)*} \\
&H_{ij}^\star = \Omega_{ij}^{(2,4)*}/\Omega_{ij}^{(2,2)*} \qquad I_{ij}^\star = \Omega_{ij}^{(1,5)*}/\Omega_{ij}^{(1,1)*} \\
&K_{ij}^\star = [\tfrac{35}{4} - 3B_{ij}^\star - 6C_{ij}^\star]/(5 - 6C_{ij}^\star)
\end{aligned}\right\} \qquad \text{(A5.2.4)}$$

A5.2.2. The viscosity of a pure gas

First-order approximation

$$[\eta]_1 = \frac{5}{16}\left(\frac{mkT}{\pi}\right)^{\frac{1}{2}} \frac{1}{\sigma^2 \Omega^{(2,2)*}(T^*)}. \qquad \text{(A5.2.5)}$$

Second-order Chapman–Cowling approximation

$$[\eta]_2 = [\eta]_1 f_\eta^{(2)} \qquad \text{(A5.2.6)}$$

$$f_\eta^{(2)} = 1 + (H^{01})^2/[H^{00}H^{11} - (H^{01})^2] \qquad \text{(A5.2.7)}$$

where

$$H^{00} = 1/[\eta]_1$$

$$H^{01} = \frac{1}{[\eta]_1}\left[\frac{7}{4} - 2E^\star\right]$$

$$H^{11} = \frac{1}{[\eta]_1}\left[\frac{301}{48} - 7E^\star + 5H^\star\right].$$

Third-order Chapman–Cowling approximation

$$[\eta]_3 = [\eta]_1 f_\eta^{(3)} \qquad \text{(A5.2.8)}$$

$$\begin{aligned}
f_\eta^{(3)} = f_\eta^{(2)} + H^{00}(H^{01}H^{12} - H^{11}H^{02})^2/\{[H^{00}H^{11} - (H^{01})^2] \\
\times [H^{00}H^{11}H^{22} + 2H^{01}H^{02}H^{12} - H^{00}(H^{12})^2 - H^{11}(H^{02})^2 - H^{22}(H^{01})^2]\}
\end{aligned} \qquad \text{(A5.2.9)}$$

† The symbol $B_{ij}^\star$ which occurs here is not to be confused with the reduced virial coefficient defined elsewhere in the book. We employ the notation here to be consistent with earlier work.[7]

where, in addition to previously defined quantities,

$$H^{02}=\frac{1}{[\eta]_1}\left[\frac{63}{32}-\frac{9}{2}E^\star+\frac{5}{2}H^\star\right]$$

$$H^{12}=\frac{1}{[\eta]_1}\left[\frac{1365}{128}-\frac{321}{16}E^\star+\frac{125}{8}H^\star-\frac{15\Omega^{(2,5)*}}{2\Omega^{(2,2)*}}\right]$$

$$H^{22}=\frac{1}{[\eta]_1}\left[\frac{25137}{1024}-\frac{1755}{32}E^\star+\frac{1905}{32}H^\star-\frac{135}{4}\frac{\Omega^{(2,5)*}}{\Omega^{(2,2)*}}+\frac{105}{8}\frac{\Omega^{(2,6)*}}{\Omega^{(2,2)*}}+\frac{3\Omega^{(4,4)*}}{\Omega^{(2,2)*}}\right].$$

Second-order Kihara approximation

$$[\eta]_2^{\mathrm{K}}=[\eta]_1 f_\eta^{\mathrm{K}(2)} \tag{A5.2.10}$$

where

$$f_\eta^{\mathrm{K}(2)}=1+\tfrac{3}{49}[4E^\star-\tfrac{7}{2}]^2. \tag{A5.2.11}$$

A5.2.3. The thermal conductivity of a pure gas

First-order approximation

$$[\lambda]_1=\frac{75}{64}\left(\frac{k^3T}{m\pi}\right)^{\frac{1}{2}}\frac{1}{\sigma^2\Omega^{(2,2)*}(T^*)}. \tag{A5.2.12}$$

Second-order Chapman–Cowling approximation

$$[\lambda]_2=[\lambda]_1 f_\lambda^{(2)} \tag{A5.2.13}$$

where

$$f_\lambda^{(2)}=1+\frac{(L^{12})^2}{L^{11}L^{22}-(L^{12})^2} \tag{A5.2.14}$$

and

$$L^{11}=\frac{1}{[\lambda]_1}$$

$$L^{12}=\frac{1}{[\lambda]_1}\left[\frac{7}{4}-2E^\star\right]$$

$$L^{22}=\frac{1}{[\lambda]_1}\left[\frac{77}{16}-7E^\star+5H^\star\right]$$

Third-order Chapman–Cowling approximation

$$[\lambda]_3=[\lambda]_1 f_\lambda^{(3)} \tag{A5.2.15}$$

where

$$f_\lambda^{(3)} = f_\lambda^{(2)} + L^{11}(L^{12}L^{23} - L^{22}L^{13})^2/\{[L^{11}L^{22} - (L^{12})^2] \times [L^{11}L^{22}L^{33} + 2L^{12}L^{13}L^{23} - L^{11}(L^{23})^2 - L^{22}(L^{13})^2 - L^{33}(L^{12})^2]\} \quad \text{(A5.2.16)}$$

and

$$L^{13} = \frac{1}{[\lambda]_1}\left[\frac{63}{32} - \frac{9}{2}E^\star + \frac{5}{2}H^\star\right]$$

$$L^{23} = \frac{1}{[\lambda]_1}\left[\frac{945}{128} - \frac{261}{16}E^\star + \frac{125}{8}H^\star - \frac{15}{2}\frac{\Omega^{(2,5)*}}{\Omega^{(2,2)*}}\right]$$

$$L^{33} = \frac{1}{[\lambda]_1}\left[\frac{14\,553}{1024} - \frac{1215}{32}E^\star + \frac{1565}{32}H^\star - \frac{135}{4}\frac{\Omega^{(2,5)*}}{\Omega^{(2,2)*}} + \frac{105}{8}\frac{\Omega^{(2,6)*}}{\Omega^{(2,2)*}} + \frac{\Omega^{(4,4)*}}{\Omega^{(2,2)*}}\right]$$

Second-order Kihara approximation

$$[\lambda]_2^{\mathrm{K}} = [\lambda]_1 f_\lambda^{\mathrm{K}(2)} \quad \text{(A5.2.17)}$$

where

$$f_\lambda^{\mathrm{K}(2)} = 1 + \tfrac{2}{21}[4E^\star - \tfrac{7}{2}]^2 \quad \text{(A5.2.18)}$$

A5.2.4. The viscosity of a binary gas mixture

First-order approximation

$$[\eta_{\mathrm{mix}}]_1 = \frac{1+Z_\eta}{X_\eta + Y_\eta} \quad \text{(A5.2.19)}$$

where

$$X_\eta = \frac{x_1^2}{[\eta_1]_1} + \frac{2x_1x_2}{[\eta_{12}]_1} + \frac{x_2^2}{[\eta_2]_1}$$

$$Y_\eta = \tfrac{3}{5}A_{12}^\star\left\{\frac{x_1^2}{[\eta_1]_1}\left(\frac{m_1}{m_2}\right) + \frac{2x_1x_2}{[\eta_{12}]_1}\left[\frac{(m_1+m_2)^2}{4m_1m_2}\left(\frac{[\eta_{12}]_1^2}{[\eta_1]_1[\eta_2]_1}\right)\right] + \frac{x_2^2}{[\eta_2]_1}\left(\frac{m_2}{m_1}\right)\right\}$$

$$Z_\eta = \tfrac{3}{5}A_{12}^\star\left\{x_1^2\left(\frac{m_1}{m_2}\right) + 2x_1x_2\left[\frac{(m_1+m_2)^2}{4m_1m_2}\left(\frac{[\eta_{12}]_1}{[\eta_1]_1} + \frac{[\eta_{12}]_1}{[\eta_2]_1}\right) - 1\right] + x_2^2\left(\frac{m_2}{m_1}\right)\right\}$$

where

$$[\eta_{12}]_1 = \frac{5}{16}\left(\frac{2m_1m_2kT}{(m_1+m_2)\pi}\right)^{\frac{1}{2}} \frac{1}{\sigma_{12}^2\Omega_{12}^{(2,2)*}(T^*)}. \quad \text{(A5.2.20)}$$

The second-order approximation to the viscosity of a binary gas mixture is most conveniently presented as a special case of that for a multicomponent mixture in view of its complexity. The requisite equations are therefore included in the set below.

A5.2.5. The viscosity of a multicomponent gas mixture with ν components

First-order approximation

$$[\eta_{\text{mix}}]_1 = -\frac{\begin{vmatrix} H_{11}^{00} & H_{12}^{00} & \cdots & H_{1\nu}^{00} & x_1 \\ H_{21}^{00} & H_{22}^{00} & \cdots & H_{2\nu}^{00} & x_2 \\ \vdots & \vdots & & \vdots & \vdots \\ H_{\nu 1}^{00} & H_{\nu 2}^{00} & \cdots & H_{\nu\nu}^{00} & x_\nu \\ x_1 & x_2 & \cdots & x_\nu & 0 \end{vmatrix}}{\begin{vmatrix} H_{11}^{00} & H_{12}^{00} & \cdots & H_{1\nu}^{00} \\ H_{21}^{00} & H_{22}^{00} & \cdots & H_{2\nu}^{00} \\ \vdots & \vdots & & \vdots \\ H_{\nu 1}^{00} & H_{\nu 2}^{00} & \cdots & H_{\nu\nu}^{00} \end{vmatrix}} \tag{A5.2.21}$$

where

$$H_{ii}^{00} = \frac{x_i^2}{[\eta_i]_1} + \sum_{\substack{k=1 \\ k \neq i}}^{\nu} \frac{2x_i x_k}{[\eta_{ik}]_1} \frac{m_k}{(m_i + m_k)^2} \left[\left(\frac{5m_i}{3A_{ik}^{\star}} + m_k\right)\right]$$

$$H_{ij}^{00} = \frac{-2x_i x_j}{[\eta_{ij}]_1} \frac{m_i m_j}{(m_i + m_j)^2} \left[\left(\frac{5}{3A_{ij}^{\star}} - 1\right)\right] \qquad i \neq j$$

and

$$[\eta_i]_1 = \frac{5}{16}\left(\frac{m_i kT}{\pi}\right)^{\frac{1}{2}} \frac{1}{\sigma_{ii}^2 \Omega_{ii}^{(2,2)^*}(T^*)}$$

$$[\eta_{ij}]_1 = \frac{5}{16}\left(\frac{2m_i m_j kT}{\pi(m_i + m_j)}\right)^{\frac{1}{2}} \frac{1}{\sigma_{ij}^2 \Omega_{ij}^{(2,2)^*}(T^*)}.$$

Second-order approximation

$$[\eta_{\text{mix}}]_2 = -\frac{\begin{vmatrix} H_{11}^{00} & \cdots & H_{1\nu}^{00} & H_{11}^{01} & \cdots & H_{1\nu}^{01} & x_1 \\ \vdots & & \vdots & \vdots & & \vdots & \vdots \\ H_{1\nu}^{00} & \cdots & H_{\nu\nu}^{00} & H_{\nu 1}^{01} & \cdots & H_{\nu\nu}^{01} & x_\nu \\ H_{11}^{10} & \cdots & H_{1\nu}^{10} & H_{11}^{11} & \cdots & H_{1\nu}^{11} & 0 \\ \vdots & & \vdots & \vdots & & \vdots & \vdots \\ H_{\nu 1}^{10} & \cdots & H_{\nu\nu}^{10} & H_{\nu 1}^{11} & \cdots & H_{\nu\nu}^{11} & 0 \\ x_1 & \cdots & x_\nu & 0 & \cdots & 0 & 0 \end{vmatrix}}{\begin{vmatrix} H^{00} & \cdots & H_{\nu}^{00} & H^{01} & \cdots & H_{\nu}^{01} \\ \vdots & & \vdots & \vdots & & \vdots \\ H_{\nu 1}^{00} & \cdots & H_{\nu\nu}^{00} & H_{\nu 1}^{01} & \cdots & H_{\nu\nu}^{01} \\ H_{11}^{10} & \cdots & H_{1}^{10} & H_{11}^{11} & \cdots & H_{1\nu}^{11} \\ \vdots & & \vdots & \vdots & & \vdots \\ H_{\nu 1}^{10} & \cdots & H_{\nu\nu}^{10} & H_{\nu 1}^{11} & \cdots & H_{\nu\nu}^{11} \end{vmatrix}} \tag{A5.2.22}$$

where, in addition to previously defined quantities

$$H_{ij}^{01} = \frac{x_i x_j m_i^2 m_j}{[\eta_{ij}]_1 (m_i + m_j)^3 A_{ij}^\star} \{14C_{ij}^\star - \tfrac{35}{3} + A_{ij}^\star [7 - 8E_{ij}^\star]\} \qquad i \neq j$$

$$H_{ii}^{01} = \frac{x_i^2}{[\eta_i]_1} \left[\frac{7}{4} - 2E_{ii}^\star\right]$$

$$+ \sum_{\substack{k=1 \\ k \neq i}}^{\nu} \frac{x_i x_k m_k^2}{[\eta_{ik}]_1 A_{ik}^\star (m_i + m_k)^3} \left\{\left(\frac{35}{3} - 14C_{ij}^\star\right) m_i + (7 - 8E_{ij}^\star) A_{ij}^\star m_k\right\}$$

$$H_{ij}^{11} = -\frac{x_i x_j m_i^2 m_j^2}{[\eta_{ij}]_1 (m_i + m_j)^4 A_{ij}^\star}$$

$$\times \left\{\frac{385}{6} - 18C_{ij}^\star - 16B_{ij}^\star + 24F_{ij}^\star - \left(\frac{301}{6} - 56E_{ij}^\star + 40H_{ij}^\star\right) A_{ij}^\star\right\} \qquad i \neq j$$

$$H_{ii}^{11} = \frac{x_i^2}{[\eta_i]_1} \left\{\frac{301}{48} - 7E_{ii}^\star + 5H_{ii}^\star\right\}$$

$$+ \sum_{\substack{k=1 \\ k \neq i}}^{\nu} \frac{x_i x_k m_k}{[\eta_{ik}]_1 (m_i + m_k)^4 A_{ik}^\star} \left\{\frac{70}{3} m_i^3 + m_i m_k^2 \left(\frac{245}{6} - 18C_{ik}^\star + 24F_{ik}^\star - 16B_{ik}^\star\right)\right.$$

$$\left. + \left[\left(\frac{77}{3} + 40H_{ij}^\star\right) m_i^2 m_k + m_k^3 \left(\frac{147}{6} - 56E_{ik}^\star\right)\right] A_{ik}^\star\right\}.$$

Note:

$$H_{ij}^{pq} = H_{ji}^{qp} \qquad p, q = 0, 1, \text{ all } i, j.$$

A5.2.6. The thermal conductivity of a binary gas mixture

First-order approximation

$$[\lambda_{mix}]_1 = \frac{1 + Z_\lambda}{X_\lambda + Y_\lambda} \qquad \text{(A5.2.23)}$$

where

$$X_\lambda = \frac{x_1^2}{[\lambda_1]_1} + \frac{2x_1 x_2}{[\lambda_{12}]_1} + \frac{x_2^2}{[\lambda_2]_1}$$

$$Y_\lambda = \frac{x_1^2}{[\lambda_1]_1} U_1 + \frac{2x_1 x_2}{[\lambda_{12}]_1} U_y + \frac{x_2^2}{[\lambda_2]_1} U_2$$

$$Z_\lambda = x_1^2 U_1 + 2x_1 x_2 U_z + x_2^2 U_2$$

$$U_1 = \frac{4}{15} A_{12}^\star - \frac{1}{12}\left(\frac{12}{5} B_{12}^\star + 1\right) \frac{m_1}{m_2} + \frac{1}{2} \frac{(m_2 - m_1)^2}{m_1 m_2}$$

$$U_2 = \frac{4}{15} A_{12}^\star - \frac{1}{12}\left(\frac{12}{5} B_{12}^\star + 1\right) \frac{m_2}{m_1} + \frac{1}{2} \frac{(m_2 - m_1)^2}{m_1 m_2}$$

$$U_y = \frac{4}{15} A_{12}^{\star} \left(\frac{(m_1+m_2)^2}{4m_1m_2} \right) \frac{[\lambda_{12}]_1^2}{[\lambda_1]_1[\lambda_2]_1} - \frac{1}{2}\left(\frac{12}{5} B_{12}^{\star} + 1 \right)$$
$$- \frac{5}{32A_{12}^{\star}} \left(\frac{12}{5} B_{12}^{\star} - 5 \right) \frac{(m_1-m_2)^2}{m_1m_2}$$
$$U_z = \frac{4}{15} A_{12}^{\star} \left[\left(\frac{(m_1+m_2)^2}{4m_1m_2} \right) \left(\frac{[\lambda_{12}]_1}{[\lambda_1]_1} + \frac{[\lambda_{12}]_1}{[\lambda_2]_1} \right) - 1 \right] - \frac{1}{12}\left(\frac{12}{5} B_{12}^{\star} + 1 \right)$$
$$[\lambda_{12}]_1 = \frac{75}{64} \left(\frac{k^3(m_1+m_2)T}{2m_1m_2\pi} \right)^{\frac{1}{2}} \frac{1}{\sigma_{12}^2 \Omega_{12}^{(2,2)*}(T^*)}. \tag{A5.2.24}$$

The second-order approximation to the thermal conductivity of a binary gas mixture is best considered as a special case of the formula for a multicomponent gas mixture presented in the next section.

A5.2.7. *The thermal conductivity of a multicomponent gas mixture of ν components*

First-order approximation

$$[\lambda_{\text{mix}}]_1 = - \frac{\begin{vmatrix} L_{11}^{11} & \cdots & L_{1\nu}^{11} & x_1 \\ \vdots & & \vdots & \vdots \\ L_{\nu 1}^{11} & \cdots & L_{\nu\nu}^{11} & x_\nu \\ x_1 & \cdots & x_\nu & 0 \end{vmatrix}}{\begin{vmatrix} L_{11}^{11} & \cdots & L_{1\nu}^{11} \\ \vdots & & \vdots \\ L_{\nu 1}^{11} & \cdots & L_{\nu\nu}^{11} \end{vmatrix}} \tag{A5.2.25}$$

where

$$L_{ij}^{11} = \frac{-x_i x_j m_i m_j}{2A_{ij}^{\star}[\lambda_{ij}]_1 (m_i+m_j)^2} \left\{ \frac{55}{4} - 3B_{ij}^{\star} - 4A_{ij}^{\star} \right\} \qquad i \neq j$$
$$L_{ii}^{11} = \frac{x_i^2}{[\lambda_i]_1} + \sum_{\substack{k=1 \\ k\neq i}}^{\nu} \frac{x_i x_k}{2A_{ik}^{\star}[\lambda_{ik}]_1 (m_i+m_k)^2} \left\{ \frac{15}{2} m_i^2 + \frac{25}{4} m_k^2 - 3m_k^2 B_{ik}^{\star} + 4m_i m_k A_{ik}^{\star} \right\}$$

and

$$[\lambda_{ij}]_1 = \frac{75}{64} \left(\frac{k^3(m_i+m_j)T}{2\pi m_i m_j} \right)^{\frac{1}{2}} \frac{1}{\sigma_{ij}^2 \Omega_{ij}^{(2,2)*}}$$

Second-order approximation†

$$[\lambda_{\text{mix}}]_2 = [\lambda'_{\text{mix}}]_2 - nk \sum_{i=1}^{\nu} [k_{Ti}]_2 [D_{Ti}]_2 \tag{A5.2.26}$$

† Expressions for the third-order Chapman–Cowling approximations to the thermal conductivity of a multicomponent mixture of monatomic gases have been given by Assael, Wakeham, and Kestin.[11]

where

$$[\lambda'_{\text{max}}]_2 = \frac{\begin{vmatrix} L^{00}_{11} & \cdots & L^{00}_{1\nu} & L^{01}_{11} & \cdots & L^{01}_{1\nu} & L^{02}_{11} & \cdots & L^{02}_{1\nu} & 0 \\ \vdots & & \vdots & \vdots & & \vdots & \vdots & & \vdots & \vdots \\ L^{00}_{\nu 1} & \cdots & L^{00}_{\nu\nu} & L^{01}_{\nu 1} & \cdots & L^{01}_{\nu\nu} & L^{02}_{\nu 1} & \cdots & L^{02}_{\nu\nu} & 0 \\ L^{10}_{11} & \cdots & L^{10}_{1\nu} & L^{11}_{11} & \cdots & L^{11}_{1\nu} & L^{12}_{11} & \cdots & L^{12}_{1\nu} & x_1 \\ \vdots & & \vdots & \vdots & & \vdots & \vdots & & \vdots & \vdots \\ L^{10}_{\nu 1} & \cdots & L^{10}_{\nu\nu} & L^{11}_{\nu 1} & \cdots & L^{11}_{\nu\nu} & L^{12}_{\nu 1} & \cdots & L^{12}_{\nu\nu} & x_\nu \\ L^{20}_{11} & \cdots & L^{20}_{1\nu} & L^{21}_{11} & \cdots & L^{21}_{1\nu} & L^{22}_{11} & \cdots & L^{22}_{1\nu} & 0 \\ \vdots & & \vdots & \vdots & & \vdots & \vdots & & \vdots & \vdots \\ L^{20}_{\nu 1} & \cdots & L^{20}_{\nu\nu} & L^{21}_{\nu 1} & \cdots & L^{21}_{\nu\nu} & L^{22}_{\nu 1} & \cdots & L^{22}_{\nu\nu} & 0 \\ 0 & \cdots & 0 & x_1 & \cdots & x_\nu & 0 & \cdots & 0 & 0 \end{vmatrix}}{D} \tag{A5.2.27}$$

and the denominator D can be written

$$D = \begin{vmatrix} L^{00}_{11} & \cdots & L^{00}_{1\nu} & L^{01}_{11} & \cdots & L^{01}_{1\nu} & L^{02}_{11} & \cdots & L^{02}_{1\nu} \\ \vdots & & \vdots & \vdots & & \vdots & \vdots & & \vdots \\ L^{00}_{\nu 1} & \cdots & L^{00}_{\nu\nu} & L^{01}_{\nu 1} & \cdots & L^{01}_{\nu\nu} & L^{02}_{\nu 1} & \cdots & L^{02}_{\nu\nu} \\ L^{10}_{11} & \cdots & L^{10}_{1\nu} & L^{11}_{11} & \cdots & L^{11}_{1\nu} & L^{12}_{11} & \cdots & L^{12}_{1\nu} \\ \vdots & & \vdots & \vdots & & \vdots & \vdots & & \vdots \\ L^{10}_{\nu 1} & \cdots & L^{10}_{\nu\nu} & L^{11}_{\nu 1} & \cdots & L^{11}_{\nu 1} & L^{12}_{\nu\nu} & \cdots & L^{12}_{\nu\nu} \\ L^{20}_{11} & \cdots & L^{20}_{1\nu} & L^{21}_{11} & \cdots & L^{21}_{1\nu} & L^{22}_{11} & \cdots & L^{22}_{1\nu} \\ \vdots & & \vdots & \vdots & & \vdots & \vdots & & \vdots \\ L^{20}_{\nu 1} & \cdots & L^{20}_{\nu\nu} & L^{21}_{\nu 1} & \cdots & L^{21}_{\nu\nu} & L^{22}_{\nu 1} & \cdots & L^{22}_{\nu\nu} \end{vmatrix} \tag{A5.2.28}$$

Also, $[D_{\text{T}i}]_2$ is given as

$$[D_{\text{T}i}]_2 = \frac{2}{5kn} \frac{\begin{vmatrix} L^{00}_{11} & \cdots & L^{00}_{1\nu} & L^{01}_{11} & \cdots & L^{01}_{1\nu} & L^{02}_{11} & \cdots & L^{02}_{1\nu} & 0 \\ \vdots & & \vdots & \vdots & & \vdots & \vdots & & \vdots & \vdots \\ L^{00}_{\nu 1} & & L^{00}_{\nu\nu} & L^{01}_{\nu 1} & \cdots & L^{01}_{\nu\nu} & L^{02}_{\nu 1} & \cdots & L^{02}_{\nu\nu} & 0 \\ L^{10}_{11} & \cdots & L^{10}_{1\nu} & L^{11}_{11} & \cdots & L^{11}_{1\nu} & L^{12}_{11} & \cdots & L^{12}_{1\nu} & x_1 \\ \vdots & & \vdots & \vdots & & \vdots & \vdots & & \vdots & \vdots \\ L^{10}_{\nu 1} & \cdots & L^{10}_{\nu\nu} & L^{11}_{\nu 1} & \cdots & L^{11}_{\nu\nu} & L^{12}_{\nu 1} & \cdots & L^{12}_{\nu\nu} & x_\nu \\ L^{20}_{11} & \cdots & L^{20}_{1\nu} & L^{21}_{11} & \cdots & L^{21}_{1\nu} & L^{22}_{11} & \cdots & L^{22}_{1\nu} & 0 \\ \vdots & & \vdots & \vdots & & \vdots & \vdots & & \vdots & \vdots \\ L^{20}_{\nu 1} & \cdots & L^{20}_{\nu\nu} & L^{21}_{\nu 1} & \cdots & L^{21}_{\nu\nu} & L^{22}_{\nu 1} & \cdots & L^{22}_{\nu\nu} & 0 \\ \delta_{i1} & \cdots & \delta_{i\nu} & 0 & \cdots & 0 & 0 & \cdots & 0 & 0 \end{vmatrix}}{D} \tag{A5.2.29}$$

in which δ_{ij} is the Kronecker delta.

The terms $[k_{\mathrm{Ti}}]_2$ can be written as

$$[k_{\mathrm{Ti}}]_2 = -\frac{\begin{vmatrix} D_{11}^{1(3)} & \cdots & D_{1\nu}^{1(3)} & S^{1(3)} \\ \vdots & & \vdots & \vdots \\ D_{\nu 1}^{\nu(3)} & \cdots & D_{\nu\nu}^{\nu(3)} & S^{\nu(3)} \\ \delta_{i1} & \cdots & \delta_{i\nu} & 0 \end{vmatrix}}{\begin{vmatrix} D_{11}^{1(3)} & \cdots & D_{1\nu}^{1(3)} \\ \vdots & & \vdots \\ D_{\nu 1}^{\nu(3)} & \cdots & D_{\nu\nu}^{\nu(3)} \end{vmatrix}} \tag{A5.2.30}$$

in which

$$D_{mm}^{k(3)} = 0 \qquad m = 1, 2 \cdots \nu \qquad k = 1, 2 \cdots \nu$$

and

$$D_{ml}^{k(3)} = d_{l,0}^{k(3)} - d_{m,0}^{k(3)} \qquad l \neq m \qquad k = 1, 2, \ldots \nu$$
$$l, m = 1, 2, \ldots \nu$$

The quantities $d_{l,0}^{k(3)}$ are themselves given by the equation

$$d_{l,0}^{k(3)} = \frac{-8}{25k} \frac{\begin{vmatrix} L_{11}^{00} & \cdots & L_{1\nu}^{00} & L_{11}^{01} & \cdots & L_{1\nu}^{01} & L_{1\nu}^{02} & \cdots & L_{1\nu}^{02} & \left(\delta_{ik} - \dfrac{\rho_1}{\rho}\right) \\ \vdots & & \vdots & \vdots & & \vdots & \vdots & & \vdots & \vdots \\ L_{\nu 1}^{00} & \cdots & L_{\nu\nu}^{00} & L_{\nu 1}^{01} & \cdots & L_{\nu\nu}^{01} & L_{\nu 1}^{02} & \cdots & L_{\nu\nu}^{02} & \left(\delta_{\nu k} - \dfrac{\rho_\nu}{\rho}\right) \\ L_{11}^{10} & \cdots & L_{1\nu}^{10} & L_{11}^{11} & \cdots & L_{1\nu}^{11} & L_{11}^{12} & \cdots & L_{1\nu}^{12} & 0 \\ \vdots & & \vdots & \vdots & & \vdots & \vdots & & \vdots & \vdots \\ L_{\nu 1}^{10} & \cdots & L_{\nu\nu}^{10} & L_{\nu 1}^{11} & \cdots & L_{\nu\nu}^{11} & L_{\nu 1}^{12} & \cdots & L_{\nu\nu}^{12} & 0 \\ L_{11}^{20} & \cdots & L_{1\nu}^{20} & L_{11}^{21} & \cdots & L_{1\nu}^{21} & L_{11}^{22} & \cdots & L_{1\nu}^{22} & 0 \\ \vdots & & \vdots & \vdots & & \vdots & \vdots & & \vdots & \vdots \\ L_{\nu 1}^{20} & \cdots & L_{\nu\nu}^{20} & L_{\nu 1}^{21} & \cdots & L_{\nu\nu}^{21} & L_{\nu 1}^{22} & \cdots & L_{\nu\nu}^{22} & 0 \\ \delta_{1l} & \cdots & \delta_{\nu l} & 0 & \cdots & 0 & 0 & \cdots & 0 & 0 \end{vmatrix}}{D} \tag{A5.2.31}$$

whereas the elements $S^{j(3)}$ are given by

$$S^{j(3)} = \frac{4}{5k} \frac{\begin{vmatrix} L_{11}^{00} & \cdots & L_{1\nu}^{00} & L_{11}^{01} & \cdots & L_{1\nu}^{01} & L_{11}^{02} & \cdots & L_{1\nu}^{02} & \left(\delta_{1j} - \dfrac{\rho_1}{\rho}\right) \\ \vdots & & \vdots & \vdots & & \vdots & \vdots & & \vdots & \vdots \\ L_{\nu 1}^{00} & \cdots & L_{\nu\nu}^{00} & L_{\nu 1}^{01} & \cdots & L_{\nu\nu}^{01} & L_{\nu 1}^{02} & \cdots & L_{\nu\nu}^{02} & \left(\delta_{\nu j} - \dfrac{\rho_\nu}{\rho}\right) \\ L_{11}^{10} & \cdots & L_{1\nu}^{10} & L_{11}^{11} & \cdots & L_{1\nu}^{11} & L_{11}^{12} & \cdots & L_{1\nu}^{12} & 0 \\ \vdots & & \vdots & \vdots & & \vdots & \vdots & & \vdots & \vdots \\ L_{\nu 1}^{10} & \cdots & L_{\nu\nu}^{10} & L_{\nu 1}^{11} & \cdots & L_{\nu\nu}^{11} & L_{\nu 1}^{12} & \cdots & L_{\nu\nu}^{12} & 0 \\ L_{11}^{20} & \cdots & L_{1\nu}^{20} & L_{11}^{21} & \cdots & L_{1\nu}^{21} & L_{11}^{22} & \cdots & L_{1\nu}^{22} & 0 \\ \vdots & & \vdots & \vdots & & \vdots & \vdots & & \vdots & \vdots \\ L_{\nu 1}^{20} & \cdots & L_{\nu\nu}^{20} & L_{\nu 1}^{21} & \cdots & L_{\nu\nu}^{21} & L_{\nu 1}^{22} & \cdots & L_{\nu\nu}^{22} & 0 \\ 0 & \cdots & 0 & x_1 & \cdots & x_\nu & 0 & \cdots & 0 & 0 \end{vmatrix}}{D} \tag{A5.2.32}$$

In these expressions n denotes the number density of molecules in the gas, ρ_i the mass density of component i, and ρ the total mass density of the gas mixture.

In addition to the elements defined earlier the elements L_{ij}^{pq} are related to the collision integrals as follows:

$$L_{ii}^{00}=0$$

$$L_{ij}^{00}=\frac{-x_i x_j}{2[\lambda_{ij}]_1 A_{ij}^\star}-\sum_{\substack{k=1\\k\neq i}}^{\nu}\frac{x_i x_k m_j}{2m_i A_{ik}^\star[\lambda_{ik}]_1} \qquad i\neq j$$

$$L_{ii}^{01}=-\sum_{\substack{k=1\\k\neq i}}^{\nu}\frac{x_i x_k m_k}{4A_{ik}^\star[\lambda_{ik}]_1(m_i+m_k)}(6C_{ik}^\star-5)$$

$$L_{ij}^{01}=\frac{x_i x_j m_i}{4A_{ij}^\star[\lambda_{ij}]_1(m_i+m_j)}(6C_{ij}^\star-5) \qquad i\neq j$$

$$L_{ii}^{02}=\sum_{\substack{k=1\\k\neq i}}^{\nu}\frac{x_i x_k m_k^2}{4A_{ik}^\star[\lambda_{ik}]_1(m_i+m_k)^2}\left\{\frac{35}{4}-3B_{ik}^\star-6C_{ik}^\star\right\}$$

$$L_{ij}^{02}=\frac{-x_i x_j m_i^2}{4A_{ij}^\star[\lambda_{ij}]_1(m_i+m_j)^2}\left\{\frac{35}{4}-3B_{ij}^\star-6C_{ij}^\star\right\} \qquad i\neq j$$

$$L_{ii}^{12}=\frac{x_i^2[7-8E_{ii}^\star]}{4[\lambda_i]_1}+\sum_{\substack{k=1\\k\neq i}}^{\nu}\frac{m_k x_i x_k}{2[\lambda_{ik}]_1 A_{ik}^\star(m_i+m_k)^3}$$
$$\times\left\{m_i^2\left(\frac{105}{4}-\frac{63}{2}C_{ik}^\star\right)+m_k^2\left(\frac{175}{16}+\frac{255}{8}C_{ik}^\star-\frac{57}{4}B_{ik}^\star-30G_{ik}^\star\right)\right.$$
$$\left.+2m_i m_k(7-8E_{ik}^\star)A_{ik}^\star\right\}$$

$$L_{ij}^{12}=-\frac{x_i x_j m_i^2 m_j}{2[\lambda_{ij}]_1 A_{ij}^\star(m_i+m_j)^3}\left\{\frac{595}{16}+\frac{3}{8}C_{ij}^\star-\frac{57}{4}B_{ij}^\star-30G_{ij}^\star-2(7-8E_{ij}^\star)A_{ij}^\star\right\} \qquad i\neq j$$

$$L_{ii}^{22}=\frac{x_i^2}{[\lambda_i]_1}\left[\frac{77}{16}-7E_{ii}^\star+5H_{ii}^\star\right]$$
$$+\sum_{\substack{k=1\\k\neq i}}^{\nu}\frac{x_i x_k}{2[\lambda_{ik}]_1 A_{ik}^\star(m_i+m_k)^4}\left\{\frac{175}{8}m_i^4+m_k^4\left(\frac{1225}{64}+\frac{315}{2}C_{ik}^\star-\frac{399}{8}B_{ik}^\star-210\,G_{ik}^\star\right.\right.$$
$$\left.+90I_{ik}^\star\right)+m_i^2 m_k^2\left(\frac{735}{8}-18C_{ik}^\star-\frac{81}{2}B_{ik}^\star+24F_{ik}^\star\right)$$
$$\left.+28m_i^3 m_k A_{ik}^\star+m_i m_k^3(49-112E_{ik}^\star+80H_{ik}^\star)A_{ik}^\star\right\}$$

$$L_{ij}^{22}=\frac{-x_i x_j m_i^2 m_j^2}{2A_{ij}^\star[\lambda_{ij}]_1(m_i+m_j)^4}$$
$$\times\left\{\frac{8505}{64}+\frac{558}{4}C_{ij}^\star-\frac{723}{8}B_{ij}^\star-210G_{ij}^\star+90I_{ij}^\star+24F_{ij}^\star\right.$$
$$\left.-(77-112E_{ij}^\star+80H_{ij}^\star)A_{ij}^\star\right\} \qquad i\neq j.$$

The L_{ij}^{pq} defined in these equations obey the symmetry condition

$$L_{ij}^{pq}=L_{ji}^{qp} \qquad p,q=0,1,2$$
$$i,j=1,2,\ldots\nu \text{ except } \left\{\begin{matrix}p=q=0\\ i\neq j\end{matrix}\right\}.$$

A5.2.8. The diffusion coefficient in a binary gas mixture

First-order

$$[D_{12}]_1 = \frac{3}{16n}\left(\frac{2kT(m_1+m_2)}{\pi m_1 m_2}\right)^{\frac{1}{2}} \frac{1}{\sigma_{12}^2 \Omega^{(1,1)*}(T_{12}^*)}. \qquad \text{(A5.2.33)}$$

Second-order Chapman–Cowling approximation

$$[D_{12}]_2 = [D_{12}]_1 f_D^{(2)} \qquad \text{(A5.2.34)}$$

and

$$f_D^{(2)} = (1-\Delta_{12})^{-1}$$

where

$$\Delta_{12} = \tfrac{1}{10}(6C_{12}^\star - 5)^2 \left\{\frac{x_1^2 P_1 + x_2^2 P_2 + x_1 x_2 P_{12}}{x_1^2 Q_1 + x_2^2 Q_2 + x_1 x_2 Q_{12}}\right\}$$

$$P_1 = \frac{2m_1^2}{m_2(m_1+m_2)}\left(\frac{2m_2}{m_1+m_2}\right)^{\frac{1}{2}} \frac{\Omega_{11}^{(2,2)*}}{\Omega_{12}^{(1,1)*}}\left(\frac{\sigma_{11}}{\sigma_{12}}\right)^2$$

$$P_{12} = 15\left(\frac{m_1-m_2}{m_1+m_2}\right)^2 + \frac{8m_1m_2}{(m_1+m_2)^2} A_{12}^\star$$

$$Q_1 = \frac{2}{m_2(m_1+m_2)}\left(\frac{2m_2}{m_1+m_2}\right)^{\frac{1}{2}} \frac{\Omega_{11}^{(2,2)*}}{\Omega_{12}^{(1,1)*}}\left(\frac{\sigma_{11}}{\sigma_{12}}\right)^2$$

$$\times[(\tfrac{5}{2} - \tfrac{6}{5}B_{12}^\star)m_1^2 + 3m_2^2 + \tfrac{8}{5}m_1 m_2 A_{12}^\star]$$

$$Q_{12} = 15\left(\frac{m_1-m_2}{m_1+m_2}\right)^2\left(\frac{5}{2} - \frac{6}{5}B_{12}^\star\right) + \frac{4m_1m_2A_{12}^\star}{(m_1+m_2)^2}\left(11 - \frac{12}{5}B_{12}^\star\right)$$

$$+\frac{8}{5}\frac{(m_1+m_2)}{(m_1m_2)^{\frac{1}{2}}}\frac{\Omega_{11}^{(2,2)*}}{\Omega_{12}^{(1,1)*}}\frac{\Omega_{22}^{(2,2)*}}{\Omega_{12}^{(1,1)*}}\left(\frac{\sigma_{11}}{\sigma_{12}}\right)^2\left(\frac{\sigma_{22}}{\sigma_{12}}\right)^2$$

and expressions for P_2 and Q_2 are obtained from P_1 and Q_1 by an interchange of subscripts.

Self-diffusion

For self-diffusion the Kihara approximation scheme leads to the somewhat simpler higher order correction factor

$$f_D^{K(2)} = 1 + (6C^\star - 5)^2/(16A^\star + 40) \qquad \text{(A5.2.35)}$$

A5.2.9. The thermal diffusion factor for a binary gas mixture.
First-order Chapman–Cowling approximation.

$$[\alpha_T]_1 = (6C_{12}^\star - 5)\frac{S_1x_1 - S_2x_2}{Q_1x_1^2 + Q_2x_2^2 + Q_{12}x_1x_2} \qquad \text{(A5.2.36)}$$

where

$$S_1 = \frac{m_1}{m_2}\left(\frac{2m_2}{m_1+m_2}\right)^{\frac{1}{2}}\left[\frac{\Omega_{11}^{(2,2)*}}{\Omega_{12}^{(1,1)*}}\right]\left(\frac{\sigma_{11}}{\sigma_{12}}\right)^2 - \frac{4m_1m_2A_{12}^\star}{(m_1+m_2)^2} + \frac{15m_2(m_1-m_2)}{2(m_1+m_2)^2}$$

and S_2 is obtained from S_1 by an interchange of subscripts. Q_1, Q_2, and Q_{12} have been defined above.

Second-order Chapman–Cowling approximation. The second-order approximation to α_T is most easily obtained directly from the definition

$$[\alpha_T]_2 = [k_{T1}]_2/x_1x_2 \tag{A5.2.37}$$

with $[k_{T1}]_2$ given by equation (A5.2.30) for $\nu = 2$.

First-order Kihara approximation

$$[\alpha_T]_1^K = \frac{(6C^\star - 5)(S_1x_1 - S_2x_2)}{Q_1^Kx_1^2 + Q_2^Kx_2^2 + Q_{12}^Kx_1x_2} \tag{A5.2.38}$$

where S_1 and S_2 have been already defined and

$$Q_1^K = \frac{2}{(m_1+m_2)m_2}\left(\frac{2m_2}{m_1+m_2}\right)^{\frac{1}{2}}\frac{\Omega_{11}^{(2,2)*}}{\Omega_{12}^{(1,1)*}}\left(\frac{\sigma_{11}^2}{\sigma_{12}^2}\right)[3m_2^2 + m_1^2 + \tfrac{8}{5}m_1m_2A_{12}^\star]$$

Q_2^K is obtained from Q_1^K by an interchange of subscripts.

$$Q_{12}^K = \frac{15(m_1-m_2)^2}{(m_1+m_2)^2} + \frac{32m_1m_2}{(m_1+m_2)^2}A_{12}^\star + \frac{8}{5}\frac{(m_1+m_2)}{(m_1m_2)^{\frac{1}{2}}}\frac{\Omega_{11}^{(2,2)*}}{\Omega_{12}^{(1,1)*}}\frac{\Omega_{22}^{(2,2)*}}{\Omega_{12}^{(1,1)*}}\frac{\sigma_{11}^2\sigma_{22}^2}{\sigma_{12}^4}.$$

Expressions for higher-order approximations to the thermal diffusion factor have been given by Saxena and Joshi.[12]

A5.2.10. The isotopic thermal diffusion factor

First-order Chapman–Cowling approximation

$$[\alpha_0]_1 = \frac{15(6C^\star - 5)(2A^\star + 5)}{2A^\star(16A^\star - 12B^\star + 55)}. \tag{A5.2.39}$$

First-order Kihara approximation

$$[\alpha_0]_1^K = \frac{15(6C^\star - 5)}{16A^\star}. \tag{A5.2.40}$$

Second-order Kihara approximation.

$$[\alpha_0]_2^K = [\alpha_0]_1 f_\alpha^{K(2)} \tag{A5.2.41}$$

with $f_\alpha^{K(2)} = 1 + \delta_\alpha$ and

$$\delta_\alpha = \frac{(7-8E^\star)}{9}\left\{\left(\frac{2A^\star}{(\frac{35}{4}) + 7A^\star + 4F^\star}\right)\left[K^\star + \frac{1}{2}\left(\frac{7(5-6C^\star) + A^\star(7-8E^\star)}{5+2A^\star}\right)\right.\right.$$
$$\left.\left.\times\left(\frac{(\frac{35}{8}) + 28A^\star - 6F^\star}{21A^\star}\right)\right] - \frac{5}{7}\left[K^\star - \frac{7}{5}\left(\frac{5-6C^\star}{5+2A^\star}\right) - \frac{3(7-8E^\star)}{10}\right]\right\}.$$

Appendix A5.3 Semi-classical formulae for the transport coefficients of dilute polyatomic gases and gas mixtures

This appendix lists the available semi-classical kinetic theory formulae for some of the transport properties of dilute polyatomic gases and gas mixtures. Formulae are given which relate various experimentally accessible quantities as well as those which relate individual transport coefficients to intermolecular pair potentials. The former are useful for comparisons with experimental data, whereas the most important of the latter have been given in the text and will be discussed in Chapter 6. Where particular quantities are often not available from experimental measurements methods whereby they may be estimated are given.

It must be emphasized that none of the expressions given are any more accurate than first-order Chapman–Cowling approximations. Some are even less accurate because of additional approximations introduced to render them useful.

A.5.3.1. The viscosity of a pure polyatomic gas

Within the Mason and Monchick approximation,[22]

$$[\eta]_1 = \frac{5}{16}\frac{(\pi mkT)^{\frac{1}{2}}}{\langle\bar{\Omega}^{(2,2)}(T)\rangle} \tag{A5.3.1}$$

where

$$\langle\bar{\Omega}^{(2,2)}(T)\rangle = \frac{1}{4}\int_0^\infty\int_0^{2\pi}\int_0^\infty \{[1-\cos^2\chi]\, b\, \mathrm{d}b\, \mathrm{d}\psi\}\exp(-\gamma^2)\gamma^6\,\mathrm{d}\gamma^2 \tag{A5.3.2}$$

and χ is the deflection angle in a binary collision between two polyatomic molecules (see Chapter 4). Methods whereby $\langle\bar{\Omega}^{(2,2)}(T)\rangle$ have been evaluated for particular intermolecular pair potentials are discussed in Chapter 6.

A.5.3.2. The self-diffusion coefficient of a pure polyatomic gas[22]

Within the Mason and Monchick approximation,[22]

$$[D_{11}]_1 = \frac{3}{8n}\left(\frac{kT\pi}{m}\right)^{\frac{1}{2}}\frac{1}{\langle\bar{\Omega}^{(1,1)}(T)\rangle} \tag{A5.3.3}$$

where

$$\langle\bar{\Omega}^{(1,1)}(T)\rangle = \frac{1}{2}\int_0^\infty\int_0^{2\pi}\int_0^\infty \{[1-\cos\chi] b\, \mathrm{d}b\, \mathrm{d}\psi\}\exp(-\gamma^2)\gamma^4\,\mathrm{d}\gamma^2 \tag{A5.3.4}$$

A5.3.3. The thermal conductivity of a pure polyatomic gas

In the semi-classical theory for molecules with a single mode of internal energy

$$\frac{m[\lambda]_1}{k[\eta]_1} = \frac{5}{2}\left(\frac{3}{2}-\Delta\right) + \frac{\rho D_{\text{int}}}{[\eta]_1}\left(\frac{c_{v,\text{int}}}{k}+\Delta\right) \tag{A5.3.5}$$

where

$$\Delta = \left(\frac{2c_{v,\text{int}}}{\pi k\zeta_{\text{rot}}}\right)\left(\frac{5}{2}-\frac{\rho D_{\text{int}}}{[\eta]_1}\right)\left[1+\left(\frac{2}{\pi\zeta_{\text{rot}}}\right)\left(\frac{5}{3}\frac{c_{v,\text{int}}}{k}+\frac{\rho D_{\text{int}}}{[\eta]_1}\right)\right]^{-1}.$$

Here, $c_{v,\mathrm{int}}$ is the internal heat capacity per molecule, $\rho = nm$ is the mass density of the gas, D_{int} the diffusion coefficient for internal energy, and ζ_{rot} the collision number for relaxation of rotational energy.

For the purposes of calculation it is often sufficient to write $D_{\mathrm{int}} = [D_{11}]_1$ and to employ the relation

$$\frac{\rho[D_{11}]_1}{\eta_1} = \frac{6}{5} A_{11}^{\star} = \frac{6\langle\bar{\Omega}^{(2,2)}(T)\rangle}{5\langle\bar{\Omega}^{(1,1)}(T)\rangle} \simeq \frac{6\Omega^{(2,2)*}}{5\Omega^{(1,1)*}} \tag{A5.3.6}$$

where $\Omega^{(2,2)*}$ and $\Omega^{(1,1)*}$ may be obtained from a spherically symmetric intermolecular pair potential. The quantity ζ_{rot} may be obtained from sound absorption measurements.

A5.3.4. *The viscosity of a multicomponent polyatomic gas mixture*

Formally, within the semi-classical theory the equation for the viscosity of a polyatomic gas mixture is identical with that for a monatomic gas mixture given in Appendix A5.2. The only differences occur in the definitions of the collision integrals which arise in the equations. Because almost all of these collision integrals can be replaced by experimental quantities the calculation of the multicomponent mixture viscosity is straightforward. If experimental values for $[\eta_{ij}]_1$ are not available the relation

$$\frac{\rho[D_{ij}]_1}{[\eta_{ij}]_1} = \frac{6}{5} A_{ij}^{\star} \tag{A5.3.7}$$

may be used to obtain them from measured binary diffusion coefficients D_{ij}. The quantity $A_{ij}^{\star}$ defined by a generalization of (A5.3.6) can itself be estimated from an assumed spherically symmetric intermolecular potential.[26]

A5.3.5. *The binary diffusion coefficient of a polyatomic gas mixture*

Within the Mason and Monchick approximation[22]

$$[D_{ij}]_1 = \frac{3}{16n}\left(\frac{2\pi kT(m_1+m_2)}{m_1 m_2}\right)^{\frac{1}{2}} \frac{1}{\langle\bar{\Omega}_{ij}^{(1,1)}(T)\rangle} \tag{A5.3.8}$$

where $\langle\bar{\Omega}_{ij}^{(1,1)}(T)\rangle$ is a generalization of the collision integral of (A5.3.4) to an unlike intermolecular pair potential. (A5.3.8) is formally identical to that for a monatomic gas mixture but again the collision integral is different.

A5.3.6. *The thermal conductivity of a polyatomic gas mixture*

By applying several physically reasonable approximations to the formal semi-classical kinetic theory expressions for the thermal conductivity of a polyatomic gas mixture, Monchick, Pereira, and Mason[23] have derived the following rather cumbersome formula for a mixture of ν components

$$\lambda_{\mathrm{mix}} = \lambda_{\mathrm{HE}} + \Delta\lambda \tag{A5.3.9}$$

where

$$\lambda_{\mathrm{HE}} = \lambda_{\mathrm{mix,trans}} + \sum_{i}^{\nu} [\lambda_i - \lambda_{i,\mathrm{trans}}]\left[1 + \sum_{j\neq i} \frac{x_j D_{i\,\mathrm{int},i}}{x_i D_{i\,\mathrm{int},j}}\right]^{-1} \tag{A5.3.10}$$

is the *Hirschfelder–Eucken* expression for the mixture thermal conductivity.[42] The term $\Delta\lambda$ incorporates the explicit effects due to inelastic collisions and is given by

$$\Delta\lambda = 4\sum_i \Biggl\{ x_i \sum_\alpha \Lambda_{i\alpha} \sum_\beta (\Delta\mathscr{L}_{\alpha\beta}^{10,10}) \Bigl(\sum_\gamma \Lambda_{\beta\gamma} x_\gamma\Bigr) - 2\left(\frac{x_i}{\mathscr{L}_{ii}^{01,01}}\right) \sum_\alpha (\Delta\mathscr{L}_{i\alpha}^{10,01}) \Bigl(\sum_\beta \Lambda_{\alpha\beta} x_\beta\Bigr) - \left(\frac{x_i^2}{\mathscr{L}_{ii}^{01,01}}\right) \Biggl[\left(\frac{l_{ii}^{01,01}}{l_{ii}^{10,10}}\right)\left(\frac{\Delta l_{ii}^{10,10}}{l_{ii}^{10,10}}\right) + 2\left(\frac{\Delta l_{ii}^{10,01}}{l_{ii}^{10,10}}\right) + \left(\frac{\Delta l_{ii}^{01,01}}{l_{ii}^{01,01}}\right) - \left(\frac{\Delta\mathscr{L}_{ii}^{01,01}}{\mathscr{L}_{ii}^{01,01}}\right) \Biggr] \Biggr\}. \quad \text{(A5.3.11)}$$

In these equations x_i represents the mole fraction of species i in the mixture. The symbol $\lambda_{\text{mix,trans}}$ is the translational contribution to the thermal conductivity of the mixture, $\lambda_{i,\text{trans}}$ is the translational contribution to the thermal conductivity of species i, and λ_i is the total thermal conductivity of the pure gas i. The quantity $\lambda_{i,\text{trans}}$ is obtained from the relation

$$\lambda_{i,\text{trans}} = \tfrac{15}{4} k\eta_i/m_i \quad \text{(A5.3.12)}$$

where η_i is the viscosity of the pure gas i. The translational contribution to the mixture thermal conductivity is obtained from (A5.2.25) of Appendix A5.2 using $\lambda_{i,\text{trans}}$ for the pure gas thermal conductivities. In this calculation $[\lambda_{ij}]_1$ is obtained from the equation

$$[\lambda_{ij}]_1 = \frac{25nk}{8A_{ij}^\star}[D_{ij}]_1 \quad \text{(A5.3.13)}$$

where D_{ij} is the experimental diffusion coefficient and $A_{ij}^\star$ is estimated as before. The remaining quantity required for the evaluation of $\lambda_{\text{mix,trans}}$, $B_{ij}^\star$ can either be estimated from the assumed spherical intermolecular potential or better obtained from the temperature dependence of the binary diffusion coefficient from the relation

$$B_{ij}^\star \simeq \frac{1}{12}\left[2\left(\frac{\partial \ln D_{ij}}{\partial \ln T}\right)_P - 1\right]\left[9 - 2\left(\frac{\partial \ln D_{ij}}{\partial \ln T}\right)_P\right] + \frac{1}{3}\frac{\partial^2 \ln D_{ij}}{\partial (\ln T)^2} \quad \text{(A5.3.14)}$$

In (A5.3.10) the symbols $D_{i\,\text{int},i}$ and $D_{i\,\text{int},j}$ represent the diffusion coefficients for internal energy of species i through species i and j respectively. Frequently the best that can be done is to equate them to the mass diffusion coefficients D_{ii} and D_{ij} respectively.

Finally the elements in the inelastic term may be written,

$$\Lambda_{\alpha\beta} = \left|\begin{matrix} \mathscr{L}_{11}^{10,10} & \cdots & \mathscr{L}_{1\beta}^{10,10} & \cdots & \mathscr{L}_{1\nu}^{10,10} & 0 \\ \vdots & & \vdots & & \vdots & \vdots \\ \mathscr{L}_{\alpha 1}^{10,10} & \cdots & \mathscr{L}_{\alpha\beta}^{10,10} & \cdots & \mathscr{L}_{\alpha\nu}^{10,10} & 1 \\ \vdots & & \vdots & & \vdots & \vdots \\ \mathscr{L}_{\nu 1}^{10,10} & \cdots & \mathscr{L}_{\nu\beta}^{10,10} & \cdots & \mathscr{L}_{\nu\nu}^{10,10} & 0 \\ 0 & \cdots & 1 & \cdots & 0 & 0 \end{matrix}\right| \Bigg/ \left|\begin{matrix} \mathscr{L}_{11}^{10,10} & \cdots & \mathscr{L}_{1\nu}^{10,10} \\ \vdots & & \vdots \\ \mathscr{L}_{\nu 1}^{10,10} & \cdots & \mathscr{L}_{\nu\nu}^{10,10} \end{matrix}\right|$$

where

$$\mathscr{L}_{ii}^{10,10} = -\frac{16}{15}\left(\frac{x_i^2 m_i}{\eta_i k}\right) - \frac{16}{25}\sum_{j\neq i}^{\nu}[x_i x_j T/PD_{ij}(m_i+m_j)^2] \times[\tfrac{15}{2}m_i^2 + (\tfrac{25}{4} - 3B_{ij}^{\star})m_j^2 + 4m_i m_j A_{ij}^{\star}]$$

and

$$\mathscr{L}_{ij}^{10,10} = \frac{16}{25}\frac{x_i x_j T m_i m_j}{PD_{ij}(m_i+m_j)^2}\left[\frac{55}{4} - 3B_{ij}^{\star} - 4A_{ij}^{\star}\right] \qquad i\neq j.$$

In addition, the other elements can be written

$$\Delta\mathscr{L}_{ii}^{10,10} = \frac{-32}{9\pi}\frac{x_i^2 m_i c_{i,\mathrm{rot}}}{k^2\eta_i\zeta_{ii}} - \frac{64}{15\pi}\sum_{j\neq i}^{\nu}\frac{x_i x_j T A_{ij}^{\star} m_i m_j}{PD_{ij}(m_i+m_j)^2}\left(\frac{c_{i,\mathrm{rot}}}{k\zeta_{ij}} + \frac{c_{j,\mathrm{rot}}}{k\zeta_{ji}}\right)$$

$$\Delta\mathscr{L}_{ij}^{10,10} = \frac{-64}{15\pi}\frac{x_i x_j T A_{ij}^{\star} m_i m_j}{PD_{ij}(m_i+m_j)^2}\left(\frac{c_{i,\mathrm{rot}}}{k\zeta_{ij}} + \frac{c_{j,\mathrm{rot}}}{k\zeta_{ji}}\right) \qquad i\neq j$$

$$\mathscr{L}_{ii}^{01,01} = \frac{-4kT}{PD_{i,\mathrm{int},i}c_{i,\mathrm{int}}}\left(x_i^2 + \sum_{j\neq i}^{\nu} x_i x_j \frac{D_{i,\mathrm{int},i}}{D_{i\ \mathrm{int},j}}\right)$$

$$\Delta\mathscr{L}_{ii}^{01,01} = \frac{-8k}{\pi(c_{i,\mathrm{int}})^2}\left(\frac{x_i^2 m_i c_{i,\mathrm{rot}}}{k\eta_i\zeta_{ii}} + \frac{6}{5}\sum_{j\neq i}^{\nu} x_i x_j \frac{m_i T A_{ij}^{\star} c_{i,\mathrm{rot}}}{m_j PD_{ij}\zeta_{ij}}\right)$$

$$\Delta\mathscr{L}_{ii}^{10,01} = \frac{16}{15\pi c_{i,\mathrm{int}}}\left[\frac{5x_i^2 m_i c_{i,\mathrm{rot}}}{k\eta_i\zeta_{ii}} + 6\sum_{j\neq i}^{\nu}\frac{x_i x_j T A_{ij}^{\star} m_i c_{i,\mathrm{rot}}}{PD_{ij}(m_i+m_j)\zeta_{ij}}\right]$$

$$\Delta\mathscr{L}_{ij}^{10,01} = \frac{32 x_i x_j T A_{ij}^{\star} m_i c_{i,\mathrm{rot}}}{5\pi c_{j,\mathrm{int}} PD_{ij}(m_i+m_j)\zeta_{ji}}$$

and

$$l_{ii}^{rs,r's'} = \lim_{x_i\to 1} \mathscr{L}_{ii}^{rs,r's'} \qquad \Delta l_{ii}^{rs,r's'} = \lim_{x_i\to 1} \Delta\mathscr{L}_{ii}^{rs,r's'}.$$

In these equations the remaining undefined quantities are $c_{i,\mathrm{int}}$ the internal heat capacity per molecule of species i, $c_{i,\mathrm{rot}}$ the rotational heat capacity of species i, and ζ_{ij} which is the collision number for equilibration of rotational energy of species i by collisions with species j. The heat capacities are generally easily accessible, but frequently the collision numbers must be estimated.[23]

References

1. Resibois, P. and Dé Leener, M. *Classical kinetic theory of fluids*, Chapter IV. Wiley, New York (1977).
2. Ferziger, J. H. and Kaper, H. G. *The mathematical theory of transport processes in gases*, Chapter 13. North-Holland, Amsterdam (1972).
3. Ref. 2, Chapter 3, p. 27.
4. Ref. 2, Chapter 4, p. 72.
5. Ref. 2. Chapter 7, p. 203.
6. Chapman, S. and Cowling, T. G. *The mathematical theory of non-uniform gases*, (3rd ed). Cambridge University Press London (1970).

7. Hirschfelder, J. O., Curtiss, C. F., and Bird, R. B. *Molecular theory of gases and liquids.* Wiley, New York (1954).
8. Kihara, T. *Imperfect gases.* Asakusa Bookstore, Tokyo (1949).
9. Ref. 2, Chapter 5, p. 139.
10. Ref. 7, Chapter 8, p. 548.
11. Assael, M. J., Wakeham, W. A., and Kestin, J. *Int. J. Thermophys.* **1,** 7 (1980).
12. Saxena, S. C. and Joshi, R. K. *J. Sci. industr. Res.* **24,** 518 (1965).
13. Wigner, F. *Phys. Rev.* **40,** 749 (1932).
14. Snider, R. F. *J. chem. Phys.* **32,** 1051 (1960).
15. Waldmann, L. *Z. Naturforsch.* **13a,** 609 (1958).
16. McCourt, F. R. and Snider, R. F. *J. chem. Phys.* **41,** 3185 (1964); **43,** 2276 (1965).
17. Snider, R. F. *Physica* **78,** 387 (1974).
18. Moraal, H. and Snider, R. F. *Chem. Phys. Lett.* **9,** 401 (1971).
19. Taxman, N. *Phys. Rev.* **110,** 1235 (1958).
20. Wang Chang, C. S., Uhlenbeck, G. E., and de Boer, J. *Studies in statistical mechanics,* Vol. 2, Part C (ed. J. de Boer and G. E. Uhlenbeck). North-Holland, Amsterdam (1964).
21. Sandler, S. I. and Dahler, J. S. *J. chem. Phys.* **44,** 1229 (1966).
22. Mason, E. A. and Monchick, L. *J. chem. Phys.* **36,** 1622 (1962).
23. Monchick, L., Pereira, A. N. G., and Mason, E. A. *J. chem. Phys.* **42,** 3241 (1965).
24. Clifford, A. A., Kestin, J., and Wakeham, W. A. *Physica* **97A,** 298 (1979).
25. Sandler, S. I. *Phys. Fluids* **11,** 2549 (1968).
26. Monchick, L., Yun, K. S., and Mason, E. A. *J. chem. Phys.* **39,** 654 (1963).
27. Senftleben, H. *Phys. Z.* **31,** 822, 961 (1930).
28. Beenakker, J. J. M. and McCourt, F. R. *Ann. Rev. Phys. Chem.* **21,** 47 (1970).
29. Hulsman, H., Van Kuik, F. G., Walstra, K. W., Knaap, H. F. P., and Beenakker, J. J. M. *Physica* **57,** 501 (1972).
30. Liu, W. K., McCourt, F. R., and Köhler, W. E. *J. chem. Phys.* **71,** 2566 (1979).
31. Liu, W. K. and McCourt, F. R. *J. chem. Phys.* **71,** 3750 (1979).
32. Viehland, L. A., Mason, E. A., and Sandler, S. I. *J. chem. Phys.* **68,** 5277 (1978).
33. Coope, J. A. R. and Snider, R. F. *J. chem. Phys.* **70,** 1075 (1979).
34. Enskog, D. *Kungl. Svenska Vet.-Ak. Handl.* **63,** (4), (1922).
35. Ref. 2, Chapter 12.
36. Thorne, H. H. Quoted by Chapman, S. and Cowling, T. G. *The mathematical theory of non-uniform gases,* (2nd edn), p. 292. Cambridge University Press London (1952).
37. van Beijeren, H. and Ernst, M. J. *Physica* **70,** 225 (1973).
38. Di Pippo, R., Dorfman, J. R., Kestin, J., Khalifa, H. E., and Mason, E. A. *Physica* **86A,** 205 (1977).
39. Mason, E. A., Khalifa, H. E., Kestin, J., Di Pippo, R., and Dorfman, J. R. *Physica* **91A,** 377 (1978).
40. Dorfman, J. R. and van Beijeren, H. *Statistical mechanics Part B: time-dependent processes,* (ed. B. J. Berne), Chapter 3. Plenum, New York (1977).
41. Hoegy, W. R. and Sengers, J. V. *Phys. Rev.* **A2,** 2461 (1970).
42. Hirschfelder, J. O. *J. chem. Phys.* **26,** 282 (1957).

6

THE TRANSPORT PROPERTIES OF GASES AND INTERMOLECULAR FORCES

6.1. Introduction

The kinetic theory of dilute gases, outlined in the previous chapter, establishes that for monatomic species the transport coefficients are related to the forces between pairs of molecules. The transport coefficients, diffusion, viscosity, and thermal conductivity, quantify the difficulty of the transport of mass, momentum, and energy respectively in a gas subjected to gradients of concentration, velocity, or temperature. Because, in a dilute gas, this transport is achieved by motion of the molecules, the nature of the collisions that the molecules undergo influences the difficulty of transport. Further, because the outcome of the binary collisions is determined by the form of the intermolecular potential, this potential influences the transport coefficients in a complicated way.

Nevertheless, the fundamental kinetic theory for the transport coefficients can be formulated so as to involve only a set of well-defined *collision integrals*, in addition to molecular masses, temperature, and pressure. These collision integrals represent variously weighted, energy-averaged cross-sections for the binary encounters between molecules of the dilute gas and are defined in Appendix A5.2. A complete list of the expressions for the most important transport coefficients of monatomic gases and gas mixtures in terms of these collision integrals is given for ease of reference in the same appendix. Since it is these expressions which form the foundation for the developments of this chapter we repeat the most important of them here. However, now we write explicit formulae for the transport coefficients of a single gas in which the fundamental constants have been assigned their known values† and where the molecular properties of the gas and the macroscopic variables are expressed in the practical units of the SI system.

† In formulating these expressions the values of Boltzmann's constant, $k = 1{\cdot}380\,662 \pm 0{\cdot}000\,044 \times 10^{-23}\,\mathrm{J\,K^{-1}}$, and Avogadro's constant $N_A = 6{\cdot}022\,045 \pm 0{\cdot}000\,031 \times 10^{23}\,\mathrm{mol^{-1}}$ have been taken from the consistent set of values recommended in CODATA Bulletin No. 11. The relative molecular masses to be used in these formulae are to be based on the carbon-12 mass scale. The relative molecular mass, M, is then the ratio of the molecular mass of the species to one-twelfth of the molecular mass of carbon-12.

For the viscosity, to a first approximation,

$$[\eta]_1 = \frac{2{\cdot}6696 \times 10^{-2}(MT)^{\frac{1}{2}}}{\sigma^2\Omega^{(2,2)*}(T^*)} \tag{6.1}$$

where

$$T^* = kT/\varepsilon$$

where the viscosity η is measured in μPa s($=\mu$N s m^{-2} = 10μpoise), the temperature T in kelvin and the length parameter of the potential σ in nanometres. M is the relative molecular mass.

For the thermal conductivity of a monatomic gas to a first approximation

$$[\lambda]_1 = \frac{0.83236(T/M)^{\frac{1}{2}}}{\sigma^2\Omega^{(2,2)*}(T^*)} \tag{6.2}$$

where the thermal conductivity is measured in mW m^{-1} K^{-1}. For the self-diffusion coefficient

$$[D_1] = \frac{2{\cdot}6636 \times 10^{-5}(T^3/M)^{\frac{1}{2}}}{P\sigma^2\Omega^{(1,1)*}(T^*)} \tag{6.3}$$

where the diffusion coefficient is measured in 10^{-5} m^2 s^{-1}, and the pressure, P, in MPa ($=10^6$ N m^{-2} = 10 bar).

Higher-order approximations to the transport coefficients of a pure gas can also be derived (see Chapter 5). The results to mth order can be written as

$$[\eta]_m = [\eta]_1 f_\eta^{(m)} \tag{6.4}$$

$$[\lambda]_m = [\lambda]_1 f_\lambda^{(m)} \tag{6.5}$$

and

$$[D_{11}]_m = [D]_1 f_D^{(m)}. \tag{6.6}$$

The correction factors $f_\eta^{(m)}$, $f_\lambda^{(m)}$, and $f_D^{(m)}$ depart by only a few per cent from unity, so that the first-order formulae for the transport coefficients are remarkably accurate. Expressions for the correction factors in terms of generalized collision integrals, $\Omega^{(l,s)*}$, are collected in Appendix A5.2.

In the case of a binary mixture of species 1 and 2, the interaction quantities $[\eta_{12}]$, and $[\lambda_{12}]_1$ which enter the kinetic theory expressions (see Appendix A5.2) and the binary diffusion coefficient, $[D_{12}]_1$, can be written in the same system of units as

$$[\eta_{12}]_1 = \frac{2{\cdot}6696 \times 10^{-2}(2M_1M_2T/(M_1+M_2))^{\frac{1}{2}}}{\sigma_{12}^2\Omega_{12}^{(2,2)*}(T_{12}^*)} \tag{6.7}$$

$$[\lambda_{12}]_1 = 0{\cdot}83236\left(\frac{T(M_1+M_2)}{2M_1M_2}\right)^{\frac{1}{2}}\frac{1}{\sigma_{12}^2\Omega_{12}^{(2,2)*}(T_{12}^*)} \tag{6.8}$$

$$[D_{12}]_1 = 2{\cdot}6636 \times 10^{-5}\left(\frac{T^3(M_1+M_2)}{2M_1M_2}\right)^{\frac{1}{2}}\frac{1}{P\sigma_{12}^2\Omega_{12}^{(1,1)*}(T_{12}^*)} \tag{6.9}$$

where M_1 and M_2 are the relative molecular masses of the two species.

In the first section of this chapter the evaluation of the transport coefficients of dilute monatomic gases from a knowledge of the intermolecular potential is described. The results are used to illustrate the sensitivity of the transport coefficients to the potential. The experimental methods used for the measurement of gaseous transport coefficients are then briefly described and the results discussed. The next sections are devoted to the inverse problem of determining the intermolecular potential from transport coefficients. First, the traditional, trial-and-error methods are discussed and, secondly, the more recent, direct inversion procedures are presented. The investigation of the intermolecular forces of monatomic species by these techniques is reviewed. In the final part of the chapter, the problem of polyatomic gases is briefly discussed. It is shown why progress in these cases has been more restricted and the need for future work in this area is indicated.

6.2. The calculation of transport properties from the intermolecular potential

Appendix A5.2 and the previous section have provided the relationships between the intermolecular pair potential and the transport properties of monatomic gases and gas mixtures. These expressions can, in principle, be extended to any order of accuracy. They indicate that if the intermolecular potential is known for each interaction in a pure gas or a binary mixture the problem of calculating the transport properties of a gas reduces to the evaluation of the integrals $\Omega^{(l,s)*}(T^*)$. For the general form of realistic intermolecular potential with which we are concerned (Fig. 1.1) the integrals defined by (A5.2.1)–(A5.2.3) in Appendix A5.2 cannot be performed analytically. However, the availability of fast computers and suitable numerical algorithms enables the evaluation of the integrals to be carried out for any specified intermolecular potential. Since it is not our purpose here to discuss the details of these procedures we provide, in Appendix 12, a listing of a computer program which will carry out the necessary integrations for a given intermolecular potential. The program is based upon an algorithm developed by Barker, Fock, and Smith.[1] The listing, together with the original publication, provide sufficient information for the interested reader to obtain a complete understanding of the numerical procedure. We also provide in Appendix 2 tabulations of all the collision integrals necessary for the evaluations of the transport properties of a gas mixture to at least a second-order approximation for a series of selected intermolecular potentials. The potentials included in these tabulations have been chosen either because of their frequent use as simple models, or because they have been found to represent closely the true intermolecular potential function for the interaction between the inert gases, where these are known. Within the former group we include members of the $n-6$ family of potential functions (see Appendix 1).

$$U^*(r^*) = \frac{n}{n-6}\left(\frac{n}{6}\right)^{6/(n-6)}\{r^{*-n} - r^{*-6}\} \tag{6.10a}$$

or, equivalently

$$U^*(\bar{r}) = \left\{\frac{6}{n-6}\,\bar{r}^{-n} - \frac{n}{n-6}\,\bar{r}^{-6}\right\}. \tag{6.10b}$$

In these expressions $U^* = U/\varepsilon$, $r^* = r/\sigma$, and $\bar{r} = r/r_m$, where ε is the well-depth parameter of the potential, σ the separation at which the potential is zero, r_m the separation at which the potential is a minimum, and n is a constant. For the second group of potential functions we have chosen the $n(\bar{r})-6$ family described by an equation of the form of (6.10b) in which the repulsive exponent, n, is allowed to vary with separation according to the equation

$$n = 13 + \gamma(\bar{r} - 1). \tag{6.11}$$

This family has been found to provide a particularly good and simple representation of the intermolecular potentials of the inert gases when the additional 'shape' parameter, γ, is treated as disposable.

In order to clarify some of the later material in this chapter, it is convenient to present here the results of collision integral calculations for some model potentials. As an example of the behaviour of the most important collision integrals $\Omega^{(2,2)*}(T^*)$ and $\Omega^{(1,1)*}(T^*)$, Fig. 6.1 shows their reduced temperature dependence for the Lennard-Jones $12-6$ potential model. The integrals all lie near unity by virtue of the reduction by their rigid-sphere values and they decrease monotonically with increasing temperature. This featureless behaviour contrasts greatly with the form of the intermolecular potential function from which they were derived. The same figure includes the corresponding collision integrals for the $18-6$ potential to illustrate their dependence upon the form of the potential. The changes in the collision integrals are small by comparison with the changes in the potentials, a result which occurs because the collision integrals at a fixed temperature arise from the effects of the potential over the entire trajectories of many binary collisions, at many impact parameters and energies.

To illustrate the magnitude of the higher-order correction terms for the viscosity and thermal conductivity ((6.4)–(6.5), Appendix A5.2) Fig. 6.2 displays the reduced temperature dependence of the factors $f_\eta^{K(2)}$ and

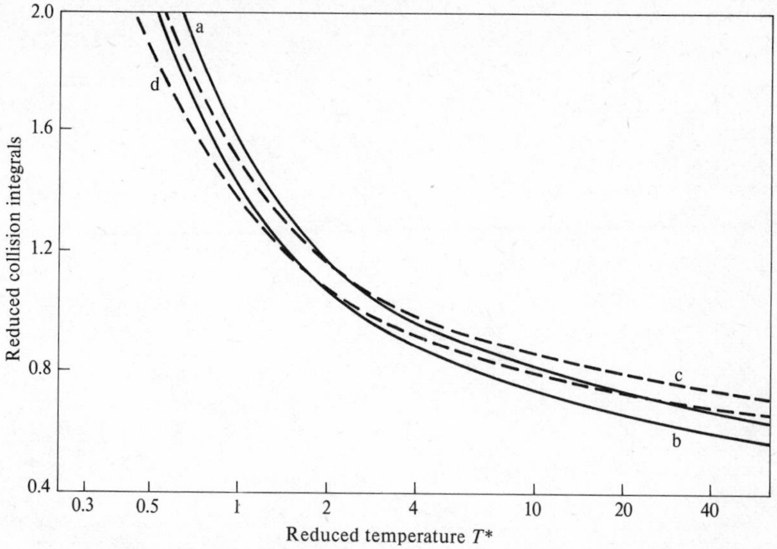

FIG. 6.1. Temperature variation of reduced collision integrals: (a) $\Omega^{(2,2)*}$ for 12−6 potential; (b) $\Omega^{(1,1)*}$ for 12−6 potential; (c) $\Omega^{(2,2)*}$ for 18−6 potential; (d) $\Omega^{(1,1)*}$ for 18−6 potential.

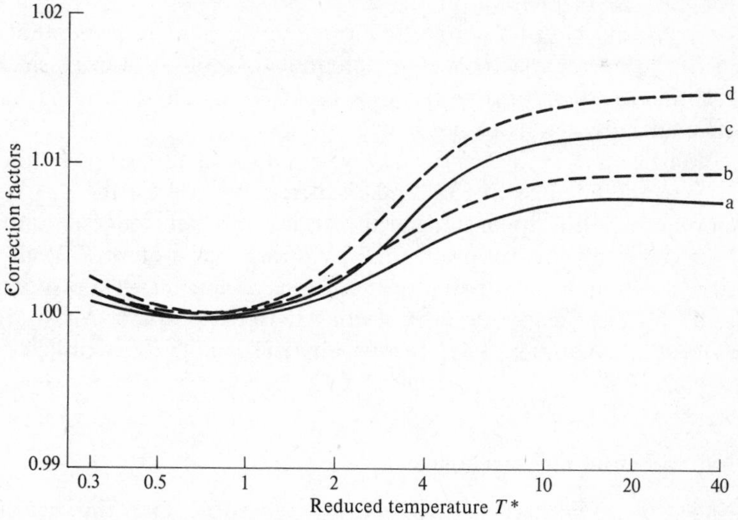

FIG. 6.2. Temperature variation of the higher-order correction factors $f_\eta^{K(2)}$ and $f_\lambda^{K(2)}$: (a) $f_\eta^{K(2)}$ for 12−6 potential: (b) $f_\eta^{K(2)}$ for 18−6 potential; (c) $f_\lambda^{K(2)}$ for 12−6 potential; (d) $f_\lambda^{K(2)}$ for 18−6 potential.

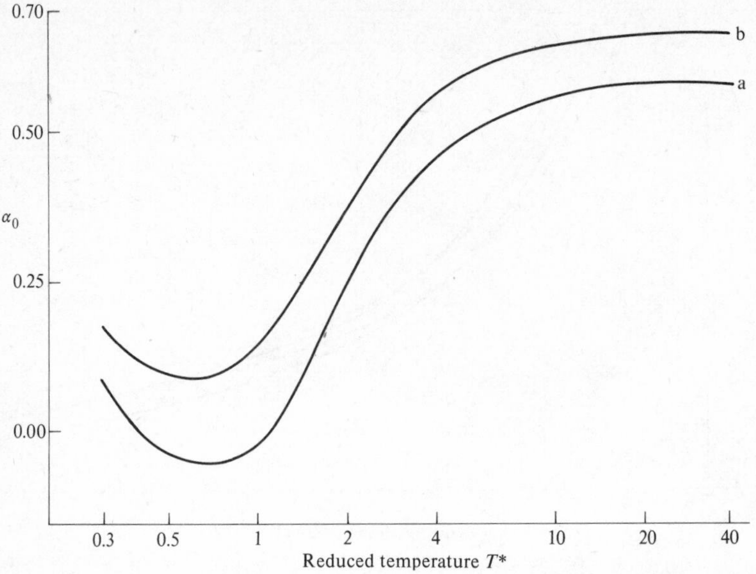

FIG. 6.3. Temperature variation of the isotopic thermal diffusion factor: (a) for the 12−6 potential; (b) for the 18−6 potential.

$f_{\lambda}^{K(2)}$ for the 12−6 potential model and the 18−6 model. It is seen that these corrections never amount to more than 2 per cent and that the change in the correction from one potential to another is even smaller. These observations will prove to be of great use in the determination of intermolecular potentials to be described later.

In contrast to Figs. 6.1 and 6.2, Fig. 6.3 displays the isotopic thermal diffusion, α_0, (defined by (5.152) and discussed in § 5.6) for the 12−6 and 18−6 potentials. It is apparent that this transport coefficient is inherently more sensitive to the form of the intermolecular potential than the viscosity, thermal conductivity, or diffusion coefficient. Thus it would appear to be, in principle, a more useful transport property for studying intermolecular potentials. The reasons why this is not so are discussed in the next section.

6.3. Experimental measurements

Since the equations of Appendix A5.2 show that the transport coefficients of gases and gas mixtures essentially depend upon combinations of the collision integrals $\Omega^{(l,s)*}$, experimental measurements of the coefficients provide a probe of the intermolecular potential through these integrals. The experimental methods for carrying out these measurements have, in most cases, been known for at least fifty years. However, only in the last ten to fifteen years have reliable measurements been made of the transport coefficients over a wide range of temperatures. The reasons for this are principally poor design of apparatus and an inadequate knowledge of the theory underlying the experimental methods.

Since the accuracy of the information which can be obtained about intermolecular forces from transport coefficient measurements is limited by the accuracy of the measurements themselves, we discuss here only the more recent experimental work of proven reliability.

6.3.1. *Viscosity*

Reliable measurements of the viscosity of pure, dilute gases and binary gas mixtures have been carried out by two different techniques in three different laboratories over a wide temperature range. Smith and his collaborators[2] have employed the familiar capillary flow method of viscometry in which the measured efflux time, τ, of a known volume of gas from a closed vessel through a small diameter circular section tube may be related to the viscosity of the gas at the temperature of the tube by the Hagen–Poiseuille equation for laminar flow.[3]

Fig. 6.4 shows a schematic diagram of the apparatus. In an experiment, gas initially contained in the front vessel, v_f, is allowed to pass through the capillary to the back vessel v_b. The pressure in the front vessel is monitored by means of the mercury level in the side arm connected to it. The time taken for the mercury level to fall from one pointer to another in this side-arm is accurately determined during the course of the flow. By using the entire series of pointers it is possible to measure the efflux time for the gas at a series of pressures upstream of the capillary during a single experimental run.

Careful design of a capillary viscometer is essential in order to obtain accurate measurements. For example, the gas density in a capillary viscometer is usually kept quite low, so that the effects of non-ideal gas behaviour are minimized. However, under these conditions 'slip' of the gas may occur at the capillary wall, because the molecules do not have zero velocity there as assumed in the continuum analysis of the flow. The slip effect increases with the mean free path of the gas and thus with decreasing gas density. The viscometer is therefore usually operated in a density range which is a compromise between the two opposing effects of gas density. The slip effect is then eliminated by using the efflux times measured at a series of pressures and extrapolating the results to infinite pressure where the slip effect is zero. Under appropriate conditions the efflux times for two different gases at the same temperature are related by

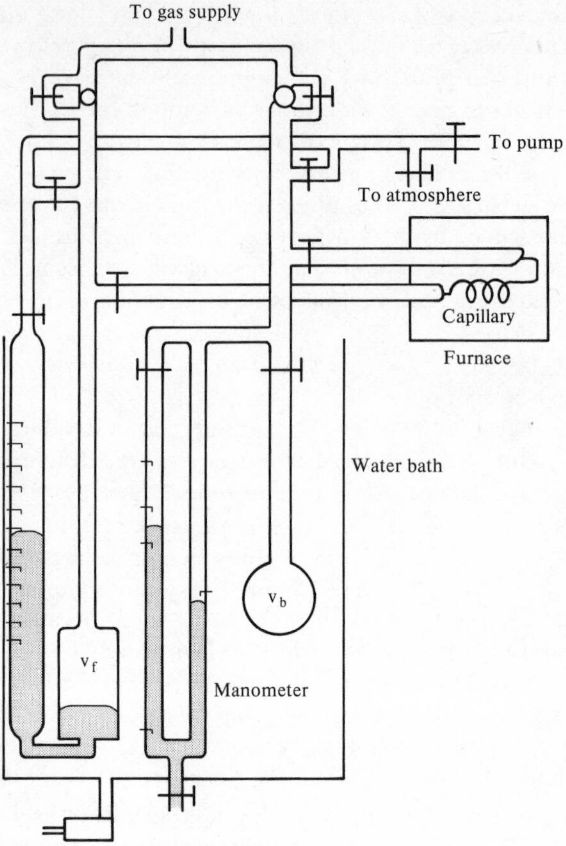

FIG. 6.4. A schematic diagram of a capillary viscometer.

the equation:

$$\frac{\tau_1}{\tau_2}=\frac{\eta_1(1+\delta)}{\eta_2} \tag{6.12}$$

where δ is a small (~1 per cent) correction term which can be estimated. For two gases at different temperatures T_1 and T_2 the ratio of efflux times is

$$\frac{\tau_1}{\tau_2}=\frac{\eta_1}{\eta_2}\frac{T_1}{T_2}[1+\delta'][1+\delta] \tag{6.13}$$

where δ' is a further small correction term. Eqns (6.12) and (6.13) together with a standard value for the viscosity of a single gas at one temperature have allowed Smith and his collaborators to measure the viscosity of several gases and gas mixtures, both monatomic and polyatomic, over the temperature range 77–1600 K, with an estimated accuracy of ±1·0 per cent. A second group of measurements have been carried out by Guevara and his collaborators over the temperature range 1100 to 2000 K also by a capillary method.[4] The estimated accuracy of these data is ±2 per cent.

Kestin and his research group[5] have employed an oscillating disc viscometer for measurement of the viscosity of nineteen pure gases and most of their binary mixtures within the temperature range 298–973 K. Fig. 6.5 contains a schematic diagram of their apparatus. In this a horizontal circular quartz disc (1) is suspended by a thin quartz strand (4) between two horizontal fixed plates (2) and (3). The disc is free to undergo damped, simple harmonic oscillations in a horizontal plane which are observed by means of a mirror attached to the disc (5). The angular

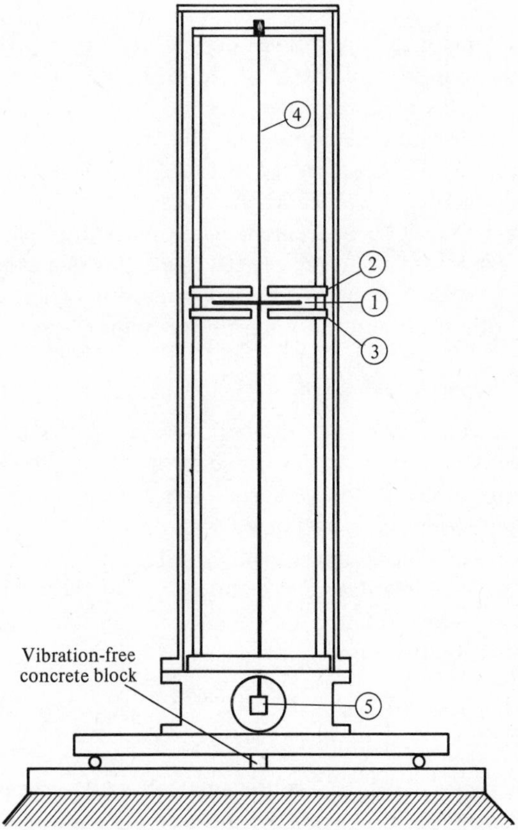

FIG. 6.5. Schematic diagram of a high-temperature oscillating-disc viscometer.

displacement θ of the disc as a function of time is given by the equation

$$\theta(t) = \theta_0 \exp\{-\pi\,\Delta(1+4t/\mathcal{T})/2\}\,.\,\sin\left(\frac{2\pi t}{\mathcal{T}}\right) \tag{6.14}$$

where Δ is the logarithmic decrement of the oscillation and $\mathcal{T}$ its period. If both Δ and $\mathcal{T}$ are measured then the viscosity of gas can be derived from the equations

$$C_N = \left\{\frac{2I}{\pi b\rho R^4}\left(\frac{\Delta\mathcal{T}_0}{\mathcal{T}} - \Delta_0\right) + \frac{2}{3}\frac{\Delta\mathcal{T}_0}{\mathcal{T}}\right\}\beta^2 + \frac{1}{45}[3\Delta^2 - 1]\left(\frac{\mathcal{T}_0}{\mathcal{T}}\right)^2\beta^4 + \frac{8}{945}\left[\frac{\Delta(\Delta^2-1)\mathcal{T}_0}{\mathcal{T}}\right]\beta^6 \tag{6.15}$$

where

$$\beta = b\left(\frac{2\pi\rho}{\eta\mathcal{T}_0}\right)^{\frac{1}{2}} \tag{6.16}$$

and C_N is a constant for the instrument depending only on its dimensions; I is the moment of inertia of the disc, R its radius, and b the spacing between the disc and the fixed plates. In addition, $\mathcal{T}_0$ and Δ_0 represent the period and logarithmic decrement of the oscillation *in vacuo*, which are also measured, and ρ the density of the gas.

A judicious choice of the characteristics of the suspension system permits absolute viscosity measurement to be made by this method. The results have an estimated accuracy of $\pm 0{\cdot}1$ per cent near room temperature falling to $\pm 0{\cdot}3$ per cent at 973 K. All three sets of viscosity data are in agreement within their mutual uncertainty bound.

6.3.2. *Thermal conductivity*

The thermal conductivity of dilute gases has been measured by a variety of techniques. The majority have been of the steady state-type wherein the temperature gradient across the gas contained in a cell of accurately known geometry is measured when a heat flux passes through it. Both parallel plate and concentric cylinder instruments have been used.[6,7] It has been demonstrated conclusively that the accuracy of such measurements is usually considerably inferior to that of viscosity by application of (5.136) for monatomic gases which reads

$$\frac{\lambda}{\eta} = \frac{15}{4}\frac{k}{m}\frac{f_\lambda^{(m)}}{f_\eta^{(m)}}$$

where again the factors $f_\lambda^{(m)}$, $f_\eta^{(m)}$ represent mth-order correction factors. This yields an essentially exact relation between the thermal conductivity and viscosity of a monatomic gas.[8]

The most likely cause of the uncertainties in thermal conductivity is natural convective heat transfer. Recently, a technique operating in a transient mode has been refined and applied over a limited temperature range to a few gases. The method consists of observing the transient temperature rise of a thin wire immersed in the gas following an essentially instantaneous initiation of a heat flux within it.[9,10] The temperature rise of the wire is monitored over a period of only one second so that convective heat transfer is eliminated.[10]

Fig. 6.6 shows a diagram of the essential part of such an apparatus. The thermal conductivity cell, which is immersed vertically in the gas, consists of a 0·15m length of platinum wire 7 μm in diameter (1) tensioned between two supports (2) and (3) by a spring (4). When a d.c. voltage is applied to the ends of the wire the current causes ohmic heating. The consequent time evolution of the temperature (and hence resistance) of the wire is determined principally by the thermal conductivity of the surrounding gas. The times at which the resistance of the wire attains certain preset values are determined by an automatic Wheatstone bridge and from a resistance–temperature calibration of the wire its temperature rise at these times determined. The theory of the instrument shows that it is possible to arrange that the measured temperature rise of the wire, ΔT, when suitably corrected, conforms to the equation

$$\Delta T = \frac{q}{4\pi\lambda} \ln\left(\frac{4\kappa t}{a^2 C}\right)$$

where q is the heat flux per unit length in the wire, a its radius, $\kappa = \lambda/\rho C_p$, is the thermal diffusivity of the gas, λ its thermal conductivity, and C is a numerical constant. Thus the thermal conductivity of the gas can be obtained from the slope of the line ΔT vs $\ln t$ given a knowledge of q.

This transient hot-wire technique has been employed for measurements of the thermal conductivity of several gases and gas mixtures near room temperature.[11] In particular, two independent sets of measurements on the inert gases have been carried out and the results are entirely consistent with the viscosity data for the same gases.[12,13] The accuracy of the experimental results is estimated as ±0·2 per cent, and the technique offers the opportunity for measurements of the thermal conductivity of monatomic and polyatomic gases over a wide range of conditions.[14]

6.3.3. *Binary diffusion coefficient*

The binary diffusion coefficient of gas mixtures has also proved an extremely difficult property to measure accurately despite the apparent simplicity of the apparatus required and the varied experimental methods employed. The most common measurement technique has been the two-bulb method in which two samples of a gas mixture of unequal

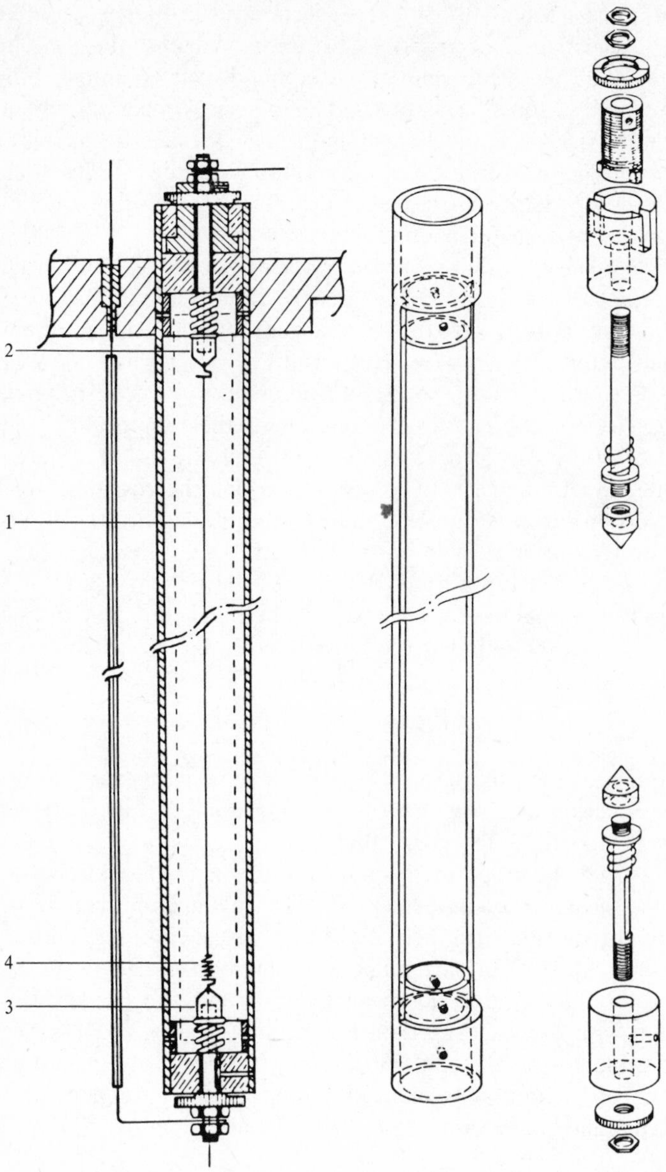

FIG. 6.6. Transient hot-wire thermal conductivity cell.

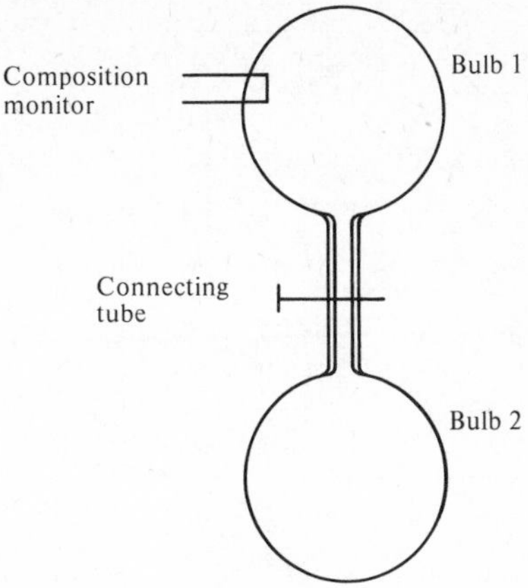

FIG. 6.7. Schematic diagram of a two-bulb diffusion apparatus.

compositions each contained initially in a separate bulb are allowed to mix through a connecting tube. Fig. 6.7 shows the apparatus schematically; the entire assembly is immersed in a thermostatted enclosure.

The approach of the system to a uniform composition following the opening of the connecting tube is monitored by measuring the composition in one bulb as a function of time. The time constant of this relaxation process is related to the geometry of the apparatus and the diffusion coefficient for the gases. For mixtures of the monatomic gases a series of measurements[15] with an accuracy comparable with that of viscosity has been carried out over the temperature range 60–400 K.

Recently Hogervost and Freudenthal[16] have developed an unusual new method for the measurement of diffusion coefficients based upon the observation of the remixing of a gas mixture subject to a d.c. discharge in a cylindrical tube after the discharge is extinguished. Results were obtained for most of the binary monatomic gas mixtures over the temperature range 300–1400 K with an accuracy of ±0·5–1 per cent. Only a few accurate measurements on polyatomic systems have been made.

6.3.4. *The thermal diffusion factor*

The phenomenon of thermal diffusion is that in which an initially uniform gas mixture subjected to a temperature gradient undergoes partial separation. Its associated transport coefficient—the thermal diffusion factor—has proved the most difficult coefficient to measure accurately. This is because the separation of the gas mixture in the temperature gradient, by thermal diffusion, is very small.

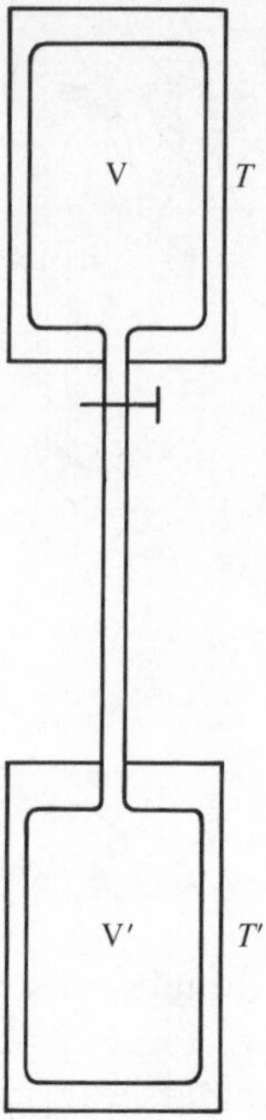

FIG. 6.8. Schematic diagram of a two-bulb thermal diffusion apparatus.

Fig. 6.8 shows a schematic diagram of the original two-bulb thermal diffusion apparatus. Two bulbs V and V′, joined by a connecting tube are maintained at different temperatures T and T' and the different mole fractions of the heavier species x_1 and x_1' in two bulbs at steady state are measured.[17] The phenomenological equation for thermal diffusion, (5.152) is

$$\frac{\partial x_1}{\partial \mathbf{r}} = -\alpha_T x_1 x_2 \frac{\partial(\ln T)}{\partial \mathbf{r}}.$$

On integration for this experimental arrangement it yields

$$\log\left\{\frac{x_1/(1-x_1)}{x_1'/(1-x_1')}\right\} = -\int_{\log T'}^{\log T} \alpha_T \, \mathrm{d}\,(\log T) \qquad (6.17)$$

for the relationship between the compositions of the mixture in the two bulbs. The thermal diffusion factor at a specific temperature may be obtained by measurement of the left side of this equation as a function of T' keeping T constant and the subsequent differentiation of these data with respect to $\log T'$. The separation, measured as a difference in mole fraction $(x_1 - x_1')$, typically amounts to 0·05 for a temperature difference of 100 K. This small value implies that the measurement of α_T with an accuracy of a few per cent requires analyses of gas mixtures far more accurate than are normally attainable.[18] Furthermore the need for differentiation of the experimental data with respect to log T' exacerbates the difficulty. In the case of isotopic mixtures this problem is even more severe since the separation is usually much smaller. Two methods have been employed in order to amplify the thermal separation. In the first, the Trennschaukel,[19] many two-bulb systems are effectively coupled in series between the same temperatures, whereas in the second, the thermal diffusion column,[20] natural convective currents are used to increase the separation arising from a given temperature gradient. Although the methods increase the separation their theory is more complicated and somewhat less certain. Consequently, the over-all accuracy of the thermal diffusion factor is not greatly improved. Nevertheless, the column method has been employed with some success by Saviron and his collaborators in recent years for studies on monatomic gas mixtures.[21] However, in general, despite the special sensitivity of the thermal diffusion factor to the intermolecular pair potential noted earlier, its use as a probe of the potential has been hampered by the difficulty of accurate measurement.

6.4. Extraction of collision integrals from experimental data

As outlined above the experimental transport coefficients of monatomic gases and gas mixtures depend in general upon a number of collision integrals. However, because of the form of the relationships it is possible to extract, with reasonable accuracy, from each source of experimental data a single collision integral characteristic of a particular pair interaction. The following discussion of this process does not cover the thermal conductivity coefficient of monatomic gases and gas mixtures since the accuracy of the experimental data available over a wide range of temperature do not warrant its inclusion at present. However, a similar analysis could be applied, since it is predominantly dependent on the same collision integral, $\Omega^{(2,2)*}$, as the viscosity (Appendix A5.2, §6.1).

Beginning with the viscosity of a pure gas, experimental measurement of $\eta_{\mathrm{exp}}(T)$ may be used to define the collision integral $\bar{\Omega}^{(2,2)}_{\mathrm{exp}}(\mathrm{T})$ by means of the equation

$$\bar{\Omega}^{(2,2)}_{\mathrm{exp}}(T)=\tfrac{5}{16}(\pi mkT)^{\frac{1}{2}}\frac{f_\eta^{(m)}}{\eta_{\mathrm{exp}}} \tag{6.18}$$

which follows from (5.124). Owing to the insensitivity of $f_\eta^{(m)}$ to the intermolecular potential function it may be calculated for any reasonable function up to third order ($m=3$). Such a value may then be used in (6.18) without adding any significant error to the evaluation of $\bar{\Omega}^{(2,2)}_{\mathrm{exp}}$.

In the case of the viscosity of a binary gas mixture the kinetic theory formulation (A5.2.17) includes contributions from the three binary interactions. However, the contributions from the like interactions which determine $[\eta_1]_1$ and $[\eta_2]_1$ may be eliminated by replacing them with experimental values for the viscosity of the pure components $[\eta_1]_{\mathrm{exp}}$ and $[\eta_2]_{\mathrm{exp}}$. The contributions from the unlike interaction occur in the collision integral ratio $A^{\star}_{12}$ and the interaction viscosity $[\eta_{12}]_1$. The collision integral ratio $A^{\star}_{12}$ is a remarkably weak function of the intermolecular potential, consequently a reliable estimate of it may be obtained from any reasonable potential. Thus mixture viscosity data may be employed[22] to determine an experimental value of the quantity $[\eta_{12}]_1$. By analogy with the pure gas a collision integral $\bar{\Omega}^{(2,2)}_{12,\mathrm{exp}}$ can be defined by

$$\bar{\Omega}^{(2,2)}_{12,\mathrm{exp}} = \tfrac{5}{16}\{2\pi m_1 m_2 kT/(m_1+m_2)\}^{\frac{1}{2}}/[\eta_{12}]_{\mathrm{exp}}. \tag{6.19}$$

This quantity contains only information relating to the unlike pair interaction. Equation (A5.2.17) is only the first-order approximation formula for the mixture viscosity, thus, it is strictly incorrect to use experimental pure gas viscosity data for $[\eta_1]$ and $[\eta_2]_1$. However, it has been shown that the effect upon the resulting value of $[\eta_{12}]_{\mathrm{exp}}$ of using the much more complicated second approximation to the viscosity is negligible.[23] In general, however, values of $\bar{\Omega}^{(2,2)}_{12,\mathrm{exp}}$ deduced in this way have a somewhat greater uncertainity (± 2 per cent) than the corresponding values of the collision integrals for the pure substances (± 1 per cent) owing to the additional approximations made in their evaluation.

A second collision integral depending on the unlike interaction potential can be obtained from binary diffusion coefficient data.[24] Since the higher-order approximation correction term $f_D^{(m)}$ is again not sensitive to the intermolecular potential, we can define a collision integral $\bar{\Omega}^{(1,1)}_{12,\mathrm{exp}}$ by means of the equation

$$\bar{\Omega}^{(1,1)}_{12,\mathrm{exp}} = \frac{3}{16n}\left[\frac{2kT(m_1+m_2)\pi}{m_1 m_2}\right]^{\frac{1}{2}} \frac{1}{D_{12}} f_D^{(m)} \tag{6.20}$$

which follows from (A5.2.29) and (A5.2.30). Clearly both $\bar{\Omega}^{(1,1)}_{12,\mathrm{exp}}$ and $\bar{\Omega}^{(2,2)}_{12,\mathrm{exp}}$ contain information about the unlike potential. In general these two quantities can be determined from the best available measurements with similar accuracies because although the extraction of $\bar{\Omega}^{(2,2)}_{12,\mathrm{exp}}$ from the viscosity data involves more approximations the original data are usually of higher accuracy.

The collision integral ratio, $A^{\star}$, is a very weak function of temperature and insensitive to the intermolecular potential. Consequently, the isotopic thermal diffusion factor, α_0, is most strongly determined by the collision integral ratio,

$$C^{\star} = \Omega^{(1,2)*}/\Omega^{(1,1)*}$$

as can be seen from (A5.2.35) and (A5.2.36). However, by a small manipulation the experimental thermal diffusion factor, $\alpha_{0,\mathrm{exp}}$, can be used to obtain a single collision integral, $\bar{\Omega}^{(1,2)}$, which does not arise in the viscosity or the diffusion coefficient. Thus we find

$$\bar{\Omega}^{(1,2)}_{\mathrm{exp}} = \left[\frac{8\alpha_{0,\mathrm{exp}}}{45 f_{\alpha}^{\mathrm{K}(2)}} + \frac{5}{6A^{\star}}\right]\bar{\Omega}^{(2,2)} \tag{6.21}$$

Here, we assume that $\bar{\Omega}^{(2,2)}$ is well known from viscosity data and make use of the fact that $f_{\alpha}^{\mathrm{K}(2)}$ as well as $A^{\star}$ is insensitive to the potential model used in its calculation.

6.5. Principle of corresponding states

Before turning to a discussion of the determination of the intermolecular potential of monatomic gases from transport coefficient data we shall give some useful results which can be derived from experimental data without a knowledge of the details of the intermolecular potential.

Let us suppose that the interaction between any pair of molecules can be described by a potential energy function of the form

$$U(r) = \varepsilon_{\mathrm{c}} F\left(\frac{r}{\sigma_{\mathrm{c}}}\right) \tag{6.22}$$

where F represents a universal function and ε_{c} and σ_{c} represent scaling parameters for energy and distance respectively which depend upon the specific interaction considered. Then (A5.2.1)–(A5.2.3) show that the collision integrals $\Omega^{(l,s)*}(T^*)$ for all potentials of this form will be universal functions of the reduced temperature ($T^* = kT/\varepsilon_{\mathrm{c}}$). This will be true, independent of the particular form of the function F. Kestin and his collaborators[8] have examined the extent to which the experimental data for the monatomic gases support this hypothesis. The most convenient way to examine the experimental data is by means of the function $\Omega_{\eta}^{*}(T^*)$ defined by the relation

$$\Omega_{\eta}^{*}(T^*) = \frac{5}{16}\left(\frac{mkT}{\pi}\right)^{\frac{1}{2}} \frac{1}{\sigma_{\mathrm{c}}^{2}\eta_{\mathrm{exp}}}. \tag{6.23}$$

Eqns (A5.2.15) and (A5.2.16) show that this function represents a combination of the collision integrals $\Omega^{(l,s)}(T^*)$.† That is

$$\Omega_{\eta}^{*}(T^*) = \Omega^{(2,2)*}(T^*)/f_{\eta}^{(m)}(T^*). \tag{6.24}$$

† The quantity, $\Omega_{\eta}^{*}(T^*)$, defined here, is exactly the same as that defined in § 5.2 in our discussion of the simple kinetic theory. It is not a single collision integral, but rather a combination of such integrals, which contains all the effect of intermolecular forces on the viscosity of a gas. The notation adopted here differs from that of ref. 8 wherein the symbol $\Omega_{22}(T^*)$ is employed for $\Omega_{\eta}^{*}(T^*)$ and $\Omega_{11}(T^*)$ for $\Omega_{D}^{*}(T^*)$.

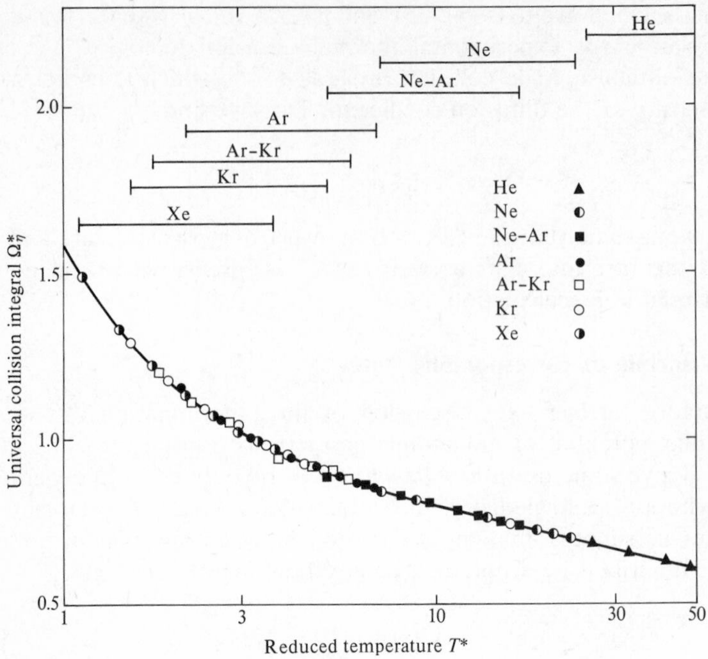

FIG. 6.9. The universal collision integral Ω^*_η.

Accordingly, if each collision integral $\Omega^{(l,s)*}$ is a universal function of T^* the $\Omega^*_\eta(T^*)$ will itself be a universal function of the reduced temperature. Fig. 6.9 shows a plot of the experimental values of $\Omega^*_\eta(T^*)$ for some of the interactions among the monatomic gases. In order to construct this plot it has been necessary to assign values to σ_c and ε_c for each interaction. However, it is not possible to determine absolute values of σ_c and ε_c solely from the requirement that the function Ω^*_η be universal. Nevertheless this requirement does permit the evaluation of the ratios of σ_c and ε_c for each interaction with respect to arbitrary reference values. In the original work[8] these reference scaling parameters were taken to be those appropriate to krypton on the basis of an approximate potential model. Here, we use as a pair of reference values our best estimates of the separation at the zero point of the potential, σ, and the depth of the minimum, ε, for the true potential of argon. These reference values, together with the scaling factor ratios for the interactions of all the monatomic gases are included in Appendix 3.

It can be seen from Fig. 6.9 that the function Ω^*_η can indeed be made universal among the interactions of the monatomic species to a high degree of accuracy (±0·5 per cent) by a suitable choice of the scaling factors σ_c and ε_c. This result suggests, but does not prove, that the intermolecular potential for the inert gases is of a nearly universal form. It is an example of the Principle of Corresponding States first encountered in Chapter 3. The function $\Omega^*_\eta(T^*)$ obtained by this procedure has been correlated by an empirical equation and both it and the analogous relation for $\Omega^*_D(T^*) \simeq \Omega^{(1,1)^*}(T^*)$ are given in Appendix 3. Other collision integrals and collision integral ratios derivable from these two functions are given in the same appendix. These correlations, together with the kinetic theory formulae of Appendix A5.2, form a very powerful interpolation and prediction scheme for the transport properties of dilute monatomic gases and their mixtures. This is because the use of the corresponding states principle allows the correlation to extend over the reduced temperature range $T^* = 1$ to 90, which greatly exceeds that available for any single substance. Futhermore, because the properties of a multicomponent gas mixture depend only upon the collision integrals for all the binary interactions in the gas, the calculation of transport property data, including the thermal conductivity, for a multicomponent mixture of monatomic components becomes straightforward using the equations of Appendix 3 and Appendix A5.2. We shall see later that further extensions of this idea to include polyatomic components are possible for some properties (§6.8).

Small departures from universality of the function Ω^*_η can be discerned on a detailed examination, and this indicates that there may be small differences between the form of the intermolecular potential functions for various monatomic interactions. However for the purposes of prediction of transport properties of gas mixtures at accessible temperatures these small discrepancies are unimportant.

6.6. Traditional methods of testing intermolecular potential functions

The application of the methods of §6.4 to the viscosity, diffusion, and thermal diffusion data for the monatomic gases and their mixtures over a range of temperatures leads to a series of values for the collision integrals $\bar{\Omega}^{(l,s)}_{\text{exp}}$ as a function of temperature for each binary interaction. Although (A5.2.1)–(A5.2.3) show that these collision integrals are related to the true intermolecular potential, it is by no means obvious how the former may be used to obtain the latter. The difficulty arises because the collision integrals are averages over the dynamics of all possible collisions at a particular temperature. Thus the collision integral at a specific temperature depends (at least formally) upon the entire intermolecular potential at all separations. The problem of inverting experimental collision integrals to obtain the potential was long thought to be impossible both in principle and in practice. Consequently, an alternative approach has been employed.

This alternative approach, which may be termed the traditional method, is a trial-and-error procedure. A potential model is assumed for the interaction of a given pair of molecules and the collision integrals $\bar{\Omega}^{(l,s)*}(T^*)$ obtained by numerical integration of (A5.2.1)–(A5.2.3). The parameters of the potential model are then varied so as to obtain the optimum agreement between the generated collision integrals $\bar{\Omega}^{(l,s)}(T)$ and the experimental values $\bar{\Omega}^{(l,s)}_{\text{exp}}(T)$. In most cases the parameters varied are σ, the separation at which the potential energy is zero and the depth of the minimum, ε. This procedure is then repeated for a series of model potentials until the smallest discrepancy between calculated and experimental collision integrals is achieved. The potential model for which this optimum agreement is secured and its associated parameters ε and σ are then taken to be a good representation of the true potential and its parameters.

When this procedure has been applied to a series of transport property measurements, the results obtained have vividly illustrated its limitations. It has frequently been observed that for a given potential model and a particular transport coefficient the optimum values of ε and σ depend upon the range of temperature studied. For example, for argon, the well depth ε/k for the Lennard-Jones $12-6$ potential model has been reported as 124 K from low-temperature viscosities and as 135 K from high-temperature viscosities. Equally, when the same potential model is used to fit data for different transport properties different values of the two parameters have been required in each case. For example, for the argon–krypton interaction the well depth of the $12-6$ potential has been estimated as 170 K from mixture viscosity data and 140 K from binary diffusion data. Finally, even the potential model selected as the optimum has varied depending on the property chosen for the test and the temperature range employed. These observations lead inevitably to the conclusion that none of the potential models used for these tests has been a good representation of the entire, true intermolecular function. The corollary to this conclusion is that the potential parameters ε and σ derived in this way are merely guides to their true values rather than determinations of them. There are however two virtues of this empirical procedure. First, it allows unsuitable intermolecular potentials to be rejected immediately, since if a particular potential does not represent all of the transport coefficients of a gas within their estimated uncertainty it cannot be the true potential. Secondly, it can provide a useful means of interpolation among experimental data for a particular gas. However, the application of these procedures, particularly to data of limited accuracy and covering a restricted temperature range, has led in the past to many estimated potentials which have since been shown to be seriously in error. Potentials obtained in this way should therefore be viewed with scepticism.

6.7. The direct determination of intermolecular forces: inversion methods

The failure of the traditional fitting procedures described above to establish unique potentials has encouraged the search for more direct methods of determining the potential from transport coefficient data. For a considerable time it was thought that such a determination was impossible. It is only in the last few years that it has been demonstrated that such *inversion* of transport property data is practicable[25] and that some insight into the reasons for its success has been gained. We shall preserve here the historical order and discuss first the practical application of inversion, deferring discussion of the basis of its validity until later. The method will be described in detail only for the viscosity, and hence for the collision integral $\bar{\Omega}^{(2,2)}_{\text{exp}}$. However analogous procedures may be used for any collision integral $\bar{\Omega}^{(l,s)}_{\text{exp}}$.

We begin by assuming an interaction that conforms to the BBMS potential function[26] (Appendix 1), with a well depth $\varepsilon/k = 142{\cdot}5$ K. It is then possible to evaluate the collision integrals $\Omega^{(l,s)*}(T^*)$ for the potential and hence generate viscosities for a hypothetical gas. For definiteness we may endow the molecules of our hypothetical gas with the mass of an argon molecule, and then the pseudo-experimental data will closely correspond to the real experimental data for argon. We suppose that the pseudo-experimental data extend over the reduced temperature range $0{\cdot}65 < T^* < 15{\cdot}0$ corresponding to the temperature range for which the viscosity is actually available. As a result we have a known (true) potential, and the corresponding pseudo-experimental viscosity data. The task of the inversion procedure is to deduce the 'true' potential from these viscosity data alone. The success of the inversion procedure can, in this case, be judged by its ability to recover the assumed 'true' potential.

The inversion procedure begins by employing the 'experimental' viscosity data to provide values of $\bar{\Omega}^{(2,2)}_{\text{exp}}(T)$ at each experimental temperature as outlined in §6.4. Next we select a well depth for the potential ε^1/k; for the moment we shall choose the 'true' value 142·5 K. We then make an initial guess at the interaction potential function and denote it by $U_0(r)$. A reasonable choice for this initial approximation is the Lennard-Jones $12-6$ function. From this 'guessed' potential function we generate the collision integrals $\Omega_0^{(2,2)*}(T^*)$ and define a characteristic, temperature-dependent length $\bar{r}_0^*$ by the equation

$$\bar{r}_0^* = \{\Omega_0^{(2,2)*}\}^{\frac{1}{2}} \tag{6.25}$$

and thence the function, $G_0(T^*)$

$$G_0(T^*) = U_0^*(\bar{r}_0^*)/T^* \tag{6.26}$$

where

$$T^* = kT/\varepsilon^1 \tag{6.27}$$

and

$$U_0^* = U_0/\varepsilon^1. \tag{6.28}$$

The dimensionless function $G(T^*)$ is known as the *inversion function*, and the subscript 0 denotes that it corresponds to the initial estimate of the potential.

The fundamental postulate of the inversion procedure is then that the series of data points $(U_1, \bar{r})$ generated by application of the equations

$$\bar{r} = \{\bar{\Omega}^{(2,2)}_{\text{exp}}(T)/\pi\}^{\frac{1}{2}} \qquad (6.29)\dagger$$

and

$$U_1 = kTG_0(T^*) \tag{6.30}$$

to the pseudo-experimental data at each temperature, constitute a closer approximation to the 'true' potential than does the original, guessed, function $U_0(r)$. As a result of these operations the new approximation to the potential $(U_1, \bar{r})$ is defined over a range of separations corresponding to the range of available viscosity data through (6.29). In order to develop an iterative scheme it is therefore necessary to extrapolate the function $U_1(r)$ to both larger and smaller separations in order that its collision integrals $\Omega_1^{(2,2)*}$ can be computed. At long range this extrapolation may be performed by means of an inverse sixth-power function

$$U(r) = C_6/r^6 \quad \text{as} \quad r \to \infty$$

which is known to be the true asymptotic limit, (Chapter 2). However, at short range an entirely empirical function has to be adopted. One such function found to be particularly suitable is

$$U(r) = A \exp(-Br^3) \quad \text{as} \quad r \to 0.$$

In any event it has been found that whatever forms of realistic extrapolation are employed at this stage, the final potential obtained, within the separation range defined by the experimental data, is essentially independent of the extrapolations employed.[25]

Once these extrapolations have been carried out the first iterate potential $U_1(r)$ may be employed to compute its corresponding collision integral $\Omega_1^{(2,2)}(T^*)$ and hence a new inversion function through equations similar to (6.25)–(6.28), viz.

$$\bar{r}_1^* = \{\Omega_1^{(2,2)*}(T^*)\}^{\frac{1}{2}} \tag{6.31}$$

$$G_1(T^*) = U_1^*(\bar{r}_1^*)/T^* \tag{6.32}$$

where $U_1^* = U_1/\varepsilon^1$.

† The factor π appears in (6.29) but not in (6.25) by virtue of the fact that $\bar{\Omega}^{(l,s)}(T) = \pi\sigma^2\Omega^{(l,s)*}(T^*)$: see § 5.6.

This new inversion function can then be employed to generate a second iterate sequence of data points (U_2, $\bar{r}$) for the 'true' potential by repeated use of (6.29) and (6.30)

$$\bar{r} = \{\bar{\Omega}^{(2,2)}_{\text{exp}}(T)/\pi\}^{\frac{1}{2}} \tag{6.33}$$

and

$$U_2(\bar{r}) = kTG_1(T_1^*). \tag{6.34}$$

The second iterate estimate for the energy at each separation is again assumed to be closer to the 'true' potential than the first and so the process may be repeated until convergence. Here, 'convergence' is in general expressed both in terms of the extent to which the collision integrals $\bar{\Omega}^{(2,2)}_m(T)$ after the mth iteration reproduce the pseudo-experimental viscosities, and in terms of the changes in the potential between steps. For this simulated case the efficacy of the method may also be judged by the extent to which the potentials obtained by inversion reproduce the potential from which the pseudo-experimental data were generated.

In order to establish the uniqueness of the potential obtained by the inversion of a given set of viscosity data it is essential to use a variety of initial estimates of the potential $U_0(r)$. In Table 6.1 and Fig. 6.10 we show the results of the application of the inversion procedure as described above. Fig. 6.10 shows the potentials obtained by inversion of the simulated argon viscosity data after three iterations beginning with each of four different initial guesses for $U_0(r)$, namely the $12-6$, $18-6$, $9-6$, and $11-6-8$ potential functions (Appendix 1). The solid line in the figure is the 'true' potential function. It can be seen that for all of the initial potentials the final inverted potential lies very close to this 'true' function. Table 6.1 provides a quantitative summary of the same result.

Table 6.1

Convergence of viscosity inversion for simulated data

	$\bar{\delta}$ per cent				Standard deviation, β per cent			
Iteration	0	1	2	3	0	1	2	3
Potential model, $U_0(r)$								
$12-6$	7·11	0·91	0·79	1·09	1·84	0·43	0·10	0·06
$9-6$	16·45	3·79	1·77	1·35	4·41	1·05	0·26	0·14
$18-6$	32·00	4·82	2·54	2·22	4·24	2·45	0·40	0·16
$11-6-8$	7·38	0·77	0·70	1·10	1·56	0·61	0·06	0·06

$\bar{\delta}$ is defined by (6.35) and β is the standard deviation of the calculated viscosities from the pseudo-experimental values. Fifty-five 'experimental' values were used in each case.

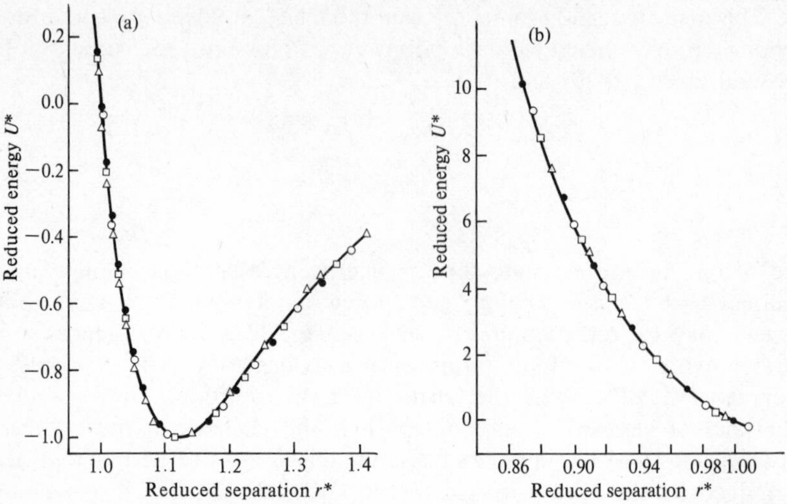

FIG. 6.10. Potentials obtained by inversion of simulated data based on the BBMS potential. The points shown are obtained from the third iteration for various initial approximate potentials. Initial approximation: (●) 9−6; (○) 11−6−8; (□) 12−6; (△) 18−6. Solid line: BBMS potential function.

Since we know the true potential $U(r_i)$ at all separations r_i we define a mean relative deviation of any iterate potential from the true value at the same separations by†

$$\bar{\delta} = \frac{100}{N} \sum_{i=1}^{N} \frac{(U_m(r_i) - U(r_i))}{U(r_i)} \tag{6.35}$$

where U_m is the potential resulting after the mth iteration and N is the number of data points used. Table 6.1 shows how this deviation decreases dramatically after one iteration for each initial potential $U_0(r)$ and rapidly converges to a value of about 1 per cent. The same table also provides the standard deviation, β, of the collision integrals $\bar{\Omega}_m^{(2,2)}(T)$, for the iterate potential $U_m(r)$, from the pseudo-experimental values taken over the whole set of data points. It can be seen that this deviation too drops substantially during the iteration process and converges to about $\pm 0{\cdot}1$ per cent in three iterations. Fig. 6.10 and Table 6.1 provide strong heuristic evidence for the uniqueness and correctness of the final intermolecular potential obtained by inversion.

In the foregoing discussion the well-depth, ε^1/k, of all of the potentials at each stage of the inversion process was chosen to be equal to the 'true' well depth. In some applications of the inversion method to real

† Because $U(r)$ passes through zero, points close to $U(r) = 0$ are excluded before the deviation is calculated so that the summation does not diverge.

transport property data this process has been possible[25] because the well depth is known from other techniques such as the spectrocopic methods discussed in Chapter 7. However, in many other cases, and in particular for the unlike interactions of the monatomic gases, this information is not yet available. Consequently an extension of the inversion procedure has been devised[27,28] to deal with these circumstances and to provide a simultaneous determination of the well depth ε/k.

The inversion process described above is found to converge eventually (in the sense prescribed earlier) whatever well depth ε^1/k is ascribed to the potential at the beginning of the process. However, owing to the manner in which the inversion is performed the rate of convergence or, equivalently, the degree of convergence after a specified number of iterations, will depend upon the relationship between the true well depth, ε/k, and the chosen value ε^1/k. This is because if the well depth chosen differs greatly from the true value then the value of T^*, kT/ε^1, calculated from a given experimental temperature T will be very different from the real value kT/ε. Consequently the value of $G(T^*)$ selected may be significantly different from the true value (Fig. 6.11). In order to achieve convergence using this incorrect choice of ε^1/k, a large number of iteration steps will be necessary. On the other hand if ε^1/k is chosen correctly only a few steps will be required to achieve convergence. Thus we see that if the inversion procedure is repeated for different well depths, ε^i/k, the correct well depth may be identified as that providing the

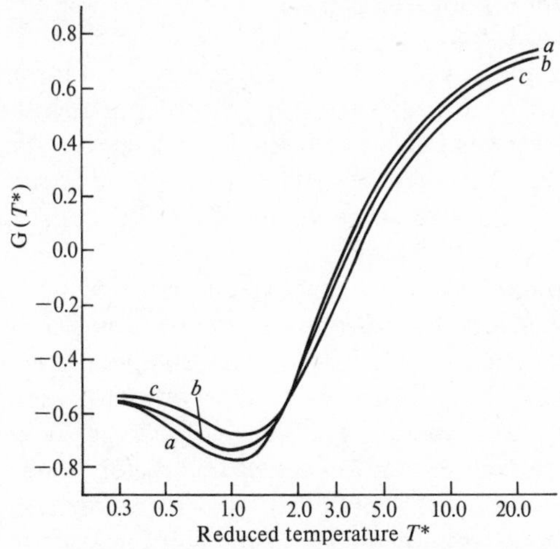

FIG. 6.11. The function $G(T^*)$ for various intermolecular potential energy functions: (a) 9−6; (b) 12−6; (c) 18−6.

most rapid convergence. The potential resulting from inversion with this well depth may then be identified with the true potential.

To illustrate the operation of this procedure we use the same synthetic viscosity data as before but now suppose that we do not know the true well depth for the interaction of two argon atoms. We therefore choose a series of values for ε^1/k in the range 130 to 150 K and perform the inversion process for each well depth using each of the three starting potentials $U_0(r)$ in turn.

In this case because the well depth is fixed at an incorrect value and remains unchanged throughout each iteration process the criterion of comparison with the 'true' potential is not useful. However, the criterion based upon reproduction of the 'experimental' viscosity data can still be employed. Eqns (6.29) and (6.30) imply a one-to-one correspondence between the collision integral at a particular temperature, $\bar{\Omega}^{(2,2)}(T)$ and a specific separation r for the intermolecular potential energy. This suggests that, when considering the well depth, the agreement with experiment for those data which determine the potential in the region of the minimum should be carefully examined. For potentials of the type of interest here the region of the potential energy minimum is defined by viscosity data at reduced temperatures around $T^* = 2$. Thus so far as the determination of the well depth is concerned, we use as a criterion of convergence the total absolute deviation of the viscosities generated by the mth iterate potential from the 'experimental' viscosities in the temperature range $T^* = 1$ to $T^* = 3$. As an example Table 6.2 includes these deviations for the case when the $12-6$ potential is used as the starting potential to invert the pseudo-experimental data for various guessed well depths. It is evident that for a well depth between $\varepsilon^1/k = 140$ and 145 K the convergence of the process is the most rapid. Thus on the basis of our earlier argument we conclude that the appropriate 'true' well depth for argon is $142{\cdot}5 \pm 5$ K. A similar result has been found to hold whichever starting potential $U_0(r)$ is used for the inversion.[27] Other convergence criteria for this well depth selection, which have some value in specific cases, have also been developed.[27–9]

In the absence of a rigorous explanation for the existence and uniqueness of this inversion procedure its application to real experimental viscosity data has been preceded by considerable heuristic testing beyond that described above.[25,27,29] In view of the body of evidence amassed in this way the practicality of the inversion method is now well established although its theoretical foundation is not completely understood.

When the procedure is applied to real experimental data the true potential is of course unknown so that the full range of tests of uniqueness must be employed in every case. It is usual also to employ several different extrapolation procedures at short and long ranges to ensure that they do not unduly contaminate the final results. Of course, the experimental error with which real experimental data are burdened inevitably results in uncertainty in the final inverted potential. However, owing to the high precision of the best viscosity data the consequent uncertainty in the final potential is small. The potentials obtained by inversion of transport property data for the pure monatomic gases and their mixtures are in good agreement with those obtained in other ways. In the main, these potentials have been obtained by inverting viscosity data, and consequently the collision integral $\bar{\Omega}^{(2,2)}(T)$. However, analogous inversion procedures can be developed for any other collision integral. For example, isotopic thermal diffusion data can be used to obtain the experimental values of the collision integral $\bar{\Omega}^{(1,2)}(T)$ through (6.21) and such data have also been inverted.[30]

The detailed results of all of these inversion procedures as applied to the interactions of monatomic species are discussed in Chapter 9 together with results from other sources. In Table 6.3 we list the ranges of separation for which the potentials of the inert gases have been obtained by transport property inversion methods. This table includes the corresponding best estimates of the separation at zero potential energy and the depth of the well obtained by these methods.

Table 6.2

Viscosity deviations in the reduced temperature range $T^* = 1$ to 3 for inversion of simulated viscosity data using the (12–6) potential as an initial guess for various potential well depths. The viscosity deviation $\Delta\eta^{1-3} = \sum(\eta_{\text{exp}} - \eta_{\text{calc}})$, where the summation extends over all data points in the range $T^* = 1$ to 3, as indicated by the superscript

Well depth $(\varepsilon/k)/K$	Iteration No.	$\frac{-\Delta\eta^{1-3}}{10^{-7}\text{ Pa s}}$	Well depth $(\varepsilon/k)/K$	Iteration No.	$\frac{-\Delta\eta^{1-3}}{10^{-7}\text{ Pa s}}$
135	1	12·37	145	1	3·49
	2	+2·99		2	−2·64
	3	+2·63		3	−1·60
	4	+3·27		4	−0·43
140	1	+7·70	150	1	−6·73
	2	+0·17		2	−5·22
	3	+0·36		3	−3·86
	4	+1·15		4	−1·93
142.5	1	4·70			
	2	−1·32			
	3	−0·64			
	4	+0·43			

Table 6.3
The ranges of separation for which intermolecular potentials among the monatomic gases have been obtained by inversion of transport property data

	Separation range		Well depth	Separation at which $U(r)=0$
System	r_{min}/nm	r_{max}/nm	(ε/k)/K	σ/nm
He–He	0·18	0·27	—	0·269
He–Ne	0·204	0·262	20	0·307
He–Ar	0·245	0·320	30	0·325
He–Kr	0·271	0·324	30	0·356
He–Xe	0·291	0·336	30	0·276
Ne–Ne	0·220	0·300	39·6	0·313
Ne–Ar	0·266	0·383	60	0·336
Ne–Kr	0·286	0·366	60	0·350
Ne–Xe	0·300	0·363	70	0·336
Ar–Ar	0·320	0·470	142·1	0·349
Ar–Kr	0·317	0·522	165	0·365
Ar–Xe	0·322	0·462	170	0·359
Kr–Kr	0·322	0·490	199·2	0·377
Kr–Xe	0·337	0·441	220	0·394
Xe–Xe	0·360	0·53	267·0	

r_{min} and r_{max} are determined from $\{\bar{\Omega}^{(2,2)}(T)/\pi\}^{\frac{1}{2}}$ at the highest and lowest temperatures respectively for which transport coefficient data are available.

6.8. The basis of the inversion procedure

Despite the considerable practical success of this inversion procedure there has been as yet no complete explanation of its mode of operation. The puzzle of the very existence of the inversion procedure is compounded by two peculiar features of the process which are essential for its success.

The first feature is the near, but not complete, universality of the inversion function $G(T^*)$ defined by (6.25)–(6.28), among a whole class of potential functions possessing repulsive and attractive branches joined by a single minimum. This behaviour is shown in Fig. 6.11 where we have plotted the inversion function for several potential functions of this type. For purely repulsive or purely attractive inverse power potentials of the form

$$U(r) = C_n/r^n$$

the inversion function G is temperature-independent. Furthermore, it is only very weakly dependent on the magnitude of n although of course it possesses the sign of C_n. Thus, at low temperatures, all the potentials employed in the construction of Fig. 6.11 lead to the same, negative, asymptotic value of G characteristic of an attractive inverse sixth-power potential, $G(T^* \rightarrow 0) = -0{\cdot}58$. At high temperatures the inversion function attains an asymptotic value characteristic of the short-range repulsive interaction for each particular potential, but this changes only slowly with the steepness of the repulsion. For example for n in the range 8 to 18, G lies in the range 0·86 to 0·82. In the intermediate temperature range the near universality of $G(T^*)$ is maintained despite the considerable differences in the potentials. In the case of either purely repulsive or purely attractive inverse power potentials of known exponent the inversion scheme is exact. By considering only high-temperature viscosity data Dymond[31] was able to make use of the near-universality of $G(T^*)$ at high temperatures to develop the forerunner of the inversion scheme described here.

The other remarkable feature of the inversion procedure is that it identifies a point on the experimental $\bar{\Omega}^{(2,2)}$ vs. T curve with a single point on the $U(r)$ function. In view of the three integrations which relate the potential over its entire range to the collision integral $\bar{\Omega}^{(2,2)}(T)$ it is very surprising that this identification should be possible.

A preliminary attempt to explain these two features of the inversion procedure and hence the entire scheme has been made[32] and is based upon a number of mathematical approximations of the various integrals linking the collision integral at a particular temperature to the potential. The essential idea of the explanation is that the predominant contribution to the collision integral at a particular temperature arises from a special subset of molecular collisions having an impact parameter near a value b_c which is related to $\bar{\Omega}^{(2,2)}$ at a particular temperature. The relative kinetic energy E of these collisions is directly related to the same temperature. For this special set of collisions it has been shown that the deflection angle is nearly a constant as the temperature is varied parametrically and that it is approximately equal to π^{-1}. For simplicity of exposition we shall here assume that this special angle is in fact identically constant and exactly equal to π^{-1} although this is not essential. On this basis the measurement of $\bar{\Omega}^{(2,2)}$ as a function of T amounts to the determination of the relationship between the special impact parameter b_c and the corresponding relative kinetic energy E, which is such that the angle of deflection for collisions with these parameters is a constant equal to π^{-1}. This is shown diagramatically in Fig. 6.12 where we plot the modulus of the deflection angle in binary collisions as a function of the impact parameter and the relative kinetic energy for a $12-6$ potential function. According to the present interpretation the experimental measurement of $\bar{\Omega}^{(2,2)}$ vs. T is equivalent to the determination of the edge of the plane

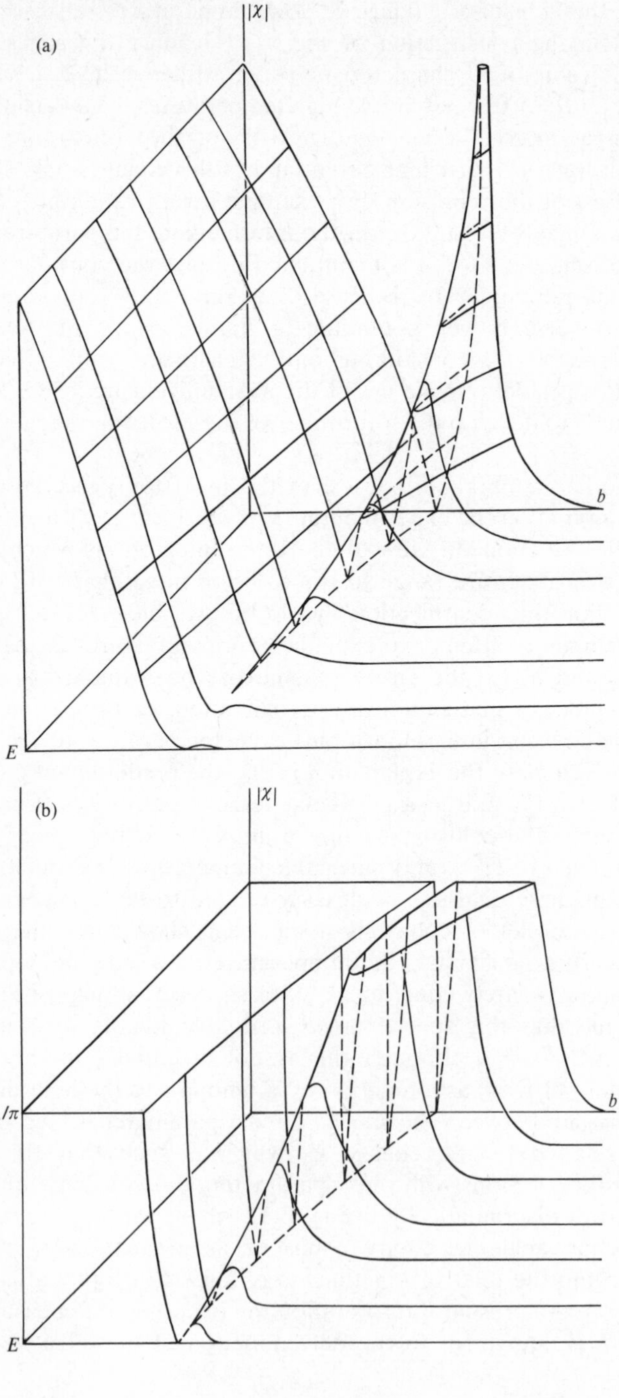

(a)
|χ|
b
E
(b)
|χ|
1/π
b
E

$|\chi| = \pi^{-1}$ parallel to the (b_c, E) plane. The determination of the intermolecular potential therefore reduces to the inversion of the (b_c, E) function for a binary collision subject to the constraint that the deflection angle be a constant. This inversion can be carried out[32] and completes the triplet of inversions in the (χ, b, E) space for a binary molecular encounter. For a discussion of the two other related inversions see Chapter 4.

The equations which result from quantifying this argument are similar to a statement of the practical inversion procedure described in the previous section. Furthermore, the near-universality of the function $G(T^*)$ is confirmed as is the identification of a single point on the potential function with an experimental $\{\bar{\Omega}^{(2,2)}, T\}$ point. Moreover, the separation $\{\bar{\Omega}^{(2,2)}(T)/\pi\}^{\frac{1}{2}}$ emerges as the natural result of the development.

The brief explanation of the inversion procedure given above is by no means complete. In particular the role of the various approximations involved has yet to be fully assessed. More importantly the explanation has been unable to establish *necessary* conditions for the successful and unique inversion of transport property data. Thus, the *sufficient* conditions discovered by the heuristic testing of the method remain the only ones available. It is to be hoped that further insight into the inversion procedure will be gained in the future in order that it may be applied to more complex systems with understanding and confidence.

6.9. Polyatomic gases

So far in this chapter the discussion has been limited to the transport properties of monatomic gases which interact through spherically symmetric potentials. Polyatomic gases on the other hand generally have orientation-dependent potentials and, moreover, possess internal energy in rotational and vibrational modes. The complications introduced into the kinetic theory of such gases and gas mixtures by these factors have been discussed in Chapter 5. First the transport of internal energy through the gas provides an additional contribution to the thermal conductivity of such gases not present for monatomic species. Secondly the occurrence of

---

FIG. 6.12. The deflection angles for a $(12-6)$ potential as a function of impact parameter, b, and relative kinetic energy, E. Measurement of the collision integral $\bar{\Omega}^{(2,2)}$ as a function of temperature is approximately the same as the determination of the outer edge of the plane parallel to the (b, E) plane at a height $|\chi| = \pi^{-1}$.

inelastic collisions, in which internal and translational energy are exchanged, can in principle affect the rate of transport of mass, momentum, and energy in the gas and hence the coefficients of diffusion, viscosity, and thermal conductivity. Finally, the collision dynamics of molecules interacting through orientation-dependent potentials are considerably more complicated than those for spherically symmetric potentials even if the collision is elastic.

In order to introduce these effects into the kinetic theory of gases rigorously a full quantum-mechanical treatment is required owing to the quantization of the internal energy of the molecules.[33,34] However, the complexity of the expressions resulting from the theory has so far inhibited their use for the interpretation of experimental data on transport properties. Instead it has been usual to employ the results of the semi-classical theory of Wang Chang and Uhlenbeck[35] or the classical theory of Taxman.[36] Although these theories are deficient in some fundamental respects (Chapter 5) the numerical consequences of the deficiencies are small for the usual transport coefficients of the gas. Even when the semi-classical and classical expressions for the transport coefficients have been employed, the complexity of the collision dynamics for polyatomic molecules has, until recently, made their exact evaluation generally impracticable. Consequently it has been necessary to introduce further approximation in order to make the calculations feasible.

First very simple molecular models have been employed such as rough spheres,[37] spherocylinders,[38,39] and loaded spheres.[40] These molecular models are not intended to be realistic representations of molecules, but they do incorporate some of the features of polyatomic molecules (inelastic collisions and non-spherical interactions). The transport coefficients for a gas composed of such molecules can be evaluated relatively simply and the results therefore provide a means of checking the accuracy of more approximate calculation methods which are applicable to more realistic intermolecular potentials. Second, physically reasonable approximations have been introduced into the classical and semi-classical kinetic theory expressions to simplify them to a level where their evaluation is only slightly more time-consuming and expensive than the monatomic case. Finally, recent developments in the treatment of binary collisions of polyatomic molecules provide the opportunity for more accurate evaluations of the transport coefficients of a polyatomic gas although only a few such calculations have yet been carried out.

Inelastic collisions in a polyatomic gas have a more profound influence on its thermal conductivity than upon the viscosity or the diffusion coefficient. The direct evaluation of the thermal conductivity of a polyatomic gas is therefore somewhat less secure than for the other two transport properties. Furthermore, it is only for the viscosity of polyatomic gases and gas mixtures that a large body of accurate experimental data are available. Consequently, we confine our discussion here to the viscosity and we shall be almost totally concerned with the forward calculation of the transport coefficient from an assumed intermolecular pair potential. It is not until the methods for carrying out this step have been perfected that serious studies of the relationship between the transport properties and the potential can be undertaken. The new developments described later in this section indicate that we have now reached this point.

6.9.1. Simple molecular models

Of the many simple molecular models employed for trial calculations of the transport properties of polyatomic gases we shall briefly discuss the rough sphere and the spherocylinder. The former introduces the effects of internal energy and inelastic collisions while retaining a spherical interaction potential, whereas the latter also introduces a non-spherical interaction.

The *rough-sphere model*[37,41] supposes that molecules are spheres which grip each other without slipping during a collision, thereby allowing transfer of rotational energy. Denoting the moment of inertia by I and the rigid sphere diameter of the molecules by d the viscosity of the gas may be written in terms of the parameter,

$$K = 4I/md^2 \quad 0 < K < \tfrac{2}{3}. \tag{6.36}$$

Here, the value $K=0$ represents a sphere with its mass concentrated at its centre, whereas the value $K=\frac{2}{3}$ represents a sphere with its mass uniformly distributed on its surface. The viscosity is given by

$$\eta = \frac{15}{8d^2}\left(\frac{mkT}{\pi}\right)^{\frac{1}{2}}\frac{(1+K)^2}{6+13K} \tag{6.37}$$

which for small K is

$$\eta = \frac{5}{16d^2}\left(\frac{mkT}{\pi}\right)^{\frac{1}{2}}\left(1-\frac{K}{6}\right). \tag{6.38}$$

The effects of internal energy upon the viscosity coefficient of the gas are therefore small since if $K=0{\cdot}1$ (quite a large value) the correction amounts to only 1·6 per cent.

Sandler and Dahler[39] have employed a *spherocylinder model* in which the molecule is a rigid, smooth circular cylinder of length L and radius d with two hemispherical end-caps of radius d. Denoting the moment of inertia about an axis normal to the cylinder axis by I their results can be expressed in terms of a geometric parameter $\beta = L/2d$ which describes the non-sphericity of the molecule and $\Delta = mL^2/8I$ which is a dimensionless moment of inertia. Their results indicate that the change in the viscosity of a gas of such molecules as β is varied is almost exactly equivalent to the change in the mean projected area of the spherocylinder as would be expected from the very simplest arguments.

6.9.2. The semi-classical and classical kinetic theories

An important result of the semi-classical and classical kinetic theories for polyatomic gases discussed in Chapter 5 is that formally the expressions for the viscosity of pure gases and gas mixtures are similar to those of a monatomic gas.[42] That is for a pure gas

$$\eta = \frac{5}{16} \frac{(mkT\pi)^{\frac{1}{2}}}{\langle \bar{\Omega}^{(2,2)}(T) \rangle} f_\eta(T) \tag{6.39}$$

where $\langle \bar{\Omega}^{(2,2)}(T) \rangle$ represents a collision integral for a polyatomic interaction and f_η includes the effects of higher-order approximations than the first in the kinetic theory solution scheme. The differences between the monatomic and polyatomic results stem from the details of the definition of the collision integral $\langle \bar{\Omega}^{(2,2)}(T) \rangle$. For a gas mixture the *interaction viscosity* $[\eta_{ij}]_1$ remains a defined quantity,[43]

$$[\eta_{ij}]_1 = \frac{5}{16} \left\{ \frac{2\pi m_i m_j kT}{m_i + m_j} \right\}^{\frac{1}{2}} \frac{1}{\langle \bar{\Omega}_{ij}^{(2,2)}(T) \rangle} \tag{6.40}$$

and may be derived from binary mixture viscosity measurements just as for a monatomic mixture with the aid of the ratio

$$A_{ij}^{\star} = \frac{\langle \bar{\Omega}^{(2,2)}(T) \rangle}{\langle \bar{\Omega}^{(1,1)}(T) \rangle}. \tag{6.41}$$

The collision integrals which enter these expressions are different from those of monatomic species owing to the possibility of inelastic collisions and the non-sphericity of the intermolecular pair potential. For example the collision integral $\langle \bar{\Omega}^{(2,2)}(T) \rangle$ is given, according to the semi-classical theory, by the equation[42]

$$\langle \bar{\Omega}^{(2,2)}(T) \rangle = \frac{1}{2} \left\{ \sum_i \exp(-\varepsilon_i) \right\}^{-2} \sum_{ijkl} \exp(-\varepsilon_i - \varepsilon_j) \times \int \{ \gamma^4 \sin^2\chi + \tfrac{1}{3}(\Delta\varepsilon)^2 - \tfrac{1}{2}(\Delta\varepsilon)^2 \sin^2\chi \} \gamma^3 \exp(-\gamma^2) I_{ij}^{kl} \sin\gamma \, d\chi \, d\psi \, d\gamma \tag{6.42}$$

In this expression I_{ij}^{kl} is the inelastic differential collision cross-section for two molecules initially in internal quantum states i and j which collide with a relative kinetic energy $kT\gamma^2$ and end up in internal states k and l. The symbol $\Delta\varepsilon$ denotes the change in the internal energy during the collision

$$\Delta\varepsilon = \varepsilon_k + \varepsilon_l - \varepsilon_i - \varepsilon_j$$

where $kT\varepsilon_i$ is the energy of the ith internal state. $I_{ij}^{kl} \sin\chi \, d\chi \, d\psi$ is the probability of scattering into the solid angle $\sin\chi \, d\chi \, d\psi$ as a result of the collision where χ is the polar deflection angle and ψ the corresponding azimuthal angle. I_{ij}^{kl} is therefore the inelastic differential cross-section for the collision which has been discussed in Chapter 4. Only recently has it become feasible to evaluate economically these differential cross-sections, and therefore the collision integrals, and few calculations have yet been carried out. The fundamental concepts associated with these new calculation methods have been discussed in Chapter 4 and their application to transport property evaluation is discussed later in this Chapter. Most work has been carried out with the aid of approximations to the formal definition of the collision integral $\langle\bar{\Omega}^{(2,2)}(T)\rangle$.

6.9.3. *The Mason and Monchick approximation*

Mason and Monchick[42] were able to show, with the aid of physically reasonable arguments, that the terms in $(\Delta\varepsilon)^2$ in (6.42) could be arranged to cancel (§ 5.9.2). Furthermore because the internal energy changes in collision for most molecules at normal temperature are small they argued that the trajectories of two colliding molecules are insignificantly distorted by the occurrence of inelastic collisions. Thus they replaced the inelastic differential scattering cross-section I_{ij}^{kl} by its elastic counterpart I_{el} for a collision between the same molecules. Using these results the semi-classical and classical expressions for the collision integral $\langle\bar{\Omega}^{(2,2)}(T)\rangle$ may be written[42]

$$\langle\bar{\Omega}^{(2,2)}(T)\rangle = \frac{1}{4}\int_0^{\infty}\int_0^{2\pi}\int_0^{\infty} \gamma^6\{(1-\cos^2\chi)b \, db \, d\psi\}\exp(-\gamma^2) \, d\gamma^2. \quad (6.43)$$

This is seen to be formally identical to the corresponding expression for monatomic gases (Appendix 5.1) but differs in that the non-spherical nature of the intermolecular potential is contained implicitly in the deflection angle χ. This is because for a collision in a non-spherically symmetric potential the molecular trajectory does not necessarily remain in a plane so that χ may depend on the azimuthal angle ψ. In view of these complexities several methods, at various levels of sophistication, have been employed to treat the transport properties of polyatomic gases and we treat them briefly in turn below.

6.9.4. Spherically-averaged potential

The formal similarity between the collision integral of (6.43) and that for monatomic gases has led to the frequent assumption of their identity. This assumption has been crudely justified by the argument that the molecules undergoing collisions in a gas have random orientations before collision and that they rotate during collision. Thus it has been assumed that the dynamics of collision taken overall are affected only by the potential averaged over all orientations of the two polyatomic molecules. This averaging procedure leads to a spherically symmetric potential for which the familiar model functions collected in Appendix 1 have been taken as suitable representations. Thus, by assuming a spherically symmetric potential model and calculating the collision integrals as for a monatomic gas the parameters of the model potential providing the best agreement with experimental viscosity data have been determined. These parameters have then often been identified with real characteristics of the true, spherically-averaged pair potential. Although this procedure is sometimes useful for the interpolation and correlation of experimental data it is of little value for the study of intermolecular forces. Indeed it has been shown that the spherical average of a non-spherical intermolecular pair potential yields appreciably different viscosities from those of a proper calculation based on the full non-spherical potential.[44]

6.9.5. The principle of corresponding states

An approach to the description of the transport properties of polyatomic gases and gas mixtures without reference to the details of the intermolecular potential is provided by the principle of corresponding states (§ 6.5). The formal similarity of (6.40) to that for monatomic gases allows the definition of $\Omega^*_\eta(T^*)$ from (6.23) to be carried over unchanged to the more complicated systems. Kestin and his collaborators[45,46] have been able to show that the functional $\Omega^*_\eta(T^*)$, which is universal among the interactions of the monatomic gases, is also universal for the interactions between several quite complicated non-polar molecules (e.g. N_2, SF_6, C_4H_{10}), some weakly polar molecules (e.g. N_2O, CCl_3F) and for the interactions of the molecules with monatomic species. Fig. 6.13 illustrates the universality by means of a plot of the experimental values of the functional $\Omega^*_\eta(T^*)$ for several polyatomic systems. In general, the temperature range of measurements on these systems is more limited than that for the monatomic species owing to the need to avoid thermal decomposition of the molecules. The figure includes as the solid line the correlation of $\Omega^*_\eta(T^*)$ obtained for the monatomic gases. The values of the scaling parameters which secure this agreement for all of the systems studied so far are included in Appendix 3. The correlations of $\Omega^*_\eta(T^*)$ and $\Omega^*_D(T^*)$

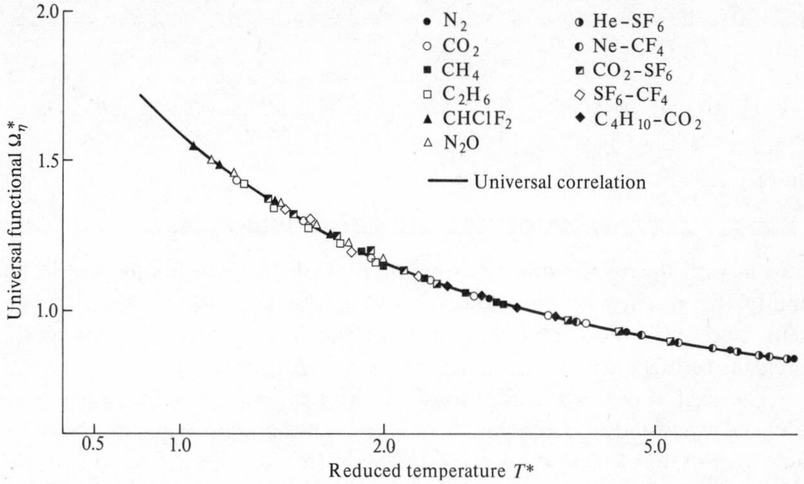

FIG. 6.13. The universal functional Ω_η^* for interactions involving polyatomic gases.

included in the same appendix together with the kinetic theory expressions of Appendices 5.2 and 5.3 permit the evaluation of the viscosity (and diffusion coefficient) of over 15 000 gas mixtures of arbitrary composition over an extremely wide range of temperature with an accuracy comparable with that of direct measurement. Thus the principle of corresponding states provides an extremely powerful correlation and prediction procedure for the properties of gases and their mixtures.

In § 6.5 it was shown that for the monatomic gases the universality of $\Omega_\eta(T^*)$ follows if all the intermolecular pair potentials can be made conformal by the choice of an energy scaling parameter ε_c and a distance scaling parameter σ_c. Clearly for the non-spherical potentials of polyatomic molecules the angular dependence of the potential makes such conformality impossible to attain between molecules of different symmetry. A full explanation of the observed universality of $\Omega_\eta^*(T^*)$ remains obscure. However it is clear that the scaling parameters ε_c and σ_c should therefore not be given the physical significance of characteristics of the true potential.[47]

6.9.6. *Monchick and Mason's treatment of polar gases*

Few attempts have so far been made to calculate the collision integrals for non-spherical intermolecular pair potentials even in the approximation where inelastic collisions are neglected. A well established model for the interaction of polar molecules is the Stockmayer potential[48] (Chapter 1),

$$U(r, \theta_1, \theta_2, \phi) = 4\varepsilon'\left\{\left(\frac{\sigma'}{r}\right)^{12} - \left(\frac{\sigma'}{r}\right)^{6}\right\} - \frac{\mu^2}{4\pi\varepsilon_0 r^3}\zeta(\theta_1, \theta_2, \phi) \qquad (6.44)$$

where

$$\zeta(\theta_1, \theta_2, \phi) = 2\cos\theta_1 \cos\theta_2 - \sin\theta_1 \sin\theta_2 \cos\phi.$$

Here θ_1 and θ_2 are the angles of inclination of the two dipoles to the line joining the centres of the molecules, ϕ is the azimuthal angle between them, and μ is the dipole moment. In the limit $\mu = 0$ the Stockmayer potential reduces to the Lennard-Jones $12-6$ potential.

The first attempt to compute the transport properties of a gas for this potential was made by Krieger[49,50] who simplified the collision dynamics by assuming that throughout every collision the two dipoles were collinear and head-to-tail, that is aligned in the configuration of maximum attractive energy with $\zeta = 2$. In this arrangement the potential is a function of r alone and the collisional problem reduces to the standard central field problem characteristic of monatomic gases.

Monchick and Mason[51] subsequently proposed a more physically reasonable approximation of the collision dynamics for the same potential and developed a practicable calculation procedure. They began the development of their method from (6.43) for the viscosity collision integral which excludes the effects of inelastic collision. First they recognized that the duration of a binary collision is approximately the same as the time for a single rotation of a polyatomic molecule.[52] Secondly, the deflection angle χ in a binary collision is determined primarily by the interaction in the neighbourhood of the closest distance of approach. The colliding molecules spend only a small fraction of the total time of a collision near this closest distance of approach and therefore the relative rotation of the two molecules in this time will be small. Consequently, Monchick and Mason assumed that each binary collision takes place at a single relative orientation of the two molecules which remains fixed throughout the encounter. It should be noted that this assumption does not imply that the molecules do not rotate relative to the laboratory during a collision but merely that they do not rotate with respect to each other. With this assumption it is clear that θ_1, θ_2, and ϕ remain fixed throughout a single binary collision so that the intermolecular potential depends only on the separation of the molecules r and can be analysed by the conventional techniques for a central field problem. It is also clear that each collision may take place at a different relative orientation for which the potential, a function of r only, is different. The collision integral $\langle\bar{\Omega}^{(2,2)}(T)\rangle$ may be shown to be an average over all orientations of the collision integral $\bar{\Omega}^{(2,2)}$

for each orientation. Mason and Monchick assumed that each relative orientation should be given equal weight in this averaging procedure so that

$$\langle \bar{\Omega}^{(2,2)}(T) \rangle = \frac{1}{8\pi} \int_{-1}^{1} \int_{-1}^{1} \int_{0}^{2\pi} \bar{\Omega}_{\omega}^{(2,2)}(T)\, \mathrm{d}\phi\, \mathrm{d}(\cos\theta_1)\, \mathrm{d}(\cos\theta_2) \quad (6.45)$$

Here $\bar{\Omega}_{\omega}^{(2,2)}(T)$ is the collision integral for the intermolecular potential $U(r)$ at the relative orientation $(\theta_1, \theta_2, \phi)$ and is therefore given by the equations of Appendix 5.2.

For the Stockmayer potential we define the reduced collision integrals by

$$\langle \Omega^{(2,2)*}(T^*) \rangle = \frac{\langle \bar{\Omega}^{(2,2)}(T) \rangle}{\pi \sigma'^2} \quad (6.46)$$

and a reduced temperature by

$$T^* = kT/\varepsilon'. \quad (6.47)$$

The viscosity of a polar gas may then be written

$$\eta = \frac{5}{16} \left(\frac{mkT}{\pi} \right)^{\frac{1}{2}} \frac{1}{\sigma'^2 \langle \Omega^{(2,2)*}(T^*) \rangle}. \quad (6.48)$$

Other collision integrals $\langle \Omega^{(l,s)*}(T^*) \rangle$ may be similarly defined and calculated. The advantage of this simplification of the collision dynamics is that collision integrals for several fixed relative orientations may be evaluated exactly as for a spherical potential and finally the averaging over all orientations carried out numerically. The computation time required is therefore only slightly more than for a spherical potential.

Monchick and Mason[51] carried out the calculation of $\langle \Omega^{(l,s)*} \rangle$ for the Stockmayer model as a function of T^* for several values of the dipole moment parameter $\delta_{max} = \mu^2/8\pi\varepsilon_0\sigma'^3\varepsilon'$. The results of similar calculations are included in Appendix 8. Plots of $\langle \Omega^{(2,2)*}(T^*) \rangle$ for several values of δ_{max} are shown in Fig. 6.14. It is apparent that the long-range dipole–dipole interaction influences the collision integral significantly only at low temperatures. For $\delta_{max} = 0$ the curve is that for the Lennard-Jones $12-6$ potential model. Furthermore the unweighted spatial average of the Stockmayer potential yields identically this potential for any value of μ since the angular-dependent term averages to zero. It is therefore clear that within this calculation scheme the spherical average of a non-spherical intermolecular potential does not lead to the same collision integrals as the full non-spherical potential.

Using the Stockmayer potential, Monchick and Mason were able to describe the available viscosity data for polar gases over their limited

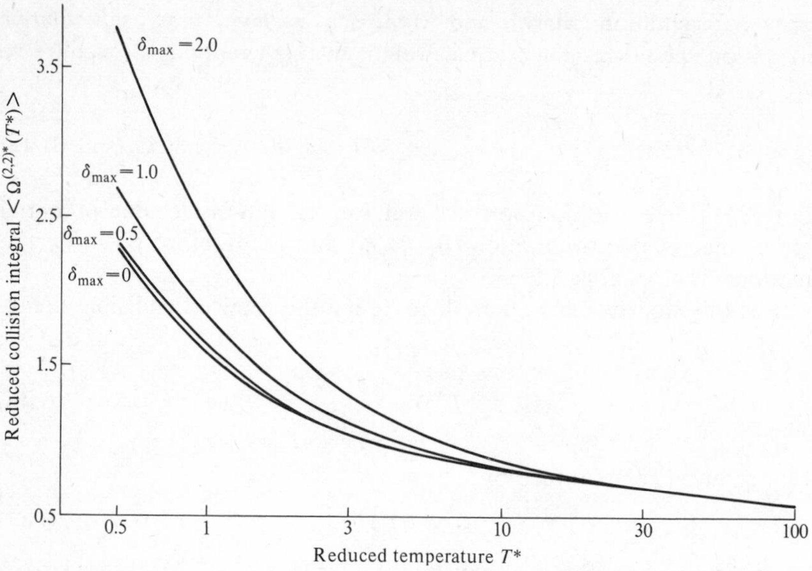

FIG. 6.14. The reduced collision integral $\langle \Omega^{(2,2)*}(T^*) \rangle$ for the Stockmayer potential.

temperature range about as well as the Lennard-Jones (12−6) potential describes the viscosity data for non-polar gases. Indeed the status of the Stockmayer potential is essentially the same as that of the (12−6) potential in that it is a model potential which incorporates some features of known interactions of molecules. However neither model should be regarded as a realistic representation of the potential.

6.9.7. *Recent developments*

The method employed by Monchick and Mason to treat the complicated dynamics of binary collisions between polyatomic molecules has the distinct advantage of simplicity and consequently is relatively rapid and inexpensive in computing time. The more complete treatments of inelastic molecular collisions have, until recently, required a very great amount of computational effort. Within the last decade there have been a number of advances in the development of approximate, but accurate methods of computing inelastic collision cross-sections. These new techniques have been discussed in Chapter 4 and here we merely indicate the ways in which they may influence the calculation of the transport properties of polyatomic gases in the future.

(a) *Quantum mechanical treatment.* As shown in Chapter 4 a rigorous quantum mechanical solution for the inelastic differential cross-section for scattering of two polyatomic molecules requires the solution of a set of coupled radial Schrödinger equations.[53] Such sets of equations have been solved for particular cases but, in general, the number of equations to be solved is so large that the computing time is prohibitive. It has therefore been necessary to introduce approximations into these coupled equations. The most attractive scheme of this type whereby differential scattering cross-sections may be evaluated and used to compute transport coefficients seems to be the infinite-order sudden (IOS) approximation.[44,54–8]

In essence the approximations introduced into the analysis of the collision dynamics are that the energy-spacing of the internal states of the molecules is small compared to the relative kinetic energy of the collision and that the differences in the centrifugal potentials for the coupled channels are small (See Chapter 4). These two approximations together turn out to be the quantum mechanical equivalent of the classical approximation of Monchick and Mason[51] that the two molecules remain in a fixed orientation relative to the line joining their centres during the collision. Interesting and important results have recently been derived with the IOS approximation. Parker and Pack[44] have examined the consequences of neglecting only the term in $(\Delta\varepsilon)$ in the semi-classical collision integrals for viscosity (6.42), and the equivalent integral for diffusion. They did not make the additional assumption of Mason and Monchick that the differential scattering cross-section may be replaced by its elastic value. In this case they have shown that the IOS result is formally identical to the procedure proposed by Monchick and Mason for non-spherical potentials. The collision integral is the equally weighted average of elastic collision integrals taken over all orientations. However, the IOS result shows that the *inelastic* contributions to the differential scattering cross-section are all automatically included by the average and not excluded as had been thought.

The IOS treatment of molecular collisions has other potential advantages for the computation of collision integrals. First it allows direct calculation of the degeneracy averaged differential cross-section I_{ij}^{kl} (Chapter 4). There is therefore no longer any necessity to approximate the semi-classical collision integrals, since a fully inelastic calculation, including the terms $\Delta\varepsilon$, can be performed. Second, the IOS approximation is most accurate when the internal energy level spacing is small compared to the kinetic energy of the collision and for high angle scattering. Fortunately, the viscosity collision integral $\langle\bar{\Omega}^{(2,2)}(T)\rangle$ is dominated by high-angle scattering and, under the conditions of most experimental viscosity measurements, the predominant mode of internal energy is rotational and the spacing of the levels is therefore very much less than kT. Consequently, the IOS approximation is most suitable under exactly the circumstances where it will find the greatest application for transport property calculations.

The only molecular interaction for which the collision integrals have so far been calculated with IOS approximation is that between a rigid rotor and an atom.[44] If we denote the angle between the axis of the rotor and the line joining the centre of the mass of the rotor to the atom by θ then the collision integral $\langle\bar{\Omega}^{(2,2)}(T)\rangle$ for this situation is given by the expressions[44]

$$\langle\bar{\Omega}^{(2,2)}(\mathrm{T})\rangle = \tfrac{1}{4}\int_0^\infty \gamma^6 \exp(-\gamma^2) Q^{(2)}(\gamma^2)\,\mathrm{d}\gamma^2 \tag{6.49}$$

where

$$Q^{(2)}(\gamma^2) = \left\{\sum_j (2j+1)\exp(-\varepsilon_j)\right\}^{-1} \sum_j (2j+1)\exp(-\varepsilon_j)$$
$$\times \sum_k \int_0^{2\pi}\int_{-1}^{1} I_j^k(\chi)\left[1-\left(\frac{\gamma'}{\gamma}\right)^2\cos^2\chi - \frac{1}{6}\left(1-\frac{\gamma'^2}{\gamma^2}\right)^2\right]\mathrm{d}(\cos\chi)\,\mathrm{d}\psi. \tag{6.50}$$

In these expressions I_j^k is the degeneracy-averaged inelastic differential cross section when the molecule is initially in internal rotational state j and ends up in state k for a scattering angle of χ. In addition $kT\gamma^2$ is the initial, relative kinetic energy of the collision and $kT\gamma'^2$ the final relative kinetic energy. The IOS approximation allows relatively simple calculation of the scattering cross-sections I_j^k (Chapter 4) and so the collision integrals may be evaluated directly.[44,58]

In fact this path to the collision integrals has rarely[58] been followed completely in any calculations. Instead it has been assumed, consistent with the IOS approximation, that $\gamma = \gamma'$, when the equation for $Q^{(2)}(\gamma^2)$ reduces to[44]

$$Q^{(2)}(\gamma^2) = \tfrac{1}{2}\int_{-1}^{1} Q^{(2)}(\gamma^2, \theta)\,\mathrm{d}(\cos\theta) \tag{6.51}$$

where

$$Q^{(2)}(\gamma^2, \theta) = 2\pi \int_{-1}^{1} (1-\cos^2\chi) I_{\mathrm{el}}(\theta, \chi)\,\mathrm{d}(\cos\chi). \tag{6.52}$$

The last formula is just that for the cross-section for elastic scattering by an intermolecular potential which is a function of r alone characterized by the relative orientation θ. Furthermore (6.51) is identical to the averaging procedure for the collision integrals employed by Monchick and Mason.[51] However, the IOS approximation shows that this formulation does not neglect inelastic collisions completely, their effect on the differential scattering cross-section being included in the $Q^{(2)}(\gamma^2, \theta)$ terms.

Parker and Pack[44] have evaluated the collision integral $\langle\bar{\Omega}^{(2,2)}(T)\rangle$ for the He–CO_2 system for a non-spherically symmetric electron gas intermolecular pair potential, according to (6.39), (6.49), (6.51), and (6.52). The potential employed can be written[44]

$$U(r, \theta) = \sum_n u_n(r) P_n(\cos\theta) \tag{6.53}$$

where $P_n(\cos\theta)$ are the Legendre polynomials. They have also performed collision integral calculations using the spherical part of this same potential $u_0(r)$. Their results indicate that differences of as much as 15 per cent arise between the collision integrals for the full $U(r, \theta)$ and those for $u_0(r)$. This finding confirms that the use of a spherically-averaged intermolecular pair potential is not justified for non-spherical molecules.

The development of the infinite-order sudden approximation for the treatment of polyatomic molecule collision dynamics will certainly encourage more systematic studies of the relationship between intermolecular potentials and the transport properties of gases containing such molecules. For example it is now possible to evaluate the collision integrals from (6.49) and (6.50) and so retain all the effects of inelastic collisions on the viscosity instead of just those contained in the differential scattering cross-section. Calculations of this nature by Maitland, Vesovic, and Wakeham[58] for the same He–CO_2 model potential used by Parker and Pack[44] indicate that the consequences of inelastic collisions can be important for all transport properties, particularly at low temperatures. The Mason–Monchick formulation appears to be a reasonable approximation for viscosity coefficients but can lead to significant errors in the case of diffusion.

(b) *The effects of electric and magnetic fields.* The effects of electric and magnetic fields upon the transport properties of polyatomic gases were discussed in Chapter 5. It has been shown theoretically that for polyatomic gases exposed to such fields a number of new transport coefficients can be identified which are intimately linked to the non-spherical nature of the intermolecular pair potential. These new transport coefficients have recently been expressed in a form which should allow their calculation for assumed potentials with the aid of, for example, the infinite-order sudden approximation.[59] Although no calculations of this type have yet been carried out there seems no doubt that the careful experimental measurements of these transport coefficients by Beenakker and his collaborators[60,61] when combined with such calculations will provide valuable information on non-spherical interactions in the future.

(c) *Other calculations.* The advent of very high-speed computers has now made possible the calculation of the transport properties of polyatomic gases through the classical theory of Taxman[36] with the aid of a suitable numerical algorithm. Evans[62] has carried out calculations of the viscosity of homonuclear diatomic molecules in the dilute gas state based on the Taxman equations. For these calculations he employed for the molecular pair interaction potential the diatomic Lennard-Jones potential

$$U=\sum_{i=1}^{2}\sum_{j=1}^{2}\varepsilon\left[\left(\frac{\sigma}{r_{ij}}\right)^{12}-\left(\frac{\sigma}{r_{ij}}\right)^{6}\right]$$

in which each atom is a Lennard-Jones interaction site. The calculations were performed for various values of the internuclear separation in the diatom. The results show that increasing anisotropy of the intermolecular potential leads to decreases in the viscosity of the gas. In view of the unrealistic nature of the intermolecular pair potential assumed no significance can be attached to the parameters ε and σ deduced from these calculations by comparison with experimental viscosity data. Calculations of the type performed by Evans are still very expensive in computing time and are therefore unlikely to be carried out routinely.

In summary, little has yet been learned about non-spherical molecular interactions from gaseous transport properties. However, several recent developments are likely to lead to increasingly sophisticated calculations of the transport coefficients of polyatomic gases. In particular, the calculation of the new transport coefficients which depend on the non-spherical part of the intermolecular potential should prove of great value to the study of intermolecular forces.

References

1. Barker, J. A., Fock, W., and Smith, F., *Phys. Fluids* **7,** 897 (1964).
2. Dawe, R. A. and Smith, E. B. *J. chem. Phys.* **52,** 693 (1970).
3. Bird, R. B., Stewart, W. E., and Lightfoot, E. N. *Transport Phenomena*, Chapter 2, Wiley, New York (1960).
4. Guevara, F. A., McInteer, B. B., and Wageman, W. E. *Phys. Fluids* **12,** 2493 (1969).
5. Kestin, J., Khalifa, H. E., Ro, S. T., and Wakeham, W. A. *Physica* **88A,** 242 (1978), and references therein. Erratum *Physica* **97A,** 466 (1979).
6. Michels, A., Sengers, J. V. and van der Guilik, P. S. *Physica* **23,** 1201 (1962).
7. Keyes, F. G. and Sandell, D. J. *J. Trans. ASME* **72,** 767 (1950).
8. Kestin, J., Ro, S. T., and Wakeham, W. A. *Physica* **58,** 165 (1972).
9. Haarman, J. W. *Physica* **52,** 605 (1971).
10. Healy, J. J., de Groot, J. J., and Kestin, J. *Physica* **82C,** 392 (1976).
11. Clifford, A. A., Kestin, J., and Wakeham, W. A. *Physica* **97A,** 287 (1979).
12. Kestin, J., Paul, R., Clifford, A. A., and Wakeham, W. A. *Physica* **100A,** 349 (1980).

13. Assael, M. J., Dix, M., Lucas, A., and Wakeham, W. A. *J. chem. Soc. Faraday Trans. I*, **77** (1981).
14. de Groot, J. J., Kestin, J., and Sookiazian, H. *Physica* **75**, 454 (1974).
15. Van Heigningen, R. J. J., Harpe, J. P., and Beenakker, J. J. M. *Physica* **38**, 1 (1968).
16. Hogervost, W. and Freudenthal, J. *Physica* **37**, 97 (1967).
17. Grew, K. E., and Ibbs, T. L. *Thermal diffusion in gases*. Cambridge University Press (1954).
18. Grew, K. E. and Wakeham, W. A. *J. Phys. B* **4**, 1548 (1971).
19. Clusius, K. and Hüber, M. *Z. Naturforsch*, **10a**, 230, 556 (1955).
20. Saxena, S. C. and Raman, S. *Rev. mod. Phys.* **34**, 252 (1962).
21. Saviron, J. M., Santamaria, C. M., Carrion, J. A., and Yarza, J. C. *J. chem. Phys.* **63**, 5318 (1975).
22. Storvick, T. R. and Mason, E. A. *J. chem. Phys.* **45**, 3752 (1966).
23. Kestin, J., Khalifa, H. E., and Wakeham, W. A. *Physica* **90A**, 215 (1978).
24. Marrero, T. S. and Mason, E. A. *J. Phys. Chem. Ref. Data* **1**, 3 (1972).
25. Gough, D. W., Maitland, G. C., and Smith, E. B. *Mol. Phys.* **24**, 151 (1972).
26. Maitland, G. C. and Smith, E. B. *Mol. Phys.* **22**, 861 (1971).
27. Maitland, G. C. and Wakeham, W. A. *Mol. Phys.* **35**, 1429 (1978).
28. Boushehri, A., Viehland, L. A., and Mason, E. A. *Chem. Phys.* **28**, 313 (1978).
29. Maitland, G. C. and Wakeham, W. A. *Mol. Phys.* **35**, 1443 (1978).
30. Clancy, P. and Smith, E. B. *Physica* **83C**, 231 (1976).
31. Dymond, J. H. *J. chem. Phys.* **49**, 3673 (1968).
32. Maitland, G. C., Mason, E. A., Viehland, L., and Wakeham, W. A. *Mol. Phys.* **36**, 797 (1978).
33. Snider, R. F. *J. Chem. Phys.* **32**, 1051 (1960).
34. Waldmann, L. *Z. Naturforsch.*, **13a**, 609 (1958).
35. Wang Chang, C. S., Uhlenbeck, G. E., and de Boer, J. *Studies in statistical mechanics* Vol. 2, Part C. (ed. J. de Boer and G. E. Uhlenbeck), North-Holland, Amsterdam (1964).
36. Taxman, N. *Phys. Rev.* **119**, 1235 (1958).
37. Pidduck, F. B. *Proc. roy. Soc. A* **101**, 101 (1922).
38. Muckenfuss, C. and Curtiss, C. F. *J. chem. Phys.* **29**, 1257 (1958).
39. Sandler, S. I. and Dahler, J. S. *J. chem. Phys.* **44**, 1229 (1966).
40. Sandler, S. I. and Dahler, J. S. *J. chem. Phys.* **43**, 1750 (1965).
41. Chapman, S. and Cowling, T. G. *The mathematical theory of non-uniform gases* (3rd edn), p. 217. Cambridge University Press (1970).
42. Mason, E. A. and Monchick, L. *J. chem. Phys.* **36**, 1622 (1962).
43. Monchick, L., Yun, K. S., and Mason, E. A. *J. chem. Phys.* **39**, 654 (1969).
44. Parker, G. A. and Pack, R. T. *J. chem. Phys.* **68**, 1585 (1978).
45. Abe, Y., Kestin, J., Khalifa, H. E., and Wakeham, W. A. *Ber. Bunsenges. Phys. Chem.* **83**, 271 (1979).
46. Kestin, J., Ro, S. T., and Wakeham, W. A. *J. chem. Phys.* **56**, 5837 (1972).
47. Bousheri, A., Viehland, L. A., and Mason, E. A. *Physica* **91A**, 424 (1978).
48. Stockmayer, W. H. *J. chem. Phys.* **9**, 398 (1941).
49. See Hirschfelder, J. O., Curtiss, C. F., and Bird, R. B. *Molecular theory of gases and liquids*, p. 597. Wiley, New York (1964).
50. Itean, F. C., Glueck, A. R., and Svehla, R. A. NASA Technical Note D-481 (1961).
51. Monchick, L. and Mason, E. A. *J. chem. Phys.* **35**, 1676 (1961).

52. Ferziger, J. H. and Kaper, H. G. *Mathematical theory of transport processes in gases*, p. 301. North-Holland, Amsterdam (1972).
53. Arthurs, A. M. and Dalgarno, A. *Proc. roy. Soc.* **A256,** 540 (1960).
54. Tsien, T. P. and Pack, R. T. *Chem. Phys. Lett.* **8,** 579 (1971).
55. Tsien, T. P., Parker, G. A., and Pack, R. T. *J. chem. Phys.* **59,** 5373 (1973).
56. Secrest, D. *J. chem. Phys.* **62,** 710 (1975).
57. Hunter, L. W. *J. chem. Phys.* **62,** 2855 (1975); Curtiss, C. F. *J. chem. Phys.* **49,** 1952 (1968).
58. Maitland, G. C., Vesovic, V., and Wakeham, W. A. *Mol. Phys.* in press (1981).
59. Liu, W. K., McCourt, F. R., and Köhler, W. E. *J. chem. Phys.* **71,** 2566 (1979).
60. Beenakker, J. J. M. and McCourt, F. R. *Ann. Rev. Phys. Chem.* **21,** 47 (1970).
61. Hulsman, H., van Kuik, F. G., Walstra, K. W., Knaap, H. F. P., and Beenakker, J. J. M. *Physica* **57,** 501 (1972).
62. Evans, D. J. *Mol. Phys.* **34,** 103 (1977).

7

SPECTROSCOPIC MEASUREMENTS

7.1. Introduction

The methods described so far for studying intermolecular forces all rely on measuring some indirect consequence of the molecular interaction. Some of these such as thermophysical properties, are macroscopic in nature, others, like molecular-beam scattering, are microscopic, but all are concerned with an average quantity involving many pair interactions between molecules moving with a range of velocities. The most direct way to study the forces between two molecules is to observe the electromagnetic spectrum of the bound dimer—or *van der Waals molecule* as it is sometimes called. Indeed the reason why the dissociation energies of common diatomic molecules such as I_2 have been known very accurately (±0·01 per cent) since the mid-1930s whereas until 1970 that of Ar_2 was uncertain by ±15 per cent is that rotation–vibration spectra of the former were readily available using relatively crude apparatus. It was not until the early 1970s that sufficiently sophisticated techniques were developed to enable the spectra of weakly interacting non-bonded species to be observed.

The existence of van der Waals dimers was first observed directly by electron diffraction[1–3] and mass spectrometry.[4,5] The main spectroscopic techniques which have been used to study the potential energy functions of these species are high-resolution ultraviolet and infrared spectroscopy, and molecular beam electric resonance spectroscopy in the microwave and radiofrequency regions. The methods give information about the vibrational and rotational energy levels of the dimer and it is through this detail that information about the intermolecular potential function is obtained. A number of other phenomena, for which the relationship between the observed property and the potential function is less direct, have also been used. These include the lineshape of absorption spectra, nuclear magnetic relaxation, Raman scattering, quadrupole relaxation, and sound absorption. Because they give the most quantitative information, most emphasis will be placed on the high-resolution techniques. In general, all the spectroscopic methods give information only about the well region of the intermolecular potential, although in the case of continuous spectra the short-range repulsive region is also probed.

We start in § 7.2 with a description of the nature of weakly-bound dimers and of the special conditions and techniques (§ 7.3) necessary to observe their high-resolution spectra. The analysis of these spectra to obtain the intermolecular potential is then considered for three classes of dimer. First, in § 7.4 we examine dimers of monatomic systems where the potential is spherically symmetric. Here, a direct formal inversion of the experimental data is possible, giving the well region of the potential. The precise information obtained from this source for the inert gases has made a significant contribution to the definition of their intermolecular potentials (§ 7.5). Second (§§ 7.6.2–7.6.4), we consider dimers involving diatomic molecules, X_2—Y, in which X_2 is able to rotate relatively freely and the potential is only weakly anisotropic. This is true only for a relatively small number of systems, typified by those involving H_2 and its isotopes. Although direct inversion of the data is not as yet possible here, in turns out that certain transitions are determined mainly by the isotropic part of the potential, whilst others also depend on the angular-dependent terms. It is this selectivity which in principle makes high-resolution spectroscopy an extremely powerful tool in studying non-spherical systems. Third, in § 7.6.5, we turn to the more common but far more complex case of non-spherical diatomic systems where the potential is highly anisotropic. This class of dimer is exemplified by Ar—HCl, the subject of much experimental and theoretical investigation, where orientations close to an end-on approach of the atom have a significantly lower energy than sideways-on, T-shaped configurations. Finally, in § 7.6.6 we summarize the information which has emerged from spectroscopic data about the 'structure' of more complex van der Waals dimers. For these systems, little detailed information about the intermolecular potentials has yet emerged. However improved resolution of the experimental data and developments in approximation procedures for calculating the energies of bound states for anisotropic potentials should combine to make spectroscopic techniques a rich source of information about the forces between polyatomic molecules over the next few years.

No attempt has been made in this chapter to cover this rapidly expanding field comprehensively. Several excellent reviews of the subject exist, notably those by Ewing[6] and Le Roy.[7,8] These have been drawn on extensively in this chapter and the reader is referred to them where more detail is required.

7.2. The origins of van der Waals molecules

A colliding pair of structureless molecules can exist in three types of state (Fig. 7.1). Effective potentials $V(r)$ of the type shown have already been discussed in § 4.2.2; the same notation is used in this chapter. In the

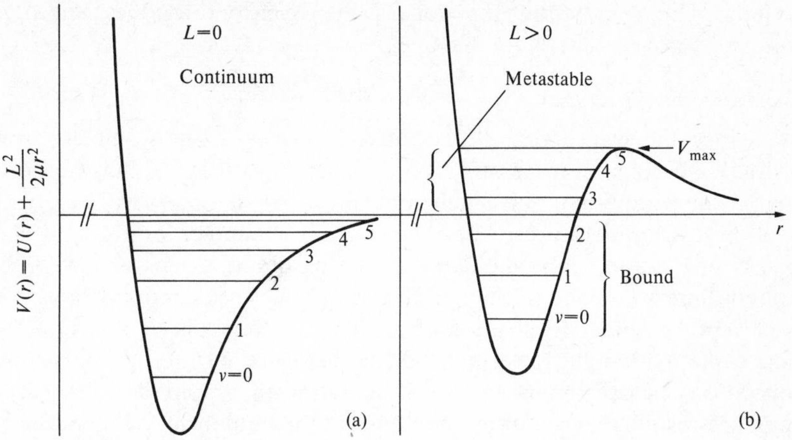

FIG. 7.1. Vibrational energy levels for a typical X_2 molecule having (a) no angular momentum ($L = 0$); (b) $L > 0$.

absence of angular momentum ($L = 0$), i.e. a non-rotating pair, molecules with energies above the dissociation limit are just *collision pairs* in *continuum* states. In the presence of a third body to remove excess kinetic energy, such a pair can drop into one of the discrete vibration–rotation energy levels in the potential well. These are *bound dimers.* (LeRoy[8] has suggested that the term *dimer* should be reserved for homonuclear pairs, such as $(H_2)_2$, referring to heteronuclear pairs, like H_2–Ar, as *bimers.* No such distinction is drawn in this chapter.)

When the approaching molecules are rotating, $L > 0$, and the effective potential $V(r)$ (see (4.10) and (4.11)) exhibits an outer maximum due to the centrifugal term. A third body may again remove excess kinetic energy to produce either bound dimers with energies below the dissociation limit or *metastable dimers* above $V(r) = 0$ but below the outer maximum in $V(r)$. These species, sometimes called *quasibound dimers,* are metastable by virtue of being able to dissociate by tunnelling through the potential barrier. Similarly, they may be formed in the absence of a third body by inward tunnelling if the colliding pair approach one another with an energy which corresponds to one of the metastable levels. A fourth situation corresponds to collisions at an energy just above V_{max}, when *resonances* can occur due to multiple orbiting followed by dissociation. For a full discussion of the orbiting phenomenon, see § 4.2.2.

The energies of the vibrational and rotational energy levels for a particular potential can be obtained by solution of the radial Schrödinger wave equation. Numerical, pertubation, variational, and WKB calculations have all been performed for a variety of model systems.[9–11] Estimates of the total number of bound vibration and rotation states, based on approximate potential functions, range from one for He_2, about 170 for Ar_2, to over 2×10^5 for CsXe.[12]

7.2.1. *Concentrations*

Since the well-depths (dissociation energies) of non-bonded dimers are typically of the order of $1\ \mathrm{kJ\ mol^{-1}}$, several orders of magnitude less than a chemical bond, only small concentrations of dimers will be present, even at low temperatures, where $kT \simeq \varepsilon$. For example, for argon at about 120 K and 1 atm, only about 0·4 per cent of the atoms are present as bound dimers.[10] Since rotational kinetic energies are comparable with the well depth ε, the concentration of metastable dimers is similar, about 0·2 per cent. Although lowering the temperature further favours dimer formation, below the normal boiling point the rapid decrease in gas pressure results in a significant decrease in the equilibrium dimer concentration. This extremely low concentration even under optimum conditions is one reason why experimental study of such dimers has proved so difficult. Their effect on bulk properties such as second virial coefficients and thermal conductivity has been investigated by Stogryn and Hirschfelder.[10]

7.2.2. *Lifetimes*

Lifetimes of metastable dimers τ can be estimated by WKB calculation procedures.[10] For Ar_2, for example, states near the top of the centrifugal barrier decay by tunnelling in about 10^{-11} s whereas well below the barrier they can exist for as long as 10^{10} s. Since at room temperature the mean time between collisions τ_c is about 10^{-10} s at 1 bar, about 10^{-7} s at 10^{-3} bar, most dimers dissociate by collision with another molecule. These calculations have been confirmed by optical pumping experiments for certain alkali metal–inert gas systems.[13]

Thus, although increasing the density of the gas increases the concentration of dimers, it also decreases the time between collisions, τ_c. This in turn increases the collisional line width $(1/\pi c\tau_c)$, so reducing the resolution of the observed spectrum. A rough criterion for the rotational fine structure of the spectrum of a van der Waals dimer to be resolved is that $(\pi c\tau_c)^{-1}$ should be less than B, the rotational constant characterizing the intermolecular motion†. Such high resolution has so far only been obtained for systems involving hydrogen, where B is relatively large, and for those involving molecules which absorb sufficiently strongly to enable

† The rotational constant $B = (\hbar^2/2I)\ \mathrm{J} \equiv h/8\pi^2 Ic\ \mathrm{cm^{-1}}$, where I is the moment of inertia of the dimer and c is the speed of light. Since the bond lengths of van der Waals dimers are much larger than normal chemical bonds, the moments of inertia are larger and the values of B correspondingly smaller than those of covalently bound molecules of similar mass.

measurements to be made at very low pressure, such as $(HCl)_2$. The advantage of the relatively new technique of molecular-beam spectroscopy is to increase the times between collision by using a rarefied beam, so increasing the lifetime of the dimers and facilitating their spectroscopic study.

7.3. Experimental techniques

7.3.1. Ultraviolet and infrared studies of dimers

Dimers consisting of inert gas atoms possess no dipole moment and consequently exhibit no electric dipole infrared spectrum. The first successful spectroscopic study of such species was performed by Tanaka and co-workers, who in a series of extremely elegant experiments observed a series of absorption bands due to Ne_2,[14] Ar_2,[15] Kr_2,[16] and Xe_2[17] in the vacuum ultraviolet in the region of the characteristic atomic line spectra of the inert gases.

Tanaka's apparatus had a path length of 6·65 m and used either helium or argon continua as background light sources. The inert gas under study was contained in a liquid nitrogen-cooled cell at a partial pressure ranging from $1{\cdot}3\times10^{-6}$ to 0·1 bar. Similar results were obtained at room temperature with an increased path length of 13 m. Exposure times for the spectra ranged from 5–80 min. Sufficient discrete band structure was observed to identify the vibrational energy level spacings of the ground state ($^1\Sigma_g^+$) and several excited electronic states. Somewhat higher resolution has been achieved for argon by Colbourn and Douglas[18] who, using a 10·6-m concave grating spectrograph at a dispersion of 0·024 nm/mm, were able to resolve rotational structure for one of the band systems.

Homonuclear molecules like N_2 and O_2 are not expected to absorb in the infrared region due to the absence of a dipole moment. However, when such molecules approach other atoms or molecules, their quadrupole moments can induce dipole moments in the other species (see § 2.2.2). The electron probability distributions of the molecules also become distorted on approach, leading to a dipole moment in the interacting pair. Since these dipoles are modulated as the molecular pair vibrates and rotates, absorption in the infrared is possible. This can be of two types:

(i) Where transitions occur in the continuum region, i.e. *collision pairs* exist, a broad diffuse absorption is observed. This is commonly referred to as *collision-induced absorption.*

(ii) Where *bound dimers* are formed and transitions occur between the discrete levels in the well, sharp absorption peaks are observed, usually superimposed on the broad background from (i).

Since both types of effect are extremely small, large path lengths (10–100 m) are required. Much of the pioneering work in this area was performed by Welsh and co-workers employing low temperatures and path lengths of 110 m for such species as $(H_2)_2$.[19] Ewing and co-workers have extended these studies by using multiple-reflection folded optics and path lengths up to 200 m ($(O_2)_2$,[20] $(N_2)_2$.[21])

Higher resolution and more detailed spectra are generally observable for dimers formed between unlike species, e.g. X_2–Y.[22–25] Effective resolutions of better than $0{\cdot}1\ \text{cm}^{-1}$ are now possible using low temperatures and phase-sensitive detection.

7.3.2. *Molecular-beam electric resonance: radiofrequency and microwave spectra*

For polar molecules, a particularly powerful technique for observing the fine structure of the spectra of van der Waals dimers is molecular-beam electric resonance[26,27] sometimes referred to as Rabi spectroscopy. The method is based on the fact that the rotational energy levels, j, of a molecule having a dipole moment are split by an electric field $\mathscr{E}$ into $j+1$ components, characterized by $|m_j|$—the well-known Stark effect. The energies of these states are increasingly perturbed as $\mathscr{E}$ increases (Fig. 7.2(a)). A molecule in state (j, m_j) can be considered to have an effective

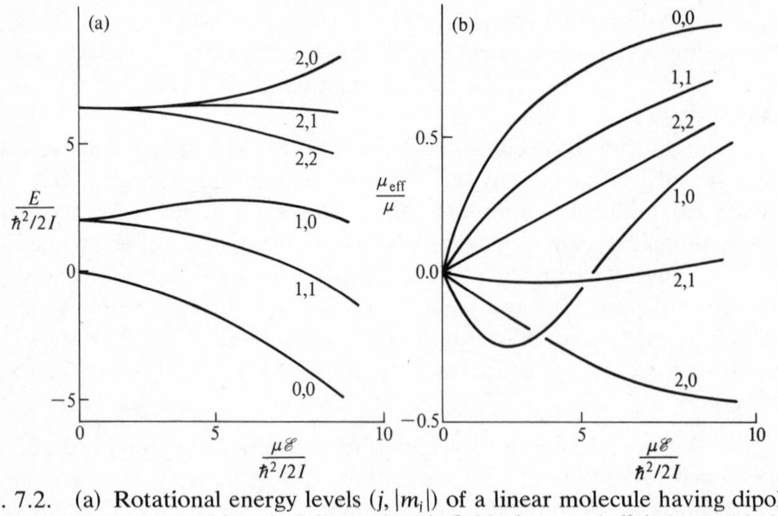

FIG. 7.2. (a) Rotational energy levels $(j, |m_j|)$ of a linear molecule having dipole moment μ, moment of inertia I, in an electric field of strength $\mathscr{E}$ (schematic). (b) Changes in μ_{eff}/μ for various $(j, |m_j|)$ states of this molecule as a function of field $\mathscr{E}$ (schematic). N.B. all nuclear spin and quadrupole hyperfine interactions have been neglected.

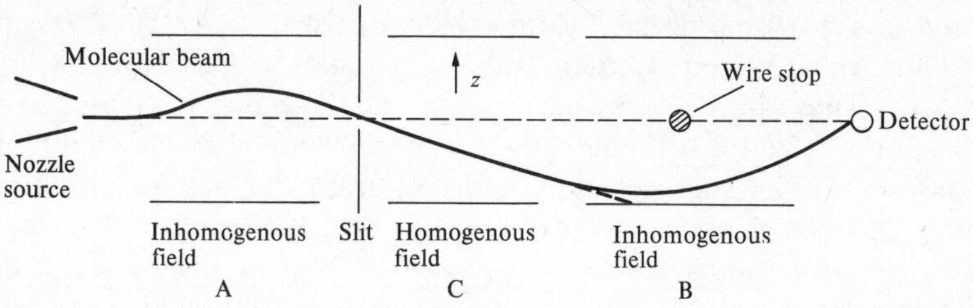

FIG. 7.3. Principle of operation of electric resonance spectrometer.

dipole moment

$$\mu_{\text{eff}}(j, m_j) = \frac{-\partial E(j, m_j)}{\partial \mathscr{E}} \quad \text{(see Fig. 7.2(b))}$$

where E is the rotational energy of the molecule. The force on such a molecule in an inhomogeneous electric field $\mathscr{E}'$ is

$$F_z = \mu_{\text{eff}}(j, m_j) \frac{\partial \mathscr{E}'}{\partial z}$$

where the field gradient is along the z-axis. Hence the molecule will be deflected by such a field to an extent which depends on μ_{eff}, i.e. on the particular (j, m_j) state occupied.

A schematic electric resonance spectrometer is shown in Fig. 7.3. A beam consisting of molecules in a mixture of states is first deflected by the defocusing field A, passes through a uniform field C, and is then re-deflected by the inhomogeneous field B. Since μ_{eff} depends on $\mathscr{E}$ in a different manner depending on the state (j, m_j), for a particular combination of fields A and B only molecules occupying a single state will be refocused at the detector. For instance, to focus state $(1, 0)$ in Fig. 7.2(b), $\mu\mathscr{E}_A/(\hbar^2/2I)$ should be less than 5, $\mu\mathscr{E}_B/(\hbar^2/2I)$ greater than 5. All other molecules will either hit the walls of the apparatus or, if μ_{eff} is very small, the wire stop placed in the direct beam. Between the two inhomogeneous fields, an oscillatory electric field is superimposed on the homogeneous field. When the oscillating field is at the correct frequency for an allowed dipole transition, some molecules absorb energy, change to a different state from the one being focused, and the intensity of the detected beam decreases. Monitoring the intensity as a function of the frequency gives the appropriate spectrum.

This technique has been applied with great success to weakly bound dimers by Klemperer and co-workers.[26–31] Beams of the dimers are produced by expanding the parent gas or gas mixture, which is pre-cooled to as low a temperature as possible, from high pressure through a supersonic nozzle. The nozzles consist of holes 30–250 μm in diameter drilled in cones sealed into the gas inlet and cooled with liquid nitrogen. They produce beams which are essentially translationally monoenergetic and whose internal energy states are characterized by low temperatures. The state-selecting and focusing fields are quadrupolar in nature, capable of potentials up to 25 kV, and the beam is detected by electron bombardment followed by mass analysis of the resulting ions. Radiation enters region C via waveguides and conventionally both radio-frequency ($\Delta j = 0$, $\Delta m_j = \pm 1$) and microwave ($\Delta j = \pm 1$, $\Delta m_j = 0$) transitions are studied. The frequency resolution of the spectrometer is in principle limited to a few kHz by the transit time through the resonance region, although at the present time, experimental line widths are about 20 kHz. Phase-sensitive detection is used to record the spectra and double resonance methods can be used to check assignments.

The information available from experiments using this technique includes the details of the rotation–vibration microwave spectrum, Stark coefficients and hence dipole moments, and detailed hyperfine structure in the radiofrequency region reflecting nuclear spin–spin, spin–rotation, and quadrupole interactions. The initial studies were analysed largely in terms of the geometry of the equilibrium or minimum energy configuration of the dimers, and some measure of the rigidity of such configurations. However more recent studies have shown considerable progress in the extraction of potential energy surfaces from the spectral data.[32] The technique should be a rich source of information on molecular interactions in the future. Details of some specific studies are given in § 7.6.

7.3.3. *Other techniques*

Several other methods have been developed for obtaining spectra of molecules formed in rarefied beams. Laser-induced fluorescence[33] uses a tunable dye laser to intersect the molecular beam and excite the primary beam molecules electronically. The resulting fluorescence is monitored as a function of exciting wavelength. The modification of the spectrum as the constituent molecule is excited to different vibrational levels contains, in principle, information on the variation of the intermolecular energy with the internal coordinates of the molecules involved. So far this technique has yielded mainly structural information (for example, on dimers involving I_2)[34,35] with little analysis in terms of potential energy surfaces being reported. An alternative to detecting the fluorescence directly is to use a cryogenic bolometer detector which is sensitive to the total energy of molecules hitting its surface.[36] Hence increases in internal energy due to excitation of the beam molecules by a radiation source, or decreases due to dissociation and subsequent out-of-beam scattering, can be detected. This method has been shown to be useful in the infrared region[36,37] and should give increasing information about potential functions as resolution of the observed spectra improves.

In some instances, continuous electronic spectra have been used to obtain quite detailed information about potential functions for excited electronic states. For instance, excited states of He_2 have been studied in this way.[38] Information about potentials can be obtained from measurements of spin-lattice relaxation times using nuclear magnetic resonance.[39] This property is particularly sensitive to the anisotropic part of the potential function[40] but extraction of quantitative information has so far been limited by the complexity of the relationships involved, and the need to use non-dilute systems in order to obtain adequate sensitivity. Raman spectroscopy (both lineshifts and line shapes) can in principle provide similar information. Such techniques are likely to gain in importance as probes of the anisotropy of potential energy functions. A brief summary of the sensitivity of measurements of this type to different regions of the intermolecular potential is given in § 7.7.

7.4. Analysis of rotation–vibration spectra to give *U*(*r*): spherical systems

7.4.1. *Introduction*

Before describing some of the spectroscopic studies of specific systems, it is worth examining in some detail what information about $U(r)$ is contained in such data in principle, and how such information is best extracted. For the analysis of spectroscopic data has suffered from similar malpractices to those perpetrated for bulk thermophysical properties i.e. the fitting of the experimental data to functions based on unrealistic forms of $U(r)$. The most common example of this has been the fitting of the vibrational energy levels $G(v)$ to the expression

$$G(v) = \omega_e(v+\tfrac{1}{2}) - \omega_e x_e(v+\tfrac{1}{2})^2 \tag{7.1}$$

where ω_e is the fundamental frequency $(\beta/\pi c)(\varepsilon/2\mu)^{\frac{1}{2}}$ and x_e is the anharmonicity constant $hc\omega_e/4\varepsilon$. Eqn (7.1) is an exact solution of the Schrödinger equation for a Morse potential

$$U(r) = \varepsilon\{(1-e^{-\beta(r-r_m)})^2 - 1\}. \tag{7.2}$$

This function is not a particularly good representation of $U(r)$ for chemically bound diatomic molecules, except in the region of the well minimum. It is even less realistic for non-bonded interactions, as has been shown by many studies covering a range of experimental properties.[41] Use of (7.1), though, automatically implies the use of this unrealistic function.

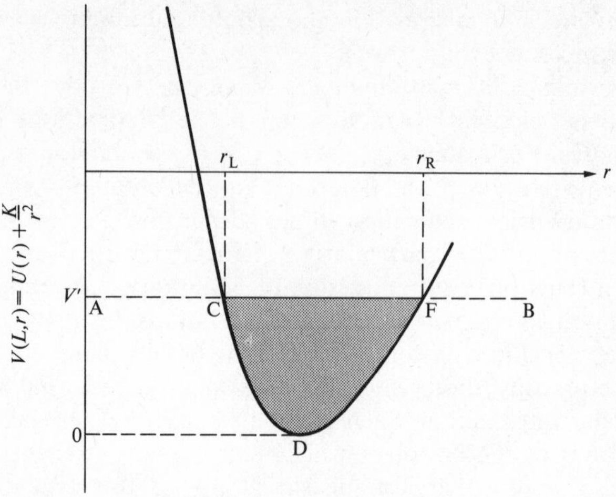

FIG. 7.4. Effective potential energy and parameters for RKR analysis. ($K = Eb^2 = L^2/2\mu$: for definition of terms see § 4.2.2.) N.B. The zero of energy is taken to be the well minimum for this analysis.

As in the case of the other properties which have been described, the most general and satisfactory method of determining $U(r)$ from spectroscopic data is by means of inversion procedures.

7.4.2. *The Rydberg–Klein–Rees inversion method*

Fig. 7.4 shows a portion of the effective potential energy curve $V(L, r)$ for a colliding pair of structureless particles.† The semi-classical Rydberg–Klein–Rees (RKR) procedure[7,42] enables the classical turning points r_L and r_R of an oscillator corresponding to a particular energy V' to be determined from a knowledge of the quantized rotational and vibrational energy level spacings from the minimum of the potential function up to the particular energy V'.

The shaded area S in Fig. 7.4, that between the lines AB and the curve CDF, can be written as

$$S = \int_{r_L}^{r_R} (V' - V)\, dr \tag{7.3}$$

where V is the effective potential energy $= U(r) + Kr^{-2}$, the latter term representing the centrifugal energy.

† Since spectroscopic studies of bound states only give information about the attractive well region of the potential, it is convenient when analyzing the data to redefine the zero of energy for $U(r)$ as the energy at the minimum in the well, $r = r_m$, rather than at $r = \infty$ (see Fig. 7.4). This procedure, which ensures that $U(r)$ is always positive, has been followed in § 7.4.2 and § 7.4.3 only.

By differentiating (7.3) with respect to V', then K, it can be seen that S is related to r_L and r_R by

$$\left(\frac{\partial S}{\partial V'}\right)_K = \int_{r_L}^{r_R} dr = r_R - r_L \tag{7.4}$$

$$\left(\frac{\partial S}{\partial K}\right)_{V'} = -\int_{r_L}^{r_R} \frac{dr}{r^2} = \frac{1}{r_R} - \frac{1}{r_L}. \tag{7.5}$$

Now let the total energy of the system be $E(I, K)$ where

(i) I is the vibrational phase integral $\oint p\, dq$, which in the first-order WKB (semi-classical) approximation is quantized according to the equation[43] (see Appendix A4.1)

$$\oint p\, dq = I = h(v + \tfrac{1}{2}),$$

v being the vibrational quantum number, p the momentum and q the conjugate coordinate,

(ii) K is the angular momentum phase integral, similarly defined, which in the JWKB framework is quantized according to[43]

$$\oint p_l\, d\theta = K = l(l+1)(\hbar^2/2\mu)$$

where l is the angular momentum quantum number, and μ the reduced mass.

The kinetic energy, T, is $E - V$ and the momentum, p, is $(2T\mu)^{\frac{1}{2}} = \{2\mu(E - V)\}^{\frac{1}{2}}$. Hence the quantization condition for I gives

$$I = h(v + \tfrac{1}{2}) = (2\mu)^{\frac{1}{2}} \oint (E - V)^{\frac{1}{2}}\, dr \tag{7.6}$$

which on subsequent differentiation with respect to E yields

$$\frac{dI}{dE} = \left(\frac{\mu}{2}\right)^{\frac{1}{2}} \oint \frac{dr}{(E - V)^{\frac{1}{2}}}. \tag{7.7}$$

The objective now is to relate the area S to the phase integrals I and K. Taking the standard integral

$$\int_b^a \left(\frac{a - x}{x - b}\right)^{\frac{1}{2}} dx = \frac{\pi}{2}(a - b),$$

let us put $a = V'$, $b = V$, and $x = E$, so that

$$\frac{\pi}{2}(V' - V) = \int_V^{V'} \left(\frac{V' - E}{E - V}\right)^{\frac{1}{2}} dE.$$

This substitution allows us to obtain an alternative expression to (7.3) for S:

$$S = \int_{r_L}^{r_R} (V' - V)\,\mathrm{d}r = \frac{2}{\pi}\int_{r_L}^{r_R} \mathrm{d}r \int_{V}^{V'} \left(\frac{V' - E}{E - V}\right)^{\frac{1}{2}} \mathrm{d}E \qquad (7.8)$$

Since

$$2\int_{r_L}^{r_R} \frac{\mathrm{d}r}{(E - V)^{\frac{1}{2}}} \equiv \oint \frac{\mathrm{d}r}{(E - V)^{\frac{1}{2}}}$$

and since the total energy, E, does not depend on r, this may be written

$$S = \frac{1}{\pi}\int_{V_0}^{V'} (V' - E)^{\frac{1}{2}} \left\{\oint \frac{\mathrm{d}r}{(E - V)^{\frac{1}{2}}}\right\} \mathrm{d}E.$$

Using (7.7) for $\mathrm{d}I/\mathrm{d}E$, the area becomes

$$S = \left(\frac{2}{\mu\pi^2}\right)^{\frac{1}{2}} \int_{V_0}^{V'} (V' - E)^{\frac{1}{2}} \frac{\mathrm{d}I}{\mathrm{d}E}\,\mathrm{d}E$$

or

$$S = \left(\frac{2}{\mu\pi^2}\right)^{\frac{1}{2}} \int_{0}^{I'} (V' - E(I, K))^{\frac{1}{2}}\,\mathrm{d}I \qquad (7.9)$$

where I' is the value of I corresponding to V' and 0 the value corresponding to the bottom of the well, V_0. Here the explicit dependence of E on I and K has been reintroduced. Differentiating (7.9) with respect to V', using (7.4), gives

$$f(V') = r_R - r_L = \left(\frac{1}{2\pi^2\mu}\right)^{\frac{1}{2}} \int_{0}^{I'} \frac{\mathrm{d}I}{\{V' - E(I, K)\}^{\frac{1}{2}}}. \qquad (7.10)$$

and with respect to K, using (7.5), gives

$$g(V') = r_L^{-1} - r_R^{-1} = \left(\frac{1}{2\pi^2\mu}\right)^{\frac{1}{2}} \int_{0}^{I'} \frac{(\partial E/\partial K)\,\mathrm{d}I}{\{V' - E(I, K)\}^{\frac{1}{2}}}. \qquad (7.11)$$

Since it is $U(r)$ which we wish to determine, (7.10) and (7.11) are usually evaluated for $l = K = 0$, when

$$\begin{aligned} V' &\equiv U(r) \\ E(I, K) &\equiv E(I, 0) \\ \left(\frac{\partial E}{\partial K}\right)_{K=0} &\equiv B_v \cdot \frac{8\pi^2\mu}{h^2} = \frac{1}{\bar{r}_v^2} \end{aligned} \qquad (7.12)$$

where B_v is the rotational constant $h^2(8\pi^2\mu\bar{r}_v^2)^{-1}$ and $\bar{r}_v$ is an effective mean separation for the level $(I, 0)$. The turning points r_R and r_L corresponding to a particular potential energy U' can therefore be found from the observed levels by evaluating (7.10) and (7.11). The function $U(r)$ may be constructed by evaluating these equations for successively higher values of I'. When applied to a set of vibrational levels with unresolved rotational fine structure—which is usually the commonest situation for van der Waals dimers—the only information available is therefore the width of the potential energy well as a function of its depth, using (7.10). Eqn (7.11) requires a knowledge of the rotational constant B_v for its evaluation, and hence both vibration and rotation fine structure need to be resolved to obtain the full potential well. Eqns (7.10) and (7.11) were first obtained by Rydberg[44] and Klein.[45]

One of the first attempts to apply the method was due to Rees[46] who attempted to fit the observed levels $E(I, K)$ to standard spectroscopic expressions such as (7.1) and hence to perform the integrations (7.10) and (7.11) analytically. His expressions give an inadequate account of anharmonicity and not surprisingly difficulties arise when sufficient data are available to calculate $U(r)$ right up to the dissociation limit where the quadratic expressions for the vibrational energy break down. To overcome these difficulties, many modifications of this procedure have been suggested.[47–49] These include fitting small groups of levels to (7.1), allowing ω_e and x_e to vary arbitrarily from one region of $U(r)$ to another, or fitting the levels to polynomial series in $(v+\frac{1}{2})$. However large numbers of terms are usually required and the series are prone to oscillation and slow convergence. Quasi-analytic procedures of this type have generally proved to be inadequate.

A much more satisfactory procedure which both produces stable solutions and places no constraints on the form of $U(r)$ is to evaluate the integrals in (7.10) and (7.11) numerically.[42] This may be done either by splitting the integrals into three parts and evaluating them part numerically and part analytically, or more conveniently by Lagrangian interpolation between the energy levels followed by Gaussian numerical integration. The numerical noise introduced using real data may be minimized by first smoothing the experimental energies before integration. A more convenient form of the equations for numerical evaluation is

$$f(U_n) = r_R - r_L = \left(\frac{h}{2\pi^2\mu c}\right)^{\frac{1}{2}} \int_0^n \frac{dx}{(U_n - U_x)^{\frac{1}{2}}} \qquad (7.13)$$

$$g(U_n) = r_L^{-1} - r_R^{-1} = \left(\frac{h}{2\pi^2 c\mu}\right)^{\frac{1}{2}} \int_0^n \frac{dx}{r_v^2(U_n - U_x)^{\frac{1}{2}}} \qquad (7.14)$$

$$= 2\left(\frac{8\pi^2 c\mu}{h}\right)^{\frac{1}{2}} \int_0^n \frac{B_x\, dx}{(U_n - U_x)^{\frac{1}{2}}}. \qquad (7.15)$$

Here $x = v + \frac{1}{2}$ and U_x, B_x are in cm^{-1}, the traditional unit for vibration–rotation transition frequencies, ($1\ \mathrm{cm}^{-1} \equiv 11{\cdot}96\ \mathrm{J\ mol}^{-1} \equiv 1{\cdot}44\ \mathrm{K}$.)

7.4.3. *Determination of dissociation energy (well depth)*

The RKR analysis gives the shape of the potential energy curve from the minimum up to the position of the last observed vibration–rotation level. In so doing it will probably define the position of the minimum, r_m, within narrow limits. However, only by determining $U(r)$ right up to the dissociation limit can an estimate be made of the well depth, ε, and even then extrapolation to the limit is usually required. The accuracy of the value determined for ε is therefore crucially dependent on the validity of the method used to perform this extrapolation from the last observed energy level.

For almost fifty years, the most commonly used method was that devised by Birge and Sponer.[50] It is based on the observation that the spacings between successive vibrational energy levels, $\Delta G_{v+\frac{1}{2}} = G(v+1) - G(v)$, usually decrease monotonically as the dissociation limit is approached. They used (7.1) to quantify this, leading to the relationship

$$\Delta G_{v+\frac{1}{2}} = \omega_e - 2x_e\omega_e(v+1). \tag{7.16}$$

This predicts that a plot of $\Delta G_{v+\frac{1}{2}}$ vs. v (a Birge-Sponer plot) should be linear and enables the dissociation energy to be obtained by adding the energy of the highest observed level ($v = v_c$) to the area under the extrapolated curve from v_c to the intercept value of v, v_{max}. However, as pointed out earlier, the use of (7.1), and hence (7.16), is equivalent to assuming a Morse potential, (7.2). The fact that such plots are commonly observed to have pronounced curvature is further evidence of the inadequacy of the Morse function, especially at large separations, and shows that large errors in the value of the dissociation energy ε may result if such a procedure is used.

In recent years more satisfactory methods have been devised based on the fact that the asymptotic functional form of $U(r)$ in the region approaching dissociation is usually known theoretically. In many cases $U(r)$ in this region can be represented by

$$U(r) = \varepsilon - \frac{C_n}{r^n} \tag{7.17}$$

where, in this section, C_n is a positive constant (see footnote on p. 396). The asymptotic form of the vibrational energy levels for this case was first devised by LeRoy and Bernstein.[51] Their method, like the RKR method, is based on the first-order WKB quantum condition for the eigenvalues of a potential $U(r)$, (7.6), which may be written

$$v+\tfrac{1}{2}=\left(\frac{8\mu}{h^2}\right)^{\frac{1}{2}}\int_{r_L}^{r_R}\{G(v)-U(r)\}^{\frac{1}{2}}\,dr. \tag{7.18}$$

Differentiation gives the density of states as

$$\frac{dv}{dG(v)}=\left(\frac{2\mu}{h^2}\right)^{\frac{1}{2}}\int_{r_L}^{r_R}\{G(v)-U(r)\}^{-\frac{1}{2}}\,dr. \tag{7.19}$$

Examination of the integrand of (7.19) for a typical model function such as a 12–6 potential shows that the most significant contributions to the integral for levels high up in the well come from the regions of r near the turning points r_R and r_L where the integrand becomes singular. Furthermore, the anharmonicity of the potential makes the contribution to the integrand from separations near r_R dominant as dissociation is approached. These observations were implemented mathematically by LeRoy and Bernstein by making the following approximations:

(i) Replace $U(r)$ in (7.19) by the asymptotic form given in (7.17);
(ii) Set $r_L=0$.

Evaluation of (7.19) then leads to[51]

$$\frac{dG(v)}{dv}=K_n\{\varepsilon-G(v)\}^{(n+2)/2n} \tag{7.20}$$

where

$$K_n=\frac{\bar{K}_n}{\mu^{\frac{1}{2}}(C_n)^{1/n}}=\frac{h}{(2\pi\mu)^{\frac{1}{2}}(C_n)^{1/n}}\left\{\frac{n\Gamma(1+1/n)}{\Gamma(\frac{1}{2}+1/n)}\right\}. \tag{7.21}$$

For the most common case of $n=6$, corresponding to the leading London dispersion term, $\bar{K}_6=59{\cdot}8301$ (when the units of energy, length, and mass are cm^{-1}, Å, and amu respectively). Integrating (7.20) gives ($n\neq 2$)

$$G(v)=\varepsilon-\{(v_{\max}-v)H_n\}^{2n/(n-2)} \tag{7.22}$$

where $H_n=\{(n-2)/2n\}K_n=\bar{H}_n/\mu^{\frac{1}{2}}(C_n)^{1/n}$. ($\bar{H}_6=19{\cdot}94336$ in cm^{-1}, Å, amu units.) For the case of $n=6$, differentiation of (7.22) gives

$$(\Delta G_{v+\frac{1}{2}})^{\frac{1}{2}}\simeq\left(\frac{dG(v+\frac{1}{2})}{dv}\right)^{\frac{1}{2}}=[3(H_6)^3]^{\frac{1}{2}}(v_{\max}-v-\tfrac{1}{2}). \tag{7.23}$$

Thus, the correct linear extrapolation in this case is $(\Delta G_{v+\frac{1}{2}})^{\frac{1}{2}}$ vs. v *not* $\Delta G_{v+\frac{1}{2}}$ vs. v. Fig. 7.5 illustrates such an extrapolation for the case of Ar_2. The large curvature in the 'true' Birge–Sponer extrapolation compared to that based on (7.16) is readily seen, as is the large error in the value of ε if the older procedure is followed. That method also underestimates the

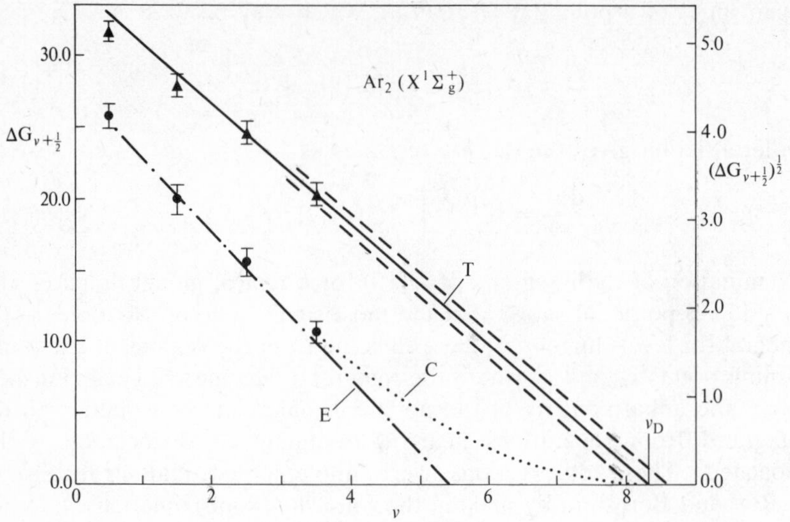

FIG. 7.5. Vibrational energy-level extrapolation for Ar_2.[52] T, C are the 'correct' extrapolations according to (7.23); E is the traditional Birge–Sponer extrapolation, (7.16). (R. J. LeRoy.[52])

total number of bound states by 50 per cent, emphasizing the strong influence that the long-range part of the potential has in determining the number of levels. The validity of (7.23) is also demonstrated and has been experimentally verified in a large number of cases.[53] This procedure of analysing energy levels near dissociation has proved of enormous value in obtaining accurate values of the well depth ε. It can also be used to determine the precise form of the long-range part of $U(r)$ to give n and C_n.

The method so far described applies in the absence of rotation ($l = K = 0$). It can be extended to vibration–rotation analysis for cases where $l \neq 0$, and to an expression for the limiting behaviour of the rotational constant B_v for levels near the dissociation limit. LeRoy[54] has obtained the expression

$$B_v = P_n(\varepsilon - G(v))^{2/n} \tag{7.24}$$

where the constant $P_n = \bar{P}_n/(\mu(C_n)^{2/n})$ and the values of $\bar{P}_n$ have been tabulated.[54] ($\bar{P}_6 = 27{\cdot}4105$ (cm^{-1}, Å, amu)). Alternatively, application of (7.22) transforms (7.24) into

$$B_v = Q_n(v_{\max} - v)^{4/(n-2)}$$

where

$$Q_n = P_n(H_n)^{4/(n-2)} = \frac{\bar{Q}_n}{\mu^{n/(n-2)} C_n^{2/(2-n)}}$$

$$Q_6 = 546{\cdot}658 \text{ (cm}^{-1}\text{, Å, amu units).}$$

Although these equations are less accurate than (7.20)–(7.23), they could be used to determine dissociation limits or potential constants to within a few per cent.

A knowledge of the long-range coefficient C_n enables a reasonable estimate of the dissociation energy ε to be determined even when the available spectroscopic data do not include levels close to the dissociation limit.[55] One method uses (7.20) in the form

$$\left[\frac{dG(v+\frac{1}{2})}{dv}\right]^{2n/(n+2)} \simeq (\Delta G_{v+\frac{1}{2}})^{2n/(n+2)} = K_n^{2n/(n+2)}\{\varepsilon - \bar{G}(v+\tfrac{1}{2})\}$$

(where $\bar{G}(v+\frac{1}{2}) = \frac{1}{2}\{G(v) + G(v+1)\}$ to interpolate graphically between the observed levels and the known limiting behaviour. Somewhat more precise values of ε can be obtained by means of a non-linear least squares fit to an expansion of the type

$$G(v) = \varepsilon - \{H_n[v_{\max} - v]\}^{2n/(n-2)}\{1 + \alpha(v_{\max} - v) + \beta(v_{\max} - v)^2 + \ldots\} \tag{7.25}$$

in which C_n, which determines H_n, and n are fixed at their known theoretical values. The result has been shown to be relatively insensitive to the precise value of C_n used or the number of terms taken in the expansion. For example, the ground-state dissociation energy of $BeAr^+$ has been determined to within one per cent using the experimental energies of a mere five of the forty bound vibrational states.[55]

7.5. Experimental results for inert gas dimers

In this section we review the spectroscopic studies of inert gas dimer systems which have been made to date, and discuss the conclusions drawn from them concerning intermolecular forces.

7.5.1. Pure components

As indicated in § 7.3, the most extensive spectroscopic studies of van der Waals dimers are those of Tanaka and co-workers on the pure inert gases. In 1970, Tanaka and Yoshino observed nine discrete band systems in the 78–108 nm region for Ar_2 and identified six vibrational levels, $v = 0$–5, for the ground state ${}^1\Sigma_g^+$ (formed from two ground state ${}^1S_0$ argon atoms).[15] The rotational structure was poorly resolved and unassigned.

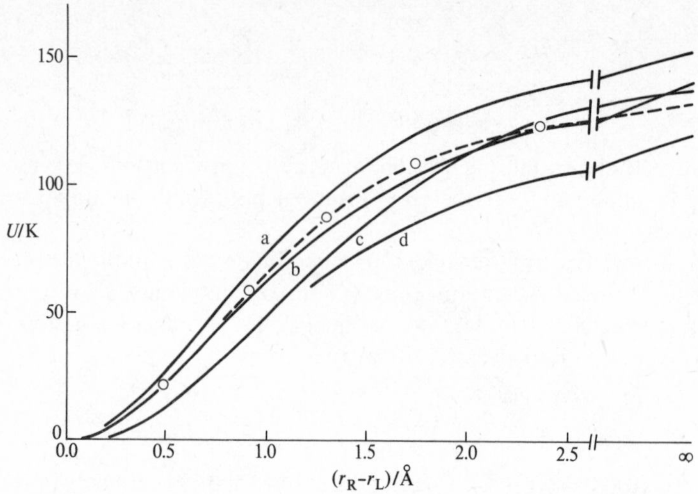

FIG. 7.6. The width $r_R - r_L$ of the argon intermolecular potential curve as a function of energy, measured from the minimum in the potential well, as determined by RKR analysis of spectroscopic data.[56] (○) RKR inversion; (---) fitted Morse curve (incorrect limit). (a) 11−6−8 function (Klein, M. and Hanley, H. J. M. *J. chem. Phys.* **53,** 4722 (1970)); (b) function of Barker, J. A. and Bobetic, M. V. *Phys. Rev.* **B2,** 4169 (1970); (c) function of Dymond, J. H. and Alder, B. J. *J. chem. Phys.* **51,** 309 (1969); (d) 12−6 potential.

This information was used in several ways to characterize $U(r)$ for argon. The method described in § 7.4.3 for determining dissociation energies from vibrational levels near dissociation was applied to the argon data by LeRoy[52] to obtain a value of $\varepsilon/k = 142{\cdot}2 \pm 7{\cdot}6$ K ($98{\cdot}7 \pm 5{\cdot}3\ \text{cm}^{-1}$) (see Fig. 7.5). Fuller use of the data was made by Maitland and Smith[56] who performed an RKR inversion to yield $f(U_n)$ which defines the width of the potential energy well as a function of its depth. This is illustrated in Fig. 7.6. Their analysis served to demonstrate the inadequacy of previously suggested forms for $U(r)$, and was used in conjunction with second virial coefficient and viscosity data to define a potential function having a well depth of 142·5 K. At essentially the same time Parson, Siska, and Lee[57] adopted the reverse approach of fitting a multiparameter equation to their differential scattering cross-section data and calculating the vibrational levels by solution of the radial wave equation. An iterative best fit to both the scattering and spectroscopic data produced a potential function with well depth 140·7 K of essentially the same shape as the Maitland–Smith (BBMS) function and one based on fitting bulk properties of gaseous, liquid, and solid argon due to Barker, Fisher, and Watts[58] ($\varepsilon/k = 142{\cdot}1$ K). Although there have been subtle refinements to the argon potential since this time (see Chapter 9), these three essentially identical functions have formed the basis of the presently accepted realistic pair potential of argon.

Colbourn and Douglas[18] have repeated Tanaka's work using a higher resolution spectrograph. They were able to resolve the rotational fine structure for one band system and so measure the energies of rotational levels associated with each of the six vibrational levels. By combining a full RKR analysis of both vibrational and rotational levels with a fit to bulk data and theoretical long-range calculations they produced a potential well, having $\varepsilon/k = 143{\cdot}3$ K ($99{\cdot}55$ cm^{-1}) and $r_m = 0{\cdot}3759$ nm, which is essentially indistinguishable from those of the three potential functions above.

Tanaka and co-workers have also studied the other homonuclear inert gas dimers. In the case of He_2,[59] the well depth is too low to allow the formation of bound states. For Ne_2, twelve band systems were observed, of which three gave information about the ground state.[14] Evidence was obtained for two stable vibrational levels and, in contrast to the argon measurements, a rotational analysis was performed on some of the $v = 0$ bands. Approximate RKR analysis of these data has been possible[60] and many attempts made to combine the spectroscopic measurements with other data for neon to determine $U(r)$. The most detailed analysis of the spectrum[61] suggests $\varepsilon/k = 41{\cdot}2$ K ($28{\cdot}6 \pm 1{\cdot}0$ cm^{-1}), and $r_0 = 0{\cdot}31$ nm, where r_0 is the mean separation in the $v = 0$ level.

For krypton and xenon, many more vibrational levels are observable and, although no rotational lines were assigned in Tanaka's work, the resulting data have enabled the potential functions to be defined within narrow limits. Nine vibrational energy level spacings were observed[16] for Kr_2; these were extrapolated by the method of LeRoy and Bernstein to give a well depth of $199{\cdot}2 \pm 0{\cdot}7$ K. This information, combined with the RKR well width inversion, gave a potential well in substantial agreement with those derived from bulk data and from differential scattering cross sections. For Xe_2, the first ten vibrational levels were observed,[17] which on extrapolation predicted that the potential well of the ground state has a depth of $281{\cdot}5$ K ($195{\cdot}5$ cm^{-1}), supporting 25 or 26 bound vibrational levels. Again partial RKR analysis to obtain the well width function, in conjunction with scattering and bulk data, enabled $U(r)$ to be closely defined. The reduced well-width plots resulting from the RKR analysis for the four heavier dimers, shown in Fig. 7.7, illustrates the near-conformal nature of the potential energy wells for this series of molecules.

7.5.2. *Mixtures*

By contrast to the homonuclear dimers, there is relatively little spectroscopic information about inert gas mixtures. Tanaka and Yoshino

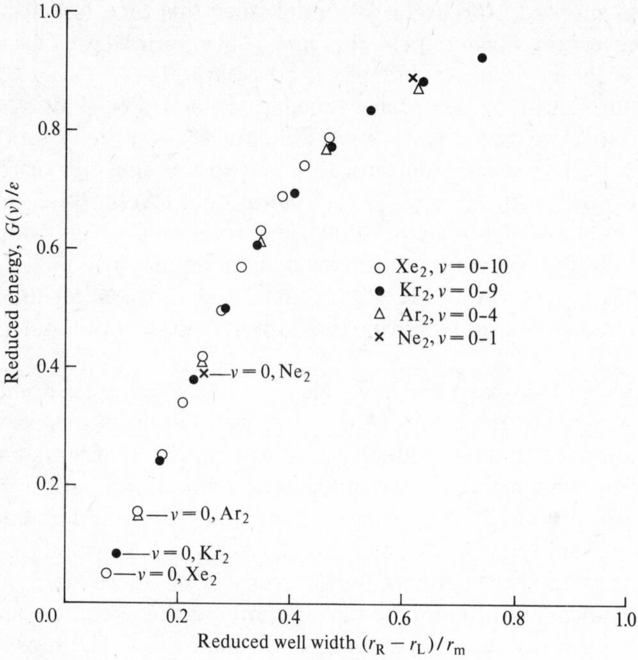

FIG. 7.7. The reduced well-width function for the inert gases Ne, Ar, Kr, and Xe, as determined by RKR analysis of spectroscopic data.[14–17]

observed six band groups in the vacuum u.v. belonging to He–Ne,[14] but all appeared to originate from just one vibrational level ($v = 0$) of the ground state. All the bands were diffuse, probably due to rotational predissociation in the $v = 0$ level, and $v > 1$ is unlikely to exist. Again for Ar–Kr,[16] they observed five groups of absorption bands, all too diffuse to assign to particular transitions. The low concentration of these mixed species obviously presents problems requiring greater instrument sensitivity and resolution before their potential energy functions may be elucidated spectroscopically.

Some light has been shed on the question of the optimum conditions for studying inert gas dimers in this way by Levine[62] He has calculated the fractional contributions to the transition moments for far-infrared and Raman spectra of bound and metastable dimers, low-energy and high-energy free particles (Fig. 7.8) using a model system comprising a 12–6 potential function and simple r-dependent dipole moment and polarizability functions for the dimer. His results for the dominant contribution to the infrared moment are illustrated in Fig. 7.8, where it can be seen that only for reduced temperatures $T^* < 0{\cdot}5$ does absorption due to dimers

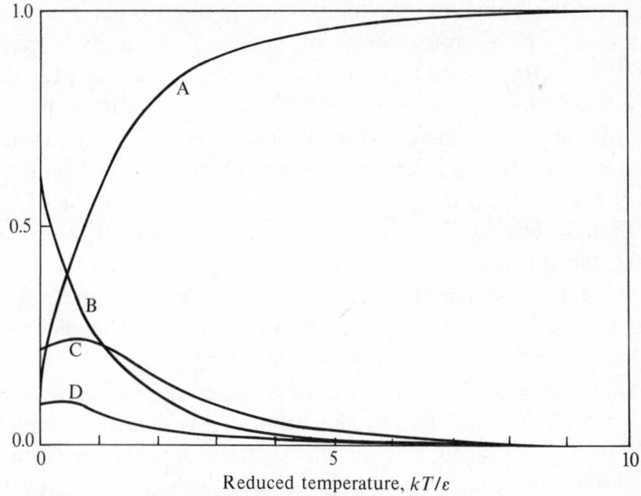

FIG. 7.8. Fractional contribution to the far-infrared transition moments for He–Ar simulation (based on ref. 62): (A) high-energy, free molecules; (B) bound dimers; (C) low-energy, free molecules; (D) metastable dimers. (ε/k = mean potential well depth.)

dominate the spectrum and by $T^* = 2$ effects are due almost entirely to free particles. Since Ar–Kr is prevented from being cooled significantly below $T^* \simeq 1$ because of condensation of the krypton the bound dimer contribution to the spectrum is not sufficiently intense to be observed. Although the application of pressure should improve the situation, it is not always easy to reconcile the optimum spectroscopic conditions with the phase diagram, especially for mixtures involving the heavier inert gases.

7.6. Analysis of spectra for non-spherical systems

7.6.1. General approach

When one or both of the interacting molecules is polyatomic, then the situation becomes far more complex. Firstly, the presence of internal degrees of freedom means that the observed rotation–vibration spectra are considerably more complicated than for the inert gas dimers. Secondly the intermolecular potential is anisotropic, depending on the relative orientation of the molecules as well as their separation. Thirdly, no formal inversion procedure exists at present by which this potential may be determined directly from the spectroscopic data.

Consequently, for these systems some assumptions must be made about the form of the intermolecular potential U and its parameters determined by fitting calculated spectra to the experimental data. Although this is obviously less satisfactory than the situation for the inert gases, it turns out that in favourable circumstances certain transitions are determined mainly by the isotropic part of the potential. This means that the evaluation of the central and anisotropic components of U is far less arbitrary than might at first be imagined, and leads to the hope that inversion of the data might eventually be possible.

The problem of calculating the discrete spectrum corresponding to the bound states of a particular anisotropic potential is essentially the same as that outlined in Chapter 4 for scattering processes in polyatomic systems. The Schrödinger wave equation is set up in the same way as described in § 4.7.6 and leads to the infinite set of close-coupled equations (4.238). The difference from the scattering problem comes in the particular boundary conditions employed in the numerical solution of the close coupled equations. The only truly bound states of van der Waals dimers occur at energies where no open channels exist, i.e. where the energies of the dimer bound states are less than the energy of the lowest bound state ($v=0$, $j=0$) of the isolated molecule(s). The appropriate boundary conditions in this case are that the wave function $F_\beta(\mathbf{r})$ obeys the relations

$$F_\beta(\mathbf{r}) \rightarrow 0, \qquad \begin{array}{l} r \rightarrow 0 \\ r \rightarrow \infty \end{array} \tag{7.26}$$

in contrast to the scattering boundary condition for $r \rightarrow \infty$, (4.240).

Dunker and Gordon[63] have developed techniques for carrying out close-coupling calculations of bound state energies. The infinite set of equations is truncated by considering only diagonal $v = v' = v''$ coupling. In the case of atom–diatomic molecule dimers, the molecule has usually been treated as a rigid rotor, with an upper limit j_{max} over which the intramolecular rotational quantum number is summed, chosen to include all open channels. Such calculations, though shown to be convergent and accurate, use extremely large amounts of computer time and are not particularly convenient for iterative parameter optimization in the fit of the potential to experimental data. Consequently, as in the case of scattering problems, approximate methods of solving the close coupling equations have been devised.

Before describing the essence of these approximations it should be pointed out that it is convenient to classify[64] dimers of the type X_2–Y according to the relative strengths of the parts of the potential which couple the radial and angular motions together. If we represent the total potential energy as the sum of an isotropic part, U_0, and an anisotropic part, U_{an}, and designate the rotational energy level spacings of the diatomic molecule X_2 and the dimer X_2–Y by $\Delta\varepsilon_{X_2}$ and $\Delta\varepsilon_{X_2Y}$ respectively, then we can identify the following types of dimer:

(1) *Weak coupling systems:* $\Delta\varepsilon_{X_2Y} \lesssim U_{an} \ll \Delta\varepsilon_{X_2}$. Here the internal rotation quantum number j remains almost a 'good' quantum number. Mixing with states associated with different orbital angular momentum quantum numbers, l, leads to small perturbations from the 'free rotor' situation. Examples of this type of system are H_2–inert gases.

(2) *Strong coupling systems:* $\Delta\varepsilon_{X_2Y} < \Delta\varepsilon_{X_2} \lesssim U_{an}$. Here, the j-states also become mixed and the most appropriate quantum number becomes that associated with the total angular momentum, $\mathbf{J}(=\mathbf{l}+\mathbf{j})$, or in some cases its projection on a body-fixed axis. Ar–HCl is the most studied system of this type.

(3) *Semi-rigid systems:* $\Delta\varepsilon_{X_2Y}$, $\Delta\varepsilon_{X_2} \ll U_{an}$. Here, the potential energy is extremely anisotropic and the dimers have well defined equilibrium structures with small-amplitude motions comparable with 'permanent', covalently bound molecules. Kr–ClF falls into this category.

The next section gives a brief account of the differences between the type of spectrum observed for these systems before we return to the approximate methods used to interpret them.

7.6.2. *Diatomic–Monatomic dimer systems*

In systems of the type X_2–Y, rotation can occur both internally (by X_2 about its centre of mass), characterized by the quantum number j, and by relative end-over-end motion of X_2 and Y, characterized by l. The total rotational motion is characterised by a quantum number $J = j+l, \ldots |j-l|$. These motions are coupled by the angular dependence of the intermolecular potential and the internal rotation is hindered by a potential energy barrier. The higher this barrier, the greater the degree of coupling between j and l and the more the J levels become shifted from the free rotation case. The way in which the energy levels of the system change as we pass from free rotation to the opposite extreme, a rigid molecule (either T-shaped or linear), is illustrated in Fig. 7.9.

The most extensively studied systems of this type are H_2–inert gases,[23–25] and these will be considered in some detail in § 7.6.3. A typical spectrum is shown in Fig. 7.10. Here the barrier to internal rotation is much smaller than the j-state spacing, coupling is weak and there is only a slight splitting of the J-levels. The H_2–Ar system is close to the centre of Fig. 7.9. This weak coupling, giving vibration–rotation frequencies close to freely rotating H_2, and the presence of few bound states due to the small well depth, make the detailed analysis of the spectrum and the

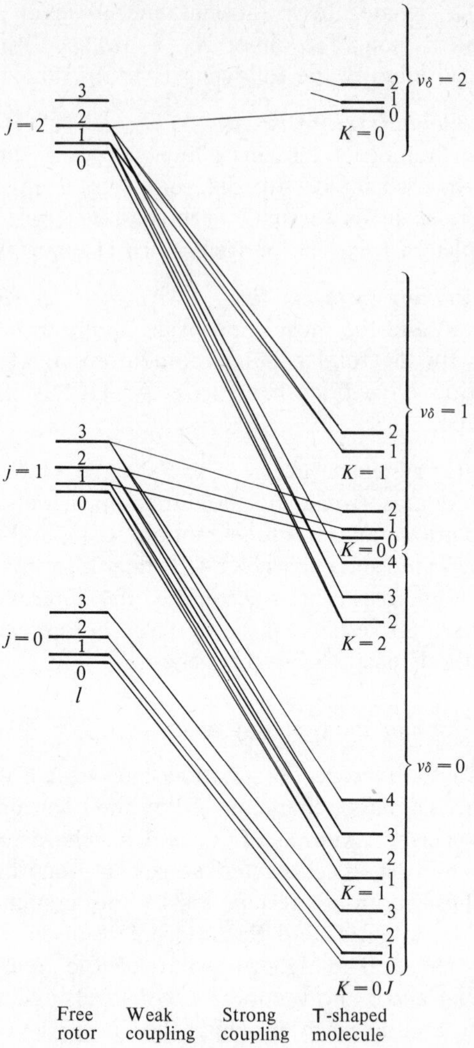

FIG. 7.9. Correlation diagram of rotatory motions of X_2–Y van der Waals molecules (Henderson, G. and Ewing, G. E.[22]). (Notice that the spacing of the l levels is much finer than that of the j levels. This is because l indicates rotation about the long X_2–Y van der Waals bond while j denotes rotation about the short X_2 chemical bond.) v_δ is the vibrational quantum number of the dimer.

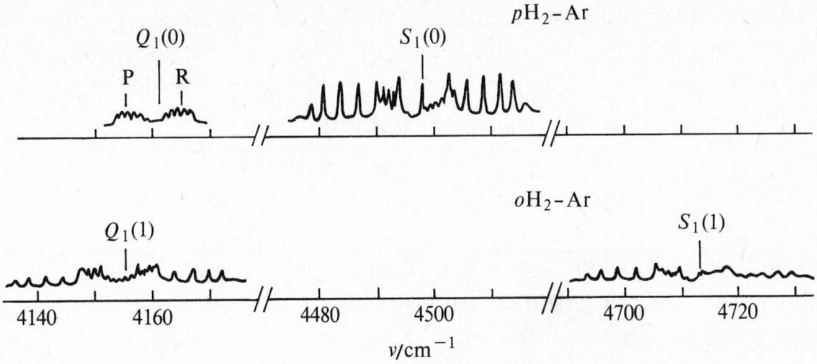

FIG. 7.10. Infrared spectrum of *ortho* and *para* hydrogen–argon mixtures at 0·5 bar, 86 K, pathlength 165 m.[24] Labels indicate free H_2 transitions.

extraction of the intermolecular potential relatively straightforward. The way in which information about the potential can be extracted from the spectra and the type of functional forms employed will be described in detail in § 7.6.3. The deeper the potential well and the stronger the coupling, the more complex the procedure of the type described in § 7.6.3 becomes.

An example of a system in which the barrier to rotation is large compared with the j-level spacing is N_2–Ar. The infrared spectrum of this dimer has been obtained at 87 K and 1 atm by Henderson and Ewing,[22] using a 122 m path length. It differs from those of H_2-inert gas systems in several ways (see Fig. 7.11): the rotation–vibration features are more closely spaced due to the larger moment of inertia of N_2; these features are shifted significantly from those of freely rotating N_2; and they are somewhat broader than in the H_2–Y case, no ΔJ hyperfine structure being resolved. The coupling between internal and dimer rotation is strong in this case, the energy level splitting being towards the right-hand side of Fig. 7.9.

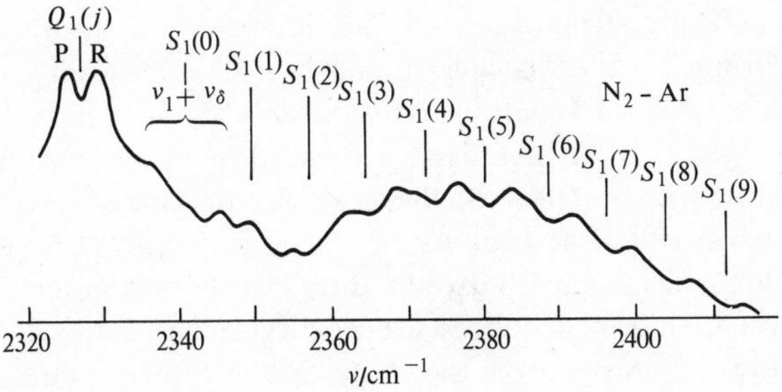

FIG. 7.11. Infrared spectrum of N_2–Ar mixture at 1 bar, 87 K, pathlength 122 m.[22] Labels indicate free N_2 transitions.

The relative lack of detail in the N_2–Ar spectrum compared with H_2–Ar means that the information about the pair potential which can be extracted is likewise less extensive. Henderson and Ewing[22] have assumed a relatively crude model for the potential:

$$U(r,\theta)=\varepsilon\left[\left(\frac{r_m}{r}\right)^{12}-2\left(\frac{r_m}{r}\right)^{6}\right]+\frac{U_R}{2}(\cos 2\theta) \qquad (7.27)$$

where θ is the angle between the centre of mass separation vector $\mathbf{r}$ and the diatomic molecule bond direction $\mathbf{R}$. This potential was used in a variational calculation of the N_2–Ar vibration–rotation energy levels for several values of the barrier height U_R neglecting both the end-over-end dimer rotation and coupling with the stretching of the N_2–Ar bond. The displacement of bands away from the band centre relative to the free N_2 bands indicates a positive U_R i.e. a T-shaped equilibrium geometry, and matching the calculated to the observed frequencies gave a value of $\bar{U}_R = 28{\cdot}8$ K (20 cm^{-1}), which must be regarded as an average value over different dimer vibrational levels. At 87 K, about 20 per cent of the dimers are locked in this semi-rigid T configuration undergoing hindered internal rotation.

The remaining parameters in (7.27) can also only be determined approximately: the geometric mean of the spectroscopic ε value for Ar_2 and a crude $12-6$ value for N_2 gave $\varepsilon_{N_2-Ar} \simeq 112$ K (78 cm^{-1}); then calculation of the P and R branch envelopes matched experiment for $r_m = 0{\cdot}39$ nm. The resulting shape of the potential is illustrated in Fig. 7.12. Although this must be regarded only as a crude and qualitative description of the true potential energy surface for N_2–Ar,[65] it clearly demonstrates how the well depth changes with orientation and the probable stabilization of the T-shaped configuration compared with the linear dimer.

The contrast between this crude analysis and that for H_2–Y (§ 7.6.4), which is still far from complete, is marked. The determination of a more realistic potential energy function would require the experimental resolution of the ΔJ-fine structure, and the inclusion of coupling between internal and dimer rotation and dimer bond vibration in its analysis, together with the use of functions considerably more complex and realistic than that used in (7.27). Ideally a direct inversion method would be developed. However, in view of the complex dependence of the bound state energies on different regions of the anisotropic potential and the variable extent of coupling between inter- and intramolecular motions which can occur, such procedures are not likely to be devised easily.

A number of dimers of the type XY–Z have been studied by molecular beam spectroscopy. For instance microwave and radiofrequency transitions for a number of isotopes of Ar–HCl indicate an

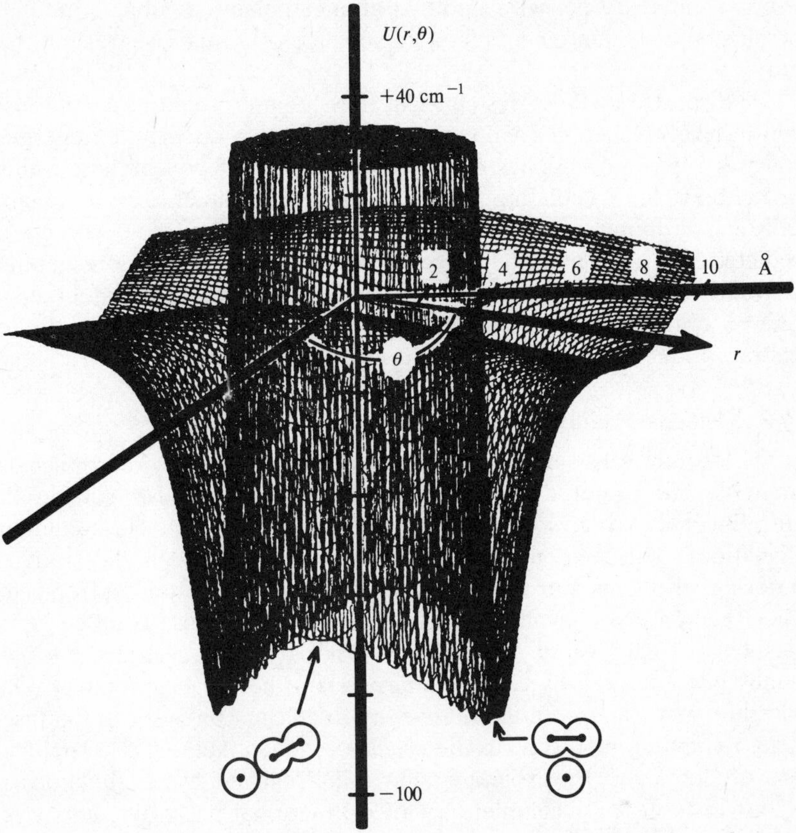

FIG. 7.12. A mapping of the N_2–Ar intermolecular potential function. A presentation of (7.27) with $\varepsilon = 0{\cdot}93$ kJ/mol (78 cm^{-1}), $r_m = 3{\cdot}9$ Å, and $U_R = 0{\cdot}24$ kJ/mol (20 cm^{-1}). Positions of the unstable linear configuration ($\theta = 0°$) and the equilibrium, stable, T-configuration ($\theta = 90°$) are shown. (G. E. Ewing.[6]).

equilibrium linear structure Ar–H–Cl with large amplitude bending excursions of the hydrogen atom from the linear structure.[28] Only J-transitions within the ground-state vibrational level of the dimer were observed (strong coupling region of Fig. 7.9), so little information about $U(r, \theta)$ can be easily deduced (but see § 7.6.5). However the vibrational frequency of the Ar–HCl bond was estimated to be 32·2 cm^{-1} compared with 25 cm^{-1} for the isoelectronic Ar_2, so presumably the potential well of the former is deeper (see § 7.6.5). By contrast, Ar–ClF is found to have the Cl atom located centrally,[65] again in an almost linear configuration with only small-amplitude bending. The existence of linear structures for these dimers is strong evidence that the dimer potentials cannot simply be represented by pairwise additive atom–atom interactions within the complex, since the latter predict a mean value of θ much nearer 90° than zero.

The spectra of these systems can thus be interpreted at a superficial level in terms of geometrical parameters and mean dissociation energies. However rapid progress has been made in the past few years in developing methods for calculating the bound states of dimers for reasonably realistic pair potentials. This enables the corresponding spectra to be predicted and by comparing these with experiment much more detailed information about the radial and angular dependence of the potential can now be obtained. We now describe these methods briefly and then illustrate how they may be applied to systems of the type X_2–Y.

7.6.3. *Methods of calculating bound-state energies*

The routine use of spectroscopic data to determine intermolecular potentials has required the development of methods for solving the Schrödinger equation which are significantly faster than close-coupling calculations. The sudden approximation methods used in the study of scattering problems introduce the simplifying assumption of replacing either the sum over j values and/or l values in the close-coupled equations by a single value, $\bar{j}$ and/or $\bar{l}$. The methods used to calculate bound-state energies, by contrast, must retain the full Hamiltonian. One procedure is to choose an alternative, more efficient basis set to that used in close-coupling methods. In the *secular equation method*,[3,66] simplifications are obtained by expanding the radial channel wave functions in (4.238) in terms of a complete set of orthonormal basis functions $\psi_i(\mathbf{r})$, converting the solution of the close-coupled equations to a matrix diagonalization problem. This method is only approximate in the same sense that practicable close-coupling calculations are approximate, in that results of arbitrary accuracy can be achieved by including enough basis functions. In fact, since it is variational in nature, it may be capable of even greater accuracy than traditional close-coupling calculations. The method will not be given in detail here; the interested reader is referred to the original papers.[8,66–8] Suffice it to say that the basis set is generated as the eigenfunctions of a basis-generating isotropic potential, $U_b(r)$. It is on the successful choice of this generating potential that the success of the method depends.

For weak-coupling systems, it has been found that a good choice is the expectation value of the spherically averaged part of the potential $U_0(r, R)$ i.e.

$$U_b(r) = \int_0^\infty \phi^*_{vj}(R)\, U_0(r, R)\, \phi_{vj}(R)\, \mathrm{d}R$$

for the diatomic molecule (X_2) state (v, j) correlating with the dimer level under consideration. (ϕ_{vj} is the vibration–rotation eigenvalue of the isolated molecule X_2 in the state (v, j)). This gives results in good agreement with close-coupling calculations using one or two orders of magnitude less computing time. For strong-coupling systems, e.g. Ar–HCl, it is found that all the bound states of an effective isotropic potential are not sufficient to form a complete basis set.[67] In this case, increasing the range of nodal behaviour on the interval to which the exact wave function is confined, by artificially introducing an infinite wall beyond some separation in the long-range tail region, has been found to be successful. Further substantial savings in computation time have been achieved by using a single effective value of l to generate the basis functions from $U_b(r)$.[8] This approximation is effective and accurate for both weak- and strong-coupling cases.

An alternative procedure, *the Born–Oppenheimer angular radial separation method*,[32,69] rests on the assumption that the vibrational/rotational motion of the diatomic molecule X_2 is much more rapid than that of the dimer bond. This separation of the various motions enables one to solve the wave equation for the intermolecular motion separately, leading to eigenvalues dependent solely on separation r. However, although this method is of a comparable speed to the secular equation method, it only gives upper and lower bounds to the eigenvalues and current versions of it are less accurate for cases where the anisotropic terms of the potential are quickly varying over the range of r that is important.[8]

Within the electronic Born–Oppenheimer approximation, the intermolecular potential does not change if isotopic substitution is made except for the effect of changes in the intramolecular bond lengths. Consequently, the spectral changes arising from mass effects provide important additional information with which to determine intermolecular potential energy functions in polyatomic systems. Since most calculations are carried out using coordinates based on the centre of mass of the dimer, the coordinate system changes on isotopic substitution. The use of isotopic data has been facilitated by the development of schemes,[70,71] for transforming expansion representations of potential functions from one centre-of-mass system to another. The specific type of potential functions used to interpret spectral data for polyatomic dimers, and the methods of analysis, will be exemplified by H_2–inert gas systems for the weakly anisotropic (coupled) case and by Ar–HCl in the case of strong coupling.

7.6.4. *Weakly anisotropic systems*

The analysis can be illustrated for the system X_2–Y as first applied to the spectra of H_2– and D_2–inert gas complexes by LeRoy and Van Kranendonk.[66] These particular polyatomic dimers have been widely

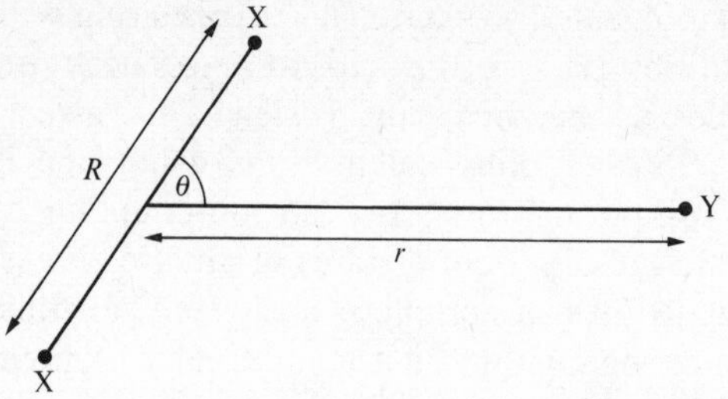

FIG. 7.13. Coordinate system for atom–diatom interaction.

studied experimentally[23–5] and have the advantage that infrared transitions which probe a significant portion of the potential energy well have been resolved. This is in contrast to some of the heavier systems considered later where existing spectroscopic data are sensitive only to the potential in the region of the minimum.

The coordinate system used is the one illustrated in Fig. 7.13. The following parameters may be defined:

(a) $\xi=(R-R_0)/R_0$ where R_0 is the expectation value of R in the ground rotation–vibration state of isolated X_2.
(b) j = internal rotational angular momentum ($\mathbf{j}$) quantum number for X_2.
(c) l = quantum number for angular momentum of relative motion of the two molecules ($\mathbf{l}$).
(d) J = total angular momentum quantum number ($\mathbf{J}=\mathbf{j}+\mathbf{l}$).
(e) v = vibrational quantum number of X_2.
(f) n = vibrational quantum number of X_2–Y, i.e. for the stretching of r.

The intermolecular potential is a function of r, ξ, and θ and may be expanded in terms of Legendre polynomials:

$$U(r,\xi,\theta)=\sum_{k=0}^{\infty} U_k(r,\xi)P_k(\cos\theta)$$
$$=U_0(r,\xi)+U_2(r,\xi)P_2(\cos\theta)+U_4(r,\xi)P_4(\cos\theta)+\ldots. \tag{7.28}$$

Eqn (7.28), involving only k-even terms, applies to dimers involving homonuclear species X_2.

The spectrum. The nomenclature used here is that adopted by Kudian and Welsh:[23] changes in j and l are denoted by the letters N, O, P, ... T, corresponding to -3, -2, $-1, \ldots, +3$, with the initial value given in brackets. Those lines arising from changes in j have the accompanying Δv as subscript. For all the examples considered here, $\Delta n = 0$ and consequently Δl changes are unsubscripted. Thus, the $T(3)$ component of the $S_1(0)$ band corresponds to

$$(v\, n\, j\, l\, J) = (0, 0, 0, 3, 3) \rightarrow (1, 0, 2, 6, 4).$$

The potential and its relation to the spectra. As far as the observed spectroscopic transitions are concerned, the generalized Born–Oppenheimer approximation enables us to consider the surface $U(r, \xi, \theta)$ as consisting of a whole range of potential energy curves for the complex, one for each pair of values of v and j. This is illustrated schematically in Fig. 7.14 for the $v = 0$, $j = 0$, and $v = 1$, $j = 0$ states of H_2–Ar. The asymptotes of these two curves are separated by the appropriate rotation–vibration spacing of the isolated molecule X_2, here H_2.

Because the wave function of a diatomic molecule in its rotational ground state ($j = 0$) is spherically symmetric, the energy levels of the complex formed by such a molecule with an inert gas atom will be mainly determined by the spherically symmetric part of U, $U_0(r, \xi)$. This is the case for both states illustrated in Fig. 7.14 and hence the transitions between them, $Q_1(0)$, give information about $U_0(r, \xi)$. For the case illustrated, only the $n = 0$ levels were observed and the rotational (l) levels have the particularly simple form shown; note that the highest levels of each curve are quasi-bound leading to broadened lines due to predissociation, which are hence a less accurate source of information about $U_0(r, \xi)$.

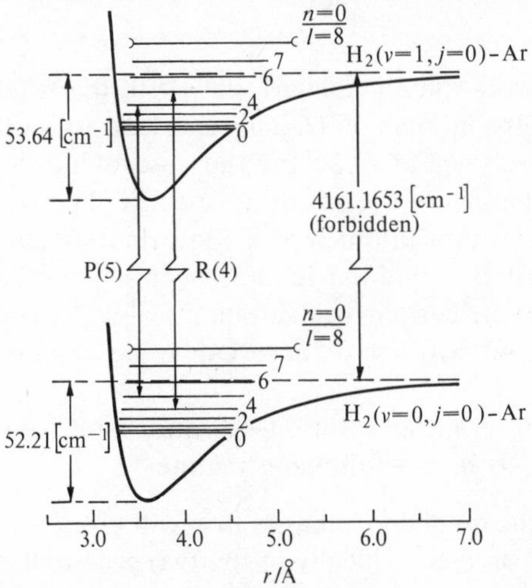

FIG. 7.14. Potential curves and energy levels for $v = 0$, $j = 0$ and $v = 1$, $j = 0$ states of H_2–Ar (schematic). (R. J. LeRoy and J. Van Kranendonk.[66])

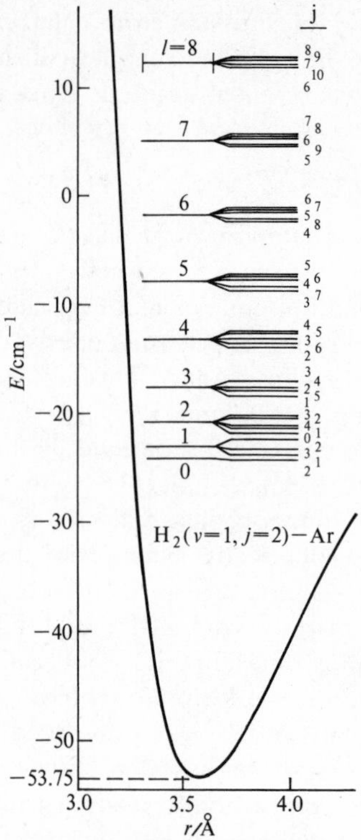

FIG. 7.15. Anistropy-induced splitting of the orbital (l) levels of a complex formed from H_2 ($v = 1, j = 2$) and Ar. (R. J. LeRoy and J. Van Kranendonk.[66])

When X_2 is rotationally excited, i.e. $j > 0$, the rotational l levels are split by the anisotropic part of U. This can lead to quite complex energy level distributions (see Fig. 7.15 for the case of $v = 1$, $j = 2$ for H_2–Ar) and large multiplicities in the resulting spectra. The consequent overlapping of lines often results in much of the information about the anisotropic part of the potential contained in the spectra being lost due to lack of resolution. However, certain transitions do only have a single J component e.g. $Q_1(1)$ and $S_1(0)$ and these can more readily be resolved and assigned.

These features of the rotation–vibration spectrum can be used to determine $U(r, \xi, \theta)$ in the following manner:

(a) Transitions involving changes in v will give information about the dependence of U on ξ. Specifically, if the two potential curves in Fig. 7.14 were identical, the average of the frequencies of the P(5) and R(4) components would be equal to the known separation of the dissociation limits of these two curves. The fact that it is not ($1{\cdot}15\ cm^{-1}$ smaller in this specific case) indicates that U depends on ξ and these differences can be used to quantify this dependence.

(b) A major source of information about the spherical part of the potential $U_0(r, \xi)$ is, as indicated above, the $Q_1(0)$ band. The relative frequencies of its components give the spacings between the complex's l rotational levels. LeRoy and Van Kranendonk[66] used these to evaluate the constants B_0, D_0 for a non-rigid rotor, which gave a satisfactory description of the H_2–X data. This type of analysis effectively defines two parameters in an assumed form for $U_0(r, \xi)$. B_0 is closely related to the length parameter σ or r_m and can usually be determined more precisely than D_0, which is related to the shape of the well, and hence ε/k. (The use of isotopes, e.g. H_2 and D_2, sometimes enables some fine tuning of the well anharmonicity to be achieved, enabling a three-parameter function to be used.) A further constraint is put on the binding energy by the onset of broadening at a particular frequency due to predissociation from the quasi-bound levels.

(c) To the same approximation as (b) applies, the energy levels for any non-zero j are only affected by terms in (7.28) up to $k \leq 2j$. In other words, the anisotropic contributions to the potential can in principle be determined by considering separate bands which are affected by successively larger numbers of the Legendre terms. This is illustrated in Fig. 7.16.

In particular, the splitting of the l levels by the potential anisotropy causes shifts in the spectral lines whose relative magnitudes are determined only by the values of j, l, and J. However, the scale of these shifts is determined by $U_2(r, \xi)$ and consideration of how this changes with l

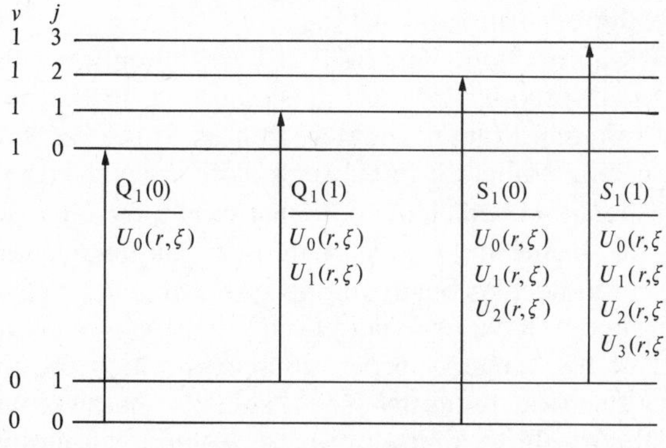

FIG. 7.16. The components for the potential $U(r, \xi, \theta)$ which largely determine particular transitions for the complex X_2–Y ($n = 0$).

gives information about the dependence of U_2 on r at both short and long ranges.

Typical model potentials. The procedure adopted by LeRoy and Van Kranendonk for the $H_2(D_2)$—inert gas systems was to assume the following simple form for the potential energy function:

$$U(r, \xi, \theta) = C_n r^{-n}(1+S_1\xi) - C_6 r^{-6}(1+S_2\xi) + [a_n C_n r^{-n}(1+S_3\xi) - a_6 C_6 r^{-6}(1+S_4\xi)]P_2(\cos\theta) \quad (7.29)$$

This is equivalent to representing the r dependence of both the isotropic part U_0 and the major anisotropic contribution U_2 by an $(n-6)$ function, for which

$$C_n = \frac{6\varepsilon r_m^n}{n-6} \quad \text{and} \quad C_6 = \frac{n\varepsilon r_m^6}{n-6}.$$

The relative strengths of the anisotropic contributions are assigned by the parameters a_n and a_6 and the ξ dependence is introduced for each r term independently using a polynomial having coefficients S_i. The truncation of these series at linear terms in ξ, and the exclusion of terms higher than $P_2(\cos\theta)$, were determined in this particular case by the quality and extent of the available spectroscopic data.

The ξ-dependence of dispersion forces is due to the variation of the X_2 polarizability and its anisotropy with ξ, and calculations show that for the lower levels of H_2 this variation is almost linear. This to some extent justifies the truncation of the series of the attractive terms, irrespective of the quality of the experimental data. The procedure followed was to evaluate the energy levels for this potential by solution of the Schrödinger wave equation using a secular equation method, to predict the resulting spectrum and to determine the parameters of (7.29) by a non-linear least squares fit to the experimental data.

The potential function given in (7.29) has a number of disadvantages. Because the spectroscopic data are most sensitive to the potential in the region of the minimum (approximately limited to the range of r given by the classical turning points for the observed dimer levels), the values of the long-range dispersion coefficients obtained by a free fit to the experimental data lie far from the correct values,[72,73] known from theoretical calculations and molecular beam studies (see Table 7.1 below). Using the known long-range behaviour as additional input can lead to a significant improvement of the range of separations over which the fitted function should reflect the true potential (see Table 7.2 below). Also, although perturbation theory shows that a power series in ξ is a natural expansion for the long-range energy, it is not appropriate at smaller separations. Consequently this function is not sufficiently flexible and large correla-

Table 7.1

A comparison of the potential parameters of the isotropic part of a generalized n − 6 potential (7.29) for H_2–inert gas interactions[66]

System	Range†	n	(ε/k)/K	r_m/nm	C_6/C_6^{theory} [73]	Method
H_2–Ne	$0{\cdot}3 \leq r \leq 0{\cdot}51$	15	37·1	0·329	1·39	Spectroscopic
	Well	12	33·7	0·33	1·53	Integral cross-sections
	Repulsion	15	45·8	0·315	1·31	Diff. cross-sections
H_2–Ar	$0{\cdot}32 \leq r \leq 0{\cdot}52$	12	75·1	0·3557	1·56	Spectroscopic
	Well	12	74·0	0·333	1·03	Integral cross-sections
	Well	12	69·1	0·317	0·72	Integral cross-sections
	Repulsion	12	91·2	0·344	1·55	Diff. cross-sections
H_2–Kr	$0{\cdot}34 \leq r \leq 0{\cdot}51$	12	87·3	0·3701	1·62	Spectroscopic
	Well	12	79·0	0·365	1·34	Integral cross-sections
	Well	12	71·9	0·335	0·73	Integral cross-sections
	Repulsion	12	109·2	0·361	1·74	Diff. cross-sections
H_2–Xe	$0{\cdot}36 \leq r \leq 0{\cdot}54$	13	96·4	0·3921	1·63	Spectroscopic
	Well	12	86·2	0·39	1·52	Integral cross-sections
	Well	12	77·7	0·353	0·75	Integral cross-sections

† Region of potential function to which measurements are sensitive.

tions occur between the attractive and repulsive coefficients a_i, S_i when it is fitted to experimental data. One alternative representation that has been used is to retain the power series expansion:

$$U(r, \xi, \theta) = \sum_{n=0}^{\infty} \sum_{k=0}^{\infty} \xi^k P_n(\cos\theta) U_{nk}(r) \tag{7.30}$$

but to treat each radial function $U_{nk}(r)$ as an independent function having its own characteristic shape parameters r_m^{nk} and ε^{nk}. Further improvement may be brought about by using a more realistic form than the $n-6$ function for the radial potential $U_{nk}(r)$. One function used by LeRoy and co-workers[67] has the form of the Buckingham–Corner potential (see Appendix 1):

$$U_{nk}(r) = A_{nk} \exp(-\beta_{nk} r) - [C_6^{nk}/r^6 + C_8^{nk}/r^8] D(r) \tag{7.31}$$

where the damping function $D(r)$ is

$$\begin{aligned} D(r) &= \exp(-a[br_m^{00}/r - 1]^c) \qquad && r \leq br_m^{00} \\ &= 1 && r > br_m^{00} \end{aligned}$$

Table 7.2

(a) Parameters for H_2–Ar potential energy functions of the form (7.30) determined from spectroscopic data[67]

Parameter	Potential function (LJ)$_1$	(HFD)$_1$	(BC)$_1$	(BC)$_3$
ε^{00}/K	75·07(0·11)	73·04	73·21(0·71)	73·25(0·72)
r_m^{00}/nm	0·355 63(0·000 46)	0·357 88	0·357 37(0·000 53)	0·357 27(0·000 52)
β^{00}/nm^{-1}	—	36·91	36·91(1·03)	36·10(0·99)
C_6^{00}/K nm^6	—	0·196 4	0·196 4	0·193 7
ε^{01}/K	43·8(1·7)	44·9	44·35(2·7)	49·2(1·2)
r_m^{01}/nm	0·383 3(0·00 26)	0·385 5	0·385 2(0·002 3)	0·376 9(0·002 6)
C_6^{01}/K nm^6	—	0·255 5	0·255 5	0·166 8
ε^{02}/K	—	—	—	2·742
r_m^{02}/nm	—	—	—	0·435 75
C_6^{02}/K nm^6	—	—	—	0·004 39
C_6^{03}/K nm^6	—	—	—	−0·036 39
ε^{20}/K	9·89(1·02)	10·5	10·57(1·30)	8·24(0·66)
r_m^{20}/nm	0·382 1(0·005 9)	0·380 5	0·381 3(0·004 5)	0·374 3(0·004 9)
C_6^{20}/K nm^6	—	0·021 01	0·021 01	0·019 4
ε^{21}/K	5·77	6·47	6·41	31·1(14·0)
r_m^{21}/nm	0·4118	0·4099	0·4109	0·394 87
C_6^{21}/K nm^6	—	0·085 31	0·073 0	0·042 6
ε^{22}/K	—	—	—	20·036
r_m^{22}/nm	—	—	—	0·402 67
C_6^{22}/K nm^6	—	—	—	0·008 215
C_6^{23}/K nm^6	—	—	—	−0·014 969
Region of validity (nm)	$0·32 \leqslant r \leqslant 0·52$ $0·06 \leqslant R \leqslant 0·10$	$0·32 \leqslant r$ $0·06 \leqslant R \leqslant 0·10$	$0·32 \leqslant r$ $0·06 \leqslant R \leqslant 0·10$	$0·32 \leqslant r$ $0 \leqslant R < 0·19$

Potential code: LJ = 12−6 radial function in (7.30); BC = Buckingham–Corner radial function in (7.30), $a = 4$, $b = 1$, $c = 3$; HFD = Buckingham–Corner radial function in (7.30), $a = 1$, $b = 1·25$, $c = 1·9$. $n = 0, 2$ in all cases; subscript on potential code is the maximum value of k in (7.30). $\beta^{nk} = \beta^{00}$ in all cases. Figures in parentheses indicate 95 per cent confident limit uncertainties. Recommended function is (BC)$_3$.

and

$$A_{nk} = \frac{\exp(\beta_{nk} r_m^{nk})(8\varepsilon^{nk} - 2C_6^{nk}/(r_m^{nk})^6)}{(\beta_{nk} r_m^{nk} - 8)}$$

$$C_8^{nk} = \frac{[\beta_{nk} r_m^{nk} \varepsilon^{nk} + (6 - \beta_{nk} r_m^{nk}) C_6^{nk}/(r_m^{nk})^6](r_m^{nk})^8}{(\beta_{nk} r_m^{nk} - 8)}$$

r_m^{nk} and ε^{nk} are the position and depth of the minimum in $U_{nk}(r)$ if $D(r) = 1$ for all r.

Table 7.2 (*Continued*)

(*b*) *Parameters for H_2–Kr, Xe potential energy functions of the form* (7.30) *determined from spectroscopic data*[68]

Parameter	H_2–Kr, $(BC)_1$	H_2–Xe, $(BC)_1$
ε^{00}/K	84·63(0·52)	94·29(0·65)
r_m^{00}/nm	0·371 92(0·000 38)	0·393 44(0·000 44)
β^{00}/nm^{-1}	34·62(0·70)	36·68(0·96)
$C_6^{00}/K\ nm^6$	0·279 6	0·440 6
ε^{01}/K	53·21(1·38)	76·35(0·99)
r_m^{01}/nm	0·391 5(0·002 4)	0·404 3(0·002 5)
$C_6^{01}/K\ nm^6$	0·363 7	0·619 2
ε^{20}/K	12·74(0·62)	14·41(0·33)
r_m^{20}/nm	0·386 0(0·005 2)	0·398 0(0·007 8)
$C_6^{20}/K\ nm^6$	0·030 2	0·051 4
ε^{21}/K	27·79(12·10)	27·5(7·5)
r_m^{21}/nm	0·406 3	0·409 0
$C_6^{21}/K\ nm^6$	0·105 0	0·178 6

For code and symbols, see footnote to Table 7.2(a).

This formulation has the additional advantage that it can incorporate theoretical values[72] of the dispersion constants C_6^{nk}. An isotropic potential of this form using damping parameters $a = 1$, $b = 1{\cdot}25$, $c = 1{\cdot}9$ has also been used by Hepburn, Scoles, and Penco[74] to successfully describe inert gas interactions (the HFD function). Here then is an example of work on spherical systems guiding the choice of potential form in non-spherical molecules.

The potentials obtained for H_2–Ar by LeRoy *et al.*[8,66,67] by fitting functions of the general form of (7.29) and (7.30), (7.31) to the available spectroscopic data are summarized in Tables 7.1 and 7.2 and in Fig. 7.17. Three forms were used for $U_{nk}(r)$: the 12−6 function, the Buckingham–Corner function (BC) with $a = 4$, $b = 1$, and $c = 3$ and the HFD function. The potentials obtained using the latter two functions are essentially identical when only terms having $k = 0, 1$ are included. Since the functional forms of the two potentials, in particular their damping functions $D(r)$, are significantly different this agreement is encouraging. These potentials are undoubtedly a significant improvement on the earlier attempts listed in Table 7.1. However, it is dangerous to assume on the basis of such a limited study that the potentials and parameters obtained for H_2–Ar in this way are now completely model-independent. Experience with the inert gases has shown that the successful determination of potentials by multi-parameter fitting procedures requires both very flexible functional forms for the potential and fitting to many properties,

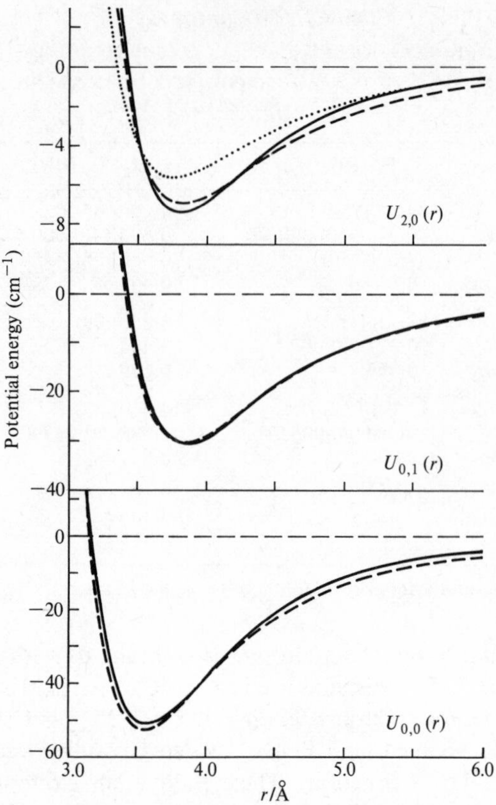

FIG. 7.17. Illustration of H_2–Ar potential energy functions listed in Table 7.2 for various values of n and k. (———) $(BC)_1$,[67] ≃ $(HFD)_1$;[67] (- - - -) $(LJ)_1$;[67] (· · · · · ·) LJ (beams).[78] (Based on R. J. LeRoy, J. S. Carley, and J. E. Grabenstetter.[67]).

sensitive to different parts of the potential, simultaneously. Although equations of the type (7.30) and (7.31) are undoubtedly more flexible than earlier attempts, it is by no means certain that they avoid undue bias of one part of the potential by another. Indeed there is some evidence that use of a Legendre polynomial expansion imposes unrealistic restrictions on the potential at intermediate and small separations.[33,67] One way of avoiding these restrictions is to build up a potential energy surface by using an independent radial function to describe each of a large number of closely spaced relative orientations (θ), and to interpolate between them numerically for intermediate orientations. Such an approach has been used to interpret molecular beam resonance spectra of atom–diatom dimers, using the simple $n(\bar{r})-6$ potential as the radial function at each θ.[75]

Apart from the theoretical dispersion coefficients, no non-spectroscopic data were used in the evaluation of the BC and HFD H_2–Ar potentials. However there is some indirect evidence that they may not be too sensitive to the particular functional form chosen. The potentials have been compared with differential and integral scattering cross-section measurements and shown to be a better over-all fit to scattering and spectroscopic data together then previously proposed functions.[67] Furthermore, the isotropic potential energy functions determined from both differential[76] and integral[77] scattering cross-sections are in good agreement with the vibrationally-averaged spherical parts of these functions. Introduction of an additional constraint, that in the limit of $R \rightarrow 0$ the potential should reduce to that corresponding to He–Ar, has produced additional refinement to the potential at small values of ξ.[8] The U_{20} term determined in this way is in good agreement with state-selected integral scattering cross-section measurements,[78] in contrast to the earlier potentials illustrated in Fig. 7.17. This version of the H_2–Ar potential, $(BC)_3$, which contains terms up to $k = 3$ and incorporates the known limiting behaviour at small and large separations as well as the information contained in the spectroscopic data about the shape of the bottom of the potential energy well, is also summarized in Table 7.2 and illustrated in Fig. 7.18. It is probably the most realistic of the H_2–Ar potential energy functions so far proposed. The H_2–Kr and H_2–Xe functions[68] given in Table 7.2(b) represent a similar improvement over the $n-6$ functions of Table 7.1.

Several general points emerge from this study. First, the spectroscopic data obviously only give information about the potential over a limited range of r: in principle covering the whole of the well region, but often in practice confined to the lower region near the minimum. Consequently any refined study of these interactions requires supplementation from theoretical calculations (especially at long range) and molecular-beam scattering data. Second, just as in the case of the inert-gas interactions, any analysis of this type is model-dependent. This situation can only be improved in the short term by the use of a wide range of properties and, in the long term, by the development of more direct forms of analysis than the multi-parameter fitting procedure used in these initial studies. The deficiencies in the potentials given in Table (7.1) are obvious from a comparison with theoretical C_6 calculations and molecular beam experiments. Those given in Table (7.2) are a significant improvement. However, the principles outlined in this section illustrate how detailed information about the complex potential energy surface is in principle contained within the spectroscopic data. They should form the basis of more comprehensive analyses and be capable of extension to some of the more complex systems discussed in the next section when data of greater resolution become available.

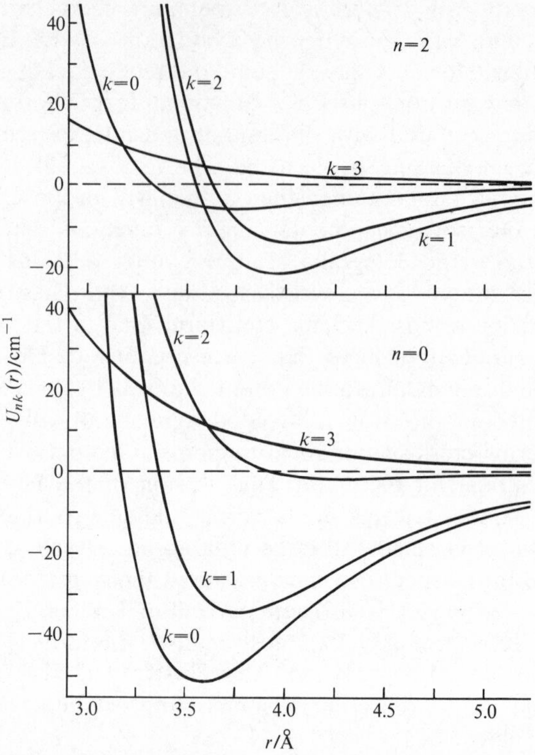

FIG. 7.18. Comparison of shapes of U_{nk} for H_2–Ar potential energy function $(BC)_3$ of form (7.39) recommended by LeRoy.[8]

This section has considered the information about the intermolecular potential contained in transitions between bound states. It is worth pointing out that the shapes of predissociated bands, arising from transitions between metastable levels which can tunnel through the centrifugal barrier, are particularly sensitive to the form of $U(r, \theta)$. They could thus be of considerable value in the future, especially for dimers having a shallow well containing only a few bound states.

7.6.5. Strongly anisotropic systems

The system of this type which has received most attention is Ar–HCl. In a series of molecular beam electric resonance studies,[25,29] Klemperer and co-workers have resolved and assigned 14 microwave and radiofrequency transitions for a number of isotopic combinations of this dimer. In the initial studies interpretation of the data was solely in structural terms, a large amplitude-bending linear structure Ar–H–Cl being indicated as the most stable configuration (see § 7.6.2). The gradual acquisition of more information about the bound states, together with nuclear quadrupole coupling constants and dipole moments for the dimer (giving $\langle\cos^2\theta\rangle$ and $\langle\cos\theta\rangle$ for the observed states, respectively) has produced an adequate body of data against which to compare the predictions of potential energy functions.

The most successful analysis to date has been carried out using the Born–Oppenheimer angular radial separation method.[31,32] The spectroscopically derived data used were the rotational constants, B_0, and the centrifugal distortion coefficients, D_0, for the ground vibration–rotation state of each of four isotopes of Ar–HCl ($H^{35,37}Cl$, $D^{35,37}Cl$), and the angles θ_1, θ_2 given by

$$\theta_1=\cos^{-1}(\langle\cos\theta\rangle)$$
$$\theta_2=\cos^{-1}(\langle\cos^2\theta\rangle^{\frac{1}{2}})$$

where θ is the angle between $\mathbf{r}$ and $\mathbf{R}$. The potential function used had the form

$$U(r,\theta)=\sum_{n=0}^{N} U_n(r)P_n(\cos\theta)$$

with $N=2$ being adequate to describe the results. The radial functions $U_n(r)$ all took the form of an exponential-inverse power function:

$$U_n(r)=\frac{\varepsilon_n}{\alpha_n-s}(s\mathrm{e}^{\alpha_n(1-r/r_\mathrm{m})}-\alpha_n(r_\mathrm{m}/r)^s).$$

The value of either ε_0 or α_0 was fixed on the basis of previous proposed potentials, molecular-beam scattering experiments, and infrared absorption studies and two six-parameter fits were carried out, resulting in the potential energy functions summarized in Table 7.3 and Fig. 7.19. It can be seen that despite being significantly different in terms of absolute energy, the relative shapes of the two functions in the region of the potential minimum are remarkably similar. This is not surprising in view of the extremely low energies ($<1\ \mathrm{cm}^{-1}$) of the observed levels compared to the well depth (133–$420\ \mathrm{cm}^{-1}$ here). This at once illustrates both the strength of the microwave technique for determining the shape of intermolecular potentials in the region of the minimum, and the need to supplement such data with a wide range of other measurements to improve the knowledge of the upper regions of the well and determine the absolute scale.

In this particular case, the isotropic part of surface II was found to reproduce the elastic differential cross-sections of Ar–HCl much better than surface I, which suggests that II is a more realistic function for this system. However, further refinement of the potential requires the use of

Table 7.3

Potential parameters for Ar–HCl functions I and II determined from microwave and radiofrequency spectra[32]

Parameter		Function I	Function II
$n=0$	$(\varepsilon_0/k)/K$	239·0	191·5
	r_{m0}/nm	0·385	0·381
	α_0	13·5	20·0
	s	6	6
$n=1$	$(\varepsilon_1/k)/K$	42·3	59·3
	r_{m1}/nm	0·426	0·4259
	α_1	10·1	10·36
	s	7	7
$n=2$	$(\varepsilon_2/k)/K$	49·8	42·9
	r_{m2}/nm	0·426	0·4259
	α_2	10·1	10·36
	s	6	6

spectroscopic data more sensitive to the angular variations of the potential energy, such as excitation of the bending mode of the dimer.[79] Relaxation phenomena and linewidth studies of the type described in §§ 7.3.3 and 7.7 should also prove useful probes of the anisotropy of this type of system.

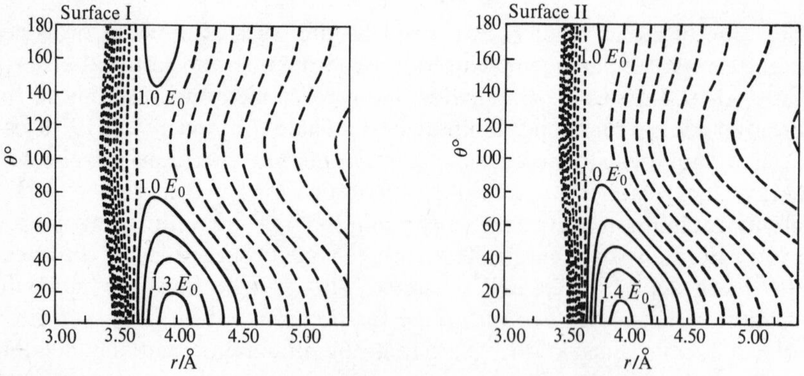

FIG. 7.19. Two best-fit contour diagrams for Ar–HCl potential functions determined from spectroscopic data. (Surface I): Solid contours outline the classically allowed region. Contour intervals are one-tenth the ground-state energy of $-160{\cdot}6\ cm^{-1}$. $\theta = 0°$ corresponds to the Ar–H–Cl configuration. $(\varepsilon/k)_0 = 239$ K. (Surface II): Solid contours outline the classically allowed region. Contour intervals are one-tenth the ground-state energy of $-128{\cdot}13\ cm^{-1}$. $(\varepsilon/k)_0 = 191{\cdot}5$ K. (S. L. Holmgren, M. Waldman, and W. Klemperer.[32])

7.6.6. *Dimers of diatomic and polyatomic molecules*

Many dimers of polyatomic molecules have now been studied either in the infrared or by molecular-beam electric resonance spectroscopy. Apart from the limited number of cases already described, however, none of the spectra obtained have been sufficiently detailed to determine the intermolecular potential energy function, even crudely. Conclusions have therefore been confined to information about minimum energy configurations, well depths and dissociation energies, and 'mean' bond lengths. These data for some of the dimers studied are summarized in Table 7.4. It is reasonable to expect that higher-resolution spectra and their more detailed analysis will in time yield a detailed mapping of the intermolecular potentials.

7.7. Sensitivity of alternative spectroscopic and scattering processes to the intermolecular potential

It was mentioned in § 7.3.3 that there were a number of spectroscopic methods which do not give discrete, high-resolution band structures which nevertheless can give useful information about the intermolecular potential. In planning experimental studies it is useful to know which particular aspects of a potential function are most sensitively tested by particular properties. An extremely useful guide to the relation between low-resolution spectroscopic measurements and anisotropic intermolecular potential functions is the work of Neilsen and Gordon.[40] In a study of molecular collisions for an atom–diatom system, they assumed that the translational motion could be treated classically and the rotational motion by quantum mechanics (semi-classical approximation) and that the relative translational motion was independent of the rotational motion (fixed classical path approximation). They solved the Schrödinger equation for the S-matrix for a single collision, averaged over all initial conditions, and then averaged over successive collisions to calculate various correlation functions to which experimental quantities can be related. For instance, the Fourier transform of the molecular dipole auto-correlation function, $\langle \boldsymbol{\mu}(0) \,.\, \boldsymbol{\mu}(t) \rangle$, is related to the rotational dipole absorption line shapes in the far-infrared. By identifying how different types of collision, which were sensitive to different parts of a potential energy surface, influenced the correlation function for different experiments, the way in which each experimental quantity depends on the potential was clarified.

Neilsen and Gordon[40] suggest how the specific information about the potential contained in a particular measurement may be extracted and have performed extensive calculations for Ar–HCl which emphasize the property–potential relations. The effect of an average collision at a given energy and impact parameter on the time evolution of a particular

Table 7.4

Structure of dimers of diatomic and polyatomic molecules studied by spectroscopic methods

Dimer	Approximate dissociation energy (kJ mol^{-1})	Minimum energy configuration	Equilibrium separation (nm)	References	Comments
$(H_2)_2$	$D_e \simeq 0{\cdot}3$; $D_0 \simeq 0{\cdot}03$	Nearly free rotor	$r_m \simeq 0{\cdot}30$ $\langle r \rangle_0 \simeq 0{\cdot}44$	19, 80	17 K, 110 m path length, One bound level, $n = 0$.
$(O_2)_2$	0·9	Rectangular	$0{\cdot}35 \pm 0{\cdot}02$	7, 20	Similar to Ar_2, Ar–N_2
$(N_2)_2$		Floppy T-shape	0·37	21	Barrier to internal rotation $\simeq 15$–$30\ cm^{-1}$
N_2–Ar	0·93	Semi-rigid T-shape	$0{\cdot}39 \pm 0{\cdot}02$	22	Barrier to internal rotation $\simeq 20\ cm^{-1}$
$(NO)_2$	6·7	Rigid, rectangular *cis* $N\hat{N}O \simeq 90°$	$0{\cdot}22 \pm 0{\cdot}02$	81, 82	Strong bond due to sharing of electrons
$(CO_2)_2$		T-shaped (?)		83	Large quadrupole moment of CO_2.
$(HF)_2$	20	Semi-rigid, non-linear 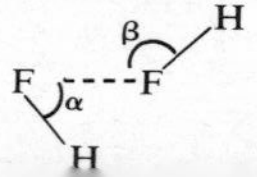	$\langle r \rangle_{F-F} = 0{\cdot}2783$	26, 84, 85	Barrier for tunnelling between equivalent dimer configurations $\simeq 6$ kJ mol^{-1}; $\mu = 2{\cdot}987 \pm 0{\cdot}003$ D.

$(HCl)_2$	8·9			86	
$(H_2O)_2$	20	Very prolate, almost symmetric, rotor; *trans* linear. 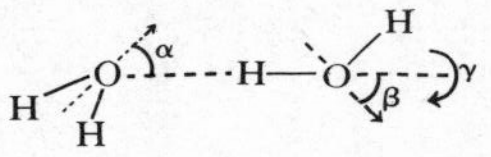 $\alpha = 58 \pm 6°$; $\beta = -51 \pm 6°$; $\gamma < 30°$	$\langle r \rangle_{O-O} = 0{\cdot}2976 \pm 0{\cdot}0010$	87	Linear hydrogen bond to within $1 \pm 6°$
ArClF		Almost linear Ar–Cl–F $Ar\hat{C}lF = 168{\cdot}9°$	$\langle r \rangle_{Ar-Cl} = 0{\cdot}3330$ $r_m = 0{\cdot}3286$	65	$\mu = 1{\cdot}053$ D
KrClF		Almost linear Kr–Cl–F $Kr\hat{C}lF = 169{\cdot}93°$	$\langle r \rangle_{Kr-Cl} = 0{\cdot}3388$	88, 89	Semi-rigid complex
HFClF		Slightly asymmetric prolate top; F–Cl–F co-linear; $H\hat{F}Cl = 55°$.	$\langle r \rangle_{F-Cl} = 0{\cdot}276$	85	Similarities with $(HF)_2$
ArOCS		Slightly asymmetric prolate top; T-shaped; $Ar\hat{C}S = 98{\cdot}11°$	$\langle r \rangle_{Ar-C} = 0{\cdot}3578$ $\langle r \rangle_{Ar-O} = 0{\cdot}3601$ $\langle r \rangle_{Ar-S} = 0{\cdot}4101$	90	
ArHF		Almost linear; large amplitude H motion	$\langle r \rangle_{Ar-F} = 0{\cdot}354$	89, 91	Similar to ArHCl

dynamic variable is given by the so-called sigma matrix, which is derived from the S-matrices for the contributing individual collisions. The study shows that sigma matrix elements for low j-states depend primarily on the attractive part of the anisotropic potential and those for high j-states depend mainly on the repulsive branch. It is this separate dependence which underlies the disparate sampling of the potential function by different kinds of experiment. For dipole absorption spectra for instance, the widths of lines involving low j values are sensitive to both the attractive and repulsive branches, whereas high j linewidths depend mainly on the repulsive branch alone. The line shifts for low j values depend primarily on the attractive potential, particularly the part having P_2 symmetry. Since high and low energy collisions depend on different regions of the potential, measurements of dipole linewidths at different temperatures would provide additional independent information. The line shifts, particularly the first, show a much different dependence on the variations in the potential parameters and so also provide independent information. The study shows that no additional information about the potential can be obtained by studying line shapes to higher densities.

Proton magnetic resonance relaxation times (spin–rotation) weight high j states more strongly than a simple population weighting. They are sensitive to a mixture of low and high j elements of the sigma matrix, and hence to the anisotropic potential over a fairly wide range of separations. Measurements at several temperatures in this case give little independent information about the potential. For the Raman spectrum, it appears that no additional information about the anisotropy of the potential is available compared with that contained in the dipole absorption spectrum. In contrast to the proton magnetic relaxation time, quadrupole relaxation times (of, for example, the chlorine nucleus) depend mainly on the reorientation of the low j states, and hence on the attractive anisotropy of the potential. Because these low j states have cross-sections very sensitive to energy, measurements of quadrupole relaxation time at different temperatures should provide independent information. Sound absorption measurements complement all the above experiments since rotational relaxation depends only on inelastic collisions which are much more closely related to the repulsive part of the potential. Experiments over a range of temperatures should provide important information about the anisotropy of the short-range part of the potential function.

It can be seen that by combining the results of different absorption experiments, information about the anisotropy of intermolecular potentials over a wide range of separation can in principle be obtained. The particular procedures for calculating the various properties discussed above are given by Neilsen and Gordon.[40]

7.8. Concluding remarks. This account of the role of spectroscopic measurements in the determination of intermolecular potential energy functions has highlighted the rapid progress made in this area since the late-1960s, initially on spherically symmetric systems and more recently on anisotropic potentials. The level of detail and discrimination provided by these techniques is unparalleled. However, it should always be borne in mind that the spectral information is usually sensitive to a limited range of separations and relative orientations and that most rapid progress towards determining large zones of complex potential energy surfaces will be made by combining information from a range of complementary techniques. Molecular-beam scattering cross-sections have a particularly important role to play here. Experience with spherical systems has shown that reliance on fixed functional forms for potential energy functions, even highly flexible ones, can be hazardous unless a wide range of data and different functions are combined. The problems of correlations between parameters are even more marked for polyatomic systems. It remains to be seen whether the progress made in developing approximate procedures for direct calculation of bound state energies from anisotropic potential functions will lead to the development of successful unambiguous inversion procedures like those in common usage for molecules having spherical symmetry.

References

1. Raoult, B., Farges, J., and Rouault, M. *C. R. Acad. Sci. Paris* **267B,** 942 (1968).
2. Audit, P. *J. Phys.* (*Paris*) **30,** 192 (1969).
3. Hilderbrandt, R. L. and Bonham, R. A. *Ann. Rev. Phys. Chem.* **22,** 279 (1971).
4. Leckenby, R. E. and Robbins, E. J. *Proc. roy. Soc.* **291A,** 389 (1966).
5. Milne, T. A. and Greene, F. T. *J. chem. Phys.* **47,** 4095 (1967).
6. Ewing, G. E. *Angew. Chem. Internat. Edit.* **11,** 486 (1972); Ewing, G. E. *Acc. Chem. Res.* **8,** 185 (1975).
7. LeRoy, R. J. In *Semiclassical methods in molecular scattering and spectroscopy* (ed. M. S. Child), Chapter 3, pp. 109–26. Reidel, Dordrecht (1980).
8. LeRoy, R. J. and Carley, J. S. *Advan. chem. Phys* **42,** 353 (1980).
9. Cashion, J. K. *J. chem. Phys.* **48,** 94 (1968).
10. Stogryn, D. E. and Hirschfelder, J. O. *J. chem. Phys.* **31,** 1531 (1959); **33,** 942 (1960).
11. Chen, C. T. and Present, R. D. *J. chem. Phys.* **54,** 3645 (1971).
12. Mahan, G. D. *J. chem. Phys.* **52,** 258 (1970).
13. Bouchiat, M. and Pottier, L. *J. Phys.* (*Paris*) **33,** 213 (1972).
14. Tanaka, Y. and Yoshino, K. *J. chem. Phys.* **57,** 2964 (1972).
15. Tanaka, Y. and Yoshino, K. *J. chem. Phys.* **53,** 2012 (1970).
16. Tanaka, Y., Yoshino, K., and Freeman, D. E. *J. chem. Phys.* **59,** 5160 (1973).

17. Freeman, D. E., Yoshino, K., and Tanaka, Y. *J. chem. Phys.* **61,** 4880 (1974).
18. Colbourn, E. A. and Douglas, A. E. *J. chem. Phys.* **65,** 1741 (1976).
19. McKellar, A. R. W. and Welsh, H. L. *Can. J. Phys.* **52,** 1082 (1974).
20. Long, C. A. and Ewing, G. E. *J. chem. Phys.* **58,** 4824 (1973).
21. Long, C. A., Henderson, G., and Ewing, G. E. *Chem. Phys.* **2,** 485 (1973).
22. Henderson, G. and Ewing, G. E. *Mol. Phys.* **27,** 903 (1974).
23. Kudian, A. and Welsh, J. L. *Can. J. Phys.* **49,** 230 (1971).
24. McKellar, A. R. W. and Welsh, H. L. *J. chem. Phys.* **55,** 595 (1971).
25. McKellar, A. R. W. and Welsh, H. L. *Can. J. Phys.* **50,** 1458 (1972).
26. Dyke, T. R., Howard, B. J., and Klemperer, W. *J. chem. Phys.* **56,** 2442 (1972).
27. Kaiser, E. W. *J. chem. Phys.* **53,** 1686 (1970).
28. Novick, S. E., Davies, P., Harris, S. J., and Klemperer, W. *J. chem. Phys.* **59,** 2273 (1973).
29. Novick, S. E., Janda, K. C., Holmgren, S. L., Waldman, M. and Klemperer, W. *J. chem. Phys.* **65,** 1114 (1976).
30. Harris, S. J., Novick, S. E., Klemperer, W., and Falconer, W. E. *J. chem. Phys.* **61,** 193 (1974).
31. Novick, S. E., Janda, K. C., and Klemperer, W. *J. chem. Phys.* **65,** 5115 (1976).
32. Holmgren, S. L., Waldman, M., and Klemperer, W. *J. chem. Phys.* **67,** 4414 (1977); **69,** 1661 (1978).
33. Smalley, R. E., Wharton, L., and Levy, D. H. *Acc. Chem. Res.* **10,** 139 (1977).
34. Kim, M. S., Smalley, R. E., Wharton, L., and Levy, D. H., *J. chem. Phys.* **65,** 1216 (1976).
35. Kubiak, G., Fitch, P. S. H., Wharton, L., and Levy, D. H. *J. chem. Phys.* **68,** 4477 (1978).
36. Gough, T. E., Miller, R. E., and Scoles, G. *J. chem. Phys.* **69,** 1588 (1978).
37. Gough, T. E., Miller, R. E., and Scoles, G. *J. Mol. Spectrosc.* **72,** 124 (1978).
38. Sando, K. M. *Mol. Phys.* **23,** 413 (1972).
39. Leonardi-Cattolica, A. M., Prins, K. O., and Waugh, J. S. *J. chem. Phys.* **54,** 769 (1971).
40. Neilsen, W. B. and Gordon, R. G. *J. chem. Phys.* **58,** 4131, 4149 (1973).
41. E.g. The fit of the vibrational energy levels for Ar_2 to a Morse curve in ref. 15, and the subsequent demonstration that this leads to a well depth ε too low by 10 per cent by Maitland, G. C. and Smith, E. B. *Mol. Phys.* **22,** 861 (1971); see also ref. 52.
42. Richards, W. G. and Barrow, R. F. *Trans. Faraday Soc.* **60,** 797 (1964); *Proc. Phys. Soc.* **83,** 1045 (1964).
43. Fröman, N. and Fröman, P. *JWKB approximation.* North-Holland, Amsterdam, (1965).
44. Rydberg, R. *Z. Phys.* **73,** 376 (1931).
45. Klein, O. *Z. Phys.* **76,** 226 (1932).
46. Rees, A. G. L. *Proc. Phys. Soc.* **59,** 998 (1947).
47. Vanderslice, J. T., Mason, E. A., Maisch, W. G. and Lippincott, E. *J. Mol. Spectros.* **3,** 17 (1959); **5,** 83 (1960).
48. Vanderslice, J. T. Mason, E. A., and Maisch, W. G. *J. chem. Phys.* **32,** 515 (1960).
49. Weissman, S., Vanderslice, J. T., and Battino, R. *J. chem. Phys.* **39,** 2226 (1963).

50. Birge, R. T. and Sponer, H. *Phys. Rev.* **28,** 259 (1926).
51. LeRoy, R. J. and Bernstein, R. B. *Chem. Phys. Lett.* **5,** 42 (1970); *J. chem. Phys.* **52,** 3869 (1970).
52. LeRoy, R. J. *J. chem. Phys.* **57,** 573 (1972).
53. LeRoy, R. J. In *Molecular Spectroscopy* 1 (ed. R. F. Barrow, D. A. Long, and D. J. Millen), p. 113. Specialist Periodical Report of the Chemical Society (1973).
54. LeRoy, R. J. *Can. J. Phys.* **50,** 953 (1972).
55. LeRoy, R. J. and Lam, W.-H. *Chem. Phys. Lett.* **71,** 544 (1980).
56. Maitland, G. C. and Smith, E. B. *Mol. Phys.* **22,** 861 (1971).
57. Parson, J. M., Siska, P. E., and Lee, Y. T. *J. chem. Phys.* **56,** 1511 (1972).
58. Barker, J. A., Fisher, R. A., and Watts, R. O. *Mol. Phys.* **21,** 657 (1971).
59. Tanaka, Y. and Yoshino, K. *J. chem. Phys.* **50,** 3087 (1969).
60. Maitland, G. C. *Mol. Phys.* **26,** 513 (1973).
61. LeRoy, R. J., Klein, M. L., and McGee, I. K. *Mol. Phys.*, **28,** 587 (1974).
62. Levine, H. B. *J. chem. Phys.* **56,** 2455 (1972).
63. Dunker, A. M. and Gordon, R. G. *J. chem. Phys.* **64,** 4984 (1976).
64. Ewing, G. E. *Can. J. Phys.* **54,** 487 (1976).
65. Pattengill, M. D. and Bernstein, R. B. *J. chem. Phys.* **65,** 4007 (1976) and references therein.
66. LeRoy, R. J. and Van Kranendonk, J. *J. chem. Phys.* **61,** 4750 (1974).
67. LeRoy, R. J., Carley, J. S., and Grabenstetter, J. E. *Faraday Discuss. Chem. Soc.* **62,** 169 (1977).
68. Carley, J. S. *Faraday Discuss Chem. Soc.* **62,** 303 (1977).
69. Levine, R. D., Johnson, B. R., Muckerman, J. T., and Bernstein, R. B. *J. chem. Phys.* **49,** 56 (1968).
70. Kreek, H. and LeRoy, R. J. *J. chem. Phys.* **63,** 338 (1975).
71. Liu, W.-K., Grabenstetter, J. E., LeRoy, R. J., and McCourt, F. R. *J. chem. Phys.* **68,** 5028 (1978).
72. Langhoff, P. W., Gordon, R. G., and Karplus, M., *J. chem. Phys.* **55,** 2126 (1971).
73. Victor, G. A. and Dalgarno, A. *J. chem. Phys.* **53,** 1316 (1970).
74. Hepburn, J., Scoles, G., and Penco, R. *Chem. Phys. Lett.* **36,** 451 (1975).
75. Hutson, J. M., Barton, A. E., Langridge-Smith, P. R. R., and Howard, B. J., *Chem. Phys. Lett.* **73,** 218 (1980).
76. Rulis, A. M., Smith, K. M., and Scoles, G. *Can. J. Phys.* **56,** 753 (1978).
77. Toennies, J. P. Welz, W., and Wolf, G. *J. chem. Phys.*, **71,** 614 (1979).
78. Zandee, L. and Reuss, J. *Chem. Phys.* **26,** 345 (1977).
79. Boom, E. W., Frenkel, D., and van der Elsken, J. *J. chem. Phys.* **66,** 1826 (1977).
80. Gordon, R. G. and Cashion, J. K. *J. chem. Phys.* **44,** 1190 (1966).
81. Billingsley, J. and Callear, A. *Trans. Faraday Soc.* **67,** 589 (1971).
82. Dinerman, C. E. and Ewing, G. E. *J. chem. Phys.* **53,** 626 (1970); **54,** 3660 (1971).
83. Mannik, L., Stryland, J. C., and Welsh, H. L. *Can. J. Phys.* **49,** 3056 (1971); Brigot, N., Odiot, S., Walmsley, S. H., and Whitten, J. L. *Chem. Phys. Lett.* **49,** 157 (1977).
84. Klemperer, W. *Ber. Bunsenges. Phys. Chem.* **78,** 128 (1974).
85. Novick, S. E., Janda, K. C., and Klemperer, W. *J. chem. Phys.* **65,** 5115 (1976).
86. Rank, D. H., Sitaram, P., Glickman, W. A., and Wiggins, T. A. *J. chem. Phys.*

39, 2673 (1963).
87. Dyke, T. R. and Muenter, J. S. *J. chem. Phys.* **57,** 5011 (1972); **60,** 2929 (1974); Dyke, T. R., Mack, K. M., and Muenter, J. S. *J. chem. Phys.* **66,** 498 (1977).
88. Novick, S. E., Harris, S. J., Janda, K. C., and Klemperer, W. *Can. J. Phys.* **53,** 2007 (1975).
89. Klemperer, W. *Faraday Discuss. Chem. Soc.* **62,** 179 (1977).
90. Harris, S. J., Janda, K. C., Novick, S. E., and Klemperer, W. *J. chem. Phys.* **63,** 881 (1975).
91. Harris, S. J., Novick, S. E., and Klemperer, W. *J. chem. Phys.* **60,** 3208 (1974).

8
CONDENSED PHASES

8.1. Introduction

The solid and liquid states of matter both have much higher densities than dilute gases, and a molecule in such a condensed phase is at all times close to several other molecules. The total intermolecular energy is consequently large, similar in magnitude to the kinetic energy, and the properties of these states depend strongly on the nature of the intermolecular forces. Provided that the relations between the forces and the bulk properties are known it is clear that the study of solids and liquids can be a valuable source of information about intermolecular interactions.

Solid-state data such as low-temperature lattice energies and nearest-neighbour distances were extensively used in the early development of model potential functions by Lennard-Jones and others,[1,2] and in recent years these and other properties of solids have played a major part in the development of accurate pair potentials for the inert gases.[3] Until recently, the statistical mechanics of liquids was not sufficiently developed, but substantial advances in the last few years have made it possible to use some equilibrium properties of atomic liquids in the refinement of pair potential functions.[4]

It is impossible in a work of this length to do justice to the subjects of solid-state and liquid-state physics, but for the reasons above they cannot be entirely neglected. We have therefore chosen to survey briefly the relevant parts of these subjects, giving references to works in which they are covered in greater depth, and to emphasize those areas which have in the past given useful information about molecular interactions. Since the primary study of this book is the nature of the intermolecular pair potential, and since the condensed phases are, as we shall see, complicated by the effects of non-additivity of the intermolecular energy, we feel that this approach is justified at the present time. With improved understanding of non-additivity and with the further development of statistical mechanical theory, particularly for time-dependent properties, it is likely that the importance of the solid and liquid states in the study of intermolecular interactions will greatly increase.

8.2. Many-body forces

It has been pointed out in earlier chapters that when several molecules are simultaneously close together their total energy is not exactly equal to the sum of the pair energies which would be observed if each pair were isolated. The presence of other nearby molecules perturbs the interactions and the energy of a group of N molecules must be written

$$U_N = \sum_{j>i} U_{ij} + \sum\sum_{i>j>k} U_{ijk} + \dots \tag{8.1}$$

Most of the energy, U_N, arises from the first term, the sum of the pair energies, but the contributions from three-body and possibly from four-body terms may not be negligible at higher densities. Early studies of the lattice properties of the inert gases assumed that non-additive many-body forces were unimportant and hence calculations were based on the assumption of pairwise additivity of the intermolecular energy.[1,2] It is now known that the over-all non-additive contribution to the lattice energy is positive,[5] so that an analysis based on pairwise-additivity suggested a shallower intermolecular potential well than the true value. The conclusions of early studies which falsely favoured such potentials as the Lennard-Jones 12−6 model resulted in part from the neglect of non-additivity, and it is now generally recognized that the three-body non-additive terms at least cannot be neglected.

The most important many-body contributions are those for which the pair separations are near to r_m, leading to large negative pair energies, and hence an associated favourable Boltzmann factor, $\exp(-U_N/kT)$. In circumstances where one or more of the pair separations is small, the pairwise additive contribution which dominates the total energy is large and positive, so that the probability of observing such a state is small. Thus, even though the many-body repulsive energy may be large, the low probability of the associated configurations reduces the importance of these states. Studies of many-body contributions have therefore generally concentrated on long-range terms, and on the three-body contributions, which appear to be dominant.

The most widely studied three-body term is the Axilrod–Teller triple-dipole energy discussed in § 2.2.4. This is the first term in the three-body corrections to the dispersion energy, and may be represented for monatomic systems by[7]

$$U_{(\mathrm{DDD})_3} = \frac{\nu_{123}(1+3\cos\theta_1\cos\theta_2\cos\theta_3)}{r_{12}^3 r_{23}^3 r_{13}^3}. \tag{8.2}$$

The coefficient ν_{123} may be calculated quite accurately using sum rule methods for simple systems.[8] In cases where accurate results are not available it may be estimated using the relation

$$\nu = -\tfrac{3}{4}\alpha C_6 \tag{8.3}$$

where α is the polarizability and C_6 is the leading coefficient in the dispersion energy. The sign of the triple-dipole energy depends on the geometry of the triangle formed by the three molecules, of internal angles θ_1, θ_2, θ_3. For acute triangles the triple-dipole energy is positive, while for most obtuse triangles it is negative. Hence near-linear arrays of molecules are stabilized by this effect, while most triangular arrangements are destabilized. For most molecular configurations appropriate to solids and liquids the net energy from the triple-dipole correction is positive.

Other long-range terms also contribute to the non-additive three-body energy.[9,10] These are analogous to the higher terms in two-body dispersion energy, e.g. the instantaneous dipole-induced quadrupole and instantaneous quadrupole-induced dipole energies which lead to the energy C_8/r^8. (See § 2.2.3). The principal three-body terms may be represented $(DDD)_3$ (the Axilrod–Teller term), $(DDQ)_3$, $(DQQ)_3$, $(QQQ)_3$, $(DDO)_3$ where D, Q, O represent dipole, quadrupole, and octopole contributions, and the subscript, 3, indicates that these energies are calculated in the third-order of perturbation theory. The terms including quadrupole contributions fall off more rapidly with distance than the $(DDD)_3$ energy, $(DDQ)_3$ as r^{-11}, $(DQQ)_3$ as r^{-13} and $(QQQ)_3$ as r^{-15}. It has been shown in some cases that these higher terms may make significant contributions to the total energy,[5] but they have not been very widely used.

All of these calculations are valid only in the limit of large separations, just as are the dispersion terms in the two-body energy. The perturbation theory used in the calculation of the non-additive dispersion energy is not valid when the pair separations are near to r_m which is the region of major interest in the condensed phases. However quantum mechanical calculations using procedures more appropriate to these separations do not lead to convenient analytical results,[11] and it has been common to employ the Axilrod–Teller correction for all ranges of pair separations. The justification for this procedure can only be based on its convenience and apparent success. Non-additive contributions for the repulsive regions have also been estimated from simplified models and used in calculations of third virial coefficients.[12] For this property, considerable cancellation occurred between the Axilrod–Teller and the repulsive corrections. For the solid and liquid states there is some reason to believe that because of the low probability of repulsive configurations arising (as discussed above) the triple-dipole energy comes fortuitously close not only to the total long-range many-body energy, but also to the total non-additive many-body contribution from all separations. Although this provides a convenient simplified working assumption, its basis remains insecure.

In general, the effect of non-additivity depends on the property under study, and will be discussed below as different properties are considered. In particular cases it is sometimes possible to separate the contributions to a property into pair and many-body non-additive terms. The latter may then be estimated and the experimental data modified to yield values for effectively 'pairwise-additive' properties.

In summary, although quantum mechanical calculations have given valuable insight into the nature and magnitude of many-body intermolecular interactions, there remain significant uncertainties about these terms which may in some circumstances limit the amount of information about intermolecular pair potentials which can be derived from studies of the condensed phases.

8.3. The solid state

8.3.1. Introduction

The solid state has traditionally played an important role in attempts to determine intermolecular forces. Because of its highly ordered structure, the ideal solid, like the perfect gas, represents a state of matter for which essentially exact statistical mechanical calculations may be performed. Although, with the recognition of the importance of non-additivity, solid-state properties must be used with more caution, there are still many properties of molecular solids which may be used as sources of information about intermolecular interactions. The simplest model of a solid,[13] consisting of molecules fixed on an infinite array of regularly ordered lattice sites, provides a basis for calculating several solid state properties in terms of the intermolecular potential. Thus, for example, the lattice energy may be readily calculated for a particular type of lattice and a model potential, making suitable assumptions about non-additivity, and it is found that the lattice energy depends very directly on the well depth. The experimentally accessible sublimation energy at 0 K leads to a value for the lattice energy, and hence may be used to determine the well depth for a model potential. In the same way, the nearest-neighbour distance at 0 K is simply related to the value of r at the potential minimum, r_m, and may be used to evaluate this parameter.

A more realistic model of the solid state recognizes the existence of molecular vibrations about the lattice sites. The vibrations are quantized, with energy quanta known as phonons, and because of zero-point energy the molecules vibrate even at 0 K. This zero-point energy must be taken into consideration when calculating the lattice energy and nearest-neighbour separation.[14] The details of the lattice vibrations depend on the force field experienced by the molecules, and studies of properties which depend on these lattice vibrations can thus be used to obtain further information about the shape of the potential. For example, low-temperature heat capacities depend largely on the second derivative of the pair potential near to the minimum, and the low-temperature coefficient of thermal expansion can give information about the third derivative in the same region.[6] Properties at higher temperatures depend on the range and distribution of the lattice vibration frequencies. These may be calculated if a potential model is assumed and the lattice structure known. Properties such as heat capacities may then be calculated as a function of temperature. Such calculations have been found to provide useful discriminants between various proposed model potential functions.[15] More information about molecular vibrations can be derived from inelastic neutron scattering studies, which lead to 'phonon dispersion curves'. These describe the frequency dependence of the speed of elastic waves, similar to sound waves, in a solid and although they have not been used directly in establishing the form of potential functions it has been shown that modern potential functions are able to reproduce the experimental data for simple molecules quite accurately.[16]

8.3.2. Static lattice properties

Solid lattices are characterized by a very high degree of structural regularity. For example, a solid which has the face-centered cubic structure, characteristic of the heavier inert gases and illustrated in Fig. 8.1, will have successive shells of neighbours at characteristic multiples of the nearest-neighbour distance, *a*, from a central molecule.[2] This is illustrated in Table 8.1. The lattice parameter, *d*, is equal to the side length of the unit cell, and for this structure $d = \sqrt{2}a$.

If the intermolecular pair potential may be written in the form

$$U(r) = \varepsilon \sum_n C_n(\sigma/r)^n \tag{8.4}$$

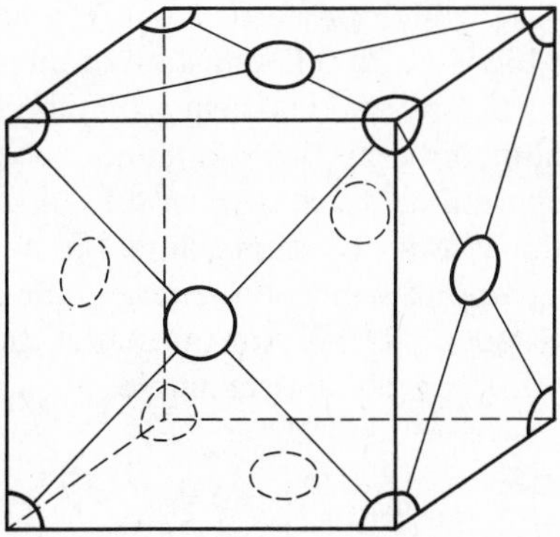

FIG. 8.1. Unit cell of the face-centered cubic lattice.

Table 8.1

Neighbour shells for a f.c.c. lattice[2]

Shell number	Distance from central molecule	No. of molecules in shell
1	a	12
2	$\sqrt{2}a$	6
3	$\sqrt{3}a$	24
4	$2a$	12
5	$\sqrt{5}a$	24

and pairwise additivity is assumed, the static lattice energy, ϕ, of N molecules may be expressed as a function of the reduced nearest-neighbour distance, a/σ:

$$\phi(a/\sigma) = \frac{N\varepsilon}{2}\sum_n C_n\left(\frac{\sigma}{a}\right)^n\left[12+6\left(\frac{1}{\sqrt{2}}\right)^n+24\left(\frac{1}{\sqrt{3}}\right)^n+\cdots\right]. \tag{8.5}$$

The factor $N/2$ in (8.5) prevents double counting of each pair contribution. For realistic powers of n in (8.4) the sums over successive shells in (8.5) converge rapidly. For most potentials the first shell contributes about 90 per cent of the total energy, using observed values of a. The total static lattice energy may be expressed in the form

$$\phi(a/\sigma) = \frac{N\varepsilon}{2}\sum C_n L_n\left(\frac{\sigma}{a}\right)^n. \tag{8.6}$$

The quantities L_n are known as *lattice sums*. For the f.c.c. lattice it is seen that they are defined by the expression in square brackets in (8.5). They have been evaluated for $4<n<20$ for all common types of simple lattice,[1,2] and are presented in Appendix 7. $\phi(a/\sigma)$ is the pairwise-additive contribution to the total lattice energy, and has the general appearance shown in Fig. 8.2. At 0 K the state of minimum free energy is the state of minimum energy, and classically may be located by differentiating (8.6). The value of a at the minimum is the low-temperature nearest-neighbour separation, a_0, and the value of $\phi(a_0)$ is the equilibrium lattice energy, ϕ_0. Within this approximation, neglecting many-body contributions and zero-point vibration energy, the lattice energy and nearest-neighbour distance at 0 K for a model potential are simple multiples of ε and σ respectively. For example, for a $12-6$ potential and a f.c.c. lattice,

$$\phi_0 = -8{\cdot}160N\varepsilon$$

$$a_0 = 1{\cdot}090_2\,\sigma.$$

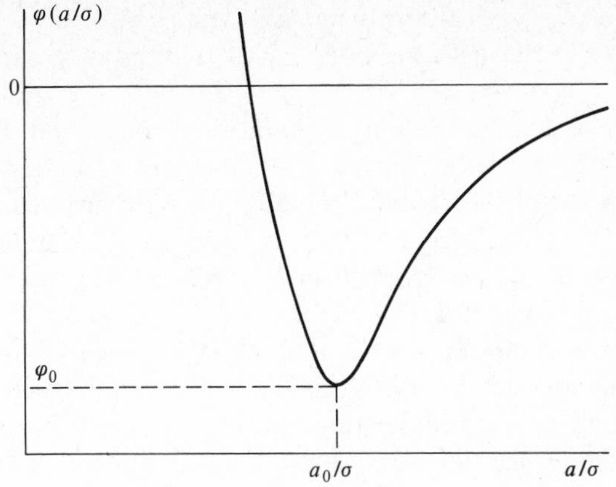

FIG. 8.2. Lattice energy, ϕ, as a function of the reduced nearest-neighbour distance, a/σ.

It is clearly a straightforward matter to use these relationships in conjunction with experimental data to determine ε and σ for a chosen model potential. In general the nearest-neighbour distance, a_0, is found to be near to r_m, the position of the minimum in the pair potential. The value of ϕ_0 is evidently determined primarily by the value of the pair potential energy at the nearest-neighbour separation, $U(a_0)$, since, as mentioned above, about 90 per cent of the total static lattice energy is derived from the nearest-neighbour interactions. Hence such data may be expected to yield very directly information about the pair potential near to the well minimum. Since the experimental uncertainty in the sublimation energy of the rare gases is generally less than 0·5 per cent, and that in a_0 is a remarkable $1{\cdot}5\times10^{-3}$ per cent for argon, the value of such calculations in the study of intermolecular interations is evident.

One slightly surprising result of such calculations is that for any reasonable pair potential function so far studied the hexagonal close-packed structure has a greater lattice energy than the face-centered cubic structure,[2] and would be predicted to be the stable form for the inert gases. As has been mentioned, this is not the case in practice for the heavier inert gases. The energy difference between the two structures is extremely small, about 0·01 per cent of the total in all cases from Ne to Xe, and several explanations have been proposed to account for this discrepancy.[16] The numbers and distances of neighbours from a central molecule are the same for the first and second neighbours in both the h.c.p. and f.c.c. lattices, and differences only arise in the third and subsequent neighbour shells. It is possible to 'tailor' the shape of the pair potential in the region of the third neighbours,[18] in order to stabilize the f.c.c. structure, but there seems to be no other reason to support such an action.

One possible explanation of this anomaly has been sought in the contribution to the lattice energy of non-additive many-body energy terms. The Axilrod–Teller triple-dipole energy was first considered, and its contribution to the lattice energy was determined by summing over all the triangles of molecules appropriate to the crystal structures.[19] Although this energy falls of as r^{-9}, the lattice sums for both f.c.c. and h.c.p. structures are less rapidly convergent than the pair sums, since the numbers of triangles to be considered increase rapidly with r. However the sums are known and show that there is a small energy difference between the values for the two lattices, which favours the f.c.c. form. Nevertheless it seems that the size of this correction is not sufficient to account for the observed stability of the f.c.c. lattice. Other non-additive three-body terms corresponding to the higher-order contributions to the dispersion energy have also been considered.[5] These fall off more quickly than r^{-9} and the lattice sums are more rapidly convergent. The differences for the two lattice structures again favour the f.c.c. structure for the $(DDQ)_3$, $(DDO)_3$, and $(QQQ)_3$ non-additive terms, but are still too small to overcome the difference between the pairwise lattice sums. It seems then that these non-additive dispersion contributions cannot provide the whole explanation for the experimentally observed structure, and recent calculations of non-additive short-range interactions come to the same conclusion.

An explanation which seems both qualitatively and quantitatively satisfactory has been proposed fairly recently.[17,20] It is noted that the twelve nearest-neighbour atoms are disposed about the central atom with different symmetries in the two lattice structures. The magnitude of the two-body dispersion energy depends on the energies of the accessible excited states of the atoms involved, as discussed in § 2.2.3(b), and those states involving atomic d orbitals are perturbed by the neighbouring atoms to different extents in the two structures. This leads to an energy difference favouring the f.c.c. lattice of a magnitude which increases from Ne to Xe, and which is about 12–25 times greater than the difference in the pairwise lattice sums. For He the permitted dipole transitions do not involve d-orbitals, and this effect will not occur. The observed lattice structure for He is of the h.c.p. type.

In spite of the failure of the long-range, non-additive contributions to account for the observed structures of rare-gas solids, calculations which were intended to resolve this problem have nevertheless shown that the size of the non-additive contributions to the lattice energy are quite large, and cannot be neglected when using lattice properties to establish the nature of pair potentials.[6,15] For example, the total three-body contribution to the lattice energy of argon is about 10 per cent of the total and for xenon it has risen to 13 per cent. Most of this contribution arises from the $(DDD)_3$ energy with the $(DDQ)_3$ amounting to about 2 per cent of the total energy, and the other terms being much smaller.

The size of these contributions forces us to consider whether higher-order contributions may also be significant. Carrying the perturbation calculation to higher order yields expressions for the fourth-order triple-dipole energy, $(DDD)_4$. The associated contribution to the f.c.c. lattice energy is found to be negative, in contrast to the other three-body energies.[5,10] For argon and krypton it seems that the $(DDD)_4$ lattice energy almost exactly cancels the total contribution from $(DDQ)_3$, $(DQQ)_3$ and $(QQQ)_3$ energies,[6] and the practice of neglecting all many body terms other than the Axilrod–Teller contribution gains some partial justification. In the case of xenon the cancellation is less complete.

8.3.3. Lattice vibrations

Each of the N atoms in a crystal lattice undergoes a complex three-dimensional vibration about its lattice site. Since each atom interacts with many neighbours the motions of the atoms are strongly coupled and it is not possible to resolve the motion of separate atoms. However the problem may be overcome by performing a normal-mode analysis in which the vibrations of the lattice are decomposed into $3N$ independent normal modes of vibration.[13,21,22] Each normal mode represents a vibration of all the atoms, having a characteristic frequency and the total lattice vibration is a superposition of these $3N$ decoupled vibrations. This approach is based on the assumption that the forces between pairs of atoms are proportional to the displacement from their equilibrium separation, which gives rise to simple harmonic motion. For the heavier inert gases at low temperatures the amplitude of the vibrations is fairly small, and the assumption of harmonic vibrations is a good first approximation. Corrections for anharmonicity may be introduced as perturbations.[23]

In molecular solids, two classes of vibration may be distinguished. The first resembles the vibrations within the isolated molecules and the second, usually at lower frequencies, represent vibrations of the whole molecule about its lattice site. We confine ourselves here to the simpler case of monatomic solids.

If the vibrational frequencies are known, the solid may be regarded for many purposes as an assembly of $3N$ independent quantum mechanical simple harmonic oscillators, of frequencies ν_1 to ν_{3N}. The vibrational quanta, $h\nu_i$, associated with the normal modes, are known as phonons. The thermodynamic properties of the solid may then be calculated using the well known statistical mechanical results for harmonic oscillators.[22]

For example, the average thermal energy of an oscillator of frequency ν at temperature T is

$$\bar{E} = \frac{h\nu}{\exp(h\nu/kT) - 1}. \tag{8.7}$$

The internal energy of the solid may be written as the sum over $3N$ such terms,

$$\begin{aligned} U - U^0 &= \sum_{i=1}^{3N} \frac{h\nu_i}{\exp(h\nu_i/kT) - 1} \\ &= kT \sum_{i=1}^{3N} \frac{x_i}{e^{x_i} - 1} \end{aligned} \tag{8.8}$$

where U^0 is the zero-point energy of the system, and we have written $x_i = h\nu_i/kT$. The heat capacity, C_v, may readily be calculated from (8.8), and is

$$C_v = k \sum_{i=1}^{3N} \frac{x_i^2 \,.\, e^{x_i}}{(e^{x_i} - 1)^2}. \tag{8.9}$$

In order to calculate the internal energy or heat capacity a knowledge of the distribution of the frequencies of the normal modes, ν_i, is required. This may be expressed in terms of a distribution function, $f(\nu)$, defined such that in the frequency range ν to $\nu + d\nu$ there are $f(\nu)\,d\nu$ frequencies.

In view of the great complexity of the actual lattice vibrations much work has been carried out using approximations to the true distributions of vibration frequencies. The simplest of these is due to Einstein, and assumes that all the frequencies have the same value, ν_E, which is characteristic of the substance. A more realistic approximation was suggested by Debye.[22] The vibrations may be analysed in terms of the standing waves which can be established in the solid. Both longitudinal and transverse waves occur, of which the former closely resemble sound waves. Debye assumed that the distribution of frequencies for both types of wave in the solid would have the same form as that for sound waves in an isotropic continuous medium, for which

$$f(\nu) = c \,.\, \nu^2 \tag{8.10}$$

Since the total number of normal modes must be $3N$, there is an upper limit to the permitted frequencies, which we write ν_{max}. Hence

$$\int_0^{\nu_{max}} f(\nu)\,d\nu = 3\,N \tag{8.11}$$

and $c = 9N/\nu_{max}^3$. For values of $\nu > \nu_{max}$, $f(\nu) = 0$. The existence of this maximum frequency is related to the non-continuous atomic nature of the solid. This imposes a limit on the smallest possible value of the wavelength, which is of the order of the distance between the atoms in the lattice. Another characteristic of this model is that the velocity of propagation of the waves, which may be different for longitudinal and transverse waves, is independent of the frequency.

Eqns (8.8) and (8.9) may now be rewritten in the form of integrals over ν, using the frequency distribution, $f(\nu)$. Hence

$$C_v = k\int_0^{\nu_{max}} \frac{(h\nu/kT)^2 e^{h\nu/kT}}{(e^{h\nu/kT}-1)^2}\frac{9N}{\nu_{max}^3}\nu^2\,d\nu. \tag{8.12}$$

It proves convenient to define a characteristic temperature, the Debye temperature, θ_D, in terms of ν_{max}

$$\theta_D = \frac{h\nu_{max}}{k}. \tag{8.13}$$

Eqn (8.12) may then be expressed as an integral over x, ($x = h\nu/kT$), to give

$$C_v = 9Nk\left(\frac{T}{\theta_D}\right)^3 \int_0^{\theta_D/T} \frac{x^4 e^x\,dx}{(e^x-1)^2} \tag{8.14}$$

The integral must be evaluated numerically, and we see that the reduced heat capacity, C_v/Nk, is a function only of the reduced temperature, T/θ_D, in this model. Tables of the heat capacity and other thermal properties, are readily available.[24] In order to apply the Debye model to a particular substance we see that it is only necessary to know an appropriate value for a single characteristic parameter, the Debye temperature, θ_D. When this is known, the value of T/θ_D is calculated at the temperature of interest, and the value of the desired property obtained from tables. The calculation of the zero-point vibrational energy, U^0, in this approximation is particularly simple. It is equal to $9Nk\theta_D/8$.

The Debye approximation gives quite an accurate description of the heat capacity of monatomic solids and, in particular, correctly predicts that at low temperature, $C_v \propto T^3$. However, it is not an exact theory, as may be seen if experimental data for a given property, such as C_v, are taken for several temperatures, and the value of θ_D needed to achieve exact agreement with experiment is determined for each temperature. If the Debye theory were exact, this 'experimental' θ_D, which we may denote $\bar{\theta}_D$, would have a constant value. The actual variation of $\bar{\theta}_D$ for argon at low temperatures is shown in Fig. 8.3.[21] Such graphs provide a convenient means of representing heat capacities, since the absolute

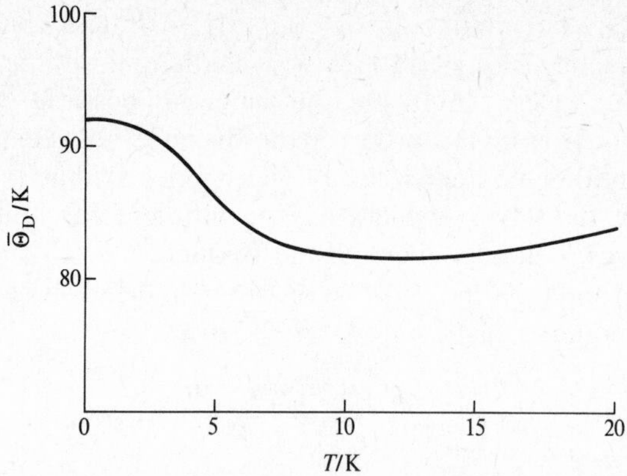

FIG. 8.3. Low-temperature behaviour of θ_D for argon.

values of C_v change markedly over the range of temperature whereas $\bar{\theta}_D$ varies only slightly. Despite these limitations of the Debye model, it has proved to offer a valuable means of making quite reliable estimates of the thermal properties of simple solids, and has been very widely used.

More exact calculations of lattice frequencies are based on the method of lattice dynamics, developed by Born and von Kármán.[25] The normal modes are plane waves of wave length λ characterized by a wave vector, $\mathbf{q}$, whose magnitude is $2\pi/\lambda$. For each $\mathbf{q}$ there are 3 independent normal modes, of different polarizations (i.e. in which the atoms move in different directions). For cubic lattices, the wave vectors which are perpendicular to the 100, 110, and 111 planes have one mode in which the atomic movements are parallel to $\mathbf{q}$ (longitudinal waves) and two for which the motion is perpendicular to $\mathbf{q}$ (transverse waves). For the 100 and 111 cases, the two transverse waves are degenerate.

For a particular crystal structure and a given value of the wave vector, it is possible, by solving the appropriate equations of motion, to determine the frequencies of the three associated normal modes if the pair potential energy function and atomic masses are known.[26] They depend on values of d^2U/dR^2 at the appropriate neighbour separations, since the force constants of simple harmonic oscillators depend on the second derivative of the potential energy. Only a limited number of wave vectors needs to be considered, owing to the restrictions on the minimum distinguishable wavelength imposed by the atomic nature of the solid. The frequency distribution of the whole solid may then be estimated by sampling a large number of values of the wave vector in the physically significant range, and calculating the associated frequencies. These may then be sorted into order, and the resulting histogram smoothed to give an approximation to the frequency distribution function, which may be

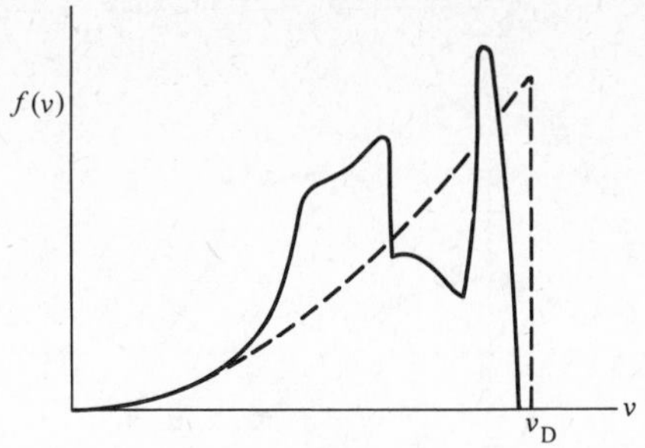

FIG. 8.4. Frequency distribution for normal modes of an f.c.c. lattice. Solid curve from lattice dynamics; dashed curve from Debye theory.

made as accurate as is necessary by sampling **q** more densely.[27] A typical frequency distribution function obtained in this way for a f.c.c. lattice is shown in Fig. 8.4, together with the corresponding Debye approximation. It is seen that the true distribution function has a much more complex appearance than the Debye function, but that the latter is accurate at low frequencies (long wavelengths) when the continuum approximation might be expected to be appropriate.

Using the accurate frequency distribution established in this way, the thermodynamic properties of a solid may be calculated as integrals over the frequency distribution of the appropriate simple harmonic oscillator contributions. The low-temperature heat capacity may thus be determined, and calculations using realistic potentials have been shown to lead to $\bar{\theta}_D$ vs. T graphs similar in shape to those found experimentally. The zero-point lattice energy may also be calculated using the exact frequency distribution

$$U_0 = \int_0^\infty \frac{h\nu}{2} f(\nu)\,d\nu. \tag{8.15}$$

Calculations using realistic pair potentials have suggested that the zero-point energy calculated in this way is less than that based on the Debye model, using the 0 K limiting value of θ_D, by around 10 per cent.[15]

An important difference between the Debye model and the lattice dynamics approach is that no prior assumptions need be made in the latter about the velocity of propagation of the waves. It is found that this depends quite considerably on the frequency and on the direction of propagation and the polarization of the waves. The relationship between the frequency, ν, and wavevector, **q**, defines the velocity of sound at this frequency, and in the direction defined by **q**, and is known as the phonon dispersion curve. For cubic lattices these are generally measured for wave

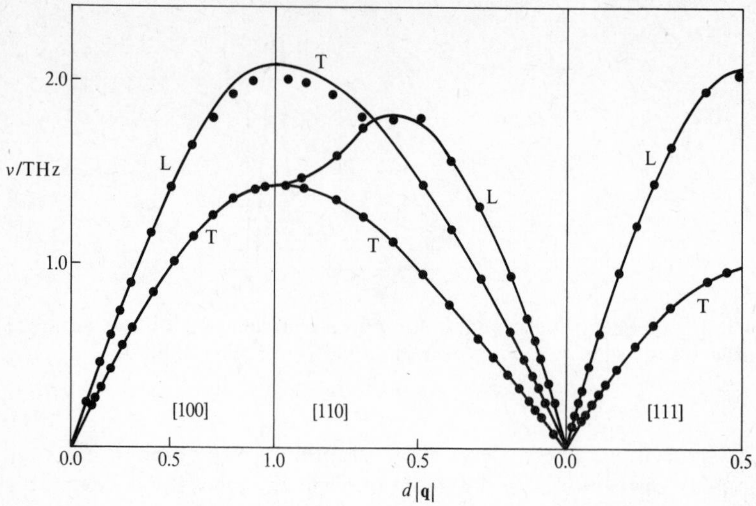

FIG. 8.5. Phonon dispersion curves for argon at 10 K. Frequencies for longitudinal (L) and transverse (T) modes in the 100, 110, and 111 directions are expressed as functions of the reduced wave vector, $d\,|\mathbf{q}|$, where d is the cubic lattice parameter.

vectors perpendicular to the 100, 110, and 111 planes. These curves may be studied experimentally by the inelastic scattering of thermal neutrons, in which the quantized lattice vibrations gain or lose phonons.[28,29] In order to establish the dispersion relations it is necessary to find the energy gained or lost by the scattered neutrons as a function of the change in the wave vector of the neutrons. A typical set of experimental results is shown in Fig. 8.5, together with theoretical curves based on a realistic intermolecular potential.

8.3.4. *Lattice vibrations: higher-order effects*

So far the discussion of lattice vibrations has been based on the assumption that these may be regarded as simple harmonic. However there is considerable evidence that this assumption is not valid. Simple harmonic motion follows from the assumption that the restoring forces between atoms are proportional to their displacement from the equilibrium separation, and hence that the potential energy curve for a pair of atoms is quadratic, having the symmetrical appearance shown in Fig. 8.6(a). For such a curve, the mean separation, $\bar{r}$, is independent of the energy. In the anharmonic case, where the curve is unsymmetrical, such as that in Fig. 8.6(b), the value of $\bar{r}$ increases as the energy is raised. Since

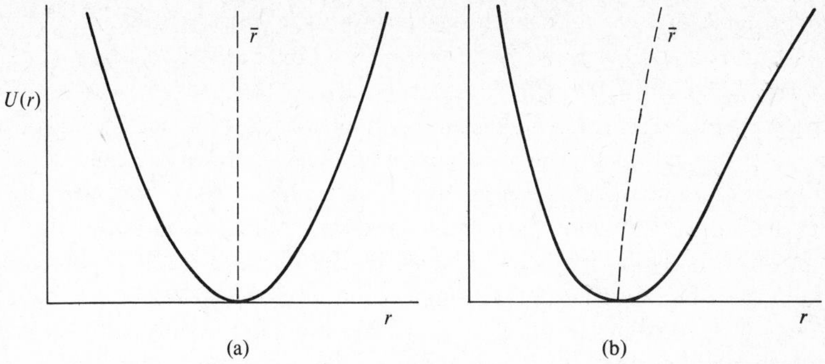

FIG. 8.6. (a) Potential function for harmonic motion; (b) function giving anharmonic behaviour. $\bar{r}$ is the mean separation.

the mean energy of a system of oscillators rises as the temperature is increased, the experimental observation that crystals expand on heating indicates that anharmonic effects are not negligible.[13] Provided that the amplitude of the oscillations is fairly small, perhaps less than 5–6 per cent of the nearest-neighbour separation, the effects of anharmonicity may be treated as a small perturbation of the harmonic results.[23] The Helmholtz free energy, from which other thermodynamic properties may be derived, can be expressed as a series

$$A = \phi + A_2 + A_3 + A_4 + \ldots$$

in which ϕ is the static lattice energy, A_2 is the harmonic free energy, and A_3, $A_4 \ldots$ are anharmonic terms. The latter may be calculated as lattice sums[23,30] and depend on the higher derivatives (up to $\mathrm{d}^4U/\mathrm{d}r^4$) of the pair potential energy function at the nearest-neighbour separation. Other thermodynamic properties may be derived from the free energy using standard relations. This approach is not applicable to light molecules such as neon, for which the amplitude of the lattice vibrations is too large, but it is suitable for the heavier inert gases at temperatures up to about one third of the melting point.

The inclusion of anharmonic terms has a significant effect on calculated values of the heat capacity, even at the lowest temperatures. For argon,[15] when anharmonic corrections were made to lattice dynamics calculations of C_v at low temperatures, the derived values of $\bar{\theta}_D$ at 0 K obtained by fitting the results were found to exceed those obtained in the harmonic approximation by about 3–4 per cent. The effects for krypton and xenon are smaller, but still significant.

In the calculation of lattice frequencies and related properties, the contribution due to non-additive many-body interactions may be included, at the cost of some increase in the length of the calculations. The importance of these contributions depends on the property under investigation.[15] For example, inclusion of the triple dipole energy in calculations for argon raised the 0 K value of $\bar{\theta}_D$ by about 0·2 per cent, but increased the zero-point energy by 0·7 per cent. This difference suggests that the low-frequency transverse modes, which dominate the low-temperature specific heat, are insensitive to the non-additive three-body terms. The higher-frequency longitudinal modes, however, which are important in the zero-point energy calculations, are significantly increased by three-body terms. In general, the effects of three-body contributions are less than those resulting from corrections for anharmonicity.

8.3.5. Experimental measurements

The properties of the solid inert gases have been the subject of extensive investigation and despite the considerable experimental difficulties of such work reliable data are available for most of the properties of primary interest in the study of intermolecular interactions. A very important aspect of this work is the method of sample preparation. The form of a solidified inert gas, whether polycrystalline or a single crystal, is extremely sensitive to the way in which it has been prepared and handled.[31] In general crystals produced directly from the vapour phase are likely to contain more defects than those produced by freezing the liquid, but even in the latter case great care must be taken and special techniques adopted if large crystals of good quality are needed. For some properties, e.g. thermal conductivity or optical properties, the effects of crystal defects on the quantity measured may be quite marked and large, well-grown crystals are required. Other measurements can make use of polycrystalline samples, for example X-ray diffraction studies using the Debye–Scherer technique. Fortunately, most of the thermodynamic properties of major interest in this work seem to be relatively insensitive to lattice defects, although the lattice parameter, d, defined in § 8.3.2. appears to be affected by imperfections.

The determination of thermodynamic properties has been based chiefly on the results of low-temperature calorimetry.[32] The sublimation energy of the solid at 0 K may be calculated using heat-capacity data measured from the lowest accessible temperatures up to the triple point, together with enthalpies of fusion and of vaporization at the triple point. Provided that the deviations of the vapour from ideality are known, these results may be combined to give the enthalpy of sublimation over the whole temperature range, and hence the limiting value at absolute zero. Alternatively measurements of the vapour pressure of the solid as a function of temperature may be used to obtain the enthalpy of sublimation directly, from the Clapeyron equation,

$$\frac{\mathrm{d}P_s}{\mathrm{d}T} = \frac{\Delta H_{\mathrm{sub}}}{T\,\Delta V_{\mathrm{sub}}} \tag{8.16}$$

where P_s is the saturated vapour pressure at temperature T. ΔH_{sub} and ΔV_{sub} are the changes in enthalpy and volume on sublimation.

The low-temperature measurements of heat capacities[33] may be used to determine the behaviour of the 'experimental Debye temperature' $\bar{\theta}_D$ (see (8.13), (8.14), and the subsequent discussion), but in order to establish reliable values of the limit of this quantity at the absolute zero it appears to be necessary to work at rather low temperatures, say below $\theta_D/50$, or about 2 K for argon. The experimental quantity most generally studied is the heat capacity at constant pressure, C_p. This can yield values of the heat capacity C_v, by use of the thermodynamic identity

$$C_p - C_v = \beta^2 T/\rho\chi_T \tag{8.17}$$

provided that the coefficients of thermal expansion, β, the density, ρ, and isothermal compressibility, χ_T are known. At low temperatures this correction is small, and does not present serious problems but at higher temperatures uncertainties in this term may limit the accuracy of the derived heat capacities, C_v.

The lattice parameter, d, may be determined in several ways.[31] X-ray diffraction studies have been extensively used to measure this quantity over a range of temperatures. Powder photographs have been widely used, although measurements on single crystals may offer a substantial increase in accuracy. An alternative route uses measurements of bulk density, ρ, of a solid sample, which is related to d through the equation

$$\rho = \frac{4M}{N_A\,\mathrm{d}^3} \tag{8.18}$$

where M is the molar mass and N_A is Avogadro's constant. The density may be determined by measuring the amount of gas which, when solidified, fills a bulb of known volume. This technique is preferred at temperatures near to the triple point, where instabilities in the structure may complicate X-ray measurements. The coefficient of thermal expansion can be derived from the temperature dependence of the lattice parameter. At temperatures near to the triple point the presence of vacancies in the crystal may significantly affect this quantity and comparisons with theoretical results should take account of this.

In recent years coherent inelastic scattering of thermal neutrons has been used to probe the vibrations of inert gas solids leading to the phonon dispersion relations.[28,29] Incident neutrons of wave vector $\mathbf{K}_0$, whose magnitude, $K_0 = 2\pi/\lambda$, and whose energy $E_0 = h^2K_0^2/2m_n$, where m_n

is the mass of a neutron, are scattered through an angle ϕ and emerge with wave vector $\mathbf{K}_1$ and energy E_1. Transfer of momentum and energy to the lattice is possible and may be studied by monitoring the changes in the properties of the neutron beam. The measurements are carried out using a 'triple-axis crystal spectrometer' in which the energy and wave vector of the incident and emergent neutron beams are determined by the conditions required for Bragg reflexion from suitable crystals which act as monochromator and analyser. In the approach most commonly used for rare-gas solids, the change in wave vector, $\mathbf{Q}$, $(\mathbf{Q}=\mathbf{K}_0-\mathbf{K}_1)$ is kept constant while the energy transfer, E_0-E_1, is stepped through a series of values. A peak occurs in the scattered neutron intensity when the energy transfer, E_0-E_1, is equal to the energy of a phonon, $h\nu_i$, associated with a normal mode $\mathbf{q}$ of wave vector, $\mathbf{q}_i$, which is equal to $\mathbf{Q}$. All the rare gases have been studied using this technique, and the phonon dispersion relationships for the principal symmetry axes of the crystals have been determined.[29]

8.3.6. *Results for the inert gases*

The experimental data for low-temperature lattice properties of argon played a vital part in the modern development of pair potentials. In a seminal paper,[34] Barker and Pompe showed that a wide range of model potentials for argon could fit virial coefficient data, high-energy molecular beam results, and theoretical estimates of the dispersion energy coefficients to a similar degree of accuracy, although the predicted dilute gas viscosities all differed systematically from the experimental results available at that time. However, when lattice energies and nearest-neighbour distances were calculated, including the Axilrod–Teller triple-dipole three-body term, these potentials gave quite different results, and comparison with the experimental values made it possible to discriminate between the proposed functions. (The viscosity data used were later shown to be in error.)

In subsequent work,[15] Bobetic and Barker used the best potential of Barker and Pompe in a lattice dynamics calculation of the low-temperature specific heat and thermal expansion of argon. Non-additive three-body dispersion forces were again incorporated in these calculations. The agreement between calculated and experimental values of $\bar{\theta}_D$ in the range 0–12 K was good, but not perfect. Further modifications were made to the potential, to improve this fit and also that for the lattice energy, which had previously been slightly in error. The $\bar{\theta}_D$ results proved to be sensitive to the value chosen for the collision diameter, σ at which the pair potential energy is zero. The final potential resulting from this work is close to the presently accepted function for argon. It was then used to calculate phonon dispersion curves and elastic constants for argon

Table 8.2

Non-additive contributions to the lattice energies of the heavier inert gases

Gas	Cohesive energy	Non-additive energies $(DDD)_3$	$(DDQ)_3$	$(DQQ)_3$	$(QQQ)_3$	$(DDD)_4$
Ar	−7733±40	580	121	18	1·2	−127
Kr	−11 158±50	1004	220	34	2·4	−281
Xe	−15 840±90	1598	374	62	4·6	−637

All values in kJ mol^{-1}. Experimental data from Crawford.[32] Non-additive energies estimated by Doran and Zucker,[5] as quoted by Barker.[6]

near to absolute zero.[35] The phonon dispersion curves were found to be in good, though not perfect, agreement with the experimental data.

Similar use has been made of solid-state data in establishing potential functions for krypton and xenon.[36] In the latter case three-body contributions including the quadrupole terms, $(DDQ)_3$, $(DQQ)_3$, and $(QQQ)_3$ were used, as was also the fourth-order triple dipole $(DDD)_4$ contribution. These results, and those for argon, are summarized in Table 8.2, and it is seen that considerable, though not complete, cancellation occurs between the $(DDD)_4$ term and those involving quadrupole interactions. The uncertainties in the latter figures may be quite large. Such calculations have clearly revealed the importance of including non-additive contributions in the calculations of lattice energies for simple solids.

8.3.7. More complex molecules

Future studies of intermolecular forces between more complex molecules may be expected to draw extensively on solid-state results. Since in the solid state molecules are essentially fixed in specific relative orientations, it is clear that many solid-state properties should be a sensitive probe of the anisotropy of molecular interactions.[37] Moreover, in such systems, additional effects such as phase transitions between different crystal structures are often observed.[38] The accurate prediction of this behaviour should prove to be a further sensitive test of the orientation dependence of intermolecular forces, and should give valuable insight into the roles of multipole interactions and of molecular shape. At higher temperatures, approaching the triple point where corrections for anharmonicity may become large, the theoretical approaches described earlier may not be satisfactory. Other, procedures more appropriate to these conditions have been investigated,[39] and Monte Carlo simulations (similar to those used extensively in studies of the liquid state, and described briefly in § 8.4.3) have been found to be suitable. Successful use has also been made of cell models,[39] of the type formerly applied to the liquid state, in which a single 'wanderer' molecule is considered to move in the (anharmonic) potential due to its neighbours which are regarded as fixed on their lattice sites. Such calculations make much smaller demands on computing facilities than the Monte Carlo simulations.

At the present time, studies of molecular solids seem to have been restricted to the use of highly simplified model potentials, such as the diatomic Lennard-Jones potential,[39,40] or other simple orientation-dependent models.[41] Pairwise additivity has usually been assumed in these exploratory calculations, although it is likely that the effects of non-additivity will be of comparable magnitude to those in the inert gases.[42] There seems to be no doubt that improved calculations will be performed in the near future as quantitative studies for diatomic and polyatomic systems lead to superior potentials for such molecules.

8.4. The liquid state

The liquid state has not in the past been generally regarded as a valuable source of information about intermolecular forces, despite the vital role played by such interactions in determining the properties and even the very existence of liquids. Only in recent years has the statistical mechanics of the liquid state reached a development at which valuable information about molecular interactions may be derived from liquid properties. Even so studies based on the liquid state have not been used in the primary development of potential energy functions, but have been used to good effect in the closing stages of the refinement of intermolecular potentials for the inert gases.

The connection between molecular interactions and the structure and properties of liquids, which is provided by statistical mechanics, is very complicated, so that it seems inappropriate here to give more than a brief account of such theories. However, extensive descriptions may be found both in a recent review by J. A. Barker and D. Henderson,[4] and in the book *Theory of simple liquids* by J. P. Hansen and I. R. McDonald.[43]

8.4.1. *Liquid structure*

The structural irregularity of the liquid state is the major reason why the development of the statistical mechanics of liquids has lagged behind that of solids and gases. In spite of the high density of liquids, they do not have the periodic structure characteristic of solids. On the other hand the same high density means that the statistical mechanical procedures developed for gases at low and moderate pressures cannot be used. Liquid structure is characterized by a degree of molecular ordering at short separations, but no long-range order. The description of the structure may conveniently be carried out for spherical molecules using the radial distribution function, $g(r)$, which describes the probability of finding another molecule at a distance r from a central reference molecule.[4,43] For a system of N molecules in a total volume V, the number of molecules, $n(r)$, in the element of volume, $4\pi r^2\,dr$, at distance r from a central molecule is given by

$$n(r)=\frac{N}{V}\cdot g(r)\,.\,4\pi r^2\,.\,dr. \tag{8.19}$$

The total number of pairs of molecules whose separations lie in the range r to $r+dr$, is then related to $g(r)$ by the expression

$$n'(r)=\frac{N^2}{2V}\cdot g(r)\,.\,4\pi r^2\,.\,dr. \tag{8.20}$$

For a system in which the molecules interact only through a pairwise additive intermolecular potential, $U(r)$, the total intermolecular energy may then be readily calculated

$$U_N=\frac{2\pi N^2}{V}\int_0^\infty U(r)\,.\,g(r)\,.\,r^2\,dr \tag{8.21}$$

and the pressure may be obtained using the virial theorem of Clausius, (which is outlined in Appendix A8.1).

$$\frac{PV}{NkT}=1-\frac{2\pi N}{3kTV}\int_0^\infty g(r)\,.\,\frac{dU(r)}{dr}\,.\,r^3\,.\,dr. \tag{8.22}$$

Other functions may be obtained from these by standard thermodynamic manipulations.

The above expressions are applicable only if the intermolecular potential energy is assumed to be pairwise additive. If non-additive interactions are to be dealt with, then higher-order distribution functions must also be considered.[43,44] The most important of these is the triplet distribution function, $g^{(3)}(r_1, r_2, r_3)$, which is related to the probability that three molecules will be found simultaneously in a triangular configuration, of sides r_1, r_2, and r_3. If only three-body non-additive interactions are important (as seems usually to be the case), then a knowledge of the triplet function, together with the radial distribution function, is sufficient to determine the equilibrium properties of the system.

However, the principal structural information is contained in the radial distribution function, and the typical appearance of $g(r)$ for a simple liquid is shown in Fig. 8.7(i). The definition of $g(r)$ is such that for a random distribution of non-interacting molecules, its value would be unity at all separations. It is seen that for a liquid $g(r)$ is zero for small values of r, where, owing to the large positive values of the intermolecular

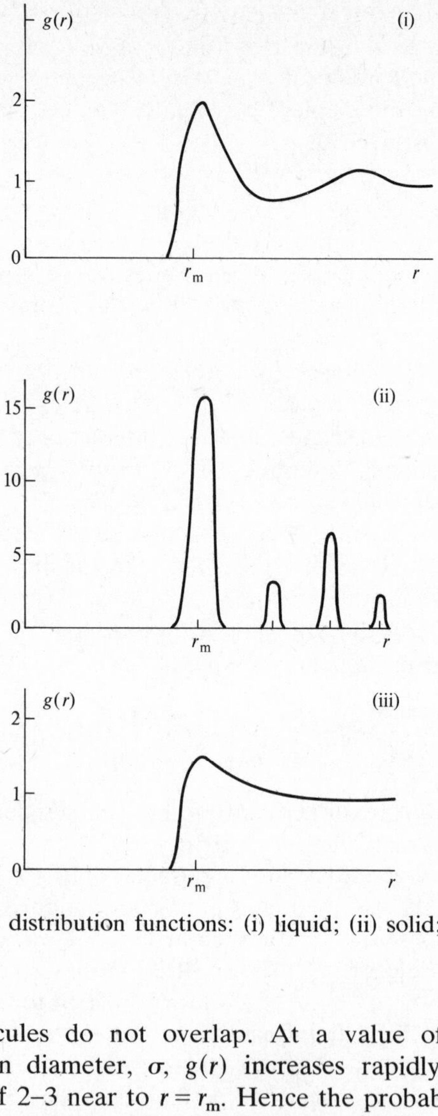

FIG. 8.7. Radial distribution functions: (i) liquid; (ii) solid; (iii) dilute gas.

energy the molecules do not overlap. At a value of r near to the molecular collision diameter, σ, $g(r)$ increases rapidly, and reaches a maximum value of 2–3 near to $r = r_m$. Hence the probability of finding a molecule at such a distance from a central molecule is several times bigger than that for a random distribution. At larger separations, $g(r)$ then falls away to a minimum, followed by a smaller maximum, and then oscillates about the limiting long-range value of unity, showing that at long distances the influence of the central molecule has died away and that there is no long-range order. Also shown in Fig. 8.7 are the corresponding results for a simple solid, and for a gas at a moderate density. The regular, long-range order of the solid is clearly shown, in contrast to the short-range order characteristic of the liquid. For the gas, $g(r)$ has a value of unity at almost all separations, with deviations at small r determined chiefly by the Boltzmann factor, $\exp(-U(r)/kT)$, based on the energy of a pair of molecules.

8.4.2. Distribution function theories of liquids

One of the most attractive approaches to the theory of liquids is that in which an attempt is made to calculate distribution functions from first principles, having assumed a suitable form for the intermolecular potential.[4,43] Applying the general principles of statistical mechanics it is possible to relate the radial distribution function, $g(r_{12})$, to the configurational intermolecular energy, U_N, through the equation

$$g(r_{12}) = \frac{1}{(N-2)!Q_N\rho^2}\int\ldots\int \exp(-U_N/kT)\,d\mathbf{r}_3\ldots d\mathbf{r}_N. \qquad (8.23)$$

Q_N is the configuration integral introduced in § 3.5, and ρ is the number density, N/V. It is seen that the direct calculations of $g(r_{12})$ involve the integration of the Boltzmann factor, $\exp(-U_N/kT)$, over all possible configurations of the molecules 3 to N while holding the two molecules of interest fixed. Clearly this procedure is not very practicable. However, it is possible, by differentiation of (8.23) with respect to $\mathbf{r}_1$, the position of molecule 1, to derive an exact and relatively simple equation relating $g(r_{12})$ to the intermolecular pair potential and to the three-body (triplet) distribution function, $g^{(3)}$. Similar linking equations can also be derived between other n-body and $(n+1)$-body distribution functions. Since, presumably, less is known about the triplet function than about the two-body function, this development may seem to be of little value. However, if an approximate expression for $g^{(3)}$ can be introduced, the equation may be solved to yield values of the radial distribution function. The best known approximation, the superposition approximation, due to Kirkwood,[45] expresses the probability of observing a configuration of three molecules at separations r_{12}, r_{23}, and r_{13} as the product of the probabilities of observing three separate pairs

$$g^{(3)}(r_{12}, r_{23}, r_{13}) \simeq g(r_{12})g(r_{23})g(r_{13}). \qquad (8.24)$$

There have been numerous calculations of distribution functions using this approach, for several forms of intermolecular potential function. Early calculations were made for hard-sphere systems[46] and for molecules interacting through a Lennard-Jones $12-6$ potential.[47] The results were promising, giving radial distribution functions which were qualitatively correct, and leading to equations of state which were accurate at low densities. However, results for higher densities were less satisfactory, and more recent developments in this area have generally been based on a rather different approach, and using two further types of distribution or correlation function.

The *total correlation* function, $h(r)$, is defined

$$h(r) = g(r) - 1 \tag{8.25}$$

and represents the fluctuation from the mean density due to a central molecule. As r becomes large, $h(r)$ tends to zero. Ornstein and Zernike[48] reasoned that the correlation, $h(r_{12})$ between molecules 1 and 2 could be considered in two parts, the *direct* influence of 1 on 2, described by the *direct correlation* function, $c(r_{12})$, and an indirect correlation, which comprised the direct correlation of molecule 1 on a further molecule, 3, which in turn exerted its total influence, described by $h(r_{23})$, on molecule 2. On integrating over all positions for molecule 3 the defining relationship for $c(r)$ results

$$h(r_{12}) = c(r_{12}) + \rho \int c(r_{13}) h(r_{23})\, d\mathbf{r}_3. \tag{8.26}$$

The direct correlation function is expected to be a short-ranged function, with a range similar to that of the intermolecular potential. A comparison between the total and direct correlation functions for a simple liquid is shown in Fig. 8.8, where it is seen that $c(r)$ is a simpler, smoother, and shorter-ranged function than $h(r)$. The equation of state may be obtained from $c(r)$ using the compressibility equation of Ornstein and Zernike[48]

$$\frac{1}{kT}\left(\frac{\partial P}{\partial \rho}\right) = 1 - 4\pi\rho \int_0^\infty c(r) r^2\, dr. \tag{8.27}$$

Since $c(r)$ and $h(r)$ are connected by (8.26), there are two independent routes to the equation of state from the correlation functions, using either the virial theorem expression, (8.22), or the compressibility equation. Exact values of $c(r)$ and $h(r)$ will yield identical results, but approximate values will give rise to inconsistent equations of state, commonly called the compressibility equation and the pressure equation.

Approximate equations may be written for $c(r)$, expressing it in terms of $g(r)$, $h(r)$, and $U(r)$, which, in conjunction with (8.26), give a pair of simultaneous equations for c and h which lead to an integral equation for $g(r)$. Two such relations which have been widely studied are the so-called hypernetted chain approximation [HNC][4,43,49]

$$c(r) = h(r) - \ln g(r) - \frac{U(r)}{kT} \tag{8.28}$$

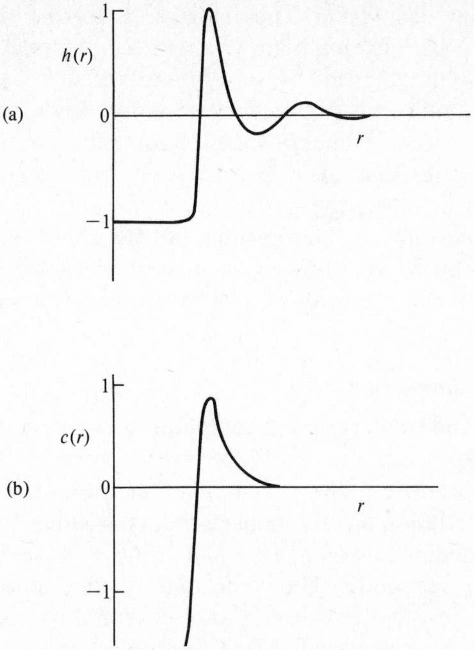

FIG. 8.8. Correlation functions for liquids: (a) the total correlation function, $h(r)$; (b) the direct correlation function, $c(r)$.

and the Percus–Yevick approximation [PY][4,43,50]

$$c(r) = g(r)\left(1 - \exp\left(\frac{U(r)}{kT}\right)\right). \tag{8.29}$$

Results have been obtained using both these approximations for several intermolecular pair potentials, notably the hard-sphere potential and the $12-6$ potential.[4,43] In both cases the resulting equations of state are found to be more accurate than those based on the superposition approximation. Nevertheless they are not exact, as is revealed for example, in the inconsistent results obtained when the equations of state are calculated from the compressibility equation and from the virial theorem.

Until recently it was not possible to assess with confidence the success of approximate statistical approaches, such as those described above. Since the calculations were performed using model potentials, such as the $12-6$ potential, the distribution functions obtained could not profitably be compared with experimental values, as the true pair potential was unknown, and there were additional uncertainties caused by non-additivity. In recent years these problems have been largely overcome thanks to the development of computer simulation techniques. These procedures have made it possible to calculate essentially exact values of the equilibrium properties of dense fluids whose molecules interact through potential energy functions specified by the investigator. (Transport properties may also be calculated, but, though there are no systematic errors in these, the results are at present generally subject to rather large uncertainty.) The results of these computer simulations should be thought of as 'quasi-experimental' data for model systems, which may then be used to test approximate statistical mechanical theories.

8.4.3. Computer simulation

Two rather different types of computer simulation procedures have been quite widely used.[4,43,51–3] These are known as the (Metropolis) Monte Carlo procedure[52] and the method of molecular dynamics.[53] In both types of calculation a cell is considered, containing a defined number of molecules, usually between 100 and 1000, whose interactions are specified by the investigator. The coordinates of the molecules are stored within the computer, and enable the total potential energy of any configuration, U_N, to be calculated. (The molecules are commonly placed initially on the sites of a crystal lattice, although this is not necessary.) When calculating the energy of a given configuration the 'minimum image' convention is generally adopted. The cell is considered to be surrounded by identical replicas of itself on each side (26 replicas in a three-dimensional case). The two-dimensional case is illustrated in Fig. 8.9. The distance between any pair of molecules i and j in the cell is compared with those between i and all the images of j in the surrounding cells, and the interaction energy is based on the shortest distance so found. A maximum separation, or 'cut-off distance', not greater than half the side length of the cell is usually employed, to prevent multiple counting of interactions.

In the Monte Carlo method the coordinates of the N molecules within the cell are changed in such a way that the probability of occurrence of a given configuration is proportional to the Boltzmann factor, $\exp(-U_N/kT)$ associated with it. This is done generally by changing the coordinates of the molecules one at a time, by adding a small randomly chosen amount to the coordinate of a given molecule. If such a change were to take the molecule through an edge of the central cell, its image in one of the neighbouring cells would have entered at the opposite side, and this molecule would then take the place of the molecule which left. The total number of molecules in the cell is thus constant, through the application of these 'periodic boundary conditions'.

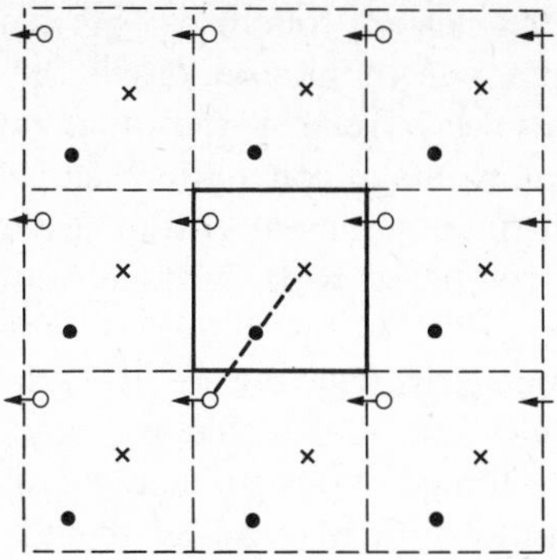

FIG. 8.9. Periodic boundary conditions and the minimum image convention. The cell under investigation is bounded by solid lines. Its surrounding images are shown with dashed lines. Three molecules are indicated, ○, ●, and ×. In calculating the interaction energy of any molecule, e.g. ●, the nearest images of the other molecules are used, as shown by the dashed lines. If a molecule leaves the cell, e.g. ○, its image enters through the opposite face.

The energy change produced by this tentative shift, ΔU, is then calculated. If the move has left the energy unchanged, or has lowered it, the new configuration is accepted. If the energy has been increased by the move the new configuration is accepted in a proportion of the cases which is given by the factor $\exp(-\Delta U/kT)$. Very many moves, sometimes millions, are made in this way, producing a series of configurations whose probability of occurrence is proportional to the Boltzmann factor, $\exp(-U_N/kT)$, and which can be analysed to give the properties of the system.[52] For example, the intermolecular energy is known at each stage, and its unweighted average over a large number of configurations may be calculated, giving directly the configurational internal energy. Similarly, the virial function, $\sum_{i>j} r_{ij}\, \mathrm{d}U(r_{ij})/\mathrm{d}r_{ij}$, may be averaged and the pressure calculated using the virial theorem. An analysis of the distance between molecules and the numbers of pairs in various ranges can give values for the radial distribution function. In addition, it is possible also to investigate quantities which are not measurable in real experiments.

It is important to note that the movement of the molecules in a Monte Carlo simulation has no physical significance, and is in no way related to the true molecular movement. It only provides a method for the sampling of configuration space over which the properties of the system must be averaged. This situation may be contrasted with that which occurs in the Molecular Dynamics simulation method.[43] A basic cell and its images are again established, but now, starting from an initial configuration in which the total (potential and kinetic) energy is defined, the motions of the molecules are followed. This is achieved by numerical integration of the equations of motion of all the molecules. For each molecule the forces resulting from all the other molecules are summed, again using the minimum image convention, and the new positions and velocities after a small time increment τ are calculated. The time step, τ, is chosen to be small compared with the time between collisions, and is typically of the order 10^{-14} s for simple systems. This procedure is repeated for many time steps, typically 10^3–10^4, during which period the motions of the molecules and their collisions may be observed. At each step in the time chain, contributions to averages similar to those studied in the Monte Carlo procedure may be evaluated. The partition of the total energy between kinetic and potential energy can also be noted, and the distribution of molecular velocities can be studied. In contrast to Monte Carlo techniques the molecular dynamics method enables the transport properties of liquids to be calculated. One way in which these may be obtained is by evaluation of the appropriate auto-correlation functions (see § 8.4.5). For instance, the diffusion coefficient is directly related to the velocity auto-correlation function, which describes the average relationship between the velocities of a given molecule at various time intervals. An alternative procedure is to perturb the system from equilibrium by imposing a gradient in a suitable macroscopic variable, and to determine the transport coefficients from the calculated fluxes, using the appropriate definitions of the transport coefficients.[54] Such methods are referred to as non-equilibrium molecular dynamics.

Since molecular dynamics yields both equilibrium and transport properties, it is in a sense a more powerful technique than the Monte Carlo method. However, it appears that results of comparable accuracy (for equilibrium properties) obtained by either method take very similar amounts of computer time for equivalent situations, and there are some circumstances in which the Monte Carlo method is more convenient. It should be noted that the statistical mechanical ensembles usually studied by the two procedures differ. The Monte Carlo approach most commonly used gives results for a canonical ensemble (i.e. constant N, V, T) while the molecular dynamics method yields data for the microcanonical ensemble (constant N, V, E). The temperature must be evaluated in this case from the distribution of energy, and is observed to fluctuate during the course of a run. For studies in which a fixed temperature is desirable, the Monte Carlo method has obvious advantages. Although the results of Monte Carlo and molecular dynamics simulations generally agree quite well, some small discrepancies have been noted.[55]

For both techniques, the demands on computing facilities are considerable. Typical Monte Carlo runs may extend over 10^5-10^6 configurations, while molecular dynamics runs of 10^3–10^5 time steps are typical.

However, extensive studies have now been reported for several model systems, and the properties of dense fluids of hard spheres and of Lennard-Jones molecules are quite well determined.[4]

The availability of exact simulation results both for hard-sphere fluids and for systems of Lennard-Jones molecules has been of great importance in the development and testing of theories of the liquid state. Thus for example the Percus–Yevick theory has been found to yield accurate results both for the equation of state and for the radial distribution functions of hard sphere fluids. Some of the most successful theories, whose development relied extensively on the availability of simulation results, are the thermodynamic perturbation theories, to which we turn in the next section.

It is in the establishment of these exact results for model systems that the main value of these computer simulations lies. Such results should not be regarded as simulations of any real substance. Where attempts are made to simulate real systems it is essential to use a totally realistic intermolecular potential function, and the effects of non-additive many-body interactions cannot be neglected.

8.4.4. Perturbation theories

Perturbation theories of the fluid state may be thought of as modern versions of van der Waals theory.[56] The essential physical principle underlying these theories is the separation of the roles of intermolecular attractive and repulsive forces. It is postulated that the structure of simple dense fluids is chiefly determined by the repulsive intermolecular interactions, and that the attractive forces maintain the high density, but do not otherwise have much effect on the structure, as described by the radial distribution function, $g(r)$. Both hard-sphere and Lennard-Jones liquids have similar radial distribution functions, which suggests that the effects of the repulsive interactions may be modelled with reasonable accuracy using a hard sphere system of a suitable diameter. The properties of hard-sphere systems are well known, both from computer simulation studies and from the Percus–Yevick equation.

In the formal development of perturbation theories, the effects of changes in the form of the intermolecular potential on the properties of a system of molecules are studied.[57] For pairwise-additive systems the pair potential energy function is written as the sum of terms, a reference potential, $U_0(r)$, plus a perturbation, $U_1(r)$.

$$U(r) = U_0(r) + U_1(r). \tag{8.30}$$

The properties of the system of molecules interacting through $U_0(r)$ are assumed to be known, and those of the perturbed systems are expressed in terms of $U_1(r)$ and those of the reference system. Thus the configurational integral Q_N, introduced in Chapter 3, may be written

$$Q_N = \frac{1}{N!}\int \ldots \int \exp\left[-\frac{1}{kT}\sum_{i>j} U_0(r_{ij}) + U_1(r_{ij})\right] \mathbf{dr}_1 \ldots \mathbf{dr}_N. \quad (8.31)$$

The exponential may be expanded in powers of U_1/kT and Q_N expressed as averages of U_1 and its powers taken over the reference system. In particular, given that

$$A = -kT \ln Z_N \quad (8.32)$$

where Z_N is the canonical partition function, and

$$Z_N = Q_N Z_1^N$$

the free energy, A, may be written[4]

$$\frac{A}{NkT} = \frac{A_0}{NkT} + \frac{2\pi N}{VkT}\int_0^\infty U_1(r) \,.\, g_0(r) \,.\, r^2 \,\mathrm{d}r + \text{higher terms}. \quad (8.33)$$

In this equation, A_0 is the Helmholtz free energy of the reference system, and $g_0(r)$ is the radial distribution function for this unperturbed system. The integral over $g_0(r)$ represents the contribution of the perturbation energy assuming that the structure of the system is identical to that of the reference system. The subsequent terms, of order $(1/kT)^2$, etc., involve integrals over the three- and four-body distribution functions of the reference system. The equation of state may be obtained using the thermodynamic relation

$$P = -(\partial A/\partial V)_T$$

which gives

$$P = P_0 + \frac{2\pi N}{3}\frac{\partial}{\partial V}\left[\frac{1}{V}\int U_1(r) \,.\, g_0(r) \,.\, r^2 \,\mathrm{d}r \ldots\right] + \text{higher terms} \quad (8.34)$$

where P_0 is the pressure of the reference system.

If the perturbation expansion is truncated after the first-order term, it is seen that the total pressure is equal to that of the reference system, P_0, plus a correction term arising from the perturbation energy, integrated over the structure of the reference fluid. Writing the van der Waals equation of state in the form

$$P = \frac{N_A kT}{\tilde{V} - b} - \frac{a}{\tilde{V}^2} \quad (8.35)$$

and associating the reference potential $U_0(r)$ with the repulsive interactions, and the perturbation energy with attractive interactions makes the analogy between these two approaches clearer.

The first-order attractive term represents the effect of the mean field due to the perturbation energy. The higher-order terms describe the contributions associated with fluctuations about the mean field. In view of the nature of the expansions for A and P as series in $1/kT$, it was originally suggested that the perturbation approach would be suitable for the high-temperature equation of state of dense gases,[58] where higher terms than the first would be small because of the factors $(1/kT)^2$, etc. In fact this theory has been applied with considerable success to most of the liquid range from the triple point to temperatures close to the critical.[59] The success of this approach has rested on the correct choice of the division of $U(r)$ into reference and perturbation potentials. By a suitable division of $U(r)$, the magnitude of the higher-order terms can be made very small, thus achieving rapid convergence of the series, even at low temperatures.

Most perturbation calculations have attempted to calculate the properties of a liquid whose molecules interacted through a model pair potential, usually the Lennard-Jones function. In the earliest studies,[58] $U_0(r)$ was taken to be a hard-sphere potential, with the hard-sphere diameter, d, slightly less than the Lennard-Jones collision diameter, σ, but with its value chosen in a rather arbitrary manner. For all values of r greater than σ, $U_1(r)$ was taken to be the Lennard-Jones potential. This division of the potential is illustrated in Fig. 8.10(a) and may be expressed

$$\begin{aligned} U_0(r) &= \infty, \qquad r \leqslant d \\ &= 0, \qquad r > d \\ U_1(r) &= 0, \qquad r \leqslant \sigma \\ U_1(r) &= 4\varepsilon\left\{\left(\frac{\sigma}{r}\right)^{12} - \left(\frac{\sigma}{r}\right)^{6}\right\} \qquad r > \sigma. \end{aligned} \tag{8.36}$$

The first-order perturbation theory was used and the results were compared with high-temperature PVT data for simple gases, since very few results were available for the Lennard-Jones fluid at that time. It was found that the results were fairly satisfactory at supercritical temperatures although sensitive to the choice of the hard-sphere diameter, d.[58] Significant progress was made only after the development of more reliable and firmly based methods for choosing d.

Such methods were first formulated in studies of the properties of systems of molecules interacting through simple inverse power repulsive potential, $U(r) = C_n/r^n$, which may be expressed in terms of a hard-sphere model, with a temperature-dependent core size, and the statistical

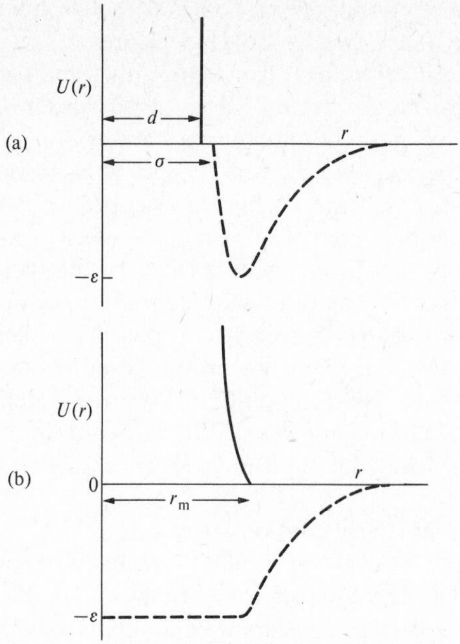

FIG. 8.10. Division of the total potential into reference and perturbation components. (———) $U_0(r)$, reference potential; (- - - -) $U_1(r)$, perturbation. (a) Barker–Henderson; (b) Weeks–Chandler–Andersen.

mechanics of such systems have been investigated in some detail.[60] A perturbation expansion of the configurational integral in powers of the 'steepness parameter' $1/n$ about the hard-sphere value, $1/n = 0$, leads to expressions for the core size as a function of n and the temperature. In the important studies of Barker and Henderson,[59,61] a similar perturbation treatment of the repulsive interactions was combined with the attractive perturbation method described previously in a double perturbation expansion. The effective hard-sphere diameter was chosen in such a way as to make the first-order repulsive perturbation term zero, and was defined by the equation

$$\mathrm{d} = -\int_0^{\sigma} \{\exp(-U(r)/kT) - 1\}\,\mathrm{d}r. \tag{8.37}$$

This relationship gives a hard core size which decreases with increasing temperature, but is independent of the density. Barker and Henderson were able to calculate the first- and second- order attractive perturbation terms using hard-sphere computer simulation results for the distribution functions of the reference system. Calculations were made for Lennard-Jones 12−6 systems, and the results were in very satisfactory agreement with data obtained by direct simulation of such systems. The second-order term, though small, was found to be necessary to achieve the excellent level of agreement. A straightforward numerical procedure by which this term could be introduced without the explicit use of high-order distribution functions was later described by Barker and Henderson.[61]

An alternative very successful perturbation theory was proposed by Weeks, Chandler, and Andersen,[62] and is based on a novel choice of reference and perturbation potentials. This is illustrated in Fig. 8.10(b), and may be written

$$\begin{aligned} U_0(r) &= U(r)+\varepsilon, & r &\leq r_m \\ U_0(r) &= 0, & r &> r_m \\ U_1(r) &= -\varepsilon, & r &\leq r_m \\ U_1(r) &= U(r), & r &> r_m. \end{aligned} \tag{8.38}$$

This division of potentials assigns the whole of the *repulsive* region of $U(r)$ to the role of reference potential and determinant of structure, rather than just the *positive* portion of $U(r)$, as has been more usual. A consequence of this division is that the perturbation energy, $U_1(r)$, is now a very smoothly varying function of r, and this seems to have the useful effect of reducing the higher-order fluctuation terms, giving a very accurate equation of state even when restricted to a first-order treatment. However a penalty is incurred also, in that the properties of the reference fluid are not now so simple as those of the hard-sphere fluid nor as well established. Weeks *et al.*[62] were able to propose a relation between their reference fluid and hard-sphere fluid, and used this to make calculations for the Lennard-Jones fluid which proved to be approximately as accurate in first-order as those based on the complete Barker–Henderson perturbation treatment.

Both of these perturbation theories provide remarkably accurate and reasonably simple procedures by which the property of fluids of spherical molecules, interacting through pairwise-additive intermolecular potentials, may be determined. They have also given valuable insights into the relationships between intermolecular forces and liquid structure which have led to useful developments in related fields.[56]

8.4.5. *Transport properties of liquids*

The difficulties which are encountered in deriving a comprehensive theory for the transport properties of dense fluids have been mentioned in § 5.10. In view of these problems it is not surprising that the accurate calculation of any transport property of a liquid, knowing its intermolecular potential function, is not yet possible, except by computer simulation.

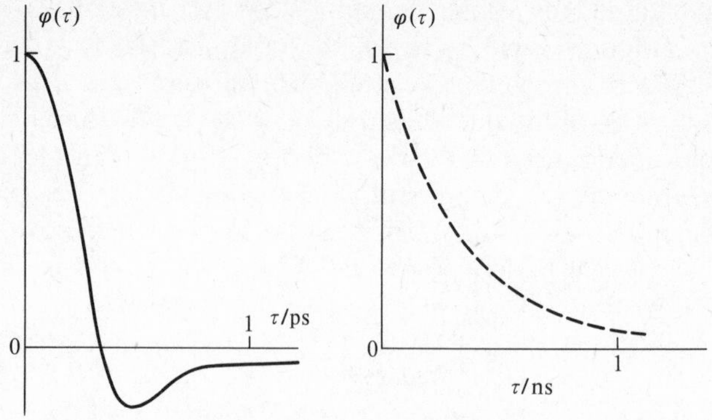

FIG. 8.11. Normalized velocity auto-correlation functions: (———) liquid; (- - - -) dilute gas. Note that the time scales differ.

Recent work in this field has concentrated on methods which relate the transport properties to time-correlation functions for dynamical variables.[63,64] Such correlation functions measure the degree to which the value of a dynamical variable, X, at a time $t+\tau$, is related to the value of the dynamical variable, Y, at time t. If X and Y are the same variable, the correlation function is called an auto-correlation function. Thus, the velocity auto-correlation function, $\phi(\tau)$, may be defined

$$\phi(\tau) = \langle \mathbf{v}(0) \cdot \mathbf{v}(\tau) \rangle \tag{8.39}$$

where $\mathbf{v}(t)$ is the velocity of a specified particle at time t, and the brackets denote an ensemble average. The typical behaviour of $\phi(\tau)$ in a dilute gas and in a simple liquid is illustrated in Fig. 8.11. In a gas, the molecule 'forgets' its previous velocity after a few collisions, and $\phi(\tau)$ drops monotonically to zero over this period. In the liquid, the molecule on average undergoes a reversal of direction, following a collision with a neighbour and this negative correlation is quite long lived. The mean square distance travelled by a molecule in a given time may be expressed as an integral over $\phi(\tau)$, and hence the velocity auto-correlation function can be related to the coefficient of self-diffusion, D (which was discussed in Chapter 5.)

$$D = \tfrac{1}{3} \int_0^\infty \phi(\tau)\, d\tau. \tag{8.40}$$

All the transport properties of liquids may be expressed in the form of integrals over the appropriate correlation functions. Properties such as self-diffusion, sheer and bulk viscosity, and thermal conductivity, which relate conjugate fluxes and forces, depend on auto-correlation functions.

A property such as thermal diffusion, which occurs as a result of coupling between fluxes, depends on cross-correlation functions. An accurate knowledge of the correlation functions thus leads directly to the transport properties of liquids.

However, no direct methods exist for the evaluation of these correlation functions for dense fluids. They may in principle be calculated in the course of a computer simulation using molecular dynamics,[65] and this approach has been proved practically feasible in the case of self-diffusion. The analysis of data to give the auto-correlation functions can be very time-consuming, and the statistical uncertainties in calculations of other properties are generally rather large. Calculations have so far been restricted to simple potential functions, such as the hard-sphere[65] or Lennard-Jones models,[66] and no extensive studies of the effects of modest changes in the form of the intermolecular potential have yet been performed. In recent years there has been some progress in the calculation of correlation functions from first principles, but so far these methods too have been restricted to simple model potentials.

For these reasons, the transport properties of liquids have so far contributed little to our knowledge of intermolecular forces. As faster computers become available it is likely that direct simulation methods will be successfully applied to a wider range of transport properties and intermolecular potentials. These may then provide a valuable additional source of information about the role of the intermolecular potential in determining the transport coefficients of liquids.

8.4.6. Experimental measurements for simple liquids

The importance of the liquefied inert gases as the simplest liquid systems has long been recognized and there are data for both their structural and thermodynamic properties over wide ranges of temperature and density. This is particularly true in the case of argon.

All measurements on the liquid inert gases involve the use of cryogenic techniques and considerable attention must be paid to the maintenance and accurate measurement of temperature. Measurements of *PVT* data may be obtained quite directly. In the simplest method,[67] a quantity of the substance under study may be trapped in a cell of known volume, and the temperature and applied pressure measured. The amount of substance can be established by *PVT* measurements at room temperature and low pressure. In an alternative approach,[68] the amount of substance was established by weighing a specially designed cryostat assembly. The weight of a given volume of sample was established by carrying out two sets of measurements over the same range of temperature and pressure, in one of which a part of the cell was occupied by an iron cylinder of known volume. The difference in sample weights gave the mass of sample whose volume was equal to that of the cylinder. In this way possible errors arising from pressure-induced changes in the volume of the sample cell were avoided.

An important property used in the calculation of thermodynamic properties is the saturation vapour pressure of a liquid which, given accurate temperature and pressure measurement, may be readily studied. From the slope of the vapour pressure, P_s, against temperature the enthalpy of vaporization may be established using the Clapeyron equation

$$\frac{dP_s}{dT} = \frac{\Delta H_{vap}}{T \Delta V_{vap}}. \tag{8.41}$$

Alternatively ΔH_{vap} may be directly measured using the standard procedures of adiabatic calorimetry. Knowing the enthalpy of vaporization along the liquid–vapour coexistence curve, and using data for gas non-ideality, the configurational energy of the liquid may be calculated.

Structural information, represented by $g(r)$, may be obtained using X-ray or neutron scattering experiments.[69] When monochromatic incident radiation is used, interference patterns occur in the diffracted radiation owing to molecular correlations, which may be used to construct the radial distribution function. The experimental difficulties in such work are quite severe, but the general features of the experimental results are now quite well defined. At constant density the effects of changes in temperature are quite small, but the result of changing the density at constant temperature may be considerable.[70]

Another property which can give useful information about the intermolecular potential function is the isotopic separation factor.[71] When an isotopic liquid mixture is in equilibrium with its vapour, the isotopic compositions of the two phases differ. If the vapour phase mole fractions of isotopes a and b are y_a and y_b, and those of the liquid are x_a and x_b, the isotopic separation factor, α, may be defined

$$\alpha = \frac{y_a/y_b}{x_a/x_b}. \tag{8.42}$$

The composition ratios may be measured using mass spectrometry.

Measurements of the transport properties of simple liquids have so far generally concentrated on the shear viscosity, and few accurate results have been reported for other properties. As no satisfactory analytical theories exist which can incorporate realistic intermolecular potential functions, and the calculation of transport properties using computer simulation is, apart from diffusion, both time-consuming and inaccurate, these properties have as yet contributed little to the study of intermolecular forces.

8.4.7. Liquid-state properites and intermolecular forces

Following the development of reliable theories of the liquid state it is now possible to assess the relationships between the equilibrium properties of liquids and intermolecular potential energy functions.

The major equilibrium properties of liquids which have been found useful are the pressure and internal energy as functions of temperature and density. Although a great deal of work has been done on the structural properties of simple liquids, using diffraction techniques to obtain the pair distribution functions, these are now known to be insensitive to the details of the intermolecular pair potential. However, they are needed to calculate the pressure and energy for a given potential model. At low densities, for imperfect gases, $g(r)$ is directly related to the pair potential, $U(r)$, by the equation[4]

$$g(r) = \exp(-U(r)/kT). \tag{8.43}$$

If accurate measurements of $g(r)$ could be made for dilute gases a direct inversion procedure giving $U(r)$ would be available, and this would constitute a most attractive route to the pair potential energy function. Unfortunately the experimental difficulties involved in accurate measurements at low pressures are very severe, and this direct approach has not yet been successfully exploited. However a closely related method based on results at higher densities has been developed and successfully applied to gaseous argon.[72]

For liquids interacting through a pairwise-additive potential energy function, the pressure may be calculated from (8.22) provided that $g(r)$ is known. The integration for the pressure contains a positive and a negative region, corresponding to repulsive and attractive forces respectively. These contributions largely cancel in the calculation of the total pressure, and the extent of the cancellation is found to depend sensitively on the shape of the pair potential near to the collision diameter, σ. Calculation of liquid pressures near to the coexistence curve, where the net pressure is small, have been found to provide a useful test of the details of the pair potential in this region.[73] The configurational internal energy on the other hand is largely determined by the intermolecular potential well, and depends on the over-all shape of this part of the potential.

Calculations of these liquid-state properties using accurate pair potentials derived from other studies show clearly that the assumption of pairwise additivity is ill founded.[73] The pressure and internal energy differ systematically from the experimental values, and it must be concluded that many-body interactions contribute significantly to these properties. Attempts to incorporate such effects into calculations have generally been restricted to the Axilrod–Teller triple-dipole term. In order to calculate the effects of three-body forces a knowledge of the three-body distribution functions is required. These are not well characterized in detail, but may be estimated using the superposition approximation (8.24) which is probably sufficiently accurate for this purpose. Alternatively, the three-body contributions may be incorporated explicitly in computer simulations. If only the triple-dipole contribution is included the contributions to the energy and to the pressure are simply related.[44] The over-all three-body contributions both to the pressure and to the energy are found to be positive. The cancellation found in the pairwise-additive calculation of the pressure does not occur with the three-body term, which may itself be similar in magnitude to the total pairwise-additive pressure. Non-additive contributions to the internal energy are about 5 per cent of the pairwise energy near to the coexistence curve.

Further information about the potential, its curvature in the region of the well, may be inferred from studies of the isotopic separation factors, introduced in § 8.4.6, and defined in (8.42). The isotopic separation results from the effects of mass differences on the quantized energy levels in the liquid. Provided that the quantum effects are small, α may be related to the isotopic masses m_a and m_b and the liquid and vapour state configurational energy, U_N, by the equation[71]

$$\ln \alpha = \frac{h^2}{96\pi k^2 T^2}\left(\frac{1}{m_a}-\frac{1}{m_b}\right)[\langle \nabla^2 U_N^{\text{liq}}\rangle - \langle \nabla^2 U_N^{\text{vap}}\rangle] \tag{8.44}$$

where the brackets, $\langle\ \rangle$, denote ensemble averages and ∇^2 is the Laplacian operator. The quantity $\nabla^2 U_N^{\text{vap}}$ is small, and may be readily evaluated so that $T^2 \ln \alpha$ is a measure of the average value of $\nabla^2 U_N$ in the liquid phase. For pairwise-additive potentials, U_N^{liq} may be written in terms of the pair potential energy function, $U(r)$, and the radial distribution function, $g(r)$. Then we may write

$$\langle \nabla^2 U_N^{\text{liq}}\rangle = \frac{4\pi N}{V}\int_0^\infty g(r)\frac{\mathrm{d}}{\mathrm{d}r}\left[r^2\,\mathrm{d}\frac{U(r)}{\mathrm{d}r}\right].\,\mathrm{d}r. \tag{8.45}$$

When the form of the integrand is studied using realistic distribution functions and pair potentials, it is found to be sharply peaked in the region of $1{\cdot}1\,\sigma$, and to be almost zero below $0{\cdot}9\,\sigma$ and above $1{\cdot}25\,\sigma$.[71] Measurements of α may thus give useful information on the curvature of the potential in this range of r, provided that $g(r)$ values are available, and the effects of many-body forces may be taken into consideration. It should perhaps be pointed out that the experimental error in α is moderately large (about 3–5 per cent in the case of argon) and that further uncertainties are introduced by the use of experimental results for $g(r)$ and in correcting for the effects of three-body forces. Nevertheless calculations of α have provided a useful test[74] of proposed potentials for Ne, and have also been applied (assuming pairwise additivity) to nitrogen.[75]

8.4.8. Results for argon

Early computer simulation studies using $12-6$ potentials and assuming pairwise additivity gave results which were often compared with experimental data for argon.[76–8] Quite good agreement with *PVT* data was usually found, but as the deficiencies of the $12-6$ potential became clearer it became apparent that little significance could be attached to this comparison. Other simulation studies using pair potentials much nearer to those presently accepted gave results which were significantly poorer than the values from the $12-6$ potential.[79] However it was pointed out by Barker *et al.* that the neglect of many-body interactions in these calculations could have serious consequences. They performed further simulations using pair potentials established on the basis of gas virial coefficients and low-temperature solid-state data, and including corrections for the Axilrod–Teller triple-dipole three-body energy.[73] These led to results which agreed with the experimental data for argon about as well as had the $12-6$ values. Some systematic discrepancies were still apparent though, especially in the pressure, and Barker and his co-workers were able to modify their pair potential to remove these, while still retaining good fits to the other properties on which the potentials had been based. The resulting potential gave radial distribution functions in good agreement with experimental values for argon, though it has been noted[4] that the $12-6$ potential assuming pairwise additivity also leads to an equally accurate description of these data.

Another significant aspect of these calculations for argon was the inclusion of quantum effects in the evaluation of the internal energy and pressure. For argon these may be estimated satisfactorily using the semi-classical Wigner expansion, similar to that described in Chapter 3 for second virial coefficients. For the internal energy the quantum corrections were fairly small, perhaps 1 per cent of the total, but for the pressure the quantum corrections were quite large, perhaps 10 per cent of the three-body contribution. As discussed in the previous section, the corrections resulting from the triple-dipole non-additive energy were large, and the importance of including such contributions in calculations of liquid state properties was reinforced by these studies.

8.4.9. More complex molecules

In recent years considerable interest has been shown in the properties of simple molecular fluids, particularly those consisting of diatomic and linear triatomic molecules.[80,81] There have been many computer simulations both of model systems with hard potentials, such as hard dumb-bells or hard spherocylinders, and also of more realistic models, mostly using potential functions based on a Lennard-Jones atom–atom interaction, such as the diatomic model introduced in Chapters 1 and 3. In addition the effects of permanent electrical multipoles have been studied. These Lennard-Jones-based potential energy functions are claimed to provide adequate pairwise-additive *effective* potential functions for such simple systems. Nevertheless it is clear from the studies of the inert gases that the effects of non-additive intermolecular interactions are likely to be important in molecular liquids also. Hence such pairwise-additive functions are best regarded as models, rather than as accurate representations of the interactions of real molecules. As model systems they have been valuable in providing insight into the effects of molecular shape on the structure and properties of simple molecular systems.

An important development in the study of molecular liquids has been the development of a very successful theory which appears to yield quite accurate structural information.[82,83] For non-spherical molecules the radial distribution function becomes a function of the relative orientation of the molecules, as well as of their separation. It is consequently harder to represent, and in the reference interaction site model, attention is focused on an alternative structural description. With each molecule are associated two or more force centres, or interaction sites. The total intermolecular energy is calculated by summing the contributions from pairs of sites on different molecules, and the structure is expressed in terms of site–site distribution functions, which depend only on the site separations. The latter are consequently more easily represented than the orientation-dependent radial distribution functions, though several types of site–site function may be needed for a complete description of the structure. Chandler and co-workers[82,83] were able to devise an approximate integral equation for the site–site distribution functions, in the case when the individual site interactions were of the hard-sphere type. This integral equation is closely related to the Percus–Yevick equation for hard spheres, and appears to give very reasonable results for a range of molecular liquids. Once again it seems that repulsive intermolecular interactions, modelled by a hard-sphere potential, are largely responsible for the structure even in these more complex liquids. The detailed part played by electrostatic interactions between permanent dipoles, quadrupoles, etc. are as yet not fully understood, although this is an area of considerable research activity at the present time.

Substantial progress in our understanding of more complex liquids can only come after more accurate pair potential functions are available. Even for simple diatomic molecules the development of such potentials is still at an early stage, and until more precise information has been obtained from microscopic and low-density macroscopic properties the quantitative study of these liquids will not be possible. Since the generally free rotation of molecules in liquids leads to extensive averaging of the orientation-dependent attractive contributions to the intermolecular energy, it seems probable that studies of molecular liquids will play only a minor role in the characterization of intermolecular potentials for more complex molecules.

8.4.10. Liquid mixtures

An area in which the effects of intermolecular interactions are of major importance is the study of liquid mixtures. The changes in thermodynamic properties which accompany the formation of a liquid mixture from its components have been investigated for many decades, and there have been numerous attempts to relate these to the molecular processes taking place. Most commonly the mixing is studied under conditions of constant temperature and pressure, and the results are presented in terms of thermodynamic excess mixing functions. When n moles of a mixture are formed from its pure components, the associated change in an extensive thermodynamic property, X, may be represented $\Delta_{\mathrm{mix}}X$. In an *ideal* mixture, these changes may be simple related to the composition of the mixture represented in terms of the mole fractions, x_i, of the components. Specifically,

$$\left.\begin{aligned} &\Delta^{\mathrm{id}}_{\mathrm{mix}}V = \Delta^{\mathrm{id}}_{\mathrm{mix}}H = 0 \\ &\Delta^{\mathrm{id}}_{\mathrm{mix}}G = -T\Delta^{\mathrm{id}}_{\mathrm{mix}}S = nRT\sum_i x_i \ln x_i. \end{aligned}\right\} \tag{8.46}$$

Such an ideal mixture obeys Raoult's law (assuming that the vapour behaves as a perfect gas mixture) but, in practice, this behaviour is rarely, if ever, observed. More commonly the mixing process is accompanied by a volume change, and the evolution or absorption of heat, and the mixture shows deviations from Raoult's law. The excess functions describing such changes are defined in terms of the observed mixing functions and those for an ideal mixture of the same composition. Thus

$$X^{\mathrm{E}} = \Delta_{\mathrm{mix}}X - \Delta^{\mathrm{id}}_{\mathrm{mix}}X \tag{8.47}$$

and the excess functions differ from the mixing functions only in the cases of the free energy and entropy. If a mixture is formed with a positive excess free energy it is said to show positive deviations from ideality, and is thermodynamically less stable than the corresponding ideal mixture. These mixtures display positive deviations from Raoult's law and, in extreme cases, show characteristic behaviour, such as limited miscibility of the components, or the formation of azeotropes boiling at a minimum temperature. Mixtures which have negative excess free energies show negative deviations from ideality, and, when the deviations are large, may form azeotropes boiling at a maximum temperature.

It is clear that the excess thermodynamic functions reflect the differences between the intermolecular interactions in the pure components and in the mixture, and also the structural changes which accompany mixing. Evidently the interactions in mixtures include those between unlike molecules, as well as between molecules of the same type, and studies of liquid mixtures may thus be expected to provide a valuable source of information on the interactions between unlike molecules. In the simplest theoretical model, the intermolecular potentials in a binary mixture may be characterized in terms of the associated well depths, $\varepsilon_{11}, \varepsilon_{22}$, and ε_{12}. The difference, w,

$$w = \varepsilon_{12} - \tfrac{1}{2}(\varepsilon_{11} + \varepsilon_{22}) \tag{8.48}$$

then gives a measure of the relative strength of the like and unlike interactions. Positive values of w indicate relatively strong unlike interactions, associated with negative deviations from ideality, and vice versa. Using this analysis, major deviations from ideality, for example those associated with 'chemical' changes such as hydrogen-bonding, may be qualitatively related to the intermolecular interactions. More quantitative studies are necessary in the absence of such specific effects and can only be carried out using quite elaborate theoretical approaches.

Although exact theories for liquid mixtures are not available, the emphasis in this area is on relating the properties of mixtures to those of the pure components, since it is the changes which occur on mixing which are of primary interest. In a widely studied approach, the one-fluid model, the properties of a binary mixture are taken to be those of a hypothetical pure liquid. It is usual to assume that this hypothetical fluid and the two pure components have conformal intermolecular potentials, and so satisfy the principle of corresponding states. The intermolecular potential for the hypothetical fluid is calculated as some composition-dependent average of the potentials for the pure component and for the unlike interactions. Thus in the random mixing approximation, the mixture potential function, $U_m(r)$, may be written

$$U_m(r) = \varepsilon_m \phi\left(\frac{r}{\sigma_m}\right) = x_1^2 U_{11}(r) + 2x_1x_2U_{12}(r) + x_2^2 U_{22}(r) \tag{8.49}$$

where U_{11} and U_{22} represent the potentials for pure components 1 and 2, and U_{12} is that for unlike interactions. The Lorentz–Berthelot combining rules (see Chapter 3) have been widely used to estimate U_{12}. For any reasonable form of $\phi(r/\sigma)$, this equation leads to expressions for the characteristic parameters, ε_m and σ_m, for the hypothetical fluid, which involve all of the six characteristic parameters for the like and unlike interactions. The properties of the mixture may then be related to those of the pure fluids, using the principle of corresponding states.

For mixtures whose molecules differ substantially in size this approach incorrectly predicts large positive contributions to the excess thermodynamic-mixing functions.[85] The successful applications of perturbation theories to pure liquids demonstrated the importance of representing correctly the structural consequencies of repulsive intermolecular forces and this the random-mixing models were unable to do. Computer simulations for mixtures of hard spheres of different sizes showed that for this important model system the random-mixing approximation gave very poor results. The development of improved mixture theories able to describe accurately the properties of mixture whose components differ in size has been stimulated by the availability of computer simulation results for hard-sphere mixtures and also for mixtures of Lennard-Jones molecules.[86–8] These have been used to good effect in assessing the success of alternative approaches to systems in which the molecular interactions are well defined, and generally reliable theories are now available for mixtures whose molecules are more or less spherical.

One of the simplest such theories, the 'van der Waals one-fluid' model[89] closely resembles the original extension to mixtures by van der Waals of his equation of state. In its modern form the one-fluid approach is again followed, but the mixture parameters ε_m and σ_m in (8.49) are derived from the relations

$$\begin{aligned} \varepsilon_m\sigma_m^3 &= x_1^2\varepsilon_{11}\sigma_{11}^3 + 2x_1x_2\varepsilon_{12}\sigma_{12}^3 + x_2^2\varepsilon_{22}\sigma_{22}^3 \\ \sigma_m^3 &= x_1^2\sigma_{11}^3 + 2x_1x_1\sigma_{12}^3 + x_2^2\sigma_{22}^3. \end{aligned} \tag{8.50}$$

Using these prescriptions, the one-fluid model is now able to cope quite well with mixtures of simple liquids, even when the molecules differ substantially in size. This theory has been extensively tested using results for mixtures of Lennard-Jones molecules obtained from computer simulations,[88] and now provides a generally reliable framework for calculating the mixing properties of simple liquids. When applied to real mixtures of simple liquids this approach has yielded useful information about the interactions between unlike molecules, and on the adequacy of the Lorentz–Berthelot combining rules. In particular, it has been demonstrated[89] that the geometric mean rule,

$$\varepsilon_{12} = (\varepsilon_{11}\varepsilon_{22})^{\frac{1}{2}}$$

generally overestimates ε_{12}. The extent of the discrepancy can be expressed in terms of a parameter k_{12} defined by

$$\varepsilon_{12} = (1 - k_{12})(\varepsilon_{11}\varepsilon_{22})^{\frac{1}{2}}.$$

Table 8.3
Deviations from geometric mean rule

	$(1-k_{12})$
$C_2H_6+C_3H_8$	1·00
$Ar+O_2$	0·99
N_2+O_2	0·99
$CO+CH_4$	0·99
$CH_4+C_2H_6$	0·99
$C_6H_6+c\text{-}C_6H_{12}$	0·97
N_2+CH_4	0·97
$CH_4+C_3H_8$	0·97
CH_4+CF_4	0·92
$CO_2+C_2H_2$	0·91

Taken from Bett, K. E., Rowlinson, J. S., and Saville, G. *Thermodynamics for chemical engineers*, p. 477. Athlone Press, London (1975).

For similar molecules the deviation from the geometric mean is very small but in some cases may be as much as 10 per cent. Some typical values of $(1-k_{12})$ estimated from the study of liquid mixtures are given in Table 8.3. These figures are subject to an uncertainty of at least 0·01 since, as explained above, the theory used to obtain them is not exact.

Other related theories, more directly based on the perturbation theories of pure liquids described earlier,[90,91] have also been developed, and tested in the same way. Information about the interactions of unlike molecules may thus be derived from thermodynamic data. The intermolecular potentials for Ar–Kr and Kr–Xe have been studied in this way,[92] using an extension of the Barker–Henderson perturbation theory to mixtures, and incorporating corrections for non-additive three-body forces and quantum corrections. Such calculations may well provide a valuable source of information on unlike-molecule interactions in the future.

Appendix A8.1. The virial theorem of Clausius

The classical virial theorem was originally deduced by Clausius and applied by him, and by van der Waals, to the description of imperfect gases. Consider a system of N molecules contained in a box of volume V. Newton's second law of motion may be applied to a given molecule, i, in the form

$$m_i \frac{d^2 x_i}{dt^2} = F_{x_i} \tag{A8.1}$$

where F_{x_i} is the x-component of the force acting on molecule i. Multiplying by x_i gives

$$x_i F_{x_i} = m_i x_i \frac{d^2 x_i}{dt^2} = \frac{d}{dt}\left(m_i x_i \frac{dx_i}{dt}\right) - m_i \left(\frac{dx_i}{dt}\right)^2. \tag{A8.2}$$

Summing over all the molecules, and taking the average over time we obtain

$$\overline{\sum x_i F_{x_i}} = \overline{\sum \frac{d}{dt}(m_i x_i u_{x_i})} - \overline{\sum m_i (u_{x_i})^2} \tag{A8.3}$$

where u_{x_i} is the x-component of the velocity.

The first term on the right-hand side of this equation may be written explicitly as a time average

$$\overline{\sum \frac{d}{dt}(m_i x_i u_{x_i})} = \frac{1}{\tau}\int_0^\tau \frac{d}{dt}(m_i x_i u_{x_i})\, dt$$

$$= \frac{1}{\tau}\{(m_i x_i u_{x_i})_{t=\tau} - (m_i x_i u_{x_i})_{t=0}\}.$$

Since $m_i x_i u_{x_i}$ must remain finite, and τ may tend to an infinitely large value, this average must be zero. Hence we may write

$$\overline{\sum x_i F_{x_i}} = -\overline{\sum m_i u_{x_i}^2} = -NkT \tag{A8.4}$$

where we have introduced the familiar result that the mean kinetic energy of N molecules in one dimension is $\frac{1}{2}NkT$. Adding in the corresponding results for the y- and z-directions we obtain

$$\overline{\sum x_i F_{x_i} + y_i F_{y_i} + z_i F_{z_i}} = \overline{\sum \mathbf{r}_i \cdot \mathbf{F}_i} = -3NkT. \tag{A8.5}$$

The left side of this equation is the *virial* of Clausius, which may be resolved into two contributions. The forces on the molecules are of two types, those due to intermolecular interactions and those resulting from collisions with the walls. The latter may readily be shown to contribute an amount $-3PV$ to the total virial, and hence we may write

$$-3PV + \overline{\sum \mathbf{r}_i \cdot \mathbf{F}'_i} = -3NkT$$

or

$$PV = NkT + \tfrac{1}{3}\overline{\sum \mathbf{r}_i \cdot \mathbf{F}'_i} \tag{A8.6}$$

where $\mathbf{F}'_i$ represents the force on molecule i resulting from molecular interactions. In general, $\mathbf{F}'_i$ may be related to the intermolecular energy of the system, U_N, by the relation

$$\mathbf{F}'_i = \nabla_i U_N$$

and hence

$$\frac{PV}{NkT}-1=-\frac{1}{3NkT}\left\langle\sum_{i=1}^{N}\mathbf{r}_i\,.\,\boldsymbol{\nabla}_i U_N\right\rangle \tag{A8.7}$$

where we have replaced the time average by the corresponding ensemble average. For spherical molecules interacting through a pairwise-additive potential function, $U(r)$, we may write

$$U_N=\sum_{j>i}U_{ij}(r_{ij})$$

and the virial is related to $U(r)$ by the equation

$$\sum_{i=1}^{N}\mathbf{r}_i\,.\,\boldsymbol{\nabla}_i U_N\equiv\sum_{j>i}r_{ij}\frac{\mathrm{d}U(r_{ij})}{\mathrm{d}r_{ij}}. \tag{A8.8}$$

The connection between the intermolecular pair potential and the equation of state may thus be written

$$\frac{PV}{NkT}-1=-\frac{1}{3NkT}\left\langle\sum_{j>1}r_{ij}\frac{\mathrm{d}U(r_{ij})}{\mathrm{d}r_{ij}}\right\rangle \tag{A8.9}$$

and, since the number of pairs of molecules with separations between r and $r+\mathrm{d}r$ is equal to $(N^2/2V)\ 4\pi r^2 g(r)\,\mathrm{d}r$, we obtain, finally,

$$\frac{PV}{NkT}-1=-\frac{2\pi}{3kT}\frac{N}{V}\int_0^\infty r^3 g(r)\frac{\mathrm{d}U(r)}{\mathrm{d}r}\,\mathrm{d}r. \tag{A8.10}$$

References

1. Lennard-Jones, J. E. and Ingham, A. E. *Proc. roy. Soc.* **A107,** 636 (1925).
2. Kihara, T. and Koba, S. *J. Phys. Soc. Jpn* **7,** 348 (1952).
3. Klein, M. L. and Venables, J. A. (Eds.). *Rare gas solids.* Academic Press, London: Vol. I (1976); Vol. II (1977).
4. Barker, J. A. and Henderson, D. *Rev. mod. Phys.* **48,** 587 (1976).
5. Doran, M. B. and Zucker, I. J. *J. Phys.* **C4,** 307 (1971).
6. Barker, J. A. Ref. 3, Chapter 4.
7. Axilrod, B. M. and Teller, E. *J. chem. Phys.* **11,** 299 (1943).
8. Bell, R. J. and Kingston, A. E. *Proc. Phys. Soc.* **88,** 901 (1966).
9. Bell, R. J. *J. Phys.* **B3,** 751 (1970).
10. Bell, R. J. and Zucker, I. J. Ref. 3, Chapter 2.
11. Williams, D. R., Schaad, L. J., and Murrell, J. N. *J. chem. Phys.*, **47,** 4916 (1967).
12. Sherwood, A. E., De Rocco, A. G., and Mason, E. A. *J. chem. Phys.* **44,** 2984 (1966).
13. Kittel, C. *Introduction to solid state physics.* Wiley, New York (1971).
14. Corner, J. *Trans. Faraday Soc.* **35,** 711 (1939); **44,** 914 (1948).
15. Bobetic, M. V. and Barker, J. A. *Phys. Rev.* **B2,** 4169 (1970).
16. Powell, B. M. and Dolling, G. Ref. 3, Chapter 15.
17. Niebel, K. F. and Venables, J. A. Ref. 3, Chapter 9.
18. Alder, B. J. and Paulson, R. H. *J. chem. Phys.* **43,** 4172 (1965).

19. Axilrod, B. M. *J. chem. Phys.* **19,** 724 (1951).
20. Niebel, K. F. and Venables, J. A. *Proc. roy. Soc.* **A336,** 365 (1974).
21. Dobbs, E. R. and Jones, G. O. *Rep. Progr. Phys.* **20,** 516 (1957).
22. Hill, T. L. *Introduction to statistical thermodynamics*, Chapter 5. Addison-Wesley, Reading, Massachusetts (1960).
23. Klein, M. L., Horton, G. K., and Feldman, J. L. *Phys. Rev.* **184,** 968 (1969).
24. Pitzer, K. S. *Quantum chemistry*, Appendix 19. Prentice-Hall, Englewood Cliffs, New Jersey (1953).
25. Born, M. and Huang, K. *Dynamical theory of crystal lattices.* Clarendon Press, Oxford (1954).
26. Klein, M. L. and Koehler, T. R. Ref. 3, Chapter 6.
27. Maradudin, A. A., Montroll, E. W., and Weiss, G. H. In *Solid state physics* (ed. F. Seitz and D. Turnbull), Suppl. 3. Academic Press, New York (1963).
28. Squires, G. L. *Contemp. Phys.* **17,** 411 (1976).
29. Powell, B. M. and Dolling, G. Ref. 3, Chapter 15.
30. Feldman, J. L. and Horton, G. K. *Proc. Phys. Soc.* **92,** 227 (1967).
31. Venables, J. A. and Smith, B. L. Ref. 3, Chapter 10.
32. Crawford, R. K. Ref. 3, Chapter 11.
33. Korpiun, P. and Lubscher, E. Ref. 3, Chapter 12.
34. Barker, J. A. and Pompe, A. *Aust. J. Chem.* **21,** 1683 (1968).
35. Barker, J. A., Klein, M. L., and Bobetic, M. V. *Phys. Rev.* **B2,** 4176 (1970).
36. Barker, J. A., Watts, R. O., Lee, J. K., Schafer, T. P., and Lee, Y. T. *J. chem. Phys.* **61,** 3081 (1974).
37. Raich, J. C. and Gillis, N. S. *J. chem. Phys.* **66,** 846 (1977).
38. Parsonage, N. G. and Staveley, L. A. K. *Disorder in crystals.* Clarendon Press, Oxford (1978).
39. Gibbons, T. G. and Klein, M. L. *J. chem. Phys.* **60,** 112 (1974).
40. English, C. A. and Venables, J. A. *Proc. roy. Soc.* **A340,** 57 (1974).
41. MacRury, T. B., Steele, W. A., and Berne, B. J. *J. chem. Phys.* **64,** 1288 (1976).
42. Monson, P. A. and Rigby, M. *Mol. Phys.*, **39,** 1163 (1980).
43. Hansen, J. P. and McDonald, I. R. *Theory of simple liquids.* Academic Press, London (1976).
44. Rowlinson, J. S. *Liquids and liquid mixtures* (2nd edn). Butterworth, London (1969).
45. Kirkwood, J. G. *J. chem. Phys.* **3,** 300 (1935).
46. Kirkwood, J. G., Maun, E. K., and Alder, B. J. *J. chem. Phys.* **18,** 1040 (1950).
47. Kirkwood, J. G., Lewinson, V. A., and Alder, B. J. *J. chem. Phys.* **20,** 929 (1952).
48. Ornstein, L. S. and Zernike, F. *Proc. Acad. Sci. Amsterdam* **17,** 793 (1914).
49. van Leeuwen, J. M. J., Groeneveld, J., and de Boer, J. *Physica* **25,** 792 (1959).
50. Percus, J. K. and Yevick, G. J. *Phys. Rev.* **110,** 1 (1958).
51. McDonald, I. R. and Singer, K. *Quart. Rev.* **24,** 238 (1970).
52. Wood, W. W. In *The physics of simple liquids* (ed. H. N. V. Temperley, J. S. Rowlinson, and G. S. Rushbrooke), Chapter 5. North-Holland, Amsterdam (1968).
53. Alder, B. J. and Hoover, W. G. Ref. 52, Chapter 4.
54. Gosling, E. M., McDonald, I. R., and Singer, K. *Mol. Phys.* **26,** 1475 (1973).
55. G. Saville, personal communication.

56. Rigby, M. *Quart. Rev.* **24,** 416 (1970).
57. Zwanzig, R. W. *J. chem. Phys.* **22,** 1420 (1954).
58. Smith, E. B. and Alder, B. J. *J. chem. Phys.* **30,** 1190 (1959).
59. Barker, J. A. and Henderson, D. *J. chem. Phys.* **47,** 4714 (1967).
60. Rowlinson, J. S. *Mol. Phys.* **8,** 107 (1964).
61. Barker, J. A. and Henderson, D. *Ann. Rev. Phys. Chem.* **23,** 439 (1972).
62. Weeks, J. D., Chandler, D., and Andersen, H. C. *J. chem. Phys.* **54,** 5237 (1971); **55,** 5422 (1971).
63. Berne, B. J. and Forster, D. *Ann. Rev. Phys. Chem.* **22,** 563 (1971).
64. March, N. H. and Tosi, M. P. *Atomic dynamics in liquids*, Chapter 4. Macmillan, London (1976).
65. Alder, B. J., Gass, D., and Wainwright T. E. *J. chem. Phys.* **53,** 3813 (1970).
66. Levesque, D. and Verlet, L. *Phys. Rev.* **A2,** 2514 (1970).
67. Streett, W. B. and Staveley, L. A. K. *J. chem. Phys.* **50,** 2302 (1969).
68. Crawford, R. K. and Daniels, W. B. *J. chem. Phys.* **50,** 3171 (1969).
69. Pings, C. J. In *The physics of simple liquids* (ed. H. N. V. Temperley, J. S. Rowlinson, and G. S. Rushbrooke), Chapter 10. North-Holland, Amsterdam (1968).
70. Mikolaj, P. G. and Pings, C. J. *Discuss. Faraday Soc.* **43,** 89 (1967).
71. Casanova, G. and Levi, A. In *The physics of simple liquids* (ed. H. N. V. Temperley, J. S. Rowlinson, and G. S. Rushbrooke), Chapter 8. North-Holland, Amsterdam (1968).
72. Harnicky, J. F., Reamer, H. H., and Pings, C. J. *J. chem. Phys.* **64,** 4592 (1976).
73. Barker, J. A., Fisher, R. A., and Watts, R. O. *Mol. Phys.* **21,** 657 (1971).
74. Thompson, S. M. *Mol. Phys.* **32,** 721 (1976).
75. Thompson, S. M., Tildesley, D. J., and Streett, W. B. *Mol. Phys.* **32,** 711 (1976).
76. Wood, W. W. and Parker, F. R. *J. chem. Phys.* **27,** 720 (1957).
77. Verlet, L. *Phys. Rev.* **159,** 98 (1967).
78. McDonald, I. R. and Singer, K. *J. chem. Phys.* **47,** 4766 (1967); **50,** 2308 (1969).
79. McDonald, I. R. and Woodcock, L. V. *J. Phys.* **C3,** 722 (1970).
80. Streett, W. B. and Gubbins, K. E. *Ann. Rev. Phys. Chem.* **28,** 373 (1977).
81. Chandler, D. *Ann. Rev. Phys. Chem.* **29,** 441 (1978).
82. Chandler, D. and Andersen, H. C. *J. chem. Phys.* **57,** 1930 (1972).
83. Lowden, L. J. and Chandler, D. *J. chem. Phys.* **61,** 5228 (1974).
84. Scott, R. L. and Fenby, D. V. *Ann. Rev. Phys. Chem.* **20,** 111 (1969).
85. Prigogine, I. *The molecular theory of solutions*. North-Holland, Amsterdam (1957).
86. Smith, E. B. and Lea, K. R. *Trans. Faraday Soc.* **59**, 1535 (1963).
87. Alder, B. J. *J. chem. Phys.* **40,** 2724 (1964).
88. McDonald, I. R. *Ann. Rep. Chem. Soc.* **69A,** 75 (1972).
89. Leland, T. W., Rowlinson, J. S., and Sather, G. A. *Trans. Faraday Soc.* **64,** 1447 (1968).
90. Leonard, P. J., Henderson, D., and Barker, J. A. *Trans. Faraday Soc.* **66,** 2439 (1970).
91. Lee, L. L. and Levesque, D. *Mol. Phys.* **26,** 1351 (1973).
92. Lee, J. K., Henderson, D., and Barker, J. A. *Mol. Phys.* **29,** 429 (1975).

9

INTERMOLECULAR FORCES: THE PRESENT POSITION

9.1. Introduction

In this chapter we summarize the present state of our knowledge of intermolecular forces, outline the problems that remain to be solved, and draw attention to the lessons that can be learnt from the successes and failures of the past.

The elucidation of the forces between even the simplest molecules has proved a very difficult problem. Although over the years many supposedly definitive potentials have been proposed, it is only very recently that functions of reasonable accuracy have been obtained and then only for a few monatomic systems. These successes have been due to new techniques which have only come to fruition since the early 1970s. In the following section we review the state of knowledge prior to this period in order to set the recent advances in their context. Then we outline the new techniques on which the present achievements in understanding the interactions of monatomic species are based.

The bulk of the chapter is devoted to examining the pair potentials that have been proposed for simple substances in recent years. The accuracy with which a potential needs to be known depends on the purpose for which it is required. We have, therefore, attempted to classify these functions into a number of categories according to their quality. At the highest level are those which could be used to evaluate properties with an accuracy comparable to that obtainable by experiment. Only for very few interactions has this standard been attained. In other cases less securely based functions have been proposed. These may prove useful in some circumstances and we have tried to emphasize the limited context in which they may be properly employed.

The concluding sections of the chapter address themselves to more general questions. Perhaps the most topical is how can the successes achieved with monatomic species be extended to polyatomic molecules? Many workers have favoured representing complex molecules by a number of centres each of which interacts with similar centres on other molecules according to a simple potential function such as those due to Lennard-Jones. In other cases the attractive interactions of non-spherical polar molecules have been expressed in terms of the multipole expansion. The circumstances in which such models may be usefully employed have yet to be elucidated.

Two other general questions are also raised. First, in what circumstances is it useful to assume that the potential energy function of simple substances have the same shape and differ only in the scaling parameters in length and energy? This is the principle of conformality which if generally true would reduce the problem of intermolecular forces to almost trivial proportions. Unfortunately its applicability is severely limited. A second question is how far can the interactions of unlike molecules be expressed in terms of the two like interactions. Traditionally workers have sought to find relations between the scaling parameters, for example $\sigma_{12} = (\sigma_{11} + \sigma_{22})/2$ and $\varepsilon_{12} = (\varepsilon_{11}\varepsilon_{22})^{\frac{1}{2}}$. These relations, called combining rules, are now generally accepted to be inadequate. More sophisticated rules of this type have been developed but as all implicitly assume conformality their scope is severely restricted.

Finally we look at the lessons that can be learnt from the past. The clearest is that often rather more critical judgement and rather less facile optimism would lead to more secure progress in the field of intermolecular forces.

9.2. The Lennard-Jones era

In the introductory chapter we followed the historical development of the understanding of intermolecular forces as far as the recognition of the dispersion effect by London in 1930. Now we outline the way in which the subject developed through the period 1930–65, which we have called the Lennard-Jones era, so that the more recent advances in the subject may be placed in the proper context.

Lennard-Jones, as early as 1924 had adopted semi-realistic potential functions of the form

$$U(r) = \frac{A}{r^n} - \frac{B}{r^m}$$

first advocated by Mie. These functions included terms which gave a representation of both the steep repulsive branch and the long-range attraction. Following the elucidation of dispersion forces it became customary to assign the exponent of the attractive branch the value six, thus further enhancing the physical soundness of the model. Until the advent of the Lennard-Jones functions almost all work had been carried out with hard-sphere models which failed to incorporate attractive forces. Though qualitatively adequate for certain transport properties they did not provide a basis for the elucidation of other properties. The powerful attack on the relationship between the properties of matter and intermolecular forces carried out by Lennard-Jones and his collaborators using his semi-realistic potential functions set the pattern for the development of the subject for three decades. The intellectual momentum of the work of Lennard-Jones and others in this period was such as to carry ideas that were brilliant and farsighted for the 1930s into a time when they positively inhibited real progress.

During this period the relationship between intermolecular energies and the bulk properties of matter were developed largely on the basis of two beliefs.

(a) That the intermolecular pair potential energy function could be adequately represented by a simple analytical expression. The Lennard-Jones $n-6$ functions were the most commonly employed and most often n was given the value twelve.

(b) That the intermolecular energy of a number of mutually interacting molecules could be taken to be the sum of the pair interaction energies in the system

$$U_N = \tfrac{1}{2}\sum_{i=1}^{N}\sum_{\substack{j=1\\ i\neq j}}^{N} U(r_{ij})$$

Although evidence of the limitations of these assumptions became available over this period most workers believed the approximations to be adequate. They appeared to give a satisfactory basis for the interpretation of most of the experimental data then available for simple substances in both the gaseous and solid states. The parameters for the $12-6$ potential function for a wide range of substances were tabulated (as in Appendix 1 of Hirschfelder, Curtiss, and Bird)[1] and the functions they defined became part of the physicists' and chemists' armoury.

A number of factors led to the apparent success of the $12-6$ function. It could account for the second virial data of simple substances provided these were available over only a limited temperature range as was usually the case. An analysis of the second virial coefficient data for Ne and Ar by Corner[2] in 1948 led to the conclusion that the appropriate exponent for the repulsive forces in an $n-6$ potential was very close to 12. The potential function defined in this way from second virial coefficient data at relatively high reduced temperatures gave a most satisfactory account of solid state properties when pairwise additivity was assumed. Secondly, when the viscosity data available at that time were fitted to the $12-6$ function, the parameters obtained were in reasonable agreement with those derived from the analysis of second virial coefficient data.

Perhaps the most striking success of the period and certainly the culminating triumph of the Lennard-Jones potential was the Monte Carlo calculation of Wood and Parker[3] in 1957. Using a $12-6$ potential function for argon with parameters selected to fit dilute gas data, and assuming pairwise additivity they attempted to calculate the properties of argon at high densities at a temperature about twice the critical. This temperature was selected to enable comparison with the experimental compressibility data of Bridgman[4] and Michels *et al.*[5] The agreement with experimental data at moderate pressures is excellent but a discrepancy with the data of Bridgman at pressures above 1000 MPa was noted and, though the authors were commendably cautious, it seemed natural to attribute this to probable inaccuracy in the experimental results. The Monte Carlo results also permitted a critical evaluation of the current models of the liquid state showing conclusively the inadequacies of the lattice theories. But it was the prediction of the properties of argon, supposedly more accurately than the experimentalists could measure that caught the imagination.

The triumph of theory was to be short-lived. It soon became clear that both the assumptions on which the success was based were incorrect. Indeed there had already existed a considerable number of facts which were inconsistent with the assumptions stated above. In the first place the calculation of the value of C_6 for argon from the $12-6$ function with the usual parameters gave $C_6 = -4\varepsilon\sigma^6 = -120 \times 10^{-79}$ J m^6 over a factor of two greater than the value estimated by London.[6] It is true that this high value was in accord with the variational calculations of Kirkwood and Müller (-129×10^{-79} J m^6) which related C_6 to diamagnetic susceptibility,[7] but it was inconsistent with all other estimates of C_6. Furthermore, as second virial coefficients at low temperatures became available[8] the $12-6$ potential was found to be incapable of reproducing them and as more accurate measurements were carried out in the 1960s the inadequacies of this function came to be accepted by those with a specialist interest in intermolecular forces. However, Lennard-Jones parameters are still being evaluated and papers claiming to show how successful the function is in the interpretation of various properties are common in the literature.

It is interesting to consider the factors that have led to the surprisingly slow development of superior potential energy functions. The first is the complicating nature of the non-additive contributions in the solid state.[9] The net effect of non-additive contributions in the solid state is to reduce the lattice energy. Thus for argon we find that the lattice energy is reduced by some 7 per cent (by about 600 J mol^{-1} to 7733 J mol^{-1}). When the Lennard-Jones $12-6$ function was used to calculate the lattice energy of the solid inert gas, assuming pairwise additivity, with parameters based on fits to second virial coefficient data, the agreement was not unreasonable. It is now recognized that this agreement was fortuitous and derived in part from the neglect of non-additivity and in part from the use of an incorrect pair potential well depth.

Table 9.1

Comparison of the spectroscopically-determined parameters for I_2 with the range of intermolecular parameters proposed for Ar_2 prior to 1970

	I–I	Ar–Ar
r_m/nm	0·2666±0·0002 (±0·1 per cent)	0·383±0·010 (±2 per cent)
$10^{21}\varepsilon$/J	246·99±0·02 (±0·01 per cent)	1·90±0·30 (±15 per cent)

A second factor which hampered an early resolution of the problem was the use of inaccurate high-temperature gas viscosity data.[10] Many of these data stemmed, either directly or indirectly from the work of Trautz,[11] and, for most of the large number of substances he investigated, his results are now known to be some 8 per cent high around 1000 K. The results obtained using these data indicated very steep repulsive forces and led to the development of new simple potential functions of which the best known is the Kihara Core function (Appendix 1). Though many attempts were made to improve on the 12−6 potential none was able to reconcile the transport properties and the second virial coefficients. Some of these functions proposed prior to 1970 are listed in Table 9.2. Their diversity is spectacular and may be compared with the corresponding uncertainty in the parameters for the intramolecular potential function for I_2 (Table 9.1). The maximum attractive energy for I_2 is roughly 100 times greater than that of Ar_2 yet it was known with higher *absolute* accuracy.

Such was the unhappy position that prevailed before about 1970. Since that time a dramatic advance in our knowledge of intermolecular potential energy functions, particularly for monatomic species, has taken place.

9.3. The basis of recent advances

These advances in our understanding have been based on a number of factors all of which have been discussed in detail earlier in this work. They may be summarized under seven headings.

(a) *New experimental measurements of bulk properties.* The importance of the new measurements of low-temperature second virial coefficients has already been mentioned.[8] Equally important was the redetermination of the high-temperature viscosities of the inert gases about 1970.[10] The new measurements showed that the earlier data were incorrect by up to 8 per cent despite the fact that measurements from more than one laboratory had tended to provide independent support for the older values. The new values did much to refute the apparent incompatibility of gas-transport data with other properties which had hindered the search for acceptable potential energy functions.

(b) *Improved calculation of intermolecular energy.* The most important achievement of calculation has been the estimation of the dispersion force coefficients which determine the long-range forces. The use of experimental information such as refractive indices, polarizabilities, and photoionization cross-sections in conjunction with the oscillator strength sum rules has led[12] to values of C_6 accurate to ± 3 per cent. The values obtained generally confirm the original estimates of London. In addition very reasonable estimates of C_8 and C_{10} have been made.[13]

At shorter range variational procedures have been used and the 64-configuration *ab initio* calculation of the intermolecular forces of helium by Phillipson[14] is of high accuracy. For larger molecules such configuration interaction calculations become prohibitively lengthy. Hartree–Fock calculations for Ne_2 and Ar_2, even though they make no allowance for electron correlation and give an upper limit to the interaction energies, have proved useful in defining the repulsive branch.[15] Statistical models of the atom have been used to compute repulsive forces with some success.[16,17] In general except for the case of He_2 where significant progress is being made towards the calculation of the complete potential function, there is no immediate prospect of calculating an accurate potential.

(c) *Spectroscopic measurements.* The recent measurements of properties more directly related to the intermolecular potential energy than the thermophysical properties have been of considerable value, in particular spectroscopic measurements on van der Waals dimers of the inert gases. The vibrational fine structure of the vacuum u.v. absorption of the argon and krypton dimers[18] has been used to determine the vibrational levels. A knowledge of these levels in the absence of rotational fine structure is not sufficient to define the potential energy function completely but can be utilized in the Rydberg–Klein–Rees (RKR) analysis to obtain the width of the potential energy well as a function of its depth.[20] The spectroscopic data have also been used to estimate the depth of the potential energy wells for the inert gases. Three different analyses give values for argon in close accord 142·1,[19] 142·5,[20] and 140·0 K.[21] This independent and accurate determination of the well depth has been of great value, enabling other properties to be used to define the complete intermolecular potential.

More recently the rotational fine structure of the Ar_2 molecule has been observed and analysed[22] and in conjunction with second virial coefficient data this was used to determine a potential energy function for argon (see Chapter 7).

(d) *Molecular beams.* Technical advances in the design and construction of apparatus have greatly increased the importance of this technique. The information obtained from supersonic nozzle molecular beams has been of great value (Chapter 4). The high intensities and narrow velocity distributions now available with this technique have enabled high-resolution measurements of the differential scattering cross-sections for the inert gases to be made. Lee and his co-workers[21] have observed oscillations arising from quantum effects and the indistinguishability of the scattered atoms in addition to the rainbow effects. Where these data are available inversion procedures may be applied and $U(r)$ can be directly determined (see Chapter 4). Unfortunately for the inert gases it has not been possible to resolve all these features in order to determine unique potential energy functions, though the potential energy functions which were found by fitting the data were in good agreement with what are now known to be the correct functions.

(e) *Non-pairwise additivity.* Corrections for non-pairwise additivity must be included when calculating those properties which depend on the simultaneous interaction of more than two molecules. It is now generally agreed that the Axilrod–Teller triple-dipole term[23] represents the dominant contribution to the long-range non-additive energy. Considerable success has been achieved in interpreting the properties of both solid and liquid inert gases using this single correction term alone, though it is possible that this success is due to the cancellation of higher terms (Chapter 8).

(f) *Data inversion methods.* Methods have recently been developed whereby the specific information about $U(r)$ contained in experimental measurements is determined directly without assuming a form for the potential energy function. They eliminate the need for the unsatisfactory trial-and-error methods that have been traditionally employed.

Of particular value is the development of inversion techniques for bulk properties which are readily available for a wide variety of substances. There are now reliable inversion methods for second virial coefficient data[24,25] and the dilute gas transport coefficients.[25] As described above the RKR inversion method has been successfully applied to the spectroscopic data for the inert gas dimers and considerable advances have been made in the inversion of molecular beam data.

(g) *Computer calculation of properties.* The availability of electronic computers has enabled thermophysical and other properties to be readily evaluated using a given potential energy function. The calculations are not restricted to analytical functions. The wealth of information that has resulted from such calculations has been of considerable value. It has led to considerable insight into how various regions of the potential energy function contribute to different properties. It has also provided a vast pool of simulated data which can be used to test theories. The inversion techniques described above have been developed and validated using such data. More sophisticated use of computers using the Monte Carlo or molecular dynamics techniques has enabled condensed phases to be simulated. The comparison of the results of such calculations using the best potential energy functions for argon[19] with the corresponding experimental data have confirmed that the Axilrod–Teller non-additivity term appears to be an adequate representation of the many-body contribution to intermolecular energy.

9.4. Potential energy functions

We shall now review the present state of our information about the pair potential energy functions of molecules. We have chosen to classify the potentials that have been proposed for various substances into four categories.

(a) *Class I.* In the first group (I) we place those functions which are considered quantitatively accurate. They have usually been derived from extensive investigations of a wide range of properties, both macroscopic and microscopic, and are capable of reproducing both the properties used in their determination and *other* properties to within the estimated experimental uncertainties. In this category we include the best modern potentials for Ar_2, Kr_2, Ne_2, He_2, Ar–Kr. The functions for He–Ar, He–Kr, and He–Xe are probably of this quality. A function proposed for Xe_2 is included in this category with some reservations, for though a wide range of data is available and the techniques for determining this function are applicable, it is not possible to reconcile all the thermophysical data.

(b) *Class II.* The second category of functions contains those functions that have been determined by methods that are expected to give reliable results but where the potential has not been extensively tested on a wide range of properties. Usually the functions have been determined using either microscopic or bulk properties but not both. In particular, functions obtained from the inversion of molecular beam cross sections fall into this category. These include the mercury–alkali metal and the inert gas–alkali metal interactions. Many inert gas cross-interactions, Kr–Xe, Ar–Xe, Ne–Xe, Ne–Kr, Ne–Ar, and He–Ne, where the inversion of thermophysical data has been the prime source of information are designated Class II.

(c) *Class III.* The third class of potential functions comprises those which result from serious attempts to describe the interactions of non-spherical polyatomic molecules. As yet there is no clear indication as to how the problem of representing the angular-dependent potential function can be best tackled. A possible approach for the attractive forces is to combine spherically symmetric dispersion forces with electrostatic contributions arising from asymmetric charge distributions. It is often convenient to represent such interactions by means of an expansion in terms of orthogonal functions such as Legendre polynomials or spherical harmonics. For many properties the anisotropy of the repulsive forces must also be considered. However the most convenient representation of this anisotropy has not yet been established.

A particularly simple means of modelling these effects employs multi-centre potential models. One such model which has been extensively investigated in recent years is the multicentre $12-6$ function (see Table 9.11). Though the inclusion of the arbitrary $12-6$ function (which fails even for the monatomic species) suggests that only limited quantitative success might be expected, this approach has many promising qualitative features.

(d) *Class IV.* The fourth category comprises the determinations not of full potential energy functions but merely the parameters that enable a model potential function to best fit selected data. Thus, for instance, thermophysical or scattering data determined over a limited range of conditions may only permit the parameters ε and σ of a best fit to, say, the $12-6$ potential to be determined. Indeed this was the universal practice until recent years and many best parameter fits are still published even for the inert gases where the $12-6$ and most other simple functions are known to be inadequate. At best this procedure offers a way of smoothing or of modestly extrapolating the data at hand. To invest such potentials with more value than this can lead to confusion.

9.5. Accurate potential functions (Class I)

9.5.1. Argon

The form of the intermolecular potential energy function for argon has been the Holy Grail of workers in the field of molecular interactions. Argon has the important attributes of spherical symmetry, almost classical behaviour (in the sense that only minor quantum corrections are necessary to account for its properties) and of being commonly available. The first serious attempts to characterize the function followed the calculation of the second virial coefficients for various $n-m$ functions in 1924,[26] and the evaluation of the transport properties for such functions in the late

1940s.[27]† In general for the data available at that time reasonably consistent parameters were obtained for the fit to the 12 − 6 function. ε/k lay in the range 120–125 K and $\sigma \sim 0{\cdot}340$ nm. As more, and better data became available increasing suspicions about the 12 − 6 functions were expressed. The reasons for its demise have been outlined above though it is important to note that the early, so-called, improvements offered little advantage. Some of these are listed in Table 9.2. (The functional forms are given in Appendix 1.) The one feature they have in common is that they are unable to account for the known properties of argon, (thermophysical, scattering, and spectroscopic). A function which represented an important departure from the existing practice of assuming a simple analytical form was that introduced in 1960 by Guggenheim and McGlashan.[28] They used an expansion about a parabolic form at the potential minimum and fitted the coefficients largely on the basis of a pairwise-additive analysis of solid-state data. The long-range behaviour was based on the best available calculations of the dispersion force coefficient C_6. The resulting function is not in accord with modern estimates but the method of approach, in particular the rejection of oversimplified functional forms, set the pattern for the best of future investigations. In the mid-1960s Munn and F. J. Smith[29] used the flexible Boys–Shavitt function [A1.11].‡ Their approach deserved more success but was limited by the need to accommodate data that was subsequently shown to be inaccurate. A similar attempt, using a rather unpromising polynomial function, by Dymond, Rigby, and E. B. Smith foundered on the same grounds.[30] Dymond and Alder[31] carried the piecemeal construction of the potential energy function to its logical conclusion. By independently varying sections of the potential and evaluating the resulting thermophysical properties they were able to identify the properties that were sensitive to small differences in the various portions of the potential function. Using this information they constructed a numerical function. This important contribution represented the most systematic attack on the problem and was sadly to fail through attempting to fit too closely data that was incorrect and, more seriously, by assuming pairwise additivity.

Perhaps the most important step towards elucidating the argon energy function was due to Barker and Pompe in 1968.[32] Their attempt was based on four principles

(i) The use of a highly flexible functional form containing 9 terms [A1.13];

† The viscosity of molecules obeying the 8 − 4 function had been calculated by H. R. Hassé and W. R. Cook, *Phil. Mag.* **3,** 977 (1927).

‡ The functional forms of all the potentials covered in this chapter are given in Appendix 1. They are referenced in the text as [A1.*n*], where *n* is the relevant section number in Appendix 1.

(ii) The use of the best calculated values of the dispersion force coefficient and molecular-beam determinations of the repulsive forces;
(iii) The use of the Axilrod–Teller term to allow for non-additive contributions to the energy in condensed phases;
(iv) The neglect of all gas transport coefficient data.

These assumptions, especially the bold neglect of the transport data then available, which emanated from more than one source and was highly consistent, can now be seen to be both judicious and effective. More recent viscosity measurements differ by up to 8 per cent from the data then available and are in much closer accord with the Barker–Pompe function. Barker and Bobetic further revised the function but the next significant advance came in 1971. If the problem of the intermolecular potential energy function of argon can be said to be 'solved' then it was in this year that the crucial advances were made. Barker, Fisher, and Watts[19] produced a refined version of the early Barker functions by the careful fitting of a wide variety of properties. Particular emphasis was placed on condensed phase properties such as the lattice parameter, lattice energy, and Debye temperature. The liquid phase properties were calculated by computer simulation (assuming the Axilrod–Teller term gave an adequate account of non-additivity). The function also fitted the high-energy beam data and the second virial coefficients and used the best calculated dispersion force coefficients. In the same year Maitland and Smith[20] also, independently, introduced an improved version of the early Barker functions using quite different arguments. They inverted the spectroscopic data of Tanaka and Yoshino[18] using the RKR analysis to obtain the precise shape of the potential bowl. With this information and from a fit to second virial coefficient data they advocated a function which though slightly different in functional form to that of Barker, Fisher, and Watts was numerically almost identical (Fig. 9.1.). The correspondence of the two functions drawn from quite different sources suggested that the problem of the interaction of argon was solved. In the following year Lee and co-workers,[21] introduced a potential function of yet another functional form based on scattering data, second virial coefficients and the spectroscopic data which was again in good agreement with the 1971 functions (Fig. 9.1).

Subsequently the work has taken two distinct directions. Firstly some workers, in particular Aziz,[33] have tried to refine the earlier functions further. Some success has been attained and results of this work represent the best available function for argon. However perhaps a word of caution should be sounded. The refinements depend to a large extent on the critical assessment of the available data—which results to accept or reject, or even which compilation of smoothed data to use. To make any substantial progress beyond the 1971 functions probably requires more

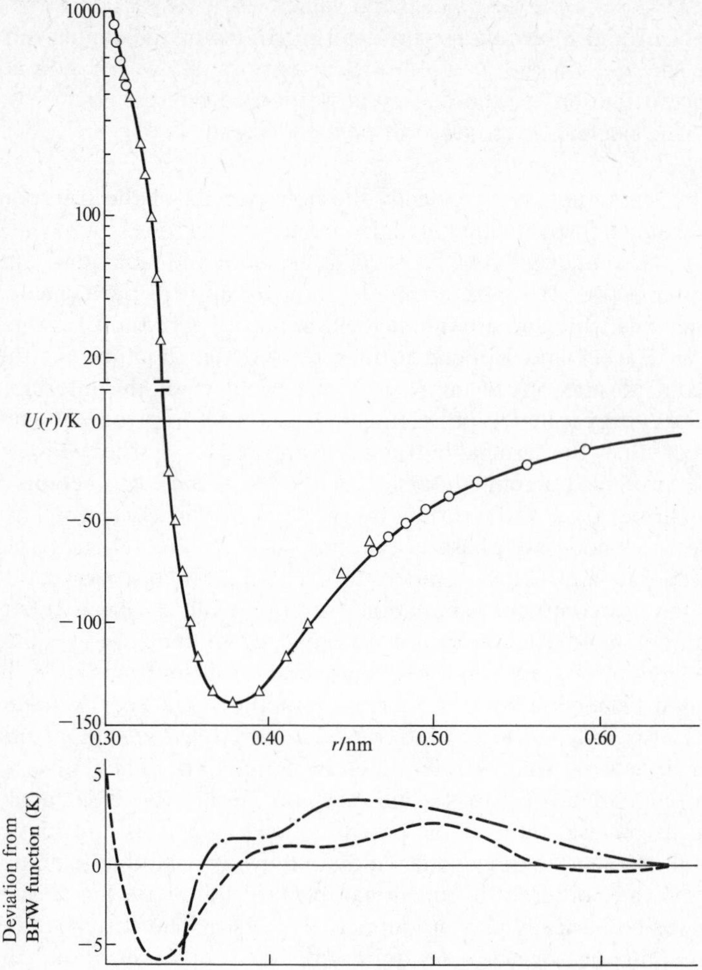

FIG. 9.1. The intermolecular pair potential energy function for argon: (——) accurate functions;[19,20,21] (△) inversion of gas viscosities;[25] (○) inversion of second virial coefficients.[25] Deviations from Barker–Fisher–Watts function;[19] (– – – –) BBMS;[20] (– · – · – · –) Lee *et al.*[21] (Maitland, G. C. and Smith, E. B. *Chem. Soc. Rev.* **2,** 181 (1973).)

accurate data. Put in a positive manner—the best functions for argon fit most of the available data within experimental error. Where discrepancies occur they are more likely to be due to errors in the data than to deficiencies of the potential function.

The second direction in which research has been pursued is in the search for simple (usually three-parameter) functions. Though analytical simplicity is less important with the advent of fast computers the most recent functions have nevertheless been very complicated and in many cases have non-physical features. It is thus desirable to be able to express results more simply. A number of simple functions have been investigated which appear to give an accurate account of thermophysical properties. A good example of such a function is the 11−6−8 function of Hanley and Klein[34] (see Table 9.2). Another such function is the core potential of Smith and Thakkar.[35] These functions are however different in shape from the modern potentials, having wells that are too deep and narrow and are thus inconsistent with the spectroscopic data. One functional form which appears to be able to fit both the thermophysical, beam and spectroscopic data is the $n(\bar{r})-6$ function[36] [A1.16]. This is the traditional Lennard-Jones $n-6$ function but with n allowed to vary with r. One form found suitable for the inert gases is $n = 13+\gamma(r/r_m - 1)$ where γ is the

Table 9.2

Pair potential functions for argon

Potential	Ref.	Year	(ε/k)/K	10σ/nm	$10r_m$/nm
LJ (12−6)	5	1949	119·8	3·405	3·822
LJ (12−6)	92	1962	125·2	3·405	3·822
Exp−6	98	1954	123·2	3·437	3·866
Exp−6	99	1955	131·49	3·384	3·784
Exp−6	100	1964	152	3·302	—
Kihara	101	1964	142·9	3·363	3·734
Guggenheim–McGlashan	28	1960	137·5	3·165	3·812
Munn–Smith	29	1965	153	3·31	—
Dymond–Rigby–Smith	30	1965	147	3·28	—
Morse	102	1961	144·8	3·386	3·855
Barker–Pompe	32	1968	147·7	3·341	3·756
Dymond–Alder	31	1969	138·2	3·280	—
11−6−8	34	1970	153	3·290	3·674
Barker–Bobetic	103	1970	140·2	3·367	—
Barker–Fisher–Watts (BFW)†	19	1971	142·1	3·361	3·761
Maitland–Smith (BBMS)†	20	1971	142·5	3·355	3·76
Parson–Siska–Lee (MSV III)†	21	1972	140·7	3·345	3·76
Morse–6–Hybrid Thakkar Smith (MIST)	35	1972	152·4	3·271	3·617
Maitland–Smith $n(\bar{r})-6$ (MS)†	36	1973	142·1	—	3·76
Aziz (AMS)†	104	1976	141·55	—	3·756
Alrichs (HFD II)	105	1976	142·9	—	3·76
Colbourn–Douglas	106	1976	143·2	—	3·75
Karnicky–Reamer–Pings	107	1976	146·3	3·389	3·86
Aziz–Chen (HFD-C)†	33	1977	143·224	—	3·759

† An adequate over-all representation of the properties of argon is given by potentials marked †. For these functions parameters are given in Table 9.2(a).

Table 9.2(a)

Potential parameters for argon

BFW: For functional form see Appendix 1, no. 13.		
Parameters:	$C_6^* = -1{\cdot}107\,27$	$A_0 = 0{\cdot}277\,83$
	$C_8^* = -0{\cdot}169\,71$	$A_1 = -4{\cdot}504\,31$
	$C_{10}^* = -0{\cdot}013\,61$	$A_2 = -8{\cdot}331\,22$
	$a = 12{\cdot}5$	$A_3 = -25{\cdot}2696$
	$\delta = 0{\cdot}01$	$A_4 = -102{\cdot}0195$
		$A_5 = -113{\cdot}25$
BBMS: See Appendix 1, no. 14.		
	$C_6^* = -1{\cdot}119\,76$	$A_0 = 0{\cdot}292\,14$
	$C_8^* = 0{\cdot}1715\,51$	$A_1 = -4{\cdot}414\,58$
	$C_{10}^* = -0{\cdot}0137\,48$	$A_2 = -7{\cdot}701\,82$
	$\alpha' = 0{\cdot}025$	$A_3 = -31{\cdot}9293$
	$\delta = 0{\cdot}01$	$A_4 = -136{\cdot}026$
	$\alpha = 12{\cdot}5$	$A_5 = -151{\cdot}0$
MSV III: See Appendix 1, no. 15.		
	$C_6^* = -1{\cdot}180$	$\beta = \beta' = 6{\cdot}279$
	$C_8^* = -0{\cdot}6118$	$b_1 = -0{\cdot}7$
	$C_{10}^* = 0$	$b_2 = 1{\cdot}8337$
	$x_1 = 1{\cdot}126\,36$	$b_3 = -4{\cdot}5740$
	$x_2 = 1{\cdot}400$	$b_4 = 4{\cdot}3667$
MS: See Appendix 1, no. 16.		
	$\gamma = 9$	
AMS: See Appendix 1, no. 16.		
	$\gamma = 7{\cdot}5$	
HFD-C: See Appendix 1, no. 21		
	$C_6^* = -1{\cdot}091\,425\,4$	$\alpha = 16{\cdot}345\,655$
	$C_8^* = -0{\cdot}600\,259\,5$	$A = 0{\cdot}950\,2720 \times 10^7$
	$C_{10}^* = -0{\cdot}370\,011\,3$	$D = 1{\cdot}4$
		$\gamma = 2{\cdot}0$

third adjustable parameter of the potential function. The best parameters for this function for argon has been given by Aziz[33] who selected a value of 7·5 for γ. This potential is perfectly adequate for both thermophysical and spectroscopic data and it was claimed, gave a better account of total cross section data than other potentials investigated.

The comparison of the experimental data and those calculated from various potential energy functions is given in Table 9.3. Such tables illustrate how the available data may be used as a guide to the adequacy of proposed potential energy functions. It can be seen that as the functions are refined a wider range of properties are accommodated and, as noted above, the best predictions lie, on the whole, within the estimated experimental errors.

Table 9.3

R.m.s deviation of experimental and calculated properties of argon (*based largely on data in ref. 104*)

Property	Potential energy functions							
	HFDC[115]	AMS[104]	MS[36]	BFW[19]	MSVIII[21]	BBMS[20]	BP[32]	11–6–8[34]
ΔB (NBS)/$cm^3\ mol^{-1}$ [108]	—	1·76	1·72	1·64	4·57	2·09	1·29	1·45
$\Delta\eta$ (MS)/(μP)[109]	3·131	3·84	7·31	6·13	16·17	4·93	17·52	1·55
$\Delta\eta$ (K)/(μP)[110]	—	0·96	1·41	2·74	5·73	2·74	4·42	1·14
$\Delta\lambda$ [111]	0·0256	0·143	0·151	0·0278	0·382	0·0216	0·158	0·249
$\Delta\alpha_T(\times 10^3)$[112]	0·996	0·837	1·289	1·826	1·892	—	2·640	0·430
ΔQ[113]	0·012	0·009	0·010	0·011	0·018	0·010	—	—
$\Delta G_{\frac{1}{2}}/(cm^{-1})$[114,104] (Expt. value = 25·71 ± 0·86)	25·67	25·04	25·12	25·22	25·50	25·32	26·35	27·38
$\Delta G_{\frac{3}{2}}/(cm^{-1})$[114,104] (Expt. value = 19·99 ± 1·05)	20·47	19·97	20·11	19·82	20·14	20·10	21·39	21·71

The letters in brackets after each property eg. $\Delta\eta$ (MS) in this and subsequent tables of this form codify the source of the data.
ΔB (NBS) are the r.m.s. deviations of the second virial coefficients of Levelt–Sengers *et al.*[108] from those calculated from the various potential functions. The lower values indicate a better performance by the function.
$\Delta\eta$ are the viscosity deviations ($\mu P \equiv 10^{-7}\ Kg\ m^{-1}\ s^{-1}$). At 300 K, η is approximately 230 μP.
$\Delta\lambda$ are the thermal conductivity deviations. At 330 K. λ is approximately 20 $mW\ m^{-1}\ K^{-1}$.
$\Delta\alpha$ is the deviation in the coefficient of thermal diffusion. The value of α at 273 K is approximately 0·013.
ΔQ is the deviation in the total cross-section.
$\Delta G_{\frac{1}{2}}$ and $\Delta G_{\frac{3}{2}}$ are *not* deviations but the first and second vibrational level spacings. The experimental values are given in parentheses.

9.5.2. *Krypton*

The early history of the investigation of the interaction between krypton atoms followed much the same course as that for argon and will not be discussed here. The serious investigations of the problem followed the spectroscopic examination of the vibrational levels of the Kr_2 dimer by Tanaka and Yoshino.[18] Nine vibrational levels for the $^1\Sigma_g^+$ ground state were observed. Buck, Dondi, Valbusa, Klein, and Scoles[37] used these data in conjunction with differential cross-section data measured over a wide energy range to define a potential of the Morse–Spline–van der Waals form [A1.15]. This potential represented an improvement over the early beam potentials. Gough, Smith, and Maitland[38] used an RKR analysis of the spectroscopic data to define the well width–depth function, together with an estimate of the well depth obtained by the method of Le Roy,[39] to obtain a numerical potential. This was found to be quite closely reproduced by the $n(\bar{r})-6$ function [A1.16] with $\gamma = 10$. Barker, Watts, Lee, Schafer, and Lee[40] fitted a very wide range of properties (second virial coefficients, u.v. spectra, gas transport data, solid-state data and differential cross-sections) to a flexible potential which was a refined version of the Barker–Pompe form [A1.13a].

The functions of Gough *et al.* (GMS) and Buck *et al.* were very similar. The Barker *et al.* function was similar in the well region but had a

Table 9.4

R.m.s. deviations of experimental and calculated properties of krypton (based largely on the results of calculations by R. A. Aziz.[43])

Property	Potential energy functions			
	GMS[38]	Barker *et al.*[40]	Buck *et al.*[56]	Aziz[43]
(ε/k)/K	199·2	201·9	200	199·9
$10r_m$/nm	4·02	4·0067	4·03	4·012
ΔB (B)/cm³ mol⁻¹ [116]	2·44	3·22	1·50	0·81
ΔB (NBS)/cm³ mol⁻¹ [108]	2·43	5·81	4·53	4·92
$\Delta\eta$ (KRW)/μP [110]	1·87	1·57	2·87	2·85
$\Delta\eta$ (CS)/μP [117]	0·99	0·37	1·31	0·70
$\Delta\eta$ (DS)/μP[10]	7·35	6·13	6·81	4·55
$\Delta\eta$ (MS)/μP [109]	5·17	3·01	5·07	2·88
$\Delta\eta$ (H)/μP [118]	4·08	2·06	4·21	2·96
ΔD (WD)/cm² s⁻¹ [119]	0·0009	0·0040	0·0007	0·0029
$\Delta\lambda$ (H)/mW m⁻¹ K⁻¹ [111]	0·0186	0·0117	0·0517	0·0265
$\Delta\lambda$ (SAX)/mW m⁻¹ K⁻¹ [120]	0·590	0·748	0·577	0·724
$\Delta\alpha_T \times 10^3$/(TW)[112]	0·453	0·817	0·436	0·595
Spectroscopic data (cm⁻¹)[18]	0·256	0·354	0·357	0·259
ΔQ (arbitrary units)[122]	0·0050	0·0090	0·0055	0·0054
$\Delta\sigma$ (arbitrary units)[121]	91	62	47	68

Definitions as in the footnote to Table 9.3. The entry labelled spectroscopic data represents the r.m.s. deviations of the fit to the vibrational energy levels. ΔD are the r.m.s. deviations in the diffusion coefficient. $\Delta\sigma$ are the r.m.s. deviations in the differential cross-sections.

rather steeper repulsive wall. Aziz and Kocay[41] investigated the degree to which the functions were capable of reproducing the known properties of krypton. All three were successful, the functions of GMS and Buck *et al.* proving slightly superior for thermal diffusion and total cross-sections (Table 9.3). They also examined the Morse–Thakkar–Smith form[42] [A1.19] which was found to be less successful. Subsequently Aziz[43] proposed a HFD function [A1.21] based on the functional form that fitted the results of Hartree–Fock calculations and known dispersion force coefficients. We conclude that for krypton, like argon, the pair potential energy function is well defined. The assessment of the merits of different potentials in the manner illustrated in Table 9.4 depends crucially on the assessment of the quality of various sets of data that are not always in good mutual agreement. This judgement of the data must, of necessity, be somewhat subjective. Further progress will require a better knowledge of the properties of krypton. Again it is of interest to note that the very complicated functions offer little advantage over the simple three-parameter $n(\bar{r})-6$ function.

9.5.3. *Helium*

Much theoretical and experimental effort has gone into determining the pair potential energy function for helium. In principle helium should be the simplest case as it is the most amenable to direct quantum mechanical calculation (see Chapter 2). In practice the potential function is less well established than that of argon and krypton.

An assessment of the performance of a number of proposed functions is given in Table 9.5 which is based on the analysis of Aziz *et al.*[44] It

Table 9.5

R.m.s. percentage deviation of experimental and calculated properties of helium (based on a private communication from R. A. Aziz and ref. 44)

	Potential energy functions						
Property	Beck[45]	Bruch & McGee[46]	Lee *et al.*[125] ESMSV	de Boer[126] LJ (12−6)	Aziz *et al.*[44] HFD	Ahlrichs *et al.*[105] HFD	Aziz *et al.*[44] MS
(ε/k)/K	10·37	10·748	10.57	10.22	10.8	10.6	10·9
$10r_m$/nm	2·969	3·023 8	2·97	2·556	2·967 3	2·979	2·976
$\Delta\eta$ (MS)/μP [109]	2·93	2·77	1·66	4·39	0·26	1·24	1·03
$\Delta\eta$ (CS)/μP [117]	1·17	1·36	0·80	2·44	0·60	1·72	0·53
$\Delta\eta$ (DS)/μP [10]	3·78	3·38	2·60	3·21	0·72	0·81	0·47
$\Delta\lambda$ (H)/mW m^{-1} K^{-1} [111]	2·49	0·75	1·49	1·32	0·24	0·79	0·15
$\Delta\lambda$ (S)/mW m^{-1} K^{-1} [123]	4·75	4·77	3·22	6·07	1·73	1·69	2·07
ΔB (G)/cm^3 mol^{-1} [124]	0·280	0·219	0·122	0·343	0·038	0·221	0·099

Definitions as in footnote to Table 9.3.
Absolute r.m.s. deviations are given for second virial coefficients.

shows that the traditional Lennard-Jones function based on a fit to second virial coefficients is as unsuccessful as for the other inert gases. More recent attempts, due to Beck[45] and Bruch and McGee[46] have difficulty in reconciling the high-temperature viscosity data. The potentials based on molecular-beam data due to Lee and co-workers (only the best of which is shown)[21] show some improvement and the most recent function due to Aziz[44] [A1.21] accommodates the available data very well. Again a simple potential of the $n(\bar{r})-6$ form, where $n(\bar{r})=12+6(\bar{r}-1)$, appears to do nearly as well as the more complicated functions.

9.5.4. Neon

The potential function of neon has been the subject of some controversy as a number of different potentials based on a different weighting of properties have been proposed. Goldman and Klein[47] paid particular attention to solid-state data and used a Morse–Spline–van der Waals functional form [A1.15]. The function of Farrar *et al.*[48] took account of differential cross-section determinations at low energy. Maitland[49] analysed the spectroscopic data using the RKR analysis to find the shape of the well. Fitting other data he proposed a double Morse–van der Waals potential. LeRoy, Klein, and McGee[50] suggested that the Maitland

Table 9.6

R.m.s. deviations of measured and calculated properties of neon (based on data calculated in ref. 51)

	Potential energy functions			
	AMS[51]	MS[36]	MMV[49]	ESMV III[48]
Property	$n(\bar{r})-6$, $\gamma=5$	$n(\bar{r})-6$, $\gamma=2$	Double Morse	Exp–Morse–van der Waals
(ε/k)/K	41·186	39·6	39·6	42·01
$10r_m$/nm	3·0739	3·07	3·07	3·101
ΔB (NBS)/cm^3 mol^{-1} [108]	0·27	0·29	0·42	0·47
$\Delta\eta$ (KRW)/μP [110]	1·66	6·48	1·84	2·19
$\Delta\eta$ (MS)/μP [109]	2·70	7·88	5·83	2·35
ΔD/cm^2 s^{-1} [127]	0·093	0·0069	0·153	0·094
$\Delta\lambda$ (SW)/mW m^{-1} K^{-1} [128]	0·59	2·68	0·39	1·00
$\Delta\lambda$ (JS)/mW m^{-1} K^{-1} [129]	3·49	1·55	5·61	2·87
$\Delta\lambda$ (H)/mW m^{-1} K^{-1} [111]	0·57	0·33	0·33	0·82
$\Delta\alpha_T(\times 10^3)$ [130]	0·413	1·331	1·172	0·406
$\Delta G_{\frac{1}{2}}$/cm^{-1} [18] (expt. value = 13·7)	13·34	12·68	12·97	14·07
D_0/cm^{-1} [50] (expt. value = 16·3)	16·48	15·67	16·34	16·75

Definitions as in footnote to Table 9.3. D_0 is the ground-state dissociation energy.

function was inconsistent with the spectroscopic data from which it was derived. They found the Goldman–Klein function best for the well shape. However this function gives a poor account of many thermophysical properties especially the gas phase viscosities. On the whole the functions of Farrar *et al.* and Maitland seem to give a better over-all account of these properties. Aziz[51] has proposed a potential of the $n(\bar{r})-6$ form which appears to be superior to all the earlier, and considerably more complicated, functions (see Table 9.6).

9.5.5. *Xenon*

A definitive analysis of the potential energy function of this gas has yet to be undertaken. There appear to be inconsistencies between the data for different properties which has hampered progress. Although we have included it in Class I the functions discussed may well be of lower quality than those for the other inert gases.

Barker *et al.*[40] have proposed a potential function for xenon–xenon interactions based on the analysis of a wide range of thermophysical and scattering properties. The best potential has the modified Barker form, [A1.13(a)] with the parameters

(ε/k)/K	$10r_m$/nm	10σ/nm	A_0	A_1	A_2	A_3	A_4	A_5
281·0	4·3623	3·890	0·2402	−4·8169	−10·9	−25·0	−50·7	−200·0
α	C_6^*	C_8^*	C_{10}^*	δ	P	Q	α'	
12·5	−1·0544	−0·166 60	−0·0323	0·01	59·3	71·7	12·5	

Discrepancies between the prediction of this potential and solid-state data appear to have been removed by more recent measurements of the cohesive energy of crystalline xenon. The authors also propose in a footnote a modified potential function containing extra terms to make the potential compatible with the spectroscopic data of Freeman, Yoshino, and Tanaka.[52] However, the results of the application of this function to bulk properties were not given.

9.5.6. *Argon–krypton*

This system has been the subject of fairly intensive investigation in recent years and is currently the best understood mixed interaction. Lee, Henderson, and Barker[53] derived a function using the thermodynamic data for liquid mixtures. Gough, Smith, and Maitland[54] obtained a function using data inversion methods on thermophysical properties. Maitland and Wakeham[55] have produced a further potential on the same basis. Buck *et al.*[56] have produced a new potential function by the semi-classical inversion of differential cross-section data for Ar–Kr. This

Table 9.7

R.m.s. deviation of experimental and calculated properties of argon–krypton mixtures (taken largely from ref. 57)

	Potential energy functions				
	Aziz *et al.*[57]	BHPS[138]	GMSM[140]	LHB[53]	MW[55]
(ε/k)/K	167·5	167·1	165·0	163·9	165
$10r_m$/nm	3·88	3·88	3·90	3·88	3·88
ΔB (B)/cm^3 mol^{-1} [131]	0·35	0·50	2·52	0·75	4·43
ΔB (S)/cm^3 mol^{-1} [132]	1·19	1·30	2·12	1·29	2·91
$\Delta\eta$ (KM)/μP [89,133]	3·82	3·20	1·73	0·70	0·80
$\Delta\eta$ (S)/μP [134]	3·24	10·92	5·74	8·35	5·92
ΔD (H)/cm^2 s^{-1} [135]	0·0382	0·0649	0·0490	0·0657	0·0307
ΔD (VH)/cm^2 s^{-1} [136]	0·000 99	0·0032	0·0014	0·0024	0·0014
ΔD (D)/cm^2 s^{-1} [137]	0·000 07	0·0021	0·000 15	0·0011	0·002
$\Delta\sigma$ (B) (arbitrary units)[138]	4·32	3·53	9·01	13·3	8·86
ΔQ (L) (arbitrary units)[113,139]	1·6	1·2	1·7	2·5	

$\Delta\sigma$ is r.m.s. deviation of differential cross-section. Other definitions as in Table 9.3.

function has been further refined by Aziz *et al.*[57] to produce a pair potential that seems superior in all respects to those that preceded it. The comparison of the experimental properties and those calculated on the basis of the various functions is given in Table 9.7. These potential energy function determinations indicate that for Ar–Kr, $\varepsilon/k = 166 \pm 1$ K, a fact that could, in principle, be used to test the so-called combining rules which relate features of the unlike interaction to those of the like interactions (see § 9.9). However, argon and krypton are sufficiently similar in molecular size and maximum interaction energy that for this system it is not possible to draw any conclusions about the combining rules. The common combining rules for ε/k all give values in the range 166–169 K.[54]

9.5.7. Helium–heavier inert gas interactions

The potential functions that represent the interactions of helium with the heavier inert gases are of particular interest as helium is sufficiently different from the other gases in the strength of its intermolecular forces to provide a good test of theories of unlike interactions. In 1973 Chen, Siska, and Lee[58] reported potential functions based on differential cross-sections measured in molecular-beam experiments. The functions for He–Ar, He–Kr, and He–Xe had virtually equal well depths. More recent measurements[59] of differential cross-sections have suggested that the well depths reported by Chen *et al.* are roughly 20 per cent too shallow but the approximate equality of the values for the three systems is confirmed.

Table 9.8

Comparison of calculated and experimental properties for mixtures of helium and the heavier inert gases (taken largely from ref. 60)

		ε/k / K	$10r_m$ / nm	10σ / nm	$\Delta\sigma$ ($T \sim 80$ K)[60]	$\Delta\sigma$ ($T \sim 300$ K)[60]	$\Delta B(T)$[131,132] / $cm^3\ mol^{-1}$	$\Delta\eta$(K)[89,141] / μP	$\Delta\eta$(S)[134,142] / μP	ΔD(H)[143] / $cm^2\ s^{-1}$	ΔD(B)[136] / $cm^2\ s^{-1}$
He–Ar	Lee[58]	24·2	3·54	3·09	1·31	1·16	1·18	5·8	2·2	0·28	0·012
	MS ($\gamma = 4$)[55]	29·4	3·43	3·07	1·39	1·07	0·69	2·7	1·4	0·04	0·005
	HFDl ($\alpha = 4$)[60]	29·4	3·50	3·12	1·15	1·17	1·10	3·4	3·6	0·15	0·10
He–Kr	Lee[58]	24·7	3·75	3·26	0·57	1·56	0·43	6·3	—	0·16	0·028
	MS ($\gamma = 4$)[55]	30·4	3·64	3·25	0·76	1·58	0·71	0·9	—	0·01	0·006
	HFDl ($\alpha = 3·9$)[60]	29·1	3·70	3·31	0·63	1·74	2·23	3·0	—	0·03	0·005
He–Xe	Lee[58]	25·2	4·15	3·61	1·16	0·99	1·57	4·7	—	0·02	0·005
	MS ($\gamma = 5$)[55]	27·4	3·98	3·56	0·65	0·66	0·73	1·3	—	0·01	0·004
	HFDl ($\alpha = 3·93$)[60]	28·4	3·93	3·53	0·62	0·76	1·05	1·4	—	0·02	0·005

$\Delta\sigma$ are the r.m.s. deviations of the differential cross-sections. σ is the collision diameter. Other definitions as in the footnote to Table 9.3.

K. M. Smith *et al.*[60] determined potential functions from gas phase properties and showed that they gave a very good account of the scattering data. They reported two useful functions, the HFD form and the $n(\bar{r})-6$, and gave suitable parameters (see Table 9.8). These potentials are in reasonably good agreement with the repulsive branches obtained by Maitland and Wakeham[55] from the inversion of the transport properties of the gaseous mixtures.

9.6. Class II potential functions

These are the results of serious attempts to obtain accurate potential functions. They are not included in category I because they have been obtained from only a limited type of data or because, if more than one type of data has been used, the potential is not capable of reconciling all the available results.

9.6.1. Kr–Xe, Ar–Xe, Ne–Xe, Ne–Ar, Ne–Kr

Ng, Lee, and Barker[61] determined the best fit to the Morse–Spline–van der Waals for Ne–Kr and Ne–Xe and to the exponential Spline–Morse–Spline–van der Waals (ESMSV) [A1.15] for Ne–Ar using low-energy differential cross-sections. They found that the unlike interactions had narrower wells than those of symmetric inert gas pairs. The values of ε and r_m they reported were: for Ne–Ar 71·9 K and 0·34 nm; for Ne–Kr 74·5 K and 0·358 nm; and for Ne–Xe 75·0 K and 0·375 nm.

Maitland and Wakeham[55] have obtained potential functions for all the unlike inert gas interactions by the inversion of gas viscosity and diffusion data. The functions were tested on thermal diffusion and second virial coefficient data and found satisfactory. Their conclusions for some interactions such as Ar–Kr, have been discussed earlier. The potentials they obtained can be represented by the $n(\bar{r})-6$ function with the parameters given in Appendix 10.

9.6.2. Alkali metal–mercury systems

These systems have been investigated by molecular-beam techniques. The differential cross-sections display rainbow structure with supernumary rainbows and superimposed rapid oscillations. These features enable the potential function to be obtained by the direct inversion methods of Buck[62] and are the potentials best established from scattering data. They are believed to be accurate and we only include them in Class II as they derive from one type of data only. The parameters are given in Table 9.9.

The potential functions for the heavier systems are essentially conformal whereas the Li–Hg system has a quite different shape. In all cases

Table 9.9
Potential parameters for alkali-metal–mercury systems.

	(ε/k)/K	$10r_m$/nm	Ref.
Li–Hg	1253±35	3·00±0·02	144
Na–Hg	638±20	4·72±0·03	145
K–Hg	603±20	4·91±0·03	146
Cs–Hg	580±17	5·09±0·02	146

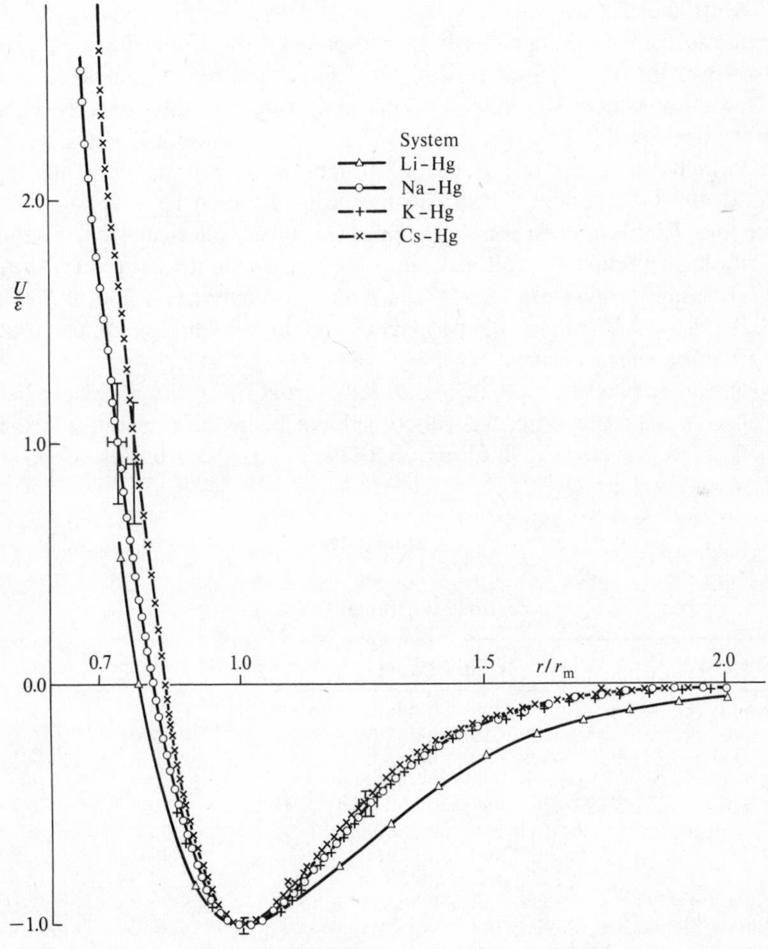

FIG. 9.2. The intermolecular potential energy functions for the alkali metal–mercury interactions. (Reproduced from Buck, U. *Advan. chem. Phys.* **30,** 313 (1975).)

the potential functions are quite different from the 12−6 function. The repulsion is softer whereas the lower part of the well is wider and the function goes to zero more rapidly as r goes to infinity (Fig. 9.2).

9.7. Class III potential functions

This class comprises serious attempts to tackle the problems of polyatomic molecules.

9.7.1. Methane

Methane has traditionally been treated as a simple spherically symmetric molecule and is probably the polyatomic molecule that could most reasonably be represented in this way. The most recent function of this type was obtained by inversion of gas viscosity data and is closely represented by the 20−6 function.[63] Table 9.10 summarizes the success with which the proposed potential functions reproduce viscosities and second virial coefficient data. The function obtained by inversion,[63] the Dymond, Rigby, and Smith[30] potential, and the spherical-shell potential of Snook and Spurling[64] all give an adequate account of the data though the ε/k values range from 182–232 and the σ values from 0·356–0·369 nm. Clearly a wider range of properties must be employed to specify the function more accurately.

Recently two attempts to include the structure of the molecule in the development of the potential function have been made using a site–site model to represent the interaction between two molecules. One such

Table 9.10

Deviation of measured and calculated properties of methane (taken largely from ref. 63)

Potential	σ/nm	(ε/k)/K	r_m/nm	$\Delta\eta$/per cent	ΔB/cm^3 mol^{-1}	ref.
Numerical	0·3559	217	0·3867	0·44	3·2	63
20−6	0·3559	217	0·3879	0·78	2·8	63
28−7	0·3398	310	0·3630	3·55	7·6	147
Kihara $\gamma = 0·2$	0·3620	204·3	0·3975	2·12	12·3	100
DRS	0·369	181·6	0·427	2·06	2·7	30
Spherical shell	0·3535	232	0·3840	1·29	5·0	64
11−6−8	0·3680	168	0·4101	0·72	27·1	148
18−12−6	0·373	185·5	0·410	5·69	2·98	149
Kihara $\gamma = 0·2$	0·3509	213	0·3853	6·4	13·2	150
Barker–Bobetic	0·4407	164·5	0·4926	31·1	49·2	151

$\Delta\eta$ represents the percentage deviation of the calculated viscosities from the experimental data. ΔB is the deviation in the second virial coefficient calculations. See ref. 63 for the sources of the experimental data.

attempt was based on a LJ (12 − 6) function[65] using the following values of the site–site interaction parameters.

	$(\varepsilon/k)/\mathrm{K}$	$10\sigma/\mathrm{nm}$
C–C	51·198	3·350
C–H	23·798	1·995
H–H	8·631	2·813

The C–H bond length was assigned the value 0·1110 nm. An alternative formulation by the same group[66] uses the function

$$U_{\alpha\beta}(r_{\alpha\beta}) = A_{\alpha\beta}\exp(b_{\alpha\beta}\,.\,r_{\alpha\beta}) + C_{\alpha\beta}r_{\alpha\beta}^{-6}$$

with the following parameters where $\varepsilon/k = 142{\cdot}87$ K and $\sigma = 0{\cdot}401$ nm.

	$A^*_{\alpha\beta}$	$b^*_{\alpha\beta}$	$C^*_{\alpha\beta}$
C–C	218 024	−14·436	−0·427 79
C–H	38 744	−14·717	−0·108 43
H–H	9260	−14·997	−0·027 36

$A^*_{\alpha\beta} = A_{\alpha\beta}/\varepsilon$; $C^*_{\alpha\beta} = C_{\alpha\beta}/\varepsilon\sigma^6$; $b^*_{\alpha\beta} = b_{\alpha\beta}/\sigma$.

The latter function has been used in simulation calculations and if pairwise additivity is assumed gives a good account of liquid methane data.

9.7.2. Hydrogen

The simplest molecules whose deviations from spherical symmetry must be explicitly incorporated into the intermolecular potential are the homonuclear diatomics, such as H_2 or N_2. Even for these simple non-spherical molecules the additional complications of shape have limited the quantitative characterization of the potentials, though considerable progress has been made in a few cases.

The computationally attractive case of hydrogen has been the subject of many theoretical studies and a survey of *ab initio* results obtained for the short range potential has been given.[159] The anisotropy of the long-range dispersion energy has been the subject of several investigations[160] and the total long-range electrostatic and dispersion interaction is now quite well defined. Configuration interaction calculations have been reported covering the whole separation range,[161–3] and the choice of suitable methods of representing the potential surface by analytical functions has been considered. It seems desirable that the form chosen for the complete potential should permit independent representations of the anisotropy for the short-range and for the components of the long-range potential, to avoid unphysical correlations between these terms. For hydrogen the anisotropy of the potential is found to be small, and appears to arise largely from the quadrupole–quadrupole interaction.[162]

There have also been many experimental studies which provide further information about this potential. The spectrum of the $(H_2)_2$ van der Waals molecule has been reported[164] and both differential[165,166] and integral[167] scattering cross-sections have also been measured for H_2 and D_2. The potentials inferred from these studies[165–9] have generally been based on spherical approximations, but with improved methods of analysis and with more detailed experimental measurements it is likely that this potential will be well characterized in the near future.

9.7.3. *Other diatomic molecules*

The potential function of nitrogen has been the subject of a number of investigations. Detailed quantum mechanical studies of the long range interactions[170] and their anisotropy[171] have been reported. The short range and intermediate regions have been investigated using *ab initio* calculations.[172] The anisotropy of this molecule is greater than that of hydrogen and the importance of using an accurate representation of the orientational dependence of the dispersion energy in calculating virial coefficients has been emphasized.[173] Spectroscopic data on the van der Waals dimer have not yet yielded detailed information about the potential but a T-shaped equilibrium structure has been proposed,[174] which is consistent with the nearest neighbour orientation in the low-temperature crystal structure. However an alternative 'staggered-parallel' conformation has also been suggested for the dimer.[175] Some high energy molecular beam scattering data have been reported for N_2–N_2[176] and attempts to fit the short-range potential using both central and anisotropic site–site potentials have been made.

Despite these developments, a complete and accurate nitrogen potential is not yet available, and various more approximate models have been used. For example, recent liquid-state computer simulations have required potential energy functions that could account for at least some of the features of condensed phase structures. This requirement has been met by the use of the di-Lennard-Jones potential function to represent the interactions of diatomic molecules. The function may be represented†

$$U = \sum_{\alpha=1}^{2} \sum_{\beta=1}^{2} 4\varepsilon_{\alpha\beta}\left\{\left(\frac{\sigma_{\alpha\beta}}{r_{\alpha\beta}}\right)^{12} - \left(\frac{\sigma_{\alpha\beta}}{r_{\alpha\beta}}\right)^{6}\right\}$$

† This form of the potential, in which $\varepsilon_{\alpha\beta}$ represents the well-depth for the $\alpha\beta$ site interaction, differs from that used in Chapter 3 and Appendix 8 which omit the factor 4. Both forms have been widely used.

where the summation includes the interaction of each of the two Lennard-Jones centres of one molecule with similar centres in a neighbouring molecule. It is uncertain how far these functions are a useful approximation to the true potential energy functions. Nevertheless for nitrogen Powles and Cheung proposed a function from liquid simulation studies which was later shown to give a good account of the second virial coefficients of the gas[67]—an independent test of some severity. The inclusion of a point quadrupole moment gave slightly worse agreement with the second virial coefficients. The value of the di-Lennard-Jones model has yet to be established. The results for liquid simulation are promising but as yet the same potential function has not proved adequate for interpreting the properties of all three states of matter. Furthermore, no attempt is usually made to allow for non-pairwise-additive contribution in the simulation calculation so that the proposed functions must be regarded as 'effective potentials', with all the limitations that this concept entails.

Our knowledge of the true potential functions of other diatomic molecules is very incomplete. The equilibrium geometries of several van der Waals dimers have been determined spectroscopically and are discussed in Chapter 7. The expansion coefficients for the dispersion energy have been calculated in some cases[170] and estimates have also been made of the anisotropic contributions to the long-range energy. Although extensive data are available for the dilute gas transport properties and second virial coefficients, methods have yet to be developed to enable effective use of this information to be made.

Singer *et al.*[68] have given di-Lennard-Jones parameters for a number of diatomic molecules and CO_2 treated as a pseudo-diatomic which have been used in liquid-state simulations. These and other parameters determined for a number of diatomic systems are given in Table 9.11. Not

Table 9.11

The di-Lennard-Jones potential

	(ε/k)/K	10σ/nm	l/σ	$10^{26}\,Q$/e.s.u.	Ref.
N_2	36·95	3·310	0·3292	—	67, 152
	43·99	3·341	0·3292	—	153
	35·32	3·314	0·3292	−1·48	154
O_2	44·5	3·09	0·3292		155
F_2	52·8	2·825	0·505		68
Cl_2	173·5	3·353	0·608		68
Br_2	257·2	2·538	0·630		68
CO_2	163·6	2·989	0·793		68

l is the internuclear separation within the molecule, and Q the quadrupole moment.

surprisingly, many simple spherically symmetric functions have also been used to represent the interaction of diatomic molecules. Some of these are included in the list of Class IV functions in §9.8.

9.7.4. *Inert-gas–diatomic-molecule interactions*

The interaction of a spherically symmetric inert-gas molecule with a diatomic molecule represents the simplest case of an anisotropic interaction. Even here, however, few systems have been studied in detail. The simplest example, that of He–H_2, has been the subject of both experimental and extensive theoretical investigation. Earlier *ab initio*[177] calculations indicated an anisotropy greater than that given by perturbation calculations of the long range dispersion energy.[178] However this discrepancy has not been observed in more recent *ab initio* work.[179] Shafer and Gordon[180] have proposed a potential function on the basis of a detailed study of spectral line shapes and rotational relaxation which was in accord with the experimental data available. They expressed the energy

$$U(r, \theta) = U_0(r) + U_2(r)P_2(\cos\theta)$$

where $P_2(\cos\theta)$ is the second Legendre polynomial. $U_0(r)$ was fitted to a Morse–Spline–van der Waals form [A1.15] with the following parameters:

$$\varepsilon/k = 15{\cdot}5\ \text{K}$$

$$r_m = 0{\cdot}338\ 15\ \text{nm} \qquad r_1 = 0{\cdot}338\ 15\ \text{nm}$$

$$r_2 = 0{\cdot}405\ \text{nm} \qquad \beta = 6{\cdot}160$$

$$C_8 = -41\ \text{a.u.} = -15 \times 10^{-100}\ \text{J m}^8$$

$$C_6 = -4{\cdot}01\ \text{a.u.} = -3{\cdot}82 \times 10^{-79}\ \text{J m}^6$$

$$b_1 = -1{\cdot}00 \qquad b_2 = 2{\cdot}636 \qquad b_3 = 13{\cdot}31 \qquad b_4 = -70{\cdot}50.$$

$U_2(r)$ was of similar form:

$$U_2(r) = \alpha_{SR}\{\exp[2\beta(1 - r/r_m)] - 2\gamma\exp[\beta(1 - r/r_m)]\} \qquad r \leq r_1$$

$$U_2(r) = \alpha_{LR}(+C_8/r^8 + C_6/r^6) \qquad r > r_2$$

In the range $r_1 > r > r_2$ the $U(r)$ was fitted to a cubic polynomial as for $U_0(r)$. The best values of the anisotropy parameters were found to be $\alpha_{SR} = 0{\cdot}262$, $\gamma = 0{\cdot}4$, and $\alpha_{LR} = 0{\cdot}105$.

For the interaction Ar–H_2 the rotational motion of the hydrogen molecule is only weakly coupled to the over-all dimer rotation in the sense that the rotation of the hydrogen molecule is not unduly restricted by the intermolecular interaction. This facilitates the investigation of this interaction by spectroscopic means by greatly simplifying the analysis of the data. The potential function devised which is believed to be of high accuracy is probably the only anisotropic function which is known with any degree of certainty. A full account of this system is given in Chapter 7 (§7.6.4) and will not be repeated here.

In the system Ar–HCl there is strong coupling between the rotational motion of the diatomic molecule and that of the dimer. Although this system has been extensively studied by molecular beam resonance spectroscopy techniques the strong coupling makes the interpretation of the experimental results more difficult and the resulting potential functions have not reached the same degree of sophistication as for the Ar–H_2 system. Nevertheless rapid progress is being made and the results of recent investigations are discussed in Chapter 7 (§7.6.5).

Serious attempts have been made to determine the potential energy surface for the Ar–N_2 interaction, which also represents a case of relatively strong rotational coupling. Electron-gas calculations by Kim[181] indicated that the model due to Kistemaker and de Vries[182] was to be preferred to that of Pattengill *et al.*[183] which was found to be insufficiently anisotropic. An analysis of the inelastic rotational cross sections was in keeping with this view.[184]

9.7.5. Benzene

Perhaps the most ambitious attempt to evaluate the intermolecular forces of a large molecule is the study of benzene by Evans and Watts.[69] They found that it was necessary to use a multi-centre model and represented the molecule by six sites situated at the vertices of a regular hexagon. Each of the sites interacted with the sites on neighbouring molecules. The site–site potential function included an exponential repulsive term and a long-range term such that the known dispersion energy was partitioned equally between the sites. A simplified version of the model in which the site–site interactions were represented by the Lennard-Jones 12 – 6 function ($\varepsilon/k = 77$ K and $\sigma = 0{\cdot}35$ nm) was found to give an adequate account of the second virial coefficients of benzene and the predicted unit cell volume in the solid state to within 15 per cent. Attempts have been made to calculate the transport properties of gaseous benzene using this model.[70]

9.7.6. Water

Because of its obvious importance a great deal of work has been carried out on the molecular interactions of water. A definitive potential energy function has yet to be established but we outline some of the more significant attempts that have been made to tackle this challenging problem.

Water is perhaps unique among simple molecules in that much of our present knowledge of the intermolecular potential energy function has come from *ab initio* quantum mechanical calculations, and fits to thermophysical data have played a less important part in the development of potential functions than is usual. This is of course a consequence of the hydrogen-bonding which dominates the interaction between water molecules, and which gives rise to a much stronger interaction than is usual for small molecules. (The maximum well depth for pairs of water molecules is believed to be around 20 kJ mol^{-1}, in contrast to the figure of about 0·3 kJ mol^{-1} for the isoelectronic Ne molecule.) The relatively coarse electronic changes which occur when two water molecules come together are just about within the range of present-day computers, whereas the more subtle electron correlation effects which give rise to the dispersion energy cannot yet be directly calculated with accuracy.

Much of the intense interest shown in water comes from a desire to understand the properties of liquid water, both in its pure state and as a solvent.[71,72] Computer simulations have been extensively used for this purpose in recent years giving valuable insight into both static and dynamic properties and there have been several attempts to develop model intermolecular potentials suitable for use in such applications. In almost all cases, pairwise additivity of the intermolecular energy has been assumed in these simulations (despite strong contra-indications from quantum mechanical calculations[73]) so that the models which have emerged are best regarded as effective pair potentials. Their relationship to the true many-body potential energy functions has been analysed by Stillinger.[74] Since these models are not true pair potentials they have not been extensively tested against experimental two-body data, although calculations of second virial coefficients have been used in the choice of adjustable parameters. There is undoubtedly an element of inconsistency in taking potential energy functions based on quantum mechanical calculations for isolated pairs of molecules, and using these as effective pair potentials for dense fluid studies, but it seems that this is the best that can be done at the present time.

An important characteristic of water interactions which has been revealed both by structural and by quantum mechanical investigations is a preference for 'tetrahedral' arrangements of hydrogen atoms, both H-bonded and covalently bonded, about the oxygen atoms. Spectroscopic studies[75] and quantum mechanical investigations have confirmed that the lowest energy structure of the dimer is as shown in Fig. 9.3. The hydrogen bond is linear within the error of the results ($\alpha \sim 0$) and the angle θ is near to the value (54°) for a tetrahedral distribution of hydrogens. In order to develop a potential energy function which favours this structure, several workers have adopted a model in which positive and negative

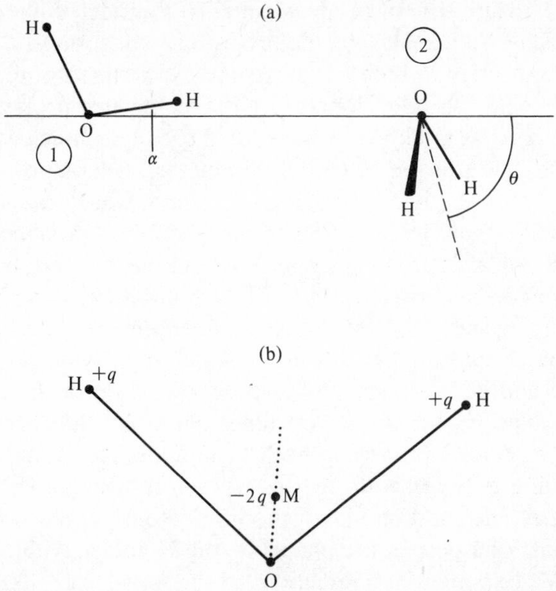

FIG. 9.3. (a) The structure of the water dimer. Molecule 1 lies in the plane of the paper. (b) The centres of force for the interactions of water molecules in the model proposed by Clementi *et al.*[72]

point charges are distributed about the oxygen atom. The resultant Coulombic interactions favour the desired geometry, and are added to a central intermolecular potential which is often based on the oxygen atom.

An early water potential developed by Rowlinson,[76] based on studies of the lattice energy and structure and on second virial coefficients, is of this type. To a central spherical Lennard-Jones potential were added positive point charges on the hydrogen atoms, and two compensating negative charges were arranged about the oxygen atom. This model was used in the earliest simulations of liquid water by Barker and Watts,[77] who used the Monte Carlo method to calculate pair distribution functions and thermodynamic properties which were in fair agreement with experiment. An attempt was also made to calculate the dielectric properties of water using this model.[78]

An extensive series of molecular dynamics simulations has been performed by Rahman and Stillinger,[79–81] using a similar type of pair potential. In its earliest form, as developed by Ben-Naim and Stillinger,[82] this consisted of a central Lennard-Jones potential, plus four tetrahedrally arranged point charges at 0·1 nm from the oxygen nucleus. A switching function was used to eliminate the Coulomb potential when the charge separations approached zero. (In Rowlinson's model a central hard core was used to avoid the same problem.) This model gave encouraging results for many static and dynamic properties, but showed a tendency to favour an excessively tetrahedral structure. A modified potential, known as the ST2 model, was developed[81] to overcome this problem. The two negative charges were drawn nearer to the oxygen nucleus, and some changes were made in the potential parameters. Simulations carried out using this potential yielded results for liquid water in good general agreement with experiment, showing some of the anomalies which are characteristic of water. Further simulations using the Monte Carlo method have recently been reported for this potential, and the importance of carrying out very long runs has been emphasized.[83]

Other potential functions for use in simulations have been developed by Clementi and co-workers.[72,84,85] These are based on fits to extensive *ab initio* calculations for the water dimer, and take the form of sums of site–site interactions, comprising both Coulombic terms and exponential functions acting between four centres in each molecule. These are illustrated in Fig. 9.3(b), and consist of the three atoms, and a site, M, on the symmetry axis. Charges $+q$ are placed on the H nuclei, with a compensating $-2q$ at M. Exponential interactions of the form $a_{ij}\exp(-b_{ij}r_{ij})$ operate between pairs of atoms on different molecules. In early work,[84] only repulsive atom–atom interactions were included, but in a later version attractive terms were added.[85] Monte Carlo calculations using the latter potential gave structural data in generally good agreement with experiment,[86] but the internal energy calculated for liquid water at 25°C was significantly less negative than the experimental value. It was suggested that a non-additive three-body energy would account for this discrepancy, though having little effect on the structural properties. It should be noted that for water the net contribution from the non-additive three-body energy appears to be negative, in contrast to the positive term which is characteristic of simple liquids.

Watts[87] has proposed an atom–atom potential for water which has been shown to be consistent with spectroscopic, molecular beam, second virial coefficient, and solid state data.

9.8. Class IV potential functions

In view of the complexity of the potential energy functions that are necessary to account for the behaviour of the inert gases it can be seen that the value of simple analytical functions is likely to be severely limited for more complex molecules. Nevertheless considerable energy has been expended on finding, for a wide range of substances, the parameters for very simple functions that result in the best fit to available thermophysical data. The function that has been most often used is that due to Lennard-Jones. Some sets of recommended parameters are given in Table 9.12.

Table 9.12

Parameters for crude potential energy functions

Sub-stance	12−6 [100]		12−6 [156]		exp−6 [100]			Kihara [100]			m−6−8 [148]			
	10σ/nm	(ε/k)/K	10σ/nm	(ε/k)/K	α	10σ/nm	(ε/k)/K	σ^*	10σ/nm	(ε/k)/K	m	γ	10σ/nm	(ε/k)/K
Ne			2·72	47·0										
Ar	3·504	117·7	3·336	141·2	18	3·302	152·0	0·125	3·314	147·2	11	3	3·292	153
Kr	3·827	164·0	3·575	191·4	18	3·551	213·9	0·150	3·521	215·6	11	3	3·509	216
Xe	4·099	222·2	3·924	257·4	22	3·823	326·0	0·175	3·878	298·8	11	3	3·841	295
N_2	3·745	95·2	3·613	103·0	30	3·445	160·2	0·25	3·526	136·2	12	2	3·54	118
O_2			3·362	128·8							10	1	3·437	113
CO_2	4·328	198·2	3·753	246·1				0·65	3·711	441·7	14	1	3·68	282
CH_4	3·783	148·9	3·706	159·7	24	3·566	255·5	0·20	3·620	204·3	11	3	3·68	168
CF_4	4·744	151·5			300	4·153	403·6	0·50	4·319	289·7				

† For the form of the exp−6 function see Appendix 1.7, for Kihara see Appendix 1.6, and for $m-6-8$ see Appendix 1.18.

The Morse potential and the Kihara potential have also been widely advocated. More recently the $m-6-8$ function has been shown to give a reasonable account of the thermophysical data for a wide range of substances.[34] Where a direct comparison with accurately known potential energy functions can be made, as for the inert gases, the simple functions are seen to be a poor representation of intermolecular energy, and it is probably not fruitful to examine all the functions that have been advocated or to attempt to assess the various best-fit parameters. In general we may find Class IV functions useful in interpolating when fitted to thermophysical data available over only a narrow range of experimental conditions but their predictive value is almost worthless. With this in mind we will devote no further attention to the topic.

9.9. General features of potential energy functions

The elucidation of accurate potential energy functions for a limited number of interactions has allowed some of the assumptions commonly made about such functions to be critically tested. Two assumptions of particular interest in this context are conformality and the combining rules for unlike interactions.

9.9.1. Conformality

A large amount of work in intermolecular forces has been based on the assumption that the potential energy functions of substantial groups of molecules are conformal. By conformal we mean that when the functions are reduced by two characteristic parameters (usually σ or r_m for separation and ε for energy) they all have the same shape. Since molecules whose potential energy functions are conformal may be shown (in normal circumstances) to adhere to the principle of corresponding states (see Chapters 3 and 6), the widespread applicability of this principle is indirect evidence that conformality may be a useful approximation. The inert gases follow the principle of corresponding states closely and it is a reasonable approximation for many simple non-polar substances.[88–90] The potential functions for Ar–Ar, Ar–Kr, and Kr–Kr, all of which are accurately known, are closely conformal.[54] However there is some evidence from molecular-beam studies that even within the inert gas series distinct differences in the shapes of the potential functions may be noted. Furthermore, the reduced value of the dispersion force coefficient C_6 is 30 per cent higher for He than Xe or Kr.[88] When we leave the inert gas series for other monatomic systems the differences in shape of the potential become more dramatic. One example quoted earlier was that of the Li–Hg potential function which appeared to be quite unlike that found for the other alkali metal–mercury systems (Fig. 9.2).

The results of these investigations indicate that, on the whole, the assumption of conformality is not generally a valid approximation. If no direct evidence is available concerning the shape of the potential energy function for a particular system, it is probably wiser when attempting to estimate gas-phase thermophysical properties, to apply the principle of corresponding states directly to the properties than to calculate properties by invoking conformality of the potentials. The corresponding states parameters and functions of Kestin, Ro, and Wakeham[88] are tabulated in Appendix 3 to facilitate such an approach.

9.9.2. *Combining rules*

When it is necessary to calculate the properties of mixtures a knowledge of the interactions between the unlike pairs is required in addition to those between the like pairs of molecules. Consequently much effort has been expended in trying to characterize the potential energy functions for the unlike-pair interactions in terms of those of the like. The most usual practice has been to attempt to express the parameters of the cross-interaction in terms of those of the like interactions. The restrictive assumption of the conformality of the functions is usually implicit in this procedure and the limitations of this assumption have been discussed above. The empirical rules which relate the parameters are called *combining rules.*

The most traditional rules are the Lorentz rule for collision diameters

$$\sigma_{12} = \frac{\sigma_{11} + \sigma_{22}}{2}$$

which is exact for hard spheres, and the geometric mean rule for well depths

$$\varepsilon_{12} = \sqrt{(\varepsilon_{11}\varepsilon_{22})}$$

often called the Berthelot rule. This second rule (expressed in terms of polarizabilities) can be seen to arise from the London formula for the dispersion energy (see § 2.2.5).

Neither rule is exact though the first appears to be a useful approximation in many circumstances. The second rule tends to overestimate the strength of the interaction between unlike molecules. In the study of mixtures it has become common practice to introduce an empirical parameter, k_{12}, which quantifies departures from the geometric mean rule for ε. Thus (see Table 8.3)

$$\varepsilon_{12} = (\varepsilon_{11}\varepsilon_{22})^{\frac{1}{2}}(1 - k_{12}).$$

Sometimes an equivalent parameter which measures departures from the Lorentz rule is found to be useful.

Many attempts have been made to devise superior combining rules based upon both empirical and theoretical considerations. Thus if U^{dis} as represented by the London expression is identified with the attractive term of the Lennard-Jones (12−6) function, $U(r) = -4\varepsilon(\sigma/r)^6$ and if additivity of collision diameters is assumed then factorization leads directly to the Hudson–McCoubrey rule,[91]

$$\varepsilon_{12} = (\varepsilon_{11}\varepsilon_{22})^{\frac{1}{2}} \frac{2(I_1 I_2)^{\frac{1}{2}}}{I_1 + I_2} \cdot \left\{\frac{2(\sigma_{11}\sigma_{22})^{\frac{1}{2}}}{(\sigma_{11} + \sigma_{22})}\right\}^6$$

where I_i here represents the ionization potential of molecule i. This is in general superior to the geometric mean rule but is still far from adequate again tending to overestimate ε_{12}. A rule which predicts even more dramatic departures from the geometric mean values is that of Fender and Halsey,[92] who advocated a harmonic mean ε_{12}

$$\varepsilon_{12} = \frac{2\varepsilon_{11}\varepsilon_{22}}{(\varepsilon_{11} + \varepsilon_{22})}.$$

This rule tends to predict that the well depth ε_{12} will be determined very largely by the component with the weaker intermolecular forces.

The experimental data which can be used to obtain values of σ_{12} and ε_{12} which can provide a test of the combining rules is severely limited. One source of information is from differential cross-sections determined in molecular beam experiments. Thus Chen, Siska, and Lee[58] investigating the interactions of helium with the heavier inert gases found that ε_{12} was almost independent of which of the heavier atoms were involved (see Table 9.13). More recent experiments have not been in complete agreement with this work indicating higher values of ε_{12} but the approximate constancy of this parameter was again observed.[60] Indeed one set of observations suggested that ε_{12} for He–Xe was slightly less than that for He–Ar despite the fact that the well-depth parameter for xenon is roughly twice that for argon. The beam results are supported by the results obtained from the inversion of gas viscosity data by Maitland and Wakeham.[55] These results, taken together provide clear evidence for the inadequacy of the geometric mean rule for ε_{12}. The rule of Fender and Halsey appears to give the correct trend though for the interactions involving helium the absolute values appear too low. Van den Biesen *et al.*[93] studied the glory structure in the total cross-sections of the mixed inert gas interactions (other than those involving He) and found like Maitland and Wakeham[55] that the Fender and Halsey rule gave reasonable agreement (Table 9.14).

Recently a number of more sophisticated combining rules have been developed by Sikora[94] and others on the basis that the repulsion of closed

Table 9.13

Test of combining rules on the helium–heavier inert gas systems

	(ε_{12}/k)/K								$10r_{m12}$/nm			
	Expt. Chen *et al.*[58]	Expt. Smith *et al.*[60]	Expt. Hogervorst[143]	Geom. mean	Kong and Chakrabarty[97]	Fender and Halsey[92]	Hudson and McCoubrey[91]	Sikora[94]	Expt. Chen *et al.*[58]	Expt. Smith *et al.*[60]	Arith. mean	Kong and Chakrabarty[97]
He–Ar	24·2	30·2	40±3	38·6	33·4	19·6	35·5	27·9	3·54	3·46	3·36	3·43
He–Kr	24·7	30·2		45·2	36·4	20·0		28·3	3·75	3·67	3·49	3·59
He–Xe	29·4	28·0	49±4	54·3	37·4	20·2	43·4	27·0	4·15	3·95	3·67	3·84

Table 9.14

Test of combining rules with heavier inert gas mixtures

	Transport properties[55]			Total cross-section[93]		Combining rules†	
	σ/nm	r_m/nm	(ε/k)/K	r_m/nm	(ε/k)/K	Fender–Halsey (ε/k)/K	Geometric mean (ε/k)/K
Ne–Ar	0·313	0·352	60±5	0·343	67	62	75
Ne–Kr	0·336	0·376	60±10	0·358	72	67	89
Ne–Xe	0·347	0·388	70±10	0·374	75	70	106
Xe–Ar	0·364	0·413	170±5	0·406	187	189	200
Xe–Kr	0·377	0·423	220±5	0·418	231	234	237

† Calculated assuming ε/k for Ne = 40, Ar = 142, Kr = 200, and Xe = 281 K.

shell species is largely due to deformation energy. This leads to the expression for the total energy

$$U_{12}(r_1+r_2)=\tfrac{1}{2}\{U_{11}(2r_1)+U_{22}(2r_2)\}$$

where $2r_1$ and $2r_2$ are the separations at which the slopes of the potential energy curves of the like species are equal. A number of rules deriving from this concept have been proposed by Sikora,[94] Smith,[95] Kong,[96] and Kong and Chakrabarty[97] which combine the atomic distortion model with a simple representation (Lennard-Jones 12−6, exp−6, or Morse) of the over-all potential function. Thus Kong gives for the Lennard-Jones 12−6 function

$$\varepsilon_{12}\sigma_{12}^{12}=\left(\frac{\varepsilon_{11}\sigma_{11}^{12}}{2^{13}}\right)\left[1+\left(\frac{\varepsilon_{22}\sigma_{22}^{12}}{\varepsilon_{11}\sigma_{11}^{12}}\right)^{\frac{1}{13}}\right]^{13}.$$

$$\varepsilon_{12}\sigma_{12}^{6}=(\varepsilon_{11}\sigma_{11}^{6}\varepsilon_{22}\sigma_{22}^{6})^{\frac{1}{2}}.$$

Unfortunately the precise testing of these rules has been hampered by the considerable uncertainties in the true values of ε_{12}. Thus, as discussed above, in the case of the helium–argon interaction estimates of this parameter have varied from 24 to 40 K. However recent reliable studies of the helium–heavier inert gas interactions have indicated a value close to 30 K for this parameter and have suggested that the atomic distortion rules represent a distinct advance. Perhaps the best of the rules in this context is the original one proposed by Sikora[94]

$$\varepsilon_{12}=(\varepsilon_{11}\varepsilon_{22})^{\frac{1}{2}}g(I)f(V)$$

where $g(I)=4I/\{(1+I)^2\}$, $I=I_2/I_1$, the ratio of the ionization potentials, and

$$f(V)=\frac{2^{13}V^{\frac{1}{2}}}{(1+V^{\frac{1}{13}})^{13}},\qquad V=\frac{\varepsilon_{22}\sigma_{22}^{12}}{\varepsilon_{11}\sigma_{11}^{12}}$$

$$\sigma_{12}=2^{-\frac{13}{12}}(\sigma_{11}^{\frac{12}{13}}+\sigma_{22}^{\frac{12}{13}})^{\frac{13}{12}}.$$

The applications of some of the more modern rules are given in Table 9.13. On the whole the new rules do better than the older prescriptions (with the possible exception of the Fender–Halsey rule). However, there are still considerable uncertainties in the experimental data and truly critical tests of the combining rules cannot yet be performed.

We must conclude that combining rules do not offer, at the present time, a secure basis for the understanding of mixtures. The proper approach to the calculation of the properties of mixtures of dilute gases is through a knowledge of all the pair interactions involved. For dense fluid mixtures a knowledge of many body non-additive energies may also be needed. This information can only be obtained by appropriate experiments on binary mixtures. It cannot be confidently inferred from the properties of the pure components.

9.10. Concluding remarks

The successful determination of the potentials for the inert gases has proved a slow and tortuous process, and even now these functions are not known quantitatively in all cases. From these studies, however, a method of approach has emerged which should ensure more fruitful progress in future efforts to determine intermolecular forces. Often the most satisfactory and direct procedure is to build up the potential in a piece-wise manner by combining data from a variety of sources. A direct approach using inversion of both microscopic and macroscopic experimental data complemented by quantum mechanical calculations is a particularly satisfactory method of obtaining the potential over-all desired separations. Where inversion or direct calculation is not possible or gives an incomplete potential, it is important that analytical functions of sufficient flexibility should be used. A fit to experimental data should then be achieved using as extensive a range of properties as possible, over a very wide temperature range. The important requirement for any proposed potential, irrespective of its method of determination, is that it should fit all available data within experimental error. In these procedures, ideally only pair (dilute-gas) properties should be used, leaving condensed-phase properties as tests of non-additive multibody, rather than of pair, energies. Where solid- and liquid-state data must be used, the Axilrod–Teller triple-dipole term appears to be an adequate effective non-additive multibody term for monatomic systems. However, the extent to which such an assumption is valid for polyatomic molecules has yet to be tested.

In addition to these principles, several additional factors unique to polyatomic molecules have emerged from the relatively few detailed studies which have so far been carried out. Inversion methods, even for atom–diatom systems, do not appear to be particularly promising at the present time. However, the benefits resulting from such procedures probably mean that continued efforts will be made to develop them for complex molecules. Meanwhile the functional forms chosen to represent potentials for polyatomic molecules must be sufficiently flexible that unrealistic constraints are not put on one region of the anisotropic potential by another. For instance the deficiency of the Legendre polynomial expansion in this respect, where the long-range anisotropic terms can in certain circumstances make the short-range potential physically unrealistic, is well established. Alternative formulations, the most extreme of which is the use of a purely numerical multi-dimensional potential, must be used when such problems arise. Theoretical calculations of the anisotropic dispersion coefficients are becoming available[170,157] and, in conjunction with short range *ab initio* calculations,[163,158] should constrain the asymptotic behaviour of proposed functions as they did for monatomic molecules.

Spectroscopic studies are likely to prove very important probes of the well region, although present molecular beam resonance methods only give detailed information on the region close to the minimum. Molecular beam scattering using state selection, which enables state-to-state differential cross-sections to be determined, should also be a sensitive probe of potential anisotropy. In the area of macroscopic properties, field-dependent gas transport coefficients are another, as yet untapped source of information about the anisotropy of potentials. Promising as these techniques are, however, the difficulties inherent in the accurate determination of the full angular dependence of potential functions for polyatomic molecules has sometimes led to the use of effective spherical potentials. The dangers of using such effective spherical potentials to calculate any property other than that from which they have been determined cannot be overemphasized. Although such potentials may eventually prove of some value, the extent to which the angle-averaging procedure varies from property to property has so far been little studied.

The intermolecular pair potential energy function is clearly one of the most powerful concepts of chemical physics. In principle, much of the behaviour of molecular systems is implicitly defined by these functions, the contributions of non-additive many-body effects representing only a minor, albeit sometimes very significant, perturbation. The potential function is thus a way of presenting a great deal of information in a very compact form. However, despite their central role in molecular theory, they have been accurately determined only for a relatively small number of molecules and there has been much uncritical work. Inspection of the literature might lead one to suppose that for every set of thermo-physical measurements, however fragmentary, there corresponds a potential energy function worthy of being placed on record. All too often single properties measured over a few degrees near room temperature are fitted to a simple model potential energy function that has not been shown to be adequate for any known species. If there is one important lesson to be learnt from the success of recent years with monatomic species, it is that intermolecular potential energy functions are of value only if determined with high accuracy and used with judgement.

References

1. Hirschfelder, J. O., Curtiss, C. F., and Bird, R. B. *Molecular theory of gases and liquids.* Wiley, New York (1954).

2. Corner, J. *Trans. Faraday Soc.* **44,** 914 (1948).
3. Wood, W. W. and Parker, F. R. *J. chem. Phys.* **27,** 720 (1957).
4. Bridgman, P. W. *Proc. Amer. Acad. Sci.* **70,** 1 (1935).
5. Michels, A., Wijker, Hub., and Wijker, Hk. *Physica* **15,** 627 (1949).
6. London, F. *Trans. Faraday Soc.* **33,** 8 (1937); *Z. phys. Chem.* **B11,** 222 (1930); Margenau, H. *Rev. mod. Phys.* **11,** 1 (1939).
7. Kirkwood, J. G. *Physik Z.* **33,** 57 (1932); Müller, A. *Proc. roy. Soc.* **A154,** 624 (1936).
8. Fender, B. E. F. and Halsey, G. D. *J. chem. Phys.* **36,** 1881 (1962); Byrne, M. A., Jones, M. R., and Staveley, L. A. K. *Trans. Faraday Soc.* **64,** 1747 (1968); Weir, R. D., Wynn Jones, I., Rowlinson, J. S., and Saville, G. *Trans. Faraday Soc.* **63,** 1320 (1967).
9. Axilrod, B. M. *J. chem. Phys.* **17,** 1349 (1949).
10. Dawe, R. A. and Smith, E. B. *J. chem. Phys.* **52,** 693 (1970).
11. For example, the results of Trautz, M. and Zink, R. *Ann. Phys.* **7,** 427 (1930) for argon.
12. Starkschall, G. and Gordon, R. G. *J. chem. Phys.* **54,** 663 (1971).
13. Starkschall, G. and Gordon, R. G. *J. chem. Phys.* **56,** 2801 (1972).
14. Phillipson, P. E. *Phys. Rev.* **125,** 1981 (1962).
15. Gilbert, T. L. and Wahl, A. C. *J. chem. Phys.* **47,** 3425 (1967).
16. Abrahamson, A. A. *Phys. Rev.* **130,** 693 (1963).
17. Gordon, R. G. and Kim, Y. S. *J. chem. Phys.* **56,** 3122 (1972).
18. Tanaka, Y. and Yoshino, K. *J. chem. Phys.* **53,** 2012 (1970); Tanaka, Y., Yoshino, K., and Freeman, D. E. *J. chem. Phys.* **59,** 5160 (1973).
19. Barker, J. A., Fisher, R. A., and Watts, R. O. *Mol. Phys.* **21,** 657 (1971).
20. Maitland, G. C. and Smith, E. B. *Mol. Phys.* **22,** 861 (1971).
21. Siska, P. E., Parson, J. M., Schafer, T. P., and Lee, Y. T. *J. chem. Phys.* **55,** 5762 (1971); Parson, J. M., Siska, P. E., and Lee, Y. T. *J. chem. Phys.* **56,** 1511 (1972).
22. Colbourn, E. A. and Douglas, A. E. *J. chem. Phys.* **65,** 1741 (1976).
23. Axilrod, B. M. and Teller, E. *J. chem. Phys.* **11,** 299 (1943).
24. Crawford, F. W., Harris, E. J. R., and Smith, E. B. *Mol. Phys.* **37,** 1323 (1970); Cox, H. E., Crawford, F. W., Smith, E. B., and Tindell, A. R. *Mol. Phys.* **40,** 705 (1980).
25. Maitland, G. C. and Smith, E. B. Seventh Symposium on Thermophysical Properties. *Amer. Soc. Mech. Eng.*, 412 (1977).
26. (Lennard)-Jones, J. E. *Proc. roy. Soc.* **A106,** 463 (1924).
27. Kihara, T. and Kotani, M. *Proc. Phys. Math. Soc. Jpn* **24,** 76 (1942); **25,** 602 (1943); de Boer, J. and van Kranendonk, J. *Physica* **14,** 442 (1948); Rowlinson, J. S. *J. chem. Phys.* **17,** 101 (1949); Hirschfelder, J. O., Bird, R. B., and Spotz, E. L. *J. Chem. Phys.* **16,** 968 (1948); *Chem. Rev.* **44,** 205 (1949); Grew, K. E. *J. Chem. Phys.* **18,** 149 (1950).
28. Guggenheim, E. A. and McGlashan, M. L. *Proc. roy. Soc.* **A255,** 456 (1960).
29. Munn, R. J. and Smith, F. J. *J. chem. Phys.* **43,** 3998 (1965).
30. Dymond, J. H., Rigby, M., and Smith, E. B. *J. chem. Phys.* **42,** 2801 (1965).
31. Dymond, J. H. and Alder, B. J. *J. chem. Phys.* **51,** 309 (1969).
32. Barker, J. A. and Pompe, A. *Austral. J. Chem.* **21,** 1683 (1968).
33. Aziz, R. A. and Chen, H. H. *J. chem. Phys.* **67,** 5719 (1977).
34. Klein, M. and Hanley, H. J. M. *J. chem. Phys.* **53,** 4722 (1970).
35. Thakkar, A. J. and Smith, V. H. *Chem. Phys. Lett.* **24,** 157 (1974).
36. Maitland, G. C. and Smith, E. B. *Chem. Phys. Lett.* **22,** 443 (1973).
37. Buck, U., Dondi, M. G., Valbusa, U., Klein, M. L., and Scoles, G. *Phys. Rev.* **A8,** 2409 (1973).

38. Gough, D. W., Smith, E. B., and Maitland, G. C., *Mol. Phys.* **27,** 867 (1974).
39. LeRoy, R. J. *J. chem. Phys.* **57,** 573 (1973).
40. Barker, J. A., Watts, R. O., Lee, J. K., Schafer, T. P., and Lee, Y. T. *J. chem. Phys.* **61,** 3081 (1974); Barker, J. A., Klein, M. L., and Bobetic, M. V. *I.B.M. J. Res. and Dev.* **20,** 222 (1976).
41. Aziz, R. A. and Kocay, W. L. *Mol. Phys.* **30,** 857 (1975).
42. Thakkar, A. J. and Smith, V. H. *Chem. Phys. Lett.* **24,** 157 (1974).
43. Aziz, R. A. *Mol. Phys.* **38,** 177 (1979).
44. Aziz, R. A., Nain, V. P. S., Carley, J. S., Taylor, W. L., and McConville, G. T. *J. chem. Phys.* **70,** 4330 (1979).
45. Beck, D. E. *Mol. Phys.* **14,** 311 (1968); **15,** 332 (1968); *J. chem. Phys.* **50,** 541 (1969).
46. Bruch, L. W. and McGee, I. J. *J. chem. Phys.* **46,** 2959 (1967).
47. Goldman, V. V. and Klein, M. L. *J. low temp. Phys.* **12,** 101 (1973).
48. Farrar, J. M., Lee, Y. T., Goldman, V. V., and Klein, M. L. *Chem. Phys. Lett.* **19,** 359 (1973).
49. Maitland, G. C. *Mol. Phys.* **26,** 513 (1973).
50. LeRoy, R. J., Klein, M. L., and McGee, I. J. *Mol. Phys.* **28,** 587 (1974).
51. Aziz, R. A. *Chem. Phys. Lett.* **40,** 57 (1976).
52. Freeman, D. E., Yoshino, K., and Tanaka, Y. *J. chem. Phys.* **61,** 4880 (1974).
53. Lee, J. K., Henderson, D., and Barker, J. A. *Mol. Phys.* **29,** 429 (1975).
54. Gough, D. W., Smith, E. B., and Maitland, G. C. *Mol. Phys.* **27,** 867 (1974).
55. Maitland, G. C. and Wakeham, W. A. *Mol. Phys.* **35,** 1443 (1978).
56. Buck, U., Dondi, M. G., Valbusa, U., Klein, M. L., and Scoles, G. *Phys. Rev.* **A8,** 2409 (1973).
57. Aziz, R. A., Presley, J., Buck, U., and Schleusener, J. *J. chem. Phys.* **70,** 4737 (1979).
58. Chen, C. H., Siska, P. E., and Lee, Y. T. *J. chem. Phys.* **59,** 601 (1973).
59. Smith, K. M., Rulis, A. M., Scoles, G., Aziz, R. A., and Duquette, G. *J. chem. Phys.* **63,** 2250 (1975).
60. Smith, K. M., Rulis, A. M., Scoles, G., Aziz, R. A., and Nain, V. *J. chem. Phys.* **67,** 152 (1977).
61. Ng, C. Y., Lee, Y. T., and Barker, J. A. *J. chem. Phys.* **61,**1996 (1974).
62. Buck, U. *J. chem. Phys.* **54,** 1923, 1929 (1971).
63. Matthews, G. P. and Smith, E. B. *Mol. Phys.* **32,** 1719 (1976).
64. Snook, I. K. and Spurling, T. H. *J. Chem. Soc. Faraday Trans. II* **68,** 1359 (1972).
65. Murad, S. and Gubbins, K. E. Computer modelling of Matter. *A.C.S Sympos. Ser.* **86,** 62 (1978).
66. Murad, S., Evans, D., Gubbins, K. E., Streett, W. B., and Tildesley, D. J. *Mol. Phys.* **37,** 725 (1979).
67. Cheung, P. S. Y. and Powles, J. G. *Mol. Phys.* **30,** 921 (1975).
68. Singer, K., Taylor, A., and Singer, J. V. L. *Mol. Phys.* **33,** 1757 (1976).
69. Evans, D. J. and Watts, R. O. *Mol. Phys.* **29,** 777 (1975); **31,** 83 (1976); **32,** 93 (1976).
70. Evans, D. J. and Watts, R. O. *Mol. Phys.* **32,** 995 (1976).
71. Stillinger, F. H. *Advan. chem. Phys.* **31,** 2 (1975).
72. Clementi, E. Liquid water structure. In *Lecture notes in chemistry*, Vol. **2,** Springer Verlag, Berlin (1976).

73. Hankins, D., Moskowitz, J. W., and Stillinger, F. H. *J. chem. Phys.* **53,** 4544 (1970); **59,** 995 (1973).
74. Stillinger, F. H. *J. chem. Phys.* **57,** 1780 (1972).
75. Dyke, T. R., Mack, K. M., and Muenter, J. S. *J. chem. Phys.* **66,** 498 (1977).
76. Rowlinson, J. S. *Trans. Faraday Soc.* **47,** 120 (1951).
77. Barker, J. A. and Watts, R. O. *Chem. Phys. Lett.* **3,** 144 (1969).
78. Watts, R. O. *Mol. Phys.* **28,** 1069 (1974).
79. Rahman, A. and Stillinger, F. H. *J. chem. Phys.* **55,** 3336 (1971); Nienhuis, G. and Deutch, J. **55,** 4213 (1971).
80. Stillinger, F. H. and Rahman, A. *J. chem. Phys.* **57,** 1281 (1972).
81. Stillinger, F. H. and Rahman, A. *J. chem. Phys.* **60,** 1545 (1974); **68,** 666 (1978).
82. Ben-Naim, A. and Stillinger, F. H. *Water and aqueous solutions* (ed. R. A. Horne), Chapter 8. Wiley-Interscience, New York (1972).
83. Rao, M., Pangali, C., and Berne, B. J. *Mol. Phys.* **37,** 1773 (1979).
84. Popkie, H., Kistenmacher, H., and Clementi, E. *J. chem. Phys.* **59,** 1325 (1973).
85. Matsuoka, O., Yoshimine, M., and Clementi, E. *J. chem. Phys.* **64,** 1351 (1976).
86. Lie, G. C., Clementi, E., and Yoshimine, M., *J. chem. Phys.* **64,** 2314 (1976).
87. Watts, R. O. *Chem. Phys.* **26,** 367 (1977).
88. Kestin, J., Ro, S. T., and Wakeham, W. A. *Physica* **58,** 165 (1972).
89. Kestin, J. and Mason, E. A. *AIP Conf. Proc.* **11,** 137 (1973).
90. Boushehri, A, Viehland, L. A., and Mason, E. A. *Physica* **91A**, 424 (1978).
91. Hudson, G. H. and McCoubrey, J. C. *Trans. Faraday Soc.* **56,** 761 (1960).
92. Fender, B. E. F. and Halsey Jr., G. D. *J. chem. Phys.* **36,** 1881 (1962).
93. Van den Biesen, J. J. H., Stokvis, F. A., van Veen, E. H., and van den Meijdenberg, C. J. N. *Physica,* **100A,** 375 (1980).
94. Sikora, P. T. *J. Phys.* **B3,** 1475 (1970).
95. Smith, F. T. *Phys. Rev.* **A5,** 1708 (1972).
96. Kong, C. L. *J. chem. Phys.* **59,** 968 (1973).
97. Kong, C. L. and Chakrabarty, M. R. *J. phys. Chem.* **77,** 2668 (1973).
98. Mason, E. A. and Rice, W. E. *J. chem. Phys.* **22,** 843 (1954).
99. Whalley, E. and Schneider, W. G. *J. chem. Phys.* **23,** 1644 (1955).
100. Sherwood, A. E. and Prausnitz, J. M. *J. chem. Phys.* **41,** 429 (1964).
101. Barker, J. A., Fock, W., and Smith, F. *Phys. Fluids* **7,** 897 (1964).
102. Konowalow, D. D. and Hirschfelder, J. O. *Phys. Fluids* **4,** 629 (1961).
103. Bobetic, M. V. and Barker, J. A. *Phys. Rev.* **B2,** 4169 (1970).
104. Aziz, R. A. *J. chem. Phys.* **65,** 490 (1976).
105. Ahlrichs, R., Penco, P., and Scoles, G. *Chem. Phys.* **19,** 119 (1976).
106. Colbourn, E. A. and Douglas, A. E. *J. chem. Phys.* **65,** 1741 (1976).
107. Karnicky, J. F., Reamer, H. H., and Pings, C. J. *J. Chem. Phys.* **64,** 4592 (1976).
108. Levelt-Sengers, J. M. H., Klein, M., and Gallagher, J. S. Report AEDC-TR-71-39. National Bureau of Standards, Washington, DC (1971).
109. Maitland, G. C. and Smith, E. B. *J. chem. Engng Data* **17,** 150 (1972).
110. Kestin, J., Ro, S. T., and Wakeham, W. A. *J. chem. Phys.* **56,** 4119 (1972).
111. Haarman, J. W. *A.I.P. Conference Proceedings*, no. 11, p. 193 (1973).
112. Taken from ref. 104. Data of Taylor, W. L. and Weissman, S. *J. chem. Phys.* **59,** 1190 (1973) and Taylor, W. L. *J. chem. Phys.* **62,** 3837 (1975).

113. Linse, C. A., van den Biesen, J. J. H., van Veen, E. H., and van den Meijdenberg, C. J. N. *Physica* **99A,** 166 (1979).
114. Tanaka, Y. and Yoshino, K. *J. chem. Phys.* **53,** 2012 (1970).
115. Aziz, R. A. and Chen, H. H. *J. chem. Phys.* **67,** 5719 (1977).
116. Brewer, J. Report No. 67-2795. Air Force Office of Scientific Research (1967).
117. Clarke, A. G. and Smith, E. B. *J. chem. Phys.* **51,** 4156 (1969).
118. Hanley, H. J. M. *J. phys. Chem. Ref. Data* **2,** 619 (1974).
119. Weissman, S. and Dubro, G. A. *Phys. Fluids* **13,** 2689 (1970).
120. Jain, P. C. and Saxena, S. C. *J. Phys. E* **7,** 1023 (1974); *J. chem. Phys.* **63,** 5052 (1975).
121. Buck, U. See ref. 43.
122. van den Meijdenberg, C. J. N. See ref. 43.
123. Jody, B. J., Jain, P. C., and Saxena, S. C. *J. heat Transfer, Trans. ASME* **97c,** 605 (1975).
124. Gammon, B. E. *J. chem. Phys.* **64,** 2556 (1976).
125. Burgmans, A. L. J., Farrar, J. M., and Lee, Y. T. *J. chem. Phys.* **64,** 1345 (1976).
126. de Boer, J. and Michels, A. *Physica* **5,** 945 (1938).
127. Weissman, S. *Phys. Fluids* **16,** 1425 (1973).
128. Springer, G. S. and Wingeier, E. W. *J. chem. Phys.* **59,** 2747 (1973).
129. Jain, P. C. and Saxena, S. C. *Chem. Phys. Lett.* **28,** 454 (1974).
130. Cunha, M. A. and Laranjeira, M. F. *Physica* **57,** 306 (1972); Laranjeira, M. F. and Cunha, M. A. *Part. Phys.* **4,** 281 (1966).
131. Brewer, J. Report no. AD 633–448. US National Bureau of Standards (1967).
132. Schramm, B., Schmiedel, H., Gehrmann, R., and Bartl, R. *Ber. Bunsenges., Phys. Chem.* **81,** 316 (1977).
133. Kestin, J., Wakeham, W., and Watanabe, K. *J. chem. Phys.* **53,** 3773 (1970).
134. Maitland, G. C. and Smith, E. B. *J. Chem. Soc. Faraday Trans. I* **70,** 1191 (1974); Gough, D. W., Matthews, G. P., and Smith, E. B. *J. Chem. Soc. Faraday Trans. I* **72,** 645 (1976).
135. Hogervorst, W. *Physica* **51,** 59 (1971).
136. van Heijningen, R. J. J., Harpe, J. P., and Beenakker, J. J. M. *Physica* **38,** 1 (1968).
137. Arora, P. S., Carson, P. J., and Dunlop, P. J. *Chem. Phys. Lett.* **54,** 117 (1978).
138. Buck, U., Huisken, F., Pauly, H., and Schleusener, J. *J. chem. Phys.* **68,** 3334 (1978).
139. Linse, C. A. Thesis. University of Leiden (1977).
140. Gough, D. W., Matthews, G. P., Smith, E. B., and Maitland, G. C. *Mol. Phys.* **29,** 1759 (1975).
141. Kalelkar, A. S. and Kestin, J. *J. chem. Phys.* **52,** 4248 (1970).
142. Maitland, G. C. and Smith, E. B. *J. Chem. Soc. Faraday Trans. I* **70,** 1191 (1974).
143. Hogervorst, W. *Physica* **51,** 59 (1971).
144. Buck, U., Hoppe, H. O., Huisken, F., and Pauly, H. *Discuss. Faraday Soc.* **55,** 185 (1973); *J. chem. Phys.* **60,** 4925 (1974).
145. Buck, U. and Pauly, H. *J. chem. Phys.* **54,** 1929 (1971).
146. Buck, U., Kick, M., and Pauly, H. *J. chem. Phys.* **56,** 3391 (1972).
147. Hamann, S. D. and Lambert, J. A. *Aust. J. Chem.* **7,** 1 (1954).
148. Hanley, H. J. M. and Klein, M. *J. phys. Chem.* **76,** 1743 (1972).

149. Patel, H. R., Viswanath, D. S., and Seshadri, D. N. *Proceedings of the Sixth Symposium on Thermophysical Properties.* American Society of Mechanical Engineers (1973).
150. Pope, G. A., Chappelear, P. S., and Kobayashi, R. *J. chem. Phys.* **59,** 423 (1973).
151. Gibbons, R. M. *Cryogenics* **13,** 658 (1973).
152. Powles, J. G. and Gubbins, K. E. *Chem. Phys. Lett.* **38,** 405 (1976).
153. Barojas, J., Levesque, D., and Quentrec, B. *Phys. Rev.* **A7,** 1092 (1973).
154. Cheung, P. S. Y. and Powles, J. G. *Mol. Phys.* **32,** 1383 (1976).
155. Cheung, P. S. Y. and Powles, J. G. Unpublished work.
156. Clifford, A. A., Gray, P., and Platts, N. *J. Chem. Soc. Faraday Trans. I* **73,** 381 (1977).
157. Mulder, F., van Dijk, G., and van der Avoird, A. *Mol. Phys.* **39,** 407 (1980); Langhoff, P. W., Gordon, R. G., and Karplus, M. *J. chem. Phys.* **55,** 2126 (1971).
158. Ree, F. H. and Winter, N. W. *J. chem. Phys.* **73,** 322 (1980).
159. McMahan, A. K., Beck, H., and Krumhansl, J. A. *Phys. Rev.* **A9,** 1852 (1974).
160. Mulder, F., van der Avoird, A., and Wormer, P. E. S. *Mol. Phys.* **37,** 159 (1979) and references contained therein.
161. Jaszunski, M., Kochanski, E., and Siegbahn, P. *Mol. Phys.* **33,** 139 (1977).
162. Gallup, G. A. *Mol. Phys.* **33,** 943 (1977); *J. chem. Phys.* **66,** 2252 (1977).
163. Price, S. L. and Stone, A. J. *Mol. Phys.* **40,** 805 (1980).
164. McKellar, A. R. W. and Welsh, H. L. *Can. J. Phys.* **52,** 1082 (1974).
165. Farrar, J. M. and Lee, Y. T. *J. chem. Phys.* **57,** 5492 (1972).
166. Dondi, M. G., Valbusa, U., and Scoles, G. *Chem. Phys. Lett.* **17,** 137 (1972).
167. Gengenbach, R., Hahn, C., Schrader, W., and Toennies, J. P. *Theor. chim. Acta* **34,** 199 (1974).
168. Butz, H. P., Feltgen, R., Pauly, H., and Vehmeyer, H. *Z. Phys.* **247,** 70 (1971).
169. Gordon, R. G. and Cashion, J. K. *J. chem. Phys.* **44,** 1190 (1966).
170. Zeiss, G. D. and Meath, W. J. *Mol. Phys.* **33,** 1155 (1977).
171. Mulder, F., van Dijk, G., and van der Avoird, A. *Mol. Phys.* **39,** 407 (1980).
172. Berns, R. M. and van der Avoird, A. *J. chem. Phys.* In press (1980); Ree, F. H. and Winter, N. W. *J. chem. Phys.* **73,** 332 (1980).
173. Evans, D. J. *Mol. Phys.* **33,** 979 (1977).
174. Long, C. A., Henderson, G., and Ewing, G. E. *Chem. Phys.* **2,** 485 (1973).
175. Sakai, K., Koide, A., and Kihara, T. *Chem. Phys. Lett.* **47,** 416 (1977).
176. Leonas, V. B., Sermyagin, A. V., and Kamyshov, N. V. *Chem. Phys. Lett.* **8,** 282 (1971).
177. Tsapline, B. and Kutzelnigg, W. *Chem. Phys. Lett.* **23,** 173 (1973).
178. Victor, C. A. and Dalgarno, A. *J. chem. Phys.* **53,** 1316 (1970); Langhoff, P. W., Gordon, R. G., and Karplus, M. *J. chem. Phys.* **55,** 2126 (1971).
179. Geurts, P. J. M., Wormer, P. E. S., and van der Avoird, A. *Chem. Phys. Lett.* **35,** 444 (1975).
180. Shafer, R. and Gordon, R. G. *J. chem. Phys.* **58,** 5422 (1973).
181. Kim, Y. S. *J. chem. Phys.* **68,** 5001 (1978).
182. Kistemaker, P. G. and de Vries, A. E. *Chem. Phys.* **7,** 371 (1975).
183. Pattengill, M. D., La Budde, R. A., Bernstein, R. B., and Curtiss, C. F. *J. chem. Phys.* **55,** 5517 (1971).
184. Pattengill, M. D. and Bernstein, R. B. *J. chem. Phys.* **65,** 4007 (1976).

APPENDIX 1

INTERMOLECULAR PAIR POTENTIAL MODELS

We define $U^*(r) = U(r)/\varepsilon$, $r/r_m = \bar{r}$, $C_n^* = (C_n/\varepsilon r_m^n)$, and $U(\sigma) = 0$. Here C_n are the dispersion coefficients which take negative values for attractive potentials.

1. Hard-sphere potential

$$U(r) = \infty \qquad r < d$$

$$U(r) = 0 \qquad r \geqslant d.$$

The simplest model of a molecular interaction with a long and illustrious history.

2. Square well potential (Fig. A1)

$$U(r) = \infty \qquad r < \sigma$$

$$U(r) = -\varepsilon \qquad R\sigma \geqslant r \geqslant \sigma$$

$$U(r) = 0 \qquad r > R\sigma$$

R defines the width of the well, and ε its depth.

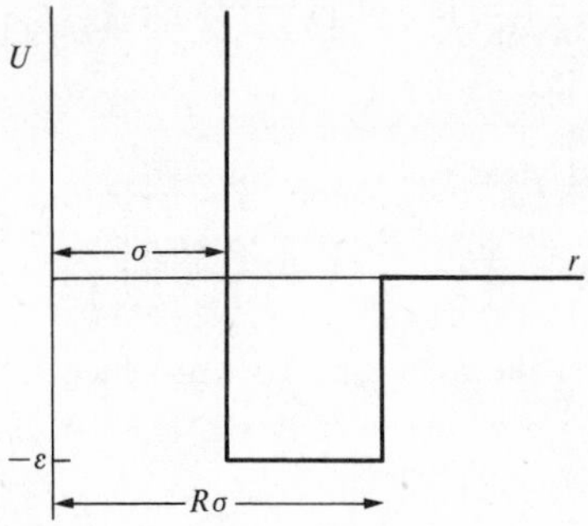

Fig. A1

3. Inverse-power repulsive potentials

$$U(r) = C_n r^{-n}$$

When $n = 4$ this is known as the Maxwell potential.

4. Sutherland potential

$$U(r) = \infty \qquad r < \sigma$$

$$U(r) = -C_m r^{-m} \qquad r \geqslant \sigma$$

The model is of hard spheres which attract each other according to an inverse power law. m is usually taken to be 6. Though apparently more simple the potential offers few advantages over the Lennard-Jones form.

5. *Lennard-Jones* $(n-m)$ *function*

$$U^*(r)=\left[\left(\frac{m}{n-m}\right)\bar{r}^{-n}-\left(\frac{n}{n-m}\right)\bar{r}^{-m}\right]$$

where, for this function,

$$\sigma=r_m\left(\frac{m}{n}\right)^{1/(n-m)}.$$

The most common form is the $(12-6)$ function

$$U^*(r)=[\bar{r}^{-12}-2\bar{r}^{-6}]$$

or

$$U^*(r)=4\left[\left(\frac{\sigma}{r}\right)^{12}-\left(\frac{\sigma}{r}\right)^{6}\right]$$

(Lennard-Jones, J. E. *Proc. roy. Soc.* **106A,** 463 (1924)). The function was first suggested by Mie in 1903.

6. *Kihara spherical core potential*

This has three adjustable parameters

$$U^*(r)=\left(\frac{n}{n-m}\right)\left(\frac{n}{m}\right)^{m/(n-m)}\left[\left(\frac{\sigma-2a}{r-2a}\right)^{n}-\left(\frac{\sigma-2a}{r-2a}\right)^{m}\right] \qquad r>2a$$

$$U^*(r)=\infty \qquad r<2a.$$

If n is taken as 12 and m as 6

$$U^*(r)=4\left\{\left(\frac{\sigma-2a}{r-2a}\right)^{12}-\left(\frac{\sigma-2a}{r-2a}\right)^{6}\right\}, \qquad r>2a$$

where a is the radius of the molecular core at which $U^*(r)=\infty$ (Kihara, T. *Rev. mod. Phys.* **25,** 831 (1953)).

7. *Exp* -6 *Potential*

$$U^*(r)=\frac{\varepsilon}{(1-6/\alpha)}\left\{\frac{6}{\alpha}\,e^{\alpha(1-\bar{r})}-\bar{r}^{-6}\right\}$$

where α is a parameter which measures the steepness of the repulsive energy. It is often assigned a value of 14 or 15. The function has a spurious maximum at low values of r.

8. *Exp*$-6-8$ *potential (Buckingham–Corner potential)*

$$U^*(r)=be^{-\alpha\bar{r}}-c[\bar{r}^{-6}+\beta\bar{r}^{-8}] \qquad \bar{r}\geq 1$$

$$=be^{-\alpha\bar{r}}-c[\bar{r}^{-6}+\beta\bar{r}^{-8}]e^{4(1-\bar{r})^3} \qquad \bar{r}<1$$

where $b=(6+8\beta)e^{\alpha}/[\alpha(1+\beta)-(6+8\beta)]$ and $c=\alpha/[\alpha(1+\beta)-(6+8\beta)]$. (Buckingham, R. A. and Corner, J. *Proc. roy. Soc.* **189A,** 118 (1947).)

9. Morse potential

$$U^*(r)=e^{(-2c/\sigma)(r-r_m)}-2e^{(-c/\sigma)(r-r_m)}$$

used most often to represent the *intra*molecular potential energy function of diatomic molecules. Konowalow and Hirschfelder and others have applied the function to the intermolecular interactions of simple molecules.

The parameter c may be related to the curvature of $U^*(r)$ at its minimum

$$c^2=\frac{1}{2}\left(\frac{\sigma^2}{\varepsilon}\right)\left(\frac{d^2U}{dr^2}\right)_{r=r_m}$$

(Morse, P. M. *Phys. Rev.* **34,** 57 (1929); Konowalow, D. D. and Hirschfelder, J. O. *Phys. Fluids* **4,** 629 (1961)).

10. Guggenheim and McGlashan

Based on a study of the solid state and other properties of argon

$$U^*(r)=\infty \qquad r<\sigma=0{\cdot}3165\ \text{nm}$$

$$=-1-\frac{k}{\varepsilon}(\bar{r}-1)^2-\frac{\alpha}{\varepsilon}(\bar{r}-1)^3+\frac{\beta}{\varepsilon}(\bar{r}-1)^4 \qquad 0{\cdot}36\leqslant r\leqslant 0{\cdot}415\ \text{nm}$$

$$=\lambda\bar{r}^{-6} \qquad r\geqslant 0{\cdot}54\ \text{nm}$$

For the regions $\sigma\leqslant r\leqslant 0{\cdot}36$ nm and $0{\cdot}415\leqslant r\leqslant 0{\cdot}54$ nm, a free-hand curve was drawn. (Guggenheim, E. A. and McGlashan, M. L. *Proc. roy Soc.* **A255,** 456 (1960).)

11. Boys and Shavitt function

$$U^*(r)=\frac{4}{\{(r/\sigma)^2+B^2\}^3}\sum_{i=0}^{\infty}C_{2i}\left\{\left(\frac{r}{\sigma}\right)^{2i}e^{A[1-(r/\sigma)^2]}-1\right\}.$$

A, B, C_0, C_2, C_4, etc. are adjustable constants. Munn investigated the use of this function for the inert gases. (Boys, S. F. and Shavitt, I. *Nature* **178,** 1340 (1956); Munn, R. J. *J. chem Phys.* **40,** 1439 (1964).)

12. Dymond–Rigby–Smith potential

$$U^*(r)=[0{\cdot}331\bar{r}^{-28}-1{\cdot}2584\bar{r}^{-24}+2{\cdot}07151\bar{r}^{-18}-1{\cdot}74452\bar{r}^{-8}-0{\cdot}39959\bar{r}^{-6}].$$

This potential is of historical interest only for the inert gases but could be useful for polyatomic molecules. (Dymond, J. H., Rigby, M., and Smith, E. B. *J. chem. Phys.* **42,** 2801 (1965).)

13. Barker–Pompe potential

$$U^*(r)=\exp[\alpha(1-\bar{r})]\sum_{i=0}^{n}A_i(\bar{r}-1)^i+\sum_{j=0}^{2}C^*_{2j+6}/(\delta+\bar{r}^{2j+6}).$$

This is a most important functional form that in the form devised by Barker, Fisher, and Watts played an important role in establishing the intermolecular potential for Ar. δ is a small non-physical parameter. (Barker, J. A. and Pompe, A. *Aust. J. Chem.* **21,** 1683 (1968); Barker, J. A., Fisher, R. A., and Watts, R. O. *Mol. Phys.* **21,** 657 (1971).)

13a. Barker (1974)

An additional term

$$\Delta U(r) = [P(\bar{r}-1)^4 + Q(\bar{r}-1)^5]\exp[\alpha'(1-\bar{r})] \qquad \bar{r}>1$$
$$= 0 \qquad \bar{r}<1$$

was added to the above function. (Barker, J. A., Watts, R. O., Lee, J. K., Schafer, T. P., and Lee, Y. T. *J. chem. Phys.* **61,** 3081 (1974).)

14. Barker–Bobetic–Maitland–Smith (*BBMS*) *potential*

$$U^*(r) = U^* \text{ (Barker–Pompe)} + \alpha' \exp\{-50(\bar{r}-1{\cdot}33)^2\}.$$

Maitland and Smith modified a version of the Barker–Pompe potential (13) due to Barker and Bobetic to bring it into line with spectroscopic data. This is another important potential for Ar. (Maitland, G. C. and Smith, E. B. *Mol. Phys.* **22,** 861 (1971).)

15. Parson–Siska–Lee potential (*Morse–Spline–van der Waals potential*)

$$\begin{aligned} U^*(r) &= \exp\{-2\beta(\bar{r}-1)\} - 2\exp\{-\beta'(\bar{r}-1)\} && 0 \leqslant \bar{r} \leqslant x_1 \\ &= b_1 + (\bar{r}-x_1)\{b_2 + (\bar{r}-x_2)\}[b_3 + (\bar{r}-x_1)b_4] && x_1 \leqslant \bar{r} \leqslant x_2 \\ &= C_6^*\bar{r}^{-6} + C_8^*\bar{r}^{-8} + C_{10}^*\bar{r}^{-10} && x_2 \leqslant \bar{r} \leqslant \infty. \end{aligned}$$

This potential which is based largely on molecular beam data gives a good representation of the argon interaction. (Parson, J. M., Siska, P. E., and Lee, Y. T. *J. chem. Phys.* **56,** 1511 (1972).)

16. $n(\bar{r})-6$ potential (*Maitland–Smith potential*)

$$U^*(r) = \left\{\left(\frac{6}{n-6}\right)\bar{r}^{-n} - \left(\frac{n}{n-6}\right)\bar{r}^{-6}\right\}.$$

n is allowed to vary with $r/r_{\rm m}$ (r^* in common notation, $\bar{r}$ in this appendix). The recommended form is

$$n = 13{\cdot}0 + \gamma(\bar{r}-1).$$

This simple three-parameter function gives a remarkably accurate representation of the inert gas potential functions. (Maitland, G. C. and Smith, E. B. *Chem Phys. Lett.* **22,** 443 (1973).)

17. Smith–Thakkar potential

$$U^*(r) = \left\{\frac{(r_{\rm m}^2-d^2)^6}{(r^2-d^2)^6} - 2\frac{(r_{\rm m}^2-d^2)^3}{(r^2-d^2)^3}\right\}, \qquad r>d$$
$$= \infty, \qquad r \leqslant d.$$

This is a simple three-parameter equation which is a modification of the spherical core Kihara type. (Smith, V. H. and Thakkar, A. J. *Chem. Phys. Lett.* **17,** 274 (1972).)

18. m − 6 − 8 potential function

$$U^*(r) = \frac{1}{(m-6)}\{(6+2\gamma)\bar{r}^{-m} - [m-\gamma(m-8)]\bar{r}^{-6}\} - \gamma\bar{r}^{-8}.$$

(Klein, M. and Hanley, H. J. M. *J. chem. Phys.* **53,** 4722 (1970).)

19. MIST (Morse–Hermite interpolation–Smith–Thakkar) potential

$$\begin{aligned} U^*(r) &= 4(Y^2 - Y) && 0 \leqslant r \leqslant R_1 \\ &= \frac{1}{\varepsilon}\sum_{k=1}^{6} a_k(r-R_1)^{[k/2]}(r-R_2)^{[(k-1)/2]}, && R_1 \leqslant r \leqslant R_2 \\ &= C_6 r^{-6}\left(1 - \frac{C_8}{nC_6 r^2}\right)^{-n} && R_2 \leqslant r < \infty \end{aligned}$$

where $Y = \exp[C(1 - r/\sigma)]$, $n = 20/29$, and $[p]$ is the largest integer $\leqslant p$. (Thakkar, A. J. and Smith, V. H. *Chem. Phys. Lett.* **24,** 157 (1974).)

20. Hartree–Fock HFD

$$U(r) = A\exp(-\alpha\bar{r}) + (C_6^*\bar{r}^{-6} + C_8^*\bar{r}^{-8} + C_{10}^*\bar{r}^{-10})F(\bar{r})$$

where

$$\begin{aligned} F(\bar{r}) &= \exp\left\{-\left[\frac{D}{\bar{r}} - 1\right]^2\right\} && \bar{r} < D \\ &= 1, && \bar{r} \geqslant D. \end{aligned}$$

D was given the value 1·28. (Ahlrichs, R., Penco, R., and Scoles, G. *Chem. Phys.* **19,** 119 (1976).)

21. HFD–Aziz potential

As above but with the repulsive term replaced by

$$A\bar{r}^{\gamma}\exp(-\alpha\bar{r}).$$

Preferred by Aziz and Chen for argon. (Aziz, R. A. and Chen, H. H. *J. chem. Phys.* **67,** 5719 (1977).)

22. Morse–6–hybrid

$$U(r) = 4(y^2 - y) \qquad r \leqslant q$$

$$U(r) = C_6 r^{-6} \qquad r \geqslant q$$

where $y = \exp(C(1 - (r/\sigma))$ and q is the point at which the two functions are joined. (Konowalow, D. D. and Zakheim, D. S. *J. Chem. Phys.* **57,** 4375 (1972).)

23. Stockmayer potential

$$U(r, \theta_1\theta_2, \phi) = 4\varepsilon'\left[\left(\frac{\sigma'}{r}\right)^{12} - \left(\frac{\sigma'}{r}\right)^{6}\right] - \frac{\mu_1\mu_2}{4\pi\varepsilon_0 r^3}\zeta(\theta_1, \theta_2, \phi).$$

This combines a Lennard–Jones 12 − 6 function with a dipole–dipole contribution.

$$\zeta(\theta_1, \theta_2, \phi) = 2\cos\theta_1\cos\theta_2 - \sin\theta_1\sin\theta_2\cos\phi$$

where θ_1, θ_2, ϕ define the relative orientation of the two dipoles.

24. The spherical-shell potential

This potential, employed to represent the interaction of globular molecules, is supposed to arise from the interaction of two spherical surfaces of diameter d which have a uniform distribution of Lennard-Jones 12 − 6 interaction sites

$$U^*(r) = \frac{(3x_m P_m^{*(4)} + P_m^{*(3)})P^{(9)} - (9x_m P_m^{*(10)} + P_m^{*(9)})P^{(3)}}{(9P_m^{*(3)}P_m^{*(10)} - 3P_m^{*(4)}P_m^{*(9)})x}$$

where

$$P^{(N)} = (r+d)^{-N} - 2r^{-N} + (r-d)^{-N}$$

$$P_m^{*(N)} = P^{(N)}(r_m)d^N$$

$$x = r/d, \qquad x_m = r_m/d.$$

(DeRocco, A. G. and Hoover, W. G. *J. chem. Phys.* **36,** 916 (1962).)

25. Double morse potential

$$U^*(r) = \exp[-2\beta(\bar{r}-1)] - 2\exp[-\beta(\bar{r}-1)]$$

where

$$\beta = \beta_1 \qquad 0 < r \leq r_m$$

$$\beta = \beta_2 \qquad r_m < r \leq r_c$$

$$U^*(r) = C_6^*\bar{r}^{-6} + C_8^*\bar{r}^{-8} + C_{10}^*\bar{r}^{-10} \qquad r_c < r < \infty$$

(Maitland, G. C. *Mol. Phys.* **26,** 513 (1973).)

26. n − 6 − 4 Potential function

A model for ion–neutral molecule interactions:

$$U^*(r) = \frac{n}{n(3+\gamma) - 12(1+\gamma)}\left\{\frac{12}{n}(1+\gamma)\bar{r}^{-n} - 4\gamma\bar{r}^{-6} - 3(1-\gamma)\bar{r}^{-4}\right\}.$$

(Viehland, L. A., Mason, E. A., Morrison, W. F., and Flannery, M. R. *Atomic Data and Nuclear Data Tables* **16,** 495 (1975).)

27. *Simons–Parr–Finlan modified Dunham expansion*

$$\frac{U(\bar{r})}{\varepsilon} = -1 + b_0(1-\bar{r})^2\left[1 + \sum_{n=1}^{N} b_n(1-\bar{r})^n\right] \quad \text{for} \quad r < r_f$$

$$\frac{U(\bar{r})}{\varepsilon} = C_6^*\bar{r}^{-6} + C_8^*\bar{r}^{-8} + C_{10}^*\bar{r}^{-10} \quad \text{for} \quad r \geqslant r_f$$

The joining point r_f is given by $U(r_f) = -0{\cdot}05\varepsilon$.
(Simons, G., Parr, R. G., and Finlan, J. M. *J. chem. Phys.* **59,** 3229 (1973); Bickes, R. W. and Bernstein, R. B. *Chem. Phys. Lett*, **26,** 457 (1974), *J. Chem. Phys.* **66,** 2408 (1977).)

APPENDIX 2

COLLISION INTEGRALS AND SECOND VIRIAL COEFFICIENTS FOR $n-6$ AND $n(\bar{r})-6$ POTENTIALS

Tables A2.1, A2.2, and A2.3 contain listings of the reduced collision integrals $\Omega^{(l,s)*}$ as functions of T^* for the three members of the $(n-6)$ potential family with $n = 9$, 12, 18, respectively. Tables A2.4–A2.12 list the collision integrals for the $n(\bar{r})-6$ family of potentials (Appendix 1) for $\gamma = 2$ to 10.

The reduced collision integrals have been calculated according to the equations

$$\chi = \pi - 2b^* \int_{r^*}^{\infty} \frac{\mathrm{d}r^*/r^{*2}}{\sqrt{\left\{1 - \dfrac{b^{*2}}{r^{*2}} - \dfrac{U^*}{E^*}\right\}}}$$

where $U^* = U(r)/\varepsilon$ and $r^* = r/\sigma$.

$$Q^{(l)*}(E^*) = 2\left[1 - \frac{1+(-1)^l}{2(1+l)}\right]^{-1} \int_0^{\infty} (1 - \cos^l \chi) b^* \, \mathrm{d}b^*$$

$$\Omega^{(l,s)*}(T^*) = \{(s+1)!\, T^{*s+2}\}^{-1} \int_0^{\infty} Q^{(l)*}(E^*) \mathrm{e}^{-E^*/T^*} E^{*s+1} \, \mathrm{d}E^*$$

with $T^* = kT/\varepsilon$.

The collision integrals listed have an uncertainty of $\pm 0{\cdot}1$ per cent.

Tables A2.13 to A2.15 contain listings of the classical, reduced second virial coefficient for the three members of the $n-6$ potential family as a function of reduced temperature T^*. Tables A2.16 to A2.24 contain similar data for the $n(\bar{r})-6$ family for $\gamma = 2$ to $\gamma = 10$.

The reduced, classical second virial coefficient is given by the expression

$$B^*(T^*) = 3\int_0^{\infty} \{1 - \exp(-U^*/T^*)\} r^{*2} \, \mathrm{d}r^*.$$

The second virial coefficient of the gas $B(T)$ is obtained from $B^*(T^*)$ with the aid of the equation

$$B(T) = b_0 B^*(T^*)$$

where $b_0 = \frac{2}{3}\pi N_A \sigma^3$, or in practical units $b_0 = 1261{\cdot}3\,\sigma^3$ where σ is measured in nanometers and b_0 in $\mathrm{cm}^3/\mathrm{mol}$.

The zero-pressure Joule–Thomson coefficient μ^0 can be obtained from the expression (Chapter 3)

$$(\mu^0 C_p^0)^* = \frac{\mu^0 C_p^0}{b_0} = T^* \frac{\mathrm{d}B^*}{\mathrm{d}T^*} - B^*.$$

Tables A2.13 to 2.24 include $T^*(\mathrm{d}B^*/\mathrm{d}T^*)$ and $(\mu^0 C_p^0)^*$ for completeness.

Listed in the same tables are the first two quantum corrections to the reduced second virial coefficient in the form of the coefficients B^*_{q1} and B^*_{q2} of the series (Chapter 3)

$$B^*_{tot} = B^* + \Lambda^{*2} B^*_{q1} + \Lambda^{*4} B^*_{q2} + \ldots .$$

Here B^*_{tot} is the total reduced second virial coefficient of the gas and Λ^* is the de Boer parameter defined as

$$\Lambda^* = h/\sigma(m\varepsilon)^{\frac{1}{2}}.$$

In practical units,

$$\Lambda^* = 4{\cdot}376/\sigma(M(\varepsilon/k))^{\frac{1}{2}}$$

where σ is measured in nanometres, M is the relative molecular mass, and (ε/k) is in kelvin. $B^*_{q1}(T^*)$ and $B^*_{q2}(T^*)$ are related to the functions $B_{q1}(T)$ and $B_{q2}(T)$ defined in Chapter 3 by the equations

$$B^*_{q1}(T^*) = B_{q1}(T)/b_0$$

and

$$B^*_{q2}(T^*) = B_{q2}(T)/b_0.$$

Table A2.1

Collision integrals for the 9 − 6 potential

T^*	$\Omega^{(1,1)*}$	$\Omega^{(1,2)*}$	$\Omega^{(1,3)*}$	$\Omega^{(1,4)*}$	$\Omega^{(1,5)*}$	$\Omega^{(2,2)*}$	$\Omega^{(2,3)*}$	$\Omega^{(2,4)*}$	$\Omega^{(2,5)*}$
.1	4.7379	4.1626	3.7689	3.4706	3.2271	4.7452	4.3409	4.0472	4.3569
.2	3.6201	3.1028	2.7178	2.4054	2.1462	3.7498	3.4228	3.1669	3.2528
.3	2.9929	2.4811	2.1075	1.8278	1.6198	3.2144	2.8742	2.5883	2.6441
.4	2.5599	2.0726	1.7407	1.5126	1.3554	2.8128	2.4574	2.1691	2.2469
.5	2.2456	1.7965	1.5109	1.3267	1.2051	2.4963	2.1455	1.8777	1.9692
.6	2.0109	1.6031	1.3584	1.2074	1.1099	2.2463	1.9135	1.6738	1.7668
.7	1.8313	1.4627	1.2517	1.1252	1.0444	2.0481	1.7393	1.5274	1.6149
.8	1.6907	1.3573	1.1735	1.0653	.9962	1.8898	1.6061	1.4190	1.4978
.9	1.5784	1.2759	1.1140	1.0196	.9592	1.7618	1.5022	1.3364	1.4056
1.0	1.4870	1.2115	1.0671	.9835	.9295	1.6571	1.4196	1.2717	1.3316
1.2	1.3484	1.1161	.9981	.9295	.8845	1.4977	1.2978	1.1775	1.2208
1.4	1.2487	1.0491	.9492	.8906	.8513	1.3836	1.2129	1.1124	1.1424
1.6	1.1739	.9993	.9125	.8607	.8253	1.2986	1.1507	1.0645	1.0843
1.8	1.1157	.9607	.8835	.8367	.8040	1.2332	1.1032	1.0277	1.0395
2.0	1.0692	.9298	.8598	.8167	.7862	1.1815	1.0656	.9982	1.0038
2.5	.9854	.8731	.8153	.7783	.7511	1.0896	.9983	.9444	.9394
3.0	.9288	.8337	.7832	.7498	.7247	1.0289	.9529	.9069	.8956
3.5	.8876	.8041	.7584	.7272	.7035	.9853	.9195	.8785	.8633
4.0	.8559	.7806	.7381	.7086	.6859	.9521	.8933	.8557	.8380
4.5	.8304	.7612	.7211	.6929	.6709	.9256	.8719	.8368	.8174
5.0	.8093	.7447	.7065	.6792	.6578	.9038	.8539	.8206	.8000
6.0	.7758	.7178	.6822	.6563	.6357	.8694	.8247	.7939	.7719
7.0	.7500	.6964	.6626	.6376	.6177	.8429	.8016	.7724	.7497
8.0	.7292	.6787	.6461	.6218	.6024	.8215	.7825	.7544	.7313
9.0	.7118	.6636	.6320	.6083	.5893	.8035	.7662	.7390	.7158
10.0	.6968	.6504	.6196	.5964	.5777	.7881	.7521	.7255	.7022
12.0	.6722	.6284	.5988	.5763	.5581	.7625	.7283	.7026	.6795
14.0	.6524	.6104	.5816	.5597	.5420	.7418	.7087	.6837	.6609
16.0	.6358	.5952	.5672	.5457	.5284	.7244	.6922	.6677	.6452
18.0	.6217	.5821	.5546	.5336	.5166	.7094	.6779	.6538	.6317
20.0	.6093	.5707	.5436	.5229	.5062	.6962	.6653	.6416	.6197
25.0	.5840	.5470	.5210	.5010	.4848	.6691	.6392	.6161	.5950
30.0	.5641	.5283	.5030	.4836	.4679	.6476	.6184	.5958	.5754
35.0	.5477	.5129	.4882	.4692	.4539	.6298	.6012	.5791	.5592
40.0	.5339	.4998	.4756	.4571	.4420	.6146	.5865	.5648	.5454
50.0	.5115	.4786	.4553	.4373	.4228	.5899	.5627	.5416	.5229
60.0	.4938	.4618	.4391	.4217	.4077	.5703	.5437	.5231	.5050
80.0	.4668	.4363	.4147	.3981	.3847	.5403	.5147	.4951	.4779
100.0	.4468	.4173	.3965	.3805	.3677	.5179	.4932	.4742	.4577

Table A2.2

Collision integrals for the 12 − 6 *potential*

T^*	$\Omega^{(1,1)*}$	$\Omega^{(1,2)*}$	$\Omega^{(1,3)*}$	$\Omega^{(1,4)*}$	$\Omega^{(1,5)*}$	$\Omega^{(2,2)*}$	$\Omega^{(2,3)*}$	$\Omega^{(2,4)*}$	$\Omega^{(2,5)*}$
.1	4.0209	3.5556	3.2384	2.9995	2.8072	4.1104	3.7738	3.5306	3.7478
.2	3.1269	2.7219	2.4281	2.1930	1.9964	3.2902	3.0253	2.8219	2.8661
.3	2.6414	2.2475	1.9595	1.7381	1.5672	2.8630	2.5948	2.3708	2.3915
.4	2.3067	1.9273	1.6621	1.4725	1.3367	2.5463	2.2648	2.0333	2.0766
.5	2.0598	1.7033	1.4679	1.3098	1.2021	2.2935	2.0109	1.7898	1.8509
.6	1.8718	1.5420	1.3356	1.2033	1.1158	2.0903	1.8173	1.6148	1.6827
.7	1.7254	1.4226	1.2415	1.1293	1.0562	1.9264	1.6689	1.4868	1.5542
.8	1.6091	1.3316	1.1718	1.0750	1.0123	1.7935	1.5537	1.3909	1.4538
.9	1.5151	1.2605	1.1184	1.0335	.9786	1.6847	1.4629	1.3171	1.3739
1.0	1.4379	1.2037	1.0762	1.0007	.9516	1.5948	1.3900	1.2590	1.3091
1.2	1.3192	1.1191	1.0137	.9516	.9107	1.4564	1.2815	1.1740	1.2113
1.4	1.2329	1.0591	.9695	.9163	.8807	1.3561	1.2054	1.1150	1.1415
1.6	1.1675	1.0144	.9361	.8893	.8572	1.2808	1.1494	1.0717	1.0895
1.8	1.1163	.9796	.9099	.8676	.8382	1.2226	1.1065	1.0383	1.0492
2.0	1.0752	.9516	.8885	.8496	.8222	1.1763	1.0726	1.0117	1.0171
2.5	1.0006	.9005	.8484	.8151	.7908	1.0939	1.0119	.9633	.9591
3.0	.9501	.8651	.8196	.7896	.7672	1.0393	.9711	.9298	.9198
3.5	.9132	.8384	.7973	.7695	.7483	1.0002	.9411	.9045	.8909
4.0	.8848	.8173	.7793	.7529	.7326	.9704	.9178	.8843	.8682
4.5	.8619	.7999	.7641	.7388	.7192	.9467	.8987	.8675	.8498
5.0	.8430	.7852	.7510	.7266	.7075	.9273	.8827	.8532	.8343
6.0	.8131	.7611	.7293	.7062	.6879	.8966	.8569	.8297	.8093
7.0	.7900	.7420	.7118	.6895	.6717	.8731	.8365	.8107	.7895
8.0	.7714	.7262	.6971	.6754	.6581	.8541	.8197	.7949	.7733
9.0	.7558	.7127	.6845	.6633	.6463	.8383	.8053	.7814	.7595
10.0	.7424	.7010	.6735	.6527	.6360	.8246	.7929	.7695	.7475
12.0	.7204	.6813	.6548	.6346	.6184	.8021	.7719	.7494	.7274
14.0	.7027	.6652	.6395	.6198	.6039	.7838	.7548	.7329	.7110
16.0	.6879	.6516	.6265	.6072	.5916	.7685	.7403	.7188	.6971
18.0	.6753	.6399	.6152	.5963	.5809	.7554	.7277	.7066	.6851
20.0	.6642	.6296	.6053	.5866	.5715	.7438	.7167	.6958	.6745
25.0	.6415	.6083	.5848	.5667	.5520	.7200	.6937	.6734	.6526
30.0	.6236	.5914	.5685	.5508	.5365	.7010	.6753	.6554	.6352
35.0	.6089	.5774	.5550	.5377	.5237	.6853	.6601	.6405	.6207
40.0	.5964	.5655	.5436	.5266	.5128	.6720	.6471	.6278	.6084
50.0	.5761	.5462	.5249	.5084	.4950	.6501	.6258	.6070	.5882
60.0	.5599	.5308	.5100	.4939	.4809	.6326	.6088	.5903	.5720
80.0	.5353	.5073	.4873	.4718	.4592	.6056	.5825	.5646	.5471
100.0	.5169	.4897	.4703	.4552	.4430	.5852	.5627	.5452	.5282

Table A2.3

Collision integrals for the 18 − 6 *potential*

T^*	$\Omega^{(1,1)*}$	$\Omega^{(1,2)*}$	$\Omega^{(1,3)*}$	$\Omega^{(1,4)*}$	$\Omega^{(1,5)*}$	$\Omega^{(2,2)*}$	$\Omega^{(2,3)*}$	$\Omega^{(2,4)*}$	$\Omega^{(2,5)*}$
.1	3.4455	3.0570	2.7946	2.5988	2.4426	3.5909	3.2996	3.0887	3.2419
.2	2.7085	2.3818	2.1525	1.9750	1.8301	2.8836	2.6590	2.4913	2.5275
.3	2.3247	2.0192	1.8017	1.6349	1.5033	2.5312	2.3166	2.1436	2.1555
.4	2.0671	1.7764	1.5735	1.4248	1.3144	2.2819	2.0626	1.8837	1.9092
.5	1.8776	1.6031	1.4186	1.2901	1.1994	2.0851	1.8645	1.6900	1.7304
.6	1.7321	1.4753	1.3097	1.1995	1.1244	1.9256	1.7099	1.5464	1.5951
.7	1.6174	1.3786	1.2307	1.1356	1.0722	1.7953	1.5890	1.4391	1.4900
.8	1.5252	1.3037	1.1713	1.0884	1.0338	1.6881	1.4935	1.3575	1.4069
.9	1.4498	1.2445	1.1254	1.0523	1.0043	1.5993	1.4172	1.2941	1.3401
1.0	1.3872	1.1967	1.0890	1.0236	.9809	1.5251	1.3554	1.2439	1.2855
1.2	1.2900	1.1248	1.0348	.9810	.9455	1.4096	1.2625	1.1700	1.2023
1.4	1.2183	1.0734	.9965	.9504	.9197	1.3248	1.1968	1.1185	1.1426
1.6	1.1636	1.0350	.9676	.9272	.8997	1.2607	1.1482	1.0807	1.0977
1.8	1.1205	1.0050	.9450	.9086	.8835	1.2107	1.1110	1.0517	1.0630
2.0	1.0857	.9810	.9267	.8933	.8700	1.1709	1.0815	1.0286	1.0353
2.5	1.0223	.9371	.8925	.8642	.8438	1.0997	1.0289	.9870	.9853
3.0	.9793	.9068	.8681	.8429	.8242	1.0526	.9938	.9584	.9517
3.5	.9479	.8842	.8495	.8262	.8087	1.0189	.9683	.9372	.9270
4.0	.9237	.8664	.8344	.8125	.7958	.9935	.9485	.9204	.9079
4.5	.9043	.8518	.8218	.8009	.7848	.9733	.9326	.9065	.8924
5.0	.8883	.8395	.8110	.7909	.7753	.9568	.9193	.8948	.8795
6.0	.8631	.8195	.7932	.7742	.7592	.9311	.8980	.8756	.8588
7.0	.8438	.8037	.7788	.7606	.7461	.9116	.8813	.8603	.8426
8.0	.8283	.7907	.7668	.7491	.7350	.8959	.8676	.8475	.8293
9.0	.8153	.7796	.7565	.7392	.7254	.8829	.8559	.8366	.8181
10.0	.8043	.7700	.7475	.7305	.7169	.8718	.8459	.8270	.8083
12.0	.7861	.7539	.7323	.7158	.7026	.8534	.8290	.8109	.7920
14.0	.7716	.7408	.7198	.7037	.6907	.8387	.8152	.7976	.7787
16.0	.7594	.7297	.7091	.6934	.6806	.8263	.8036	.7863	.7674
18.0	.7491	.7201	.6999	.6844	.6718	.8157	.7935	.7765	.7577
20.0	.7400	.7117	.6918	.6765	.6641	.8064	.7846	.7678	.7491
25.0	.7214	.6942	.6750	.6601	.6480	.7873	.7662	.7499	.7314
30.0	.7068	.6804	.6616	.6470	.6352	.7721	.7515	.7355	.7173
35.0	.6947	.6689	.6505	.6362	.6245	.7595	.7393	.7236	.7056
40.0	.6845	.6591	.6410	.6269	.6155	.7488	.7289	.7134	.6956
50.0	.6677	.6432	.6255	.6118	.6006	.7312	.7118	.6966	.6792
60.0	.6544	.6304	.6131	.5996	.5886	.7172	.6981	.6832	.6661
80.0	.6340	.6108	.5940	.5810	.5703	.6955	.6769	.6625	.6459
100.0	.6187	.5960	.5796	.5669	.5565	.6791	.6610	.6468	.6306

Table A2.4

Collision integrals for the $n(\bar{r})-6$ potential

$\gamma = 2{\cdot}0$; $\sigma/r_m = 0{\cdot}8945$

T^*	$\Omega^{(1,1)*}$	$\Omega^{(1,2)*}$	$\Omega^{(1,3)*}$	$\Omega^{(1,4)*}$	$\Omega^{(1,5)*}$	$\Omega^{(2,2)*}$	$\Omega^{(2,3)*}$	$\Omega^{(2,4)*}$	$\Omega^{(2,5)*}$
.1	3.7619	3.3495	3.0696	2.8616	2.6968	3.8877	3.5758	3.3468	3.5450
.2	2.9755	2.6207	2.3615	2.1494	1.9673	3.1199	2.8748	2.6917	2.7609
.3	2.5463	2.1935	1.9286	1.7201	1.5564	2.7329	2.4939	2.2963	2.3246
.4	2.2431	1.8954	1.6458	1.4641	1.3324	2.4510	2.1998	1.9909	2.0308
.5	2.0150	1.6832	1.4588	1.3058	1.2008	2.2240	1.9682	1.7645	1.8187
.6	1.8391	1.5288	1.3305	1.2018	1.1162	2.0388	1.7884	1.5989	1.6598
.7	1.7008	1.4137	1.2388	1.1293	1.0576	1.8877	1.6488	1.4765	1.5376
.8	1.5903	1.3257	1.1707	1.0760	1.0144	1.7638	1.5393	1.3840	1.4417
.9	1.5004	1.2566	1.1184	1.0352	.9812	1.6616	1.4524	1.3126	1.3650
1.0	1.4263	1.2014	1.0770	1.0030	.9547	1.5765	1.3824	1.2561	1.3026
1.2	1.3120	1.1187	1.0157	.9547	.9144	1.4447	1.2775	1.1732	1.2080
1.4	1.2285	1.0600	.9722	.9200	.8848	1.3485	1.2035	1.1155	1.1402
1.6	1.1650	1.0162	.9394	.8933	.8617	1.2759	1.1489	1.0730	1.0895
1.8	1.1152	.9820	.9136	.8719	.8429	1.2196	1.1070	1.0404	1.0501
2.0	1.0751	.9546	.8925	.8542	.8271	1.1746	1.0738	1.0143	1.0187
2.5	1.0023	.9043	.8530	.8201	.7961	1.0944	1.0144	.9667	.9619
3.0	.9528	.8693	.8245	.7949	.7727	1.0411	.9743	.9338	.9233
3.5	.9166	.8431	.8025	.7749	.7539	1.0027	.9449	.9089	.8948
4.0	.8886	.8222	.7846	.7584	.7383	.9736	.9219	.8890	.8725
4.5	.8661	.8050	.7695	.7444	.7249	.9503	.9032	.8725	.8543
5.0	.8474	.7904	.7565	.7323	.7133	.9312	.8875	.8584	.8390
6.0	.8179	.7666	.7350	.7119	.6936	.9011	.8620	.8351	.8143
7.0	.7951	.7476	.7175	.6952	.6775	.8779	.8419	.8164	.7948
8.0	.7766	.7318	.7028	.6812	.6638	.8592	.8252	.8008	.7787
9.0	.7611	.7184	.6902	.6690	.6520	.8436	.8110	.7873	.7650
10.0	.7479	.7066	.6792	.6584	.6416	.8301	.7987	.7755	.7530
12.0	.7260	.6870	.6605	.6403	.6240	.8078	.7779	.7555	.7330
14.0	.7083	.6709	.6451	.6254	.6094	.7897	.7609	.7390	.7167
16.0	.6935	.6573	.6320	.6127	.5970	.7745	.7464	.7250	.7028
18.0	.6809	.6455	.6207	.6017	.5862	.7614	.7339	.7128	.6908
20.0	.6698	.6351	.6108	.5920	.5767	.7499	.7229	.7020	.6802
25.0	.6471	.6137	.5901	.5718	.5569	.7261	.6999	.6796	.6583
30.0	.6291	.5967	.5736	.5557	.5412	.7072	.6815	.6615	.6408
35.0	.6143	.5826	.5599	.5424	.5281	.6915	.6662	.6465	.6262
40.0	.6017	.5706	.5483	.5310	.5170	.6781	.6532	.6337	.6137
50.0	.5812	.5510	.5293	.5125	.4988	.6561	.6317	.6126	.5933
60.0	.5648	.5353	.5141	.4977	.4843	.6385	.6145	.5958	.5769
80.0	.5398	.5114	.4909	.4751	.4622	.6112	.5879	.5697	.5517
100.0	.5210	.4934	.4735	.4581	.4456	.5907	.5678	.5500	.5325

Table A2.5

Collision integrals for the $n(\bar{r})-6$ potential

$\gamma = 3{\cdot}0;\ \sigma/r_m = 0{\cdot}8940$

T^*	$\Omega^{(1,1)*}$	$\Omega^{(1,2)*}$	$\Omega^{(1,3)*}$	$\Omega^{(1,4)*}$	$\Omega^{(1,5)*}$	$\Omega^{(2,2)*}$	$\Omega^{(2,3)*}$	$\Omega^{(2,4)*}$	$\Omega^{(2,5)*}$
.1	3.7174	3.3135	3.0395	2.8354	2.6729	3.8450	3.5414	3.3190	3.5092
.2	2.9463	2.5980	2.3434	2.1354	1.9570	3.0987	2.8596	2.6798	2.7404
.3	2.5250	2.1786	1.9186	1.7136	1.5520	2.7192	2.4838	2.2885	2.3116
.4	2.2272	1.8856	1.6398	1.4601	1.3294	2.4411	2.1927	1.9858	2.0221
.5	2.0029	1.6763	1.4545	1.3027	1.1981	2.2165	1.9631	1.7609	1.8124
.6	1.8296	1.5235	1.3270	1.1990	1.1135	2.0329	1.7845	1.5962	1.6550
.7	1.6931	1.4094	1.2357	1.1264	1.0547	1.8829	1.6456	1.4741	1.5336
.8	1.5837	1.3219	1.1677	1.0731	1.0115	1.7598	1.5366	1.3819	1.4382
.9	1.4948	1.2532	1.1155	1.0323	.9782	1.6581	1.4500	1.3105	1.3618
1.0	1.4213	1.1981	1.0741	.9999	.9515	1.5735	1.3801	1.2541	1.2996
1.2	1.3078	1.1155	1.0127	.9515	.9111	1.4421	1.2753	1.1712	1.2052
1.4	1.2246	1.0569	.9690	.9166	.8813	1.3461	1.2014	1.1134	1.1375
1.6	1.1614	1.0130	.9361	.8898	.8580	1.2737	1.1468	1.0709	1.0867
1.8	1.1117	.9788	.9102	.8683	.8391	1.2174	1.1049	1.0382	1.0473
2.0	1.0717	.9512	.8890	.8504	.8231	1.1725	1.0717	1.0120	1.0159
2.5	.9988	.9007	.8492	.8160	.7918	1.0921	1.0121	.9643	.9589
3.0	.9493	.8656	.8205	.7905	.7681	1.0387	.9719	.9311	.9201
3.5	.9129	.8391	.7982	.7704	.7491	1.0003	.9423	.9060	.8914
4.0	.8848	.8181	.7801	.7537	.7333	.9710	.9191	.8859	.8690
4.5	.8622	.8007	.7649	.7395	.7197	.9476	.9002	.8692	.8506
5.0	.8434	.7860	.7517	.7271	.7078	.9284	.8843	.8550	.8352
6.0	.8136	.7618	.7299	.7064	.6879	.8980	.8586	.8314	.8102
7.0	.7906	.7426	.7121	.6895	.6715	.8747	.8382	.8125	.7904
8.0	.7719	.7266	.6972	.6752	.6575	.8558	.8214	.7966	.7741
9.0	.7563	.7130	.6844	.6628	.6455	.8399	.8070	.7830	.7602
10.0	.7428	.7011	.6731	.6520	.6349	.8263	.7945	.7710	.7481
12.0	.7206	.6811	.6541	.6335	.6168	.8037	.7734	.7507	.7278
14.0	.7027	.6647	.6384	.6182	.6019	.7854	.7561	.7339	.7111
16.0	.6877	.6508	.6251	.6052	.5892	.7699	.7414	.7196	.6970
18.0	.6748	.6388	.6135	.5940	.5781	.7566	.7287	.7072	.6847
20.0	.6635	.6282	.6033	.5840	.5684	.7449	.7174	.6961	.6739
25.0	.6403	.6063	.5821	.5634	.5481	.7207	.6939	.6732	.6515
30.0	.6220	.5888	.5652	.5469	.5320	.7014	.6752	.6547	.6336
35.0	.6068	.5744	.5512	.5332	.5186	.6853	.6595	.6394	.6187
40.0	.5939	.5621	.5393	.5216	.5072	.6716	.6462	.6263	.6059
50.0	.5729	.5420	.5198	.5026	.4886	.6491	.6242	.6047	.5850
60.0	.5562	.5260	.5043	.4874	.4737	.6311	.6066	.5874	.5682
80.0	.5306	.5014	.4805	.4642	.4510	.6032	.5794	.5607	.5423
100.0	.5113	.4830	.4626	.4468	.4340	.5822	.5588	.5405	.5228

Table A2.6

Collision integrals for the n(r̄) − 6 potential

$\gamma = 4{\cdot}0;\ \sigma/r_m = 0{\cdot}8936$

T^*	$\Omega^{(1,1)*}$	$\Omega^{(1,2)*}$	$\Omega^{(1,3)*}$	$\Omega^{(1,4)*}$	$\Omega^{(1,5)*}$	$\Omega^{(2,2)*}$	$\Omega^{(2,3)*}$	$\Omega^{(2,4)*}$	$\Omega^{(2,5)*}$
.1	3.6866	3.2886	3.0195	2.8194	2.6601	3.8154	3.5161	3.2972	3.4837
.2	2.9284	2.5863	2.3360	2.1311	1.9549	3.0810	2.8461	2.6696	2.7273
.3	2.5142	2.1731	1.9163	1.7130	1.5522	2.7081	2.4766	2.2841	2.3045
.4	2.2205	1.8831	1.6393	1.4603	1.3297	2.4342	2.1892	1.9844	2.0181
.5	1.9986	1.6751	1.4545	1.3029	1.1981	2.2123	1.9617	1.7610	1.8103
.6	1.8267	1.5230	1.3270	1.1989	1.1132	2.0305	1.7842	1.5968	1.6538
.7	1.6911	1.4091	1.2356	1.1262	1.0542	1.8816	1.6459	1.4750	1.5331
.8	1.5823	1.3216	1.1675	1.0727	1.0108	1.7592	1.5372	1.3828	1.4379
.9	1.4937	1.2529	1.1151	1.0317	.9773	1.6580	1.4507	1.3114	1.3617
1.0	1.4205	1.1977	1.0736	.9991	.9504	1.5737	1.3809	1.2549	1.2996
1.2	1.3071	1.1150	1.0119	.9504	.9097	1.4426	1.2761	1.1718	1.2051
1.4	1.2240	1.0562	.9680	.9152	.8796	1.3467	1.2020	1.1138	1.1372
1.6	1.1607	1.0121	.9349	.8882	.8561	1.2742	1.1473	1.0711	1.0864
1.8	1.1109	.9777	.9087	.8665	.8369	1.2178	1.1052	1.0382	1.0468
2.0	1.0708	.9500	.8873	.8484	.8208	1.1728	1.0718	1.0119	1.0152
2.5	.9977	.8991	.8471	.8136	.7890	1.0923	1.0119	.9638	.9579
3.0	.9478	.8636	.8181	.7877	.7649	1.0386	.9714	.9304	.9188
3.5	.9112	.8369	.7955	.7672	.7456	1.0000	.9416	.9050	.8899
4.0	.8829	.8156	.7771	.7503	.7295	.9705	.9182	.8847	.8672
4.5	.8601	.7980	.7616	.7358	.7157	.9470	.8992	.8678	.8486
5.0	.8411	.7830	.7483	.7233	.7036	.9275	.8831	.8534	.8329
6.0	.8109	.7585	.7260	.7022	.6833	.8969	.8570	.8295	.8076
7.0	.7876	.7390	.7080	.6849	.6665	.8733	.8364	.8102	.7876
8.0	.7687	.7227	.6928	.6703	.6523	.8541	.8192	.7941	.7710
9.0	.7528	.7088	.6797	.6577	.6400	.8380	.8046	.7801	.7568
10.0	.7391	.6967	.6682	.6466	.6291	.8242	.7919	.7679	.7445
12.0	.7165	.6763	.6487	.6277	.6106	.8012	.7704	.7472	.7238
14.0	.6983	.6595	.6327	.6120	.5953	.7825	.7527	.7301	.7067
16.0	.6830	.6453	.6190	.5987	.5822	.7668	.7377	.7155	.6923
18.0	.6698	.6330	.6072	.5872	.5709	.7532	.7247	.7027	.6798
20.0	.6583	.6222	.5967	.5769	.5609	.7413	.7132	.6915	.6688
25.0	.6346	.5998	.5749	.5557	.5400	.7166	.6892	.6679	.6458
30.0	.6158	.5819	.5576	.5387	.5234	.6968	.6700	.6491	.6274
35.0	.6003	.5670	.5432	.5247	.5097	.6804	.6540	.6333	.6121
40.0	.5871	.5544	.5310	.5128	.4980	.6664	.6403	.6198	.5990
50.0	.5655	.5337	.5109	.4932	.4788	.6433	.6177	.5977	.5776
60.0	.5483	.5173	.4949	.4776	.4635	.6248	.5996	.5799	.5603
80.0	.5220	.4920	.4705	.4537	.4402	.5962	.5716	.5525	.5337
100.0	.5022	.4730	.4521	.4358	.4226	.5745	.5505	.5317	.5136

Table A2.7

Collision integrals for the $n(\bar{r})$-potential

$\gamma = 5{\cdot}0;\ \sigma/r_m = 0{\cdot}8931$

T^*	$\Omega^{(1,1)*}$	$\Omega^{(1,2)*}$	$\Omega^{(1,3)*}$	$\Omega^{(1,4)*}$	$\Omega^{(1,5)*}$	$\Omega^{(2,2)*}$	$\Omega^{(2,3)*}$	$\Omega^{(2,4)*}$	$\Omega^{(2,5)*}$
.1	3.6492	3.2570	2.9911	2.7925	2.6337	3.7821	3.4858	3.2684	3.4543
.2	2.9001	2.5618	2.3151	2.1144	1.9426	3.0539	2.8217	2.6485	2.7097
.3	2.4916	2.1561	1.9045	1.7053	1.5473	2.6873	2.4605	2.2725	2.2930
.4	2.2031	1.8718	1.6323	1.4559	1.3265	2.4190	2.1790	1.9778	2.0102
.5	1.9851	1.6671	1.4497	1.2995	1.1952	2.2013	1.9549	1.7567	1.8044
.6	1.8161	1.5170	1.3232	1.1958	1.1102	2.0223	1.7794	1.5937	1.6492
.7	1.6824	1.4042	1.2322	1.1231	1.0512	1.8752	1.6422	1.4725	1.5291
.8	1.5750	1.3174	1.1643	1.0696	1.0076	1.7541	1.5342	1.3805	1.4345
.9	1.4874	1.2490	1.1119	1.0285	.9740	1.6537	1.4480	1.3092	1.3585
1.0	1.4148	1.1941	1.0704	.9959	.9471	1.5700	1.3784	1.2528	1.2965
1.2	1.3024	1.1116	1.0086	.9470	.9061	1.4396	1.2739	1.1697	1.2022
1.4	1.2197	1.0528	.9646	.9116	.8759	1.3441	1.1999	1.1117	1.1344
1.6	1.1567	1.0087	.9313	.8844	.8522	1.2718	1.1451	1.0689	1.0835
1.8	1.1071	.9742	.9050	.8626	.8328	1.2155	1.1030	1.0359	1.0439
2.0	1.0671	.9464	.8836	.8444	.8165	1.1705	1.0695	1.0094	1.0122
2.5	.9940	.8953	.8430	.8092	.7844	1.0899	1.0094	.9611	.9547
3.0	.9440	.8596	.8137	.7831	.7600	1.0362	.9688	.9275	.9154
3.5	.9073	.8326	.7909	.7623	.7404	.9974	.9387	.9020	.8862
4.0	.8788	.8111	.7723	.7451	.7240	.9678	.9152	.8815	.8633
4.5	.8559	.7933	.7566	.7304	.7100	.9441	.8960	.8644	.8446
5.0	.8367	.7782	.7430	.7177	.6977	.9246	.8798	.8498	.8288
6.0	.8063	.7534	.7204	.6962	.6769	.8937	.8535	.8257	.8032
7.0	.7827	.7335	.7021	.6786	.6598	.8699	.8326	.8062	.7829
8.0	.7636	.7170	.6866	.6637	.6453	.8505	.8153	.7898	.7660
9.0	.7475	.7029	.6732	.6508	.6327	.8343	.8005	.7756	.7517
10.0	.7336	.6905	.6615	.6394	.6216	.8203	.7875	.7632	.7391
12.0	.7107	.6697	.6416	.6201	.6026	.7970	.7657	.7421	.7180
14.0	.6921	.6526	.6252	.6041	.5869	.7780	.7477	.7246	.7007
16.0	.6765	.6381	.6112	.5904	.5735	.7620	.7324	.7096	.6859
18.0	.6631	.6255	.5991	.5786	.5619	.7482	.7191	.6966	.6732
20.0	.6513	.6145	.5883	.5681	.5516	.7360	.7073	.6851	.6619
25.0	.6271	.5915	.5660	.5463	.5302	.7107	.6827	.6609	.6383
30.0	.6078	.5731	.5482	.5289	.5131	.6905	.6630	.6416	.6195
35.0	.5919	.5579	.5334	.5145	.4990	.6737	.6466	.6254	.6038
40.0	.5784	.5449	.5209	.5022	.4869	.6593	.6325	.6116	.5903
50.0	.5563	.5237	.5003	.4820	.4672	.6356	.6093	.5888	.5682
60.0	.5387	.5068	.4838	.4660	.4515	.6166	.5907	.5705	.5505
80.0	.5116	.4808	.4586	.4414	.4274	.5872	.5620	.5423	.5232
100.0	.4913	.4613	.4397	.4229	.4093	.5649	.5402	.5210	.5026

Table A2.8

Collision integrals for the $n(\bar{r})$-potential

$\gamma = 6{\cdot}0$; $\sigma/r_m = 0{\cdot}8926$

T^*	$\Omega^{(1,1)*}$	$\Omega^{(1,2)*}$	$\Omega^{(1,3)*}$	$\Omega^{(1,4)*}$	$\Omega^{(1,5)*}$	$\Omega^{(2,2)*}$	$\Omega^{(2,3)*}$	$\Omega^{(2,4)*}$	$\Omega^{(2,5)*}$
.1	3.6271	3.2398	2.9772	2.7806	2.6231	3.7628	3.4689	3.2535	3.4378
.2	2.8866	2.5519	2.3079	2.1095	1.9396	3.0415	2.8118	2.6406	2.6990
.3	2.4825	2.1506	1.9016	1.7042	1.5471	2.6790	2.4549	2.2689	2.2865
.4	2.1970	1.8689	1.6313	1.4557	1.3265	2.4137	2.1762	1.9767	2.0065
.5	1.9810	1.6656	1.4493	1.2994	1.1950	2.1980	1.9538	1.7568	1.8024
.6	1.8132	1.5161	1.3229	1.1956	1.1098	2.0204	1.7791	1.5944	1.6480
.7	1.6803	1.4036	1.2319	1.1227	1.0505	1.8743	1.6425	1.4733	1.5285
.8	1.5734	1.3169	1.1638	1.0689	1.0067	1.7537	1.5347	1.3814	1.4341
.9	1.4861	1.2485	1.1113	1.0276	.9728	1.6538	1.4487	1.3101	1.3583
1.0	1.4138	1.1935	1.0696	.9948	.9457	1.5702	1.3791	1.2536	1.2963
1.2	1.3015	1.1109	1.0075	.9456	.9044	1.4401	1.2745	1.1703	1.2020
1.4	1.2188	1.0518	.9633	.9100	.8739	1.3447	1.2004	1.1121	1.1340
1.6	1.1558	1.0075	.9298	.8825	.8499	1.2723	1.1455	1.0691	1.0830
1.8	1.1061	.9728	.9033	.8604	.8303	1.2160	1.1032	1.0359	1.0432
2.0	1.0659	.9449	.8816	.8420	.8138	1.1709	1.0696	1.0093	1.0114
2.5	.9925	.8933	.8406	.8064	.7812	1.0900	1.0092	.9606	.9535
3.0	.9423	.8573	.8109	.7799	.7565	1.0360	.9682	.9266	.9139
3.5	.9053	.8300	.7878	.7588	.7365	.9970	.9380	.9008	.8844
4.0	.8766	.8083	.7689	.7413	.7198	.9672	.9142	.8801	.8613
4.5	.8534	.7902	.7529	.7263	.7055	.9433	.8948	.8628	.8423
5.0	.8341	.7749	.7391	.7133	.6929	.9236	.8783	.8479	.8263
6.0	.8033	.7496	.7161	.6914	.6717	.8923	.8517	.8234	.8003
7.0	.7794	.7294	.6973	.6734	.6541	.8682	.8304	.8035	.7796
8.0	.7599	.7126	.6815	.6581	.6392	.8486	.8128	.7868	.7624
9.0	.7435	.6981	.6679	.6449	.6263	.8321	.7977	.7724	.7478
10.0	.7294	.6855	.6558	.6332	.6149	.8178	.7845	.7597	.7350
12.0	.7060	.6642	.6355	.6134	.5955	.7941	.7622	.7381	.7134
14.0	.6870	.6467	.6186	.5969	.5793	.7747	.7439	.7202	.6957
16.0	.6711	.6318	.6043	.5829	.5656	.7584	.7282	.7050	.6806
18.0	.6574	.6189	.5918	.5707	.5536	.7443	.7146	.6916	.6675
20.0	.6453	.6076	.5808	.5599	.5430	.7319	.7025	.6798	.6559
25.0	.6205	.5839	.5578	.5375	.5209	.7060	.6774	.6550	.6318
30.0	.6007	.5651	.5395	.5196	.5033	.6853	.6571	.6351	.6124
35.0	.5844	.5494	.5242	.5047	.4887	.6680	.6402	.6184	.5962
40.0	.5705	.5361	.5113	.4920	.4762	.6533	.6257	.6041	.5824
50.0	.5478	.5142	.4900	.4712	.4559	.6289	.6018	.5806	.5596
60.0	.5296	.4967	.4730	.4546	.4396	.6093	.5827	.5618	.5414
80.0	.5017	.4699	.4470	.4292	.4147	.5790	.5530	.5326	.5132
100.0	.4807	.4497	.4274	.4101	.3960	.5560	.5305	.5106	.4919

Table A2.9

Collision integrals for the $n(\bar{r})-6$ potential

$\gamma = 7{\cdot}0$; $\sigma/r_m = 0{\cdot}8920$

T^*	$\Omega^{(1,1)*}$	$\Omega^{(1,2)*}$	$\Omega^{(1,3)*}$	$\Omega^{(1,4)*}$	$\Omega^{(1,5)*}$	$\Omega^{(2,2)*}$	$\Omega^{(2,3)*}$	$\Omega^{(2,4)*}$	$\Omega^{(2,5)*}$
.1	3.5956	3.2142	2.9554	2.7614	2.6060	3.7382	3.4469	3.2339	3.4128
.2	2.8659	2.5359	2.2954	2.0998	1.9323	3.0247	2.7978	2.6285	2.6826
.3	2.4674	2.1402	1.8943	1.6990	1.5432	2.6664	2.4449	2.2610	2.2756
.4	2.1857	1.8617	1.6265	1.4521	1.3235	2.4041	2.1691	1.9716	1.9988
.5	1.9722	1.6602	1.4455	1.2963	1.1920	2.1906	1.9486	1.7532	1.7966
.6	1.8061	1.5117	1.3196	1.1926	1.1067	2.0146	1.7752	1.5915	1.6434
.7	1.6743	1.3997	1.2287	1.1196	1.0473	1.8695	1.6392	1.4708	1.5246
.8	1.5682	1.3133	1.1607	1.0657	1.0034	1.7497	1.5319	1.3791	1.4306
.9	1.4814	1.2451	1.1081	1.0243	.9694	1.6502	1.4461	1.3078	1.3550
1.0	1.4094	1.1902	1.0664	.9914	.9421	1.5671	1.3767	1.2513	1.2932
1.2	1.2975	1.1075	1.0041	.9420	.9006	1.4374	1.2722	1.1680	1.1990
1.4	1.2151	1.0484	.9597	.9062	.8699	1.3421	1.1982	1.1098	1.1310
1.6	1.1521	1.0040	.9260	.8786	.8457	1.2699	1.1432	1.0667	1.0799
1.8	1.1024	.9692	.8994	.8563	.8259	1.2136	1.1009	1.0334	1.0401
2.0	1.0623	.9411	.8776	.8377	.8092	1.1685	1.0672	1.0068	1.0082
2.5	.9887	.8893	.8362	.8017	.7762	1.0876	1.0066	.9578	.9501
3.0	.9384	.8530	.8062	.7748	.7510	1.0335	.9655	.9236	.9102
3.5	.9012	.8255	.7828	.7533	.7307	.9943	.9350	.8975	.8805
4.0	.8723	.8034	.7636	.7355	.7137	.9643	.9110	.8765	.8571
4.5	.8489	.7852	.7474	.7203	.6990	.9402	.8914	.8590	.8379
5.0	.8294	.7696	.7333	.7070	.6862	.9204	.8748	.8440	.8217
6.0	.7982	.7439	.7098	.6846	.6645	.8888	.8478	.8191	.7953
7.0	.7740	.7234	.6907	.6663	.6466	.8644	.8262	.7989	.7743
8.0	.7543	.7062	.6746	.6507	.6314	.8446	.8083	.7819	.7569
9.0	.7376	.6915	.6606	.6371	.6181	.8278	.7930	.7673	.7419
10.0	.7233	.6786	.6483	.6252	.6064	.8134	.7796	.7543	.7289
12.0	.6994	.6569	.6275	.6048	.5864	.7893	.7569	.7323	.7069
14.0	.6801	.6389	.6102	.5880	.5699	.7696	.7381	.7140	.6888
16.0	.6638	.6237	.5955	.5736	.5557	.7529	.7221	.6983	.6733
18.0	.6498	.6105	.5826	.5610	.5433	.7385	.7082	.6846	.6599
20.0	.6375	.5988	.5713	.5499	.5324	.7258	.6958	.6724	.6480
25.0	.6120	.5746	.5477	.5267	.5096	.6993	.6700	.6470	.6232
30.0	.5918	.5551	.5288	.5082	.4914	.6781	.6492	.6265	.6033
35.0	.5750	.5390	.5131	.4928	.4763	.6603	.6318	.6093	.5867
40.0	.5607	.5252	.4996	.4797	.4635	.6452	.6169	.5946	.5724
50.0	.5372	.5027	.4777	.4582	.4424	.6201	.5923	.5704	.5490
60.0	.5185	.4846	.4601	.4411	.4256	.5999	.5725	.5510	.5302
80.0	.4897	.4569	.4332	.4148	.3999	.5687	.5419	.5209	.5012
100.0	.4680	.4360	.4130	.3951	.3806	.5450	.5187	.4982	.4793

Table A2.10

Collision integrals for the $n(\bar{r})-6$ potential
$\gamma = 8{\cdot}0;\ \sigma/r_m = 0{\cdot}8915$

T^*	$\Omega^{(1,1)*}$	$\Omega^{(1,2)*}$	$\Omega^{(1,3)*}$	$\Omega^{(1,4)*}$	$\Omega^{(1,5)*}$	$\Omega^{(2,2)*}$	$\Omega^{(2,3)*}$	$\Omega^{(2,4)*}$	$\Omega^{(2,5)*}$
.1	3.5781	3.1989	2.9422	2.7500	2.5962	3.7244	3.4347	3.2227	3.3975
.2	2.8540	2.5273	2.2895	2.0962	1.9305	3.0143	2.7885	2.6204	2.6742
.3	2.4597	2.1359	1.8923	1.6984	1.5433	2.6585	2.4389	2.2570	2.2709
.4	2.1807	1.8596	1.6258	1.4521	1.3235	2.3987	2.1660	1.9702	1.9962
.5	1.9689	1.6591	1.4453	1.2961	1.1917	2.1871	1.9472	1.7531	1.7951
.6	1.8038	1.5110	1.3193	1.1922	1.1061	2.0125	1.7748	1.5920	1.6425
.7	1.6727	1.3992	1.2283	1.1189	1.0463	1.8683	1.6394	1.4715	1.5240
.8	1.5669	1.3128	1.1601	1.0649	1.0022	1.7491	1.5323	1.3798	1.4302
.9	1.4803	1.2445	1.1073	1.0232	.9680	1.6500	1.4467	1.3085	1.3547
1.0	1.4085	1.1895	1.0654	.9901	.9405	1.5672	1.3773	1.2519	1.2929
1.2	1.2966	1.1066	1.0029	.9404	.8986	1.4377	1.2727	1.1684	1.1986
1.4	1.2142	1.0473	.9581	.9042	.8676	1.3425	1.1985	1.1099	1.1305
1.6	1.1511	1.0026	.9242	.8763	.8431	1.2702	1.1434	1.0667	1.0792
1.8	1.1013	.9676	.8973	.8538	.8231	1.2138	1.1010	1.0332	1.0392
2.0	1.0610	.9393	.8753	.8350	.8061	1.1687	1.0671	1.0064	1.0071
2.5	.9871	.8871	.8334	.7984	.7726	1.0875	1.0062	.9571	.9486
3.0	.9363	.8504	.8030	.7711	.7469	1.0331	.9647	.9225	.9083
3.5	.8989	.8225	.7792	.7493	.7262	.9937	.9339	.8961	.8783
4.0	.8697	.8001	.7597	.7311	.7089	.9634	.9097	.8749	.8547
4.5	.8460	.7816	.7432	.7156	.6939	.9392	.8898	.8571	.8352
5.0	.8263	.7657	.7289	.7020	.6808	.9191	.8730	.8418	.8188
6.0	.7947	.7396	.7049	.6791	.6585	.8872	.8456	.8165	.7919
7.0	.7701	.7187	.6853	.6603	.6401	.8625	.8237	.7959	.7706
8.0	.7500	.7011	.6688	.6442	.6244	.8423	.8054	.7785	.7527
9.0	.7331	.6861	.6544	.6303	.6108	.8252	.7898	.7635	.7375
10.0	.7185	.6729	.6418	.6180	.5987	.8105	.7760	.7502	.7242
12.0	.6941	.6505	.6204	.5971	.5781	.7859	.7528	.7276	.7016
14.0	.6743	.6321	.6026	.5797	.5610	.7658	.7336	.7089	.6830
16.0	.6576	.6165	.5874	.5648	.5463	.7488	.7172	.6928	.6671
18.0	.6432	.6029	.5741	.5518	.5335	.7340	.7029	.6787	.6533
20.0	.6306	.5908	.5624	.5403	.5222	.7209	.6902	.6662	.6411
25.0	.6044	.5658	.5380	.5164	.4986	.6938	.6636	.6400	.6155
30.0	.5835	.5457	.5184	.4972	.4798	.6719	.6422	.6188	.5950
35.0	.5662	.5290	.5022	.4812	.4641	.6537	.6242	.6010	.5778
40.0	.5514	.5148	.4883	.4676	.4508	.6380	.6088	.5858	.5631
50.0	.5272	.4914	.4655	.4453	.4289	.6121	.5834	.5607	.5388
60.0	.5078	.4727	.4473	.4275	.4114	.5913	.5629	.5405	.5194
80.0	.4780	.4439	.4193	.4002	.3846	.5589	.5311	.5093	.4893
100.0	.4554	.4222	.3983	.3797	.3645	.5343	.5070	.4857	.4666

Table A2.11

Collision integrals for the $n(\bar{r})-6$ potential

$\gamma = 9{\cdot}0;\ \sigma/r_m = 0{\cdot}8910$

T^*	$\Omega^{(1,1)*}$	$\Omega^{(1,2)*}$	$\Omega^{(1,3)*}$	$\Omega^{(1,4)*}$	$\Omega^{(1,5)*}$	$\Omega^{(2,2)*}$	$\Omega^{(2,3)*}$	$\Omega^{(2,4)*}$	$\Omega^{(2,5)*}$
.1	3.5543	3.1782	2.9238	2.7332	2.5806	3.7039	3.4165	3.2065	3.3764
.2	2.8364	2.5127	2.2773	2.0862	1.9226	3.0006	2.7770	2.6105	2.6587
.3	2.4460	2.1255	1.8847	1.6928	1.5390	2.6481	2.4305	2.2501	2.2600
.4	2.1699	1.8521	1.6206	1.4481	1.3201	2.3905	2.1597	1.9655	1.9884
.5	1.9603	1.6533	1.4411	1.2927	1.1885	2.1806	1.9425	1.7496	1.7892
.6	1.7966	1.5062	1.3156	1.1888	1.1028	2.0072	1.7710	1.5891	1.6377
.7	1.6665	1.3949	1.2248	1.1156	1.0429	1.8638	1.6362	1.4689	1.5199
.8	1.5614	1.3088	1.1566	1.0614	.9986	1.7452	1.5295	1.3774	1.4265
.9	1.4753	1.2408	1.1038	1.0197	.9642	1.6466	1.4441	1.3062	1.3512
1.0	1.4038	1.1858	1.0619	.9864	.9366	1.5640	1.3748	1.2496	1.2895
1.2	1.2924	1.1029	.9991	.9364	.8944	1.4349	1.2704	1.1661	1.1953
1.4	1.2101	1.0435	.9542	.9000	.8631	1.3399	1.1961	1.1075	1.1272
1.6	1.1471	.9987	.9201	.8719	.8384	1.2677	1.1410	1.0641	1.0759
1.8	1.0973	.9636	.8930	.8491	.8181	1.2113	1.0984	1.0305	1.0358
2.0	1.0570	.9352	.8707	.8301	.8009	1.1662	1.0645	1.0035	1.0035
2.5	.9829	.8826	.8285	.7931	.7668	1.0848	1.0033	.9539	.9447
3.0	.9320	.8455	.7977	.7653	.7407	1.0302	.9615	.9190	.9042
3.5	.8942	.8173	.7735	.7431	.7197	.9906	.9305	.8924	.8739
4.0	.8648	.7947	.7537	.7246	.7019	.9602	.9061	.8709	.8500
4.5	.8410	.7759	.7369	.7088	.6866	.9358	.8860	.8529	.8303
5.0	.8210	.7598	.7223	.6949	.6732	.9155	.8690	.8375	.8136
6.0	.7891	.7332	.6978	.6715	.6504	.8833	.8412	.8118	.7864
7.0	.7641	.7118	.6778	.6522	.6315	.8583	.8190	.7908	.7646
8.0	.7437	.6939	.6609	.6357	.6154	.8378	.8005	.7731	.7465
9.0	.7264	.6785	.6462	.6215	.6013	.8206	.7845	.7578	.7309
10.0	.7115	.6650	.6332	.6088	.5889	.8056	.7705	.7442	.7173
12.0	.6867	.6422	.6112	.5872	.5677	.7805	.7468	.7211	.6942
14.0	.6664	.6233	.5929	.5693	.5500	.7600	.7271	.7018	.6752
16.0	.6493	.6072	.5772	.5539	.5349	.7426	.7103	.6852	.6589
18.0	.6346	.5931	.5636	.5406	.5217	.7275	.6956	.6707	.6447
20.0	.6216	.5808	.5515	.5287	.5100	.7141	.6825	.6578	.6321
25.0	.5946	.5549	.5263	.5039	.4856	.6862	.6552	.6308	.6057
30.0	.5731	.5342	.5060	.4840	.4660	.6637	.6330	.6089	.5845
35.0	.5553	.5170	.4892	.4674	.4497	.6449	.6145	.5905	.5668
40.0	.5400	.5022	.4748	.4533	.4358	.6287	.5985	.5748	.5516
50.0	.5150	.4780	.4511	.4301	.4130	.6020	.5722	.5487	.5265
60.0	.4949	.4585	.4322	.4116	.3948	.5804	.5509	.5278	.5063
80.0	.4640	.4286	.4031	.3832	.3670	.5468	.5180	.4953	.4751
100.0	.4406	.4061	.3812	.3618	.3461	.5213	.4929	.4706	.4515

Table A2.12

Collision integrals for the $n(\bar{r})-6$ potential

$\gamma = 10{\cdot}0;\ \sigma/r_m = 0{\cdot}8904$

T^*	$\Omega^{(1,1)*}$	$\Omega^{(1,2)*}$	$\Omega^{(1,3)*}$	$\Omega^{(1,4)*}$	$\Omega^{(1,5)*}$	$\Omega^{(2,2)*}$	$\Omega^{(2,3)*}$	$\Omega^{(2,4)*}$	$\Omega^{(2,5)*}$
.1	3.5405	3.1660	2.9131	2.7239	2.5728	3.6942	3.4077	3.1981	3.3649
.2	2.8271	2.5060	2.2731	2.0839	1.9216	2.9917	2.7683	2.6024	2.6537
.3	2.4401	2.1224	1.8834	1.6925	1.5390	2.6405	2.4244	2.2458	2.2574
.4	2.1663	1.8507	1.6202	1.4480	1.3199	2.3851	2.1563	1.9639	1.9868
.5	1.9579	1.6526	1.4408	1.2923	1.1879	2.1770	1.9409	1.7493	1.7882
.6	1.7950	1.5057	1.3152	1.1882	1.1019	2.0049	1.7704	1.5894	1.6371
.7	1.6653	1.3944	1.2242	1.1147	1.0417	1.8625	1.6362	1.4695	1.5195
.8	1.5604	1.3082	1.1558	1.0602	.9971	1.7446	1.5298	1.3780	1.4261
.9	1.4744	1.2400	1.1028	1.0183	.9625	1.6463	1.4445	1.3068	1.3508
1.0	1.4029	1.1850	1.0606	.9848	.9346	1.5640	1.3753	1.2501	1.2891
1.2	1.2915	1.1018	.9975	.9344	.8921	1.4352	1.2708	1.1663	1.1948
1.4	1.2091	1.0421	.9523	.8977	.8604	1.3402	1.1964	1.1075	1.1265
1.6	1.1459	.9970	.9179	.8692	.8354	1.2680	1.1411	1.0640	1.0749
1.8	1.0959	.9617	.8905	.8462	.8148	1.2115	1.0984	1.0302	1.0346
2.0	1.0554	.9330	.8680	.8269	.7973	1.1662	1.0643	1.0030	1.0022
2.5	.9809	.8799	.8253	.7893	.7626	1.0846	1.0027	.9530	.9429
3.0	.9296	.8424	.7940	.7611	.7360	1.0297	.9607	.9178	.9020
3.5	.8915	.8138	.7694	.7384	.7144	.9899	.9294	.8908	.8714
4.0	.8618	.7909	.7492	.7195	.6962	.9592	.9046	.8690	.8471
4.5	.8376	.7717	.7320	.7033	.6806	.9345	.8843	.8507	.8271
5.0	.8174	.7553	.7170	.6890	.6667	.9141	.8670	.8349	.8102
6.0	.7850	.7282	.6920	.6650	.6433	.8814	.8387	.8087	.7824
7.0	.7595	.7063	.6714	.6451	.6238	.8560	.8161	.7873	.7602
8.0	.7387	.6879	.6540	.6282	.6072	.8352	.7971	.7692	.7416
9.0	.7211	.6721	.6389	.6134	.5926	.8176	.7808	.7534	.7257
10.0	.7058	.6582	.6255	.6003	.5798	.8023	.7665	.7395	.7117
12.0	.6804	.6347	.6028	.5780	.5578	.7767	.7422	.7157	.6880
14.0	.6596	.6152	.5838	.5594	.5394	.7557	.7219	.6959	.6683
16.0	.6420	.5986	.5676	.5435	.5237	.7378	.7046	.6788	.6515
18.0	.6268	.5841	.5535	.5296	.5099	.7223	.6894	.6638	.6369
20.0	.6134	.5712	.5409	.5172	.4978	.7085	.6760	.6504	.6238
25.0	.5855	.5444	.5147	.4914	.4723	.6797	.6476	.6223	.5966
30.0	.5633	.5229	.4936	.4706	.4518	.6564	.6247	.5996	.5745
35.0	.5447	.5050	.4760	.4534	.4348	.6369	.6054	.5805	.5561
40.0	.5289	.4896	.4610	.4386	.4203	.6201	.5888	.5641	.5403
50.0	.5028	.4643	.4363	.4144	.3965	.5923	.5613	.5369	.5141
60.0	.4819	.4440	.4165	.3950	.3775	.5698	.5392	.5149	.4931
80.0	.4497	.4128	.3861	.3653	.3483	.5348	.5046	.4809	.4605
100.0	.4253	.3892	.3632	.3429	.3265	.5081	.4783	.4549	.4356

Table A2.13

The classical second virial coefficient, the quantum corrections and the Joule–Thomson coefficient for the (9 – 6) potential

T^*	B^*	$T^*(dB^*/dT^*)$	$(\mu^0 C_p^0)^*$	B_1^*	B_2^*
.3	-.35023043E+2	.950346E+2	.130058E+3	.217346E+2	-.209450E+2
.4	-.17526941E+2	.376858E+2	.552127E+2	.608019E+1	-.361971E+1
.5	-.11195490E+2	.211303E+2	.323258E+2	.263000E+1	-.108732E+1
.6	-.80431751E+1	.140746E+2	.221178E+2	.143507E+1	-.442698E+0
.7	-.61799861E+1	.103484E+2	.165284E+2	.901283E+0	-.217773E+0
.8	-.49562299E+1	.809966E+1	.130559E+2	.620962E+0	-.121708E+0
.9	-.40934657E+1	.661463E+1	.107081E+2	.456445E+0	-.745155E-1
1.0	-.34536372E+1	.556883E+1	.902247E+1	.351787E+0	-.488360E-1
1.2	-.25699817E+1	.420384E+1	.677382E+1	.230868E+0	-.242911E-1
1.4	-.19900750E+1	.335827E+1	.534834E+1	.165893E+0	-.138559E-1
1.6	-.15812733E+1	.278566E+1	.436694E+1	.126658E+0	-.868468E-2
1.8	-.12782065E+1	.237319E+1	.365140E+1	.100952E+0	-.582930E-2
2.0	-.10449740E+1	.206232E+1	.310729E+1	.830705E-1	-.412084E-2
2.5	-.64588818E+0	.154202E+1	.218791E+1	.561715E-1	-.203197E-2
3.0	-.39520418E+0	.122089E+1	.161610E+1	.415672E-1	-.116889E-2
3.5	-.22434083E+0	.100309E+1	.122743E+1	.325793E-1	-.743706E-3
4.0	-.10119508E+0	.845691E+0	.946886E+0	.265657E-1	-.507877E-3
4.5	-.87634201E-2	.726653E+0	.735416E+0	.222953E-1	-.365433E-3
5.0	.62786902E-1	.633484E+0	.570697E+0	.191249E-1	-.273692E-3
6.0	.16541191E+0	.497078E+0	.331666E+0	.147643E-1	-.167801E-3
7.0	.23449971E+0	.402063E+0	.167563E+0	.119313E-1	-.112069E-3
8.0	.28340156E+0	.332123E+0	.487212E-1	.995746E-2	-.795332E-4
9.0	.31929450E+0	.278523E+0	-.407714E-1	.851076E-2	-.590550E-4
10.0	.34636506E+0	.236165E+0	-.110200E+0	.740908E-2	-.454092E-4
12.0	.38352470E+0	.173558E+0	-.209967E+0	.584983E-2	-.290313E-4
14.0	.40679474E+0	.129599E+0	-.277196E+0	.480552E-2	-.200222E-4
16.0	.42187814E+0	.971154E-1	-.324763E+0	.406113E-2	-.145785E-4
18.0	.43181689E+0	.721919E-1	-.359625E+0	.350577E-2	-.110555E-4
20.0	.43836574E+0	.525081E-1	-.385858E+0	.307677E-2	-.865292E-5
25.0	.44604697E+0	.178033E-1	-.428244E+0	.234023E-2	-.518584E-5
30.0	.44717507E+0	-.463316E-2	-.451808E+0	.187605E-2	-.343516E-5
35.0	.44522639E+0	-.201702E-1	-.465397E+0	.155856E-2	-.243497E-5
40.0	.44175855E+0	-.314587E-1	-.473217E+0	.132864E-2	-.181238E-5
50.0	.43297624E+0	-.465271E-1	-.479503E+0	.101946E-2	-.111238E-5
60.0	.42360366E+0	-.558895E-1	-.479493E+0	.822295E-3	-.749915E-6
80.0	.40590467E+0	-.663731E-1	-.472278E+0	.587171E-3	-.405536E-6
100.0	.39046628E+0	-.716350E-1	-.462101E+0	.452951E-3	-.253146E-6

Table A2.14

The classical second virial coefficient, the quantum corrections and the Joule–Thomson coefficient for the (12 − 6) potential

T^*	B^*	$T^*(dB^*/dT^*)$	$(\mu^0 C_p^0)^*$	B_1^*	B_2^*
.3	-.27880563E+2	.766072E+2	.104488E+3	.249583E+2	-.339213E+2
.4	-.13798823E+2	.302671E+2	.440659E+2	.701989E+1	-.590611E+1
.5	-.87201972E+1	.169237E+2	.256439E+2	.305161E+1	-.178579E+1
.6	-.61979650E+1	.112488E+2	.174468E+2	.167256E+1	-.731344E+0
.7	-.47100326E+1	.825711E+1	.129671E+2	.105461E+1	-.361667E+0
.8	-.37342215E+1	.645413E+1	.101884E+2	.729172E+0	-.203100E+0
.9	-.30471111E+1	.526491E+1	.831202E+1	.537685E+0	-.124897E+0
1.0	-.25380786E+1	.442826E+1	.696634E+1	.415582E+0	-.821893E-1
1.2	-.18359470E+1	.333749E+1	.517343E+1	.274090E+0	-.411812E-1
1.4	-.13758463E+1	.266262E+1	.403846E+1	.197775E+0	-.236416E-1
1.6	-.10519103E+1	.220602E+1	.325793E+1	.151545E+0	-.149033E-1
1.8	-.81203208E+0	.187733E+1	.268936E+1	.121171E+0	-.100551E-1
2.0	-.62762426E+0	.162972E+1	.225734E+1	.999901E-1	-.714164E-2
2.5	-.31261261E+0	.121553E+1	.152814E+1	.680158E-1	-.355749E-2
3.0	-.11523336E+0	.960002E+0	.107524E+1	.505769E-1	-.206388E-2
3.5	.18957395E-1	.786714E+0	.767756E+0	.398041E-1	-.132272E-2
4.0	.11541735E+0	.661483E+0	.546065E+0	.325731E-1	-.909040E-3
4.5	.18761812E+0	.566754E+0	.379136E+0	.274237E-1	-.657774E-3
5.0	.24334383E+0	.492595E+0	.249251E+0	.235910E-1	-.495136E-3
6.0	.32290464E+0	.383972E+0	.610673E-1	.183028E-1	-.306238E-3
7.0	.37608869E+0	.308256E+0	-.678324E-1	.148538E-1	-.206045E-3
8.0	.41343415E+0	.252480E+0	-.160954E+0	.124425E-1	-.147163E-3
9.0	.44059801E+0	.209701E+0	-.230897E+0	.106700E-1	-.109886E-3
10.0	.46087543E+0	.175867E+0	-.285009E+0	.931649E-2	-.849179E-4
12.0	.48822405E+0	.125799E+0	-.362425E+0	.739427E-2	-.547573E-4
14.0	.50482619E+0	.905886E-1	-.414238E+0	.610140E-2	-.380359E-4
16.0	.51514039E+0	.645289E-1	-.450611E+0	.517644E-2	-.278645E-4
18.0	.52153603E+0	.445031E-1	-.477033E+0	.448410E-2	-.212437E-4
20.0	.52537428E+0	.286637E-1	-.496711E+0	.394773E-2	-.167056E-4
25.0	.52851692E+0	.669120E-3	-.527848E+0	.302287E-2	-.101102E-4
30.0	.52692551E+0	-.174929E-1	-.544418E+0	.243674E-2	-.674935E-5
35.0	.52322681E+0	-.301125E-1	-.553339E+0	.203394E-2	-.481487E-5
40.0	.51857505E+0	-.393116E-1	-.557887E+0	.174105E-2	-.360319E-5
50.0	.50836146E+0	-.516478E-1	-.560009E+0	.134520E-2	-.223096E-5
60.0	.49821263E+0	-.593621E-1	-.557575E+0	.109126E-2	-.151435E-5
80.0	.47979011E+0	-.680819E-1	-.547872E+0	.786339E-3	-.827385E-6
100.0	.46406948E+0	-.725244E-1	-.536594E+0	.610910E-3	-.520358E-6

Table A2.15

The classical second virial coefficient, the quantum corrections and the Joule–Thomson coefficient for the (18 − 6) potential

T^*	B^*	$T^*(\mathrm{d}B^*/\mathrm{d}T^*)$	$(\mu^0 C_p^0)^*$	B_1^*	B_2^*
.3	−.21205394E+2	.586192E+2	.798246E+2	.307143E+2	−.685518E+2
.4	−.10416672E+2	.232297E+2	.336464E+2	.872562E+1	−.121379E+2
.5	−.65146867E+1	.130194E+2	.195341E+2	.382715E+1	−.372392E+1
.6	−.45726686E+1	.866942E+1	.132421E+2	.211417E+1	−.154499E+1
.7	−.34251515E+1	.637264E+1	.979779E+1	.134228E+1	−.772971E+0
.8	−.26716375E+1	.498666E+1	.765830E+1	.933724E+0	−.438650E+0
.9	−.21405238E+1	.407144E+1	.621197E+1	.692244E+0	−.272348E+0
1.0	−.17467399E+1	.342693E+1	.517367E+1	.537633E+0	−.180812E+0
1.2	−.12030830E+1	.258565E+1	.378873E+1	.357546E+0	−.920341E−1
1.4	−.84649140E+0	.206442E+1	.291092E+1	.259795E+0	−.535631E−1
1.6	−.59526035E+0	.171139E+1	.230665E+1	.200255E+0	−.341780E−1
1.8	−.40912771E+0	.145701E+1	.186614E+1	.160953E+0	−.233118E−1
2.0	−.26598388E+0	.126524E+1	.153123E+1	.133430E+0	−.167201E−1
2.5	−.21362966E−1	.944119E+0	.965482E+0	.916399E−1	−.850518E−2
3.0	.13195695E+0	.745746E+0	.613789E+0	.686745E−1	−.502039E−2
3.5	.23619566E+0	.611075E+0	.374879E+0	.543995E−1	−.326465E−2
4.0	.31111161E+0	.513662E+0	.202550E+0	.447671E−1	−.227197E−2
4.5	.36716699E+0	.439914E+0	.727471E−1	.378761E−1	−.166227E−2
5.0	.41040998E+0	.382135E+0	−.282752E−1	.327261E−1	−.126359E−2
6.0	.47208880E+0	.297414E+0	−.174675E+0	.255835E−1	−.794629E−3
7.0	.51324572E+0	.238275E+0	−.274971E+0	.208959E−1	−.542104E−3
8.0	.54207861E+0	.194650E+0	−.347428E+0	.176011E−1	−.391791E−3
9.0	.56298965E+0	.161146E+0	−.401843E+0	.151675E−1	−.295532E−3
10.0	.57854361E+0	.134612E+0	−.443932E+0	.133015E−1	−.230410E−3
12.0	.59938230E+0	.952694E−1	−.504113E+0	.106371E−1	−.150822E−3
14.0	.61187084E+0	.675256E−1	−.544345E+0	.883331E−2	−.106063E−3
16.0	.61948012E+0	.469357E−1	−.572544E+0	.753558E−2	−.784922E−4
18.0	.62405422E+0	.310690E−1	−.592985E+0	.655945E−2	−.603615E−4
20.0	.62665211E+0	.184837E−1	−.608168E+0	.579989E−2	−.478303E−4
25.0	.62818213E+0	−.386653E−2	−.632049E+0	.448186E−2	−.293969E−4
30.0	.62610069E+0	−.184749E−1	−.644576E+0	.363974E−2	−.198513E−4
35.0	.62244150E+0	−.287037E−1	−.651145E+0	.305710E−2	−.142958E−4
40.0	.61809395E+0	−.362197E−1	−.654314E+0	.263098E−2	−.107826E−4
50.0	.60882296E+0	−.464213E−1	−.655244E+0	.205100E−2	−.675446E−5
60.0	.59974454E+0	−.529173E−1	−.652662E+0	.167588E−2	−.462642E−5
80.0	.58336298E+0	−.604763E−1	−.643839E+0	.122128E−2	−.256037E−5
100.0	.56939063E+0	−.645243E−1	−.633915E+0	.957021E−3	−.162373E−5

Table A2.16

The classical second virial coefficient, the quantum corrections and the Joule–Thomson coefficient for the $n(\bar{r})-6$ potential, $\gamma = 2{\cdot}0$

T^*	B^*	$T^*(dB^*/dT^*)$	$(\mu^0 C_p^0)^*$	B_1^*	B_2^*
.3	-.25875772E+2	.717893E+2	.976651E+2	.322450E+2	-.602511E+2
.4	-.12704342E+2	.282364E+2	.409408E+2	.906327E+1	-.104973E+2
.5	-.79740404E+1	.157325E+2	.237066E+2	.393759E+1	-.317543E+1
.6	-.56323717E+1	.104283E+2	.160607E+2	.215704E+1	-.130084E+1
.7	-.42544162E+1	.763812E+1	.118925E+2	.135946E+1	-.643421E+0
.8	-.33525317E+1	.595973E+1	.931226E+1	.939540E+0	-.361361E+0
.9	-.27185121E+1	.485444E+1	.757295E+1	.692529E+0	-.222228E+0
1.0	-.22494536E+1	.407790E+1	.632735E+1	.535061E+0	-.146234E+0
1.2	-.16035130E+1	.306713E+1	.467064E+1	.352645E+0	-.732575E-1
1.4	-.11810219E+1	.244288E+1	.362390E+1	.254298E+0	-.420428E-1
1.6	-.88402190E+0	.202112E+1	.290514E+1	.194743E+0	-.264923E-1
1.8	-.66438292E+0	.171783E+1	.238221E+1	.155629E+0	-.178656E-1
2.0	-.49573662E+0	.148955E+1	.198528E+1	.128360E+0	-.126822E-1
2.5	-.20816858E+0	.110809E+1	.131626E+1	.872167E-1	-.630817E-2
3.0	-.28442442E-1	.873001E+0	.901444E+0	.647912E-1	-.365392E-2
3.5	.93447845E-1	.713680E+0	.620232E+0	.509461E-1	-.233796E-2
4.0	.18085080E+0	.598590E+0	.417739E+0	.416579E-1	-.160415E-2
4.5	.24610746E+0	.511554E+0	.265446E+0	.350467E-1	-.115887E-2
5.0	.29634157E+0	.443427E+0	.147085E+0	.301284E-1	-.870944E-3
6.0	.36777239E+0	.343647E+0	-.241251E-1	.233462E-1	-.537008E-3
7.0	.41522927E+0	.274095E+0	-.141135E+0	.189263E-1	-.360248E-3
8.0	.44832343E+0	.222853E+0	-.225471E+0	.158385E-1	-.256577E-3
9.0	.47220668E+0	.183546E+0	-.288661E+0	.135701E-1	-.191074E-3
10.0	.48987555E+0	.152452E+0	-.337424E+0	.118391E-1	-.147283E-3
12.0	.51333611E+0	.106427E+0	-.406909E+0	.938261E-2	-.945219E-4
14.0	.52717651E+0	.740480E-1	-.453128E+0	.773211E-2	-.653728E-4
16.0	.53542491E+0	.500765E-1	-.485348E+0	.655236E-2	-.476995E-4
18.0	.54021459E+0	.316505E-1	-.508564E+0	.567007E-2	-.362309E-4
20.0	.54276649E+0	.170732E-1	-.525693E+0	.498705E-2	-.283926E-4
25.0	.54358146E+0	-.869681E-2	-.552278E+0	.381071E-2	-.170504E-4
30.0	.54041836E+0	-.254180E-1	-.565836E+0	.306635E-2	-.113058E-4
35.0	.53557759E+0	-.370350E-1	-.572613E+0	.255552E-2	-.801730E-5
40.0	.53005174E+0	-.455002E-1	-.575552E+0	.218451E-2	-.596759E-5
50.0	.51857311E+0	-.568416E-1	-.575415E+0	.168385E-2	-.366056E-5
60.0	.50753748E+0	-.639196E-1	-.571457E+0	.136326E-2	-.246509E-5
80.0	.48791886E+0	-.718846E-1	-.559803E+0	.979166E-3	-.132937E-5
100.0	.47139925E+0	-.759028E-1	-.547302E+0	.758751E-3	-.827324E-6

Table A2.17

The classical second virial coefficient, the quantum corrections and the Joule–Thomson coefficient for the $n(\bar{r})-6$ potential, $\gamma = 3{\cdot}0$

T^*	B^*	$T^*(dB^*/dT^*)$	$(\mu^0 C_p^0)^*$	B_1^*	B_2^*
.3	-.25721383E+2	.715492E+2	.972705E+2	.321709E+2	-.595072E+2
.4	-.12602751E+2	.280973E+2	.407000E+2	.903109E+1	-.103312E+2
.5	-.78983832E+1	.156358E+2	.235342E+2	.391921E+1	-.311533E+1
.6	-.55721442E+1	.103544E+2	.159266E+2	.214484E+1	-.127257E+1
.7	-.42044396E+1	.757837E+1	.117828E+2	.135056E+1	-.627782E+0
.8	-.33098711E+1	.590955E+1	.921942E+1	.932644E+0	-.351721E+0
.9	-.26813426E+1	.481117E+1	.749251E+1	.686948E+0	-.215809E+0
1.0	-.22165622E+1	.403983E+1	.625639E+1	.530399E+0	-.141709E+0
1.2	-.15768668E+1	.303637E+1	.461324E+1	.349164E+0	-.707150E-1
1.4	-.11587263E+1	.241702E+1	.357575E+1	.251533E+0	-.404418E-1
1.6	-.86493972E+0	.199875E+1	.286369E+1	.192454E+0	-.254024E-1
1.8	-.64777742E+0	.169807E+1	.234585E+1	.153677E+0	-.170803E-1
2.0	-.48110266E+0	.147183E+1	.195293E+1	.126658E+0	-.120919E-1
2.5	-.19707740E+0	.109389E+1	.129097E+1	.859248E-1	-.597827E-2
3.0	-.19726734E-1	.861057E+0	.880783E+0	.637472E-1	-.344475E-2
3.5	.10044603E+0	.703294E+0	.602848E+0	.500678E-1	-.219395E-2
4.0	.18654020E+0	.589344E+0	.402804E+0	.408981E-1	-.149910E-2
4.5	.25076020E+0	.503177E+0	.252417E+0	.343760E-1	-.107890E-2
5.0	.30014849E+0	.435734E+0	.135585E+0	.295271E-1	-.808040E-3
6.0	.37027156E+0	.336960E+0	-.333119E-1	.228465E-1	-.495152E-3
7.0	.41675318E+0	.268109E+0	-.148645E+0	.184974E-1	-.330376E-3
8.0	.44908336E+0	.217383E+0	-.231700E+0	.154619E-1	-.234170E-3
9.0	.47234619E+0	.178472E+0	-.293875E+0	.132337E-1	-.173630E-3
10.0	.48949726E+0	.147690E+0	-.341807E+0	.115348E-1	-.133306E-3
12.0	.51213324E+0	.102128E+0	-.410005E+0	.912627E-2	-.849514E-4
14.0	.52533663E+0	.700738E-1	-.455263E+0	.751001E-2	-.583950E-4
16.0	.53307085E+0	.463438E-1	-.486727E+0	.635602E-2	-.423778E-4
18.0	.53743210E+0	.281046E-1	-.509328E+0	.549384E-2	-.320323E-4
20.0	.53961840E+0	.136766E-1	-.525942E+0	.482701E-2	-.249915E-4
25.0	.53970657E+0	-.118239E-1	-.551530E+0	.368008E-2	-.148648E-4
30.0	.53599044E+0	-.283623E-1	-.564353E+0	.295562E-2	-.977771E-5
35.0	.53070636E+0	-.398451E-1	-.570551E+0	.245920E-2	-.688581E-5
40.0	.52481233E+0	-.482060E-1	-.573018E+0	.209914E-2	-.509422E-5
50.0	.51274759E+0	-.593925E-1	-.572140E+0	.161407E-2	-.309253E-5
60.0	.50125735E+0	-.663572E-1	-.567615E+0	.130409E-2	-.206462E-5
80.0	.48096124E+0	-.741596E-1	-.555121E+0	.933563E-3	-.109796E-5
100.0	.46394724E+0	-.780599E-1	-.542007E+0	.721513E-3	-.675794E-6

Table A2.18

The classical second virial coefficient, the quantum corrections and the Joule–Thomson coefficient for the $n(\bar{r})-6$ potential, $\gamma = 4{\cdot}0$

T^*	B^*	$T^*(dB^*/dT^*)$	$(\mu^0 C_p^0)^*$	B_1^*	B_2^*
.3	-.25581929E+2	.713177E+2	.968996E+2	.320825E+2	-.587402E+2
.4	-.12513138E+2	.279687E+2	.404818E+2	.899476E+1	-.101609E+2
.5	-.78324899E+1	.155483E+2	.233808E+2	.389898E+1	-.305391E+1
.6	-.55201281E+1	.102884E+2	.158085E+2	.213159E+1	-.124373E+1
.7	-.41615484E+1	.752539E+1	.116869E+2	.134099E+1	-.611857E+0
.8	-.32734479E+1	.586531E+1	.913876E+1	.925277E+0	-.341916E+0
.9	-.26497505E+1	.477317E+1	.742292E+1	.681013E+0	-.209287E+0
1.0	-.21887208E+1	.400650E+1	.619522E+1	.525458E+0	-.137115E+0
1.2	-.15544815E+1	.300956E+1	.456404E+1	.345493E+0	-.681368E-1
1.4	-.11401303E+1	.239452E+1	.353465E+1	.248628E+0	-.388199E-1
1.6	-.84913786E+0	.197931E+1	.282845E+1	.190054E+0	-.242990E-1
1.8	-.63412730E+0	.168092E+1	.231504E+1	.151634E+0	-.162860E-1
2.0	-.46916505E+0	.145644E+1	.192560E+1	.124880E+0	-.114951E-1
2.5	-.18821878E+0	.108154E+1	.126976E+1	.845787E-1	-.564511E-2
3.0	-.12936713E-1	.850641E+0	.863577E+0	.626612E-1	-.323371E-2
3.5	.10573613E+0	.694208E+0	.588472E+0	.491552E-1	-.204878E-2
4.0	.19068354E+0	.581228E+0	.390544E+0	.401094E-1	-.139327E-2
4.5	.25399200E+0	.495798E+0	.241806E+0	.336803E-1	-.998390E-3
5.0	.30263381E+0	.428932E+0	.126298E+0	.289039E-1	-.744747E-3
6.0	.37159644E+0	.331002E+0	-.405942E-1	.223289E-1	-.453079E-3
7.0	.41720630E+0	.262738E+0	-.154468E+0	.180535E-1	-.300377E-3
8.0	.44884876E+0	.212442E+0	-.236406E+0	.150723E-1	-.211685E-3
9.0	.47154940E+0	.173859E+0	-.297691E+0	.128860E-1	-.156137E-3
10.0	.48822834E+0	.143336E+0	-.344893E+0	.112204E-1	-.119299E-3
12.0	.51010619E+0	.981530E-1	-.411953E+0	.886156E-2	-.753711E-4
14.0	.52271796E+0	.663650E-1	-.456353E+0	.728078E-2	-.514171E-4
16.0	.52997041E+0	.428319E-1	-.487139E+0	.615345E-2	-.370606E-4
18.0	.53392714E+0	.247445E-1	-.509183E+0	.531211E-2	-.278405E-4
20.0	.53576592E+0	.104375E-1	-.525328E+0	.466202E-2	-.215982E-4
25.0	.53515656E+0	-.148443E-1	-.550001E+0	.354550E-2	-.126875E-4
30.0	.53090364E+0	-.312342E-1	-.562138E+0	.284160E-2	-.825733E-5
35.0	.52518548E+0	-.426072E-1	-.567793E+0	.236006E-2	-.576122E-5
40.0	.51892846E+0	-.508821E-1	-.569811E+0	.201131E-2	-.422695E-5
50.0	.50628131E+0	-.619389E-1	-.568220E+0	.154231E-2	-.252934E-5
60.0	.49433575E+0	-.688067E-1	-.563142E+0	.124327E-2	-.166804E-5
80.0	.47335582E+0	-.764655E-1	-.549821E+0	.886733E-3	-.869249E-6
100.0	.45583932E+0	-.802579E-1	-.536097E+0	.683301E-3	-.526239E-6

Table A2.19

The classical second virial coefficient, the quantum corrections and the Joule–Thomson coefficient for the $n(\bar{r})-6$ potential, $\gamma=5{\cdot}0$

T^*	B^*	$T^*(dB^*/dT^*)$	$(\mu^0 C_p^0)^*$	B_1^*	B_2^*
.3	-.25469137E+2	.711364E+2	.966056E+2	.320015E+2	-.579918E+2
.4	-.12439977E+2	.278651E+2	.403050E+2	.896033E+1	-.999372E+1
.5	-.77785408E+1	.154771E+2	.232556E+2	.387951E+1	-.299338E+1
.6	-.54774979E+1	.102344E+2	.157119E+2	.211873E+1	-.121525E+1
.7	-.41263865E+1	.748198E+1	.116084E+2	.133165E+1	-.596100E+0
.8	-.32435909E+1	.582900E+1	.907259E+1	.918058E+0	-.332201E+0
.9	-.26238620E+1	.474193E+1	.736579E+1	.675181E+0	-.202819E+0
1.0	-.21659175E+1	.397907E+1	.614498E+1	.520591E+0	-.132555E+0
1.2	-.15361741E+1	.298743E+1	.452361E+1	.341864E+0	-.655745E-1
1.4	-.11249528E+1	.237591E+1	.350086E+1	.245748E+0	-.372066E-1
1.6	-.83627406E+0	.196319E+1	.279946E+1	.187671E+0	-.232008E-1
1.8	-.62304988E+0	.166665E+1	.228970E+1	.149602E+0	-.154949E-1
2.0	-.45951318E+0	.144361E+1	.190312E+1	.123108E+0	-.109005E-1
2.5	-.18114204E+0	.107118E+1	.125232E+1	.832344E-1	-.531299E-2
3.0	-.76027372E-2	.841831E+0	.849434E+0	.615746E-1	-.302326E-2
3.5	.10979705E+0	.686467E+0	.576670E+0	.482408E-1	-.190397E-2
4.0	.19376395E+0	.574262E+0	.380498E+0	.393182E-1	-.128770E-2
4.5	.25628733E+0	.489419E+0	.233131E+0	.329818E-1	-.918072E-3
5.0	.30428172E+0	.423011E+0	.118729E+0	.282776E-1	-.681604E-3
6.0	.37222793E+0	.325749E+0	-.464791E-1	.218082E-1	-.411108E-3
7.0	.41706452E+0	.257945E+0	-.159120E+0	.176064E-1	-.270453E-3
8.0	.44808990E+0	.207984E+0	-.240106E+0	.146796E-1	-.189259E-3
9.0	.47028076E+0	.169655E+0	-.300626E+0	.125353E-1	-.138694E-3
10.0	.48652745E+0	.139331E+0	-.347197E+0	.109031E-1	-.105334E-3
12.0	.50770254E+0	.944388E-1	-.413264E+0	.859421E-2	-.658228E-4
14.0	.51975779E+0	.628528E-1	-.456905E+0	.704911E-2	-.444648E-4
16.0	.52655142E+0	.394683E-1	-.487083E+0	.594863E-2	-.317645E-4
18.0	.53011885E+0	.214952E-1	-.508624E+0	.512827E-2	-.236666E-4
20.0	.53162014E+0	.727909E-2	-.524341E+0	.449506E-2	-.182204E-4
25.0	.53032495E+0	-.178387E-1	-.548164E+0	.340922E-2	-.105215E-4
30.0	.52553659E+0	-.341161E-1	-.559653E+0	.272610E-2	-.674561E-5
35.0	.51938076E+0	-.454051E-1	-.564786E+0	.225961E-2	-.464353E-5
40.0	.51275465E+0	-.536131E-1	-.566368E+0	.192229E-2	-.336535E-5
50.0	.49950982E+0	-.645661E-1	-.564076E+0	.146958E-2	-.197020E-5
60.0	.48709265E+0	-.713533E-1	-.558446E+0	.118161E-2	-.127451E-5
80.0	.46539813E+0	-.788867E-1	-.544285E+0	.839251E-3	-.642474E-6
100.0	.44735238E+0	-.825797E-1	-.529932E+0	.644558E-3	-.378033E-6

Table A2.20

The classical second virial coefficients, the quantum corrections and the Joule–Thomson coefficient for the $n(\bar{r})-6$ potential, $\gamma=6{\cdot}0$

T^*	B^*	$T^*(dB^*/dT^*)$	$(\mu^0 C_p^0)^*$	B_1^*	B_2^*
.3	-.25370803E+2	.709770E+2	.963478E+2	.319171E+2	-.572408E+2
.4	-.12376487E+2	.277741E+2	.401505E+2	.892480E+1	-.982588E+1
.5	-.77318994E+1	.154147E+2	.231466E+2	.385950E+1	-.293261E+1
.6	-.54407557E+1	.101873E+2	.156280E+2	.210554E+1	-.118664E+1
.7	-.40961634E+1	.744412E+1	.115403E+2	.132209E+1	-.580271E+0
.8	-.32179925E+1	.579736E+1	.901535E+1	.910674E+0	-.322441E+0
.9	-.26017206E+1	.471473E+1	.731645E+1	.669219E+0	-.196320E+0
1.0	-.21464624E+1	.395518E+1	.610165E+1	.515618E+0	-.127972E+0
1.2	-.15206335E+1	.296816E+1	.448880E+1	.338158E+0	-.629995E-1
1.4	-.11121390E+1	.235968E+1	.347182E+1	.242808E+0	-.355850E-1
1.6	-.82547841E+0	.194912E+1	.277460E+1	.185238E+0	-.220969E-1
1.8	-.61381485E+0	.165418E+1	.226800E+1	.147527E+0	-.146998E-1
2.0	-.45152586E+0	.143237E+1	.188389E+1	.121300E+0	-.103028E-1
2.5	-.17541666E+0	.106204E+1	.123746E+1	.818618E-1	-.497914E-2
3.0	-.34172130E-2	.834022E+0	.837439E+0	.604650E-1	-.281171E-2
3.5	.11285074E+0	.679562E+0	.566711E+0	.473069E-1	-.175841E-2
4.0	.19594043E+0	.568010E+0	.372070E+0	.385101E-1	-.118159E-2
4.5	.25775711E+0	.483659E+0	.225901E+0	.322682E-1	-.837340E-3
5.0	.30516518E+0	.417633E+0	.112468E+0	.276377E-1	-.618139E-3
6.0	.37218324E+0	.320924E+0	-.512593E-1	.212761E-1	-.368928E-3
7.0	.41630617E+0	.253498E+0	-.162808E+0	.171494E-1	-.240382E-3
8.0	.44675652E+0	.203811E+0	-.242946E+0	.142782E-1	-.166726E-3
9.0	.46846822E+0	.165687E+0	-.302781E+0	.121768E-1	-.121168E-3
10.0	.48430545E+0	.135522E+0	-.348783E+0	.105786E-1	-.913044E-4
12.0	.50480813E+0	.908620E-1	-.413946E+0	.832079E-2	-.562315E-4
14.0	.51632465E+0	.594350E-1	-.456890E+0	.681217E-2	-.374820E-4
16.0	.52266983E+0	.361664E-1	-.486503E+0	.573914E-2	-.264457E-4
18.0	.52585368E+0	.182818E-1	-.507572E+0	.494023E-2	-.194751E-4
20.0	.52702017E+0	.413573E-2	-.522884E+0	.432429E-2	-.148285E-4
25.0	.52503814E+0	-.208558E-1	-.545894E+0	.326985E-2	-.834669E-5
30.0	.51970779E+0	-.370463E-1	-.556754E+0	.260798E-2	-.522777E-5
35.0	.51310543E+0	-.482693E-1	-.561375E+0	.215689E-2	-.352130E-5
40.0	.50610045E+0	-.564238E-1	-.562524E+0	.183127E-2	-.250021E-5
50.0	.49223805E+0	-.672911E-1	-.559529E+0	.139523E-2	-.140866E-5
60.0	.47933036E+0	-.740089E-1	-.553339E+0	.111860E-2	-.879220E-6
80.0	.45688804E+0	-.814286E-1	-.538317E+0	.790767E-3	-.414555E-6
100.0	.43828554E+0	-.850267E-1	-.523312E+0	.605021E-3	-.228991E-6

Table A2.21

The classical second virial coefficients, the quantum corrections and the Joule–Thomson coefficient for the n(r̄)−6 potential, γ = 7·0

T^*	B^*	$T^*(dB^*/dT^*)$	$(\mu^0 C_p^0)^*$	B_1^*	B_2^*
.3	-.25292942E+2	.708600E+2	.961530E+2	.318402E+2	-.565071E+2
.4	-.12325054E+2	.277030E+2	.400280E+2	.889114E+1	-.966084E+1
.5	-.76937878E+1	.153649E+2	.230587E+2	.384023E+1	-.287262E+1
.6	-.54106060E+1	.101491E+2	.155597E+2	.209273E+1	-.115833E+1
.7	-.40713056E+1	.741327E+1	.114846E+2	.131275E+1	-.564576E+0
.8	-.31969113E+1	.577145E+1	.896836E+1	.903423E+0	-.312749E+0
.9	-.25834750E+1	.469237E+1	.727585E+1	.663346E+0	-.189859E+0
1.0	-.21304282E+1	.393549E+1	.606592E+1	.510707E+0	-.123412E+0
1.2	-.15078359E+1	.295219E+1	.446002E+1	.334484E+0	-.604332E-1
1.4	-.11016081E+1	.234616E+1	.344777E+1	.239885E+0	-.339673E-1
1.6	-.81663326E+0	.193734E+1	.275397E+1	.182814E+0	-.209946E-1
1.8	-.60627954E+0	.164369E+1	.224997E+1	.145457E+0	-.139053E-1
2.0	-.44504265E+0	.142287E+1	.186791E+1	.119492E+0	-.970521E-2
2.5	-.17085537E+0	.105424E+1	.122509E+1	.804867E-1	-.464496E-2
3.0	-.17902859E-3	.827274E+0	.827453E+0	.593512E-1	-.259978E-2
3.5	.11510589E+0	.673532E+0	.558426E+0	.463680E-1	-.161249E-2
4.0	.19742584E+0	.562496E+0	.365070E+0	.376966E-1	-.107516E-2
4.5	.25861637E+0	.478530E+0	.219914E+0	.315491E-1	-.756331E-3
5.0	.30550017E+0	.412803E+0	.107302E+0	.269924E-1	-.554430E-3
6.0	.37167800E+0	.316520E+0	-.551578E-1	.207387E-1	-.326556E-3
7.0	.41514503E+0	.249382E+0	-.165763E+0	.166874E-1	-.210159E-3
8.0	.44505985E+0	.199900E+0	-.245160E+0	.138721E-1	-.144067E-3
9.0	.46632009E+0	.161928E+0	-.304392E+0	.118138E-1	-.103536E-3
10.0	.48176758E+0	.131880E+0	-.349887E+0	.102500E-1	-.771842E-4
12.0	.50162199E+0	.873860E-1	-.414236E+0	.804363E-2	-.465713E-4
14.0	.51261158E+0	.560707E-1	-.456541E+0	.657182E-2	-.304442E-4
16.0	.51851301E+0	.328820E-1	-.485631E+0	.552653E-2	-.210818E-4
18.0	.52131364E+0	.150576E-1	-.506256E+0	.474933E-2	-.152456E-4
20.0	.52214296E+0	.958744E-3	-.521184E+0	.415086E-2	-.114041E-4
25.0	.51946179E+0	-.239480E-1	-.543410E+0	.312823E-2	-.614817E-5
30.0	.51357316E+0	-.400794E-1	-.553653E+0	.248792E-2	-.369160E-5
35.0	.50650684E+0	-.512562E-1	-.557763E+0	.205246E-2	-.238427E-5
40.0	.49910564E+0	-.593716E-1	-.558477E+0	.173874E-2	-.162274E-5
50.0	.48459281E+0	-.701725E-1	-.554765E+0	.131964E-2	-.838008E-6
60.0	.47116499E+0	-.768324E-1	-.547997E+0	.105455E-2	-.476779E-6
80.0	.44792475E+0	-.841496E-1	-.532074E+0	.741496E-3	-.181745E-6
100.0	.42872509E+0	-.876559E-1	-.516381E+0	.564864E-3	-.762915E-7

Table A2.22

The classical second virial coefficients, the quantum corrections and the Joule–Thomson coefficient for the $n(\bar{r})-6$ potential, $\gamma = 8{\cdot}0$

T^*	B^*	$T^*(dB^*/dT^*)$	$(\mu^0 C_p^0)^*$	B_1^*	B_2^*
.3	-.25216694E+2	.707356E+2	.959523E+2	.317493E+2	-.557501E+2
.4	-.12276094E+2	.276316E+2	.399077E+2	.885337E+1	-.949154E+1
.5	-.76580332E+1	.153160E+2	.229741E+2	.381912E+1	-.281128E+1
.6	-.53825969E+1	.101123E+2	.154949E+2	.207887E+1	-.112944E+1
.7	-.40483892E+1	.738371E+1	.114321E+2	.130272E+1	-.548582E+0
.8	-.31776046E+1	.574676E+1	.892437E+1	.895694E+0	-.302883E+0
.9	-.25668660E+1	.467115E+1	.723801E+1	.657111E+0	-.183286E+0
1.0	-.21159163E+1	.391684E+1	.603276E+1	.505509E+0	-.118777E+0
1.2	-.14963839E+1	.293710E+1	.443349E+1	.330614E+0	-.578265E-1
1.4	-.10922939E+1	.233342E+1	.342571E+1	.236815E+0	-.323249E-1
1.6	-.80890754E+0	.192624E+1	.273514E+1	.180274E+0	-.198760E-1
1.8	-.59978781E+0	.163380E+1	.223359E+1	.143291E+0	-.130991E-1
2.0	-.43954250E+0	.141391E+1	.185345E+1	.117604E+0	-.909901E-2
2.5	-.16716947E+0	.104685E+1	.121402E+1	.790528E-1	-.430604E-2
3.0	.22536871E-2	.820869E+0	.818615E+0	.581914E-1	-.238484E-2
3.5	.11660378E+0	.667786E+0	.551182E+0	.453914E-1	-.146449E-2
4.0	.19818863E+0	.557218E+0	.359030E+0	.368511E-1	-.967188E-3
4.5	.25877860E+0	.473601E+0	.214822E+0	.308022E-1	-.674136E-3
5.0	.30515749E+0	.408141E+0	.102983E+0	.263224E-1	-.489775E-3
6.0	.37052137E+0	.312237E+0	-.582841E-1	.201812E-1	-.283535E-3
7.0	.41334833E+0	.245351E+0	-.167997E+0	.162085E-1	-.179455E-3
8.0	.44273711E+0	.196046E+0	-.246691E+0	.134513E-1	-.121036E-3
9.0	.46355129E+0	.158204E+0	-.305348E+0	.114378E-1	-.856055E-4
10.0	.47861170E+0	.128254E+0	-.350357E+0	.990969E-2	-.628163E-4
12.0	.49781803E+0	.838965E-1	-.413921E+0	.775676E-2	-.367306E-4
14.0	.50827688E+0	.526706E-1	-.455606E+0	.632318E-2	-.232671E-4
16.0	.51372859E+0	.295443E-1	-.484184E+0	.530668E-2	-.156057E-4
18.0	.51613887E+0	.117658E-1	-.504373E+0	.455199E-2	-.109230E-4
20.0	.51662328E+0	-.229756E-2	-.518921E+0	.397164E-2	-.790053E-5
25.0	.51322273E+0	-.271416E-1	-.540364E+0	.298199E-2	-.389311E-5
30.0	.50675589E+0	-.432292E-1	-.549985E+0	.236402E-2	-.211207E-5
35.0	.49920674E+0	-.543707E-1	-.553577E+0	.194477E-2	-.121240E-5
40.0	.49139170E+0	-.624553E-1	-.553847E+0	.164336E-2	-.716359E-6
50.0	.47619686E+0	-.732004E-1	-.549397E+0	.124181E-2	-.246037E-6
60.0	.46222162E+0	-.798083E-1	-.542030E+0	.988662E-3	-.576373E-7
80.0	.43813895E+0	-.870267E-1	-.525166E+0	.690897E-3	.625064E-7
100.0	.41830745E+0	-.904392E-1	-.508747E+0	.523687E-3	.849675E-7

Table A2.23

The classical second virial coefficient, the quantum corrections and the Joule–Thomson coefficient for the $n(\bar{r})-6$ potential, $\gamma = 9{\cdot}0$

T^*	B^*	$T^*(dB^*/dT^*)$	$(\mu^0 C_p^0)^*$	B_1^*	B_2^*
.3	-.25149203E+2	.706256E+2	.957748E+2	.316550E+2	-.549902E+2
.4	-.12232803E+2	.275681E+2	.398009E+2	.881449E+1	-.932149E+1
.5	-.76264956E+1	.152726E+2	.228991E+2	.379744E+1	-.274963E+1
.6	-.53579571E+1	.100794E+2	.154374E+2	.206467E+1	-.110039E+1
.7	-.40282860E+1	.735737E+1	.113857E+2	.129246E+1	-.532498E+0
.8	-.31607181E+1	.572475E+1	.888547E+1	.887786E+0	-.292959E+0
.9	-.25523849E+1	.465221E+1	.720459E+1	.650735E+0	-.176673E+0
1.0	-.21033064E+1	.390019E+1	.600350E+1	.500196E+0	-.114111E+0
1.2	-.14865082E+1	.292361E+1	.441012E+1	.326658E+0	-.552015E-1
1.4	-.10843331E+1	.232199E+1	.340632E+1	.233678E+0	-.306702E-1
1.6	-.80237344E+0	.191625E+1	.271862E+1	.177678E+0	-.187484E-1
1.8	-.59436551E+0	.162488E+1	.221924E+1	.141077E+0	-.122862E-1
2.0	-.43501652E+0	.140580E+1	.184081E+1	.115673E+0	-.848746E-2
2.5	-.16429289E+0	.104011E+1	.120440E+1	.775860E-1	-.396378E-2
3.0	.39808429E-2	.814967E+0	.810986E+0	.570046E-1	-.216758E-2
3.5	.11746606E+0	.662446E+0	.544979E+0	.443917E-1	-.131476E-2
4.0	.19836513E+0	.552275E+0	.353909E+0	.359853E-1	-.857866E-3
4.5	.25839045E+0	.468949E+0	.210558E+0	.300372E-1	-.590844E-3
5.0	.30429111E+0	.403709E+0	.994183E-1	.256360E-1	-.424205E-3
6.0	.36887649E+0	.308114E+0	-.607621E-1	.196099E-1	-.239838E-3
7.0	.41108399E+0	.241428E+0	-.169656E+0	.157175E-1	-.148225E-3
8.0	.43995848E+0	.192260E+0	-.247698E+0	.130197E-1	-.975757E-4
9.0	.46033275E+0	.154516E+0	-.305817E+0	.110522E-1	-.673161E-4
10.0	.47500851E+0	.124638E+0	-.350370E+0	.956063E-2	-.481418E-4
12.0	.49356477E+0	.803761E-1	-.413189E+0	.746247E-2	-.266544E-4
14.0	.50348568E+0	.492087E-1	-.454277E+0	.606807E-2	-.159004E-4
16.0	.50847779E+0	.261208E-1	-.482357E+0	.508109E-2	-.997178E-5
18.0	.51048647E+0	.836894E-2	-.502118E+0	.434950E-2	-.646585E-5
20.0	.51061404E+0	-.567488E-2	-.516289E+0	.378776E-2	-.428003E-5
25.0	.50646371E+0	-.304859E-1	-.536950E+0	.283198E-2	-.155069E-5
30.0	.49938928E+0	-.465500E-1	-.545939E+0	.223697E-2	-.463355E-6
35.0	.49132981E+0	-.576708E-1	-.549001E+0	.183436E-2	.164243E-7
40.0	.48307543E+0	-.657349E-1	-.548810E+0	.154562E-2	.238250E-6
50.0	.46715325E+0	-.764377E-1	-.543591E+0	.116211E-2	.382557E-6
60.0	.45259172E+0	-.830006E-1	-.535592E+0	.921255E-3	.390819E-6
80.0	.42760361E+0	-.901236E-1	-.517727E+0	.639224E-3	.327409E-6
100.0	.40709156E+0	-.934383E-1	-.500530E+0	.481708E-3	.261591E-6

Table A2.24

The classical second virial coefficient, the quantum corrections and the Joule–Thomson coefficient for the $n(\bar{r})-6$ potential, $\gamma = 10{\cdot}0$

T^*	B^*	$T^*(dB^*/dT^*)$	$(\mu^0 C_p^0)^*$	B_1^*	B_2^*
.3	-.25097841E+2	.705524E+2	.956503E+2	.315681E+2	-.542464E+2
.4	-.12198545E+2	.275209E+2	.397194E+2	.877743E+1	-.915394E+1
.5	-.76011649E+1	.152389E+2	.228401E+2	.377647E+1	-.268865E+1
.6	-.53380229E+1	.100535E+2	.153915E+2	.205081E+1	-.107158E+1
.7	-.40119597E+1	.733629E+1	.113482E+2	.128239E+1	-.516510E+0
.8	-.31469780E+1	.570698E+1	.885395E+1	.879990E+0	-.283077E+0
.9	-.25405945E+1	.463681E+1	.717740E+1	.644430E+0	-.170079E+0
1.0	-.20930428E+1	.388656E+1	.597961E+1	.494928E+0	-.109453E+0
1.2	-.14784935E+1	.291245E+1	.439095E+1	.322721E+0	-.525764E-1
1.4	-.10779087E+1	.231245E+1	.339036E+1	.230546E+0	-.290129E-1
1.6	-.79714433E+0	.190785E+1	.270499E+1	.175081E+0	-.176176E-1
1.8	-.59007549E+0	.161731E+1	.220739E+1	.138858E+0	-.114700E-1
2.0	-.43148954E+0	.139886E+1	.183035E+1	.113735E+0	-.787280E-2
2.5	-.16218766E+0	.103424E+1	.119643E+1	.761100E-1	-.361900E-2
3.0	.50779075E-2	.809747E+0	.804669E+0	.558078E-1	-.194830E-2
3.5	.11779237E+0	.657650E+0	.539857E+0	.433819E-1	-.116337E-2
4.0	.19807141E+0	.547773E+0	.349702E+0	.351098E-1	-.747157E-3
4.5	.25757950E+0	.464660E+0	.207081E+0	.292627E-1	-.506366E-3
5.0	.30303686E+0	.399579E+0	.965417E-1	.249405E-1	-.357605E-3
6.0	.36688955E+0	.304196E+0	-.626934E-1	.190301E-1	-.195335E-3
7.0	.40850359E+0	.237640E+0	-.170863E+0	.152187E-1	-.116337E-3
8.0	.43687811E+0	.188556E+0	-.248322E+0	.125810E-1	-.735644E-4
9.0	.45681942E+0	.150866E+0	-.305953E+0	.106598E-1	-.485536E-4
10.0	.47111267E+0	.121025E+0	-.350088E+0	.920524E-2	-.330542E-4
12.0	.48901432E+0	.768042E-1	-.412210E+0	.716256E-2	-.162505E-4
14.0	.49838618E+0	.456549E-1	-.452731E+0	.580793E-2	-.826357E-5
16.0	.50290427E+0	.225737E-1	-.480331E+0	.485096E-2	-.410894E-5
18.0	.50449527E+0	.482312E-2	-.499672E+0	.414287E-2	-.181058E-5
20.0	.50424920E+0	-.922199E-2	-.513471E+0	.360006E-2	-.485280E-6
25.0	.49930674E+0	-.340386E-1	-.533345E+0	.267881E-2	.924840E-6
30.0	.49158426E+0	-.501055E-1	-.541690E+0	.210723E-2	.129237E-5
35.0	.48297679E+0	-.612242E-1	-.544201E+0	.172165E-2	.133429E-5
40.0	.47424833E+0	-.692814E-1	-.543530E+0	.144587E-2	.126878E-5
50.0	.45753729E+0	-.799585E-1	-.537496E+0	.108084E-2	.106925E-5
60.0	.44233697E+0	-.864844E-1	-.528821E+0	.852577E-3	.885762E-6
80.0	.41635901E+0	-.935137E-1	-.509873E+0	.586683E-3	.623525E-6
100.0	.39510188E+0	-.967206E-1	-.491822E+0	.439065E-3	.456784E-6

APPENDIX 3

THE EXTENDED LAW OF CORRESPONDING STATES CORRELATION OF THERMOPHYSICAL PROPERTIES[1,2]

The correlation is based upon the hypothesis that, to a good degree of accuracy, the intermolecular potentials for a number of pair interactions can be rendered conformal by the choice of two scaling factors ε_c and σ_c characteristic of each interaction. From this hypothesis it follows that all the collision integrals for this universal potential are universal functions of a reduced temperature, $T^* = kT/\varepsilon_c$. The combinations of these universal collision integrals necessary for the evaluation of the viscosity, diffusion coefficient, and thermal conductivity of monatomic gases and gas mixtures at low density and the viscosity and diffusion coefficient of polyatomic gas mixtures have been correlated empirically. These functionals of the supposed conformal potential are defined in the terms of the collision integrals of Chapter 5 as

$$\Omega_\eta^* = \Omega^{(2,2)*}/f_\eta \tag{A3.1}$$

$$\Omega_D^* = \Omega^{(1,1)*} \tag{A3.2}$$

$$A^\star = \Omega^{(2,2)*}/\Omega^{(1,1)*} \tag{A3.3}$$

$$B^\star = \frac{5\Omega^{(1,2)*} - 4\Omega^{(1,3)*}}{\Omega^{(1,1)*}} \tag{A3.4}$$

$$\xi = f_\lambda/f_\eta. \tag{A3.5}$$

The reduced second virial coefficient, B^*, is also a functional of the pair potential function (Chapter 3) and for the monatomic gases can be included in the correlation scheme. The correlations of these quantities as functions of the reduced temperature are given below. These correlations differ from those of the original references since the scaling factors have been arbitrarily chosen so that ε_c and σ_c for argon are consistent with recent best estimates of the potential well depth ε_{Ar} and the separation at zero potential energy σ_{Ar} for this species.[3]

$$\Omega_\eta^* = \exp\{0{\cdot}46649 - 0{\cdot}57015 \ln T^* + 0{\cdot}19164(\ln T^*)^2 - 0{\cdot}03708(\ln T^*)^3 + 0{\cdot}00241(\ln T^*)^4\} \qquad 1 < T^* < 90 \tag{A3.6}$$

$$\Omega_D^* = \exp\{0{\cdot}348 - 0{\cdot}459 \ln T^* + 0{\cdot}095(\ln T^*)^2 - 0{\cdot}010(\ln T^*)^3\} \qquad 1 < T^* < 25 \tag{A3.7}$$

$$A^\star = \exp\{0{\cdot}1281 - 0{\cdot}1108 \ln T^* + 0{\cdot}0962(\ln T^*)^2 - 0{\cdot}0271(\ln T^*)^3\} + 0{\cdot}0024(\ln T^*)^4\} \qquad 1 < T^* < 25 \tag{A3.8}$$

$$B^\star = \exp\{0{\cdot}1789 - 0{\cdot}1233 \ln T^* + 0{\cdot}0558(\ln T^*)^2 - 0{\cdot}0074(\ln T^*)^3\} \qquad 1 < T^* < 25 \tag{A3.9}$$

$$\xi = 1 + 0{\cdot}0042[1 - \exp 0{\cdot}33(1 - T^*)] \qquad 1 < T^* < 90 \tag{A3.10}$$

Table A3.1

The scaling factor ratio $p = \varepsilon_c/\varepsilon_{Ar}$

	He	Ne	Ar	Kr	Xe	N_2	O_2	CO_2	CH_4	CF_4	SF_6	C_2H_6	C_3H_8	n-C_4H_{10}	i-C_4H_{10}	C_2H_4	N_2O	CCl_3F	$CHClF_2$
He	0·07350	0·1375	0·3596	0·2092	0·1115	0·4187	0·5820	0·3275	0·2156	0·2576	0·3200	—	—	—	—	—	—	—	—
Ne		0·2967	0·4147	0·5716	0·5143	0·4587	0·6563	0·5630	0·5733	0·5487	0·6905	—	—	—	—	—	—	—	—
Ar			1·0000	1·1860	1·2512	0·8484	0·9470	1·1058	1·2282	1·0151	1·2331	—	—	—	—	—	—	—	—
Kr				1·3758	1·4927	0·9081	1·1003	1·4289	1·1739	1·1514	1·3834	—	—	—	—	—	—	—	—
Xe					1·8571	—	—	—	—	—	—	—	—	—	—	—	—	—	—
N_2						0·7361	0·7785	1·0678	0·9190	0·8911	1·1450	1·0761	1·1777	1·2428	—	—	—	—	—
O_2							0·8919	1·0999	—	0·9637	1·2049	—	—	—	—	—	—	—	—
CO_2								1·7325	1·3412	1·2853	1·5530	1·6242	1·7278	1·8436	—	—	—	—	—
CH_4									1·1395	1·0210	1·1105	1·7787	1·5624	1·7929	—	—	—	—	—
CF_4										1·1052	1·3676	1·2428	—	—	—	—	—	—	—
SF_6											1·4671	1·4882	—	—	—	—	—	—	—
C_2H_6												1·7082	1·8944	1·8957	—	—	—	—	—
C_3H_8													1·8964	2·0012	—	—	—	—	—
n-C_4H_{10}														2·0168	2·0018	—	—	—	—
i-C_4H_{10}															1·8423	—	—	—	—
C_2H_4																1·7250	—	—	—
N_2O																	1·8840	—	—
CCl_3F																		1·8885	—
$CHClF_2$																			2·0005

— denotes no investigation.

Table A3.2

The scaling factor ratio $s = \sigma_c / \sigma_{Ar}$

	He	Ne	Ar	Kr	Xe	N_2	O_2	CO_2	CH_4	CF_4	SF_6	C_2H_6	C_3H_8	n-C_4H_{10}	i-C_4H_{10}	C_2H_4	N_2O	CCl_3F	$CHClF_2$
He	0·7767	0·8034	0·8824	0·9815	1·1121	0·9043	0·8514	0·9809	0·9918	1·1000	1·1756	—	—	—	—	—	—	—	—
Ne		0·8225	0·9392	0·9605	1·0343	0·9250	0·9031	0·9985	0·9696	1·1000	1·1741	—	—	—	—	—	—	—	—
Ar			1·0000	1·0334	1·0957	1·0389	1·0040	1·0757	1·0270	1·1844	1·2689	—	—	—	—	—	—	—	—
Kr				1·0690	1·1319	1·0863	1·0356	1·0845	1·0997	1·2221	1·3105	—	—	—	—	—	—	—	—
Xe					1·1692	—	—	—	—	—	—	—	—	—	—	—	—	—	—
N_2						1·0842	1·0495	1·1060	1·0905	1·2249	1·2966	1·1936	1·2774	1·3488	—	—	—	—	—
O_2							1·0097	1·0744	—	1·1917	1·2701	—	—	—	—	—	—	—	—
CO_2								1·1252	1·1127	1·2440	1·3212	1·2005	1·2844	1·3458	—	—	—	—	—
CH_4									1·1106	1·2525	1·3616	1·1872	1·2853	1·3285	—	—	—	—	—
CF_4										1·3668	1·4452	1·3431	—	—	—	—	—	—	—
SF_6											1·5679	1·4345	—	—	—	—	—	—	—
C_2H_6												1·3048	1·3795	1·4658	—	—	—	—	—
C_3H_8													1·4901	1·5555	—	—	—	—	—
n-C_4H_{10}														1·6496	1·6466	—	—	—	—
i-C_4H_{10}															1·6803	—	—	—	—
C_2H_4																1·2151	—	—	—
N_2O																	1·1054	—	—
CCl_3F																		1·7186	—
$CHClF_2$																			1·3871

$$B^* = (T^*)^{-\frac{1}{2}}\{-0{\cdot}7175 + 0{\cdot}2377 \ln T^* + 0{\cdot}50172(\ln T^*)^2 - 0{\cdot}1026(\ln T^*)^3 + 0{\cdot}0068(\ln T^*)^4\}\exp\left(\frac{1{\cdot}0582}{T^*}\right) \qquad 0{\cdot}5 < T^* < 130 \tag{A3.11}$$

where

$$T^* = kT/\varepsilon_c. \tag{A3.12}$$

These correlations are to be used in conjunction with the equations of Appendix 5.2. In Tables A3.1 and A3.2 we provide values of the scaling parameter ratios.

$$p = \varepsilon_c/\varepsilon_{Ar} \quad \text{and} \quad s = \sigma_c/\sigma_{Ar} \tag{A3.13}$$

for all the interactions so far studied in this way. The reference values of ε_{Ar} and σ_{Ar} are taken to be

$$\varepsilon_{Ar}/k = 141{\cdot}6\ \text{K} \qquad \sigma_{Ar} = 0{\cdot}3350\ \text{nm}. \tag{A3.14}$$

The accuracy of the correlation scheme is estimated as ±0·5 per cent for viscosity, ±2 per cent for the diffusion coefficient, and ±0·5 per cent for the thermal conductivity of monatomic gases and gas mixtures. The correlation cannot be directly applied to the thermal conductivity of polyatomic gases or mixtures containing polyatomic components.

References

1. Kestin, J., Ro, S. T., and Wakeham, W. A. *Physica* **58,** 165 (1972).
2. Kestin, J. and Mason, E. A. *AIP Conf. Proc.* **11,** 137 (1973).
3. Aziz, R. A. *J. chem. Phys.* **65,** 490 (1976).

APPENDIX 4

RECOMMENDED VALUES FOR THERMOPHYSICAL PROPERTIES OF SOME REPRESENTATIVE GASES

In this appendix we provide recommended values of the thermophysical properties of selected gases. The viscosity and the second virial coefficients available from direct measurements have been assessed and values at convenient temperature intervals are listed. It is intended that these data should provide a standard against which proposed intermolecular pair potentials for these molecules should be judged.

To this end the five gases have been chosen to represent various classes of molecules. Thus, helium represents a monatomic gas for which quantum effects are significant whereas argon represents a classical monatomic gas. Hydrogen is a diatomic gas for which quantum effects are important, and nitrogen represents a classical diatomic gas. Finally carbon tetrafluoride is an example of a globular molecule.

Recommended data. Table A4.1 lists values for the viscosity which have been obtained by smoothing the most reliable experimental data available.[1–5] For helium, argon, and nitrogen the estimated uncertainty in the values is ±0·5 per cent in the temperature range 300–1000 K and ±1·5 per cent outside of this range. For carbon tetrafluoride the uncertainty in the range 300–600 K is ±0·5 per cent, whereas it rises to ±1·5 per cent outside of it. For hydrogen there have been fewer measurements of equivalent accuracy and we have listed the values arising from critical evaluation of all the experimental data.[6] The uncertainty for the hydrogen viscosities is ±5 per cent below 80 K and ±1·5 per cent above that temperature.

Table A4.2 includes second virial coefficients for the same five gases, also taken from a critical compilation.[7] In this case the estimated uncertainties are included in the table.

Calibration data for the viscosity and thermal conductivity of gases. In Table A4.3 we provide data for the zero-density viscosity and thermal conductivity of some common gases at a single temperature. The zero-density viscosity, η_0, is the limiting value of the viscosity in the expansion

$$\eta(T, \rho) = \eta_0(T) + a_1(T)\rho + a_2(T)\rho^2 + \ldots$$

as $\rho \to 0$, and the zero-density thermal conductivity, λ_0, is the value of the thermal conductivity in the same limit of the expansion

$$\lambda(T, \rho) = \lambda_0(T) + b_1(T)\rho + b_2(T)\rho^2 + \ldots.$$

These two expansions are based on the theoretical description of transport coefficients in dense gases.[8] Higher-order terms in the expansions are not simple powers of the density, but for the present purposes this is unimportant.

The data given for the viscosity and thermal conductivity in the subsequent

Table A4.1

Recommended viscosity data for some representative gases at a pressure of 1 bar

Temperature T/K	Viscosity: $\eta/(\mu \mathrm{Pa\,s})$				
	He	Ar	H_2	N_2	CF_4
20	—	—	1·02	—	—
40	—	—	2·07	—	—
60	—	—	2·85	—	—
80	—	—	3·53	—	—
100	—	—	4·14	—	—
120	10·9	9·78	4·72	8·10	9·78
150	12·6	12·0		10·0	12·1
200	15·2	15·9	6·75	12·9	14·8
250	17·6	19·5		15·5	17·40
300	19·94	22·74	8·95	17·88	22·36
400	24·37	28·85	10·9	22·21	26·50
500	28·40	34·17	12·7	26·00	30·35
600	32·23	39·08	14·4	29·48	33·9
700	36·01	43·46	16·1	32·70	37·0
800	39·50	47·69	17·6	35·85	40·0
900	42·97	51·65	19·1	38·75	42·8
1000	46·18	55·43	20·5	41·40	45·4
1100	49·1	58·9	21·9	43·9	—
1200	52·1	62·4	23·3	46·3	—
1300	54·9	65·7	24·6	48·7	—
1400	57·8	68·9	25·9	51·1	—
1500	60·6	72·0	27·1	53·4	—
1600	63·3	75·1	28·3	55·6	—

tables are intended to serve as values with which an apparatus designed for the measurement of these properties in a relative manner can be calibrated. Consequently, we list the most accurately determined values of the viscosity and thermal conductivity at zero density and a single temperature together with the first density coefficients $a_1(T)$ and $b_1(T)$ for each transport coefficient. In addition, we provide the linear temperature correction for each transport coefficient at the reference temperature. All the values reported for the zero density transport coefficients are believed to be accurate to $\pm 0{\cdot}2$ per cent. The uncertainties in the density coefficients and temperature derivatives are of course larger. In the case of the monatomic gases the values reported for the thermal conductivity and viscosity which have been determined by direct measurement are internally consistent within their mutual uncertainty.

Table A4.2

Second virial coefficients for some representative gases

Helium		Argon		Hydrogen		Nitrogen		Carbon tetrafluoride	
T/K	B/cm^3 mol^{-1}	T/K	B/cm^3 mol^{-1}	T/K	B/cm^3 mol^{-1}	T/K	B/cm^3 mol^{-1}	T/K	B/cm^3 mol^{-1}
2·0	−174±8	85	−276±5	15	−230±5	75	−275±8	225	−172·5±1
2·5	−134±5	90	−251±3	17	−191±5	80	−243±7	250	−137·5±1
3·0	−109±2	95	−225±3	19	−162±5	90	−197±5	275	−109·0±0·5
3·5	−92·6±2	100	−202·5±2	22	−132±5	100	−160±3	300	−87·0±0·5
4·0	−80·2±2	125	−183·5±1	25	−110±3	110	−132±2	325	−69·0±0·5
5·0	−62·7±1	150	−154·5±1	30	−82±3	125	−104±2	350	−55·0±0·5
7·0	−40·9±1	200	−123·0±1	40	−52±2	150	−71·5±2	400	−32·0±0·5
10·0	−23·1±1	250	−84·7±1	50	−33±2	200	−35·2±1	450	−16·0±0·5
15·0	−10·8±1	300	−15·5±0·5	75	−12±1	250	−16·2±1	500	−4·0±0·5
20·0	−3·4±0·5	400	−1·0±0·5	100	−1·9±1	300	−4·2±0·5	600	14·0±0·5
30·0	2·5±0·5	500	7·0±0·5	150	7·1±0·5	400	9·0±0·5	700	25·0±0·1
50·0	7·4±0·5	600	12·0±0·5	200	11·3±0·5	500	16·9±0·5	800	33·0±0·1
100·0	11·7±0·5	700	15·0±1	300	14·8±0·5	600	21·3±0·5		
200·0	12·1±0·5	800	17·7±1	400	15·2±0·5	700	24·0±0·5		
400·0	11·2±0·5	900	20·0±1						
700·0	10·1±0·5	1000	22·0±1						

Table A4.3

(a) Viscosity

Gas	Temperature, T_r (°C)	η_0 (μPa s)	a_1 (nPa s m^3/kg)	$(\partial\eta_0/\partial T)T_r$ (μPa s/K)	Ref.
He	25	19·86	−10·64	0·045	1, 9
Ne	25	31·75	4·28	0·072	1, 9
Ar	25	22·60	11·10	0·064	1, 9
Kr	25	25·36	8·14	0·079	1, 9
Xe	25	23·05	6·17	0·079	9, 10
H_2	25	8·908	14·82	0·022	9, 10
N_2	25	17·78	10·56	0·047	9
O_2	25	20·61	9·98	0·055	9, 10
CO_2	25	14·91	1·23	0·046	9, 10
CH_4	25	11·07	17·76	0·033	10, 11

(b) Thermal conductivity

Gas	Temperature, T_r (°C)	λ_0 (mW/mK)	b_1 (μW m^2/K kg)	$(\partial\lambda_0/\partial T)T_r$ (mW/mK^2)	Ref.
He	27·5	155·9	291	0·350	12
Ne	27·5	49·45	34	0·111	12
Ar	27·5	17·74	21·6	0·050	12
Kr	27·5	9·496	8·7	0·029	12
Xe	27·5	5·509	6·0	0·019	12
H_2	27·5	188·8	94·7	0·475	14
N_2	27·5	26·03	36·2	0·063	13
CO_2	27·5	16·85	19	0·085	13
CH_4	27·5	34·93	74	0·092	13

References

1. Kestin, J., Ro, S. T., and Wakeham, W. A. *J. chem. Phys.* **56,** 4119 (1972).
2. Kestin, J., Ro, S. T. and Wakeham, W. A. *J. chem. Phys.* **56,** 5837 (1972).
3. Kestin, J., Khalifa, H. E., and Wakeham, W. A. *J. chem. Phys.* **67,** 4254 (1977).
4. Maitland, G. C. and Smith, E. B. *J. Chem. Soc. Faraday Trans. I* **70,** 1191 (1974).
5. Gough, D. W., Matthews, G. P., and Smith, E. B. *J. Chem. Soc. Faraday Trans. I,* **72,** 645 (1976).
6. Maitland, G. C. and Smith, E. B. *J. chem. engng Data* **17,** 150 (1972).
7. Dymond, J. H. and Smith, E. B. *The virial coefficients of pure gases and mixtures.* Clarendon Press, Oxford (1980).
8. Dorfman, J. R. and Cohen, E. G. D. *J. math. Phys.* **8,** 282 (1967).
9. Kestin, J. and Leidenfrost, W. *Physica* **25,** 1033 (1959).
10. Kestin, J., Ro, S. T., and Wakeham, W. A. *Trans. Faraday Soc.* **67,** 2308 (1971).

11. Kestin, J. and Yata, J. *J. chem. Phys.* **49,** 4780 (1968).
12. Kestin, J., Paul, R., Clifford, A. A., and Wakeham, W. A. *Physica* **100A,** 349 (1980).
13. Clifford, A. A., Kestin, J., and Wakeham, W. A. *Physica* **97A,** 287 (1979).
14. Clifford, A. A., Kestin, J., and Wakeham, W. A. *Ber. Bunsenges. Phys. Chem.* **84,** 9 (1980).

APPENDIX 5

CHARACTERISTIC PARAMETERS OF SOME SIMPLE SUBSTANCES

	T_c/K	V_c/cm^3 mol^{-1}	P_c/bar	T_B/K	B_0/cm^3 mol^{-1}	ω†
^{4}He	5·19	57·3	2·27			
Ne	44·40	41·7	27·6	127	9·0	
Ar	150·8	74·9	48·7	410	18·3	−0·002
Kr	209·4	91·2	55·0	577	22·3	−0·002
Xe	289·7	118	58·4	775	29·7	0·002
CH_4	190·5	99	46·1	509·3	25·2	0·013
CF_4	227·6	140	37·4	518·1	48·3	0·173
SF_6	318·7	198	37·6	693	72·0	0·209
$C(CH_3)_4$	433·7	303	32·0			0·195
N_2	126·2	89·5	33·9	327	22·8	0·040
O_2	154·6	73·4	50·4	405	18·7	0·021
N_2O	309·6	97·4	72·5			0·145
CO_2	304·2	94·0	73·8	713	30·2	0·225
C_2H_6	305·4	148	48·8			0·105
C_3H_8	369·8	203	42·5			0·152
C_2H_4	282·4	129	50·3			0·085

The critical parameters are taken from (1) Mathews, J. F. *Chem. Rev.* **72,** 71 (1972) (Inorganic substances); (2) Kudchadker, A. P., Alani, G. H., and Zwolinski, B. J. *Chem. Rev.* **68,** 659 (1968) (Organic compounds).

† The parameter ω is Pitzer's acentric factor which is described in Appendix 6.

APPENDIX 6

PITZER'S ACENTRIC FACTOR: CORRESPONDING STATES FOR SECOND VIRIAL COEFFICIENTS

Pitzer[1] showed that the deviations of complex molecules from the simple corresponding states behaviour shown by monatomic fluids could be correlated using a single additional parameter, the acentric factor, ω. This may be defined

$$\omega = -\log(P_s/P_c) - 1{\cdot}000$$

where P_s is the saturated vapour pressure of the substance at $T = 0{\cdot}700\ T_c$, and P_c is the critical pressure. This definition is chosen to make $\omega \sim 0$ for Ar, Kr, and Xe. Values for a number of substances are given in Appendix 5.

In the extended principle of corresponding states the reduced properties of a fluid are taken to be functions of three variables, the reduced temperature, T_R $(= T/T_c)$, reduced pressure, P_R $(= P/P_c)$ and the acentric factor ω. For second virial coefficients the analytical expressions given below were developed and were found to reproduce the results for many substances quite accurately. It should be noted however that the recent low-temperature (below $0{\cdot}6\ T_c$) measurements for the rare gases are not well reproduced.

$$\frac{BP_c}{RT_c} = (0{\cdot}1445 + 0{\cdot}073\omega) - (0{\cdot}330 - 0{\cdot}46\omega)T_R^{-1} - (0{\cdot}1385 + 0{\cdot}50\omega)T_R^{-2}$$
$$- (0{\cdot}0121 + 0{\cdot}097\omega)T_R^{-3} - 0{\cdot}0073\omega T_R^{-8}$$

References

1. Pitzer, K. S. and Curl, R. F., Jr. *J. Am. Chem. Soc.* **79,** 2369 (1957).

APPENDIX 7

LATTICE SUMS, L_n, FOR SOME CUBIC LATTICES

	L_n		
n	Face-centred (c.c.p.)	Hexagonal (h.c.p.)	Body-centred (b.c.c.)
4	25·3383	25·3391	22·6387
5	16·9675	16·9684	14·7585
6	14·4539	14·4549	12·2533
7	13·3594	13·3603	11·0542
8	12·8019	12·8028	10·3550
9	12·4925	12·4933	9·8945
10	12·3113	12·3119	9·5640
11	12·2009	12·2014	9·3133
12	12·1319	12·1323	9·1142
13	12·0877	12·0880	8·9518
14	12·0590	12·0592	8·8167
15	12·0400	12·0402	8·7030
16	12·0274	12·0275	8·6062
17	12·0198	12·0199	8·5236
18	12·0130	12·0131	8·4525
19	12·0093	12·0094	8·3914
20	12·0063	12·0063	8·3386

(c.c.p, b.c.c) Lennard-Jones, J. E. and Ingham, A. E. *Proc. roy Soc.* **A107,** 636 (1925); (h.c.p.) Barron, T. H. K. and Domb, C. *Proc. roy Soc.* **A227,** 447 (1955).

APPENDIX 8

COLLISION INTEGRALS AND SECOND VIRIAL COEFFICIENTS FOR NON-SPHERICAL INTERMOLECULAR POTENTIAL MODELS

The Stockmayer potential. The Stockmayer potential, which is a model for the interaction of two permanent dipolar molecules, may be written

$$U(r)=4\varepsilon'\left\{\left(\frac{\sigma'}{r}\right)^{12}-\left(\frac{\sigma'}{r}\right)^{6}\right\}-\frac{\mu^2}{4\pi\varepsilon_0 r^3}\zeta(\theta_1,\theta_2,\phi)$$

where $\zeta(\theta_1,\theta_2,\phi)=2\cos\theta_1\cos\theta_2-\sin\theta_1\sin\theta_2\cos\phi$. Here, μ is the dipole moment of the molecule, σ' a characteristic length for the potential, and ε' a characteristic energy. The angles θ_1 and θ_2 are the angles of inclination of the axes of the dipoles to the line joining their centres, whereas ϕ is the azimuthal angle between them.

The potential can also be written in the form

$$U(r)=4\varepsilon'\left\{\left(\frac{\sigma'}{r}\right)^{12}-\left(\frac{\sigma'}{r}\right)^{6}-\delta\left(\frac{\sigma'}{r}\right)^{3}\right\}$$

where

$$\delta=\frac{\mu^{*2}}{16\pi\varepsilon_0}\zeta(\theta_1,\theta_2,\phi)$$

and

$$\mu^*=\mu/(\varepsilon'\sigma'^3)^{\frac{1}{2}}.$$

Table A8.1 contains the reduced collision integrals $\langle\Omega^{(1,1)^*}(T^*)\rangle$ and $\langle\Omega^{(2,2)^*}(T^*)\rangle$ computed for this potential model according to the scheme proposed by Monchick and Mason.[1] The collision integrals have been reduced with respect to σ' and the temperature with respect to ε' so that

$$\langle\Omega^{(l,s)^*}(T^*)\rangle=\frac{\langle\bar{\Omega}^{(l,s)}(T)\rangle}{\pi\sigma'^2}$$

where

$$T^*=kT/\varepsilon'.$$

The collision integrals are tabulated for four values of the parameter δ_{max}, which is the maximum value that can be taken by δ,

$$\delta_{max}=\mu^{*2}/8\pi\varepsilon_0$$

and is therefore a measure of the dipole–dipole interaction energy.

Table A8.1

The collision integrals for the Stockmayer potential

T^* \ δ_{max}	0·25		0·50		1·0		2·0	
	$\langle\Omega^{(1,1)*}\rangle$	$\langle\Omega^{(2,2)*}\rangle$	$\langle\Omega^{(1,1)*}\rangle$	$\langle\Omega^{(2,2)*}\rangle$	$\langle\Omega^{(1,1)*}\rangle$	$\langle\Omega^{(2,2)*}\rangle$	$\langle\Omega^{(1,1)*}\rangle$	$\langle\Omega^{(2,2)*}\rangle$
0·10	4·002	4·266	4·655	4·833	6·454	6·729	9·824	10·34
0·20	3·164	3·305	3·355	3·516	4·198	4·433	6·225	6·637
0·30	2·657	2·836	2·770	2·936	3·319	3·511	4·785	5·126
0·40	2·320	2·522	2·402	2·586	2·812	3·044	3·972	4·282
0·50	2·073	2·277	2·140	2·329	2·472	2·665	3·437	3·727
0·60	1·885	2·081	1·944	2·130	2·225	2·417	3·054	3·329
0·70	1·738	1·924	1·791	1·970	2·036	2·225	2·763	3·028
0·80	1·622	1·795	1·670	1·840	1·886	2·070	2·535	2·788
0·90	1·527	1·689	1·572	1·733	1·765	1·944	2·349	2·596
1·00	1·450	1·601	1·490	1·644	1·665	1·838	2·196	2·435
1·20	1·330	1·465	1·364	1·504	1·509	1·670	1·956	2·181
1·40	1·242	1·365	1·272	1·400	1·394	1·544	1·777	1·989
1·60	1·176	1·289	1·202	1·321	1·306	1·447	1·639	1·838
1·80	1·124	1·231	1·146	1·259	1·237	1·370	1·530	1·718
2·00	1·082	1·184	1·102	1·209	1·181	1·307	1·441	1·618
2·50	1·005	1·100	1·020	1·119	1·080	1·193	1·278	1·435
3·00	0·9538	1·044	0·9656	1·059	1·012	1·117	1·168	1·310
3·50	0·9162	1·004	0·9256	1·016	0·9626	1·062	1·090	1·220
4·00	0·8871	0·9732	0·8948	0·9830	0·9252	1·021	1·031	1·153
5·00	0·8446	0·9291	0·8501	0·9360	0·8716	0·9628	0·9483	1·058
6·00	0·8142	0·8979	0·8183	0·9030	0·8344	0·9230	0·8927	0·9955
7·00	0·7908	0·8741	0·7940	0·8780	0·8066	0·8935	0·8526	0·9505
8·00	0·7720	0·8549	0·7745	0·8580	0·7846	0·8703	0·8219	0·9164
9·00	0·7562	0·8388	0·7584	0·8414	0·7667	0·8515	0·7976	0·8895
10·00	0·7428	0·8251	0·7446	0·8273	0·7515	0·8356	0·7776	0·8676
12·00	0·7206	0·8024	0·7220	0·8039	0·7271	0·8101	0·7464	0·8337
14·00	0·7029	0·7840	0·7039	0·7852	0·7078	0·7899	0·7228	0·8081
16·00	0·6880	0·7687	0·6888	0·7696	0·6919	0·7733	0·7040	0·7878
18·00	0·6753	0·7554	0·6760	0·7562	0·6785	0·7592	0·6884	0·7711
20·00	0·6642	0·7438	0·6648	0·7445	0·6669	0·7470	0·6752	0·7569
25·00	0·6415	0·7200	0·6418	0·7204	0·6433	0·7221	0·6490	0·7289
30·00	0·6236	0·7011	0·6239	0·7014	0·6249	0·7026	0·6291	0·7076
35·00	0·6089	0·6855	0·6091	0·6858	0·6099	0·6867	0·6131	0·6905
40·00	0·5964	0·6724	0·5966	0·6726	0·5972	0·6733	0·5998	0·6762
50·00	0·5763	0·6510	0·5764	0·6512	0·5768	0·6516	0·5785	0·6534
75·00	0·5415	0·6141	0·5416	0·6143	0·5418	0·6147	0·5424	0·6148
100·00	0·5181	0·5889	0·5182	0·5894	0·5184	0·5903	0·5186	0·5895

Table A8.2 contains reduced second virial coefficients for the Stockmayer potential computed according to the expansion[2]

$$B^*(T^*) = \left(\frac{4}{T^*}\right)^{\frac{1}{4}}\left\{\Gamma\left(\frac{3}{4}\right) - \frac{1}{4}\sum_{n=1}^{\infty}\sum_{k=0}^{k\leq n/2}\frac{2^{(n-k)}G_k}{(n-2k)!\,(2k)!\,(2k+1)}\right.$$
$$\left.\times\Gamma\left(\frac{2n-2k-1}{4}\right)\delta_{max}^{2k}/T^{*(n+k)/2}\right\}$$

Table A8.2

The classical virial coefficient for the Stockmayer potential, $B^*(T^*)$

$T^*\backslash\delta_{max}$	0·25	0·50	1·0	2·0
0·30	−39·1572	−102·22	—	—
0·40	−16·834 13	−29·6199	−235·0	—
0·50	−10·014 09	−14·809 93	−60·52	—
0·60	−6·895 75	−9·313 64	−26·714	−1141·0
0·70	−5·142 41	−6·582 59	−15·3703	−282·2
0·80	−4·027 43	−4·979 93	−10·217 48	−106·8
0·90	−3·258 83	−3·935 10	−7·404 11	−52·80
1·00	−2·698 14	−3·203 35	−5·671 95	−31·024
1·20	−1·936 77	−2·250 18	−3·689 83	−14·5675
1·40	−1·445 30	−1·659 24	−2·606 87	−8·6034
1·60	−1·102 76	−1·258 48	−1·932 37	−5·748 78
1·80	−0·850 94	−0·969 62	−1·475 11	−4·133 80
2·00	−0·658 40	−0·752 02	−1·146 28	−3·115 25
2·50	−0·331 61	−0·389 12	−0·627 15	−1·726 84
3·00	−0·128 18	−0·167 30	−0·327 65	−1·038 20
3·50	0·009 53	−0·018 91	−0·134 82	−0·635 67
4·00	0·108 22	0·086 55	−0·001 46	−0·375 53
4·50	0·181 94	0·164 82	0·095 54	−0·195 61
5·00	0·238 73	0·224 85	0·168 76	−0·064 98
6·00	0·319 68	0·309 98	0·270 87	0·109 69
7·00	0·373 70	0·366 50	0·337 55	0·219 04
8·00	0·411 58	0·406 02	0·383 64	0·292 48
9·00	0·439 12	0·434 67	0·416 81	0·344 29
10·00	0·459 67	0·456 02	0·441 41	0·382 20
12·00	0·487 37	0·484 78	0·474 42	0·432 59
14·00	0·504 18	0·502 25	0·494 49	0·463 21
16·00	0·514 64	0·513 13	0·507 08	0·482 72
18·00	0·521 13	0·519 92	0·515 05	0·495 49
20·00	0·525 04	0·524 04	0·520 04	0·503 95
25·00	0·528 30	0·527 63	0·524 98	0·514 31
30·00	0·526 77	0·526 29	0·524 39	0·516 75
35·00	0·523 11	0·522 75	0·521 31	0·515 54
40·00	0·518 48	0·518 20	0·517 07	0·512 55
50·00	0·508 30	0·508 11	0·507 36	0·504 34
75·00	0·484 10	0·484 01	0·483 65	0·482 19
100·00	0·464 05	0·464 00	0·463 78	0·462 91

where

$$G_k = \sum_{m=0}^{k} \frac{k!\,3^m}{(k-m)!\,(2m+1)}$$

and $B^*(T^*) = B(T)/(\frac{2}{3}\pi N_A \sigma'^3)$.

The calculations have been performed for the same values of δ_{max} as were employed in the evaluation of the collision integrals, and were checked using the alternative but equivalent formulation of Buckingham and Pople.[3] The dipole parameter, t^*, introduced in Chapter 3, and generally used in virial coefficient studies is related to δ_{max} by the equation

$$t^* = \delta_{max}/\sqrt{2}.$$

Table A8.3

Classical second virial coefficients for the diatomic 12−6 potential, $B^* = B/([2\pi/3]N_A\sigma^3)$

T^* \ R^*	0	0·1	0·2	0·3	0·4	0·5	0·6
0·40	−13·799	−12·839	−11·000	−9·342	−8·055	−7·074	−6·314
0·50	−8·720	−8·191	−7·132	−6·122	−5·298	−4·642	−4·115
0·60	−6·198	−5·846	−5·120	−4·400	−3·790	−3·287	−2·871
0·70	−4·710	−4·449	−3·899	−3·335	−2·843	−2·427	−2·074
0·80	−3·734	−3·527	−3·081	−2·613	−2·195	−1·834	−1·521
0·90	−3·047	−2·874	−2·497	−2·093	−1·725	−1·401	−1·116
1·00	−2·538	−2·389	−2·060	−1·701	−1·369	−1·072	−0·808
1·20	−1·836	−1·717	−1·450	−1·151	−0·866	−0·606	−0·369
1·40	−1·376	−1·276	−1·046	−0·784	−0·530	−0·293	−0·074
1·60	−1·052	−0·964	−0·760	−0·523	−0·290	−0·069	+0·137
1·80	−0·812	−0·733	−0·547	−0·328	−0·111	+0·098	0·294
2·00	−0·628	−0·555	−0·383	−0·178	+0·028	0·227	0·416
2·50	−0·313	−0·250	−0·101	+0·079	0·265	0·447	0·623
3·00	−0·115	−0·059	+0·075	0·241	0·413	0·585	0·752
4·00	+0·115	+0·164	0·281	0·428	0·584	0·742	0·897
5·00	0·243	0·287	0·395	0·531	0·677	0·825	0·973
6·00	0·323	0·364	0·464	0·593	0·732	0·874	1·017
8·00	0·413	0·451	0·543	0·661	0·791	0·924	1·059
10·00	0·461	0·496	0·582	0·694	0·818	0·946	1·075
15·00	0·511	0·542	0·621	0·723	0·837	0·955	1·076
20·00	0·525	0·555	0·629	0·726	0·833	0·946	1·061

Characteristic Boyle parameters

T^*_B	3·418	3·207	2·761	2·318	1·956	1·677	1·463
B^*_0	0·368	0·386	0·433	0·498	0·574	0·657	0·744
V^*_B	0·811	0·852	0·956	1·098	1·262	1·439	1·625

The diatomic Lennard-Jones (12−6) *potential.* The diatomic Lennard-Jones potential is a model for the interaction of two homonuclear diatomic molecules, and may be written

$$U(r) = \varepsilon \sum_{l=1}^{2} \sum_{m=1}^{2} \left\{ \left(\frac{\sigma}{r_{lm}}\right)^{12} - \left(\frac{\sigma}{r_{lm}}\right)^{6} \right\}.$$

Here ε and σ are characteristic parameters for the interaction between sites two of which are situated on each molecule a distance $R^*\sigma$ apart. Table A8.3 contains the reduced second virial coefficient

$$B^* = B(T)/(\tfrac{2}{3}\pi N_A\sigma^3)$$

as a function of $T^* = kT/\varepsilon$, for seven values of R^*. The table also includes the reduced Boyle temperature, T^*_B for which $B^*(T^*_B) = 0$ and the parameters (Chapter 3)

$$B^*_0 = -B^*(T^* = 0{\cdot}7\ T^*_B)$$

and

$$V_B^* = T_B^* \left(\frac{dB^*}{dT^*}\right)_{T_B^*}$$

References

1. Monchick, L. and Mason, E. A. *J. chem. Phys.* **35,** 1676 (1961).
2. Rowlinson, J. S. *Trans. Faraday Soc.* **45,** 974 (1949).
3. Buckingham, A. D. and Pople, J. A. *Trans. Faraday Soc.* **51,** 1173 (1955).

APPENDIX 9

THE INTERMOLECULAR PAIR POTENTIAL FOR ARGON†

Separation r/nm	Potential Energy $U(r)$/K	Separation r/nm	Potential Energy $U(r)$/K	Separation r/nm	Potential Energy $U(r)$/K
0·202	56 772·0	0·350	−98·40	0·452	−71·92
0·210	42 632·0	0·354	−113·57	0·456	−68·40
0·220	29 573·0	0·358	−124·88	0·460	−65·04
0·230	20 331·0	0·362	−132·97	0·470	−57·28
0·240	13 842·0	0·366	−138·40	0·480	−50·42
0·250	9321·0	0·370	−141·63	0·490	−44·37
0·260	6193·0	0·374	−143·07	0·500	−39·07
0·270	4047·0	0·376	−143·22	0·510	−34·44
0·274	3394·0	0·378	−143·05	0·520	−30·38
0·278	2835·0	0·380	−142·58	0·530	−26·84
0·282	2358·0	0·384	−140·89	0·540	−23·76
0·286	1950·0	0·388	−138·37	0·550	−21·08
0·290	1603·0	0·392	−135·20	0·560	−18·75
0·294	1309·0	0·396	−131·54	0·570	−16·70
0·298	1059·0	0·400	−127·52	0·580	−14·92
0·302	847·0	0·404	−123·24	0·590	−13·35
0·306	668·5	0·408	−118·79	0·600	−11·97
0·310	518·3	0·412	−114·24	0·620	−9·68
0·314	392·3	0·416	−109·65	0·640	−7·89
0·318	287·0	0·420	−105·07	0·660	−6·48
0·322	199·4	0·424	−100·53	0·680	−5·35
0·326	126·9	0·428	−96·08	0·700	−4·45
0·330	67·13	0·432	−91·72	0·720	−3·72
0·334	18·26	0·436	−87·48	0·740	−3·13
0·338	−21·41	0·440	−83·37	0·760	−2·64
0·342	−53·29	0·444	−79·41	0·780	−2·24
0·346	−78·60	0·448	−75·59	0·800	−1·91

† Based on the results of Aziz, R. A. and Chen, H. H. *J. chem. Phys.* **67,** 5719 (1977). $\varepsilon/k = 143{\cdot}22$K; $r_m = 0{\cdot}3759$ nm; $\sigma = 0{\cdot}335$ nm.

APPENDIX 10

PARAMETERS OF THE $n(\bar{r})-6$ POTENTIAL MODEL FOR SOME GASES

	m	γ	r_m/nm	(ε/k)/K	σ/nm	Quality	Ref.
He—He	12	6	0·2967	10·9	0·2633	I	2
He—Ne	13	8	0·3005	20	0·2690	II	9
He—Ar	13	4	0·343	29·4	0·3065	I	8
He—Kr	13	4	0·364	30·4	0·3253	I	8
He—Xe	13	5	0·398	27·4	0·3554	I	8
Ne—Ne	13	5	0·307 39	41·186	0·2745	I	3
Ne—Ar	13	9	0·343	66·9	0·3056	II	6
	13	4	0·3516	60	0·3142	II	9
	13	4	0·348	62·0	0·3110	I	10
Ne—Kr	13	9	0·358	71·8	0·3190	I	6
	13	5	0·3656	67·1	0·3265	I	10
Ne—Xe	13	9	0·374	74·5	0·3332	II	6
	13	5	0·3924	65·42	0·3504	I	10
Ar—Ar	13	7·5	0·3756	141·55	0·3350	I	4
Ar—Kr	13	9	0·3902	165	0·3477	I	7
	13	9	0·388	165	0·3457	I	6
Ar—Xe	13	10	0·406	187·4	0·3615	II	6
Kr—Kr	13	10	0·402	199·2	0·3580	I	5
Kr—Xe	13	10	0·418	231·1	0·3720	II	6
	13	8	0·4230	220	0·3771	II	9
Xe—Xe	13	11	0·436	275·3	0·3880	II	6
	13	11	0·426	281	0·3790	I/II	11

Potential energy function[1] $n = m + \gamma(r/r_m - 1)$.

References

1. Maitland, G. C. and Smith, E. B. *Chem. Phys. Lett.* **22,** 483 (1973).
2. Aziz, R. A., Nain, V. P. S., Carley, J. S., Taylor, W. L., and McConville, G. T., *J. chem. Phys.* **70,** 4330 (1979).
3. Aziz, R. A. *Chem. Phys. Lett.* **40,** 57 (1976).
4. Aziz, R. A. *J. chem. Phys.* **64,** 490 (1976).
5. Aziz, R. A. Private communication.
6. van den Biesen, J. J. H., Stokvis, F. A., van Veen, E. H., and van den Meijdenberg, C. J. N. *Physica* **100A,** 375 (1980).
7. Gough, D. W., Matthews, G. P., Smith, E. B., and Maitland, G. C., *Mol. Phys.* **29,** 1759 (1975).
8. Smith, K. M., Rulis, A. M., Scoles, G., Aziz, R. A., and Nain, V. *J. chem. Phys.* **67,** 152 (1977).
9. Maitland, G. C. and Wakeham, W. A. *Mol. Phys.* **35,** 1443 (1977).
10. Aziz, R. A. Private communication.
11. Maitland, G. C. and Smith, E. B. In *Seventh Symposium on thermophysical properties* (ed. A. Cezairlian), p. 412. Am Soc. Mech. Eng. (1977).

APPENDIX 11

FORTRAN-IV COMPUTER PROGRAM FOR THE EVALUATION OF SECOND VIRIAL COEFFICIENTS

```
C      *******************************************
C      EVALUATION OF THE SECOND VIRIAL COEFFICIENT
C      *******************************************
C
C
C
       DIMENSION X(600),PR(205,3),
      +          BTR(100),BT(100),TEMP(100),TEMPR(100),TS(100),D(20),
      +          DB(20),A(10),IHEAD(8)
       EQUIVALENCE (PR(1,1),UR(1)),(PR(1,2),DUR(1)),(PR(1,3),D2UR(1))
       COMMON AR(205),UR(205),DUR(205),D2UR(205),VP(10)
      +,M,CC,DD,E,EPSILON,BLAM,RM3,D1,D2,D3,D4,RMIN,CSIX,LONGR
      +,ISHORTR,IGAMMA,RCRIT,BIGA,BIGB,RATIO,NUM,XF,RSTAR
C
C
C
C***********************************************************************
C         **** ******* ********* ****** ****** ************* ****
C         THIS PROGRAM EVALUATES SECOND VIRIAL COEFFICIENTS, B(T),
C                       BY SIMPSON'S RULE INTEGRATION
C                       ** ********* **** ***********
C***********************************************************************
C
C
C
C--------INPUT DATA
C--------L          =NO. OF POTENTIALS FOR WHICH B(T) IS TO BE EVALUATED
C--------IHEAD      =HEADER FOR EACH POTENTIAL
C--------NUM        =NUMBER OF POTENTIAL POINTS TO BE READ IN (R,U(R))
C--------LONGR      =TYPE OF LONG RANGE EXTRAPOLATION
C        - SEE SUBROUTINE POTL
C--------ISHORTR    =TYPE OF SHORT RANGE EXTRAPOLATION
C        - SEE SUBROUTINE POTL
C--------IGAMMA     =GAMMA COEFFICIENT FOR N(R) - 6   (LONG RANGE
C        EXTRAPOLATION CALCULATION)
C--------CSIX       =C6 COEFFICIENT FOR CSIX/R**6   (LONG RANGE
C        EXTRAPOLATION CALCULATION)
C--------BIGA,BIGB =PARAMETERS IN SHORT RANGE EXTRAPOLATION PROCEDURES
C        - SEE SUBROUTINE POTL
C--------RMIN,RCRIT=LOWER AND UPPER VALUES OF R BEYOND WHICH AN
C        EXTRAPOLATION PROCEDURE MUST BE USED.  RMIN MUST BE CHOSEN
C        BETWEEN 2ND AND 3RD DATA POINTS AND RCRIT BETWEEN N-2 AND
C        N-1 DATA POINTS
C--------RATIO      =SIGMA/RM (WHERE RM=VALUE OF R FOR MINIMUM
C        FUNCTION VALUE)
C--------SIGMA      =VALUE OF R WHERE U(R) = 0
C--------EPSILON    =ABSOLUTE VALUE OF U(R)FOR MINIMUM FUNCTION VALUE
C--------BLAM       =LAMBDA (DE BOER PARAMETER)
C--------AR         =R,INTERMOLECULAR SEPARATION
C--------UR         =U(R),INTERMOLCULAR POTENTIAL ENERGY
C--------TS         =TSTAR = T/EPSILON
```

```
C--------M           =NO. OF SIMPSON'S RULE POINTS
C--------CC          =SIMPSON'S RULE - LOWER LIMIT
C--------DD          =SIMPSON'S RULE - UPPER LIMIT
C
      READ(5,1000)L
 1000 FORMAT(I2)
      WRITE(6,1002)
      WRITE(6,1001)
 1001 FORMAT(/,1X,"EVALUATION OF SECOND VIRIAL COEFFICIENTS"/)
      WRITE(6,1003)
 1002 FORMAT(1H1,"****************************************")
 1003 FORMAT(1X,"****************************************")
      N=0
    1 J=0
      NT=0
      READ(5,2001)IHEAD
      WRITE(6,2000)IHEAD
 2000 FORMAT(/8A10,//)
 2001 FORMAT(8A10)
      CALL COEFF
C
C--------TO READ IN EITHER NUMERICAL POTENTIAL FUNCTION OR COEFFICIENTS
C--------FOR ANALYTICAL POTENTIAL
C
      PI=3.1415926536
      VV=3.0/(32.0*PI*PI*SQRT(PI))
C
C--------READ IN NO. OF INTEGRATION POINTS, UPPER AND LOWER BOUNDS ON R*
C
      READ(5,5000)M,CC,DD
 5000 FORMAT(I4,2F4.2)
      WRITE(6,5400)M,CC,DD
 5400 FORMAT(1X,"INTEGRATION HAS USED",I4,1X,"POINTS IN RANGE R* =",
     +F7.2,1X,"TO",F5.1,/)
C
C--------CALCULATE RANGE AND INTERVAL FOR SIMPSONS RULE
C
      Y=DD-CC
      YY=Y/M
C
C--------SET SIMPSONS RULE COEFFICIENTS, X
C
      DO 200 I=2,M,2
      X(I)=4.0
      X(I+1)=2.0
  200 CONTINUE
      X(1)=1.0
      X(M+1)=1.0
      NT=NT+1
C
C--------READ IN NUMBER OF T* VALUES
C
      READ(5,5300)NTSTAR
 5300 FORMAT(I4)
      WRITE(6,6000)
 6000 FORMAT(5X,"T*",15X,"B*",15X,"T*.DB*/DT*",13X,"JT",13X,"B1(Q)",
     +13X,"B2(Q)",13X,"B0(Q)"/)
C
C--------READ IN T* VALUES
C
      DO 500 NT=1,NTSTAR
      READ(5,5500)TS(NT)
 5500 FORMAT(F10.6)
      EE=1.0/TS(NT)
      J=J+1
C
C--------RSTAR=R/SIGMA
```

```
C
      RSTAR=CC
      G=0.0
      H=0.0
      B=0.0
      C=0.0
      MM=M+1
C
C
      DO 300 I=1,MM
      XF=1.0/RSTAR
      CALL POTL
C
C--------CALCULATES U(R*) AND ITS DERIVATIVES FOR THIS R*
C
      XF=RSTAR**2
      V=-EE*VP(1)
      IF(-100.0-V)21,22,22
   21 Z=EXP(V)
      ZZ=Z-1.0
      GO TO 23
C
C--------IF U(R*) LARGE, SET EXP(-U(R*)/T*) = Z = 0
C
   22 Z=0.0
      ZZ=-1.0
   23 DF=X(I)
C
C--------EVALUATION OF B INTEGRAL (G) IN RANGE CC < R* < DD
C--------H=T*.DB*/DT*
C--------R=JOULE-THOMSON COEFFICIENT = T*.DB*/DT* - B*
C--------B,C,U = QUANTUM CORRECTION FACTORS
C
      G=G+XF*DF*ZZ
      H=H+DF*Z*XF*VP(1)
      FF=XF*Z*VP(2)
      B=B+DF*FF*VP(2)
      A1=Z*VP(3)
      A2=2.0*Z*VP(2)
      A3=1.1111111111*EE*RSTAR*Z*VP(2)
      A4=-0.138888888889*EE*EE*Z*VP(2)*XF
      A1=XF*A1*VP(3)
      A2=A2*VP(2)
      A3=A3*VP(2)*VP(2)
      A4=A4*VP(2)**3
      A5=A1+A2+A3+A4
      C=C+DF*A5
      RSTAR=RSTAR+YY
  300 CONTINUE
C
C
C
C--------EVALUATION OF HH, THE CONTRIBUTION TO B INTEGRAL FROM R* > DD,
C        ASSUMING U(R) = C6EFF/R**6
C
      RSTAR=RSTAR-YY
      XF=RSTAR*RSTAR
      C6EFF=XF**3*VP(1)
      HH=C6EFF*EE*(1.0/(XF*RSTAR))
C
C--------CONRIBUTION TO B INTEGRAL FROM R* < CC IS CC**3 (ASSUMES
C        U(R* < CC) = INFINITY)
C--------G IS B INTEGRAL EVALUATED FOR 0 < R* < INFINITY
```

```
C
      G=-YY*G+HH+CC**3
      H=-H*YY*EE-HH
      FF=48.0*PI*PI
      B=YY*B*EE**3/FF
      A6=1920.0*PI**4
      C=-YY*EE**4*C/A6
      R=H-G
      Q=SQRT(EE)
      U=VV*Q*EE
C
C--------PRINT OUT REDUCED VIRIAL COEFFICIENTS AND QUANTUM CORRECTIONS
C
      WRITE(6,7000)TS(NT),G,H,R,B,C,U
 7000 FORMAT(1X,F7.3,7X,E15.8,5(5X,E13.6))
C
C--------EVALUATION OF VIRIAL COEFFICIENTS (M**3/MOLE * 10**-6)
C         INCLUDING QUANTUM CORRECTIONS
C
      TEMPR(J)=TS(NT)
      TEMP(J)=TS(NT)*EPSILON
      BTR(J)=G+BLAM*B+BLAM**2*C
      BT(J)=BTR(J)*RM3
      C=0.0
  500 CONTINUE
C
C
C
      N=N+1
      WRITE(6,1100)
 1100 FORMAT(/,39X,"T*",8X,"TEMP",11X,"B*+B(QU)",7X,"B(T)",/,50X,"(K)"
     +,22X,"(CM**3/MOLE)",/)
      WRITE(6,1200)(TEMPR(I),TEMP(I),BTR(I),BT(I),I=1,J)
 1200 FORMAT(35X,F7.3,F13.2,5X,F12.6,F12.2)
C
C--------REPEAT PROGRAM FOR NEXT POTENTIAL FUNCTION
C
      IF(N-L.GT.0)GO TO 1
      WRITE(6,9100)
 9100 FORMAT(///)
      STOP
      END
C
C
C
C
      SUBROUTINE COEFF
C
C--------IF THE VALUE OF NUM IS POSITIVE, READ IN NUMERICAL
C--------POTENTIAL VALUES.  IF IT IS NEGATIVE READ IN COEFFICIENTS
C--------FOR ANALYTICAL EQUATION FOR POTENTIAL.
C--------SPECIFIC ANALYTIC POTENTIAL USED HERE IS A MIE-LENNARD-JONES
C--------(N-M) FUNCTION.
C
C
      COMMON AR(205),UR(205),DUR(205),D2UR(205),VP(10)
     +,M,CC,DD,E,EPSILON,BLAM,RM3,D1,D2,D3,D4,RMIN,CSIX,LONGR
     +,ISHORTR,IGAMMA,RCRIT,BIGA,BIGB,RATIO,NUM,XF,RSTAR
      READ(5,2500)NUM
 2500 FORMAT(I6)
      READ(5,2700) SIGMA,EPSILON,BLAM
 2700 FORMAT(3F10.4)
C
C--------CALCULATE RM3 ....RM3=2/3*PI*N*SIGMA**3
```

```
C
      RM3=SIGMA**3*1.26163
      IF(NUM.GE.0) GO TO 30
      READ(5,2600)D1,D2,D3,D4
 2600 FORMAT(4E15.6)
      RETURN
   30 READ(5,3000)LONGR,ISHORTR,IGAMMA
 3000 FORMAT(3I6)
      READ(5,3100)BIGA,BIGB,RMIN,RCRIT,RATIO,CSIX
 3100 FORMAT(6F10.4)
      READ(5,4000)(AR(JV),UR(JV),JV=1,NUM)
 4000 FORMAT(2F15.8)
C
C--------CALCULATE SLOPES AT MID POINTS OF GIVEN VALUES
C--------STORE IN DUR(I)
C
      DUR(1)=(UR(2)-UR(1))/(AR(2)-AR(1))
      DO 100 JV=2,NUM-1
      A9=(UR(JV+1)-UR(JV))/(AR(JV+1)-AR(JV))
      B=(UR(JV)-UR(JV-1))/(AR(JV)-AR(JV-1))
      DUR(JV)=(A9+B)*0.5
  100 CONTINUE
      DUR(NUM)=(UR(NUM)-UR(NUM-1))/(AR(NUM)-AR(NUM-1))
C
C--------CALCULATE SECOND DERIVATIVES
C
      D2UR(1)=(DUR(2)-DUR(1))/(AR(2)-AR(1))
      DO 150 JV=2,NUM-1
      A9=(DUR(JV+1)-DUR(JV))/(AR(JV+1)-AR(JV))
      B=(DUR(JV)-DUR(JV-1))/(AR(JV)-AR(JV-1))
      D2UR(JV)=(A9+B)*0.5
  150 CONTINUE
      D2UR(NUM)=(DUR(NUM)-DUR(NUM-1))/(AR(NUM)-AR(NUM-1))
      RETURN
      END
C
C
C
C
      SUBROUTINE POTL
C
C
C--------IF THE VALUE OF NUM IS POSITIVE, INTERPOLATES NUMERICAL
C--------POTENTIAL VALUES, AND 1ST AND 2ND DERIVATIVES FOR CURRENT
C--------VALUE OF RSTAR.  IF IT IS NEGATIVE CALCULATES POTENTIAL
C--------AND 1ST AND 2ND DERIVATIVES USING ANALYTICAL FUNCTION.
C--------SPECIFIC ANALYTIC POTENTIAL USED HERE IS A MIE-LENNARD-JONES
C--------(N-M) FUNCTION.
C
C
      DIMENSION PR(205,3)
      EQUIVALENCE (PR(1,1),UR(1)),(PR(1,2),DUR(1)),(PR(1,3),D2UR(1))
      COMMON AR(205),UR(205),DUR(205),D2UR(205),VP(10)
     +,M,CC,DD,E,EPSILON,BLAM,RM3,D1,D2,D3,D4,RMIN,CSIX,LONGR
     +,ISHORTR,IGAMMA,RCRIT,BIGA,BIGB,RATIO,NUM,XF,RSTAR
      IF(NUM.GE.0)GO TO 29
C
C--------ANALYTIC POTENTIAL
C
      VP(1)=D3*XF**D1+D4*XF**D2
C
C--------1ST DERIVATIVE
C
      VP(2)=-D3*D1*XF**(D1+1.0)-D4*D2*XF**(D2+1.0)
```

```
C
C--------2ND DERIVATIVE
C
      VP(3)=D3*D1*(D1+1.0)*XF**(D1+2.0)+D4*D2*(D2+1.0)*XF**(D2+2.0)
      RETURN
   29 IF(RSTAR-RMIN)5,5,6
    5 GOTO(7,8,9)ISHORTR
C
C--------SHORT RANGE EXTRAPOLATION CALCULATION FOR R*.LE.RMIN
C--------CALCULATES U(R) AND ITS FIRST AND SECOND DERIVATIVES OF U(R)
C        WITH RESPECT TO R* WHERE R*=R/SIGMA
C--------FOR ISHORTR=1, U(R) = (A+B(R/RM1))
C--------FOR ISHORTR=2, U(R) = A EXP(-BR)
C--------FOR ISHORTR=3, U(R) = A EXP(-BR**3)
C
    7 AA=RATIO*RSTAR-1.0
      BB=EXP(-12.5*AA)
      CC=BIGA+BIGB*AA
      VP(1)=CC*BB
      VP(2)=RATIO*(-12.5*VP(1)+BIGB*BB)
      VP(3)=BIGB*BB*(1.0-2.0*RATIO**2*12.5)+VP(1)*(12.5*RATIO)**2
      RETURN
C
    8 VP(1)=EXP(BIGA+BIGB*RSTAR)
      VP(2)=BIGB*VP(1)
      VP(3)=BIGB**2*VP(1)
      RETURN
C
    9 VP(1)=EXP(BIGA+BIGB*RSTAR**3)
      VP(2)=VP(1)*BIGB*3.0*RSTAR**2
      VP(3)=VP(1)*(6.0*BIGB*RSTAR+9.0*BIGB**2*RSTAR**4)
      RETURN
C
C--------FOR RMIN < R* < RCRIT, INTERPOLATE U(R*) AND ITS FIRST TWO
C--------DERIVATIVES FROM THE NUMERICAL LISTS.
C
    6 IF(RSTAR-RCRIT)11,11,12
   11 JI=1
   13 IF(AR(JI)-RSTAR)14,15,15
   14 JI=JI+1
      GO TO 13
   15 JI=JI-1
      C3=AR(JI)
      C4=AR(JI+1)
      C5=AR(JI+2)
C
      DO 400 K=1,3
      C9=PR(JI,K)
      C1=PR(JI+1,K)
      C2=PR(JI+2,K)
      C6=C9+(C1-C9)*(RSTAR-C3)/(C4-C3)
      C7=C1+(C2-C1)*(RSTAR-C4)/(C5-C4)
      VP(K)=C6+(C7-C6)*(RSTAR-C3)/(C5-C3)
  400 CONTINUE
C
      RETURN
C
C--------FOR R* > RCRIT USE LONG RANGE EXTRAPOLATON PROCEDURE
C--------IF LONGR=1, U(R) = -CSIX/R**6
C--------IF LONGR=2, U(R) = 6R**-N/(N-6) - NR**-6/(N-6)
C        WHERE N=13 + GAMMA (R/RM - 1)
C
   12 GO TO (16,17,10)LONGR
   10 RETURN
C
   16 VP(1)=-CSIX*XF**6
      VP(2)=-6.0*VP(1)*XF
      VP(3)=42.0*VP(1)*XF**2
      RETURN
```

```
C
   17 RSTAR=RSTAR*RATIO
      IF(RSTAR-6.0)19,19,20
   20 BB=60.0
      GO TO 28
   19 BB=13.0+IGAMMA*(RSTAR-1.0)
   28 DEN=BB-6.0
      VP(1)=6.0*(1.0/RSTAR)**BB/DEN-BB*(1.0/RSTAR)**6/DEN
      WP1=6.0*IGAMMA*RSTAR/DEN**2+6.0*BB/DEN+6.0*IGAMMA*RSTAR*
     +ALOG(RSTAR)/DEN
      WP1=-WP1*(1.0/RSTAR)**(BB+1.0)
      WP2=-(1.0/RSTAR)**7*(6.0*BB/DEN+6.0*IGAMMA*RSTAR/DEN**2)
      VP(2)=WP1-WP2
      ZP1=-IGAMMA*ALOG(RSTAR)-(BB+1.0)/RSTAR+DEN**2/(-6.0*IGAMMA*
     +RSTAR-6.0*BB**2+36.0*BB-6.0*IGAMMA*RSTAR*ALOG(RSTAR)*DEN)*
     +(6.0*IGAMMA/DEN)*(-1.0/DEN+2.0*IGAMMA*RSTAR/DEN**2-1.0+BB/DEN-
     +ALOG(RSTAR)-(1.0/RSTAR)+IGAMMA*ALOG(RSTAR)/DEN)
      ZP1=ZP1*WP1
      ZP2=-7.0*WP2/RSTAR+(1.0/RSTAR)**7*(-6.0*IGAMMA*RSTAR/DEN+IGAMMA*
     +6.0*BB/DEN**2-6.0*IGAMMA/DEN**2+12.0*IGAMMA**2*RSTAR/DEN**3)
      VP(3)=ZP1-ZP2
      RSTAR=RSTAR/RATIO
      RETURN
      END
```

APPENDIX 12

FORTRAN-IV COMPUTER PROGRAM FOR THE EVALUATION OF COLLISION INTEGRALS

```
C
C
C--------THIS PROGRAM EVALUATES TRANSPORT COLLISION INTEGRALS,
C--------OMEGA(L,S), FOR A NUMERICAL OR AN ANALYTIC PAIR POTENTIAL
C--------ENERGY FUNCTION U(R), USING THE METHOD OF BARKER, FOCK AND
C--------SMITH, PHYS. FLUIDS, 7, 897 (1964).  IT WAS COMPILED AND RUN
C--------WITHOUT ERROR ON THE CDC6500 AND CYBER 174 AT IMPERIAL
C--------COLLEGE, LONDON.
C
C
C--------THE MAIN PROGRAM
C--------    (1) SETS THE VALUES OF TSTAR (= REDUCED TEMPERATURE,
C                T X BOLTZMANN'S CONSTANT/EPSILON, WHERE EPSILON IS THE
C                WELL DEPTH) AT WHICH THE OMEGA(L,S) ARE TO BE EVALUATED
C                (HERE TSTAR = 1(1)10)
C--------    (2) CALLS THE COLLISION INTEGRAL SUBROUTINES VIA
C                COLLIS(TSTAR, II)
C--------    (3) OUTPUTS THE OMEGA(L,S) VIA SUBROUTINE SHOUT(TSTAR)
C
C
C
C
      DIMENSION TSTAR(100),IHEAD(13)
      COMMON/COLL/OMEGA(3,5)
      INTEGER IHEAD
      WRITE(6,13)
      WRITE(6,12)
   12 FORMAT(35X,64H* COLLISION CROSS-SECTIONS (Q) AND COLLISION (OMEG
     +A) INTEGRALS *)
      WRITE(6,13)
   13 FORMAT(35X,64(1H*))
C
C--------READ HEADING FOR PROGRAM.
C
      READ(5,14)IHEAD
   14 FORMAT(13A10)
      WRITE(6,15)IHEAD
   15 FORMAT(//13A10)
C
C--------READ IN NUMBER OF TSTAR VALUES TO BE USED.
C
      READ(5,10) N
   10 FORMAT(I4)
      DO 5 II = 1,N
      READ(5,11)TSTAR(II)
   11 FORMAT(F5.2)
      CALL COLLIS(TSTAR,II)
      CALL SHOUT(TSTAR,II)
    5 CONTINUE
      STOP
      END
```

```
C
C--------THIS SUBROUTINE OUTPUTS THE COLLISION INTEGRALS
C--------OMEGA(L,S), L = 1 - 3, S = 1 - 5.
C
      SUBROUTINE SHOUT(TSTAR,II)
      DIMENSION TSTAR(100)
      COMMON/COLL/OMEGA(3,5)
      WRITE(6,11)TSTAR(II), (OMEGA(1,J),J=1,5),OMEGA(2,2),OMEGA(2,3),
     + OMEGA(2,4),OMEGA(3,3)
   11 FORMAT(//,48X,19HCOLLISION INTEGRALS,/,48X,19(1H*)
     + ,//,8X,2HT*,10X,3H1,1,9X,3H1,2,9X,3H1,3,9X,3H1,4,9X,
     + 3H1,5,9X,3H2,2,9X,3H2,3,9X,3H2,4,9X,3H3,3,//,10(3X,F9.4))
      RETURN
      END
C
C                  *********************************
C                  *COLLISION INTEGRAL SUBROUTINES*
C                  *********************************
C
C
C--------THESE SUBROUTINES CALCULATE THE COLLISION INTEGRALS FOR A
C--------GIVEN REDUCED TEMPERATURE AND INTERMOLECULAR POTENTIAL.
C--------THE CALCULATION METHOD AND VARIABLE NAMES ARE BASED ON THE
C--------PAPER BY BARKER ET AL. (PHYSICS OF FLUIDS, 1964, 7, 897).
C--------BRIEFLY, THE ANGLE INTEGRATION IS DONE BY THE TRAPEZIUM RULE
C--------AND THE CROSS-SECTION AND COLLISION INTEGRALS ARE OBTAINED
C--------USING A FIVE POINT GAUSS - LEGENDRE INTEGRATION TECHNIQUE.
C--------THE SUBROUTINES (AND A BRIEF DESCRIPTION) ARE
C                 COLLIS   - POINTS ALONG THE REDUCED ENERGY AXIS
C                            AT WHICH THE INTEGRAND IS TO BE EVALUATED
C                            ARE SET UP.  THE CROSS-SECTION INTEGRALS
C                            ARE EVALUATED FOR THESE ENERGIES.
C                 SOLVE    - CALCULATES THE PARAMETERS REQUIRED TO DEAL
C                            WITH THE ORBITING CONDITIONS.
C                 OMGINT   - EVALUATES THE COLLISION INTEGRALS.
C                 FIPSI    - EVALUATES THE POTENTIAL VALUE AT A W*, AND
C                            CRITICAL ENERGY FOR THE ORBITING CONDITION.
C                 CHIINT   - EVALUATES THE ANGLE INTEGRATION.
C                 COEFF    - INPUTS THE PARAMETERS NEEDED FOR THE
C                            POTENTIAL PROCEDURE FIPSI.
      SUBROUTINE COLLIS(TSTAR,II)
C
C
C--------THE VARIABLES USED IN THIS SUBROUTINE ARE
C
C--------GSTAR(J)    - POINTS ON REDUCED ENERGY AXIS.
C--------GWEIGH(J)   - GAUSS WEIGHT FOR CORRESPONDING GSTAR(J).
C--------QSTAR(L,J)  - L TH CROSS-SECTION AT CORRESPONDING GSTAR(J).
C--------WSTAR(J)    - POINTS ALONG THE REDUCED INTERMOLECULAR
C                      DISTANCE AXIS.  WSTAR = 1/(REDUCED DISTANCE).
C--------IPSI(J)     - POTENTIAL ENERGY AT A DISTANCE CORRESPONDING
C                      TO 1/WSTAR.
C--------BSTAR(J)    - POINTS ALONG REDUCED IMPACT PARAMETER AXIS.
C                      BSTAR = (RED. IMPACT PARAMETER) SQUARED.
C--------BWEIGH(J)   - GAUSS WEIGHT FOR CORRESPONDING BSTAR(J).
C--------PIVOT(J)    - GAUSS PARAMETER TO SET UP POINTS ALONG AXIS OF
C                      INTEGRATION VARIABLE.
C--------WEIGHT(I)   - THE WEIGHT APPLIED IN THE GAUSS INTEGRATION.
C--------NGPNTS      - NO. OF POINTS ON GSTAR AXIS.
C--------GLASTV      - FINAL VALUE ON GSTAR AXIS USED IN NUMERICAL
C                      INTEGRATION (USED FOR EXTRAPOLATING TAIL)
C--------NWST1       - NUMBER OF WSTAR VALUES OF SPACING WSTEP1
C--------NWST2       - NUMBER OF WSTAR VALUES OF SPACING WSTEP2
C--------WSTEP1      - SPACING OF FIRST NWST1  WSTAR VALUES
C                      (I.E. TAKES WSTAR FROM ZERO TO NWST1 * WSTEP1)
C--------WSTEP2      - SPACING OF WSTAR VALUES FROM
C                      WSTAR = (NWST1 * WSTEP1) TO (NWST1 * WSTEP1
C                      + NWST2 * WSTEP2)
```

```
C--------BRGN        - ZERO TO BRGN IS REGION OF (IMPACT DIAMETER)**2
C                      OVER WHICH CROSS-SECTION INTEGRATION IS
C                      DONE NUMERICALLY.  VALUES OUTSIDE THIS REGION
C                      CONTRIBUTE AN INSIGNIFICANT AMOUNT TO THE
C                      INTEGRAL.
C--------NBSTEP      - NO. OF 5-POINT GAUSS PANELS USED IN
C                      CROSS-SECTION INTEGRATION.
C--------GSTEP       - INITIAL GSTAR PANEL WIDTH.
C--------NGSTEP      - REPEAT NO. OF 5 POINT PANELS (USED IN
C                      GSTAR INTEGRATION) BEFORE WIDTH IS DOUBLED.
C--------NGPANL      - NO. OF TIMES PANEL WIDTH IS DOUBLED IN
C                      INTEGRATION OVER GSTAR.
C--------IFLAG       - FLAG TO INDICATE WHETHER SUBROUTINES HAVE BEEN
C                      CALLED BEFORE IN MAIN PROGRAM (IFLAG=1).  IF SO
C                      THE CROSS-SECTION INTEGRATION NEED NOT BE DONE
C                      FOR EVERY REDUCED TEMPERATURE.
C--------TSTAR       - REDUCED TEMPERAURE.
C--------RUNW        - RUNNING VARIABLE FOR WSTAR.
C--------RUNG        - RUNNING VARIABLE FOR GSTAR.
C--------GTCRIT      - GREATEST VALUE OF CRITICAL ORBITING ENERGY.
C--------WCRIT       - WSTAR VALUE AT WHICH GTCRIT OCCURS.
C--------BCRIT       - CRITICAL ORBITING IMPACT PARAMETER.
C--------NBPNTS      - NO. OF POINTS ALONG BSTAR AXIS.
C--------BRUN        - RUNNING VARIABLE FOR BSTAR.
C--------UPBSBD      - UPPER BSTAR BOUND FOR AN INTEGRATION WHERE
C                      ORBITING OCCURS.
C--------BSTEP       - BSTAR STEP SIZE WIDTH.
C--------CHI         - DEFLECTION ANGLE FOR GIVEN BSTAR AND GSTAR.
C--------BINTD1      - INTEGRAND FOR FIRST CROSS-SECTION
C--------BINTD2      - INTEGRAND FOR SECOND CROSS-SECTION.
C--------BINTD3      - INTEGRAND FOR THIRD CROSS-SECTION.
C
C
      DIMENSION GSTAR(100),GWEIGH(100),QSTAR(3,100),WSTAR(150),

     + IPSI(150),BSTAR(150),BWEIGH(150),PIVOT(5),WEIGHT(5),OMEGA
     + (3,5),TSTAR(100)
      REAL IPSI
      COMMON/BLK1/WSTAR,IPSI
      COMMON/BLK2/QSTAR,GSTAR,GWEIGH,NGPNTS,GLASTV
      DATA PIVOT/0.04691008,0.23076535,0.5,0.76923466,0.95308992/
      DATA WEIGHT/0.11846344,0.23931433,0.28444444,0.23931433,
     + 0.118463444/
C
C--------READ IN DATA FOR VARIABLES NWST1,NWST2,WSTEP1,WSTEP2,
C--------BRGN,NGSTEP,GSTEP,NGSTEP,NGPANL
C--------TYPICAL VALUES FOR RESULTS ACCURATE TO 0.1 PERCENT
C--------IN RANGE TSTAR = 0.3 - 50 ARE:
C        NWST1 = 30, NWST2 = 80, NGSTEP = 2, NGPANL = 8,
C        NBSTEP = 10, WSTEP1 = 0.025, WSTEP2 = 0.0125,
C        BRGN = 9.0, GSTEP = 0.5
C        (FOR TSTAR GREATER THAN 100, USE NGPANL = 10)
C        N.B. NBSTEP MUST BE ≤ 30, NGSTEP * NGPANL ≤ 20.
C
      READ(5,10)NWST1,NWST2,NGSTEP,NGPANL,NBSTEP
      READ(5,11)WSTEP1,WSTEP2,BRGN,GSTEP
   10 FORMAT(5I4)
   11 FORMAT(4F7.4)
      NORB = 0
      IF(II .NE. 1) GOTO 19
      CALL COEFF
C
C
C--------SET UP OF WSTAR POINTS.
```

```
C
C
      DO 1 I = 1,NWST1
    1 WSTAR(I) = FLOAT(I)*WSTEP1
      RUNW = WSTAR(NWST1)
      DO 2 I = 1,NWST2
    2 WSTAR(NWST1+I) = RUNW + FLOAT(I)*WSTEP2
      NWPNTS = NWST1+NWST2
C
C
C--------SET UP OF GSTAR POINTS.
C
C
      RUNG = 0.0
      NGPNTS = 0
      DO 3 I = 1,NGPANL
      DO 4 J = 1,NGSTEP
      DO 5 K = 1,5
      NGPNTS = NGPNTS+1
      GINCR = PIVOT(K)*GSTEP
      GSTAR(NGPNTS) = GINCR+RUNG
    5 GWEIGH(NGPNTS) = WEIGHT(K)*GSTEP
    4 RUNG = RUNG+GSTEP
    3 GSTEP = GSTEP+GSTEP
      GLASTV = RUNG
C
C
C--------EVALUATION OF IPSI(I) FOR WSTAR(I), AND OF GTCRIT.
C
C
      GTCRIT = 0.
      WCRIT = 0.
      DO 6 I = 1,NWPNTS
      CALL FIPSI(WSTAR(I),IPSI(I),CRIT)
      IF(GTCRIT .GT. CRIT) GOTO 6
      WCRIT = WSTAR(I)
      GTCRIT = CRIT
    6 CONTINUE
C
C
C--------CROSS-SECTION INTEGRATION.
C
C
C--------FOR EACH GSTAR THREE CROSS-SECTIONS ARE EVALUATED.
C
C
      DO 7 J = 1,NGPNTS
      DO 8 L = 1,3
    8 QSTAR(L,J) = 0.0
      IF(GSTAR(J) .GT. GTCRIT) GOTO 9
C
C
C--------IF GSTAR IS GREATER THAN GRSTCRIT THEN ORBITING DOES NOT
C--------OCCUR AT THIS ENERGY.
C
C
      CALL SOLVE(WCRIT,GSTAR(J),BCRIT)
      IF(BCRIT .LE. 0.0) GOTO 9
C
C
C--------SET UP OF BSTAR POINTS IF ORBITING OCCURS.
C--------INTEGRATION OVER BSTAR DONE IN THREE STAGES
C--------BSTAR = 0 - 0.9*BCRIT, 0.9 - 1.1*BCRIT, 1.1*BCRIT - BRGN
```

```
C
C
      NORB = NORB + 1
      NBPNTS = 0
      BRUN = 0.0
      UPBSBD = 0.9*BCRIT
      BSTEP = UPBSBD/FLOAT(NBSTEP)
   20 DO 21 J1 = 1,NBSTEP
      DO 12 K = 1,5
      NBPNTS = NBPNTS+1
      BINCR = BSTEP*PIVOT(K)
      BSTAR(NBPNTS) = BRUN+BINCR
   12 BWEIGH(NBPNTS) = BSTEP*WEIGHT(K)
   21 BRUN = BRUN+BSTEP
      BSTEP = 0.2*BCRIT/FLOAT(NBSTEP)
      IF(10*NBSTEP .GT. NBPNTS) GO TO 20
      BSTEP = BRGN/FLOAT(NBSTEP)
      IF(15*NBSTEP .GT. NBPNTS) GO TO 20
      GOTO 15
C
C
C--------SET UP OF BSTAR POINTS IF ORBITING DOES NOT OCCUR.
C--------INTEGRATION DONE IN ONE STAGE, BSTAR = 0 - BRGN.
C
C
    9 BRUN = 0.0
      NBPNTS = 0
      BSTEP = BRGN/FLOAT(NBSTEP)
      DO 13 J1 = 1,NBSTEP
      DO 14 K = 1,5
      NBPNTS = NBPNTS+1
      BINCR = BSTEP*PIVOT(K)
      BSTAR(NBPNTS) = BRUN+BINCR
   14 BWEIGH(NBPNTS) = WEIGHT(K)*BSTEP
   13 BRUN = BRUN+BSTEP
C
C
C--------EVALUATION OF THE CROSS-SECTION INTEGRALS.
C--------NOTE, IF CHI IS GREATER THAN ABOUT 20*PI THE AVERAGE
C--------VALUE OF THE TERM 1 - COS(CHI) **N   IS SUBSTITUTED.
C
C
   15 DO 16 J1 = 1,NBPNTS
      CHI = CHIINT(GSTAR(J),BSTAR(J1),NWPNTS)
      IF(ABS(CHI) .LT. 63.0) GOTO 17
      BINTD1 = 1.0
      BINTD2 = 0.5
      BINTD3 = 1.0
      GOTO 18
   17 COSCHI = COS(CHI)
      BINTD1 = 1.0 - COSCHI
      BINTD2 = 1.0 - COSCHI*COSCHI
      BINTD3 = 1.0 - COSCHI**3
C
C
C--------SUMMATION OF WEIGHTED INTEGRANDS.
C
C
   18 QSTAR(1,J) = QSTAR(1,J) + BINTD1*BWEIGH(J1)
      QSTAR(2,J) = QSTAR(2,J) + BINTD2*BWEIGH(J1)
      QSTAR(3,J) = QSTAR(3,J) + BINTD3*BWEIGH(J1)
   16 CONTINUE
C
C
C--------THE QSTAR(2,J) REQUIRES NORMALISATION. THE NORMALISATION
C--------TERMS FOR THE OTHER TWO QSTARS CANCEL OUT.
```

```
C
C
      QSTAR(2,J)=1.5*QSTAR(2,J)
    7 CONTINUE
      WRITE(6,30)NGPNTS
   30 FORMAT(1X,26HNO. OF KINETIC ENERGIES = ,I6)
      WRITE(6,33)NORB
   33 FORMAT(1X,28HNO. OF ORBITAL COLLISIONS = ,I4)
      WRITE(6,32)
   32 FORMAT(//74X,14HCROSS-SECTIONS,/74X,14(1H*)/)
      WRITE(6,31) (GSTAR(J),GWEIGH(J),(QSTAR(I,J),I=1,3),J=1,NGPNTS)
   31 FORMAT(//17X,4HK.E.,15X,7HWEIGHTS,15X,4HQ(1),17X,4HQ(2),17X,
     +  4HQ(3),//,5(11X,F10.5))
   19 CALL OMGINT(TSTAR,II)
      RETURN
      END
C
C
      SUBROUTINE SOLVE(WCRIT,GSTAR,BCRIT)
C
C
C--------THIS SUBROUTINE SOLVES THE EQUATION
C        CRIT(WSTAR) - GSTAR = 0
C--------FOR THE CRITICAL ORBITING IMPACT PARAMETER BCRIT.
C--------(HERE CRIT(WSTAR) = IPSI(WSTAR) - 0.5 * WSTAR * DERIV(WSTAR)
C        DERIV(WSTAR) IS THE DERIVATIVE OF IPSI W.R.T. WSTAR.)
C--------THE EQUATION IS SOLVED BY THE BISECTION METHOD.
C--------THE VARIABLES USED IN THIS SUBROUTINE ARE
C        WCRIT,GSTAR,BCRIT AS USED IN COLLIS.
C        WSTAR  - RUNNING VARIABLE ON  WSTAR AXIS.
C        WSTAR1,WSTAR2 - ENDS OF THE REGION CONTAINING THE ROOT.
C
C
      REAL IPSI
      WSTAR2 = WCRIT
      WSTAR1 = 0.
      DO 3 I1 = 1,30
      WSTAR = (WSTAR1+WSTAR2)*0.5
      CALL FIPSI(WSTAR,IPSI,CRIT)
      IF(CRIT-GSTAR)1,1,2
    1 WSTAR1 = WSTAR
      GO TO 3
    2 WSTAR2 = WSTAR
    3 CONTINUE
      BCRIT  =  (CRIT-IPSI)/(CRIT*WSTAR*WSTAR)
      RETURN
      END
C
C
      SUBROUTINE OMGINT(TSTAR,II)
C
C
C--------THIS SUBROUTINE EVALUATES THE COLLISION INTEGRAL NUMERICALLY
C--------IN THE REGION   0 .LE. GSTAR .LE. GLASTV  AND FOR EACH
C--------OMEGA(L,S) EVALUATES THE INTEGRAL IN THE REMAINING REGION
C--------( GLASTV .LT. GSTAR .LE. INFINITY ) USING THE FOLLOWING
C--------EXPRESSION WHICH RESULTS FROM INTEGRATION BY PARTS
C--------ASSUMING QSTAR(L) IS CONSTANT IN THIS REGION
C--------OMEGA(L,S)/QSTAR(L) = EXP(-A) + (SUM FROM N=1 TO S OF
C--------(A**N*EXP(-A)/N FACTORIAL))
C        WHERE A = GLASTV/TSTAR
C--------THE OTHER VARIABLES USED ARE
C        OMEGA(L,S) - REDUCED COLLISION INTEGRALS.
C        EXPENT     - EXPONENT USED TO EVALUATE THE INTEGRAND.
C        FACTOR     - NORMALIZATION FACTOR. NOTE, CORRECT FACTOR IS
C                     ONE HALF THE NORMALIZATION FACTOR USED IN THE
C                     PAPER BY BARKER ET AL.
```

```
C          FIT1,FIT2   - TERMS IN THE HIGH ENERGY PARTIAL INTEGRATION.
C          OMINTD      - OMEGA INTEGRAND.
C
C
        DIMENSION QSTAR(3,100),GWEIGH(100),GSTAR(100),TSTAR(100)
        COMMON/COLL/OMEGA(3,5)
        COMMON/BLK2/QSTAR,GSTAR,GWEIGH,NGPNTS,GLASTV
        INTEGER S,EXPENT
C
C
C--------GAUSS INTEGRATION.
C
C
        FACTOR = 1.0
        DO 1 S = 1,5
        FACTOR  =  FACTOR/FLOAT(S+1)
        EXPENT = S+1
        DO 2 L = 1,3
        OMEGA(L,S) = 0.0
        DO 3 J = 1,NGPNTS
        OMINTD = EXP(-GSTAR(J)/TSTAR(II)) * QSTAR(L,J) *
     +  (GSTAR(J)/TSTAR(II))**EXPENT * GWEIGH(J)/TSTAR(II)
        OMEGA(L,S) = OMEGA(L,S) + FACTOR*OMINTD
    3   CONTINUE
    2   CONTINUE
    1   CONTINUE
C
C
C--------APPROXIMATING THE TAIL.
C
C
        FIT1 = EXP(-GLASTV/TSTAR(II))
        FIT2 = FIT1
        DO 4 S = 1,5
        FIT1 = (GLASTV/TSTAR(II)) * (FIT1/FLOAT(S))
        FIT2 = FIT2 + FIT1
        DO 5 L = 1,3
        OMEGA(L,S) = OMEGA(L,S) + FIT2 * QSTAR(L,NGPNTS)
    5   CONTINUE
    4   CONTINUE
        RETURN
        END
C
C
        SUBROUTINE FIPSI(WSTAR,IPSI,CRIT)
C
C
C--------THIS SUBROUTINE EVALUATES THE POTENTIAL ENERGY FOR A
C--------GIVEN WSTAR SO THAT  U*(RSTAR) = IPSI(WSTAR)
C--------WHERE IPSI AND WSTAR ARE DEFINED ABOVE
C          RSTAR - INTERMOLECULAR DISTANCE REDUCED W.R.T. SIGMA.
C          RSTAR = 1/WSTAR.
C          U*(RSTAR) - REDUCED POTENTIAL FUNCTION DEFINED IN THE
C                      USUAL WAY.
C          SIGMA       - RSTAR WHERE U*(SIGMA)=0.
C--------THE SUBROUTINE ALSO EVALUATES CRIT WHICH IS USED IN THE
C--------SUBROUTINE SOLVE.
C--------CRIT       =     IPSI(WSTAR) - 0.5*DERIV(WSTAR)*WSTSAR
C--------WHERE DERIV IS THE DERIVATIVE OF IPSI W.R.T. WSTAR.
C--------(THE DERIVATIVE OF U* W.R.T. RSTAR (CALLED DRIVRS) IS
C--------EVALUATED FIRST THEN CONVERTED TO DERIV.)
C--------THE SUBROUTINE AS GIVEN HERE CAN USE ONE OF TWO ANALYTIC
C--------POTENTIALS OR A NUMERICAL POTENTIAL DEPENDING ON
C--------THE VALUE OF NPARAM.
C          NPARAM      - POSITIVE: ANALYTIC POTENTIAL.
C                   =3 - REPULSIVE AND ATTRACTIVE INVERSE POWER
C                        POTENTIAL FUNCTION.
C                   =2 - NRSTAR - 6 POTENTIAL.
```

```
C                 - NEGATIVE: NUMERIC POTENTIAL.
C             =-N - N PAIRS OF POTENTIAL POINTS (RSTAR,U*)
C--------OTHER VARIABLES USED ARE DESCRIBED IN THE APPROPRIATE
C--------SECTION.
C
C
      DIMENSION PARAM(20),DFRNCE(150),USTARN(150),RSTARN(150)
      REAL IPSI,N
      COMMON/BLK3/PARAM,NPARAM,DFRNCE,USTARN,RSTARN
      IF(NPARAM)6,3,1
    1 IF(NPARAM.NE.3) GOTO 2
C
C
C--------FIRST ANALYTIC POTENTIAL (N-6)
C
C
C--------U*(RSTAR) = C * ( (RSTAR**-A) - (RSTAR**-B) )
C--------WHERE       C = (A/(A-B)) * ( (A/B)**(B/(A-B) )
C        PARAM(1) - A     USUALLY DENOTED N
C        PARAM(2) - B     USUALLY EQUAL TO 6
C        PARAM(3) - C     NORMALIZATION FACTOR
C--------FOR THE LENNARD - JONES POTENTIAL PARAM(1) = 12
C                                         PARAM(2) = 6
C                                         PARAM(3) = 4
C
C
      IPSI = PARAM(3) * (WSTAR**PARAM(1) - WSTAR**PARAM(2))
      DERIV = PARAM(3) * (PARAM(1)*WSTAR**(PARAM(1)-1.0)
     +        - PARAM(2)*WSTAR**(PARAM(2)-1.0))
      CRIT = IPSI - 0.5*DERIV*WSTAR
      RETURN
    2 IF(NPARAM.NE.2) GOTO 3
C
C                     ***************
C
C--------SECOND ANALYTIC POTENTIAL (N(RSTAR) - 6)
C
C
C--------U*(RSTARM) = (6/(N-6)*(RSTARM**-N) - (N/(N-6)*(RSTARM**-6)
C--------WHERE RSTARM - INTERMOLECULAR DISTANCE REDUCED W.R.T. RM.
C                  RM - INTERMOLCULAR DISTANCE S.T. U*(RM) = -1
C                   N - 13 + GAMMA * (RSTARM-1)
C--------THE VARIABLES NOT GIVEN SO FAR ARE
C          PARAM(1) - SIGMA/RM
C          PARAM(2) - GAMMA          (4 .LE. GAMMA .LE. 10)
C          WSTARM   - 1/RSTARM
C
C
      WSTARM = WSTAR/PARAM(1)
      RSTARM = 1.0/WSTARM
      IF(RSTARM .LE. 6.0)GOTO 4
C
C
C--------THIS PROVIDES A VERY STEEP POTENTIAL AT SHORT RANGE.
C
C
      N = 60.0
      GOTO 5
    4 N = 13.0 + PARAM(2)*(RSTARM-1.0)
    5 DEN = N-6.0
      IPSI = 6.0*WSTARM**N/DEN - N*WSTARM**6/DEN
      DRIVRS = (6.0*N/DEN + 6.0*PARAM(2)*RSTARM/DEN**2) *
     +        (WSTARM**7 - WSTARM**(N+1.0)) - 6.0*PARAM(2)*RSTARM*
     +                ALOG(RSTARM)*WSTARM**(N+1.0)/DEN
      DERIV = (-1.0/(WSTARM*WSTARM))*DRIVRS
      CRIT = IPSI - 0.5*DERIV*WSTARM
      RETURN
```

```
C
C                           ***************
C
    3   WRITE(6,13)
   13   FORMAT(1X,"NUMERICAL POTENTIAL")
        CALL COEFF
        GOTO 1
C
C
C--------NUMERICAL POTENTIAL.
C
C
C--------THE POTENTIAL IS OBTAINED BY QUADRATIC INTERPOLATION BETWEEN
C--------THE GIVEN DATA POINTS.
C--------THE DERIVATIVE W.R.T. RSTAR IS OBTAINED BY A QUADRATIC
C--------THE POTENTIAL AND CRIT FOR RSTAR OUTSIDE THE REGION COVERED
C--------BY THE DATA POINTS ARE OBTAINED USING EXTRAPOLATION FUNCTIONS.
C--------VARIABLES NOT GIVEN SO FAR ARE
C             RSTARN(I) - INTERMOLECULAR DISTANCE REDUCED W.R.T. SIGMA
C             USTARN(I) - CORRESPONDING POTENTIAL.
C--------NOTE. THESE ARE THE DATA POINTS SUPPLIED BY THE USER.
C        ******THEY MUST BE SUPPLIED WITH RSTAR INCREASING******
C           DFRNCE(I) - THE DIFFERENCES USED TO ESTIMATE THE DERIVATIVE.
C
C
    6   NUM = -NPARAM-1
        RSTAR = 1.0/WSTAR
        IF(RSTAR .LE. RSTARN(1)) GOTO 7
        IF(RSTAR .GE. RSTARN(NUM)) GOTO 8
        DO 9 J = 1,NUM
        I = J
        IF(RSTARN(J) .GE. RSTAR) GOTO 10
    9   CONTINUE
   10   U1 = USTARN(I-1)
        U2 = USTARN(I)
        U3 = USTARN(I+1)
        R1 = RSTARN(I-1)
        R2 = RSTARN(I)
        R3 = RSTARN(I+1)
        D1 = DFRNCE(I-1)
        D2 = DFRNCE(I)
        D3 = DFRNCE(I+1)
        UF1 = U1 + (U2-U1)*(RSTAR-R1)/(R2-R1)
        UF2 = U2 + (U3-U2)*(RSTAR-R2)/(R3-R2)
        DF1 = D1 + (D2-D1)*(RSTAR-R1)/(R2-R1)
        DF2 = D2 + (D3-D2)*(RSTAR-R2)/(R3-R2)
        IPSI = UF1 + (UF2-UF1)*(RSTAR-R1)/(R3-R1)
        DRIVRS = DF1 + (DF2-DF1)*(RSTAR-R1)/(R3-R1)
        DERIV = (-RSTAR*RSTAR)*DRIVRS
        CRIT = IPSI - 0.5*DERIV*WSTAR
        RETURN
C
C
C--------SHORT RANGE EXTRAPOLATION
C--------U(R*) = A*EXP(-BR*)
C
C
C--------U*(RSTAR) = PARAM(1) * EXP(-PARAM(2)*RSTAR)
C--------PARAM(1) AND PARAM(2) OBTAINED BY FITTING FIRST TWO
C--------PAIRS OF DATA POINTS.
C
C
    7   IPSI = PARAM(1)*EXP(-PARAM(2)*RSTAR)
        CRIT = IPSI*(1.0-0.5*PARAM(2))
        RETURN
C
C
C--------LONG RANGE EXTRAPOLATION.
```

```
C---------U(R*) = CR**-6
C
C
C---------U*(RSTAR) = -PARAM(3) * RSTAR**-6
C---------PARAM(3) OBTAINED EITHER BY FITTING LAST DATUM
C---------OR  FROM LITERATURE.
C
C
    8   IPSI = - PARAM(3)*WSTAR**6
        CRIT = -2.0*IPSI
        RETURN
        END
C
C
        FUNCTION CHIINT(GSTAR,BSTAR,NWPNTS)
C
C
C---------THIS FUNCTION EVALUATES THE DEFLECTION ANGLE FOR A GIVEN
C---------BSTAR AND GSTAR. THE INTEGRATION IS DONE BY THE TRAPEZIUM RULE
C---------AND CONTINUES UNTIL EITHER 1 - THE REGION BOUNDED BY THE WSTAR
C---------POINTS PROVIDED IN THE PROGRAM IS EXCEEDED (AN UNLIKELY CASE),
C---------OR  2 - THE DENOMINATOR OF THE INTEGRAND
C---------( Y(N) ) BECOMES ZERO ( USUAL CASE ). THE WSTAR AT WHICH THIS
C---------CONDITION OCCURS IS THE RECIPROCAL OF THE DISTANCE OF CLOSEST
C---------APPROACH. THE LAST PANEL IS THEN APPROXIMATED USING REVERSE
C---------INTERPOLATION.
C---------FOR MORE DETAILS ON THE MATHEMATICS OF THIS FUNCTION SEE THE
C---------PAPER BY BARKER ET AL.
C---------VARIABLES NOT DEFINED SO FAR:
C         WINTGD - INTEGRAL PART OF THE FUNCTION.
C---------THE FUNCTION SUBPROGRAM RETURNS THE VALUE OF THE DEFLECTION
C---------ANGLE VIA THE FUNCTION NAME.
C
C
        DIMENSION Y(150),WSTAR(150),IPSI(150)
        REAL IPSI
        COMMON/BLK1/WSTAR,IPSI
        Y(1) = 1.0-BSTAR*WSTAR(1)*WSTAR(1)-IPSI(1)/GSTAR
        IF(Y(1) .GE. 0.0) GOTO 1
        WINTGD = 2.0*WSTAR(1)/(1.0-Y(1))
        GOTO 4
    1   WINTGD = 2.0*WSTAR(1)/(1.0 + SQRT(Y(1)))
        DO 2 N = 2,NWPNTS
        Y(N) = 1.0-BSTAR*WSTAR(N)*WSTAR(N)-IPSI(N)/GSTAR
        IF(Y(N) .GT. 0.0) GOTO 3
        WINTGD = WINTGD+2.0*(WSTAR(N)-WSTAR(N-1))*SQRT(Y(N-1))/
     +         (Y(N-1)-Y(N))
        GOTO 4
    3   WINTGD = WINTGD + 2.0*(WSTAR(N)-WSTAR(N-1))/(SQRT(Y(N-1))+
     +      SQRT(Y(N)))
        IF(ABS(Y(N)) .LT. 1.0E-20) GOTO 4
    2   CONTINUE
    4   CHIINT = 3.14159 - 2.0*SQRT(BSTAR)*WINTGD
        RETURN
        END
C
C
        SUBROUTINE COEFF
C
C
C---------THIS SUBROUTINE INPUTS THE PARAMETERS NECESSARY FOR THE
C---------POTENTIAL ENERGY FUNCTION FIPSI.
C---------IN THE CASE OF A NUMERICAL POTENTIAL THE SUBROUTINE
C---------ALSO EVALUATES THE DIFFERENCES BETWEEN SUCCESSIVE POINTS.
C---------ALL VARIABLES AS DESCRIBED ABOVE.
C
C
        DIMENSION PARAM(20),DFRNCE(150),USTARN(150),RSTARN(150)
        COMMON/BLK3/PARAM,NPARAM,DFRNCE,USTARN,RSTARN
```

```
C
C--------READ IN THE NUMBER OF PARAMETERS FOR ANALYTIC
C--------POTENTIAL (MAXIMUM 20) OR -(THE NUMBER OF
C--------DISTANCE,POTENTIAL PAIRS) FOR A NUMERICAL
C--------POTENTIAL (MAXIMUM 150).
C
    1   READ(5,10) NPARAM
   10   FORMAT(I4)
        IF(NPARAM) 2,3,4
C
C--------READ IN (NPARAM) PARAMETER VALUES
C
    4   READ(5,11)(PARAM(I),I=1,NPARAM)
   11   FORMAT(F9.4)
        RETURN
    3   WRITE(6,5)
    5   FORMAT(1X,"NO POTENTIALS WITH ZERO PARAMETERS",///)
        GO TO 1
    2   NN = -NPARAM
        IF(NN .GT. 150) GO TO 17
C
C--------READ IN EXTRAPOLATION PARAMETERS
C--------READ NN REDUCED DISTANCE,POTENTIAL PAIRS
C
        READ(5,12)(PARAM(I),I=1,3)
        READ(5,13)(RSTARN(I),USTARN(I),I=1,NN)
   13   FORMAT(2F9.4)
   12   FORMAT(F9.4)
        DFRNCE(1) = (USTARN(2)-USTARN(1))/(RSTARN(2)-RSTARN(1))
        NN1 = NN-1
        DO 6 I = 2,NN1
        A = (USTARN(I+1)-USTARN(I))/(RSTARN(I+1)-RSTARN(I))
        B = (USTARN(I)-USTARN(I-1))/(RSTARN(I)-RSTARN(I-1))
    6   DFRNCE(I) = (A+B)/2.0
        DFRNCE(NN) = (USTARN(NN)-USTARN(NN1))/(RSTARN(NN)-RSTARN(NN1))
        RETURN
   17   WRITE(6,14)
   14   FORMAT(1X,"NOT SUFFICIENT STORAGE FOR GREATER THAN 150 PAIRS")
        STOP
        END
```

APPENDIX 13

INVERSION FUNCTIONS FOR GAS PHASE PROPERTIES

The inversion methods that utilize the functions tabulated here are based on the use of a thermophysical property, P, to define a characteristic temperature-dependent cross-section $\pi\tilde{r}_P^2$ and hence a characteristic length $\tilde{r}_P$. The potential energy at this separation $U(\tilde{r}_P)$ is given by the equation

$$U(\tilde{r}_P) = G_P(T^*)kT$$

where $G_P(T^*)$ is the appropriate inversion function. The success of the method is due to the fact that the functions $G_P(T^*)$ are not very sensitive to the nature of the potential energy function and thus may be estimated using only approximate potential functions. A knowledge of T^* $(=kT/\varepsilon)$ and so an estimate of the well depth ε (or in practice any characteristic feature of the potential function) may be required. Accounts of the inversion techniques are given in §§ 3.11.2 and 6.7.

Viscosity

$$\eta = \frac{5}{16}\frac{(m\pi kT)^{\frac{1}{2}}}{\overline{\Omega}^{(2,2)}(T)} f_\eta. \tag{A13.1}$$

From the definition $\tilde{r}_\eta = [\overline{\Omega}^{(2,2)}/\pi]^{\frac{1}{2}}$, $\tilde{r}_\eta$ may be related to the coefficient of viscosity. Evaluating the various constants involved gives the simple equation

$$\tilde{r}_\eta = 0.16339\left[\frac{(MT)^{\frac{1}{2}} f_\eta}{\eta}\right]^{\frac{1}{2}}$$

where $\tilde{r}_\eta$ is in nanometres, M is the relative molecular mass, T the absolute temperature in K, η is in units of μPa s (μN s m^{-2}), and f_η is a factor in the range 1·000–1·008 which may be calculated using an approximate potential function. (f_η may be calculated for an $n-6$ or $n(\tilde{r})-6$ potential using the expressions of Appendix 5.2 and the collision integrals of Appendix 2. Values for the $12-6$ and $18-6$ potentials are plotted in Fig. 6.2.) Then $U(\tilde{r}_\eta) = G_\eta(T^*) \,.\, kT$. The values of G_η calculated using the Lennard-Jones $12-6$ function and the BBMS potential, a realistic function devised for argon, are given in Table A13.1. The corresponding values of f_η are given in Table A13.2.

Self-diffusion. The coefficient of self-diffusion is related to the collision integral, $\overline{\Omega}^{(1,1)}$ as shown in the equation below

$$D = \frac{3}{8P}\left[\frac{k^3T^3}{m}\right]^{\frac{1}{2}} \frac{f_D}{\Omega^{(1,1)}}. \tag{A13.2}$$

Again defining the distance $\tilde{r}_D = [\overline{\Omega}^{(1,1)}/\pi]^{\frac{1}{2}}$ we obtain an expression for $\tilde{r}_D$ in terms of D, giving

$$\tilde{r}_D = 5.161 \times 10^{-3}\left[\frac{f_D}{PD}\left(\frac{T^3}{M}\right)^{\frac{1}{2}}\right]^{\frac{1}{2}}$$

Table A13.1
Inversion functions

	BBMS				LJ(12−6)			
T^*	G_η	G_D	G_α	G_B	G_η	G_D	G_α	G_B
0·3	−0·561	−0·691	−1·062	−0·00373	−0·549	−0·679	−1·063	−0·0056
0·4	−0·566	−0·746	−1·212	−0·0254	−0·573	−0·743	−1·196	−0·0367
0·5	−0·599	−0·817	−1·293	−0·0979	−0·611	−0·806	−1·287	−0·117
0·6	−0·644	−0·874	−1·328	−0·239	−0·652	−0·858	−1·320	−0·251
0·7	−0·684	−0·908	−1·301	−0·402	−0·689	−0·893	−1·294	−0·418
0·8	−0·714	−0·923	−1·225	−0·595	−0·718	−0·910	−1·220	−0·585
0·9	−0·732	−0·920	−1·111	−0·744	−0·736	−0·908	−1·111	−0·720
1·0	−0·739	−0·902	−0·969	−0·833	−0·743	−0·892	−0·979	−0·806
1·1	−0·737	−0·869	−0·810	−0·861	−0·741	−0·861	−0·833	−0·839
1·2	−0·727	−0·824	−0·644	−0·831	−0·730	−0·819	−0·682	−0·823
1·3	−0·709	−0·770	−0·477	−0·756	−0·711	−0·768	−0·530	−0·767
1·4	−0·686	−0·706	−0·313	−0·650	−0·686	−0·711	−0·382	−0·683
1·5	−0·657	−0·635	−0·156	−0·525	−0·657	−0·649	−0·238	−0·581
1·6	−0·624	−0·566	−0·007	−0·390	−0·623	−0·548	−0·101	−0·468
1·7	−0·558	−0·490	+0·134	−0·254	−0·588	−0·517	+0·0281	−0·350
1·8	−0·550	−0·417	+0·267	−0·121	−0·551	−0·450	0·150	−0·233
1·9	−0·511	−0·340	+0·388	+0·0063	−0·512	−0·382	0·264	−0·118
2·0	−0·471	−0·267	0·502	0·127	−0·474	−0·315	0·371	−0·008
2·4	−0·316	+0·013	0·882	0·518	−0·324	−0·0626	0·733	+0·367
2·8	−0·175	0·254	1·165	0·790	−0·190	+0·159	1·009	0·643
3·0	−0·114	0·360	1·278	0·893	−0·130	0·257	1·121	0·750
3·4	−0·005	0·544	1·465	1·050	−0·025	0·430	1·310	0·918
4·0	+0·127	0·762	1·670	1·205	+0·103	0·640	1·523	1·088
4·5	0·212	0·910	1·795	1·288	0·189	0·779	1·656	1·183
5·0	0·281	1·017	1·893	1·346	0·255	0·893	1·761	1·250
6·0	0·384	1·186	2·032	1·419	0·358	1·066	1·916	1·336
8·0	0·510	1·388	2·190	1·483	0·485	1·284	2·103	1·415
10·0	0·582	1·500	2·278	1·508	0·558	1·410	2·209	1·447
12·0	0·629	1·570	2·328	1·518	0·605	1·492	2·278	1·461
15·0	0·674	1·632	2·372	1·524	0·651	1·571	2·345	1·469
20·0	0·720	1·687	2·407	1·523	0·694	1·646	2·410	1·471

where P is the pressure in MPa and D is the coefficient of self-diffusion in $10^{-5}\ \mathrm{m}^2\ \mathrm{s}^{-1}$.

The potential energy at the separation $\tilde{r}_D$ is given by the equation

$$U(\tilde{r}_D) = kT \, . \, G_D(T^*),$$

where G_D is a function similar to G_η and is tabulated in Table A13.1. The correction factor f_D is given in Table A13.2.

Thermal conductivity. Like viscosity, the coefficient of thermal conductivity, λ, of a substance with no internal degrees of freedom, is related to the collision integral, $\bar{\Omega}^{(2,2)}$. The equation is

$$\lambda = \frac{25 c_v}{32 \bar{\Omega}^{(2,2)}} \left\{ \frac{kT\pi}{m} \right\}^{\frac{1}{2}} f_\lambda = \frac{15}{4} \eta \cdot \frac{k}{M} \cdot \frac{f_\lambda}{f_\eta} \qquad \text{(A13.3)}$$

Table A13.2

Second-order Kihara correction factors for the calculation of transport properties

	12−6				BBMS (A1.14)			
T^*	f_η	f_D	f_λ	f_α	f_η	f_D	f_λ	f_α
0·3	1·000 98	1·000 20	1·001 53	1·046 83	1·001 63	1·000 77	1·001 84	1·033 49
0·5	1·000 01	1·000 03	1·000 01	1·005 21	1·000 27	1·000 06	1·000 42	1·026 59
0·7	1·000 06	1·000 06	1·000 10	0·993 50	1·000 02	1·000 01	1·000 03	0·999 30
1·0	1·000 00	1·000 01	1·000 01	1·000 36	1·000 03	1·000 09	1·000 05	0·985 51
1·5	1·000 33	1·000 58	1·000 51	0·972 03	1·000 51	1·000 72	1·000 79	0·975 74
2·0	1·001 33	1·001 06	1·002 07	0·972 77	1·001 49	1·001 69	1·002 32	0·976 80
2·4	1·002 24	1·002 56	1·003 48	0·976 21	1·002 34	1·002 45	1·003 64	0·980 17
3·0	1·003 45	1·003 71	1·005 37	0·981 81	1·003 44	1·003 41	1·005 36	0·985 76
5·0	1·005 81	1·006 01	1·009 03	0·994 33	1·005 46	1·005 11	1·008 50	0·998 78
10·0	1·007 32	1·007 65	1·011 39	1·003 94	1·006 36	1·005 77	1·009 90	1·009 96
20·0	1·007 74	1·008 17	1·012 04	1·006 89	1·005 87	1·005 08	1·009 12	1·014 93

In Chapter 5, Appendix 2, f_η is defined by (A5.2.11), f_D by (A5.2.35), f_λ by (A5.2.18), and f_α by (A5.2.41).

Since $\bar{\Omega}^{(2,2)}$ is equal to $\pi\tilde{r}_\lambda^2$, then

$$\tilde{r}_\lambda = 0{\cdot}9123\left\{\left(\frac{T}{M}\right)^{\frac{1}{2}}\frac{f_\lambda}{\lambda}\right\}^{\frac{1}{2}}$$

where λ is the coefficient of thermal conductivity in units of mW m^{-1} K^{-1}.

The potential energy may be obtained using the equation

$$U(\tilde{r}_\lambda) = kT\,.\,G_\eta(T^*).$$

The function G_η is involved, since both viscosity and thermal conductivity are related to the same collision integral. However, f_λ differs slightly from f_η and is tabulated in Table A13.2.

Thermal diffusion. One inversion method uses $\Omega^{(1,2)}$ determined from the isotopic thermal diffusion factor for monatomic species, $[\alpha_0]_1^K$. From (5.156)

$$C^* = \frac{8[\alpha_0]_1^K A^\star}{45} + \frac{5}{6}$$

and since

$$C^\star = \Omega^{(1,2)*}/\Omega^{(1,1)*} = \bar{\Omega}^{(1,2)}/\bar{\Omega}^{(1,1)}$$

$$A^\star = \Omega^{(2,2)*}/\Omega^{(1,1)*} = \bar{\Omega}^{(2,2)}/\bar{\Omega}^{(1,1)}$$

$$\bar{\Omega}^{(1,2)} = \frac{8[\alpha_0]_1^K}{45} + \frac{5}{6A^\star}\bar{\Omega}^{(2,2)}$$

$[\alpha_0]_1^K$ may be obtained from the experimental values $[\alpha_0]_{\text{expt}}$ by the relation

$$[\alpha_0]_1^K = [\alpha_0]_{\text{expt}}/f_\alpha$$

where f_α is a correction factor that is insensitive to the potential energy function. In general $\bar{\Omega}^{(2,2)}$ is well known from viscosity measurements. Hence $\bar{\Omega}^{(1,2)}$ can be determined and

$$\tilde{r}_\alpha = [\bar{\Omega}^{(1,2)}/\pi]^{\frac{1}{2}}$$

$$U(\tilde{r}_\alpha) = G_\alpha kT.$$

Again G_α and f_α are tabulated in Tables A13.1 and A13.2. [Strictly G_α should be labelled $G_{\Omega(1,2)}$].

Second virial coefficients. We define

$$\tilde{r}_B = \left[\frac{B + T\left(\frac{\mathrm{d}B}{\mathrm{d}T}\right)}{2/3\ \pi N_A}\right]^{\frac{1}{3}}$$

$$= 9{\cdot}256\left[B + T\left(\frac{\mathrm{d}B}{\mathrm{d}T}\right)\right]^{\frac{1}{3}}$$

where B is the second virial coefficient in $m^3\ mol^{-1}$ and $\tilde{r}_B$ is in nm.

Again

$$U(\tilde{r}_B) = G_B kT.$$

SUBSTANCE INDEX

INDEX